NEURAL
and
METABOLIC
CONTROL
of
MACRONUTRIENT
INTAKE

Edited by

Hans-Rudolf Berthoud, Ph.D.
Randy J. Seeley, Ph.D.

CRC Press
Taylor & Francis Group
Boca Raton London New York

CRC Press is an imprint of the
Taylor & Francis Group, an **informa** business

CRC Press
Taylor & Francis Group
6000 Broken Sound Parkway NW, Suite 300
Boca Raton, FL 33487-2742

First issued in paperback 2019

ISBN-13: 978-0-8493-2752-0 (hbk)
ISBN-13: 978-0-367-39935-1 (pbk)
Library of Congress Card Number 97-24341

Library of Congress Cataloging-in-Publication Data

Neural and metabolic control of macronutrient intake / edited by Hans
   -Rudolf Berthoud, Randy J. Seeley.
         p.   cm.
      Includes bibliographical references and index.
      ISBN 0-8493-2752-0 (alk. paper)
      1. Appetite.   2. Food preferences.   3. Nutrition--Psychological
   aspects.   4. Neuropsychology.   I. Berthoud, Hans-Rudolf.
   II. Seeley, Randy J.
      [DNLM: 1. Nutrition.   2. Dietary Carbohydrates--metabolism.
   3. Dietary Fats--metabolism.   4. Dietary Proteins--metabolism.
   5. Food Preferences--physiology.   6. Food Preferences--psychology.
   QU 145 N494 1999]
   QP136.N47   1999
   612.3′9--dc21
   DNLM/DLC
   for Library of Congress   99-29233
                                                                                 CIP

Visit the Taylor & Francis Web site at
http://www.taylorandfrancis.com

and the CRC Press Web site at
http://www.crcpress.com

NEURAL
and
METABOLIC
CONTROL
of
MACRONUTRIENT
INTAKE

Preface

Several years ago the idea for this book grew out of a conversation at the European Winter Conference on Brain Research. Over a beer, the two editors of this book were discussing the topic of macronutrient selection and the potential involvement of the brain in its control. Macronutrient selection refers to the process by which general omnivores choose to consume differing amounts of protein, carbohydrate, or fats in their diets. Considerable controversy still surrounds to what degree macronutrient content of particular foods guides ingestive behavior. While neither of us had done a tremendous amount of research on macronutrient selection, over the course of several beers we began a spirited "discussion" of this topic. Quickly, however, we realized how difficult a topic it was to discuss because there was no resource that brought together the large and diverse literature related to macronutrient selection. That is to say, our arguments rapidly devolved because we lacked a way to resolve disputes over the current state of knowledge about issues related to macronutrient selection. Neither of us possessed the breadth of knowledge (from biochemistry to human and animal behavior) that a real consideration of these issues would require. Before we left that meeting, we decided that very few individuals likely had the necessary breadth of knowledge and therefore a book that could tie together the pieces of this intricate puzzle would be a valuable resource for individuals from a broad array of backgrounds with general interests in dietary selection.

What you will find in this volume is our attempt to bring together chapters from the people who have most influenced current thinking on this topic. These individuals represent the truly interdisciplinary nature of this topic. They range from biochemists, nutritionists, psychologists, physiologists, neuroscientists, and molecular biologists. In addition to diverse backgrounds, the authors of these chapters are individuals who differ considerably on their opinions about the nature of macronutrient selection. We have gone to great lengths to put into this single volume chapters from people who clearly advocate the existence of separate control systems for the individual macronutrients and those who do not. In this way, we believe readers can evaluate these arguments for themselves. The book is arranged to provide three types of information. First, the basic background of the biochemical and physiological systems as they relate to macronutrient selection. Second, opinions and data concerning to what degree animals and humans show evidence of macronutrient selection. Third, evidence about how the central nervous system might be involved in the choices animals make among macronutrients.

Few basic research problems with this level of controversy have such important practical implications. Increased consumption of dietary fat in developed countries is closely tied to increases in heart disease, diabetes, and obesity. If separate control systems exist, it leaves open the possibility that strategies could be devised to intervene in these control systems and alter the proportion of fat in the diet. If such separate control systems do not exist, such interventions will eventually fail. Consequently, the stakes of this particular debate both for basic understanding of how the brain controls ingestive behavior and for our ability to decrease the incidence of these dangerous conditions are extremely high.

The Editors

Hans-Rudolf Berthoud was born near Zürich, Switzerland, and obtained his Diploma in Biology (1969) and Ph.D. in Neurobehavioral Biology and Physiology (1973) from the Swiss Federal Institute of Technology (ETH), Zürich. During his postdoctoral training, Dr. Berthoud continued to study the role of the hypothalamus in ingestive behavior with Gordon J. Mogenson at the University of Western Ontario in Canada, and then collaborated with Bernard Jeanrenaud and Albert Renold as "Chef de Traveaux" at the University of Geneva Medical School on projects focusing on the neural control of pancreatic secretion. In 1982 he emigrated to the U.S., where he began a long collaboration with Terry Powley at Purdue University investigating anatomy and physiology of the vagus nerve. In 1992 Dr. Berthoud was appointed Associate Professor at the Pennington Biomedical Research Center in Baton Rouge, and Louisiana State University Medical Center in New Orleans, LA, where he continued his studies on morphological, physiological, and behavioral aspects of vagal and central nervous mechanisms that control ingestive behavior and energy metabolism.

Dr. Berthoud has published over 80 papers in leading journals, including several reviews and book chapters. He is a member of the Society for Neuroscience (SN), the American Physiological Society (APS), the Society for the Study of Ingestive Behavior (SSIB), the North American Association for the Study of Obesity (NAASO), and the International Society for Autonomic Nervous System Physiology (ISAN). Dr. Berthoud is on the Editorial Board for the *American Journal of Physiology, Comparative and Regulatory Physiology*, and *Obesity Research*.

Randy J. Seeley received his undergraduate degree with honors from Grinnell College in 1989. He received his Ph.D. at the University of Pennsylvania in the Department of Psychology working with Dr. Harvey Grill. He then was a Post-Doctoral Fellow for 2 years working with Drs. Steve Woods and Dan Porte, Jr. at the University of Washington. He was also on the faculty of the Department of Psychology at the University of Washington for 2 years before moving to the University of Cincinnati in the fall of 1997. He was appointed as an Associate Professor in the Department of Psychiatry in the College of Medicine at the University of Cincinnati and a member of the Graduate Neuroscience Program, an Adjunct Associate Professor in the Department of Psychology, and Associate Director of the Obesity Research Center. Since 1993 he has coauthored over 60 papers and chapters concerning his work on the neural and metabolic control of food intake and body weight in rodent models including reviews in the *New England Journal of Medicine* and *Science*. He is on the editorial board for the *American Journal of Physiology* and a member of the Society for Neuroscience, Society for the Study of Ingestive Behavior, North American Association for the Study of Obesity, American Psychological Association, and American Psychological Society.

Contributors

Karen Ackroff
Department of Psychology
Brooklyn College of CUNY
Brooklyn, NY 11210

Bernard Beck
INSERM U. 308
Mécanismes de Régulation du Comportement
  Alimentaire
38 rue Lionnois
54000 Nancy, France

Stephen C. Benoit
Department of Psychiatry
University of Cincinnati College of Medicine
P.O. Box 670559
Cincinnati, OH 45267-0559

Hans-Rudolf Berthoud
Neurobiology of Nutrition Laboratory
Pennington Biomedical Research Center
Louisiana State University
6400 Perkins Rd.
Baton Rouge, LA 70808

Charles J. Billington
Department of Psychiatry
Veterans Administration Medical Center
Minneapolis, MN 55417

John E. Blundell
Biopsychology Group
School of Psychology
University of Leeds
Leeds, LS2 9JT, U.K.

David A. Booth
Nutritional Psychology Research Group
University of Birmingham
Birmingham B15 2TT, England

Lynda M. Brown
Department of Nutrition and Food Science
University of Maryland
Rm. 3304, Marie Mount Hall
College Park, MD 20742

T. W. Castonguay
Department of Nutrition and Food Science
University of Maryland
Marie Mount Hall
College Park, MD 20742

Mark Chavez
The Monell Chemical Senses Institute
3500 Market St.
Philadelphia, PA 19104

Jian Chen
Pennington Biomedical Research Center
Louisiana State University
6400 Perkins Rd.
Baton Rouge, LA 70808

Mihai Covasa
Department of V.C.A.P.P. and Program in
  Neuroscience
College of Veterinary Medicine
Washington State University
Pullman, WA 99164

T. L. Davidson
Department of Psychological Sciences
1364 Psychological Sciences Bldg.
Purdue University
West Lafayette, IN 47907-1364

John M. de Castro
Department of Psychology
Neuropsychology and Behavioral Neuroscience
  Program
Georgia State University
University Plaza
Atlanta, GA 30303

Ambrose A. Dunn-Meynell
Department of Neurosciences
Neurology Service, VA Medical Center
East Orange, NJ 07018

J. P. Flatt
Department of Biochemistry and Molecular
  Biology
University of Massachusetts Medical Center
Worcester, MA 01655

Louis A. Foster
Department of Anatomy and Physiology
College of Veterinary Medicine
Kansas State University
Manhattan, KS 66506

Mark I. Friedman
The Monell Chemical Senses Institute
3500 Market St.
Philadelphia, PA 19104

K. Fukagawa
Department of Internal Medicine
Oita University

Bennett G. Galef, Jr.
Department of Psychology
McMaster University
Hamilton, ON L8S 4K1 Canada

Kyriaki Gerozissis
Physiopathologie de la Nutrition
CNRS ESA 7059
Universite Paris 7, 2 Place Jussieu
75251 Paris Cedex 05, France

Dorothy W. Gietzen
Department of Anatomy, Physiology and Cell
  Biology
School of Veterinary Medicine
University of California, Davis
Davis, CA 95616

Michael J. Glass
Minnesota Obesity Center
Department of Psychology
Veterans Affairs Medical Center
Minneapolis, MN 55417

Jason C. G. Halford
University of Liverpool
Department of Psychology
Eleanor Rathbone Building
Liverpool, L6972A, U.K.

Ruth B. S. Harris
Pennington Biomedical Research Center
Louisiana State University
6400 Perkins Rd.
Baton Rouge, LA 70808

Leigh Anne Howell
Pennington Biomedical Research Center
Louisiana State University
6400 Perkins Rd.
Baton Rouge, LA 70808

Frank H. Koegler
Oregon Regional Primate Research Center
Oregon Health Sciences University
505 N.W. 185th Ave.
Beaverton, OR 97006

Wolfgang Langhans
Institute of Animal Sciences
Group of Physiology and Animal Husbandry
Swiss Federal Institute of Technology (ETH)
Zurich, 8092 Switzerland

Sarah F. Leibowitz
The Rockefeller University
1230 York Ave.
New York, NY 10021

Barry E. Levin
Department of Neurosciences
Neurology Service, 127C VA Medical Center
East Orange, NJ 07009

Allen S. Levine
Department of Medicine
Veterans Affairs Medical Center
One Veterans Dr.
Research Service (151)
Minneapolis, MN 55417

Ling Lin
Pennington Biomedical Research Center
Louisiana State University
6400 Perkins Rd.
Baton Rouge, LA 70808

Tiffany D. Mitchell
Pennington Biomedical Research Center
Louisiana State University
6400 Perkins Rd.
Baton Rouge, LA 70808

Timothy H. Moran
Department of Psychiatry and Behavioral
  Sciences
Johns Hopkins University School of Medicine
720 Rutland Ave., Ross Bldg., Rm. 618
Baltimore, MD 21205

Javier R. Morell
Department of Psychological Sciences
1364 Psychological Sciences Bldg.
Purdue University
West Lafayette, IN 47907-1364

Stylianos Nicolaïdis
Institut Europeen des Sciences du Gout et des
  Comportements Alimentaires
Universite de Bourgogne
21000 Dijon, France

Leona M. O'Reilly
Rowett Research Institute
Bucksburn, Aberdeen, Scotland AB21 9SB

Martine Orosco
Physiopathologie de la Nutrition
CNRS ESA 7059
Universite Paris 7, 2 place Jussieu case 7126
75251 Paris Cedex 05, France 75251

David Raubenheimer
Department of Zoology and University Museum
  of Natural History
University of Oxford
South Parks Rd.
Oxford, Ox1 3PS, U.K.

Christine A. Riedy
Department of Dental Public Health
University of Washington
Seattle, WA

Robert C. Ritter
Department of V.C.A.P.P. and Program in
  Neuroscience
College of Veterinary Medicine
Washington State University
Pullman, WA 99164

W. Sue Ritter
Department of Veterinary and Comparative
  Anatomy, Physiology and Pharmacology
Washington State University
Pullman, WA 99164-6520

Barbara J. Rolls
Nutrition Department
Pennsylvania State University
226 Henderson Bldg.
University Park, PA 16802

Edmund T. Rolls
University of Oxford
Department of Experimental Psychology
South Parks Rd.
Oxford OX1 3UD
England

Vanessa H. Routh
Department of Pharmacology and Physiology
New Jersey Medical School
Newark, NJ 07103

Donna H. Ryan
Pennington Biomedical Research Center
Louisiana State University
6400 Perkins Rd.
Baton Rouge, LA 70808

Anthony Sclafani
Department of Psychology
Brooklyn College and the Graduate School
City University of New York
2900 Bedford Ave.
Brooklyn, NY 11210

Randy J. Seeley
Department of Psychiatry
University of Cincinnati
P.O. Box 670559
Cincinnati, OH 45267-0559

Stephen J. Simpson
Department of Zoology and University Museum
  of Natural History
University of Oxford
South Parks Rd.
Oxford, Ox1 3PS, U.K.

Brenda K. Smith
Pennington Biomedical Research Center
Louisiana State University
6400 Perkins Rd.
Baton Rouge, LA 70808

Edward M. Stricker
Department of Neuroscience
479 Crawford Hall
University of Pittsburgh
Pittsburgh, PA 15260

R. James Stubbs
Human Nutrient Unit
Rowett Research Institute
Greenburn Rd.
Bucksburn, Aberdeen, Scotland AB21 9SB U.K.

Patrick Tso
Department of Pathology
University of Cincinnati Medical Center
231 Bethesda Avenue (ML 0529)
Cincinnati, OH 45267-0529

Louise Thibault
School of Dietetics and Human Nutrition
Macdonald Campus of McGill University
21.111 Lakeshore Rd.
Ste. Anne de Bellevue, QC H9X 3V9 Canada

Gertjan van Dijk
Department of Animal Physiology
University of Groningen
Kerklaan 30
Haren 9750AA The Netherlands

Malcolm Watford
Department of Nutritional Sciences
Cook College
Rutgers University
New Brunswick, NJ 08901

Michael Wiater
Department of Veterinary and Comparative
  Anatomy, Physiology and Pharmacology
Washington State University
P.O. Box 646520
Pullman, WA 99164-6520

Stephen C. Woods
Department of Psychiatry
University of Cincinnati Medical Center
P.O. Box 670559
Cincinnati, OH 45267-0559

David A. York
Pennington Biomedical Research Center
Louisiana State University
6400 Perkins Rd.
Baton Rouge, LA 70808

Bradley D. Youngblood
Pennington Biomedical Research Center
Louisiana State University
6400 Perkins Rd.
Baton Rouge, LA 70808

Acknowledgments

The authors would like to thank Stephen C. Woods for his input and guidance on the ideas and organization of this book. We would also like to thank Patricia James for administrative help. In particular the authors would also like to thank Becky Topmiller for her tireless efforts to keep us organized and fighting through the huge task of pulling all of these chapters together. We might have been able to do this book without her, but we know it would not have been any fun.

Contents

Section I   Evidence for macronutrient selection: basic mechanisms and strategies to achieve regulation

Section II Effects of metabolic processing on energy intake and macronutrient selection

Section III Detection of macronutrients and their metabolites

Section IV   Neural integration of sensory and metabolic information in the control of macronutrient selection

Section I

Evidence for Macronutrient Selection: Basic Mechanisms and Strategies to Achieve Regulation

1 Specific Appetites and Homeostatic Systems

Edward M. Stricker

CONTENTS

1   INTRODUCTION

Animals need a great many different nutrients for optimal growth and development. Although it is vital for good health that animals meet these nutritional needs, there are so many individual nutrients that a specific appetite for each of them would require too much of the brain's capacity for function. Perhaps for that reason, the intakes of amino acids, vitamins, and most minerals are not the subject of conscious, voluntary behavior; instead, these nutrients appear to be consumed serendipitously while animals eat food to obtain caloric nourishment. Similarly, the "consumption" of oxygen, the nutrient needed most often and in largest amounts, is usually controlled by a physiological reflex and does not normally intrude on conscious thought. On the other hand, abundant scientific research has indicated that the sensation of thirst, the specific appetite for water, exists in all vertebrate animals studied to date. The present volume considers whether specific appetites also guide the intake of carbohydrates, fats, and proteins. This introductory chapter provides a brief, focused review of thirst in order to identify general variables and principles that might add perspective when considering the existence of specific appetites for macronutrients.

2   THIRST

Vertebrate animals make two key responses when deprived of drinking water: thirst and secretion of the antidiuretic hormone, vasopressin.[1] Thus, the dehydrated animal not only seeks to obtain and ingest water, but also adaptively conserves body water and thereby minimizes further water loss in urine. Consideration of how animals detect water deficiency reveals the two dimensions in which body fluids are regulated: osmolality and volume.

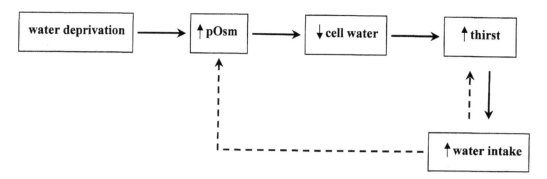

FIGURE 1   A schema for the control of osmoregulatory thirst. Water deprivation (or a NaCl load) increases plasma osmolality (pOsm), which causes water to move from cells by osmosis. The dehydration of cerebral osmoreceptor cells stimulates thirst and water intake, which restores body fluid osmolality to normal. A more rapid source of inhibitory feedback occurs early in an episode of water drinking in association with swallowing and with osmotic dilution. Solid arrows indicate stimulation; dashed arrows indicate inhibition.

2.1   OSMOREGULATION

The osmolality or concentration of body fluids refers to the ratio of osmolytes to water. When the osmolality is elevated, whether by increasing the numerator of the ratio (as with a NaCl load) or by decreasing the denominator (as with water deprivation), water movement by osmosis from body cells into extracellular fluid ensures that the osmolality in all body fluid compartments remains equal. Dehydration of certain neurons causes thirst and vasopressin secretion. Those cells appear to be located in the basal forebrain outside the blood–brain barrier, and they respond readily to small changes in the effective osmolality of circulating fluid. These "osmoreceptor cells" are not uniquely sensitive to dehydration (as photoreceptors in the retina are to light); instead, they have unique neural connections with other brain sites that control thirst and vasopressin secretion.[1,2] After destruction of the osmoreceptor cells by brain lesions, animals no longer increase thirst and secretion of vasopressin in response to increased plasma osmolality.[3,4]

Activation of cerebral osmoreceptor cells by dehydration creates an excitatory stimulus to ingest water that should persist until sufficient intake corrects the deficiency and removes the stimulus and hence the motivation to continue water consumption. Because ingested water does not get fully absorbed from the intestines into the circulation for at least 10 to 20 min, that delay before plasma osmolality is lowered substantially and the stimulus of thirst is removed should cause water to be consumed in excess. However, such overconsumption does not actually occur. Thus, some early mechanism must detect ingestion of water, or its consequences, and inhibit thirst (and secretion of vasopressin) despite continued systemic dehydration. In fact, such inhibitory signals have been shown to arise from the oropharynx, related to swallowing, and from the intestines and/or liver, related to osmotic dilution.[5,6] This simple negative feedback system in the control of osmoregulatory thirst is diagrammed in Figure 1.

2.2   VOLUME REGULATION

Animals deprived of drinking water lose water from plasma as well as from cells, and the loss of plasma volume (hypovolemia) is itself a stimulus of thirst independent of increases in plasma osmolality.[7,8] Deficits in blood volume are detected by stretch receptors (called "baroreceptors") located on the distensible walls of the inferior vena cava and right atrium.[9] When blood volumes are low, these neurons send an afferent signal to the nucleus of the solitary tract in the caudal brain

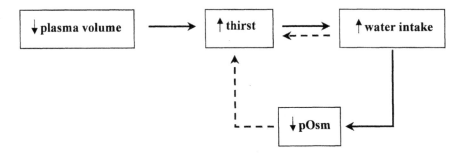

FIGURE 2   A schema for the control of hypovolemic thirst. A loss of plasma volume stimulates thirst and water intake by activation of cardiac baroreceptors and of the renin-angiotensin system. Water intake does not repair plasma volume deficits but causes osmotic dilution of body fluids, which inhibits further drinking despite continued hypovolemia and elevated blood levels of angiotensin. A more rapid source of inhibitory feedback occurs early in an episode of water drinking. Solid arrows indicate stimulation; dashed arrows indicate inhibition. Not shown is the NaCl appetite also stimulated by these signals during hypovolemia; the induced NaCl intake together with water ingestion allows restoration of the plasma volume deficits.

stem, which then communicates to the forebrain where thirst and secretion of vasopressin are stimulated. As might be expected, those responses are abolished when the cardiac receptors are destroyed, or when a small balloon is inflated locally to maintain distension of the venous walls despite hypovolemia.[9-11]

Ingested water, when absorbed, largely moves by osmosis into body cells. That outcome is desirable when thirst results from increased plasma osmolality and cellular dehydration, because it promotes osmoregulation. Increased cellular hydration is not desirable, however, when thirst results instead from loss of plasma volume, because water consumption then produces osmotic dilution without much correction of hypovolemia. Fortunately, osmotic dilution provides a potent stimulus for inhibiting thirst despite continued hypovolemia, and prevents excessive consumption of water.[12] This arrangement is diagrammed in Figure 2.

To restore plasma volume, hypovolemic animals must ingest a NaCl solution that is osmotically equivalent to plasma. Thus, when they first drink water due to thirst, they must then consume salt to restore body fluid osmolality. In fact, rats do drink water and concentrated NaCl solution in response to hypovolemia, and they do so in appropriate amounts to create the isotonic NaCl solution that is ideal for plasma volume restoration[13] (also see below).

2.3   ANGIOTENSIN

A third stimulus of thirst is angiotensin II.[14] Renin, an enzyme secreted into the blood from the kidneys in response to blood loss, initiates a cascade of steps resulting in the formation of angiotensin in blood.[1] Angiotensin is a potent pressor agent and thus is very useful in combating hypovolemia and hypotension. Angiotensin also is a potent stimulus for secretion of aldosterone, the adrenocortical steroid hormone responsible for conservation of Na^+ in urine. Angiotensin also stimulates vasopressin secretion, thirst, and salt appetite through its actions in the subfornical organ, a small structure located in the dorsal portion of the third cerebral ventricle that contains an abundance of angiotensin receptors, and, like cerebral osmoreceptors, lacks a blood–brain barrier; those stimulatory effects of circulating angiotensin are abolished by surgical destruction of the subfornical organ.[15-17] Thus, the detection of plasma volume deficits is mediated both by neural input to the brain stem and by blood-borne endocrine input into the forebrain. The fact that redundant mechanisms are used to detect the loss of plasma volume and to activate appropriate behavioral responses should not be surprising given the critical significance of adequate blood volume to life.

2.4   SUMMARY

Three signals stimulate thirst: neural input to the brain stem from cardiovascular baroreceptors, blood-borne osmolytes that affect forebrain osmoreceptors, and a blood-borne endocrine stimulus detected by angiotensin receptors in the subfornical organ. However, excitatory signals are not the only stimuli that control thirst; anticipatory processes provide early, preabsorptive inhibition of further drinking, and thirst induced by hypovolemia and angiotensin are inhibited by osmotic dilution despite the continued presence of excitatory signals. Complementary to water intake is the secretion of vasopressin which is controlled by the same excitatory and inhibitory signals that affect thirst.

Several principles emerge from this exercise that generalize to nutrients other than water and specific appetites other than thirst: (1) variables critical to organismal function are regulated (e.g., body fluid osmolality, blood volume); (2) nutrient deficits that impair such regulation generate stimuli for specific ingestive behaviors; (3) these signals may be provided by neural input from the periphery to the brain, or by blood-borne substrates or hormones detected in the brain; (4) stimuli occur early in an episode of ingestion and limit further intake; (5) these same signals stimulate complementary physiological responses that conserve the needed nutrient.

3   ANALOGY TO NACL APPETITE

The useful generality of these principles can be evaluated by considering whether they extend to another key nutrient, Na^+, and another well-studied specific appetite, the strong motivation to consume salty-tasting foods and fluids. As mentioned, plasma Na^+ concentration appears to be regulated in conjunction with osmo- and volume-regulation of body fluids. Accordingly, salt appetite arises in response to Na^+ deficits as expected;[18,19] however, it also appears unexpectedly after prolonged deprivation of dietary Na^+,[20] when animals are not Na^+ deficient because increased circulating levels of aldosterone limit urinary Na^+ loss. A brief analysis of the major, relevant issues regarding the control of NaCl appetite under these conditions is as follows.

Numerous studies indicate that multiple signals control NaCl appetite. Angiotensin appears to be one excitatory signal because salt appetite in Na^+-deficient rats can be abolished either by destruction of the subfornical organ, by bilateral nephrectomy, or by systemic administration of an angiotensin receptor blocker.[17,21,22] Aldosterone may provide another excitatory signal;[23] alternatively, aldosterone may stimulate NaCl intake by inhibiting central inhibitory systems.[24] Alterations in the osmolality of body fluids is a third stimulus, with osmotic dilution reducing inhibition of NaCl appetite and hyperosmolality increasing it.[24,25] Gastric distension provides yet another inhibitory signal.[26] These diverse signals are detected, respectively, by receptors for angiotensin and aldosterone in the brain, by osmoreceptors in the forebrain, intestines, and/or liver, and by stretch receptors in the stomach. Note that receptors in the liver and gastrointestinal tract allow early detection and modulation of NaCl intake,[27] and that secretion of aldosterone complements NaCl appetite by promoting urinary Na^+ conservation. The latter effect is clearly analogous to the complementary roles of vasopressin and thirst in response to water deficits; not analogous to the effect of vasopressin, however, is the stimulation of NaCl intake by aldosterone, whether it does so directly or indirectly. In short, general features in the control of salt appetite resemble those in the control of thirst (and, in fact, specific details in the two control systems overlap considerably).[1]

4   ANALOGY TO HUNGER

Unlike thirst, the sensation of hunger does not reflect a need for caloric nutrients because bodily cells usually are continually provided with nourishment adequate for metabolism. Also unlike thirst, hunger apparently does not have to be stimulated by discrete excitatory signals.[28] Thus, given the availability of palatable and nutritious food, animals will eat unless they have just eaten, in which

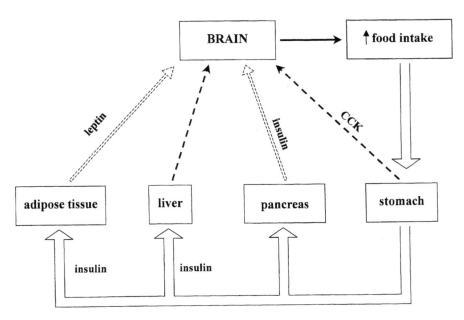

FIGURE 3   A schema for the control of food intake. The brain controls food intake; in turn, it is influenced by the taste of food, which may excite or inhibit intake (not shown). Ingested food is delivered to the stomach, where distension provides a vagal neural signal to the brain stem that inhibits further intake; the intestinal hormone cholecystokinin helps to augment this distension signal. Ingested food passes from the stomach to the intestines and then the liver, where storage of calories in the presence of insulin provides another vagal neural signal of satiety to the brain stem. Blood-borne pancreatic insulin also inhibits food intake by its direct action in the brain. Finally, postabsorptive delivery of ingested food to the white adipose tissue promotes storage of calories, in the presence of insulin, and thereby stimulates secretion of leptin; this protein provides another blood-borne endocrine signal of satiety to the brain. Not shown are other signals that may limit food intake (e.g., dehydration, toxins), and the separate sites for the vagal neural inputs (the nucleus of the solitary tract in the brain stem) and the blood-borne endocrine inputs (the arcuate nucleus of the hypothalamus). Solid arrow indicates stimulation; dashed arrows indicate neural inhibition of food intake; dashed, narrow, open arrows indicate endocrine inhibition of food intake; wide open arrows indicate pre- and postgastric flow of ingested calories.

case the inhibitory effects of satiety generated from the previous meal are still operative. This arrangement presumably evolved because, in the wild, the appearance of food is not predictable, and opportunities to eat are exploited whenever they occur. Compelling support of this perspective are laboratory findings that food deprivation does not cause a critical shortage of caloric nutrients in the circulation (in contrast to the water deficits during water deprivation), and that the spontaneous meals eaten by rats are unpredictable in size, but meal sizes do predict how long rats wait until they eat again.[29]

Several signals appear to stimulate satiety after a meal (see Figure 3). One signal results from gastric distension detected by stretch receptors on the walls of the stomach; the bigger the distension, the bigger the inhibition of eating.[30] This afferent signal projects to the nucleus of the solitary tract and the area postrema in the caudal brain stem via the gastric branch of the vagus nerve; as expected, animals eat unusually large meals when the gastric vagus is cut or its projection sites in the brain are destroyed.[31,32] Experiments in rats have identified an intestinal hormone secreted during the meal, cholecystokinin, that potentiates the gastric vagal signal and produces an exaggerated sensation of satiety related to gastric distension.[33,34] This signal reflects the bulk of gastric contents, not its nutrient contents, because there are no known detectors of calories in the stomach.[35]

The ingestion of calories, detected in the liver and possibly in the intestines, provides another satiety signal that is sent to the caudal brain stem by another branch of the vagus nerve.[36,37] This stimulus appears to develop only in the presence of insulin, a pancreatic hormone secreted when food is consumed.[38,39] Insulin also appears to promote satiety by a direct effect on the brain.[40] Thus, satiety normally is not lost as the stomach empties, because the gastric signal related to distension is replaced by the postgastric signal related to caloric delivery (except in animals with insulin-deficient diabetes mellitus).

Food intake and body weight have been shown to be closely related.[41] Changes in body weight largely reflect changes in the storage of calories in adipose tissue. Those stores are diminished during food deprivation to provide calories needed for cellular function. When food intake resumes, ingested calories are stored readily in the adipose tissue; thus, their effects on the liver are diminished, thereby reducing the duration of this postgastric satiety signal. In addition, food empties from the stomach more rapidly, thereby reducing the duration of gastric satiety signals. In consequence, food intake recurs more frequently and daily intakes are increased until the lost body weight is restored. The opposite effects occur when calorie stores in adipose tissue have been augmented during a period of excessive food intake; both gastric and postgastric satiety signals are prolonged because the rate at which calories are stored in the swollen adipocytes is decreased and food intake is diminished until the gained body weight is lost.[28]

Aside from this indirect influence of body weight on food intake, in recent years a direct inhibitory signal to the brain from adipose tissue has been discovered. This satiety stimulus is a blood-borne protein, leptin, that is secreted from adipose tissue in proportion to adipocyte size; the fatter the animal, the more leptin is secreted and the less inclination the animal has to eat.[42] Leptin receptors have been found in the hypothalamus, and when they are stimulated food intake is reduced.[43,44]

Finally, two other meal-related signals are also known to be potent inhibitors of food intake.[45] One is thirst, as might result from the presence of relatively large amounts of salt in food (see above). The other is nausea, as might result from the inadvertent intake of toxins in food. However, these two variables are not relevant to most meals, and thus they are excluded from the schema presented in Figure 3.

5   CONCLUSIONS

Despite the superficial similarity in the intakes of water and food, the mechanisms that control these ingestive behaviors differ greatly from one another. Hunger is not generated by caloric deficiencies, and food intake is not controlled according to a depletion–repletion scheme. In fact, the recruitment of stored calories during food deprivation prevents such deficiencies, and normally there are no excitatory signals for hunger other than those generated by the palatability of the food. Instead, multiple endogenous signals for inhibiting food intake have been identified. Some inhibitory signals are not specific and disrupt all ingestive behaviors (e.g., nausea, gastric distension), whereas others are specific to food ingestion (e.g., postgastric delivery of calories, insulin, leptin) or work by stimulating a competing drive (e.g., thirst).

From this perspective, the specific appetites for water and NaCl in support of water and sodium homeostasis, respectively, have many features that resemble the general appetite for food in support of caloric homeostasis. With regard to the latter, it is apparent that numerous signals can inhibit food intake, and that removal of those inhibitory stimuli will increase food intake (especially in the presence of palatable food). Some of these signals are neural and some are blood-borne; some are related to ingested and stored calories and some are not. They include preabsorptive stimuli that occur early in an episode of ingestion and influence further intake. The signals also stimulate secretion of insulin, which complements satiety by promoting the storage of calories and thereby prevents the unwanted excretion of glucose in urine. Insulin also promotes satiety, analogous to the effect of aldosterone on NaCl appetite (but not the effect of vasopressin on thirst). In each of these respects, the control of the specific appetites for water and NaCl is similar to the general hunger for caloric nutrients.

It will be interesting to consider whether succeeding chapters in this volume can provide sufficient evidence to indicate that the individual macronutrients are regulated; to support the existence of specific appetites for each of them; to identify neural and/or blood-borne substrates and hormones that excite or inhibit such ingestive behaviors, and the specific receptors that detect such signals; and to identify physiological responses that complement the specific appetites.

ACKNOWLEDGMENTS

The author's work is supported by a grant from the National Institute of Mental Health (MH-25140).

REFERENCES

1. Stricker, E. M. and Verbalis, J. G., Fluid intake and homeostasis, in *Fundamental Neuroscience*, Zigmond, M. J., Bloom, F. E., Landis, S. C., Roberts, J. L., and Squire, L. R., Eds., Academic Press, San Diego, 1999, 1091–1109.
2. Ramsay, D. J. and Thrasher, T. N., Thirst and water balance, in *Handbook of Behavioral Neurobiology 10: Neurobiology of Food and Fluid Intake,* Stricker, E. M., Ed., Plenum Press, New York, 1990, 387–419.
3. Johnson, A. K. and Buggy, J., Periventricular preoptic-hypothalamus is vital for thirst and normal water economy, *Am. J. Physiol.*, 234, R122, 1978.
4. Mangiapane, M. L., Thrasher, T. N., Keil, L. C., Simpson, J. B., and Ganong, W. F., Deficits in drinking and vasopressin secretion after lesions of nucleus medianus, *Neuroendocrinology*, 37, 73, 1983.
5. Thrasher, T. N., Nistal-Herrera, J. F., Keil, L. C., and Ramsay, D. J., Satiety and inhibition of vasopressin secretion after drinking in dehydrated dogs, *Am. J. Physiol.*, 240, E394, 1981.
6. Curtis, K.S. and Stricker, E.M., Enhanced fluid intake by rats after capsaicin treatment, *Am. J. Physiol.*, 272, R704, 1997.
7. Fitzsimons, J. T., Drinking by rats depleted of body fluid without increase in osmotic pressure, *J. Physiol. (London)*, 159, 297, 1961.
8. Stricker, E. M., Some physiological and motivational properties of the hypovolemic stimulus for thirst, *Physiol. Behav.*, 3, 379, 1968.
9. Gauer, O. H. and Henry, J. P., Circulatory basis of fluid volume control, *Physiol. Rev.*, 43, 423, 1963.
10. Zimmerman, M. B., Blaine, E. H., and Stricker, E. M., Water intake in hypovolemic sheep: effects of crushing the left atrial appendage, *Science*, 211, 489, 1981.
11. Kaufman, S., Role of right atrial receptors in the control of drinking in the rat, *J. Physiol. (London)*, 349, 389, 1984.
12. Stricker, E. M., Osmoregulation and volume regulation in rats: inhibition of hypovolemic thirst by water, *Am. J. Physiol.*, 217, 98, 1969.
13. Stricker, E. M. and Jalowiec, J. E., Restoration of intravascular fluid volume following acute hypovolemia in rats, *Am. J. Physiol.*, 218, 191, 1970.
14. Fitzsimons, J. T. and Simons, B. J., The effect on drinking in the rat of intravenous infusion of angiotensin, given alone or in combination with other stimuli of thirst, *J. Physiol. (London)*, 203, 45, 1969.
15. Ferguson, A. V. and Renaud, L. P., Systemic angiotensin acts at subfornical organ to facilitate activity of neurohypophysial neurons, *Am. J. Physiol.*, 251, R712, 1986.
16. Simpson, J. B. and Routtenberg, A., Subfornical organ: site of drinking elicitation by angiotensin II, *Science*, 181, 1172, 1973.
17. Weisinger, R. S., Denton, D. A., DiNicolantonio, R., Hards, D. K., McKinley, M. J., Oldfield, B., and Osborne, P. G., Subfornical organ lesion decreases sodium appetite in the sodium-depleted rat, *Brain Res.*, 526, 23, 1990.
18. Richter, C. P., Increased salt appetite in adrenalectomized rats, *Am. J. Physiol.*, 115, 155, 1936.
19. Jalowiec, J. E., Sodium appetite elicited by furosemide: effects of differential dietary maintenance, *Behav. Biol.*, 10, 313, 1974.
20. Stricker, E. M., Thiels, E., and Verbalis, J. G., Sodium appetite in rats after prolonged dietary sodium deprivation: a sexually dimorphic phenomenon, *Am. J. Physiol.*, 260, R1082, 1991.

21. Stricker, E. M., Vagnucci, A. H., McDonald, R. H., Jr., and Leenen, F. H. H., Renin and aldosterone secretions during hypovolemia in rats: relation to NaCl intake, *Am. J. Physiol.*, 237, R45, 1979.

22. Sakai, R. R., Nicolaidis, S., and Epstein, A. N., Salt appetite is suppressed by interference with angiotensin II and aldosterone, *Am. J. Physiol.*, 251, R762, 1986.

23. Rice, K. K. and Richter, C. P., Increased sodium chloride and water intake of normal rats treated with desoxycorticosterone acetate, *Endocrinology*, 33, 106, 1943.

24. Stricker, E. M. and Verbalis, J. G., Sodium appetite, in *Handbook of Behavioral Neurobiology*, Vol. 10, Stricker, E. M., Ed., Plenum Press, New York, 1990, 387–419.

25. Stricker, E. M. and Verbalis, J. G., Central inhibitory control of sodium appetite in rats: correlation with pituitary oxytocin secretion, *Behav. Neurosci.*, 101, 560, 1987.

26. Stellar, E., Hyman, R., and Samet, S., Gastric factors controlling water- and salt-solution drinking, *J. Comp. Physiol. Psychol.*, 47, 220, 1954.

27. Tordoff, M. G., Schulkin, J., and Friedman, M. I., Hepatic contribution to satiation of salt appetite in rats, *Am. J. Physiol.*, 251, R1095, 1986.

28. Stricker, E. M., Biological bases of hunger and satiety: therapeutic implications, *Nutr. Rev.*, 242, 333, 1984.

29. Woods, S. C. and Stricker, E. M., Central control of food intake, in *Fundamental Neuroscience*, Zigmond, M. J., Bloom, F. E., Landis, S. C., Roberts, J. L., and Squire, L. R., Eds., Academic Press, San Diego, 1999, 1111–1126.

30. Snowdon, C. T., Gastrointestinal sensory and motor control of food intake, *J. Comp. Physiol. Psychol.*, 71, 68, 1970.

31. Gonzalez, M. F. and Deutsch, J. A., Vagotomy abolishes cues of satiety produced by gastric distension, *Science*, 212, 1283, 1981.

32. Stricker, E. M., Curtis, K. S., Peacock, K., and Smith, J. C., Rats with area postrema lesions have lengthy eating and drinking bouts when fed ad libitum: implications for feedback inhibition of ingestive behavior, *Behav. Neurosci.*, 111, 624, 1997.

33. Schwartz, G. J., McHugh, P. R., and Moran, T. H., Integration of vagal afferent responses to gastric loads and cholecystokinin in rats, *Am. J. Physiol.*, 261, R64, 1991.

34. Antin, J., Gibbs, J., and Smith, G. P., Cholecystokinin interacts with pregastric food stimulation to elicit satiety in the rat, *Physiol. Behav.*, 20, 67, 1978.

35. Phillips, R. J. and Powley, T. L., Gastric volume rather than nutrient content inhibits food intake, *Am. J. Physiol.*, 271, R766, 1996.

36. Stricker, E. M., Rowland, N., Saller, C. F., and Friedman, M. I., Homeostasis during hypoglycemia: central control of adrenal secretion and peripheral control of feeding, *Science*, 196, 79, 1977.

37. Friedman, M. I., Hepatic-cerebral interactions in insulin-induced eating and gastric acid secretion, *Brain Res. Bull.*, 5(Suppl. 4), 63, 1980.

38. Booth, D. A., Some characteristics of feeding during streptozotocin-induced diabetes in the rat, *J. Comp. Physiol. Psychol.*, 80, 238, 1972.

39. Lindberg, N. O., Coburn, P. C., and Stricker, E. M., Increased feeding by rats after subdiabetogenic streptozotocin treatment: a role for insulin in satiety, *Behav. Neurosci.*, 98, 138, 1984.

40. Woods, S. C., Insulin and the brain: a mutual dependency, in *Progress in Psychobiology and Physiological Psychology*, Vol. 16, Fluharty, S. J., Morrison, A. R., Sprague, J. M., and Stellar, E., Eds., Academic Press, New York, 1995, 53–81.

41. Mayer, J., Regulation of energy intake and the body weight: the glucostatic theory and the lipostatic hypothesis, *Ann. N.Y. Acad. Sci.*, 63, 15, 1955.

42. Considine, R. V., Sinha, M. K., Heiman, M. L., Kriauciunas, A., Stephens, T. W., Nyce, M. R., Ohannesian, J. P., Maarco, C. C., McKee, L. J., Bauer, T. L., and Caro, J. F., Serum immunoreactive-leptin concentrations in normal-weight and obese humans, *New Engl. J. Med.*, 334, 292, 1996.

43. Tarataglia, L. A., Dembski, M., Weng, X., Deng, N., Culpepper, J., Devos, R., Richards, G. J., Campfield, L. A., Clark, F. T., Deeds, J., Muir, C., Sanker, S., Moriarity, A., Moore, K. J., Smutko, J. S., Mays, G. G., Woolf, E. A., Monroe, C. A., and Tepper, R. I., Identification and expression cloning of a leptin receptor OB-R, *Cell*, 83, 1, 1996.

44. Campfield, L. A., Smith, F. J., Guisez, Y., Devos, R., and Burn, P., Recombinant mouse OB protein: evidence for a peripheral signal linking adiposity to central neural networks, *Science*, 269, 546, 1995.

45. Stricker, E. M. and Verbalis, J. G., Caloric and noncaloric controls of food intake, *Brain Res. Bull.*, 27, 299, 1991.

2 Too Many Choices? A Critical Essay on Macronutrient Selection

Mark I. Friedman

CONTENTS

1 INTRODUCTION

"And now, for the main course?"

"I'd like some fish."

"O.K., what kind would you like? We have bass, cod, flounder, haddock, monk, perch, red snapper, swordfish, trout …"

"Salmon?"

"O.K. we can do that. How would you like that — baked, broiled, or poached?"

"Grilled?"

"Sure. And how would you like it done — rare, medium, or well?"

"Medium-rare."

"That comes with either baked, broiled, French fried, mashed, or scalloped potatoes."

"I'd like wild rice."

"O.K. We've got rice. And you get your choice of cole slaw, corn, lima beans, or string beans."

"I'd like a salad instead."

"We have house dressing, thousand island, French, Italian, and oil and vinegar."

"I'd like blue cheese, if you have it."

"I think we do."

"Could I have it on the side?"

"Sure."

"And could you leave out the cucumbers."

"I'll put it down. 'No cukes.'"

"And could they go easy on the onion."

"Light on onion."

"Also, could I have some bread when you get a chance?"

"Sure thing. I'll bring you a basket with our fresh-baked selection."

"Oh, I'd like olive oil for the bread instead of butter. O.K.?"

"No problem. What would you like to drink with your meal?"

"Could I see the wine list?"

As basic as it is, eating is not so simple. As the above interchange illustrates, there are many choices to be made when it comes to eating. In the least, there are two fundamental decisions to make: what to eat and, once a food is selected, how much to eat. The second decision about quantity is relatively straightforward, although the physiological mechanisms involved are quite elaborate. "How much?" is a question about amount: volume, weight, or number. On the other hand, the decision about quality — what to eat — is far more complicated because there are many determinants of selection and any number of choices. In this essay I question whether the macronutrients are among those choices.

2   DECISIONS, DECISIONS

Decisions about what to eat are influenced directly by decisions about who, where, when, and how to eat because they prescribe the choices that are presented.

When it comes to eating, the decision of "who" may take different forms. "With" whom one eats can affect food selection in several ways. In some species, selection patterns are set via the social transmission of food preferences; for example, offspring will prefer what a parent chooses to eat. For humans, social situations provide an opportunity for the intersection of individual preferences which will affect what is eaten and, at times, the type of social occasion dictates the food choices. In certain instances food is eaten by a parent and regurgitated to feed offspring. In this case, "for" whom one eats affects food selection. "Who" one eats is also a food choice if you happen to be a predator or a cannibal.

The decisions about where and when to eat also directly affect what is eaten. For humans who have a choice, deciding to eat at home, in a restaurant, to pick up "take out" food, have a picnic, or eat in the car, plane, or train will determine the array of foods available for selection. For other animals, the question of where to eat may involve the decision of whether to hoard or bury the food or eat it on the spot. If you are a small rodent that is caching food, hoarding seeds is one thing, whereas hoarding a large turnip is quite another. When food is consumed also determines what food is consumed because different foods tend to be available at different times. Nocturnal species tend to be eaten by predators that eat at night, and seasonal foods are eaten most often when they are in season. And even restaurants have their breakfast, lunch, and dinner menus.

How food is eaten affects which foods are chosen. "How" in this context is primarily a morphological issue concerning the structure and operation of the feeding apparatus and the body parts used in procuring food. An animal with a proboscis is pretty much restricted to a liquid diet. A fast moving predator will have a greater variety of prey than a slower moving predator of equal size unless, perhaps, the slow one has a large cerebral cortex that allows it to plan and set traps. Anatomical limitations can also change because of accidents, disease, and aging.

Choices of which food to eat seems nearly endless; there are many edible things in the world and there are even more inedible things that could be eaten. The human diet is, of course, the prime example of food choice considering the variety of cuisines and the inventiveness of those who prepare food. Fortunately, many decisions about what to eat are made for us through the operation and influence of natural selection, the environment, and experience. In fact, depending on what kind of animal "us" is, practically all the decisions may have been made in advance without any individual input. Take a clam, for example. Clearly, as with all sessile animals, the issue of where to eat is pretty much fixed quite early in life. When to eat can depend on the tides, and what to eat largely depends on what drifts by. But, between a clam and human, the degree and kinds of choice available to different species vary enormously.

Undoubtedly, many factors determine why a particular food is chosen when there is a choice. Accessibility is certainly involved, as is the risk or cost of acquiring it. A food that is relatively expensive, either energetically or monetarily, or is risky to obtain is less likely to be chosen if other foods are available that provide a similar or even lesser benefit. The sensory characteristics of a food and, even more so, the innate or learned hedonic response to those sensory properties may be the most important determinant of choice, at least in the short term.

It has long been thought that food is chosen because of its nutritional content.[1-4] The basic assumption behind nutrient-based selection is that particular food items are chosen because they provide a particular nutritional benefit. Intuitively, it is easy to see why foods would be selected if they relieve a nutritional deficiency, although empirically it might be questionable how easily animals, including humans, can pull this off. It is not so obvious why foods would be selected for their nutritional value in the absence of such a deficiency. It is also not clear how many different nutrients are actively selected; there are certainly many different kinds of nutrients considering the variety of vitamins, minerals, micronutrients, and macronutrients (i.e., proteins, fats, and carbohydrates). So, how many nutritional choices really have to be made?

3   ARE MACRONUTRIENTS CHOICES?

Is macronutrient content a basis for food choice? If you are talking about humans, then the answer, at least for some, is, "Absolutely yes." Many people pay a great deal of attention to the carbohydrate and, especially, fat content of the foods they eat, and this is made easy in the U.S. because of government-mandated nutritional labeling requirements. Indeed, such conscious, nutrient-based food selection is limited only by a person's motivation to track down the detailed nutrient composition of his/her foods and by the size of food labels. If, however, the question is posed for humans who are not conscious of the macronutrient food composition of their food or for other animals, the answer at present is, I believe, "Probably not."

It is important to bear in mind that the issue is not whether animals select macronutrients. Clearly, they do if given the opportunity. Rather, the issue is whether selections are made because they recognize or detect individual macronutrients for what they are. Although the notion is appealing, it seems possible that the capacity to choose foods based on their macronutrient composition could be an illusion resulting from the way experiments are performed and their results are interpreted. When it comes to decisions about what to eat, there may not be as many choices as we think.

3.1   When Is It Ever So Simple?

Laboratory experiments on macronutrient selection usually involve providing the subject (typically laboratory rats or mice) with a choice of separate macronutrients: carbohydrate, fat, and protein. Often, each macronutrient is mixed with vitamins and minerals to assure that requirements for these nutrients are met regardless of the pattern of selection among macronutrients. It is certainly reasonable to offer a choice of individual macronutrients if what one wants to study experimentally is macronutrient selection. But when is the choice ever so simple under natural conditions, under conditions in which animals evolved?

Macronutrients are frequently, if not invariably, found together along with various combinations of other nutrients such as vitamins and minerals. Foods with considerable amounts of fat typically also contain considerable amounts of protein (e.g., meat and seeds). Carbohydrates often are also provided along with protein (e.g., grains and beans). It is not difficult to see how confusing it would be for an animal to try to select foods because of their macronutrient content. Even humans with access to nutrient analyses mistakenly identify macronutrient content, for example, when the desire for high-fat desserts and candies is characterized as a carbohydrate craving.

3.2   WHO NEEDS MACRONUTRIENT SELECTION ANYWAY?

For specialist feeders, the choice of foods was made long ago by natural selection, which limited options to one food or a small number of foods that support nutritional status long enough to assure reproduction. For generalist feeders, and omnivores in particular, this evolutionary influence loses its reach. Although natural selection produced the generalist feeder and omnivore, these feeding strategies, once in place, are by their very nature influenced less by distal (evolutionary) than by proximal factors. Food choices are not that preset, nor so limited, and it is this flexibility that presumably confers an adaptive advantage on the generalist. If we just let the generalist be a generalist (also, the omnivore an omnivore), they will presumably select a variety of foods. This selection could be based on any number of factors described above other than macronutrient composition and, because of the variety, provide a nutritionally adequate diet. Under such circumstances, where is the need to select foods because of their macronutrient content?

3.3   WHAT'S SO SPECIAL ABOUT INDIVIDUAL MACRONUTRIENTS?

Presumably, animals select particular macronutrients because there is something special about them; they relieve a particular deficit, satisfy a specific need, or provide some identifiable benefit. But what makes carbohydrate, fat, and protein so special for animals with a metabolic profile like rats and humans?

All three macronutrients can be used as an energy source, which seems to minimize any advantage of selecting one over another for that purpose. Indeed, recent work suggests that changes in energy production at the level of ATP synthesis provide signals that control food intake in rats.[5] Unless the different macronutrients are utilized with different efficiencies, they should provide an equivalent benefit making them indistinguishable in terms of a feedback signal. In fact, under normal circumstances, the macronutrients are not utilized for energy production to the same degree, even though there is the capacity to do so. Animals might, therefore, be able to discriminate which macronutrient their energy is coming from, which would lead them to select a particular macronutrient at a particular time. However, even under these conditions they would be eating for energy, not for carbohydrate, fat, or protein per se.

It could be argued that, for an omnivore, simply eating for calories would guarantee a sufficient macronutrient intake, especially because these nutrients tend to be found together (at least in pairs). Indeed, the need and appetite for calories appear to predominate over that for at least one macronutrient, namely, protein: rats fed a protein-deficient diet do not increase food intake and, consequently, their intake of protein unless their energy needs are increased by putting them in the cold.[6]

There also is less need to choose among macronutrients because they are interconvertible. Fats can be synthesized from carbohydrate (and to some extent amino acids), and carbohydrate, in the form of glucose, can be synthesized from protein (amino acids) and glycerol. If the need for fat can be fulfilled by carbohydrate and the need for carbohydrate met by protein, there seems less reason for selecting one macronutient over another.

Unlike fat and carbohydrate, essential amino acids and fatty acids cannot be synthesized from other macronutrients; they must be consumed. Because meeting the requirement for these nutrients depends on the behavior of eating, it stands to reason that there should be a mechanism to assure the selection of foods with protein and essential fatty acids.

Protein is extremely important for growth in developing animals, but, at least for rats, it is questionable whether rapidly growing animals in fact select a protein source.[7] The protein requirement for an adult animal is fairly small, suggesting that a capacity for recognizing and selecting protein may not be so crucial for a mature animal. In fact, adult rats that are self-selecting from among the three macronutrients fail to consume enough protein to maintain health if they are simply exposed for a few days to the fat or carbohydrate source prior to having access to all three macronutrients. This suggests that either the ability to select protein is overridden by experience with other macronutrient sources or that the selection of high-protein foods is not based on a need for protein.

The requirement for essential fatty acids is also quite small and endogenous levels can meet that need for a relatively long time. It seems likely, given that macronutrients occur together in foods, that the need for essential fatty acids would be met if there is an adequate supply of food. Indeed, it seems more likely that an animal would die of starvation before its endogenous supply of essential fatty acids is exhausted.

4   BUT WHAT ABOUT THE DATA?

Although there may be reasons for not needing a special ability to select macronutrients, what about all the experiments that demonstrate such a capacity? After all, animals do shift their selection of macronutrients in response to a variety of experimental manipulations, and many times in a manner one might expect given the nature of the treatment. Yes, but …

4.1   ARE THEY SELECTING MACRONUTRIENTS OR FOODS?

In other words, are they selecting the macronutrient or the source of the macronutrient? In order to demonstrate empirically that a macronutrient per se is chosen, it is necessary to vary the source or form of that macronutrient. If it is fat, carbohydrate, or protein that is being selected then it should not matter what form that macronutrient takes as long as they are nutritionally equivalent. Unfortunately, there have been very few experiments on macronutrient selection that have controlled for the source of the macronutrients in this way. Indeed, when this is done the pattern of selection seems to follow form and not function; that is, it varies with the source, not the nutrient.[3,8,9] It would seem that until the source of macronutrients is varied in those situations in which macronutrient selection is thought to operate, the capacity for such selection should be taken with a grain of salt.

4.2   ARE THEY SELECTING A MACRONUTRIENT OR A MEANS TO AN END?

The notion of macronutrient selection can be considered akin to that of specific appetite.[1] From this perspective, a particular macronutrient is selected because the animal has innate, untrained knowledge of which macronutrient it needs and the means to recognize it. Presumably, the specific macronutrient is recognized for what it is by virtue of its sensory properties (i.e., taste, smell, etc.). Such a view of macronutrient selection is on dangerous ground considering that salt appetite, the *sine qua non* of specific appetites, may be less specific than previously thought both in terms of the need and what can satisfy it.[10,11] To demonstrate a specific appetite for a particular macronutrient, one must show, at the least, that the choice is made immediately without time or opportunity for postingestive feedback and without previous experience. If the chosen macronutrient is selected because it previously relieved a deficit or need, then selection is a learned behavior. In this case, the sensory properties of the macronutrient serve as conditioned stimuli, not as specific identifiers of the nutrient. Does this alone make the phenomenon of macronutrient selection uninteresting? Absolutely not. But what it does mean is that the symbol of a physiological consequence is being selected, not the macronutrient in the food cup. There is no need for the ability to recognize macronutrients per se in order to do this.

4.3   ARE THEY SELECTING A SPECIFIC MACRONUTRIENT OR WHAT THEY ALREADY LIKE?

Animals faced with a choice of foods have a history of food experiences. Even newborns, through flavors transmitted in amniotic fluid[12] or milk,[13] have a food-related "past," even if it's not exactly their own. Which foods are selected can be shaped by such prior experience,[14] and this goes for macronutrient sources as well. Under typical laboratory conditions of macronutrient selection, rats' selection patterns are dramatically affected by whether they had eaten one of the macronutrients in advance of the choice situation.[15] Familiar macronutrient sources are preferred to new ones. Which macronutrient source is selected in response to an experimental treatment can also be influenced by existing preferences; in a macronutrient choice situation, treatments that stimulate intake tend to increase the intake of the preferred macronutrient.[16,17] It seems that regardless of what determines the initial pattern of selection among macronutrient sources, subsequent selections are made within the context of existing preferences. A particular macronutrient source may be chosen, therefore, not because of the macronutrient it contains, but because it is familiar or already preferred.

4.4   IS IT SELECTION OR REJECTION?

There are limits on how much food is eaten, set usually by the need for calories. Because macro-nutrients provide these calories, changes in the intake of one macronutrient source will be offset by compensatory changes in the consumption of another in order to maintain energy intake. A change in the pattern of macronutrient selection therefore could be as much a reflection of a decrease in the intake of one macronutrient source as it is an increase of another. In other words, a macronutrient can be selected because another one is rejected, that is, selected by default. It is not obvious how to determine whether selection or rejection drives the choice of a macronutrient source; we could just as easily talk about macronutrient "rejection" as "selection." From either a theoretical or mechanistic point of view this is an important distinction. It is important even if the difference is a semantic one, because it would shape the way we think about the problem and approach it experimentally.

4.5   IS THERE A RESULT THAT DOESN'T FIT?

Validation of the concept of macronutrient selection often derives from experiments in which an experimental manipulation alters the selection pattern in a seemingly appropriate manner. The experimental induction of diabetes mellitus, for example, shifts a rat's selection pattern away from one in which carbohydrate is most preferred.[18] This seems logical because diabetic rats cannot readily utilize carbohydrate; why eat what you cannot use? However, it would also make sense if diabetic rats increased their intake of carbohydrate because one could argue that they eat more to compensate for the low utilization rate. It is not clear what the criteria for a "wrong" choice among macronutrients are; if one assumes that rats or other animals practice macronutrient selection, then a change in either direction confirms the assumption. If the shift in pattern does not make sense, then it can only be that we do not understand the mechanism underlying macronutrient selection. There appears to be no result that would refute the hypothesis that animals monitor their intake of specific macronutrients (see Galef[4]).

5 IS RESARCH ON MACRONUTRIENT SELECTION WORTHWHILE?

I have argued here that in the real world there is little need to select foods for their macronutrient content. Also, I have made a case that the apparent capacity to select macronutrients for their nutritional benefit is a product of the laboratory methods used to study it and the largely untested assumptions underlying the interpretation of the results of those experiments. In essence, I have questioned the validity of the entire research enterprise regarding macronutrient selection.

Am I also arguing that research on macronutrient selection should stop? No. At least, not yet. Before giving it up, the effort needs to be reevaluated not just on paper as done here, but in the laboratory. The model for analysis of macronutrient selection developed by Simpson and Raubenheimer[19] appears to provide a powerful tool to evaluate selection behavior and compare patterns across different conditions. However, more experiments that address the problems raised above concerning whether it is macronutrients or macronutrient sources that are selected in various situations are still sorely needed, as are studies dealing with the role of previous experience and existing preferences. It would be quite valuable to have experiments that address some of the untested assumptions that underlie interpretation of selection data. Attempts to demonstrate the capacity for macronutrient selection under ecologically relevant conditions would be extremely worthwhile.

Conducting research is like eating in the sense that it involves many decisions, not the least of which is choosing what problem to solve. The critical effort I suggest above should tell us not only whether macronutrient selection is a real phenomenon, but also whether it is a viable choice for a research problem.

ACKNOWLEDGMENTS

The author thanks Paul Breslin, Marcia Pelchat, Karen Teff, and Michael Tordoff for their helpful comments and discussions during preparation of this chapter. This work was supported in part by National Institutes of Health grants DK-53109 and DK-36339.

REFERENCES

1. Lát, J., Self-selection of dietary components, in *Handbook of Physiology,* Section 6: Alimentary Canal, 1, Control of Food and Water Intake, Code, C.F., Ed., American Physiological Society, Washington, D.C., 1967, 27.
2. Overmann, S. R., Dietary self-selection by animals, *Psychol. Bull.,* 83, 218, 1976.
3. Kanarek, R. B., Determinants of dietary self-selection in experimental animals, *Am. J. Clin. Nutr.,* 42, 940, 1985.
4. Galef, B. G., Jr., A contrarian view of the wisdom of the body as it relates to dietary self-selection, *Psychol. Rev.,* 98, 218, 1991.
5. Friedman, M. I., An energy sensor for control of energy intake, *Proc. Nutr. Soc.,* 56, 41, 1997.
6. Andik, I., Donhoffer, Sz., Farkas, M., and Schmidt, P., Ambient temperature and survival on a protein-deficient diet, *Brit. J. Nutr.,* 17, 257, 1963.
7. Galef, B. G., Jr., this volume.
8. Matsuno, A. Y. and Thibault, L., Effect of sucrose and fructose macronutrient diets on feeding behavior of rats, *Physiol. Behav.,* 58, 1277, 1995.
9. Glass, M. J., Cleary, J. P., Billington, C. J., and Levine, A. S., Role of carbohydrate type on diet selection in neuropeptide Y-stimulated rats, *Am. J. Physiol.,* 273, R2040, 1997.
10. Tordoff, M. G., The importance of calcium in the control of salt intake, *Neurosci. Biobehav. Rev.,* 20, 89, 1996.

11. Tordoff, M. G., Polyethylene glycol-induced calcium appetite, *Am. J. Physiol.*, 273, R587, 1997.
12. Mennella, J. A., Johnson, A., and Beauchamp, G. K., Garlic ingestion by pregnant women alters the odor of amniotic fluid, *Chem. Senses*, 20, 207, 1995.
13. Mennella, J. A. and Beauchamp, G. K., Maternal diet alters the sensory qualities of human milk and the nursling's behavior, *Pediatrics*, 88, 737, 1991.
14. Mennella, J. A. and Beauchamp, G. K., Early flavor experiences: research update, *Nutr. Rev.*, 56, 205, 1998.
15. Reed, D. R., Friedman, M. I., and Tordoff, M. G., Experience with a macronutrient source influences subsequent macronutrient selection, *Appetite*, 18, 223, 1992.
16. Gosnell, B. A., Krahn, D. D., and Majchrzak, M. J., The effects of morphine on diet selection are dependent upon baseline diet preferences, *Pharmacol. Biochem. Behav.*, 37, 207, 1990.
17. Smith, B. K., Berthoud, H. R., York, D. A., and Bray, G. A., Differential effects of baseline macronutrient preferences on macronutrient selection after galanin, NPY, and an overnight fast, *Peptides*, 18, 207, 1997.
18. Kanarek, R. B. and Ho, L., Patterns of nutrient selection in rats with streptozotocin-induced diabetes, *Physiol. Behav.*, 32, 639, 1984.
19. Simpson, S. J. and Raubenheimer, D., Geometric analysis of macronutrient selection in the rat, *Appetite*, 28, 201, 1997.

3 Is There a Specific Appetite for Protein?

Bennett G. Galef, Jr.

CONTENTS

1   THE WISDOM OF THE BODY

The title Walter Cannon[1] devised for his classic monograph *The Wisdom of the Body* characterized accurately the extraordinary efficacy of physiological systems in maintaining homeostatic equilibria. As Cannon[1] demonstrated in successive chapters, the body possesses internal processes that detect deviations from optimal values in the blood content of water, salt, sugar, protein, fat, calcium, and oxygen, as well as internal effectors capable of redressing any detected imbalance.

Cannon was aware of the important role of behavioral interactions with the external world in maintaining the internal milieu, and devoted an early chapter of *The Wisdom of the Body* to discussion of hunger and thirst as appetites assuring ingestion of substances needed to maintain homeostasis.

2   BEHAVIORAL HOMEOSTASIS

Current appreciation of the role of behavior in maintenance of internal homeostasis rests, however, not on the work of Cannon, but on that of Curt Richter.[2-4] It was Richter who forcefully extended the concept of homeostasis from internal physiological processes to exteroceptors and to behaviors involved in the identification, acquisition, and ingestion of nutrients needed to maintain or restore the internal milieu.

In a series of classic studies carried out during the 1930s and 1940s, Richter demonstrated that, when challenged either by artificially induced nutrient deficiencies or by homeostatic perturbations resulting from spontaneous changes in physiological state, rats would alter their patterns of food selection so as to redress any disturbance to internal homeostasis that they experienced.

The results of Richter's empirical work led him to conclude that animals (and humans) could select foods to ingest that would remedy any deficiency that they experienced in their internal reserves of each micro- and macronutrient.[2] Thus, by implication, Richter came to champion the view that animals and humans possessed: (1) interoceptors able to identify deficiencies in each of the 30 to 50 nutrients they required for health, (2) exteroceptors able to identify each of these nutrients in the external world, and (3) specific appetites that would cause them to seek out and ingest appropriate substances when interoceptors detected deficiency states.

3   SODIUM APPETITE

Richter's[2] model of "total self-regulatory functions" was, in essence, an extrapolation from his understanding of the regulation of sodium.[3] In 1936, Richter[4] had found that adrenelectomized rats, that die from sodium loss in a few days if fed only water and a standard rodent diet, will survive indefinitely if also given access to concentrated sodium solutions that intact rats find so unpalatable that they refuse to ingest them. Richter suggested that ingestion by adrenalectomized rats of concentrated sodium solutions was the result of innate systems detecting sodium deficiency, identifying sources of sodium in the external world, and motivating ingestion of sodium.

3.1   BEHAVIORAL EVIDENCE

Subsequent experiments have shown that Richter was essentially correct in his conclusion that there are motivational systems in rodents, and in mammals more generally, that enable them to detect sodium depletion and respond to such depletion behaviorally, thus restoring homeostatic equilibrium.[5,6] Further, there is evidence consistent with the view that such adaptive responses of rodents to sodium depletion are unlearned: (1) rats show enhanced ingestion of sodium salts the very first time that they are made sodium deficient,[7] and they do so within seconds of their first contact with sodium-salt solutions;[8,9] (2) sodium-deficient rats avidly and immediately ingest LiCl[10,11] that tastes like NaCl[12] but does not alleviate symptoms of sodium depletion and can have toxic effects; and (3) mice made sodium deficient for the first time begin drinking a sodium solution within 30 s of being given access to it, drink the bulk of sodium solution within 10 to 15 min of starting ingestion, and ingest an amount of sodium closely related to the extent of their sodium loss.[13] In summary, both basic systems for detection and control of ingestion of appropriate amounts of sodium seem to develop independent of prior experience of the beneficial effects of ingesting sodium when sodium deficient (for further review, see Schulkin[14]).

Of course, as is the case with any congenital motivational system, learning can increase the efficiency of appetitive[15,16] responses. For example, in a classic study of sodium appetite, Kriekhaus and Wolf[17] showed that rats that had learned while sodium replete to lever press for access to unpalatable solutions of sodium acetate or sodium phosphate would, when made sodium deficient for the first time, lever press vigorously, even if that behavior no longer resulted in access to sodium. Other rats given the same operant training, but with either potassium or calcium solutions rather than a sodium solution as a reward, made far fewer lever presses when made sodium deficient than did rats that had been trained to lever press for solutions containing sodium.

Taken together, the behavioral data clearly indicate that, when rodents are made sodium deficient, they experience an appetite for sodium salts.

3.2 PHYSIOLOGICAL EVIDENCE

Results of studies of physiological substrates of salt ingestion are consistent with the hypothesis that rodents have an inherent ability to respond adaptively to sodium depletion. For example, sodium-deficient rats exhibit a considerable decrease in excitation of sodium-responsive fibers in the chorda tympani in response to stimulation of the tongue with sodium,[18] and salt deprivation causes a marked reduction in the response of salt-sensitive cells in the nucleus tractus solitarus (NTS) as well as a substantial increase in the activity in response to salt on the tongue of cells in NTS that are normally responsive to sweet tastes.[19]

4 APPETITES FOR OTHER NUTRIENTS?

The responses of many animals to sodium depletion seem, as Richter[4] proposed, to rest on an innate, specific appetite. However, as Rozin and Schulkin[20] pointed out many decades later, it does not seem possible, as Richter[2] had also proposed, for dietary generalists (omnivores) to select a nutritionally balanced diet on the basis of specific appetites for each of the 30 to 50 macro- and micronutrients that they require.

5 DIETARY SELF-SELECTION

Results of Richter's studies of diet selection by rats given access to a cafeteria of 17 purified dietary components convinced Richter, and much of the academic community,[21] that animals had "a special appetite not only for salt and sugar, but also for protein, carbohydrate, sodium, calcium, phosphorus, potassium and the vitamins" (Richter et al.,[22] p. 744). The relatively uncritical acceptance of Richter's hypothesis concerning diet selection surely reflected, at least in part, acceptance of his assertion that laboratory demonstrations of an ability of animals to compose a nutritious diet by selecting from a cafeteria of foods only confirmed what could be deduced from the observation that animals survive in natural circumstances. "The survival of animals and humans in the wild state in which the diet had to be selected from a variety of beneficial, useless and even harmful substances is proof [emphasis added] of this ability…to make dietary selections which are conducive to normal growth and development" (Richter et al.,[22] p. 734).

6 COMMON SENSE AND EVOLUTIONARY THEORY

Evolutionary arguments contributed to acceptance of the hypothesis that animals have specific appetites for each nutrient that they require. Long before Richter began his empirical work, it had been argued that ecological pressures must have produced animals with "instinctive dietary habits" (Osborne and Mendel,[23] p. 20), with a capacity for "instinctive discernment inherited from a long line of naturally selected ancestors" shaped to select appropriate foods with "unerring precision" (Crichton-Browne,[24] p. 22). Consequently, although a few saw the importance of critically examining results of studies of dietary self-selection in animals (e.g., References 25 and 26), there was little perceived need to look closely at laboratory evidence consistent with the view that animals could self-select optimal diets. Such laboratory data demonstrated only what both evolutionary theory and naturalistic observation dictated must be true.

When rats in laboratory situations failed to self-select an adequate diet, as they often did,[21,26] such failures were interpreted[2] as the result of procedural artifacts of one kind or another: use of domesticated animals as subjects (though findings with domesticated animals were unobjectionable when those findings confirmed expectations), age-related exhaustion of an ability to self-select (though age of subjects was uncorrelated with success at self-selection),[26,27] or use of complex natural foods rather than purified food components in the cafeteria of foods offered to subjects (though this put advocates of Richter's position in the logically awkward stance of having to argue that cafeteria-feeding studies using purified nutrients provided better laboratory analogues of foraging in natural circumstances than did studies using natural foods).

7   CAFETERIA-FEEDING EXPERIMENTS AND PROTEIN INTAKE

In the cafeteria-feeding experiments that Richter and others conducted, protein intake was the major determinant of growth rate and survival. As Scott[28] pointed out in 1946 (p. 403), rats in laboratory cafeteria-feeding situations "either do or do not like casein; if they like it, they eat an average of 3 grams a day and grow well; if they do not, they eat less than 0.1 gram per day, lose weight, and die within a short period."

In the classic experiments of Richter et al.,[22] rats were provided with two sources of protein and a relatively unpalatable carbohydrate.[7] Thus, the particular foods offered for choice by Richter et al.[22] may have led to relatively high protein intake by rats that chose foods on the basis of palatability alone. Perhaps as a result of this fortuitous choice of items by the experimenter, rats which were offered Richter's cafeteria of foods[28-30] survived and grew, though, as discussed in the penultimate section of this chapter, rats did not do so well in Richter's cafeteria-feeding situation as Richter and others thought that they did.

The relevant question, of course, is not whether an experimenter can choose an array of foods to offer in a cafeteria situation that ensures that subjects' flavor preferences cause them to ingest. To the contrary, the relevant question is whether, in general, animals will detect and select protein for ingestion on the basis of need when such selection is not assured by the relative palatabilities of available foods.

It is, in fact, almost embarrassingly easy to create situations in the laboratory in which rats are not able to select an adequate diet from among alternative foods. As the number of protein-poor foods offered together with a protein-rich food increases beyond one, the probability of adaptive food choice falls off rapidly.[31] As the time between ingestion of a nutritious food and the onset of beneficial effects of its ingestion increases,[32-34] and as the relative palatability of protein-rich foods decreases,[29,35] the probability of selecting enough protein-rich food to eat falls dramatically.

Consider a recent example. Beck and Galef[36] offered weaning and adolescent rats housed alone in large (1 × 1 m) cages a choice among four foods, three of which were low in protein (4.4% protein by weight, considerably less than the 12.5% protein generally considered adequate for young rats) and one that, although protein-rich (17.5% protein by weight), was relatively unpalatable. Subjects were given access to bowls containing each of the four diets for 24 h/day and an experimenter simply weighed each rat and each food bowl daily. Over a 7-day period, 14 of 18 subjects failed to eat substantial amounts of the protein-rich diet and failed to gain weight (as they should have at their ages), and many were seriously ill at the end of the 7 days.

The contrast with the behavior of subjects suffering sodium deprivation and offered salt solutions could not be more marked. Not only did protein-deprived subjects not immediately recognize and ingest appropriate amounts of protein, they did not learn over a period of a week to select the protein-rich food from among four alternatives by sampling among the four foods available and evaluating the postingestive consequences of their ingestion.

It is relevant to theoretical issues discussed below to note that 21 of 28 other weanling rats that shared their enclosures with older rats that the authors had trained to eat the protein-rich food thrived in Beck and Galef's[36] experimental situation. The behavioral mechanisms supporting this transfer of diet preference from knowledgeable rats to their naive companions are well understood[36] and involve a tendency, especially on the part of protein-deprived rats,[37] to show an enhanced intake of any food that they have smelled on the breath of a conspecific (for reviews, see Galef[38] and Galef and Beck[39]).

8 THEORETICAL IMPLICATIONS OF FAILURES TO SELF-SELECT

The failure of rats to gain weight in rather simple cafeteria-feeding situations was demonstrated numerous times in the literature[26,39] but largely ignored in discussions of dietary self-selection, taken together with the success of naive rats living together with knowledgeable conspecifics, seriously undermines Richter et al.'s[22] assertion, quoted above, that survival of animals in nature, where diets must be selected from among alternatives of varying nutritive value, proves that they can select diets conducive to normal growth and development.

When an area is colonized by members of some generalist species, the vast majority of immigrants may well fail to find foods that meet their nutritional needs and, consequently, may perish from malnutrition. However, should one individual, for whatever reason, stumble upon or learn to select an adequate diet, it will serve as a potential model for those that come after, greatly facilitating their search for a nutritious diet. As the results of experiments by Beck and Galef[36] described above indicate, presence of such models could make inhabitable otherwise uninhabitable areas.[39] Consequently, survival of animals in nature does not provide compelling evidence that individual animals can identify needed nutrients in natural environments.

Equally important, members of any species, even a species as far ranging as *Rattus norvegicus*, are not present everywhere within their species' range. Rats can survive only in those areas that provide, in accessible form, all resources necessary for life. If some portion of the environment contains a necessary nutrient only in a form that rats cannot innately recognize or learn to eat, that portion of the environment will be devoid of rats. Consequently, existence of omnivorous species in natural environments tells us little about the range of environments in which individual species' members can select a nutritionally adequate diet.

The message here is that results of studies of dietary self-selection must be examined far more critically than has been customary because failure to find success in diet selection in the laboratory is not, as Richter believed, indicative of a failure to capture in the laboratory the important features of a naturally occurring event.

9 LEARNING OR INSTINCT IN PROTEIN SELECTION?

Even in those cases where animals do come to select adequate amounts of protein from a cafeteria of foods, they appear to do so on the basis of their ability to learn to associate ingestion of substances with their postingestive consequences rather than on the basis of an innate, specific appetite for protein. The evidence that animals are able to associate a food flavor with its postingestional consequences, either positive[40] or negative,[41] is convincing, as is evidence that animals show potentially adaptive changes in their unconditioned responses to novel foods when they become ill.[42,43] Consequently, association learning can play an important role in development of adaptive patterns of diet selection in those laboratory situations that are neither so benign that innate palatability preferences lead directly to selection of nutritionally adequate diets, nor so demanding that only the presence of a knowledgeable conspecific can direct a newcomer to ingest foods containing needed nutrients. Harris et al.[31] provided deficient subjects that were unable to learn to identify the source of a necessary nutrient (in the Harris et al. experiment, vitamin B) in a cafeteria with an opportunity to learn about the positive consequences for health of ingesting the food in the cafeteria rich in vitamin B. The result was a marked improvement in subjects' subsequent performance in the cafeteria situation in which they had previously failed to consume a nutritionally balanced diet.

Learning, rather than instinct, almost surely provides the basis for much of whatever success animals have in cafeteria-feeding situations. Indeed, evidence of an important role of learning in successful diet selection was present even in Richter's earliest data on cafeteria feeding.[22]

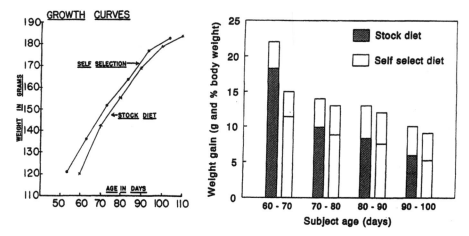

FIGURE 1 Left panel: average growth curves for 8 rats self-selecting a diet and 19 rats eating stock diet (Richter, 1942–1943, Figure 9). (From Richter, C. P., Holt, L., and Barelare, B., *Am. J. Physiol.*, 122, 734, 1938. With permission.) Right panel: mean change in body weight (total histograms) and in percent of body weight at 60 days of age (lower portion of each histogram) of rats either fed stock diet or allowed to self-select a diet over 10-day periods. Data calculated directly from the graph in the left panel.

10 RICHTER'S CLASSIC STUDY OF "TOTAL SELF-REGULATORY FUNCTIONS"

The data reported in Richter's classic papers on cafeteria feeding[2,22] seem to show that animals feeding from a cafeteria of purified ingredients select a highly nutritious diet without delay (left panel of Figure 1). The data, therefore, seem to provide support for the hypothesis that there exist specific appetites for each micro- and macronutrient present in the cafeteria and necessary for growth. However, if Richter had graphed his data differently, he might have reached a conclusion different from that which he did reach.

The left panel of Figure 1 reproduces Figure 9 from Richter's (1942–1943) paper on total self-regulatory functions.[2] It shows the average growth curves of eight rats given the opportunity to self-select a diet from a cafeteria of 17 purified ingredients, and of 19 rats maintained on the McCullom diet, a standard laboratory-rodent diet of the 1930s. The way the data are presented in the figure leads naturally to inference that rats can grow at least as fast when self-selecting a diet as when eating a stock diet, and can do so almost immediately after they are first placed in a situation where they are required to compose a diet for themselves from a multitude of ingredients. Indeed, Richter noted[2] that the cafeteria-fed rats grew at the same rate as those fed stock diet while consuming fewer calories.

The data plotted in Richter's figure are unusual in two respects: first, the ages at which rats in the two groups are weighed are offset by 5 days. This unusual procedure for data collection is important because subjects were growing rapidly throughout the experiment and subjects in the "self-selection" group that were weighed when 5 days older than subjects in the "stock-diet" group therefore seemed to be growing faster relative to subjects in the stock-diet group than they actually were. Second, members of the stock-diet group began the experiment at a considerably lower mean weight than did members of the self-selection group.

10.1 REANALYSIS OF THE DATA OF RICHTER ET AL. (1938)

Unusual features of the data in Richter's presentation can be corrected by examining the average weight gains of subjects in the two groups at comparable ages rather than their average absolute weights at different ages. As the author did not have access to Richter's raw data, he first enlarged Richter's Figure 9[2] by 250%. He then used drafting instruments to find the points of intersection of the line segments indicating the mean weights of subjects in the self-selection group on the days when they were the same ages as were subjects in the stock-diet group when they were weighed. The author then calculated the increase in body weight of subjects in the two groups during successive 10-day periods. The results are plotted in the right panel of Figure 1.

There are two important features of the replotted data. First, from 60 to 100 days of age, subjects in the self-selection group gained less weight (approximately 49 g) than did subjects in the stock-diet group (approximately 59 g). As also can be seen in Figure 1, this difference between groups in weight gain becomes more pronounced if one corrects for the roughly 10% difference in body weight of subjects assigned to the two groups when they were 60 days old. Second, and more important, the performance of subjects in the self-selection group, relative to those in the stock-diet group, was particularly poor early in the experiment and improved as the experiment progressed. Early in the experiment, when subjects were from 60 to 70 days of age, subjects assigned to the self-selection group gained only 68% as much weight as did subjects assigned to the stock-diet group. Late in the experiment, when subjects were 90 to 100 days old, the corresponding figure was 88%. Such change in adequacy of self-selection of nutrients would be expected on the hypothesis that experience with the cafeteria of foods improved the ability of subjects in the self-selection group to compose a nutritious diet. Improvement in performance over days would not be predicted on the hypothesis that subjects had specific appetites for each necessary nutrient.

In summary, although rats feeding from the particular cafeteria of foods used in Richter's laboratory did unusually well,[26] even when feeding from that cafeteria the behavioral "wisdom of the body" was not so impressive as it was long believed to have been.

11   A CAVEAT

There is one important problem with comparisons between existing evidence of sodium appetite and existing evidence of protein appetite. The experimental methods used to study behavioral homeostasis in response to sodium and to protein deficiency have been quite different. Consequently, it is possible that the different pictures of control of sodium intake and protein intake that emerge from consideration of the literature may reflect, at least in part, methodological differences rather than differences in behavioral response to a lack of sodium or protein.

Students of sodium appetite first induce a deficit in their subjects, then expose their subjects to a cafeteria containing a source of sodium and one or more other nutrients. Subjects are not made sodium deficient in the choice situation and, when they are faced with a cafeteria of foods, they have only one problem to solve to return their internal milieu to its normal state; i.e., they have to identify sodium and ingest appropriate amounts of it.

Subjects in studies of protein appetite are usually placed, while still in homeostatic equilibrium, in a situation where they have a source of protein, a source of carbohydrate, a source of fat, and one or more sources of vitamins and minerals. Thus, subjects in studies of protein appetite usually develop a protein deficit while trying to solve a more complex ingestive problem than that presented to the typical subject in a study of sodium appetite.

It is possible that rats first made protein deficient, then given access to a source of protein and one or more protein-free foods, might behave like sodium-deficient rats given access to sodium in a choice situation. For example, Deutsch and co-workers[44] found that within 1 min of first being given a choice between a novel protein-containing diet and a novel carbohydrate, protein-deprived rats ate more of some proteins (soybean, gluten, zein, yeast, etc.), but not of others (casein or lactalbumin), than did protein-replete control subjects (also Heinrich et al.[45]). As Deutsch et al.[44] discuss at some length, the partial success of protein-deprived subjects in selecting proteins is both suggestive and problematic if one wishes to argue that rats have an innate ability to recognize protein.

It would be particularly interesting to repeat, using protein rather than sodium as a reward, the classic experiment on sodium appetite by Kriekhaus and Wolf[17] described above in the paragraphs concerning behavioral evidence of sodium appetite. The point of the repetition would be to determine whether rats made protein deficient for the first time have an unlearned motivation to secure protein similar to the innate sodium appetite they exhibit in response to first experience of a sodium deficit.

Given the difficulty that rats have in learning to ingest enough protein when placed in a cafeteria-feeding situation and the ease with which sodium-deficient rats find sodium in a cafeteria, it seems likely that the performance of protein-deficient rats in the Kriekhaus and Wolf[17] experiment would differ markedly from that of sodium-deficient rats facing the same circumstances.[46] However, the necessary experiment has not, to the best of this author's knowledge, been carried out.

12   CONCLUSION

The physiological processes internal to vertebrates have an impressive ability to maintain the internal milieu within limits compatible with life. In the case of some needed substances, such as sodium and water, the unlearned behavioral responses of the animal to imbalances in internal state are equally impressive. There is strong evidence of inherent mechanisms that act to motivate animals deficient in sodium or water[47] to ingest avidly the substance they need once it is encountered. Behavioral responses elicited by depletion of protein appear far less impressive. Naive animals seem to require extended periods of trial-and-error learning or interaction with knowledgeable conspecifics in order to identify protein-rich foods, and can do so only in a fairly restricted range of environments.[21,39] Even when given weeks in which to learn to select adequate amounts of protein-rich food from among an array of foods, rats fail as often as they succeed.[21,26] There is no compelling evidence of an innate-specific appetite for protein, and the absence of such evidence suggests that mechanisms for the neural control of protein selection are unlikely to exist.

ACKNOWLEDGMENTS

I thank the Natural Sciences and Engineering Council of Canada for funds that facilitated work on this manuscript, and Ed Stricker, Jay Schulkin, and Randy Seeley for help at various stages in its preparation.

REFERENCES

1. Cannon, W. B., *The Wisdom of the Body,* Norton, New York, 1939.
2. Richter, C. P., Total self regulatory functions of animals and human beings, *Harvey Lecture Series,* 38, 63, 1942–1943.
3. Richter, C. P., Salt appetite of mammals: its dependence on instinct and metabolism, in *L'Instinct dans le Comportement des Animaux et de l'Homme,* Masson et Cie, Paris, 1956, 577.
4. Richter, C. P., Increased salt appetite in adrenalectomized rats, *Am. J. Physiol.,* 113, 155, 1936.
5. Denton, D. A., *The Hunger for Salt,* Springer-Verlag, New York, 1982.
6. Schulkin, J., *Sodium Hunger,* Cambridge University Press, Cambridge, 1991b.
7. Epstein, A. N., Oropharyngeal factors in feeding and drinking, in *Handbook of Physiology: Alimentary Canal,* Code, C. F., Ed., American Physiological Society, Washington, D.C., 1967, 197.
8. Nachman, M., Taste preferences for sodium salts by adrenalectomized rats, *J. Comp. Physiol. Psychol.,* 55, 1124, 1962.
9. Wolf, G., Innate mechanisms for regulation of sodium intake, in *Olfaction and Taste,* Pfaffman, C., Ed., Rockefeller University Press, New York, 1969.
10. Nachman, M., Taste preferences for lithium chloride by adrenalectomized rats, *Am. J. Physiol.,* 205, 219, 1963.

11. Schulkin, J., Behavior of sodium deficient rats: the search for a salty taste, *J. Comp. Physiol. Psychol.*, 96, 628, 1982.

12. Morrison, G. R., Behavioural response patterns to salt stimuli in the rat, *Can. J. Psychol.*, 21, 141, 1967.

13. Denton, D. A., McBurnie, M., Ong, F., Osborne, P., and Tarjan, E., Sodium deficiency and other physiological influence on voluntary sodium intake of BALB/c mice, *Am. J. Physiol.*, 255, R1025, 1988.

14. Schulkin, J., Innate, learned, and evolutionary factors in the hunger for salt, in *Chemical Senses*, Vol. 4, Friedman, M. I., Tordoff, M. G., and Kare, M. R., Eds., Marcel Dekker, New York, 1991a, 211.

15. Craig, W., Appetites and aversions as constituents of instincts, *Biol. Bull.*, 34, 91, 1918.

16. Tinbergen, N., *The Study of Instinct*, Clarendon Press, Oxford, 1951.

17. Kriekhaus, E. E. and Wolf, G., Acquisition of sodium by rats: interaction of innate mechanisms and latent learning, *J. Comp. Physiol. Psychol.*, 65, 197, 1968.

18. Contreras, R. J., Kosten, T., and Frank, M. E., Activity in salt taste fibers: peripheral mechanism for mediating changes in salt intake, *Chem. Senses*, 8, 275, 1984.

19. Jacobs, K. M., Mark, G. P., and Scott, T. R., Taste responses of the nucleus tractus solitarus of sodium-deprived rats, *J. Physiol.*, 406, 393, 1988.

20. Rozin, P. N. and Schulkin, J., Food selection, in *Handbook of Behavioral Neurobiology*, Vol. 10, Stricker, E. M., Ed., Plenum Press, New York, 1990, 297.

21. Galef, B. G., Jr., A contrarian view of the wisdom of the body as it relates to dietary self-selection, *Psychol. Rev.*, 98, 218, 1991.

22. Richter, C. P., Holt, L., and Barelare, B., Nutritional requirements for normal growth and reproduction in rats studied by the self-selection method, *Am. J. Physiol.*, 122, 734, 1938.

23. Osborne, T. B. and Mendel, L. B., The choice between adequate and inadequate diet, as made by rats, *J. Biol. Chem.*, 35, 19, 1918.

24. Crichton-Browne, J., *Delusions in Diet*, Funk & Wagnalls, London, 1910.

25. Anon., Self-selection of diets, *Nutr. Rev.*, 2, 199, 1944.

26. Lat, J., Self-selection of dietary components. Oropharyngeal factors in feeding and drinking, in *Handbook of Physiology: Alimentary Canal*, Code, C. F., Ed., American Physiological Society, Washington, D.C., 1967, 367.

27. Scott, E. M., Smith, S., and Verney, E., Self-selection of diet. VII. The effect of age and pregnancy on selection, *J. Nutr.*, 35, 281, 1948.

28. Scott, E. M., Self-selection of diet. I. Selection of purified components, *J. Nutr.*, 31, 397, 1946.

29. Kon, S. K., LVIII. The self-selection of food constituents by the rat, *Biochem. J.*, 25, 473, 1931.

30. Pilgrim, F. J. and Patton, R. A., Patterns of self-selection of purified dietary components by the rat, *J. Comp. Psychol.*, 40, 343, 1947.

31. Harris, L. J., Clay, J., Hargreaves, F., and Ward, A., Appetite and choice of diet. The ability of the vitamin B deficient rat to discriminate between diets containing and lacking the vitamin, *Proc. R. Soc. London Ser. B*, 113, 161, 1933.

32. Harriman, A., Provitamin A selection by vitamin A depleted rats, *J. Physiol.*, 86, 45, 1955.

33. Westoby, M., An analysis of diet selection by large generalist herbivores, *Am. Naturalist*, 108, 290, 1974.

34. Young, P. T. and Wittenborn, J., Food preferences of normal and rachitic rat, *J. Comp. Psychol.*, 30, 261, 1940.

35. Scott, E. M. and Quint, E., Self-selection of diet. IV. Appetite for protein, *J. Nutr.*, 32, 293, 1946.

36. Beck, M. and Galef, B. G., Jr., Social influences on the selection of a protein-sufficient diet by Norway rats, *J. Comp. Psychol.*, 103, 132, 1989.

37. Galef, B. G., Jr., Beck, M., and Whiskin, E. E., Protein deficiency magnifies social influence on the food choices of Norway rats, *J. Comp. Psychol.*, 105, 55, 1991.

38. Galef, B. G., Jr., Communication of information concerning distant diets in a social, central-place foraging species: *Rattus norvegicus*, in *Social Learning: Psychological and Biological Perspectives*, Zentall, T. R. and Galef, B. G., Jr., Eds., Erlbaum, Hillsdale, NJ, 1988, 119.

39. Galef, B. G., Jr. and Beck, M., Diet selection and poison avoidance by mammals individually and in social groups, in *Handbook of Behavioral Neurobiology*, Vol. 10, Stricker, E. M., Ed., Plenum Press, New York, 1990, 329.

40. Booth, D. A., Food-conditioned eating preferences and aversions with interoceptive elements: conditioned appetites and satieties, *Ann. N.Y. Acad. Sci.*, 443, 22, 1985.
41. Garcia, J. and Hankins, W. G., On the origin of food aversion paradigms, in *Learning Mechanisms in Food Selection*, Barker, L. M., Best, M. R., and Domjan, M., Eds., Baylor University Press, Waco, TX, 1977, 3.
42. Rozin, P., Specific aversions as a component of specific hungers, *J. Comp. Physiol. Psychol.*, 64, 237, 1967.
43. Rozin, P., Specific aversions and neophobia resulting from vitamin deficiency or poisoning in half-wild and domestic rats, *J. Comp. Physiol. Psychol.*, 66, 82, 1968.
44. Deutsch, J. A., Moore, B. O., and Heinrichs, C., Unlearned specific appetite for protein, *Physiol. Behav.*, 46, 619, 1989.
45. Heinrichs, S. C., Deutsch, J. A., and Moore, B. O., Olfactory self-selection of protein-containing foods, *Physiol. Behav.*, 47, 409, 1990.
46. Reed, D. R., Friedman, M. I., and Tordoff, M. G., Experience with a macronutrient source influences subsequent macronutrient selection, *Appetite*, 18, 223, 1992.
47. Stricker, E. M. and Sterritt, G. M., Osmoregulation in the newly hatched domestic chick, *Physiol. Behav.*, 2, 117, 1967.

4 Geometric Models of Macronutrient Selection

Stephen J. Simpson and David Raubenheimer

CONTENTS

1   INTRODUCTION

Nutritional regulation is a complex problem of balancing the location, selection, ingestion, and utilization of numerous chemical compounds against multiple and changing metabolic requirements. This multidimensional aspect of nutrition has presented considerable challenges to researchers, many of which have been addressed only partially or not at all. The most common approach has been to downplay or ignore the interactions among nutrients in a diet by treating nutrition as a univariate phenomenon. For example, when considering whether animals regulate their intake of a given dietary component, such as protein, it has been common to vary concentrations in the food of the nutrient of interest without taking full account of the commensurate changes which occur to other food components. Elsewhere, notably in behavioural ecology,[1,2] it is assumed that a single food property, its energy content, is biologically preeminent. Such an approach becomes problematic when there are nutrient-specific influences on feeding and food selection with nutrients exerting their effect through what they are, not how much energy they yield. While the unidimensional approach has provided valuable insights into nutritional and ecological processes, it remains to be seen how severely such insights are constrained by the failure to consider the interactions that take place among nutrient groups.

Attempts to address explicitly the multiple nutritional dimensions of foods have been made by Parks[3] and Moon and Spencer[4] using mixture theory (see Toyomitzu et al.[5] and Friggens et al.[6] for applications). Mixture theory provides a powerful means of describing foods and diets, but it does not extend easily to incorporating the amounts of the various foods (or nutrients) ingested, nor the state of the animal.[7,8] Hence, an alternative "geometric" framework, based on a state-space approach, has been developed and applied to the study of nutritional regulation by insects.[8-13] The geometrical approach enables the unification within a single model of a range of important nutritionally relevant factors including food types, food items, nutrient requirements, nutrient deficits, changes in an animal's nutritional state resulting from the ingestion of particular foods, body composition, and animal performance. The approach provides a means of integrating the study of the ingestive and postingestive processes that together produce nutritional regulation, and of placing and interpreting such mechanisms in the context of their evolutionary and developmental history and ecological circumstances.

Our aim in the present chapter is to introduce the geometric models and to show how they can deal with the particular issue of the regulation of macronutrient intake. Although derived from work on insects, we provide reinterpretations of existing data from rats and hens to illustrate the generality of the approach.

2   THE GEOMETRY OF MACRONUTRIENT SELECTION

The detailed substance of the geometric models has been developed in a series of recent papers[8,10-14] to which the reader is referred. In essence, the animal is depicted as existing within a multidimensional space where each axis represents a nutrient. Within this *nutrient space* lie points ("targets", meant nonteleologically) which represent the animal's requirements, either instantaneous or integrated over a given period during its life. The most fundamental of these in the hierarchy of nutritional regulation, termed the *nutrient target*, is the overall nutrient requirement of the animal's tissues. Part of the nutrient target represents the optimal nutrient allocation to growth and is termed the *growth target*. The remaining, nonstructural component of the nutrient target represents the optimal nutrient requirement for supplying energetic needs. Reaching the nutrient target requires food to be ingested, although liberation of reserves may also contribute. Some ingested nutrients will inevitably be lost in converting intake to meet tissue needs, so intake requirements (the *intake target*) will exceed the nutrient target to varying degrees in different nutrient dimensions. An animal that meets its intake target will, by definition, provide nutrients to its tissues at the optimal rate in the optimal concentrations and amounts.

The present nutritional state of the animal is also represented as a point in nutrient space. The relationship between this and the intake target determines which foods need to be ingested to provide the most effective route for moving toward the intake target. Foods, being fixed mixtures of various compounds, are represented as lines radiating from the origin into nutrient space at an angle which reflects the specific ratio of nutrients they contain. We have termed these food lines *rails* to capture the idea that an animal is confined to eat the ratio of nutrients that a given food contains: it can move along the rail by eating more but it cannot get off without differentially utilizing nutrients postingestively. The optimal food joins the point representing the current state of the animal to the intake target in nutrient space. By ingesting this food in sufficient quantity, the animal moves from its present state directly to the intake target, simultaneously meeting its multiple nutrient requirements.

Suboptimal foods will not allow the animal to reach the intake target, necessitating a behavioural compromise in which some required nutrients are overeaten and others undereaten relative to the intake target. The intake target can only be reached in the absence of an optimal food if two or more available suboptimal foods are *nutritionally complementary*, i.e., their "rails" bound a space that encompasses the intake target in nutrient space. In such circumstances the animal may reach its intake target by ingesting a specific combination of the two or more foods. *Diet selection* is thus the task of locating the optimal food and/or mixing complementary foods, in both cases ingesting appropriate amounts of the foods to reach the intake target.

The scheme as described to this juncture is conceptually simple but poses practical challenges when exploring nutrient selection within a nutrient space composed of several dimensions. Hence, a key experimental aim is to describe nutritional regulation in the simplest terms, both by compounding axes and removing unregulated (or indirectly regulated) nutrients from the model.[12] Macronutrients provide an obvious starting point for constructing models. They comprise the largest dry weight proportion of animals' diets and are readily compounded into three axes, one each for protein, lipid, and carbohydrate. It is important when combining axes that the ratio of constituent compounds (e.g., amino acids within protein) is close to optimal. In our research on insects, we have been fortunate in that dietary lipid is nutritionally inconsequential for the herbivorous species we examine, allowing us to focus on two dimensions, carbohydrate and protein.

3   IS INTAKE OF A MACRONUTRIENT REGULATED?

Part of the process of reducing the number of axes included in a description of nutrient selection is to discover which nutrients' intake is regulated. The geometrical framework provides a simple and powerful means of doing this, as can be explained by an hypothetical example where animals are provided with one of four simultaneous pairings of foods and consumption is measured over a given period. Let us say that these foods contain two nutrient groups A and B and the rest of the food, R, is a neutral bulking agent (e.g., cellulose in the case of our insect subjects). Foods a and b both have nutrients A and B present in a 1:5 ratio but differ in content of R such that food a has 10:50% A:% B and b has 5:25. Foods c and d have the reverse combinations, namely, 25:5 and 50:10. Animals are given a vs. c, a vs. d, b vs. c, or b vs. d. We will also say that the intake requirement for A and B over the period of the experiment is 100 mg A and 100 mg B. Figure 1 shows the outcome under four different scenarios when intake of A and B are plotted against each other.

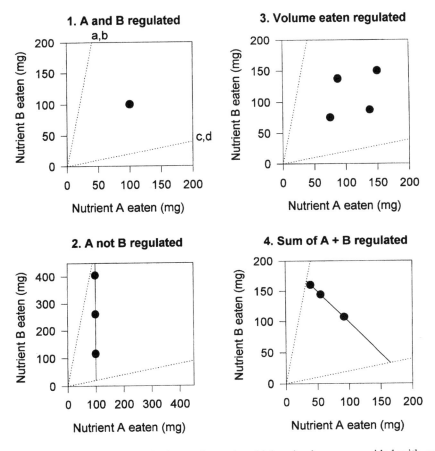

FIGURE 1   Outcomes of an hypothetical experiment in which animals were provided with one of four nutritionally complementary pairings of foods containing nutrients A and B, and a third component, R. The foods contain the following concentrations of nutrients A and B (% A:% B): food a, 10:50; food b, 5:25; food c, 50:10; and food d, 25:5. Pairings were a with c, b with d, a with d, or b with c. The resulting intake of nutrients A and B are plotted under four scenarios (see text).

Scenario 1: both A and B are regulated — Here animals have all converged upon the target point of intake. Achieving this has required that each treatment group ingest a unique combination of their two choice foods as follows (in milligrams dry weight food intake): 167 a vs. 333 c; 333 b vs. 167 d; 167 a vs. 167 d; 333 b vs. 333 c. Such an outcome provides strong evidence for orthogonal regulation of the intake of A and B, and would be strengthened further if additional complementary food choices gave the same result.

Scenario 2: nutrient A is regulated, but not B — In this case, the outcomes of the four treatments will fall along the vertical line shown in Figure 1. All treatments will ingest 100 mg of nutrient A, but the amount eaten of B will depend on other influences controlling intake. In the example shown, we have made the proportional distribution of feeding between the two paired foods a function of their relative concentrations of B. Hence, animals eat proportionally more of the intake required to reach 100 mg of A from the food with most B, as would occur if B were a nonregulated phagostimulant. This results in the following dry weight food intakes: 800 a vs. 80 c; 400 b vs. 160 d; 500 a vs. 100 d; 1000 b vs. 200 c. As a result, the last two treatments end up ingesting the same quantity of nutrient B.

Scenario 3: neither A nor B are regulated — The example shown is where volume ingested is the regulated variable, not nutrient intake. Foods are eaten randomly in the example such that intake is on average distributed equally between the two foods in a treatment, up to the critical total volume, provided by 500 mg total consumption.

Scenario 4: the sum of A and B is regulated — Here A and B are treated by the systems controlling intake as being completely interchangeable, as might occur, for instance, if energy intake is regulated, not whether that energy comes from nutrient A or B. In such cases, A and B could be combined into a single dimension. Outcomes in such a case will lie along a line with a negative slope. In the present example the slope is -1, indicating that A and B are interchangeable on a mass-for-mass basis. A slope of -2 would show that a mass unit of A yields twice as much of the regulated variable than a unit of B (e.g., as in energy derived from lipid vs. carbohydrate). In the example, we have made proportional intake of the choice foods a function of their concentration of B (as in scenario 2), leading to the following dry weight food intakes: 317 a vs. 32 c; 370 b vs. 148 d; 278 a vs. 56 d; 556 b vs. 111 c. The last two treatments end up ingesting the same quantities of both nutrients A and B.

We used an experimental design based on this hypothetical example to investigate macronutrient selection in the locust, *Locusta migratoria*, and found that locusts strongly defended their intake of protein and carbohydrate in a manner closely similar to scenario 1 (Figure 2A).[15] Salt intake was also regulated when locusts could do so independently of macronutrients (Figure 2B).[16] Similarly, when data were recast from selected examples taken from the vertebrate literature, there was evidence for target-like regulation of protein and carbohydrate intake in rats and hens (Figure 2C,D).[8,17-19]

4   MAKING COMPROMISES WHEN THE INTAKE TARGET IS UNREACHABLE

When an animal is restricted to a single, suboptimal food or else has only noncomplementary foods to select between, it is forced to overeat some nutrients and undereat others, reaching some *point of compromise* in the available nutrient space. Such a situation exposes whether and how the mechanisms regulating intake of different nutrient groups interact. The balancing of these ingestive conflicts by control mechanisms is one of the most interesting, and ignored, issues in the study of nutrition.

A simple means of exploring the interactions between regulatory systems for different nutrients is to provide animals with one of an array of suboptimal foods and to measure intake.[8,9,11,12] Collectively, the resulting points of nutrient intake across the array of food rails form a pattern that

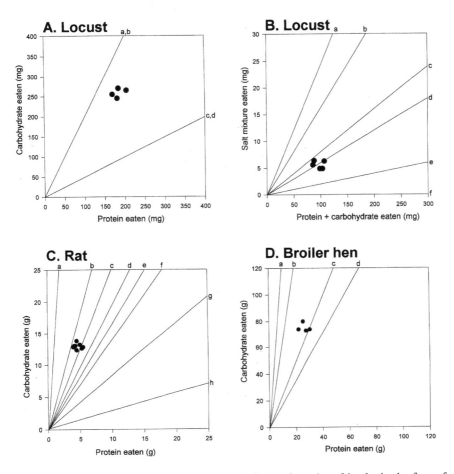

FIGURE 2 Mean values from experiments indicating defense of a point of intake in the face of variation in nutrient composition of two choice foods. (A) Data from locusts fed over 6 days of the fifth stadium on one of four pairings of synthetic foods, containing either 14:28 or 7:14%protein:%digestible carbohydrate (rails a, b) with either 28:14 or 14:7 (c, d). Foods contained identical concentrations of other nutrients, totaling 4%, and the remainder was indigestible cellulose. (Data are from Chambers et al.[15]) (B) Locusts fed synthetic foods containing a fixed concentration of protein and digestible carbohydrate (totalling 30%) and one of seven concentrations of salt mixture over the first 2 days of the fifth stadium. Food pairings were a vs. f, b vs. f, c vs. f, c vs. e, d vs. e. (Data are from Trumper and Simpson.[16]) (C) Results from adult male rats provided with pairs of foods varying in protein and carbohydrate content. Pairings were a vs. e, a vs. g, a vs. h, b vs. c, b vs. d, b vs. f, b vs. g, b vs. h. (Data are from Theall et al.[18]) (D) Defense of protein-carbohydrate intake by broiler hens provided with one of four food pairings, a vs. c, b vs. c, c vs. d, b vs. d. (Data come from Shariatmadari and Forbes[19] with information on dietary carbohydrate and lipid composition from J.M. Forbes, personal communication.) (From Raubenheimer, D. and Simpson, S. J., *Nutr. Res. Rev.*, 10, 151, 1997. With permission.)

describes a *rule of compromise* that quantifies the relationship between the mechanisms regulating intake of the nutrients concerned. Raubenheimer and Simpson[8] and Simpson and Raubenheimer[12] provide detailed discussions of the ways in which intake arrays can be described and interpreted. A key point is that such interpretations rely on there being an estimate of the position of the intake target, which provides an essential referent.

Various rules of compromise have been described to date in insects and vertebrates, a selection of two-dimensional examples of which are shown in Figures 3 to 6. Figure 3 illustrates two examples of what we have termed the *no interaction rule*. In one case,[16] locusts regulated protein and

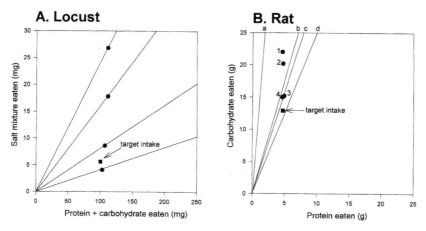

FIGURE 3 Examples of the *no interaction* rule of compromise, where the intake target cannot be reached and regulation of one nutrient is abandoned in the face of competition from another. (A) Locusts provided with foods containing suboptimal salt levels (see Figure 2B). (From Trumper and Simpson.[16]) (B) Data from Theall et al.[18] for adult male rats given food pairings a vs. b (point 1), a vs. c (point 2), b vs. c (point 3) or b vs. d (point 4). The defended target intake for protein vs. carbohydrate is shown as a square (see Figure 2C). (From Raubenheimer, D. and Simpson, S. J., *Nutr. Res. Rev.*, 10, 151, 1997. With permission.)

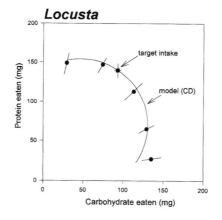

FIGURE 4 An example of the *closest distance rule*, where an animal moves as close as possible in nutrient space (under a given scaling: mass units in this case) to the intake target. Data are from the African migratory locust (*Locusta migratoria*) over the first 3 days of the fifth larval stadium. Filled circles represent mean (±SE) intake of ten insects fed synthetic foods containing the nutrients in a balance of (%protein:%digestible carbohydrate) 7:35, 14:28, 21:21, 28:14, or 35:7 (from left to right in the figure). The solid square represents the selected point of intake by locusts allowed to switch between complementary foods, while the arc shows the prediction for the CD rule. (From Raubenheimer, D. and Simpson, S. J., *Nutr. Res. Rev.*, 10, 151, 1997. With permission.)

carbohydrate intake (combined as a single axis) and let salt intake vary passively when fed single foods containing suboptimal salt levels, even though they regulated both salt and the macronutrients when allowed to switch between complementary foods (see Figure 2B). The other example comes from a study where rats were provided with pairs of foods which both contained less than the defended target ratio of protein to carbohydrate.[18] The rats maintained protein intake at the cost of

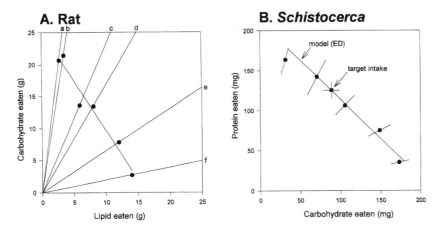

FIGURE 5 Examples of the *equal distance* rule of compromise. (A) Results from Friggens et al.'s[6] study of lactating rats provided with one of six foods (a to f) varying in protein, carbohydrate, and lipid content. When the mass ingested of carbohydrate and lipid are plotted against each other, the points align along a line with slope close to –2 (best fit regression is shown), suggesting that energy content determines the intake relationship between the two macronutrients. (B) Data are from the same experiment as in Figure 4, but for fifth-instar nymphs of the desert locust (*Schistocerca gregaria*). Unlike the grass-specialising *Locusta*, this polyphagous, desert-dwelling species maximized the combined intake of protein and carbohydrate. (Figure 5B from Raubenheimer, D. and Simpson, S. J., *Nutr. Res. Rev.*, 10, 151, 1997. With permission.)

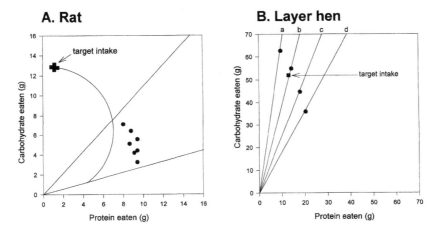

FIGURE 6 Other rules of compromise. (A) Data from Tews et al.[21] for young, protein-loaded male rats which were either allowed to choose between foods which encompassed the target intake (five treatments converge on the same intake point in the cluster of circles at the top left), or else were provided with one of seven treatments where the two foods both contained higher than the target P:C ratio (the pairings fell within the two food rails shown). In these cases, rats weighted overeating protein less strongly than undereating carbohydrate relative to the CD rule (shown by the arc). (B) Results recast from Shariatmadari and Forbes[19] for layer hens given either a single food (rails a to d) or a choice of a vs. d. Diet b allowed chickens to reach their target. When they were unable to reach the target intake they weighted protein regulation more strongly than that for carbohydrate relative to CD (where birds would have moved to the point on their food rail which lay at right angles to the target). (From Raubenheimer, D. and Simpson, S. J., *Nutr. Res. Rev.*, 10, 151, 1997. With permission.)

carbohydrate regulation, indicating that regulatory mechanisms for protein swamped those for carbohydrate under these circumstances.

A second rule of compromise is illustrated in Figure 4. This shows what we have called the *closest distance rule*, where an animal moves to the geometrically closest point on a food rail to the intake target. The animal is minimizing the sum of undereating one nutrient and overeating the other, irrespective of which of the two nutrients happens to be in excess or deficit. The example shown is for the African migratory locust, *Locusta migratoria*, fed foods varying in protein and carbohydrate content. This rule would appear to be common among insect species, even though the position of the intake target for protein and carbohydrate varies markedly with phylogeny and ecology.[11,20]

The third rule is termed the *equal distance rule* and occurs when animals move along their food rail to the point where the sum of the two nutrients is the same as at the intake target. There are two quite different explanations for such an outcome. One is that the two nutrients are interchangeable. Figure 5A plots intake of carbohydrate and lipid against each other in an experiment where lactating rats were provided with one of several foods varying in protein, carbohydrate, and lipid content.[6] In this case the intake points (plotted in grams eaten) fell along a straight line with a slope close to −2, strongly suggesting that the two macronutrients were being used to meet energy requirements (lipid providing twice the energy content of carbohydrate per mass). In contrast, the second example shown in Figure 5B requires a different explanation. Here the desert locust, *Schistocerca gregaria*, demonstrates the equal distance rule for protein and carbohydrate intake. It is known that both of these macronutrients are regulated separately by this species. Also, protein and carbohydrate are not interchangeable to a growing locust. Unlike the closely related African migratory locust which shows the closest distance rule (see Figure 4), the desert locust maximizes its intake of protein and carbohydrate up to a combined limit. The two species have closely similar intake targets for protein and carbohydrate, but the difference in their rule of compromise is explicable in terms of their respective ecologies.[8] The desert locust eats a wide range of plant species and lives in a harsh environment where food is scarce and of variable quality, while the African migratory locust lives in somewhat more equable environments and specializes in eating grasses. The generalist species is better able to capitalize on excess nutrients eaten than is the specialist.

Other rules of compromise also exist, two of which are plotted in Figure 6 for rats and layer hens. Data shown in Figure 3 indicated that rats given foods containing a less than optimal ratio of protein to carbohydrate regulated their protein but not carbohydrate intake. The example in Figure 6A, from a different experiment,[21] suggests that a different strategy is employed when foods are unbalanced in the opposite direction. Rats ingested more protein than expected under the closest distance rule, perhaps indicating deamination and respiration of excess protein to make up for limiting carbohydrate.

5   CHANGES WITH TIME: PHYSIOLOGICAL, DEVELOPMENTAL, AND EVOLUTIONARY

So far, we have discussed intake targets and rules of compromise as static features integrated over a particular period of time. In fact, both move as trajectories in time, both within the life of an individual over physiological and developmental time, and also across generations through evolutionary time.

The defended point of intake in nutrient space changes with an animal's recent nutritional experience. For example, locusts and caterpillars select a protein-rich food following a short experience (only one meal in the case of the locust) of a protein-deplete food, and show a similar preference for carbohydrate-rich food after a 4-h period of carbohydrate deprivation.[15,22-24] Figure 7 shows how previous nutritional experience influences the nature of regulatory responses to protein

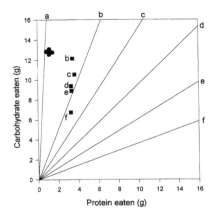

FIGURE 7 Data from Tews et al.[21] where rats were prefed for 7 to 10 days on either a food with a higher (circles) or lower (squares) than optimal ratio of protein to carbohydrate, then given a choice between food a and one of foods b to f. When protein had been in excess in the pretreatment food, rats subsequently regulated intake of both protein and carbohydrate (all five points lie on top of each other), but when protein had been limiting, they defended a higher intake of protein but largely abandoned carbohydrate regulation. (From Simpson, S. J. and Raubenheimer, D., *Appetite*, 28, 201, 1997. With permission of Academic Press, Limited, London.)

and carbohydrate when rats were pretreated on foods containing unbalanced proportions of protein-to-carbohydrate and then allowed to select between nutritionally complementary foods.[21] While rats that were pretreated on a food containing a higher than optimal protein to carbohydrate ratio subsequently defended their intake of both macronutrients, those animals pretreated on a food with low protein and excess carbohydrate defended a higher level of protein intake and abandoned carbohydrate regulation. Responses to macronutrients may change, not only with previous experience of those same nutrients, but also with micronutrient imbalances, as seen, for example, in zinc-deficient rats which ingested higher fat-to-carbohydrate ratios than did control animals.[25]

Activity levels affect the relative amounts of protein and carbohydrate eaten, with adult locusts increasing their intake of carbohydrate but not protein after a period of flight.[8] Environmental conditions, notably temperature, also influence the relative amounts of protein and carbohydrate eaten, particularly in endothermic animals.[26]

In the longer term, targets and rules of compromise move during ontogeny as animals grow, mature, reproduce, and senesce.[27-30] They also move across generations as animals are selected, either naturally or artificially, to exploit different food sources and adopt changed life-history strategies.[11]

6 REGULATORY MECHANISMS

This is not the place to consider in detail the nature of regulatory mechanisms controlling macro-nutrient selection, since other chapters in the present volume do that. We would, however, like briefly to highlight the broad conceptual issues, introduce some of our recent theoretical modeling of the design features of taste systems as they relate to nutritional regulation, and point the reader in the direction of literature which addresses the regulation of macronutrient intake in insects.

Regulating nutrient intake requires two sources of information, the first regarding the composition of the food and the second, the nutritional state of the animal. The responsiveness of an animal to a food of given composition should reflect, through feedbacks, the animal's nutritional history. Feedbacks between state and responsiveness occur on several time scales: evolutionary, developmental, and physiological. An instance of the first of these concerns is the responsiveness of the taste system to stimulation by nutrients. In most taxa, sugars, amino acids, and salt nutrients serve as phagostimulants (see Bernays and Simpson[31] for insect examples). More than this, it is

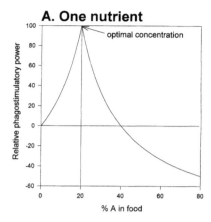
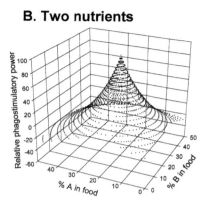

FIGURE 8 (A) Results of a mathematical model predicting from first principles (see Simpson and Raubenheimer[13]) the relationship between the concentration of a nutrient, A, and the phagostimulatory power of a food. The equation describing the relationship is $Px = ((At/(At + |At − Ax|)) − 0.5) \times 200$, where Px is the phagostimulatory power of the food, At and Ax are the proportions of A in the optimal food (20% in the hypothetical example) and food x, respectively. Subtracting 0.5 and multiplying by 200 scales Px so that a food with no A has a phagostimulatory power of zero. Negative values simply indicate that such foods are less stimulatory than one lacking A altogether. (B) Expansion of the model to include two nutrients, A and B. In this case the equation is $Px = (1/(1 + sqrt((Ax/At − 1)^2 + (Bx/Bt − 1)^2)) − 0.5) \times 200$. In the example both At and $Bt = 20\%$. (From Simpson, S. J. and Raubenheimer, D., *Entomol. Exp. Appl.*, 80, 55, 1996. With kind permission from Kluwer Academic Publishers.)

possible to predict mathematically how the phagostimulatory power of a food (the extent to which it stimulates and maintains ingestion) should vary with its nutrient composition[13,32] (Figure 8) such that an animal will (1) mix its intake of two or more complementary foods to provide the optimal concentration of nutrients in its diet, (2) eat predominantly from the optimal food if it is available, or (3) eat most of the food which is closest to being optimal if available foods are suboptimal and noncomplementary in composition. So far, data from insects and vertebrates are consistent with the mathematical model.[13,33]

The parameters of the taste system as defined by the above mathematical model are set in evolutionary time, indicating ancestrally prevailing ecological circumstances and average nutrient requirements. Predictable changes in these requirements with development may also become embodied through evolution in the design of the gustatory (and other) systems, with response properties changing as part of the default developmental programme.[28] However, such mechanisms lack the resolution to respond to shorter-term nutritional perturbations. Epigenetic transference of regulatory responses may occur across generations through maternal effects, as is seen in phase change in locusts,[34] or through *in utero* learning and postnatal social experience, as in some mammals.[35] Activity-dependent developmental changes in the taste system have been reported in locusts,[36] with the number of taste receptors changing within the life of an individual in response to the chemical complexity of the nutritional environment.

On a considerably shorter time scale, nutrient-specific feedbacks for sugars and amino acids have been described in insects. These involve direct feedbacks and learned associations. The former includes direct modulation of peripheral gustatory sensitivity by blood concentrations of amino acids and sugars, these in turn providing an instantaneous measure of an insect's nutritional state (Figure 9).[23,37,38] Such a mechanism provides a remarkably direct means of linking nutritional state to food composition. Nutrient feedbacks would appear to operate more centrally in the integrative pathways controlling feeding behaviour in vertebrates (chapters in the present volume).

Learning of various sorts — including aversion learning, learned specific appetites, and induced neophobic responses — has also been shown to play an important role in regulating food selection

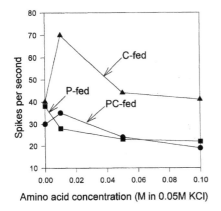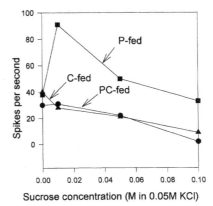

FIGURE 9 The effect of nutritional perturbations on the gustatory responsiveness of locusts to sugars and amino acids. Locusts were prefed for 4 h on one of three synthetic foods. The C food contained digestible carbohydrate but no protein; the P food contained protein but no digestible carbohydrate, while the PC food contained both nutrient groups. Data show mean electrophysiological responses from mouthpart taste hairs to stimulation with a range of amino acid and sugar solutions. Note how responsiveness has been modulated by prefeeding such that receptors respond most vigorously to the nutrients which were absent from the pretreatment food. Such responses have been shown to be mediated directly by levels of amino acids and sugars in the blood, and to be closely associated with changes in dietary selection behaviour. (Data after Simpson et al.,[23] from Simpson, S. J. and Raubenheimer, D., *Entomol. Exp. Appl.*, 80, 55, 1996. With kind permission from Kluwer Academic Publishers.)

in insects.[39-44] An impressive example is the ability of locusts to learn to associate the odour of a food with its protein content and to be attracted by odours previously paired with high-protein food, but only when in a state of protein deficit.[39] Recently, separate learned-specific appetites have been demonstrated for protein and carbohydrate when visual conditioning stimuli were used.[44]

7 THE RELATIONSHIP BETWEEN PRE- AND POSTINGESTIVE REGULATORY SYSTEMS

Whereas animals are constrained to ingest the particular ratio of nutrients present in their selected foods, they are nevertheless able to "jump rails" postingestively by differentially utilizing ingested nutrients. Hence, the growth target may be reached from a range of intake points in nutrient space. Selected examples of regulated growth in the face of different points of intake are shown for locusts, rats, and hens in Figure 10, while details of some of the mechanisms involved in insects may be found in Zanotto et al.[45-47] Discussion of methodological and statistical issues associated with quantifying postingestive regulation may be found in Raubenheimer[48] and Raubenheimer and Simpson.[10,49,50] Finally, the dynamic interactions between pre- and postingestive mechanisms regulating intake and utilization of macronutrients have recently been considered in a class of models based around the marginal value theorem.[14,51]

8 CONCLUSIONS

The geometric models introduced in this chapter aim to address the interactive nature of nutritional regulation: the interactions occurring between nutrients within foods, between regulatory systems controlling intake (selection and consumption) of different nutrients, between pre- and postingestive mechanisms, and between animals and their ecological environments. None of these can be considered in isolation without the risk of missing key insights. Within the framework, intake targets and rules of compromise provide essential referents when it comes to exploring the differences

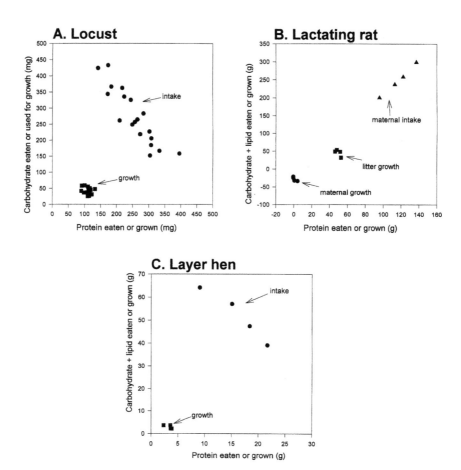

FIGURE 10   The relationship between intake and growth, showing how growth is defended in two nutrient dimensions across a range of intakes. (A) Data from Raubenheimer and Simpson[9] for locusts fed for the entire fifth stadium on 1 of 19 foods varying in protein and digestible carbohydrate content and with only trace lipids. Carbohydrate-derived growth included lipid and structural carbohydrate in cuticle. (B) Data from Friggens et al.[6] for lactating rats. Maternal intake on the four foods spanned a considerably wider range in both nutrient dimensions than did either maternal or litter growth. The maternal growth target was to maintain body protein content and to lose 20 g of body lipid, while litter growth was regulated at circa 50-g increase in both dimensions. (C) Results recast from Shariatmadari and Forbes'[19] study on layer hens. (From Raubenheimer, D. and Simpson, S. J., *Nutr. Res. Rev.*, 10, 151, 1997. With permission.)

between species, genotypes, or developmental stages, or when investigating the effects on behaviour of pharmacological or genetic manipulations. The models helped us to begin to understand the complexity and diversity of insect nutrition. Our hope is that they may prove similarly useful in studying vertebrates.

REFERENCES

1. Stephens, D. W. and Krebs, J. R., *Foraging Theory*, Princeton University Press, Princeton, NJ, 1986.
2. Hughes, R. N., Ed., *Diet Selection. An Interdisciplinary Approach to Foraging Behaviour*, Blackwell Scientific, Oxford, 1993.
3. Parks, J. R., *A Theory of Feeding and Growth of Animals*, Springer-Verlag, Berlin, 1982.
4. Moon, P. and Spencer, D. E., A geometry of nutrition, *J. Nutr.*, 104, 1535, 1974.

5. Toyomitzu, M., Kimura, S., and Tomita, Y., Response-surface analyses of the effects of dietary protein, fat and carbohydrate on feeding and growth pattern in mice from weaning to maturity, *Anim. Prod.*, 56, 251, 1993.

6. Friggens, N. C., Hay, D. E. F., and Oldham, J. D., Interactions between major nutrients in the diet and the lactational performance of rats, *Br. J. Nutr.*, 69, 59, 1993.

7. Emmans, G. C., Diet selection by animals: theory and experimental design, *Proc. Nutr. Soc.*, 50, 59, 1991.

8. Raubenheimer, D. and Simpson, S. J., Integrative models of nutrient balancing: application to insects and vertebrates, *Nutr. Res. Rev.*, 10, 151, 1997.

9. Raubenheimer, D. and Simpson, S. J., The geometry of compensatory feeding in the locust, *Anim. Behav.*, 45, 953, 1993.

10. Raubenheimer, D. and Simpson, S. J., The analysis of nutrient budgets, *Functional Ecol.*, 8, 783, 1994.

11. Simpson, S. J. and Raubenheimer, D., A multi-level analysis of feeding behaviour: the geometry of nutritional decisions, *Philos. Trans. R. Soc. London B,* 342, 381, 1993.

12. Simpson, S. J. and Raubenheimer, D., The geometric analysis of feeding and nutrition: a user's guide, *J. Insect Physiol.*, 41, 545, 1995.

13. Simpson, S. J. and Raubenheimer, D., Feeding behaviour, sensory physiology and nutrient feedback: a unifying model, *Entomol. Exp. Appl.*, 80, 55, 1996.

14. Raubenheimer, D. and Simpson, S. J., Nutrient transfer functions: the site of integration between feeding behaviour and nutritional physiology, *Chemoecology*, 8, 61, 1998.

15. Chambers, P. G., Simpson, S. J., and Raubenheimer, D., Behavioural mechanisms of nutrient balancing in *Locusta migratoria* nymphs, *Anim. Behav.*, 50, 1513, 1995.

16. Trumper, S. and Simpson, S. J., Regulation of salt intake by nymphs of *Locusta migratoria*, *J. Insect Physiol.*, 39, 857, 1993.

17. Simpson, S. J. and Raubenheimer, D., Geometric analysis of macronutrient selection in the rat, *Appetite*, 28, 201, 1997.

18. Theall, C. L., Wurtman, J. J., and Wurtman, R. J., Self-selection and regulation of protein:carbohydrate ratios in foods adult rats eat, *J. Nutr.*, 114, 711, 1984.

19. Shariatmadari, F. and Forbes, J. M., Growth and food intake responses to diets of different protein contents and a choice between diets containing two concentrations of protein in broiler and layer strains of chicken, *Br. Poultry Sci.*, 34, 959, 1993.

20. Simpson, S. J., Raubenheimer, D., and Chambers, P. G., Nutritional homeostasis, in *Regulatory Mechanisms of Insect Feeding*, Chapman, R. F. and de Boer, G., Eds., Chapman and Hall, New York, 1995, 251.

21. Tews, J. K., Repa, J. J., and Harper, A. E., Protein selection by rats adapted to high or moderately low levels of dietary protein, *Physiol. Behav.*, 51, 699, 1991.

22. Simpson, S. J., Simmonds, M. S. J., Blaney, W. M., and Jones, J. P., Compensatory dietary selection occurs in larval *Locusta migratoria* but not *Spodoptera littoralis* after a single deficient meal during *ad libitum* feeding, *Physiol. Entomol.*, 15, 235, 1990.

23. Simpson, S. J., Simmonds, M. S. J., Blaney, W. M., and James, S., Variation in chemosensitivity and the control of dietary selection behaviour in the locust, *Appetite*, 17, 141, 1991.

24. Simmonds, M. S. J., Simpson, S. J., and Blaney, W. M., Dietary selection behaviour in *Spodoptera littoralis*: the effects of conditioning diet and conditioning period on neural responsiveness and selection behaviour, *J. Exp. Biol.*, 162, 73, 1992.

25. Kennedy, K. J., Rains, T. H., and Shay, N. F., Zinc deficiency changes preferred macronutrient intake in subpopulations of Sprague-Dawley outbred rats and reduces hepatic pyruvate kinase gene expression, *J. Nutr.*, 128, 43, 1998.

26. Musten, B., Peace, D., and Anderson, G. H., Food intake regulation in the weanling rat: self-selection of protein and energy, *J. Nutr.*, 104, 563, 1974.

27. Chyb, S. and Simpson, S. J., Dietary selection in adult *Locusta migratoria* L., *Entomol. Exp. Appl.*, 56, 47, 1990.

28. Barton-Browne, L., Ontogenetic changes in feeding behaviour, in *Regulatory Mechanisms in Insect Feeding*, Chapman, R. F. and de Boer, G., Eds., Chapman and Hall, New York, 1995, 307.

29. Leibowitz, S. F., Lucas, D. J., Leibowitz, K. L., and Jhanwar, Y. S., Developmental patterns of macronutrient intake in female and male rats from weaning to maturity, *Physiol. Behav.*, 50, 1167, 1991.

30. Veyrat-Durebex, C. and Alliot, J., Changes in pattern of macronutrient intake during ageing in male and female rats, *Physiol. Behav.*, 62, 1273, 1997.
31. Bernays, E. A. and Simpson, S. J., Control of food intake, *Advances in Insect Physiology*, 16, 59, 1982.
32. Simpson, S. J., Experimental support for a model in which innate taste responses contribute to regulation of salt intake by nymphs of *Locusta migratoria*, *J. Insect Physiol.*, 40, 555, 1994.
33. Chambers, P. G., Raubenheimer, D., and Simpson, S. J., The rejection of nutritionally unbalanced foods by *Locusta migratoria*: the interaction between food nutrients and added flavours, *Physiol. Entomol.*, 22, 199, 1997.
34. McCaffery, A. R., Simpson, S. J., Islam, M. S., and Roessingh, P., A gregarizing factor present in the egg pod foam of the desert locust *Schistocerca gregaria*, *J. Exp. Biol.*, 201, 347, 1998.
35. Provenza, F. D. and Cincotta, R. P., Foraging as a self-organizational learning process: accepting adaptability at the expense of predictability, in *Diet Selection*, Hughes, R. N., Ed., Blackwell Scientific, Oxford, 1993, 78.
36. Rogers, S. and Simpson, S. J., Experience-dependant changes in the number of chemosensitive sensilla on the mouthparts of *Locusta migratoria*, *J. Exp. Biol.*, 200, 2313, 1997.
37. Simpson, S. J. and Simpson, C. L., Mechanisms controlling modulation by haemolymph amino acids of gustatory responsiveness in the locust, *J. Exp. Biol.*, 168, 269, 1992.
38. Simpson, S. J. and Raubenheimer, D., The central role of the haemolymph in the regulation of nutrient intake in insects, *Physiol. Entomol.*, 18, 395, 1993.
39. Simpson, S. J. and White, P. R., Associative learning and locust feeding: evidence for a "learned hunger" for protein, *Anim. Behav.*, 40, 506, 1990.
40. Bernays, E. A. and Raubenheimer, D., Dietary mixing in grasshoppers: changes in acceptability of different plant secondary compounds associated with low levels of dietary proteins, *J. Insect Behav.*, 4, 545, 1991.
41. Champagne, D. E. and Bernays, E. A., Phytosterol suitability as a factor mediating food aversion learning in the grasshopper *Schistocerca americana*, *Physiol. Entomol.*, 16, 391, 1991.
42. Trumper, S. and Simpson, S. J., Mechanisms regulating salt intake in fifth-instar nymphs of *Locusta migratoria*, *Physiol. Entomol.*, 19, 203, 1994.
43. Bernays, E. A., Effects of experience on feeding, in *Regulatory Mechanisms in Insect Feeding*, Chapman, R. F. and de Boer, G., Eds., Chapman and Hall, New York, 1995, 279.
44. Raubenheimer, D. and Tucker, D., Pairing of visual cues with the separate consumption of protein and carbohydrate, *Anim. Behav.*, 54, 1449, 1997.
45. Zanotto, F. P., Simpson, S. J., and Raubenheimer, D., The regulation of growth by locusts through postingestive compensation for variation in the levels of dietary protein and carbohydrate, *Physiol. Entomol.*, 18, 425, 1993.
46. Zanotto, F. P., Raubenheimer, D., and Simpson, S. J., Selective egestion of lysine by locusts fed nutritionally unbalanced foods, *J. Insect Physiol.*, 40, 259, 1994.
47. Zanotto, F. P., Gouveia, S. M., Simpson, S. J., Raubenheimer, D., and Calder, P., Nutritional homeostasis in locusts: is there a mechanism for increased energy expenditure during carbohydrate overfeeding?, *J. Exp. Biol.*, 200, 2437, 1997.
48. Raubenheimer, D., Problems with ratio analysis in nutritional studies, *Functional Ecol.*, 9, 21, 1995.
49. Raubenheimer, D. and Simpson, S. J., Constructing nutrient budgets, *Entomol. Exp. Appl.*, 77, 99, 1995.
50. Raubenheimer, D. and Simpson, S. J., Analysis of covariance: an alternative to nutritional indices, *Entomol. Exp. Appl.*, 62, 221, 1992.
51. Raubenheimer, D. and Simpson, S. J., Meeting nutrient requirements: the roles of power and efficiency, *Entomol. Exp. Appl.*, 80, 65, 1996.

5 Macronutrient Selection in Free-Feeding Humans: Evidence for Long-Term Regulation

John M. de Castro

CONTENTS

1   INTRODUCTION

Food and fluid intake in the natural environment of free-living humans are influenced by a large array of variables. These influences span the gamut from physiological and genetic variables,[1-5] to endogenous and exogenous rhythms,[6-9] to psychological,[10-14] and to socio/cultural factors.[15-27] In addition, these effects result from past experiences and training, present environmental conditions, and the expectations of future events. It is amazing, given the number and complexity of variables operative, that there can be any regulation of energy intake and expenditure at all. The fact, though, is that it is somehow regulated.

The relative stability of body weight for most people through the adult life span is a testament to the fact that ultimately, in some way, regulation occurs. In order for body weight to remain stable there must be a balancing of the expenditure of energy with intake. If expenditure exceeds intake then body weight will decrease, while if intake exceeds expenditure weight will increase. To produce an expenditure–intake equivalence, expenditure could be adjusted to match intake and/or intake could be adjusted to match expenditure. The evidence that is available suggests that, although both processes may be involved, adjustments occur primarily to the intake of total food energy. Hence, the present chapter will focus on the intake side of the equation.

Energy in food is derived from the oxidation of the three macronutrients — carbohydrate, fat, and protein — and from alcohol. The sum of the intake of these nutrients comprises the total food energy ingested. Regulation of intake might be based simply on the total food energy ingested or could involve separate regulatory signals specific to each of the component nutrients.

There has been a considerable amount of investigation in the laboratory exploring these issues. However, there are little data in regard to whether there is regulation of the intake of individual macronutrients by free-living humans. It is one thing to demonstrate that a regulation occurs for a nutrient in a tightly controlled laboratory situation where there are few other factors operating to influence intake. It is quite another to demonstrate it with normal humans immersed in the complex multifaceted environments of modern living and ingesting a heterogeneous diet which is partially controlled by the individual and partially by others.[28,29]

This chapter will discuss and present data on the question of whether normal free-living humans regulate their intake of individual macronutrients. One difficulty immediately encountered when one attempts to address such a question is how to obtain reliable and valid data on the eating habits of free-ranging humans on a continuous, 24-h-per-day basis. The data presented in the chapter were derived from a 7-day diet diary technique.[30] Hence, the chapter will begin with a discussion of the diet diary method. The next difficulty involved in answering the question of the regulation of intake of individual macronutrients is determining what is the relevant time frame. Regulation could be fairly immediate and adjust intake on a meal-to-meal basis, short-term regulation, and/or it might involve delayed adjustments spanning days or even weeks. Hence, the chapter will discuss both the short-term, meal-to-meal regulation of individual macronutrients and also the evidence for long-term, day-to-day regulation.

1.1   THE 7-DAY DIET DIARY TECHNIQUE

Much of the data presented has been obtained from normal free-living humans with a 7-day dietary diary technique.[30] We have been employing this technique for a number of years to measure the spontaneous, natural intake of free-living normal adult humans.[1-32] The participants record in a small, pocket-sized diary each meal, when it starts and ends, exactly what they eat or drink, how it was prepared, their subjective states at the beginning and again at the end of the meal, where they eat the meal, and the number and nature of other people eating with them. This is a fairly tedious task and we have found that subjects' motivation to record accurately wanes after a week. As a result, the vast majority of our studies has restricted data acquisition to a 1-week recording period. Fortunately, it has been established that a 7-day period of recording is sufficient to obtain a fairly accurate sample of an individual's usual intake.[33-38] It also provides a reliable measure, as good agreement has been found for diet diary records that were repeated after as much as 2 years.[37-41]

There is a long-standing tradition of suspicion in regard to the validity of self-reports of intake. Indeed, the 24-h recall procedure for acquiring dietary information has been shown to produce fairly inaccurate results.[35,42-46] However, the 7-day diet diary technique has been found to be reasonably valid and accurate. There is evidence that the subjects' self-reports of their intake are truthful reports of the foods they actually ingest. Surreptitious measurements of the actual amount of food consumed at meals have been found to be in close agreement with the diary records.[46-48]

The case for the validity of the technique notwithstanding, there is evidence that the amounts of food reported in diaries underestimate actual intakes by about 20%.[41,49-54] However, this is only a problem when the absolute amounts of nutrients ingested are required. In most research applications only relative amounts are looked at. In most cases, the quantities ingested by one group or condition are related to those ingested by another group or condition. The error created by underestimation would be expected to affect all subjects equally and thus would not change relative values. In addition, in a number of analyses the individual subject's intake on one occasion is compared or correlated with that same subject's intake on another occasion. Underestimation would not be expected to affect the same subject's behavior differently on different occasions. Hence, although a bona fide problem, underestimation is usually not a significant issue in interpreting most of the results of the application of the diet diary technique.

There are a number of sources of unsystematic error which affect diet diary records of intake. However, these sources of error would be expected to add to error variance and thereby tend to obscure relationships rather than produce systematic differences between intakes associated with different groups or conditions. "Random inaccuracy may lead to false negative conclusions by reducing true associations but will not generate misleading correlations" (p. 708[41]). That significant systematic and subtle relationships have been discerned in many prior projects[1-32] in spite of the error attests to the robustness of the phenomena observed and indicates that the technique is sensitive enough for most research purposes.

The diet diary self-report method that we employ has a number of features that help to insure accurate recording.[30] In particular, the subjects record every item that they either eat or drink at the time that it is consumed. This minimizes the distortion of memory and improves the reporting of details and the estimation of quantities. Also, as a check on the validity of the reports we have each subject identify two individuals with whom they will be eating during the recording period. These individuals are later contacted and asked to recall or verify what the subject ate at the meals where they were present. After over a thousand verifications, in no case has the observer reported that the subject ate something that wasn't recorded in the diary or reported that the subject did not eat a recorded food. In addition, the participants are aware that we will be contacting the people with whom they will be eating and this may increase their accuracy in recording their intake. Recently, we have added a new wrinkle. Subjects are given a camera and are asked to take a picture of the foods that they are eating prior to ingesting them. We have found that the inclusion of the pictures increases the estimates of the amounts ingested by 5 to 8%. Hence, we believe that these modifications of the technique greatly improve the accuracy of estimation and reduce the under-estimation of intake.

2   SHORT-TERM MACRONUTRIENT INTAKE REGULATION

On the short term, intake can be regulated by adjustments to the amount ingested in meals and/or to the frequency of meals. The metric of the former is the meal size, while for the latter it is the intermeal interval. In free-feeding laboratory conditions the meal size tends to predict the duration of the following interval. This is true for both lab animals[55-57] and for humans.[58] This "postprandial" relationship suggests that regulation of intake occurs via adjustments to the interval until the next meal. However, this requires that a meal be initiated at a time dictated by the amount previously ingested. In order for this to occur people must have the freedom to eat whenever they have the desire to eat.

In the natural environment of humans it is more difficult to adjust the timing of meals as the environment often requires specific scheduling of meals, e.g., breakfast timing dictated by work start times, preset lunch times, family dinner times, etc. As a result, for humans in the natural environment we have found that there is no significant relationship between the meal size and the after-meal interval.[31,32] On the other hand, there is a significant relationship between the duration of the interval prior to a meal and the size of the meal.[31,32] This so-called "preprandial" relationship suggests that under normal, everyday conditions people adjust the amount eaten in a meal based on how long it has been since they last ate; that is, they regulate by adjusting meal size to their deprivation level. The evidence indicates that this is due to the environmental constraints of the natural environment. Indeed, when humans are in an unconstrained environment where they have the freedom to eat whenever they wish, they show the "postprandial" pattern.[58] On the other hand, when lab animals were constrained to eat on a particular schedule they displayed a "preprandial" pattern.[59]

Time in and of itself cannot affect eating. Rather, it operates as a result of time-based changes in physiological processes. In particular, over time nutrients are utilized and nutrient stores are depleted. Eating then could be regulated based on the amounts required to replenish the stores. Our research has suggested that an important variable is the nutrients remaining in the stomach.[31]

Over the before-meal interval the stomach empties a proportion of the nutrients ingested at the prior meals. The amount remaining in the stomach at the time of meal ingestion appears to have a negative effect on subsequent intake: with the more present, the smaller the amount ingested in the subsequent meal.[31]

In free-living humans, the contents of the stomach cannot be easily monitored and directly measured. However, the regularity and predictability of gastric function allow for the application of a computer model to the estimation of the stomach contents. The stomach empties in a very regular and predictable fashion such that an amount proportional to the square root of the caloric content of the stomach empties every minute.[60-62] Given that the diary records provide information as to amount that was ingested and the time it was ingested, it is relatively simple to calculate how much should have emptied over the interval and how much should be remaining at the time of the meal. Using this approach we have found a significant negative correlation between the amount predicted to be in the stomach at the point of meal initiation and the meal size.[31,32,62,63] This negative correlation with meal size is found regardless of whether meal size is expressed in terms of total food energy, carbohydrate, fat, or protein intake. It also occurs regardless of whether stomach content is expressed in terms of total food energy, carbohydrate, fat, or protein intake. This negative relationship indicates that the more that is present in the stomach at the beginning of the meal, the less will be eaten in the meal — a thoroughly reasonable outcome.

The slopes of the regression lines from these analyses can provide an estimate of the magnitude of the relationship between the before-meal stomach contents and meal size. These slopes are depicted in Figure 1. Although all of the slopes are significantly negative, they do markedly differ in magnitude. In particular, the amount of protein in the stomach (horizontal hatching) has significantly larger negative slopes than either fat (slanted hatching) or carbohydrate (vertical hatching) and even considerably larger slopes than those for overall food energy in the stomach (cross-hatching). For the prediction of overall food energy intake in the meal, the average slope for protein stomach content is −1.92. This suggests that for every calorie of protein in the stomach at the beginning of the meal, the meal size is reduced by nearly 2 cal. In addition, this effect occurs regardless of how meal size is measured. This suggests that protein in the stomach has a very potent restraining effect on subsequent intake that is substantially greater than the other macronutrients.

These results were further explored by employing multiple linear regressions predicting meal size using the estimated stomach contents of the three macronutrients as the independent variables. Since the intakes of all three macronutrients are positively correlated, it is impossible to see with univariate correlations the impact of one macronutrient independent of the effects of the others. However, with multiple linear regression, it is possible to view the effect of each while taking into consideration, and mathematically holding constant, the effects of the other two. The results of these analyses are depicted in Figure 2. In this figure the beta coefficients from the multiple regression analysis are shown for four separate multiple regression predictions of meal size expressed as total food energy (first set of three bars), protein (second set), fat (third set), or carbohydrate (fourth set) intake. Regardless of the measure of meal size, the estimated before-meal protein content of the stomach (horizontal hatching) has a large, negative, and significant impact on intake. Only in the case of stomach carbohydrate (vertical hatching) prediction of the carbohydrate content of the meal does another macronutrient have a significant negative beta coefficient. In fact, carbohydrate is significantly positive in predicting fat intake, while fat (slanted hatching) is significantly positive in predicting protein intake.

These results suggest that the effectiveness of the premeal stomach contents in suppressing subsequent intake is due to the effect of protein and not fat or carbohydrate. In regard to short-term regulation, protein would appear to somehow be special. Prior intake of protein is the only macronutrient, in the present analyses, that appears to have a regulatory, negative-feedback effect on subsequent intake. This effect appears to be expressed as a result of the protein remaining in the stomach at the time of ingestion of the next meal having a potent negative impact on intake in the meal.

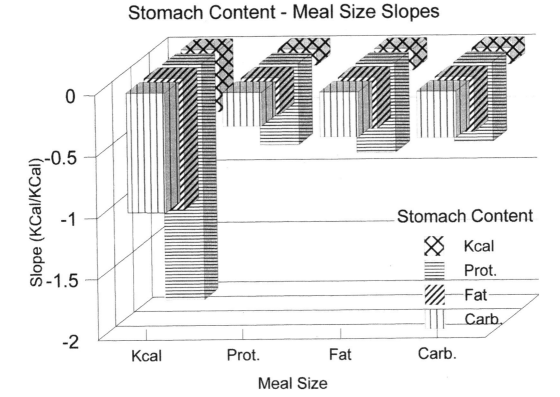

FIGURE 1 Mean slopes of the regression lines relating the estimated before-meal stomach contents of total food energy (Kcal) (left set of 3 bars), protein (left center set of 3 bars), carbohydrate (right center set of 3 bars), or fat (right set of 3 bars) and the amounts ingested in the meals of total food energy (Kcal) (cross-hatching), protein (horizontal hatching), fat (slanted hatching), and carbohydrate (vertical hatching).

There is an indication in these data that suggests that there may be specific short-term negative-feedback regulation of protein intake. The estimated before-meal stomach content of protein produced a significantly larger negative beta coefficient when predicting the meal size of protein than when predicting the meal size of overall total food energy or the meal size of either fat or carbohydrate. Hence protein in the stomach is not only the only effective macronutrient in suppressing subsequent intake, but it is also particularly more effective in suppressing subsequent protein intake. Hence protein in the stomach has a particularly potent restraining effect on subsequent intake. In addition, it affects the ingestion of protein to a greater extent than any other nutrient. However, it should be noted that these effects only account for a small proportion of the variance (<4%) in meal size. Intake is highly variable and the short-term effect of protein would appear to be a small regulatory signal operating amid a large cadre of potent nonregulatory signals.

3   LONG-TERM MACRONUTRIENT INTAKE REGULATION

All regulation must eventually be expressed as short-term regulation. However, clues as to the nature and extent of these process can be glimpsed by investigating more integrated long-term data. In this section two different views will be presented: one from twin data of the heritability of macronutrient intake and one from daily macronutrient intake effects on macronutrient intake on subsequent days.

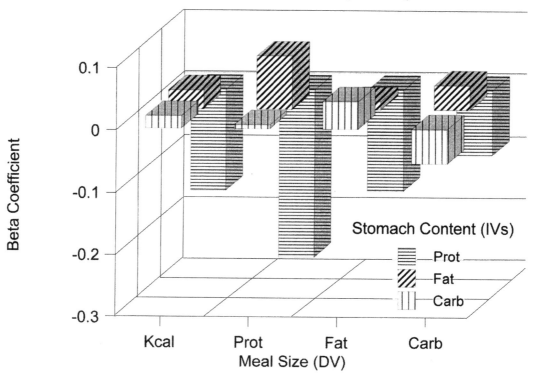

FIGURE 2 Mean beta coefficients from four multiple linear regressions predicting the amounts ingested in the meals of total food energy (Kcal) (left set of 3 bars), protein (left center set of 3 bars), carbohydrate (right center set of 3 bars), or fat (right set of 3 bars) based on the independent variables of the estimated before-meal stomach contents of protein (horizontal hatching), fat (slanted hatching), and carbohydrate (vertical hatching).

3.1 GENETIC INFLUENCES

One way to investigate the degree of long-term regulation of intake is to look at the degree to which the genes affect intake. If the intakes of specific nutrients are heritable, then it suggests that the amounts of these nutrients are in some way regulated. To investigate this hypothesis a twin study was performed to assess the influence of inheritance on the body sizes and nutrient intakes of adults. Diet diary nutrient intake data were collected from adult twins who live independently: 110 identical and 102 fraternal same-sex adult twin pairs and 53 pairs of mixed-gender fraternal twins participated.[1-4] Information regarding the overall and meal intake of nutrients was then analyzed with linear structural modeling techniques of heritability analysis. These techniques are sensitive; allow the assessment of both genetic and early, familial, environmental effects; allow for the assessment of true gender differences in genetic and environmental effects; and allow the inclusion of other variables in the models to assess where in complex interactive chains that heredity affects the behaviors.[65,66]

The models revealed significant genetic influences on body size, height, and weight. That body size is primarily determined by the genes has been previously described by a number of other investigators using both twin[67-70] and adoption[71-73] study methodologies. However, we demonstrated that not only body size, but also the amount of food energy ingested daily, as well as its macronutrient, alcohol, and water content, was significantly affected by inheritance.[1] Applying linear structural modeling analysis to these twin data revealed that 65% of the variance in daily energy

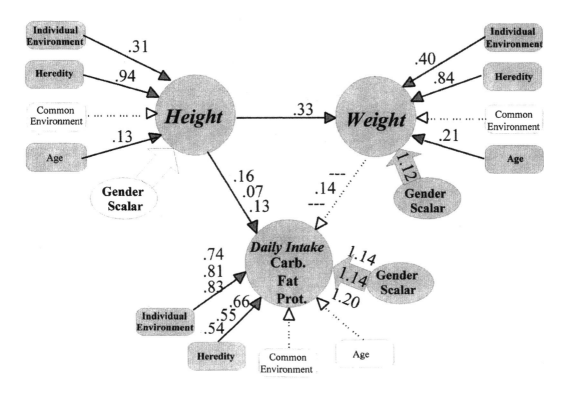

FIGURE 3 Linear structural model and the most parsimonious model fitting the twin data for height, weight, and total daily intakes of carbohydrate, fat, and protein are represented by the solid path lines. Where three path coefficients are presented, the top coefficient represents the influence of the factor on carbohydrate intake; the middle, fat; and the bottom, protein. Dotted lines represent nonsignificant paths. For all remaining parameters, removing any one leads to a statistically significant reduction in the model's account of the observations.

intake could be attributed to heredity. In addition, heredity accounted for 44% of the variance in meal frequency and 65% of the variance in average meal size. In contrast, the analysis indicated that the shared, familial environments in which the twins were raised had no significant impact on the levels or pattern of intake in adulthood.[1]

A possible alternative explanation for the apparent genetic effects on intake is that it occurs secondarily to genetic effects on body size. That is, what is inherited is body size, and body size influences the amount of intake. Hence, when amount of intake is looked at in isolation it shows significant heritability. However, by the application of a complex interactive linear structural model we were able to demonstrate that the heritability of overall daily food energy and macronutrient intake was independent of body size.[2] The results of the application of this linear structural model to the twin data are summarized in a path diagram in Figure 3. In this model the genetic and environmental effects on body size, height, and weight are considered along with the daily intakes of the macronutrients. Body size is allowed to influence macronutrient intake. So the influences of heredity, environment, gender, and age on macronutrient intake are assessed with the influence of body size already considered. Remarkably, body size only has a small influence on the daily intake of the macronutrients and accounts for less than 3% of the variance. On the other hand, heredity accounts for 44, 30, and 29% of the variance in daily intake of carbohydrate, fat, and protein, respectively. Hence, it would appear that the genes have a major impact on macronutrient intakes independent of their effects on body size.

Another possible explanation for the apparent heritabilities of macronutrient intake is that it occurs secondarily to genetic effects on overall daily intake; that is, what is inherited is daily food

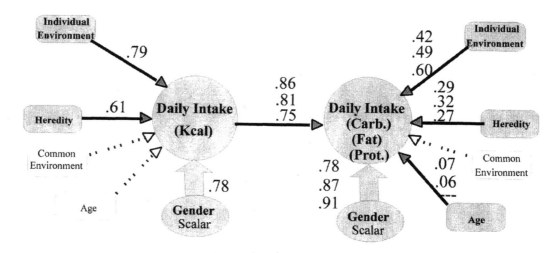

FIGURE 4 Linear structural model and the most parsimonious model fitting the twin data for total daily caloric intakes and the daily intakes of carbohydrate, fat, and protein are represented by the solid path lines. Dotted lines represent nonsignificant paths. Where three path coefficients are presented, the top coefficient represents the influence of the factor on carbohydrate intake; the middle, fat; and the bottom, protein. For all remaining parameters, removing any one leads to a statistically significant reduction in the model's account of the observations.

energy intake, and since it is composed of intake of the macronutrients, their apparent heritability may be forced. However, the application of another linear structural model demonstrated that the heritability of macronutrient intake was to some extent independent of the inheritance of the overall daily intake of food energy.[2] The results of the application of this linear structural model to the twin data are summarized in a path diagram in Figure 4. The effects of the genes account for only 8, 10, and 7% of the variance in the daily intake of carbohydrate, fat, and protein, respectively. Although these effects appear to be small it should be recalled that the overall daily intake of food energy, whose effects are removed in the model, is composed of the daily intakes of each of the macronutrients. Thus, the fact that there is still a residual effect of the genes on the daily intake of each of the macronutrients individually when their combined effects are removed is remarkable and suggests that there are genetic influences on the intakes of each of the macronutrients that are independent of genetic effects on the overall intake of food energy.

The fact of genetic influences on the amounts of macronutrients ingested implies that in some way the levels of intake of the macronutrients must be influenced or regulated. The genes are normally perceived as affecting physical structure. Indeed, inheritance is ultimately expressed on a molecular level and physical structure. However, we have obtained evidence of late that this phenotype may encode the likelihood and level of engagement in environmental variables and also in the responsiveness of the individual to environmental variables.[26] Hence, the genetic influence on intake may not be direct, but indirect, by its effects on the behavioral proclivities and responsivities of the individual.

These ideas can be seen in an analysis of the heritability of the before-meal stomach contents of the macronutrient relationships to the amount eaten in the meal.[74] An analysis of the twin data indicated that the amounts of total food energy and each of the macronutrients in the stomach at the beginning of each meal were clearly influenced by the genes. Heritability accounted for 40, 36, 41, and 36% of the variance in before-meal stomach contents of total energy, carbohydrate, fat, and protein, respectively, and 45, 48, 42, and 44% of the variance in after-meal stomach contents, respectively. Linear structural modeling indicated that the heritabilities of the stomach contents of the macronutrients were secondary to genetic influences on overall intake and meal size. But regardless of it being primary or secondary, the fact remains that the genes influence the level of macronutrients that tend to be found in the stomach before and after meals.

As reviewed above, these macronutrients in the stomach have a restraining effect on subsequent intake. This is measured by the regression assessing the relationship between the stomach contents and the size of the subsequent meal. The individual's responsiveness to stomach filling can be measured with the correlation coefficient or the slope of the regression line between the two factors. To investigate the extent to which the genes might affect an individual's responsiveness to stomach filling, a heritability analysis was performed on the twin data for the correlations and slopes. Although significant heritabilities were found for the correlations and slopes between meal size and the before-meal stomach contents of total energy, carbohydrate, and fat, the clearest and strongest relationships were present for protein in the stomach. The genes accounted for 32, 23, 18, and 26% of the variance in the correlations and 30, 28, 25, and 34% of the variance in the slopes relating before-meal stomach protein content and the amount ingested in the meal of total energy, carbohydrate, fat, and protein, respectively.

These heritabilities results suggest that the amounts of macronutrients ingested are regulated. They further suggest that the genes may act not only directly on the intakes, but also indirectly by affecting the amounts of macronutrients found in the stomach at the beginning of the meals and the individual's responsiveness to these stomach contents. Hence, macronutrient regulation can be seen to result from the operation of a number of mechanisms. These include anatomical differences and personal preferences, habits, and sensitivities.

3.2   DELAYED NEGATIVE FEEDBACK

If short-term mechanisms were responsible for regulation, then the total amount of food energy ingested in a day would be relatively constant, provided that activity levels also were relatively constant. However, this is not the case. Intake varies considerably from day to day.[75-77]

Hence, in order for regulation to occur some mechanism must be present to adjust intake during a day based on prior day's intake. The level of intake during a day must, in some way, produce negative feedback to affect intake on subsequent days. Hence, an autocorrelation between food energy intake on a day and that ingested on the next day should be present and negative. This must also be true if there is regulation of the amounts ingested of individual macronutrients. Negative autocorrelations should be present for daily intakes of each individual macronutrient. However, prior research has failed to obtain significant negative autocorrelations. In fact, the daily intake autocorrelations that have been obtained have been found to be small and predominantly positive.[76,77] That is, when a person eats a lot on one day, that same person may or may not eat a lot on the next day. "The finding challenges the belief that some short-term homeostatic mechanism exists that causes high energy intakes to be followed by low ones."[77] Hence, there is currently little understanding regarding how daily intake affects subsequent intake to produce energy balance.

To take a new look at daily intake autocorrelations, the 7-day diary intakes obtained from 324 men and 409 women were analyzed for the amounts of food energy and each of the macronutrients ingested on each of the 7 days. Autocorrelations were then calculated between intakes on 1 day and those occurring on subsequent days. The results for the mean macronutrient autocorrelations between the amounts ingested in a day and during each of the four subsequent days are presented in Figure 5. In this figure, each set of four bars represents the autocorrelation between intake on 1 day and that occurring either 1, 2, 3, or 4 days later. The autocorrelations with the 2-day lag (solid bars) were significantly stronger than with other delays. Hence it would appear that although there is a slight negative feedback that tends to restrain intake 1 day later, it is not until 2 days later that the effect is maximal. It continues on the third day, but disappears by the fourth day.

Interestingly, there appears to be clear macronutrient-specific effects. Carbohydrate intake (left set of bars) had a significantly larger negative correlation than either fat or protein with the amount of carbohydrate ingested (leftmost four bars in each set) either 1 or 2 days later. Fat intake (middle

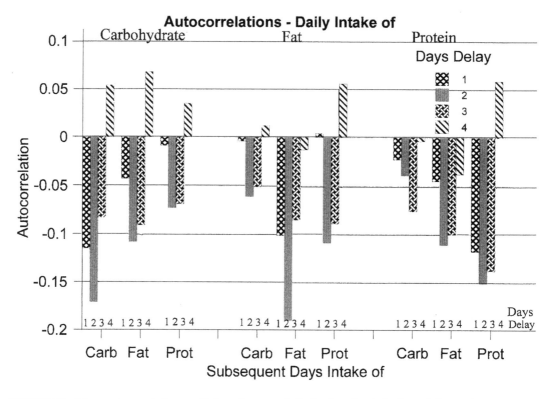

FIGURE 5 Mean autocorrelation coefficients between daily intake of carbohydrate (left set of 12 bars), fat (middle set of 12), and protein (right set of 12) and intake on subsequent days of carbohydrate (left four bars of each set), fat (middle four bars), and protein (right four bars). The first bar of each set of four represents the correlations calculated between the amounts ingested on a day and on the next day (lag 1), and 2 (lag 2), 3 (lag 3), and 4 (lag 4) days later.

set of bars) had a significantly larger negative correlation than either carbohydrate or protein with the amount of fat ingested (middle four bars in each set) either 1 or 2 days later. Similarly, protein intake (rightmost set of bars) had a significantly larger negative correlation than either carbohydrate or fat with the amount of protein ingested (right set of four bars in each set) either 1 or 2 days later. But there were no nutrient-specific effects for intake occurring 3 or 4 days later. Hence, carbohydrate appears to maximally affect carbohydrate, fat, and protein, 2 days after ingestion.

Autocorrelation assumes that the error terms from each of the variables in the correlation are themselves uncorrelated. Violations of this assumption can result in an overestimation of the proportion of the variance that is accounted for by the regression. This assumption is nearly always violated with autocorrelations. Repeated measure variables, such as daily energy intakes, fall into a pattern called a simplex structure.[78] Simplex autoregressive models, unlike autocorrelations, do not assume that the error terms are uncorrelated. The model allows the estimation of intraday path coefficients which indicate the covariation of measurement over time and innovation (ζ) within day variation, variance that is not accounted for by the prior days' intakes. These paths and innovations can be estimated with a linear structural modeling.[65,79]

The simplex autoregressive LISREL model was applied to the prediction of the amount of each macronutrient ingested on each of two subsequent days based on the days' macronutrient intakes.[79] This was a very complex model that not only allowed each macronutrient to affect intake of that macronutrient on the two subsequent days, but also allowed each macronutrient to affect the intake of the other two macronutrients on the two subsequent days. The model and the results of its application to the day-to-day intakes reported in the 7-day diet diaries are depicted in Figure 6.

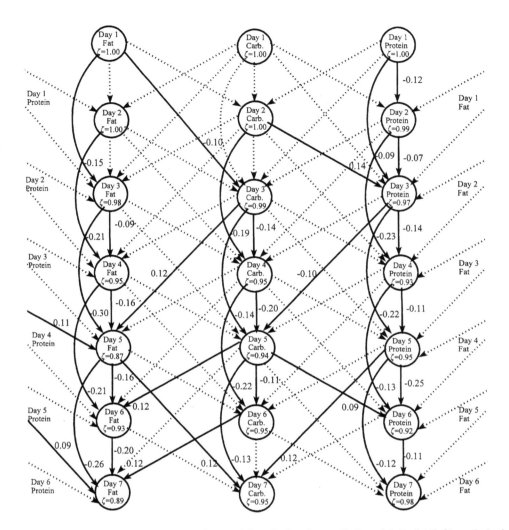

FIGURE 6   LISREL Simplex autoregressive model applied to the prediction of daily fat (left), carbohydrate (middle), and protein (right) intake. The numbers represent path coefficients produced by the analysis. Solid lines represent significant paths, while dotted lines represent nonsignificant paths. Removal of any solid-line path presented in the figure produces a significant degradation of the model's fit as assessed with a χ^2 test. ς represents variance that is not accounted for by the prior day's intakes.

The simplex linear structural modeling analysis results parallel those found with the autocorrelation analysis. In the model there are 66 possible paths, indicating an effect of a macronutrient on different macronutrients. Of these, only 11 are statistically significant, and 8 of these have a positive sign rather than the expected negative, which would be indicative of a negative feedback process. These results indicate that a particular macronutrient has very little, if any, influence on the subsequent days' intake of the other macronutrients. On the other hand, the model contains 15 paths representing each macronutrient's influence on the intake of that same macronutrient 2 days later. Of these, 14 were significant and negative, ranging from −0.09 to −0.30.

Hence, the simplex linear structural modeling analysis supports the conclusions from the autocorrelational analysis that the negative effect of ingestion of a macronutrient is greatest on the subsequent intake of that particular macronutrient 2 days later. However, because the macronutrient intakes tend to covary, the autocorrelations are only suggestive of a macronutrient specificity. The simplex analysis, on the other hand, by independently evaluating the individual paths, having extracted the covariances, provides a view of the macronutrients' independent effects. This analysis makes a strong case for the specificity of the delayed negative feedback. In this analysis, the intake of carbohydrate, with a 2-day delay, influenced carbohydrate intake only, while fat affected fat alone and protein affected protein alone.

4   DISCUSSION

The results of the studies reviewed in this chapter indicate that both short- and long-term regulation of macronutrient intake can be detected in the eating behavior of normal, free-living humans. The fact that there appears to be significant heritabilities for the intakes of the macronutrients strongly suggests that their intake must, in some way, be regulated by the physiology. The genes encode physiology. So if there is a genetic influence on a factor, it must do so by encoding a physiological substrate that produces it. However, the influence of the genes appears to be subtler than simply encoding gross anatomy. The data imply that the genes may influence behavioral tendencies and preferences and the individual's responsiveness to environmental factors. This must have a physiological substrate; but it is likely a subtle one involving nuances in the structure of the nervous system. Its net effect is to influence the behaviors expressed by the individual, and these in turn affect intake.

On the short term, regulation of macronutrient intake can be observed. In the natural environment, humans appear to adjust intake primarily by altering the amounts ingested in meals and not the frequency or timing of meals. Macronutrient regulation appears to occur as the result of a negative feedback process which affects the amounts ingested in a meal. In particular, the amount of protein remaining in the stomach at the beginning of a meal appears to have a marked suppressive effect on the amount ingested. It appears to affect the intake of all nutrients — protein more so than the others — but nonetheless they all appear to be negatively affected by the before-meal stomach content of protein.

That the negative feedback from stomach protein may be fairly nonselective is reasonable when one imagines the difficulty for the system to detect and meter the amounts of different nutrients present in a mixed heterogeneous meal. In order to adjust the amounts of the macronutrients taken in a meal, the system first has to be able to measure the amounts ingested. This is a very difficult if not impossible task to carry out based on the immediate exteroceptive stimuli (sight and odor of food) and the stimuli from the oral cavity. It would be much simpler for the system to meter the amounts of nutrients taken in later as they are processed in the digestive system. Regulation can occur by simply having whatever was taken in selectively affect the next meal. In this way, if too much protein is taken in, then there will be more protein in the stomach at the time of the next meal and this will suppress intake at that meal. This should result in less protein being ingested; but if it doesn't, then the next meal will be affected.

This short-term negative feedback from the contents of the stomach appears to be one of the processes affected by the genes. First, the data clearly suggest that the amount that an individual normally has present in the stomach at the beginning of a meal is affected by inheritance. Second, the genes appear to have another, subtler effect. They influence the degree of response that an individual has to these stomach contents. The genes then can affect the amount of the macronutrients normally ingested by an individual by influencing the stomach content negative feedback loop, both by affecting the amount in the stomach and also the individual's responsiveness to that content.

This mechanism appears to produce a degree of regulation of intake. However, it has only a modest effect and accounts for only a small fraction in the variance in meal intake. That it isn't sufficient to produce intake regulation is indicated by the large variation in day-to-day intake. If it were a major influence then intake would not vary greatly from day to day; but in fact, it does.[75-77] It appears that there is another mechanism that corrects for the errors made as a result of the insufficiency of the short-term regulatory system. It is a delayed negative feedback system that acts over days. This regulatory system operates particularly with a 2-day delay to adjust intake in response to prior intake.

What physiological mechanisms may underlie this delayed negative feedback is a mystery. The fact that the delay is long relative to the life of most physiological or biochemical events indicates that GI, plasma, and hepatic factors are unlikely intermediaries. In the case of fat, it is possible that feedback from a long-term energy storage depot (adipose tissue) may be involved. However, for carbohydrate and protein storage, half-lives are far too short and the turnover too high to allow for a long-term feedback signal that would allow adjustments after a 2-day delay. This suggests that the mechanism of the macronutrient specificity of the feedback is not simple and maybe not unitary.

Although these delayed feedback effects are statistically significant, the effect sizes are small. The largest univariate autocorrelation reported accounts for less than 4% of the total variance in daily intake, and the best linear structural modeling only accounted for 13% of the variance. Most effect sizes were smaller than these. Hence, intake on a day only accounts for a small proportion of the variance in the intake of nutrients on subsequent days. To some extent these small effect sizes are not surprising given that these data were acquired from humans in their natural environments where complex arrays of stimuli affect the individual and create variance in behavior. That single variables, e.g., negative feedback, only account for a small proportion of the variance may simply be a reflection of the fact that there are large numbers of variables operative in these environments. However, the question remains as to how negative feedback could promote regulation given that it only accounts for a small amount of the variance in intake.

Although there are a large number of relatively potent variables that may be producing much larger effects, these influences are transient.[28,29] They have large but short-term effects on intake. On the other hand, prior intake appears to feed back both at the time of the next meal and again after a delay of at least a day, and usually longer, and its influence persists for a number of days beyond. It sets a bias that promotes an adjusted overall level of intake. The bias may not be apparent as its influence is overshadowed by more potent stimuli on a momentary basis. However, its presence is persistent. Over time, it continues to influence the amount ingested, while the nonregulatory short-term influences do not persist. Over time they cancel out one another's effects. The random short-term influences average over time and result in no net effect on intake, while the effects of the persistent bias continue to influence intake, producing a cumulative net alteration of intake. In effect over time, the integral of the effects of the short-term environmental factors is zero, while the integral for the small negative feedback effects becomes substantial. It is possible that over long periods of time these processes result in regulation of intake.

REFERENCES

1. de Castro, J. M., Genetic influences on daily intake and meal patterns of humans, *Physiology and Behavior*, 53, 777, 1993.
2. de Castro, J. M., Independence of genetic influences on body size, daily intake, and meal patterns of humans, *Physiology and Behavior*, 54, 633, 1993.
3. de Castro, J. M., A twin study of genetic and environmental influences on the intake of fluids and beverages, *Physiology and Behavior*, 54(4), 677–687, 1993.
4. de Castro, J. M., Genes and environment have gender independent influences on the eating and drinking of free-living humans, *Physiology and Behavior*, 63, 385, 1998.
5. Henson, M. B., de Castro, J. M., Johnson, C. J. and Stringer, A., Food intake by brain injured humans who are in the chronic phase of recovery, *Brain Injury*, 7(2), 169, 1993.
6. de Castro, J. M., Circadian rhythms of the spontaneous meal patterns, macronutrient intake, and mood of humans, *Physiology and Behavior*, 40, 437, 1987.
7. de Castro, J. M., Seasonal rhythms of human nutrient intake and meal patterns, *Physiology and Behavior*, 50, 243, 1991.

8. de Castro, J. M., Weekly rhythms of spontaneous nutrient intake and meal pattern of humans, *Physiology and Behavior*, 50, 729, 1991.

9. de Castro, J. M. and Pearcey, S., Lunar rhythms of the meal and alcohol intake of humans, *Physiology and Behavior*, 57, 439, 1994.

10. de Castro, J. M. and Elmore, D. K., Subjective hunger relationships with meal patterns in the spontaneous feeding behavior of humans: evidence for a causal connection, *Physiology and Behavior*, 43, 159, 1988.

11. Elmore, D. K. and de Castro, J. M., Self-rated moods and hunger in relation to spontaneous eating behavior in bulimics, recovered bulimics, and normals, *International Journal of Eating Disorders*, 9(2), 179, 1990.

12. Elmore, D. K. and de Castro, J. M., Meal patterns of normal, untreated bulimic, and bulimic women, *Physiology and Behavior*, 49(1), 99, 1991.

13. de Castro, J. M., The relationship of cognitive restraint to the spontaneous food and fluid intake of free-living humans, *Physiology and Behavior*, 57(2), 287, 1995.

14. de Castro, J. M. and Goldstein, S. J., Eating attitudes and behaviors of pre- and postpubertal females: clues to the etiology of eating disorders, *Physiology and Behavior*, 58, 15, 1995.

15. de Castro, J. M. and de Castro, E. S., Spontaneous meal patterns in humans: influence of the presence of other people, *American Journal of Clinical Nutrition*, 50, 237, 1989.

16. de Castro, J. M., Social facilitation of duration and size but not rate of the spontaneous meal intake of humans, *Physiology and Behavior*, 47(6), 1129, 1990.

17. de Castro, J. M., Brewer, M., Elmore, D. K. and Orozco, S., Social facilitation of the spontaneous meal patterns of humans is independent of time, place, alcohol, or snacks, *Appetite*, 15, 89, 1990.

18. de Castro, J. M., The relationship of spontaneous macronutrient and sodium intake with fluid ingestion and thirst in humans, *Physiology and Behavior*, 49(3), 513, 1991.

19. de Castro, J. M., Social facilitation of the spontaneous meal size of humans occurs on both weekdays and weekends, *Physiology and Behavior*, 49(6), 1289, 1991.

20. de Castro, J. M. and Brewer, E. M., The amount eaten in meals by humans is a power function of the number of people present, *Physiology and Behavior*, 51, 121, 1992.

21. de Castro, J. M. and Brewer, E. M., The amount eaten in meals by humans is a power function of the number of people present, *Physiology and Behavior*, 51(1), 121, 1992.

22. Redd, E. M. and de Castro, J. M., Social facilitation of eating: effects of instructions to eat alone or with others, *Physiology and Behavior*, 52(4), 749, 1992.

23. de Castro, J. M., Family and friends produce greater social facilitation of food intake than other companions, *Physiology and Behavior*, 56(3), 445, 1994.

24. de Castro, J. M., Social facilitation and inhibition of eating, in *Not Eating Enough: Strategies to Overcome Underconsumption of Field Rations*, National Academy of Sciences, Washington D.C., 1995, 373.

25. de Castro, J. M., Bellisle, F., Feunekes, G. I. J., Dalix, A. M. and De Graaf, C., Culture and meal patterns: a comparison of the food intake of free-living American, Dutch, and French students, *Nutrition Research*, 17, 807, 1997.

26. de Castro, J. M., Inheritance of social influences on eating and drinking in humans, *Nutrition Research*, 17, 631–648, 1997.

27. de Castro, J. M., Socio-cultural determinants of meal size and frequency, *British Journal of Nutrition*, 77(Suppl. 1), S39, 1997.

28. de Castro, J. M., How can eating behavior be regulated in the complex environments of free-living humans, *Neuroscience and Biobehavioral Reviews*, 20, 119, 1996.

29. de Castro, J. M., How can energy balance be achieved by free-living human subjects?, *Proceedings of the Nutrition Society*, 56, 1, 1997.

30. de Castro, J. M., Methodology, correlational analysis, and interpretation of diet diary records of the food and fluid intakes of free-living humans, *Appetite*, 23, 179, 1994.

31. de Castro, J. M. and Kreitzman, S. N., A microregulatory analysis of spontaneous human feeding patterns, *Physiology and Behavior*, 35, 329, 1985.

32. de Castro, J. M., McCormick, J., Pedersen, M. and Kreitzman, S. N., Spontaneous human meal patterns are related to preprandial factors regardless of natural environmental constraints, *Physiology and Behavior*, 38, 25, 1986.

33. Basiotis, P. P., Welsh, S. O., Cronin, F. J., Kelsay, J. L. and Mertz, W., Number of days of food intake records required to estimate individual and group nutrient intakes with defined confidence, *Journal of Nutrition*, 117, 1638–1641, 1987.

34. Cellier, K. M. and Hankin, M. E., Studies of nutrition and pregnancy. I. Some considerations in collecting dietary information, *American Journal of Clinical Nutrition*, 13, 55–62, 1963.

35. Larkin, F. A., Metzner, H. L. and Guire, K. E., Comparison of three consecutive-day and three random-day records of dietary intake, *Journal of the American Dietetic Association*, 91(12), 1538–1542, 1991.

36. Tinker, L. F., Schneeman, B. O. and Willits, N. H., Number of weeks of 24 hour food records needed to estimate nutrient intake during a community-based clinical nutrition trial, *Journal of the American Dietetic Association*, 93(3), 332–334, 1993.

37. Heady, J. A., Diets of bank clerks: development of a method of classifying the diets of individuals for use in epidemiological studies, *Journal of the Royal Statistical Society (Ser. A)*, pt 3, 124, 336–361, 1961.

38. St. Jeor, S. T., Guthrie, H. A. and Jones, M. B., Variability of nutrient intake in a 28 day period, *Journal of the American Dietetic Association*, 83, 155–162, 1983.

39. Adleson, S. F., Some problems in collecting dietary data from individuals, *Journal of the American Dietetic Association*, 36, 453–461, 1960.

40. Block, G., A review of validations of dietary assessment methods, *American Journal of Epidemiology*, 115, 492–505, 1982.

41. Livingstone, M. B., Prentice, A. M., Strain, J. J., Coward, W. A., Black, A. E., Barker, A. E., McKenna, P. G. and Whitehead, R. G., Accuracy of weighed dietary records in studies of diet and health, *British Medical Journal*, 300, 708–712, 1990.

42. Brown, J. E., Tharp, T. M., Dahlberg-Luby, E. M., Snowdon, D. A., Ostwald, S. K., Buzzard, I. M., Rysavy, S. D. M. and Wieser, S. M. A., Videotape dietary assessment: validity, reliability, and comparison of results with 24-hour dietary recalls from elderly women in a retirement home, *Journal of the American Dietetic Association*, 90, 1675–1679, 1990.

43. Dubois, S. and Boivin, J.-F., Accuracy of telephone dietary recalls in elderly subjects, *Journal of the American Dietetic Association*, 90, 1680–1687, 1990.

44. Mullenbach, V., Kushi, L. H., Jacobson, C., Gomez-Martin, O., Prineas, R. J., Roth-Yousey, L. and Sinaiko, A. R., Comparison of 3-day food record and 24-hour recall by telephone for dietary evaluation in adolescents, *Journal of the American Dietetic Association*, 92(6), 743–745, 1992.

45. Myers, R. J., Klesges, R. C., Eck, L. H., Hanson, C. L. and Klem, M. L., Accuracy of self-reports of food intake in obese and normal-weight individuals: effects of obesity on self-reports of dietary intake in adult females, *American Journal of Clinical Nutrition*, 48, 1248–1251, 1988.

46. Krantzler, N. J., Mullen, B. J., Schultz, H. G., Grivetti, L. E., Holden, C. A. and Meiselman, H. L., The validity of telephoned diet recalls and records for assessment of individual food intake, *American Journal of Clinical Nutrition*, 36, 1234–1242 1982.

47. Eagles, J. A. and Longman, D., Reliability of alcoholics reports of food intake, *Journal of the American Dietetic Association*, 42, 136–139, 1963.

48. Gersovitz, M., Madden, J. P. and Smicikalas-Wright, H., Validity of the 24-hour dietary recall and seven-day record for group comparisons, *Journal of the American Dietetic Association*, 73, 48–55, 1978.

49. Prentice, A. M., Black, A. E., Coward, W. A., Davies, H.L., Goldberg, R. R., Murgatroyd, P. R., Ashford, J., Sawyer, M. and Whitehead, R.G., High levels of energy expenditure in obese women, *British Medical Journal*, 292, 983–987, 1986.

50. Lissner, L., Habicht, J.-P., Strupp, B. J., Levitsky, D. A., Haas, J. D. and Roe, D. A., Body composition and energy intake: do overweight women overeat and underreport?, *American Journal of Clinical Nutrition*, 49, 320–325, 1989.

51. Bandini, L. G., Schoeller, D. A., Cyr, H. N. and Dietz, W. H., Validity of reported energy intake in obese and nonobese adolescents, *American Journal of Clinical Nutrition*, 52, 421–425, 1990.

52. Mertz, W., Tsui, J. C., Judd, J. T., Reiser, S., Hallfrisch, J., Morris, E. R., Steele, P. D. and Lashley, E., What are people really eating? The relation between energy intake derived from estimated diet records and intake determined to maintain body weight, *American Journal of Clinical Nutrition*, 54, 291–296, 1991.

53. Goran, M. I. and Poehlman, E. T., Total energy expenditure and energy requirements in healthy elderly persons, *Metabolism*, 41, 744–753, 1992.

54. Livingstone, M. B., Prentice, A. M., Coward, W. A., Strain, J. J., Black, A. E. W., Davies, P. S., Stewart, C. M., Mckenna, P. G. and Whitehead, R. G., Validation of estimates of energy intake by weighted-dietary record and diet history in children and adolescents, *American Journal of Clinical Nutrition*, 56, 29–35, 1992.

55. LeMagnen, J. and Tallon, S., La periodicite spontanee de la prise d'aliments ad libitum du rat blanc, *Journal of Physiology (Paris)*, 58, 323–349, 1966.

56. LeMagnen, J. and Tallon, S., L'effect du jeune preable sur les caracteristiques temporelles de la prise d'aliments chez le rat, *Journal of Physiology (Paris)*, 60, 143–154, 1968.

57. de Castro, J. M., Meal pattern correlations: facts and artifacts, *Physiology and Behavior*, 15, 13–15, 1975.

58. Bernstein, I. L., Zimmerman, J. C., Czeisler, A. C. and Weitzman, E. D., Meal patterning in "free-running" humans, *Physiology and Behavior*, 27, 621–623, 1981.

59. de Castro, J. M., Regulatory alchemy: an attempt to make rat eating patterns humanlike and human eating patterns ratlike, *Appetite*, 7(3), 249, 1986.

60. Hopkins, A., The pattern of gastric emptying: a new view of old results, *Journal of Physiology (London)*, 182, 144–150, 1966.

61. Hunt, J. N. and Knox, M. T., Regulation of gastric emptying, in *Handbook of Physiology: Alimentary Canal*, Vol. 4, Motility, Code, C. F. and Heidel, W., Eds., American Physiological Society, Washington, D.C., 1968, pp. 1917–1935.

62. Hunt, J. N. and Stubbs, D. F., The volume and content of meals as determinants of gastric emptying, *Journal of Physiology (London)*, 245, 209–225, 1975.

63. de Castro, J. M., Macronutrient relationships with meal patterns and mood in spontaneous feeding behavior of humans, *Physiology and Behavior*, 39(5), 561–569, 1987.

64. de Castro, J. M., Physiological, environmental, and subjective determinants of food intake in humans: a meal pattern analysis, *Physiology and Behavior*, 44(4/5), 651–659, 1988.

65. Neale, M. C. and Cardon, L. R., *Methodology for Genetic Studies of Twins and Families*, Kluwer Academic Publishers, Dordrecht, the Netherlands, 1992.

66. Heath, A. C., Neale, M. C., Hewitt, J. K., Eaves, L. J., and Fulker, D. W., Testing structural equation models for twin data using LISREL, *Behavioral Genetics*, 19, 9–35, 1989.

67. Bray, G. A., *The Body Weight Regulatory System: Normal and Disturbed Mechanisms*, Chiofi, L. A., James, W. P. T. and VanItallie, T. R., eds., Raven Press, New York, 1981, 185–195.

68. Feinleib, M., Garrison, R. J., Fabsitz, R., Christuan, J. C., Hrubec, Z., Bohani, N. O., Kannel, W. B., Rosermas, R., Schwartz, J. T., and Wagner, J. O., The NHLBI twin study of cardiovascular disease risk factors methodology and summary of results, *American Journal of Epidemiology*, 106, 284–295, 1977.

69. Stunkard, A. J., Foch, T. T. and Hrubec, Z., A twin study of human obesity, *Journal of the American Medical Association*, 256, 51–54, 1986.

70. Stunkard, A. J., Harris, J. R., Pedersen, N. L. and McClearn, G. E., The body-mass index of twins who have been reared apart, *New England Journal of Medicine*, 322, 1483–1487, 1990.

71. Price, R. A., Cadoret, R. J., Stunkard, A. J. and Troughton, E., Genetic contribution to human obesity: an adoption study, *American Journal of Psychiatry*, 144, 1003–1008, 1987.

72. Sorenson, T. I. A., Price, R. A., Stunkard, A. J. and Schulsinger, F., Genetics of human obesity in adult adoptees and their biological siblings, *British Medical Journal*, 298, 87–90, 1989.

73. Stunkard, A. J., Sorenson, T. I. A., Hanis, C., Teasdale, T. W., Chakraborty, R., Schull, W. J. and Schulsinger, F., An adoption study of human obesity, *New England Journal of Medicine*, 314, 193–198, 1986.

74. de Castro, J. M., Inheritance of premeal stomach content influences on eating and drinking in free-living humans, *Physiology and Behavior*, 66, 223–232, 1999.

75. Balogh, M., Kahn, H. A. and Medalie, J. H., Random repeat 24-hour dietary recalls, *American Journal of Physiology*, 24, 304–310, 1971.

76. Morgan, K. J., Johnson, S. R. and Goungetas, G., Variability of food intakes: an analysis of a 12-day data series using persistence measures, *American Journal of Epidemiology*, 126, 326–335, 1987.

77. Tarasuk, V. and Beaton, G. H., The nature and individuality of within-subject variation in energy intake, *American Journal of Clinical Nutrition*, 54, 464–470, 1991.

78. Guttman, L., A new approach to factor analysis: the radex, in *Mathematical Thinking in Social Sciences*, Lazerfeld, P. F., Ed., Free Press, Glencoe, IL, 1954, pp. 258–349.

79. Boomsma, D. I., Martin, N. G. and Molenaar, P. C. M., Factor and simplex models for repeated measures: application to psychomotor measures of alcohol sensitivity in twins, *Behavioral Genetics*, 19, 79–96, *1989.*

80. de Castro, J. M., Prior days intake has macronutrient specific delayed negative feedback effects on the spontaneous food intake of free-living humans, *Journal of Nutrition*, 128, 61–67, 1998.

6 Macronutrient-Specific Hungers and Satieties and Their Neural Bases, Learnt from Pre- and Postingestional Effects of Eating Particular Foodstuffs

David A. Booth and Louise Thibault

CONTENTS

0-8493-2752-0/00/$0.00+$.50
© 2000 by CRC Press LLC

1   OVERVIEW

1.1   AIMS OF THIS CHAPTER

This book aims to review critically the evidence for neural mechanisms of macronutrient-specific dietary selection. Such an enterprise is precarious because the data usually interpreted as showing neural modulation of nutrient-controlled intake are confounded by sensorimotor artifacts.[1,2] The same flaw in claimed evidence for control of food intake by metabolic energy yield was identified a long time ago[3] but is also still widely neglected. This chapter therefore expounds again[4-15] the behavioral mechanisms required for an individual to achieve selection of a diet in anticipation of the nutritional effects of different foods' contents of one or more of the energy-yielding nutrients, carbohydrate, fat, or protein, and hence the experimental designs needed to get evidence for such energy nutrient ("macronutrient") selection and its neural bases.

The literature on the effects of neuroactive drugs on intake from two or three diets differing in macronutrient composition was recently reviewed.[2] No set of findings met the basic criteria for evidence that a neurotransmitter mediated protein-, carbohydrate-, or fat-specific intake. Indeed, only one potentially relevant program of research has so far provided direct neuroanatomical, pharmacological, and biochemical evidence of dietary selection specific to a nutrient: this is the conditioned reduction of the amounts eaten of diets deficient in the micronutrients within protein, i.e., certain indispensible amino acids, as detailed in Chapter 23, Gietzen's chapter on amino acid receptors, and related to Sections 2.2 and 3.2 of this chapter. Considerable progress has been made in understanding the unlearnt nutrient-specific appetite for sodium salts in rats.[16] However, Na+ is not an energy nutrient as are proteins, carbohydrates, and fats — the topic of this book.

This dearth of cogent data has arisen because many investigators appear not to have understood the logical necessities for the control of ingestion by any nutrient in the diet. There has been a general neglect of published evidence on the mechanisms in the rat and other species that achieve the identification of foods providing a required nutrient and then influence the selective intake of such foods so that a lack of that nutrient is remedied or even avoided, and perhaps also excessive intake is restrained to some degree.

As a result, a high proportion of published experiments on "macronutrient selection" are incorrectly designed. Hence, they fail to provide evidence that the macronutrients in the foods consumed had any role in determining the amounts that were in fact ingested, let alone that the manipulation of the brain or body by drug injection or other means affected the intake of the macronutrient as claimed. The results are most likely already known to be explicable as effects of the drug on the ease of consuming the foodstuffs, or on preferences for or perceptions of sensory characteristics of one or more of the foods offered.

This chapter will therefore rehearse the mechanisms necessary for any macronutrient selection to be achieved and its intake to be brought under some degree of quantitative control. However, these mechanisms cannot be investigated or even clearly thought about when words are used and theory is built in ill-conceived ways, as they continue to be. Therefore, longstanding terminological confusions and inadequate models of ingestion must first be addressed.

1.2   TERMINOLOGY

1.2.1   A Saga of Misconceptions

The research field of ingestive behavior has been plagued with tendentious and sometimes incoherent attempts to prescribe the meanings of words such as hunger, thirst, satiety, appetite, preference, palatability, regulation, and so on. These are words from ordinary language for aspects of individuals' actions and reactions. They are very useful as tags or headings for certain broad types of causal process operative in food and drink choices and intake. They should not be hijacked in the cause of labeling pieces of data regardless of what those observations show or fail to show about the mechanisms involved.

It is now well recognised in philosophy that it is a mistake to try to delimit the legitimate use of a commonsense term or to attempt to define operationally a theoretical construct in science. An even cruder fallacy is to assume that the existence of two words implies the necessity for two concepts or even the existence of two entities named by those words, e.g., satiety and satiation, hunger and appetite (for food), or palatability and (sensory) preference.

The usual mistake is to use these rough categorisations of processes controlling ingestive behavior as names for particular pieces of data, such as aspects of the intake pattern or certain wordings of appetite ratings. This presumes that the intake or rating directly and exclusively measures that particular sort of causal process. There is usually no evidence for this assumption within the investigation or even in other investigations. Only a question-begging or even illogical verbal analysis is offered in support of the terminological prescription. Indeed, there is often direct evidence that the intake measure or the rating fails to distinguish the claimed mechanism from others.

This erroneous approach is commonly advocated in terms that fly in the face of the logic of scientific investigation. Satiety, palatability, eating motivation, hunger sensations, and so on are claimed to be "concepts" defined by the size of a meal, for example, or by the word "hunger" or "pleasantness" in a rating. On the contrary, a causal process involved in ingestive behavior, in its neural or somatic underpinnings or in its social context, can only be measured by varying a particular source of influence on that process independently of other potential influences and observing some effect on an aspect of ingestion or on some symbolic reference to the eater's actions toward foods. Amounts eaten, eating rates, or rating scores by themselves can in principle tell us nothing about hungers and satieties. The effect of a drug on such a measure also tells us nothing about the neural and mental processes on which it is acting. The neuropsychopharmacological experiment has to be designed to test the effect of the drug on one causal process as distinct from any other potential explanations.[17] For example, to demonstrate that a drug decreases the pleasure of eating, and does not decrease sensory preference in an affectless manner nor increase nastiness, the administration of the drug must reduce the lip-licking response to the taste of sugar without decreasing ingestion of the sweet material or increasing the gaping response to bitter agents.[18]

1.2.1.1   Objective uses of appetite ratings and intake parameters

Hunger and thirst are still sometimes said to be sensations and yet not the sensation of an epigastric pang or of a dry mouth, respectively. Some use the term 'sensation' dualistically for a conscious product of brain activity. Others operate in reductionist mode, supposedly referring only to the CNS state itself triggered by a physiological state of peripheral tissues. These are incoherent residues of the muddled controversy between 'peripheralist' and 'centralist' approaches to motives and emotions; this was generally considered by the 1960s to have been won by the centralists, but only because of failure to consider eating and drinking behavior as causal systems within which many different peripheral sources of neural input modulate distinct patterns of neural output.

A more realistic term is the 'sense' of hunger or thirst. This is correct identification of physiological status by ratings or intake tests: a measure is shown to vary with controlled levels of energy or water deficit.

That is, no theoretical construct can be reduced to its measure. A rating cannot just be assumed to be a direct scale of some subjective state, let alone of a physiological event, whatever the investigator thinks is the meaning of the word used — a sensation, a desire, a motivation, or whatever. For example, line ratings labeled "Not at all" at one end and "Extremely" at the other are not "visual analog scales" of whatever word is above the line: they are not visual (or motor) but spatial formats. They are not analogs of anything, they simply express a person's view of his or her position between the extremes; and they do not scale anything until they have been shown to be sensitive specifically to independently controlled variation of the strength of some relationship between input and output for the rater in the circumstances of rating use.

Thus the scientific use of appetite ratings is to measure the rater's performance in discriminating between levels of separately measured influences on eating or drinking. This validated rating is indeed objective appetite, whereas the labeling of an intake parameter with some "concept" is merely subjective. That is, the correct terms are "rated appetite/satiety" (not "subjective appetite/satiety") and "bout weight" or "meal size," not an "objective measure of satiation" — even if the same parameter, meal size, were not also taken to be a measure of "palatability"!

1.2.1.2   Sating processes and sated states

This inability of an intake parameter by itself to measure any mechanism is illustrated by the mishandling of data on the discontinuation of eating, which could have as much role in dietary selection as the initiation of eating does. The size of a meal has been defined as a measure of 'satiation,' and the delay to the next meal defined as a measure of 'satiety.' This prescription commits a linguistic solecism, has grossly misleading mechanistic implications, and proves to be self-contradictory in its primary application.

The basic error in this terminology is the fallacy of misplaced concreteness — assuming that because a word is a noun there must be some entity corresponding to it. Satiety and satiation are two nouns from the one verb, to sate or (from the same root) to satiate or to satisfy an appetite. For the ingestive appetites more specific terms are commonly used, such as to quell hunger, finish a meal or fill up (with food and/or water), or to slake or quench, in the case of thirst, the appetite for watery fluids. The two noun forms differ only in 'satiation' being a gerund ('satiating' being used adjectivally): that is, satiation is the process of sating, i.e., going from a less to a more sated state, and satiety is a state or degree of being sat(iat)ed. The process of moving to more extreme states and the states between which appetite is moved during and after ingestion are the same category of phenomenon. There is no more difference in the phenomena being referred to than between the water in a river flowing past a bridge and the water currently under that bridge.

The current insistence that satiation refers to the state at the end of a meal and satiety refers to the state (or series of states?) between meals has helped to sustain the most unfortunate myth that different mechanisms are involved in ending meals from the mechanisms that sustain the suppression of appetite between meals. For example, many investigators of human eating presume that sensory and expected satiety dominate the termination of meals, while visceral and metabolic satiety are assumed to predominate in the intervals between meals. Yet, intestinal stimulation and tissue oxidation of both carbohydrate and fat in a meal can begin well before the meal ends, and memories and expectations from the sensory cues in a meal can be remembered indefinitely and certainly for several hours until the next meal. Thus the processes of satiation and a satiated state of some degree continue from soon after the start of a meal, past the end of the meal, until and beyond the rise of a desire to eat again. Indeed, the satiating influences of one meal are probably operative during the next meal under *ad libitum* conditions.

The most absurd aspect of this notion that satiation is distinct from satiety is that it is an operational definition that does not even work in its primary application. The satiating effects of foods are measured between meals by the amount of food eaten during a test meal offered at the time in question. Yet the size of that meal, which is said to measure "satiety" (after meals), by this definition is determined by something different called "satiation" (within meals).

In short, arguments about what word to use to describe what piece of data detract from the proper task of scientific investigators: measuring the diverse causal processes by which the ingestion of foods influences the subsequent dietary intake. Which oral or visceral patterns of stimulation by foods and the eating context are inhibiting appetite for what further eating choices at any particular moment is a matter for experimental designs, not prescriptive definitions. The relevant experiments have already shown, for example, that visceral signalling of a slight satiety can begin during a meal,[19] that sensorily triggered processes of satiation for an expected menu can be predicted for several hours after meal[20] and hence memories lasting days are involved in dietary selection from one meal to the next, not necessarily physiological mediation.[21]

1.2.1.3   Sensory preference and sensual pleasure

Palatability is a term that has been abused almost as badly as satiety, with as dire an effect on the design of experiments on dietary selection. Palatability has been variously (and contradictorily) supposed to be a property of the food which is invariant across all contexts (at least for one individual), or a pleasurable effect in the mind of the eater measured by any rating of "pleasantness" or "liking" and by the facial movement pattern elicited by a sweet taste, e.g., lip licking.

Neither palatability nor anything else can be in the food and not in the mind, or in the brain and not at the oral receptors. What we know is that eating is typically facilitated by the patterns of stimulation to the external senses from a well-known foodstuff. Hence the food's viewed and smelt characteristics promote its selection in anticipation and extend its ingestion once in the mouth by delivering the expected gustatory, tactile, thermal, irritative, and olfactory stimulation to the oronasal cavity. This learnt sensory facilitation of acceptance, or of preference (assuming an implicit comparison with acceptance of other foods), is an effect of the food on the mind, not a physico-chemical or ghostly entity in either sphere.

There is nothing completely invariant either in the effects of a food's sensory characteristics on an individual's appetite for eating it. The concept of palatability as a fixed position of a food in an individual's dietary preference hierarchy is plainly unrealistic. Corn flakes are preferred to ice cream at breakfast in most countries around the North Atlantic, but often vice versa at other times of day. Eaters of meat and of cakes prefer steak to chocolate gateau for the main course of a cooked meal, but the other way round for dessert in the same meal (whether or not that main course included steak or any other meat). In other words, sensory preferences can depend very heavily on context (the effective aspects of which are not identifiable from the above anecdotes). Thus, a food has no particular level of palatability to a person or animal, except perhaps in an approximate sense for a particular context of preference testing within or between meals. The concept of a fixed palatability cannot work where it is most often invoked, in supposed competition with presumed growing effects of the food on satiety within a meal.

Furthermore, ratings of pleasantness or liking cannot be taken to be measures of momentary sensory preference or sensual pleasure any more than of a food's stable palatability. The investigator may be thinking of brain process of reward but the pleasantness rating need have nothing to do with such a mechanism in some circumstances. Demonstration that a rating using a word such as "palatable" is sensitive to differences in sensory stimulation from foods does nothing to show that palatability ratings measure a thrill of pleasure[18] or are even exclusively sensitive to sensory factors in appetite. The rated "pleasantness" of a food could be strongly influenced by differences in visceral state between the beginning and end of a meal or half an hour later: the average pleasantness of several staple foods is in fact more sensitive than hunger ratings to both postingestional effects of maltodextrin and the satiating effect of eating what is believed to be a high calorie food.[22]

The same point about nonspecificity of intake or rating measures applies to temporal patterns of ingestive movements, so-called microbehavioral parameters. The only way of showing that one aspect of eating rates is sensitive to orosensory variation and not to visceral variation, and vice versa for another eating pattern, is to measure the differences in the movement parameters when oral and visceral stimulation are varied independently of each other, preferably both within the normal range of interest, e.g., within and between *ad libitum* meals on a nutritious choice of diets. This confounding of an eaten food's cephalic stimulation with its visceral actions is the basic weakness of the literature on selection between diets differing in macronutrient composition.

Therefore, even if some experiments (yet to be done) showed that ratings of "hunger pangs" or intervals between mouthfuls are more sensitive to food in the digestive tract than to orosensory differences, and that ratings of "pleasantness" or mouthful chewing rates have qualitatively reverse sensitivities, an effect of gastric loading on the rated pleasantness or chewing rates for some foods would still not show that visceral signals affect sensory preference. As in any other scientific study of potentially interacting mechanisms, the operation of each hypothesised causal process has to be measured independently: seeing a balloon rise when a blower is turned on beneath it does not provide the slightest evidence that the blower is an antigravity machine.

Also, the use of a word such as "pleasantness" or the identification of a sensorily sensitive pattern in eating movements cannot distinguish the expression of relative acceptance (preference) from the experiencing of a sensual thrill (pleasure). Much more specific measurement criteria must be satisfied to distinguish preference from pleasure.[18] That an investigator has the concept of pleasure, palatability, or indeed preference does not give the right to assume that there is an entity separate from the momentary effect on eating from stimulation of the head senses by the food. There needs to be independent evidence for such an entity. One prospect of such evidence is to distinguish between two aspects of the response patterns involved in the behavior toward foods or drinks, namely, the movements of acceptance or rejection which normally achieve ingestion (passage of the material down the throat), and movements which do not function in ingestion but have presumably served in the history of the species to elaborate the gathering of nutrients into the mouth or the expulsion of poisons from the oral cavity. It is commonly assumed that these facial movements and manual or forepaw gestures express the excitement of different valences of emotionality, i.e., pleasure or revulsion, in ways that have served genetically beneficial social functions. For example, 'yum' and 'yuck' expressions could have encouraged or discouraged the adult's feeding of an infant, the young child's joining of an older peer in feasting on a delicious food, or avoiding the site of some nasty-tasting berries, with increased reproductive success resulting. Whatever their function, there is only evidence for a process separate from ingestive motivation (possibly pleasurable sensation or emotion) if some of the facial movements can be dissociated from the movements resulting in ingestion.[18,23]

Thus, behavioral words should be used in scientific discussion in their original broad and empirical senses, and prejudicial verbal definitions avoided. If the issue is whether there is a difference between the processes that end a meal and those which delay the next bout of eating, then evidence on the specific candidate processes must be collected rather than effects on meal size and meal-to-meal intervals just being given different labels. If the processes controlling which diet is taken at the start of the meal differ between diets (despite being equated by some measure of 'palatability'), then their sensory bases must be characterised and not arbitrarily (and misleadingly) attributed to differences in macronutrient composition.

1.2.2   Hungers and Satieties, Preferences, and Aversions

1.2.2.1   Causal processes in ingestion

The fundamental point is that hunger, satiety, preference, and aversion are not *causes* of ingestion. Rather, these terms are rough labels for the facilitatory and inhibitory processes that are active during the ingestion or refusal of a food. The causes of the processes of selecting what to eat and initiating, maintaining, and terminating the eating of that material are the *sources* of hunger and its satisfaction. These include the sensory characteristics of the foods, the signals from the viscera to the brain, the social context which involves conspecifics' eating, and, in the human case, beliefs and emotions about the foods, the body, and the occasion (Figure 1).

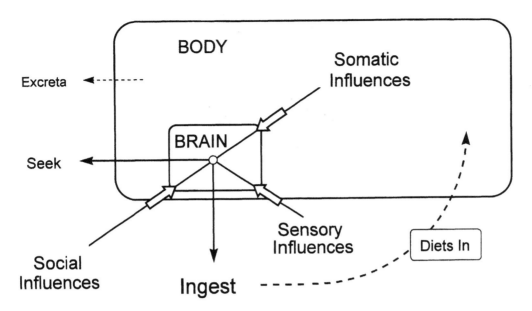

FIGURE 1   Outline of the basic mechanisms of dietary selection specific to a nutrient (whether unlearnt or learnt). The large box is the animal's body and the small compartment is neural circuitry that engineers the information processing evident within the ingestive behavior. The continuous lines are the causal processes transforming sensory and somatic information (as well as influences from conspecifics and the physical environment) into movement pattern information. The broken lines are the resultant transfer of nutritive materials into the body and its output of energy and excreta.

All that is meant by an appetite or desire for something is the observable propensity to approach and contact that sort of object or situation. An appetite differs from a taxis only in the complexity of the influences on approach and contact,[24] requiring information processing about which it is efficient to use mentalistic language, and for which sophisticated network engineering would be required as its physical basis (i.e., a brain, lacked by a chemotactic bacterium). A preference is simply a difference in degree of appetite between two examples of something of the sort desired. An aversion to or fear of something is a tendency to escape, reject, or avoid it, whether as a category of objects or situations or relatively to one example more than to another.

Specifically, ingestive appetites are motor responses to the stimulation received from pieces of material ("small objects of desire"[15]) which normally achieve entry of some of the material into the mouth and its passage down the throat. An ingestive aversion tends to yield refusal to let the material enter the mouth or, if it has got in, to result in its ejection before swallowing.

Thus, being hungry or satiated is a cognitive-behavioral state of having an appetite for food or lacking one as a result of eating, whether evident in the actual eating of some foods and/or, no less objectively, by their symbolic acceptance ("I'd like that" or "That would be pleasant to eat"). A person or an animal is hungry when ready to eat some solid or liquid food,[25] thirsty when ready to drink some watery fluids,[26] and sated when such readiness to ingest has been lost as a result of recent ingestion. There need be no hunger pangs or dry mouth nor a consciously preferred food or drink available, nor need it be judged to be the occasion or time for eating or drinking. Nevertheless, such perceptions do commonly occur, at least for some people sometimes, and they contribute to the state of hunger. However, they are not themselves the behavior or the wish to eat or drink: indeed, the epigastric pangs are sensations of hunger only because they occur when we want to eat and are unlike an ache that sometimes accompanies nausea and wanting to vomit. The individual having the appetite or aversion may at that time be aware of percepts, emotions, and conceptualised attitudes toward foods and drinks, parts of her or his body, and some eating or drinking places.

The taking of such a conscious viewpoint is, however, a highly a sophisticated performance.[15] None of these perceptual, rational, and emotional processes can be assumed to have been achieved without specific evidence on that occasion, even in a human adult showing an appetite or aversion, let alone by a young child or the member of another species selecting or rejecting a piece of food.

Since an appetite for food and/or drink may be expressed in many different ways, no single aspect of the intake pattern nor any particular rating should be "defined" as the measure of hunger or thirst or a nutrient-specific appetite. Indeed, in animal husbandry, the term appetite is widely used to refer to the *ad libitum* daily intake of a cow or chicken. Le Magnen[27,28] expounded his theoretical view of the control of amounts eaten from different diets in terms of two aspects of (ingestive) "appetite": (1) intakes in one-diet tests of different diets and (2) relative intakes in tests on two or more variants of a diet presented simultaneously. A two-stimulus intake test is generally more sensitive to sensory differences between diets than are two one-stimulus intake tests. However, this rule of thumb neglects crucial differences between single-diet and choice tests in the moment-to-moment variations in the control of ingestion which cumulate into the total amounts eaten of each diet, to which we now turn.

1.2.2.2   Dynamics of selective ingestion

When an appetite, such as the ingestive appetites of hunger or thirst, declines as it is being carried into effect, i.e., during eating or drinking, respectively, the appetite is then becoming satisfied, sated, or satiated. Since there are differential appetites (such as for foods and for fluids), there may also be differentiated satieties. There has been very little attention to the issue whether or not the hunger satiety state differs from thirst satiety, i.e., satiety for eating but not for drinking, or vice versa. We are certainly not so familiar with the plural of satiety as we are with that of appetite. One reason for this may be that we generally consume both foods and drinks at meals and so, when we are "full," we are sated for both food and drink, at least to some degree and for certain ranges of foods and drinks.

Nevertheless, the principle of multiple satieties is implied by the phenomenon dubbed "sensory-specific satiety." The effect of eating a food is becoming bored with that particular foodstuff or knowing that we have had enough of it (it remains unclear which of these, or some other effect). At least as accurate a phrasing would be "food-specific satieties": there is one for each food in which there has developed a temporary lack of interest as a result of eating that material.

As we shall see, food-specific satieties can also arise from sources other than immediately prior eating of the same food. Indeed, these other sorts of loss of interest in a food as a result of eating include the key inhibitory mechanism of macronutrient selection; this is reduced acceptance of a particular foodstuff because of an incipient excessive ingestion of a nutrient that the food is anticipated to supply, whether as a result of learning or as the inborn result of survival value of such nutrient-specific sensory satiety. These satieties specific to both the sensory characteristics and the effects of the nutrient contents of food must be distinguished from satieties specific to only the sensed qualities of the food. Switching from eating a food rich in one macronutrient to another food rich in a different macronutrient is not evidence for macronutrient-specific selection: the observation could result merely from the attractions of sensory variety. Variety might function to balance nutrient intakes but it cannot direct behavior to particular macrontrients.

As for 'satiety,' so the term 'hunger' should not be restricted to physiological signals of energy deficit or to the behavioral or brain state triggered by them, let alone to food deprivation or the ethical category of being deprived of the right to vital sustenance. Hunger is appetite for food in any sense of nutriment, by whatever processes that disposition to eat has been facilitated.

Similarly, the term 'appetite' cannot be restricted to the results of stimulation from food before it has been swallowed or from the context of eating. The appetite for a particular food or at a particular time of day may depend on internal signals as well as on external stimuli. Thus, hungers dependent on the need for sodium salts or protein amino acids are commonly known as sodium or protein appetites. Yet caloric appetite is commonly known simply as hunger. This further shows the interchangeability of the two terms. In this chapter, therefore, appetite for the energy in foods and any appetite for a particular form of macronutrient will often be called a "hunger."

Thus the term "appetite" (in an ingestive sense) means exactly the same as "hunger" except that the latter term can be used to exclude thirst, the appetite for water or drinks. This is of particular importance to the physiologists of hydromineral balance. Some nutritionists are unhappy with the idea that water is a nutrient: being the ultimate nonnutritive drink somehow excludes it from the diet. However, hunger cannot be confined to solids and thirst to fluids because there are clearly many liquid and semisolid foods, some rich in energy nutrients, minerals, and vitamins.

1.2.2.3 Variation in preference/aversion for a food

A preference for a foodstuff is a state of relatively greater acceptance of that item than of another item, as would be expressed in selection of the former item if there were a choice. In this relative sense there is at the same time a comparative aversion to the latter item. However, the term "aversion" is often used with a presumption of absolute or "active" aversion, i.e., an actual repulsion from the item, not merely a lack of interest in it. Thus, when the acceptance of a food material declines while it is being eaten to the point at which it is no longer accepted (at least temporarily), this can be regarded as a relative aversion. It is a further issue whether or not there is any aversiveness in further eating of the food at that moment of being sated with it.

Satiety, therefore, is a temporary state of relative aversion to the sensory qualities of the food(s) no longer wanted, or at least a loss of preference for them for the time being. Note that a sated eater may still rate a favorite dessert as highly palatable or as not having decreased in pleasantness at all: this illustrates that investigators cannot legislate how people must perform when using ratings. Indeed, it is likely to puzzle an assessor to keep asking how palatable or even how pleasant a foodstuff is that is being eaten. This may be why ratings using the word "palatability" are so variable[29] and the rated pleasantness of dessert does not always go down substantially when it has been finished,[30,31] and yet the pleasantness of familiar foods not recently eaten goes down very reliably when sated.[22] Assessors are left in conflict between expressing a momentary general attitude to eating (satiety, loss of interest in most foods) and stating their knowledge of their usual attitude to the particular food the question is being asked about (the somewhat unrealistic commonsense concept of a fixed palatability; see Section 1.3.1.2).

The degree of preference for a food at any given time depends entirely on what can be sensed at that moment. The material is recognised by the eater or forager as having its distinctive combination of levels of wavelengths and spatial distribution of reflected light, sniffed aroma, pressure patterns on the fingers and in the mouth, sweetness, sourness, retronasally sensed aroma chemicals, and so on. Thus, selections among foods and drinks depend entirely on their sensory identities, not on their contents of energy/nutrients or water. Indeed, people may enthusiastically consume energy-free yoghurt when hungry or drink seawater when desperately thirsty with no net yield of water to the body. Ingestive behavior will never be understood nor its normal bases elucidated effectively until experimenters manipulate and measure these sensory patterns independently of each other in their usual context of each other's pattern levels, as well as in the context of the usual range of internal states and social circumstances. Preference or aversion may vary with context, including momentary stages of nutrient deficit or excess (see Section 1.3.2).

1.3   BASIC MECHANISMS

1.3.1   Failed Models of Hunger(s)

Neglect of the immediate sensory controls of ingestion, in or out of conjunction with gastric distension and chemoreceptive or metabolic signals, has arisen historically from assumptions that particular categories of gustatory stimulation automatically function to meet nutrient requirements. The continued influence of this misleading model of the nutritional roles of appetites for foods and drinks largely accounts for lack of progress in understanding macronutrient selection and its neural bases.

*1.3.1.1   Preferences and aversions for the taste of salt
             or of sugar*

Curt Richter provided considerable evidence that rats had capacities to select foods containing particular nutrients, including sodium salts and protein vs. carbohydrate or fat.[32] The case of the hunger for salt[33] has provided a model for nutrient-specific dietary selection and regulatory intake control which has been unrealistically assumed to apply to all other forms of macro- and micro-nutrient intake. Salt appetite was easier to induce and understand than protein selection in pregnancy, for example.

It is tempting to extend the model of salt appetite to giving the taste of sugar a central role in the control of caloric intake. Indeed, such an argument may appear stronger for sugar than for salt in that sweetness congenitally elicits a strong ingestive reflex in many mammals but salt does not. The taste of sugar (and saccharin), the ingestive reflex it elicits, and the pleasure it can evoke have been taken by psychologists and neurophysiologists to be a paradigm of the mechanisms of intake control. Analysis of the rat's bout of drinking a sweet fluid has bolstered the model of "palatability" competing with "satiety," which is considered in Section 1.3.1.2.

Against this assumption, Dethier and Bodenstein[34] showed that the blowfly's responses to the taste of sugar were controlled osmotically only, not by energy metabolism, glucose-sensitive cells, or glucoreceptors. In vertebrates, too, any sugar craving during acute hyperglycemia is likely to be a learnt instrumental response to reduce distress: saccharin drinking did not increase in naive rats made hypoglycemic with insulin, while there was some increase in bar-pressing for saccharin in experienced rats.[35]

The main problem for the sweetness theory of hunger for energy or carbohydrate-specific cravings is that most of the natural sources of carbohydrate energy do not taste sweet. They provide starch, not sugars. Furthermore, sweetness is not needed to signal to a hungry forager the presence of some readily digestible energy in ripe fruit (the commonest prehistoric source of sugars): the fruit generally becomes soft and changes color when the sugar content builds. It is even more doubtful that the human race survived by being tempted to suffer bees' stings for the taste of honey.

In any case, what needs explaining is the congenital reflex to sweetness. Why is it needed to facilitate the ingestion of milk, for which there is ample tactile-motor provision? It has been suggested[13-15] that sugars are superreleasers of a counterreflex to the unlearnt rejection of the taste of alkaloids that protects the youngster from death as the result of foraging from plants beyond mother's diet. The bitter taste of these quaternary nitrogen compounds extends to the primary amine, urea, and to some essential amino acids and peptides in milk. To prevent the infant spitting out its sole source of nutrients, a modification of the bitter receptor connected to ingestive movements is required that is more sensitive to the conjunctions of carbonyl ($C=O$), amino ($N-H_2$), and C–H groups found in amino acids than to the heterocyclic nitrogen in plant alkaloids. This hypothesis is supported by recent demonstration of the genetic closeness of some bitter and sweet receptors. The lack of sweetness preference among mammalian obligate carnivores (cats) is also strong circumstantial support. In other words, the congenital liking for sweetness is a sort of protein appetite, and nothing to do with need for carbohydrates or energy.

There appears to be no solid backing for the notion that reactions to the sweet taste alone suffice to account for intake control in general, or even specifically the control of energy intake within the regulation of energy exchange. Even the taste of salt is far from a reliable mediator in regulation of sodium balance. Some sensory effects may participate in the change in preference with sodium deficit,[36] but intake is dominated by the avidity for materials tasting of salt, not by the perception of saltiness or the gustatory Na^+ receptors and their afferent neurophysiology. Weak salty tastes are sufficiently attractive (possibly via stimulation of sugar receptors and hence a slightly sweet taste) to sustain considerable need-free intake of Na^+ *ad libitum*. Furthermore, a loss of body fluids that stimulates secretion of renin and aldosterone raises the preference for quite concentrated NaCl solutions in rats,[37] and possibly in the human fetus and infant.[38,39] Notoriously, human beings in industrialised countries ingest far more sodium than they need, excreting most of it, behavior which has been blamed for chronic hypertension, though there are weaknesses in the epidemiological evidence.[40] Nevertheless, rats depleted of sodium show an increased preference for the taste of salt at all concentrations tested.[41] This increased preference for a relevant sensory characteristic in the presence of nutritional depletion is what can be defined as a specific appetite or hunger, as distinct from a mere sensory preference, on the one hand, or a generic hunger for many types of food on the other hand. Such an apparently unlearnt sodium appetite has been replicated in sheep,[33] but it remains controversial whether or not human beings show it.[42]

So it is puzzling why the salty taste and the sweet taste have so dominated thinking about the nutritional selectivity of dietary choices and the regulatory functions of ingestion.

One reason is that taste is widely regarded in research on the physiology of ingestion as the only sensory modality of importance in preference and intake. This may be an illusion, however, based on the ease of preparing a solution of a tastant, the readily visible ingestive and expressive reflexes to tastants, and narrow range of concentrations of tastants in a particular food or drink relative to the variation among types of fruit, vegetables, confectionery, etc.[43] In fact, foodstuffs are much more distinctive within the mouth in their texture and retronasal aroma than in their taste, let alone the importance before entry into the mouth of selection or rejection by visual appearance, orthonasal smell, and/or manual texture. Movements of the tongue and pharynx in response to textures and aromas are invisible, unlike the sucking or spitting to tastes, yet they are no less crucial to ingestion and to pleasure in eating.

This preconception that gustation is all important seems to have encouraged the idea that starch has a taste. However, the behavioral and neurophysiological experiments have not adequately excluded effects of the distinctive somatosensory (textural) characteristics of starch particles or of the non-Newtonian viscosity of solutions of starch hydrolysed to maltodextrin and/or ancilliary odors from the plant source of the starch.

Regarding taste as a model for ingestion has had a subtler counterproductive influence which also is at work on the possibilities of macronutrient selection. There are very few main types of gustatory receptor, each with a well-known specific ligand, e.g., glucose, sodium chloride, hydrochloric acid, and urea. This has encouraged the "labeled line" fallacy of perceptual performance — namely, that the brain can identify what is stimulating the receptors in one-to-one fashion, e.g., glucose is sweet and NaCl is salty.

An example of the unfortunate effects of this approach in research into human and rodent ingestion is the assumption that fat can be tasted, or at least has a specific texture, in the way that sugar and salt have specific tastes. This is most implausible for foods in the wild or the uncooked or cooked plain plant or animal foods that we and our commensual vermin still eat: fats contribute to very different textural qualities in meat cell walls, within muscles, and in the visible fat, and different again in milk, nuts, and so on.

People judge how much fat is in a variant of a particular foodstuff by recognising the similarity of its exact micropattern of stimulation to mechanoreceptors to a familiar standard variant, although there can be some information also in the strength of the aroma of fat from that source, e.g., of cow's milk. For example, the smoothness of the butterfat globules in milks and creams cues their fat content, not just the fluid's viscosity.[44-46] Furthermore, the smoothness of single cream may be qualitatively different from that of whole milk and not just a higher level of the smoothness parameters of the cream in the milk.[45]

Animal laboratory experiments commonly use isolated preparations of the macronutrients. Pure fat in liquid or solid form feels oily or greasy. That texture of bulk fat cannot be mimicked by any digestible carbohydrate or protein. So any dietary selection paradigm which includes a high fat choice is inevitably confounded by oral sensorimotor factors. Results from such a choice therefore should never be described as providing evidence for fat selection or fat intake — merely for dietary intake which involves a greasy texture.

Rats appear to love very soft and smooth foods such as oils and greasy pastes. An analogy in people may be the contribution of the melting of cocoa butter on the tongue to the addictive qualities of chocolate.[47]

Among solid foods, a diet which readily crumbles but gives particles which are hard appears to be attractive to rodents and human beings. Rats certainly prefer friable chow pellets to hard blocks or floury powders. It may be the texture and size that contribute to the rat's liking for whole grains of wheat. People, too, prefer their cookies to be crunchy and their potato chips to be crisp. All these textures arise from interactions of fat with a matrix of starch (and sugar in the case of cookies).

Thus we need to understand the mechanoreceptor stimulation patterns that comprise these textures of fatty and other foods. When we understand better the neural pathways of sensory preferences for fat, it may become possible to disconfound sensory from nutritional actions of triglycerides sufficiently to establish a neural basis for fat-specific dietary selection.

1.3.1.2   Fixed palatabilities and satiating powers inherent
to foods

The physiologists' original model of the control of the amount eaten in a bout was of "palatability" driving intake and a growing "satiety" state increasingly inhibiting intake. This model is currently undergoing a revival under the influence of the nonmechanistic, "concept"-prescribing approach to intakes and ratings criticised above (Section 1.2.1). Nevertheless, long ago it was comprehensively debunked theoretically[4-8,48] and refuted empirically.[49,50] Both the sensory facilitation from a food (Section 1.2.1.3) and its satiating effect[20] may vary markedly with context.

Sensory facilitation is not inherent in a particular sensory characteristic either — the level of a sensory factor is crucial. The operation of unlearnt brainstem reflexes gives the taste of sugars (or of sour or bitter substances) an increasing effect with increasing concentration. Nevertheless, the sensory influences facilitating (or inhibiting) intake and selection do not generally increase in effect with increasing intensity throughout the range. In a familiar and (vaguely speaking) palatable food or drink, the most ingestive facilitation comes from a particular intensity of sweetness, as well as of aroma, feel, color, etc. (and even of bitterness in coffee or sourness in an orange); this is the result of learning.[6,15,51,52] This learnt peak in sensory preference generally escapes the notice of experimenters because either they test the same levels of the same sensory characteristics as they (unintentionally) trained the animal to like, or they elicit the innate sweetness acceptance reflex increasingly by a series of rising sugar concentrations.

The learnt preference peak is also neglected because experimenters do not use experimental designs and data analyses that detect the stimulus generalisation decrement at concentrations both above and below the learnt level of sweetness as well as of sweet/sour mixtures, aromas, and all the other sensory qualities of a familiar food.

Thus it is highly unrealistic to suppose that the amount eaten in a meal of a sweet fluid results from the palatability of sweetness being gated out by postingestional inhibitory effects. There is no preference for sweetness as such in familiar foodstuffs: the preference is for the particular level of sweetness learnt to be in that material, whether or not that maximum sensory motivation contributes more or less to appetite in different parts of the meal. Indeed, rats learn a relative aversion to a particular level of sweetness (strong or weak) associated with osmotic aftereffects of ingesting hypertonic sugar solutions.[51] Thus, as acknowledged by workers on meal-size control using 0.8 M sucrose or sweetened condensed milk in rats,[53] an aversion is conditioned to that level of sweetness during familiarisation with the diet and testing regimen. The conditioned aversion sharply reduces selection of the diet[51] and the initial rate of eating.[28] It may reduce the acceptance of the diet to such a low level that merely sensory satiety is sufficient to terminate the meal.

However, when the gastrointestinal osmotic effects of this diet are reduced by withdrawing or diverting much of it out of the digestive tract, the learnt aversion to that level of sweetness is extinguished.[53] Also, low levels of sugar still reaching the intestine despite the open gastric fistula will condition a preference to that level of sweetness.[51,54] In those circumstances, there may be no aversive consequences of continuing to ingest the sugary fluid, even those that an excess of a normal food is liable to provoke. Hence, the meal may be resumed again and again, as soon as a transiently accumulating boredom wears off each time.

These effects of concentrated sugars are very different from those of concentrated maltodextrin and starch detailed in Section 2.1.2. It is therefore false to claim that "concentrated carbohydrate" only produces this learnt aversion from the start of the meal:[55] this effect is peculiar to hypertonic test diets such as concentrated sugar solutions,[51] whereas dilute sugar solutions and concentrated maltodextrin sols are hypotonic.[49,51] A learnt aversion specific to the end of a meal (Section 2.1.2) requires a much milder punishment, generated by digestion of concentrated starch products to maltose in the duodenal lumen and/or glucose in the wall of the small intestine.[49]

1.3.1.3 Regulation of nutritional status through control of ingestive behavior

Conceptual confusion concerning physiological regulation, i.e., the stabilisation of the internal environment against challenges from the external environment, has also affected strategies for investigating macronutrient selection.

Demonstration of a regulatory capacity, i.e., evidence for defense of a physiological state by behavioral change or biochemical adaptation, raises two sorts of theoretical questions. These two categories of scientific issue are not reducible to one or other sort of issue, neither are they contradictory nor incompatible in any way, as some have argued.

One set of issues is the advantage that was gained by the gene pool of the species in some past ecology by possession of that regulatory capacity: some evidence can be gained on that by quantitative functional analysis of selective fitness of the contemporary species in a range of environments plausibly related to its evolutionary history; this area of mathematical zoology must be distinguished from "just-so stories" told by psychologists or physiologists on the basis of some laboratory data or field observations.

Coevolution of species and ecologies has produced individual animals with currently operative causal mechanisms that achieve functions which were of selective advantage at some time in the past. Hence, the control of ingestion cannot be understood solely in functional terms, although a narrowly mechanistic analysis is also incomplete. Thus, the distinction between behaviorally causal mechanisms and evolutionarily regulatory functions is crucial to interpreting evidence on the neural control of macronutrient selection by the individual, as distinct from control of nutrient selection by inclusive fitness of the species.

In short, the phrase "regulation of intake" confuses what is regulated with one of the means of that regulation. Rather, the control of intake plays a part in physiological regulation, as does the control of expenditure and excretion.

Such regulation is served by mechanisms controlling the amounts eaten of particular foods in such a way that a nutritional deficit is rectified and an excess is restrained. Quantitative control of dietary intake of the energy content of all the macronutrients could, in principle, be a powerful stabiliser of a physiological state or process such as hepatic energy production, neuronal glucose uptake, or adipose tissue fat content. If the appetite for foods in general (i.e., hunger in the everyday sense of the word) mainly serves to control energy intake, then the total amount of food eaten on an occasion must be influenced by sensory or immediate postingestional effects of those foods in a manner which anticipates the energy eventually yielded by that meal. These anticipatory mechanisms seem to allow for the average energy density of the diet with considerable accuracy since an individual's long-term energy intake can be constant within a few percent (although this argument has been challenged[56]). These mechanisms could hardly operate without having some sensitivity to differences in energy density between types of food. Short-term overeating of high-fat and other energy-dense foods is strong evidence for active control of the volume of food in accord with usual average energy density, contrary to its characterisation as passive overconsumption of fat. Human foods have only recently become labelled with their energy contents, and then only packaged food items. Our ancestors must have had ways of anticipating the energy yield from foods without being told. Certainly laboratory rats cannot read the food composition tables which investigators use to report their eating behaviour in terms of energy intake.

Thus, this field's fundamental notion of control of intake to regulate energy balance, so far from being a coherent theory, merely sets up an array of very difficult mechanistic puzzles.

Even more problematic is the notion of separate control of the intake of different macronutrients. The claim that protein intake is regulated separately from carbohydrate intake[57] is confused because much of the carbon in protein is used by adults for energy, just like the carbon in starches and sugars, as well as in fats. What influences protein intake more directly and potently than carbohydrate intake is its effects on nitrogen metabolism, particularly on the balance of essential amino acids.

The first evidence for protein-specific dietary selection that helped to regulate nitrogen exchange came from rats given a choice between a protein-free but otherwise complete diet and either a concentrated protein diet or a dilute one:[58] the rats on the dilute protein diet whose protein intake fell below about 10% of energy increased their intake of that diet, while the rats on the concentrated diet decreased their intake, at least until liver enzymes had adapted to the high protein intake.

When such tests were repeated with discrete periods of access to arbitrarily flavored diets, rats deprived of protein (and all other nutrients) acquired in one or two trials a preference for the flavor of the richer protein diet,[59] which was expressed when they were again deprived. When the appropriate controls were run a decade later, this protein-conditioned sensory preference indeed proved to depend on an internal state set up by a recent deficit in intake of protein in particular: the preference was not expressed after gastric administration of hydrolysed protein, but appeared in full strength after intubation of an equicaloric dose of carbohydrate;[60,61] see also Table 1.

TABLE 1
Learnt Carbohydrate-Specific Appetite in the Rat.

Rat number	As trained, not repleted		Carbohydrate-repleted		Protein-repleted	
	Intake (ml)	Preference	Intake (ml)	Preference	Intake (ml)	Preference
615	20.2	0.48	14.2	0.56	15.0	0.63
617	10.6	0.81	15.4	0.39	11.6	0.75
618	14.4	0.72	12.0	0.60	11.2	0.69
619	6.4	0.63	7.4	0.25	5.8	0.71
Mean	12.9	0.66[a]	12.2	0.45	10.9	0.69[b]
S.D.	5.9	0.14	3.5	0.16	3.8	0.05

Note: Total fluid intake was from two odors and preference was for the odor previously conditioned during food deprivation by aftereffects of carbohydrate infusion, as the ratio to total intake, in food-deprived rats after gastric infusion of saline or equicaloric doses of carbohydrate or protein (N = 4). (The mean preference ratios were 0.68[b] after saline and 0.48 after carbohydrate infusion in a previous experiment on six rats without the protein infusion test but otherwise identical.) Thus the carbohydrate-paired odor and carbohydrate-deprivation state selectively facilitated intake, whereas the odor and protein deprivation did not.

[a] $P < 0.05$.

[b] $P < 0.005$ that mean preference ratio is not different from 0.5 by (t test).

From Baker, B.J., Ph.D. thesis, University of Birmingham, 1988.

The notion of separate controls of carbohydrate and fat intakes is the most problematic of all. Oxidation does not distinguish the carbon from fat from that from carbohydrate. There have been claims that the body regulates glucose stores in hepatic glycogen separately from fat stores in adipose tissue and that dietary selection contributes to the distinct carbohydrate and fat balances. However, the evidence for separate control of fat and carbohydrate intakes is thin at best (in both senses of thin).

1.3.2   Anticipatory Chemospecific Control of Dietary Selection

1.3.2.1   Conditioning of sensory control of dietary selection and intake

Aside from all those detailed difficulties, the older physiological models have in common one fatal flaw. They approach the eating of food as the intake of energy or even the ingestion of different energy nutrients, such as glucostatic starch, without specifying how the energy or particular macronutrient is identified during dietary choices. The control of fat intake has been vested recently in a bodily fat "balance," but the behavioral mechanisms remain to be specified. None of these theories is based on any identified physiological processes or even plausibly proposed mechanisms by which the energy, glucose, or fat yield of a meal could be measured while the eater is selecting foods for that occasion and deciding how much to eat of each material. It is therefore misleading to report food intakes in energy units or weights or proportions of macronutrients. The only legitimate measure for studies of the control of intake is the volume (or its surrogate, weight) of each particular foodstuff consumed, which can be roughly identified by the eater, visually, by eating movements and in the stomach.

This point was repeatedly stated by Jacques Le Magnen from 1955 onward,[27,62] and within the American literature since 1967.[63,64] Thus, he argued from the time those behaviorally nonmechanistic regulatory models were first advocated that the control of dietary intake by the caloric contents of foods cannot be immediate but must be indirect. His argument was that there is not enough time before the end of the meal for that energy to be metabolised following absorption of the digestion products of macronutrients ingested in the foods. Hence, he proposed, the postprandial metabolic effects of macronutrients have to be predicted somehow from the diet as it is sensed during eating by receptors in the stomach and the oronasal cavity (in the case of the albino laboratory rat, at least, which is often almost blind).

Le Magnen pointed out that there was no consistent sensory cue to the energy content of a natural food, unlike the salty taste which Richter had shown was better liked when the body signaled a lack of sodium salts to the eater/drinker. Hence, he argued, to the extent that contemporaneous influences on selecting amounts of foods to eat do reflect their yield of energy and other nutrition, this immediate control must be established over the longer term by alterations in the operation of those orosensory and gastric controls to allow for the effects of several hours of digestion and assimilation of the meal.

As a pointer to the nature of these adjustments in oral and gastric control of intake, Le Magnen[27] cited the then recent demonstration that wild rats could learn to anticipate the effects of poison in a bait and become frightened of eating that particular bait,[65] a paradigm that was later widely exploited by John Garcia and colleagues. So Le Magnen postulated the conditioning of selective appetites among foods according to the postprandial consequences for energy metabolism.

Thus, in his experiments from 1955 to 1963,[27,28] Le Magnen was concerned to demonstrate that rats could learn to select foodstuffs to which arbitrary sensory cues had been added, and to eat greater amounts of them if those cues predicted a good supply of energy to the tissues and an effective reduction in physiological signals triggering renewed appetite for food. He indeed showed that a variety of manipulations, thought to augment hunger, conditioned aversions to whatever sensory cues were incorporated in the immediately preceding meal, and thereby reduced intake of foods having the flavors that predicted such longer-term undersatiety.

It emerged later, however, that radioisotopic carbon (i.e., dietary fuel) in the mouth of the rat can reach the liver and the brain within 1 to 2 min,[19] even when a fistula in the wall of the stomach is open (thus invalidating the use of the term "sham feeding" for that preparation[66] (D.A. Booth and D. Barber, unpublished data, 1986). Thus, in some circumstances at least, a 'sample' of the first few mouthfuls of a meal may be circulating and metabolised before the end of that bout of eating, even when its duration is brief as an *ad libitum* meal usually is in the rat. Indeed, this initial rise in rate of absorption of energy substrates is in principle predictive of the subsequent peak of absorption,[7,8,67] if not of the whole course of postprandial delivery of digestion products to parenteral metabolism.

Thus, it may not be necessary to learn about the nutritional consequences of the whole period of digestion and absorption of a meal. The initial stimulation of chemoreceptors in the small intestine and/or the liver or brain could provide an adequate anticipation of the energy and other nutrient contents of a meal.

1.3.2.2   Learnt sensory aversions vs. oral metering in meal size control

Another limitation of Le Magnen's demonstrations of anticipatory adjustments of orosensory control of intake (which indeed continue to this day in some experiments on ingestive learning in rats) was that the postprandial manipulations to condition sensory control of selection and intake were unlikely to have mimicked the normal effects of digestion and absorption, either over the whole period of the meal's assimilation or in the period shortly following cessation of eating. He injected concentrated solutions of glucose, supraphysiological doses of insulin, or intake-suppressant doses of amphetamine, or diluted the diet with kaolin. Hypertonic sugar solutions in the human digestive tract can cause nausea and expand uncomfortably when injected under the skin or into the peritoneal cavity.[68] Those adverse osmotic effects were known to be powerful enough to suppress food intake in starved rats[69] and later proved[51] to condition aversion even to the strongly sweet taste of a concentrated sugar solution or of added saccharin. It was also found later that a wide variety of drugs condition aversion to taste or smells sensed before their injection, including amphetamine (e.g., Reference 70). Hypoglycemic doses of insulin were found to be similarly aversive.[71] So there is no reason to think that the conditioning of selective rejection and reduced intake by hypertonic glucose, insulin, or amphetamine is specific to these agents' putative effects on appetite for food. The difficulty of eating and/or the nasty texture of kaolin can by themselves condition aversion, even when the energy density of the diet is maintained with fat.[3]

Indeed, later workers who did not cite Le Magnen's work found that the conditioning of aversions or the reinforcement of avoidance was the major preventive process against ingestion of solids or liquids which are poisonous[72] or nutrient-deficient,[73] rather than serving as an anticipatory mechanism in selection among nutritious diets and the amounts eaten of them. Usually the poisoned or deficient animal would eat nothing of the averted diet ever again[72] and would show an indiscriminate interest in all other foods, including even novel ones about which the healthy animal is hesitant.[73]

A reduction in the malaise of nutrient deficiency has only weak preference conditioning and learnt intake-increasing effects.[74] However, the importance of the existing observations has been exaggerated because the animals continue to suffer aftereffects of chronic deficiency. There has been no test of the learning from a rapid and complete cure of the chronic pathology created by these vitamin or mineral deficiencies or poisons. When an aversion-conditioning nutrient deficiency is corrected over a period of minutes, there is a strong olfactory preference conditioning effect.[75] So it is a myth that the only ingestive conditioning is aversive (and that the aversions are only to taste) or that there is no substantial preference conditioning by the repair of nutritional deficiencies.

In addition, in healthy rats that were only mildly food deprived or were feeding freely, strong conditioning of sensory preferences was demonstrated by the postingestional effects of dilute glucose and other hypotonic carbohydrate preparations,[51] and of balanced amino acid mixtures and high-quality protein.[59,76] These glucose-conditioned preferences were sufficiently strong to make a rat like to drink all its sugar from a barely tasting dilute solution rather than from a simultaneously presented, highly sweet concentrated solution. In retrospect, this glucose-conditioned preference on top of the unlearnt preference for intense sweetness may help to account for the fact that a rat will drink three times its own weight each day of a mixture of saccharin and dilute glucose.[77] Sclafani and colleagues[64] have since demonstrated that the flavor-preference conditioning effect of glucose is virtually inextinguishable under their experimental conditions and is entirely dependent on postingestional effects of the glucose, not any sweet taste.

However, the strength of these acquired aversions and preferences, far from solving the problem of nutrition-anticipatory dietary selection, actually creates a serious difficulty.

Aversion conditioning causes refusal of the averted material. Preference conditioning causes exclusive selection of the now-liked material. These extremes are not the phenomena to be explained. The issue is how the size of a substantial bout of eating is controlled in anticipation of the energy yield or how a nutritious 'cocktail' of diets varying in macronutrient content can be mixed into a meal. There has to be some mechanism which has the effect of "oral metering," even if the volume passing through the mouth is not literally measured, as was proposed for thirst[78] and has been argued for hunger, too.[79]

Even worse, the directions of these learnt responses are counterregulatory for the required graded controls of intake. The intake of a nutrient-dilute diet should increase to provide the needed amount, whereas deficiency conditions decrease in intake. The intake of a nutrient-rich diet should decrease to limit the risks from excess, but the more the carbohydrate in a meal (without aversive osmotic effects), the greater the conditioned increase in intake.

These are both crude paradoxes about nutritionally conditioned sensory aversions and preferences. The key problem about learnt anticipation of nutrition is different again, however. It also appears to be sufficiently subtle or difficult for its resolution[48] to have been generally misunderstood or ignored by workers on the learnt control of the sizes and macronutrient compositions of meals and in discussions endeavouring to weasel out of the above two paradoxes. Le Magnen[27,62] pointed out that his experiments measured two distinct expressions of appetite for food — bout size on a single food and relative intakes in a choice between foods. He found far from complete rejection of one odor and exclusive acceptance of the other after training; these are weaker than maximum learnt responses and so provide an obvious way out of the first pair of paradoxes. However, the aversive responses induced by the agents he used were not obviously regulatory control effects: decreased intake in anticipation of concentrated glucose or hunger reduction by amphetamine looks superficially functional, but less intake after dilution of the diet or hunger induction by insulin is paradoxical from a regulatory point of view.

The solution to all three paradoxes is not to seek regulatory control of amounts consumed solely in the nutritional conditioning of preference/aversion to the sensory qualities of the foods. The logic is the same as realising that "palatability" (sensory preferences of whatever origin) and sensory-specific satiety by themselves are incapable of explaining how meal sizes are adjusted to allow for anticipated nutrient yields. Purely sensory preferences can explain the selection of the food to eat and the initial and continued enthusiasm of eating, but they cannot explain adjustment of the cessation of eating to the nutritional consequences of the meal. The inhibition of intake later in the meal has to be adjusted *separately* from the facilitation of intake at the start of the meal. The only conceivable way to separate the control by one and the same sensory stimulation at two ends of the meal is to make the responses to those sensory characteristics contingent on some effect which develops during the meal. For this separation to serve regulatory functions, it should provide predictive information in addition to that provided by the sensory cues. The most immediate change is the volume in the stomach. This provides information on the amount eaten, which is not directly available from the food's stimulation within the mouth nor is it always readily visible. The amount in the stomach also provides information about the meal as a whole. Early parts of the meal could stimulate intestinal receptors and even begin to be digested, absorbed, and circulated to nutrient detectors in other tissues. Nevertheless, these signals can only provide information on nutrient composition and perhaps concentrations. Total yields cannot be predicted without conjunction of this chemical information with information on meal volume.

Thus the acquired sensory inhibition of eating in regulatory anticipation of nutritional consequences has to be triggered simultaneously by meal volume-related gastric stimulation at least, whether or not other postoral signals are also incorporated in this satiated state. This learnt response has to supercede the learnt and unlearnt facilitation of eating from the head senses and visceral signals. To achieve regulatory control of meal size, it is not necessary that the acquired liking for a food be contingent on an empty stomach, but such learning is liable to occur if the mechanism exists for making an acquired aversion for a food contingent on a partly filled stomach. Furthermore, as we shall see in Sections 2.1 and 2.2, such combination of sensory facilitation and/or inhibition with internal signals requires chemically specific sensitivities, unlike gastric distension, in order to provide a mechanism for nutrient-specific dietary selection.

When Le Magnen's results on acquired anticipatory appetites were examined from this point of view, it was evident that he had obtained no evidence for separate inhibitory sensory control of the termination of meals. The learnt differentiation in meal sizes in single-diet tests was always in the same direction as the learnt differentiation of intake in the two-diet tests (where diets were distinguished only by the added odor, predictive of different nutritional consequences of eating the training meals having those odors). Furthermore, when he presented group mean cumulative intakes at single-odour meals after training with postprandial amphetamine,[80] the ratio of initial rates of intakes was the same as the ratio of meal sizes. This suggested that the conditioned aversion that presumably determined the initial eating rate was also the sole determinant of the point of cessation of eating.

The criterion for evidence of anticipatory regulatory control of meal size was therefore proposed[48,49] to be a dissociation of the initial flavor preference from a terminating aversion to the same flavor. This would be evidence for a conditioned satiety response to the combination of food flavor and gastric filling. It would be the behaviorally simplest mechanism to establish oral metering which limited the intake of nutrient-rich foods. Alternatively, and equally relevant to regulatory control by nutrient yield, an initial flavor aversion could be coupled with a preference toward the end of the meal for the same flavour. This would provide a mechanism for increasing the intake of nutrient-poor (but not totally deficient) foods without countering a functionally desirable initial preference for richer alternatives.

This criterion of anticipatory control of meal size was first met by rats learning from the effects of consuming concentrated suspensions of starch.[48] The odor (or texture) paired with a dilute suspension was consumed at a similar rate to the odor of concentrated starch, perhaps because of a sufficiently strong carbohydrate-conditioned preference. However, the learnt meal sizes on the odor of concentrated starch were substantially smaller than those on the odor added to the dilute starch.

This initial demonstration of a learnt sensory-visceral satiety state interpreted grouped cumulative intake curves as providing direct evidence of increasing inhibition of intake toward the end of the meal, having the flavor anticipating the effects of concentrated starch. However, this appearance of a slowing of eating rate is an artifact of averaging across rats which stop eating at different times after the start of the meal. Slowing is not reliably seen in individual meals unless the meal is large and concentrated.[7] However, the follow-up paper[49] used brief, two-stimulus tests at the start and end of test meals and clearly confirmed the interpretation that there was a switch from relative preference to relative aversion for the odor paired with the richer diet. It should also be noted that this is evidence for an aversively conditioning effect of the rich diet on the flavor with the full stomach, but not necessarily of the conditioning of an actual aversion. All that is necessary to terminate eating is a conditioned loss of all sensory facilitation of eating ("preference") when the stomach is moderately full.

2   LEARNT NUTRIENT-SPECIFIC HUNGERS AND SATIETIES

2.1   FROM PREFERENCE/AVERSION TO HUNGER/SATIETY

2.1.1   Preference and Hunger

We have seen that sodium appetite is not just a liking for the taste of salt. It is not just an increase in liking for salty tastes, either, as in the permanent increase in need-free intake of salt after rats[37] and people[38] have been depleted of fluid, triggering the circulation of aldosterone and angiotensin to the brain, which increases this avidity for sodium salts. The full form of sodium appetite, properly so called, is an increase in preference for the taste of sodium salts when the body is depleted of sodium ions.

Sodium appetite also is not just bodily sodium deficit, nor even the (extra) intake of sodium salts during sodium deficit. The internal state and its recognition in behavior are only operating in sodium appetite when the preference for salty tastes is also increased.

In other words, sodium appetite is an increase in ingestion in response specifically to the conjunction of the taste of salt and a bodily deficit of sodium.

This illustrates the crucial distinction between a preference for a food and a specific hunger. Both the preference and the hunger may be for the same sensorily limited range of foods. However, the preference is independent of bodily state, whereas a specific hunger depends on an internal state of incipient or actual deficit. It follows that the observation of food preference, e.g., selection of one diet more than of others, is not by itself evidence that there is currently an internal state which would make it biologically appropriate to eat that foodstuff. A hunger is a food preference that does depend on a bodily need. Indeed, the specific hunger is defined by what the need is for — sodium, protein, or whatever (including hunger for energy, as opposed to thirst for water). There is a distinction here between behavior and function, or between antecedents and consequences, the blurring of which has caused great confusion. A hunger is not defined by the nutrition delivered by its exercise, but by the need for nutrition which arouses the hunger. Hence, of course, if the need state has been combined with the nutrient-containing food's sensory characteristics, e.g., by the conditioning or reinforcing action of the ingestion of that food, then indulging the hunger will normally anticipate correctly the resultant meeting of the need.

Equally, a hunger is not just a detected metabolic state; it is in addition a preference for certain sensory characteristics of foods. If the hunger is functional, then foods having those sensory qualities will generally provide the needed nutrient, as salty-tasting natural materials generally contain substantial amounts of sodium salts; indeed, the pure salty taste (without bitter tastes, also) is a marker for high proportions of sodium salts relative to potassium, calcium, iron, and so on.

Thus, acquired nutrient-specific hungers cannot be just learnt preferences, nor can they be just metabolic memories.[81] Learnt hungers and satieties are metabolic expectancies[6] controlled by both food sensory memories and digestive state ("metabolic") memories (see Chapter 14). Acquired hungers are ingestion-facilitatory responses to learnt conjunctions of food stimuli and bodily signals which have gained integrated or configural facilitatory control of ingestion. When the learnt internal state is present but the learnt food sensory characteristics are not, then intake and preference are not increased by the hunger mechanism (though they might be increased across many foods by general arousal). When the learnt sensory characteristics are present but not the learnt need state, then there also is no hunger to express in increased intake or selection preference.

2.1.2   Aversion and Satiety

The converse of all the above follows exactly for aversions and satieties, so long as satiation is understood simply as the suppression of appetite by obtaining its objects. Satiety is not simply aversion; it is aversion — or at least lack of preference — contingent on fulness of the stomach or some other state of incipient repletion. Satiety is not simply the response to postingestional signals of repletion; it is a sensorily contingent rejection response: very few states of satiety preclude acceptance of every sort of food. A five-course meal still leaves room for coffee and liqueurs (if that is a style of dining you are used to).

Thus, learnt satieties are ingestion-inhibitory (relative aversion) results of stimulation by particular conjunctions of food cues and bodily signals. After learning from a meal with a carbohydrate-rich start, a food flavor that is preferred near the beginning of a meal becomes rejected after the stomach has become moderately full as a result of eating or intubation of nonnutritive fluid.[82] This is not a general or even unlearnt effect of gastric distension on the attractions of food. The reverse happens after learning from a meal on dilute carbohydrate: the food flavor is highly preferred at the end of the meal;[48,49,82] this way round has been called learnt desatiation[11] and may contribute to the large meals that rats learn to take on sugary drinks when much of the fluid is diverted from the duodenum through an open fistula in the stomach or by withdrawal.[83,84]

The learnt eating-suppressant combination of stimulation patterns to the head senses and to the viscera (and perhaps to the brain itself) is not merely the sum of the effects of food cues' postingestional signals. The external and internal stimulation no longer acts separately but is merged by learning into a distinctive complex of its own, i.e., a configural stimulus or a holistic *gestalt*. The clearest evidence for this comes from experiments which contrast learnt appetite and learnt satiety from the same food cue in the empty and full states.[10,49,82]

This evidence for conditioned satiety thus directly refutes the notion, still commonly deployed (Section 1.3.1.2), that the buildup of satiety subtracts from a fixed palatability of the food to bring a meal to its end. The preference for one food over another is not fixed. That relative preference/aversion does not just decrease to neutrality, either; what is learnt is a reversal of preferences as the meal proceeds and the internal element on the configural control changes from emptiness to fullness. The concepts of palatability and satiety controlling the amount ingested have to be replaced by consideration of the internal-state-dependent changes in momentary relative preference/aversion for the sensory characteristics of the ingestate.[85,86]

This dramatic switch from liking to disliking a flavor (or at least to losing all interest in it) has provided key insights into the physiological mechanisms involved in the learning. These insights come from seeking to explain the extraordinary fact that the same training manipulation induces both the preference early in a meal and the aversion late in a meal, even to the same flavor or texture if only one food is presented throughout the meal. The manipulation producing these opposite effects is the inclusion of readily digested starch in sufficient concentration in the early part of the training meals. The first point to note is that concentrated sugars condition aversion throughout the meal, whereas concentrated maltodextrin (partial hydrolysate of starch, Ross Laboratories brand name Polycose) conditions preference at the start and aversion at the end; therefore, the aversion seen throughout the meal from $0.8\,M$ sucrose[53] is irrelevant to the existence of conditioned satiety, and sucrose and maltodextrin should not be lumped together as "carbohydrate" in this context.[55] The key further finding is that the maltodextrin must be ingested early in the meal; if it is eaten in the latter part of the meal, then preference is conditioned to the food throughout the meal. The only obvious learning mechanism that could explain this pattern of results is that concentrated maltodextrin induces a mildly aversive effect after several minutes' delay;[22,49] the delay is long enough for maltodextrin eaten early in the meal not to inhibit eating directly while the first training meal is continuing, but short enough to come promptly after the end of the meal. A weakly aversive effect can only condition stimuli shortly beforehand, i.e., dessert cues, and if (as in the rat experiments) these are the same flavor throughout the meal, the food cue has to be compounded with the internal state distinctive to the end of the meal in order to elicit relative aversion to replace the preference observed early in the meal. The strong preference-conditioning effect of glucose from the maltodextrin will act longer on cues presented previously, thus generating a learnt facilitation of intake of the flavor with an empty stomach, i.e., the conditioned preference early in the meal. When the maltodextrin is eaten too late in the meal, then its aversive aftereffect comes too late after the end of the meal for such a weak punishment to condition food stimuli removed by the termination of eating that much longer in the past. Hence, with maltodextrin late in the meal, only the strong preference-conditioning effect can operate on the dessert flavor, as it also does on the even earlier first-course flavor to produce a counterregulatory increase in meal size.[22]

This nutritionally regulatory form of food-specific satiety set up by aversive conditioning in previous meals is totally different from the habituation-like, food-specific satiety induced by immediately preceding exposure to the food, which cannot control the intake of a nutrient but only encourage variety in the diet. Learnt sensory-visceral satiety depends on visceral state, whereas exposure-induced satiety does not. Exposure-induced, food-specific satiety depends on very recent stimulation by that satiated food's sensory characteristics, whereas sensory-visceral satiety is evoked immediately when the relevant food is first presented within or after a meal.

The learnt sensory-visceral satiety also is not confined within meals. Several of the demonstrations, in rats[49] and people,[15,86] involved tests up to 20 min after the end of a meal. This provides another example of the invalidity of the assumption that different mechanisms bring a meal to an end than those which delay a return to eating. While the signals from moderate repletion are present, they can contribute to the inhibitory configuration with the cues from foods whose conjunction with fulness has been followed promptly by a mildly aversive consequence.

Behavior theory raises complex issues about this form of learnt control of ingestion,[9,10,48,49] which remain unanswered and may indeed be unanswerable in the form posed, not just for learnt satieties but for all ingestion. Eating may be stimulus-elicited, i.e., a classically conditioned response or, in ordinary language, an involuntary reaction. It may be a discriminative or occasioned instrumental response, i.e., a reinforced operant, traditionally conceived as intentional activity. Natural animal activity probably has both elements. Therefore, this evidence for learnt control of behavior by internal stimuli should not be ignored by those providing other evidence within instrumental paradigms[81] just because it was generated within experimental designs having more affinity to classical conditioning and has usually been discussed in those terms, e.g., "conditioned satiety."[48]

The practical implications of learnt satiety have also been neglected. Readily digested carbohydrate in sufficient concentrations to produce transiently excessive duodenal stimulation will reduce the attractions of desserts on the menu in which the carbohydrate-dense food is included early in the meal. This, then, is a use of energy-dense foods which could help weight control, as well as a reason for including some sugar and/or refined starches in main courses. The efficacy of this potential less fattening habit has yet to be tested in the field, but successful dieters learn satiety better than nondieters do and unsuccessful dieters learn worse than nondieters, not acquiring the sensory-visceral configuration but learning either only the food cues or only the body cues.[15,86] If children were offered easily digested carbohydrate-dense foods in their main courses and allowed to self-regulate, this could help to prevent the onset of obesity.[87] Sugar in baked beans may be a good thing!

2.2   DEPENDENCE OF LEARNT HUNGERS AND SATIETIES ON NUTRITIONALLY SPECIFIC STATES

If the learnt sensory preference/aversion depends only on emptiness or fullness of the stomach, then the resulting dietary selection cannot be specific to a state of need for a particular nutrient. Only if the deficit or repletion state is nutrient-specific is there a mechanism set up by the learning which restricts the selection or rejection of a diet to circumstances where there are physiological signals of incipient need or excess.

Control of ingestion by configurations of food cues and visceral signals can certainly be learnt with simply distension signals from the digestive tract.[82] So the use of starch products (digested to glucose) in those experiments should not be taken to imply that a carbohydrate-specific hunger or satiety has been learnt. Differences in learnt preferences between stages in a meal or hungry and full states certainly cannot establish the nature of the internal stimulation controlling intake.[88] The interpretation of experiments that measure only half of the full set of comparisons must remain open.[89,90]

There is only one way to test the nutritional specificity of the learnt selection or rejection of a food cue — that is, to compare the learnt response between the internal state related to the nutrient in question and an internal state related to another nutrient. Very few such experiments have yet to be carried out. They require not only comprehension and implementation of the training and testing paradigm for learnt food-viscera configurations, but also the physiological and biochemical manipulations to distinguish between internal states.

This has been achieved in several published experiments demonstrating a learnt protein-specific appetite in rats[60,61] and people.[91] Protein lack is compared with protein repletion by intubation or feeding with protein-free or protein-rich preparations, equating energy contents by adding extra carbohydrate to the protein-free preparation, and using a nutrient-free load as a further control.

In one such series of experiments on protein appetite in rats, the test conditions were reversed and carbohydrate-free and carbohydrate-rich gastric infusions were given, made equicaloric using protein. The results were positive (Table 1), providing the only evidence known to the present authors for a carbohydrate-specific appetite. This result also addresses the issue of asymmetrical effects of protein and carbohydrate loads:[90] the choice tests disconfound differences in effects of macronutrients on total intake from learnt differences in macronutrient-specific, need-contingent, sensory selection.

We now turn to results of studies of the effects of drug administration on dietary selection that may provide incidental evidence for nutritionally specific sensory preference/aversion, i.e., macronutrient hunger or satiety.

3   EVIDENCE FOR NEURAL CONTROL OF MACRONUTRIENT SELECTION

3.1   CRITERION FOR POSITIVE EVIDENCE OF MACRONUTRIENT INTAKE

3.1.1   Nutrient Composition of Food Intake Is Not Nutrient Intake

What the eater ingests is the perceived foodstuff. All that the investigator observes is the intake of that material (plus some noningestatory fixed action patterns and/or — in the human case — some linguistic communications, either of which may express processes also controlling the ingestion). It is not justifiable to take the intake (or any of the symbolism) as evidence in itself of ingestive behaviour toward the macronutrients that happen to be in that food. The reporting of dietarily selective intakes as "macronutrient intake" is therefore to be deplored in this context.

Even the well-nigh universal practice of reporting dietary intake in energy units is grossly misleading in the context of mechanisms controlling ingestion and the bidirectional defense (regulation) of physiological states (not set points[5,92]) related to energy balance.

If dietary intake that includes fat can be shown to contribute to a balancing of the fat content of the body, this is still not evidence that the ingestion is controlled by the nutritional effects of the fat in the foods; fat intake could be completely uncontrolled and the balancing achieved purely by metabolic mechanisms.

3.1.2   Nutrient Commonality across Sensory Diversity

It is technically impracticable to make protein with the texture of starch or to give starch the texture of protein. It is not feasible, either, to match or mask the tastes of the fully hydrolysed ("elemental") starch and protein, i.e., glucose and amino acids. Therefore, choices between dextrin and casein will always be confounded by sensory effects, and even by sensorimotor differences in handling the 'cement' or 'glue' produced in the mouth from dextrin or casein, respectively, on wetting by the rat's abundant salivation.

Therefore, the minimum requirement for evidence of a particular neural basis for dietary selection for the nutritive effects of carbohydrate or protein is to observe qualitatively the same effects of a manipulation such as administering a centrally acting drug on intakes in a choice between dextrin and casein, and a choice between two sensorily totally different forms of carbohydrate and protein. Obviously the comparison contrast should be no worse confounded by differential (un)palatability than is choice between dextrin and casein. Baker and Booth[93] used soluble forms of the two macronutrients, i.e., calcium caseinate and maltodextrin (with chalk to equate Ca contents).

Two or three examples of these quite crude variations in the test diets would at least put scientific life into an hypothesis of neurally based protein sparing or carbohydrate selection. Then it would be worthwhile investing in more critical approaches such as gastric infusion and sensorily arbitrary diets.

It is rather more difficult to meet the evidential requirements in the case of fat selection and a neural basis for it. The intensely hydrophobic character of triglycerides makes their bulk form as oil or grease impossible to mimic in texture, and their dispersed form as oil-in-water emulsion (cream) or fat in protein gels (cheese) or starch matrices (baked products) much more difficult to match or mask than is generally appreciated.[94] As well as their oily texture, fats carry strong and distinctive aromas with them, quite different from those in maltodextrin or casein hydrolysate or the macromolecular forms, even when proteins or starches are purified by use of organic solvents etc. A comparison of a solid high-fat diet with an aqueous emulsion for the fatty choice with solid or liquid protein- or carbohydrate-rich diets is an essential first step. The smoothness of texture in common between solid and liquid would remain a serious concern, however. Again, the only sound way forward involves bypassing the sensory characteristics of fat by relating realistic forms and doses of gastric administration to the oral ingestion of fat-free diets which are learnt to be preferred when there is a deficit in the usual effects of recent consumption of fat.

In short, for there to begin to be evidence that the intake is controlled by the nutrient content of the diet, the choice between two or more diets has to be in a constant pattern across conditions of dietary formulation that differ markedly in sensory characteristics. Only then does the possibility arise that the drug is acting on some nutrient appetite pathways rather than on sensorimotor pathways.[95,96]

3.1.3   Application of the Criterion

These requirements for evidence of intake which is selective for one or more macronutrients have recently been reviewed in detail and applied to the research literature on effects of administering drugs on dietary selection in rats.[2] The claims for catecholaminergic or opioid involvement in protein intake and peptidergic involvement in carbohydrate intake were not substantiated, largely because casein was the only protein preparation included in the test diets in all three series of experiments.

The few experiments testing the effects of drug administration on human dietary selection were critiqued earlier by a more cognitive application of the same basic criterion of disconfounding sensory differences from nutrient composition.[97] These criticisms apply independently of the points that the allegedly high-carbohydrate foods were, in fact, high-fat foods, and that the protein levels even in those "junk" foods were too high to permit the effects of plasma tryptophan uptake into the brain observed in rats fed on a protein-free diet. There is no evidence for a protein-sparing effect of the markedly serotoninergic appetite-suppressant d-fenfluramine; rather, when dieters are faced with the opportunity to eat while their appetite for food is reduced by a drug they believe to have such effects, they will try to select items in modest amounts that they feel like eating from foods which they believe to be nutritious, such as ham sandwiches. These items happen to be higher in protein. No evidence was collected on whether or not these research participants believed that these foods were high in protein; if they did know that the ham sandwiches were high in protein, their selection is not evidence of a brain mechanism designed to spare protein as distinct from neural mechanisms for learning of a vast range of material contained in formal and informal education.

There is now clear evidence for a learnt protein-specific hunger in human adults[91] as in rats (reviewed in Section 2.2). Nevertheless, there are as yet no data on the issue whether or not preferences for frequently eaten high-protein foods are based at all on this learnt hunger. It is doubtful that usual eating patterns involve meals low enough in protein content or long enough intervals between meals to generate the physiological state of incipient protein deficit which is necessary for a protein-specific food preference, i.e., genuine protein selection, uninformed by nutrition education and unbiased by social pressures.

3.2   PROTEIN HUNGER AND OPIATES

The evidence opens the possibility that the opioids are involved in protein selection. However, the literature does not demonstrate opioid involvement in protein intake by a constant pattern of selection in a choice of two or more diets formulated on differing principles of sensory character. For example, in a choice between diets having a 4:1 carbohydrate/fat ratio and a 4:1 protein/fat ratio, the reported diet-specific, deprivation-dependent facilitation of protein intake by morphine[98] might be an effect of morphine on sensory satiety rather than an augmentation of protein appetite. In a choice among casein, sucrose/corn starch/dextrin, and lard diets, low doses (2 and 2.5 mg/kg) of morphine caused an increased caloric intake from the casein diet during the following 2 h,[99,100] whereas in a choice among casein, corn starch/dextrin/sucrose, and hydrogenated fat/safflower oil diets, higher doses (20 and 40 mg/kg) did not.[101] Replicating the increase in protein intake on a different free-feeding dietary paradigm (e.g., a choice between two diets) and with proteins other than casein with IP morphine in a dose range of 2 to 2.5 mg/kg would be evidence for excitatory involvement of opioid transmission in protein-specific dietary selection under *ad libitum* conditions.

There are opposite effects of morphine on casein diet intake during restricted feeding, which could be a different sort of evidence for endorphinergic modulation of protein appetite. However, this finding has not been supported by the effects of an antagonist. Furthermore, gnawing has been proposed to contribute to effects of morphine on feeding.[102,103] Since the texture of casein may affect gnawing, it is possible that sensorimotor constraints from the diet interacted with deprivation status to induce the observed changes in selection of this protein preparation.

3.3   PUTATIVE CARBOHYDRATE HUNGER OR SATIETY
AND 5HT OR NE

A variety of starch and sugar preparations have consistently provided evidence that rats' intake of carbohydrate is suppressed while casein intake is relatively spared after systemic administration of the largely serotonergic drug d-fenfluramine[104-106] or the injection of serotonin itself (5HT) into the region of the paraventricular nucleus of the anterior hypothalamus (PVN). These observations were made in a choice among macronutrients early in the dark phase of a 12-h/12-h light/dark cycle and at the end of the light phase.[107-110] This 'dusk' period is when rats' gastric emptying speeds up[110] and so their meal frequency increases.[111] Also, at this time a majority of rats increase their intake of carbohydrate-rich diets relative to protein-rich and fat-rich diets.[112] An hypothesis that these two phenomena are connected is supported by the observation that 5HT injection into the PVN region does not suppress carbohydrate-rich diet intake selectively in the middle or late dark period.[108]

If serotonergic activity in the PVN region does counter the early nighttime rise in carbohydrate intake, there are two theoretical possibilities: 5HT there may suppress either a rise in appetite specifically for carbohydrate-rich diets or a release from carbohydrate-specific satiety. Suppression of a carbohydrate-need-dependent preference for the sensory qualities of carbohydrate-yielding diets is unlikely because no evidence was found for the involvement of serotoninergic activity in carbohydrate-conditioned preference for odor[96] or texture.[93,113] This leaves the possibility that PVN 5HT is involved in a reduction in meal size by a relative sensory aversion which has been coupled by learning to high levels of glucose released from the earliest stages of digestion — a carbohydrate-specific conditioned satiety. As noted in Section 2.2 above, the known mechanisms of carbohydrate-conditioned satiety are not necessarily specific to carbohydrate or any nutrient, but the possibility of conditioned meal termination dependent on a configuration of food cues and excessive carbohydrate digestion has yet to be tested. Clearly, such experiments should be done in rats showing the crespicular rise in carbohydrate intake. If learnt glucose-specific satiety can be demonstrated, then the effects of d-fenfluramine on its performance should be tested. Positive results would provide the first example of some understanding of a neural basis for a macronutrient-specific satiety (or hunger).

Injection of norepinephrine (NE) close to the lateral PVN, or other manipulations increasing alpha-2 activity, increases the size of meals or even elicits a meal during the same period, shortly before and in the early period of darkness. The evidence is that this noradrenergic facilitation of intake arises from blockade of conditioned satiation.[95] This learnt reduction in meal size is the rejection of a particular foodstuff when the stomach has been filled to an extent which has in the past been followed by an aversive effect of digesting concentrated starch.[22,49] However, only mechanical distension of the upper digestive tract is necessary as the internal cue for reduction of intake of food providing the external cue.[82] The learnt reduction in meal size blocked by PVN NE may therefore not be carbohydrate-specific, even though it was induced by an aversive effect of starch. The critical experiments remain to be done.

3.4   IN SEARCH OF EVIDENCE FOR A FAT-SPECIFIC HUNGER

High-fat diets differ in sensory characteristics from diets containing carbohydrate or protein. Isolated fat can have a glossy sheen which is seldom found in carbohydrate and protein. Emulsified fat makes dairy products look milky and feel smooth. Fat in foodstuffs made of starch and/or sugar provides distinctive crispness or crunchiness. These textures arise from the hydrophobicity specific to fats which so far is impossible to mimic with nonfatty materials, as noted in Section 3.1.2. High-fat diets and pure oil are sensorily attractive to rats, probably because of their fluidity and smoothness, and such strong preferences might override the control by sensory-postingestional integration that is required for nutrient-specific intake.

High-fat diets are typically diluted with cellulose in order to equate the energy density to high-carbohydrate and high-protein diets. This gives the fat-containing diets further textural characteristics that cannot be mimicked by carbohydrate or protein preparations.

It has been reported that the source of dietary fat affects the intake of high-fat diet in a choice between diets.[114] This specificity to the character of a particular type of fat proved that there was no fat-specific selection, i.e., no control of intake by effects that different fat preparations share.

The elucidation of neural mechanisms is further complicated because neurotransmitters such as 5-HT and DA are involved in brainstem sensorimotor circuits essential to tactile reflexes. Thus, changes in textural preferences or sensorimotor coordination could be mediating the effects on intake of high-fat diets observed after administration of serotonergic and catecholaminergic agents via routes that permit the drug to act throughout the brain. Indeed, the first reported effects of systemic administration of dl-fenfluramine on intakes in dietary choice proved to be related to the textures of the diets, not to their nutrient contents. The effect of dl-fenfluramine on selection between pairs of three sizes of chow crumbs containing dextrin or maltodextrin and casein or calcium caseinate (insoluble or soluble powder of each nutrient) was to reduce the intake preference for a coarser over a finer crumb, with no consistent effect of the carbohydrate or protein added.[96] The drug therefore affects how large-scale dietary textures are managed by the rat. Hence, it could change the effects on intake or preference of the pasty texture of the lard or vegetable oil diets and in ways that differ from changes in mastication of the powdery textures of casein and of carbohydrate diets including cornstarch or dextrin. Since 5-HT systems normally act to inhibit mastication,[115] the effect of fenfluramine could have been to reduce intake of the diet which was the most laborious to eat. Until the suppression of fat intake by fenfluramine is replicated with macronutrient diets not differing in texture in ways that are confounded with nutrient contents, the neural systems on which the drug acts cannot be safely implicated in the dietary selection specifically of fat.

Suppression of intake of lard or vegetable fat (safflower oil) was a common effect of dl- and d-fenfluramine.[106,109,116] However, the actions of d-fenfluramine also resulted in a decrease in caloric intake from carbohydrate,[104-106] and a carbohydrate intake suppressive effect of d-fenfluramine was seen also where fat was not available, i.e., in choice between diets having a high level of cornstarch or casein.[104,105] Thus, suppression of fat intake is not a consistent effect of fenfluramine across varied conditions. Therefore, there is no evidence that it acts on neural pathways involved in selection of fat, even if a fat-specific appetite indeed exists.

Morphine was shown to increase the intake of the diet that rats ate most of in a choice — that is, when treated with morphine, more of a cornstarch/dextrin/sucrose mixture was eaten by rats that preferred this diet, and more vegetable shortening/safflower oil intake occurred in rats that favored that diet.[117] Such effects of a drug on baseline preference could be mediated by intensification of the sensory signal or affective reactions that drive the baseline and so have nothing to do with nutrient-specific appetite pathways.

Further work is needed to identify the mechanisms controlling the greater intake of the fat mixtures (as well as protein-rich diet) which was observed at the end of the dark period in rats that predominantly took carbohydrate mixtures around the start of dark.[112] This late-night selection might be the unmasking of a basic sensory preference for fat which was obscured by a carbohydrate-specific preference earlier in the dark phase. Even if this shift in selection was shown to be an appetite of fat or a satiety for carbohydrate, any drug-increased selection of fat-rich diet late in the dark phase would similarly have to be analysed for nutritional or sensory control. At the very least, the drug effect would have to be shown to be independent of the form of fat in the diet, e.g., paste, emulsion, and oil.

3.5   PROTOCOL FOR EVIDENCE OF MACRONUTRIENT SELECTION

It should be clear from the above that no evidence for nutrient selection, let alone for a physiological process involved in nutrient-specific intake, can be obtained without data comparing at least two versions of each macronutrient that are totally distinct sensorily. A set of diets has been proposed[2] that provides three-way sensory contrasts for carbohydrate, fat, and protein. These are two kinds of solid diets, based on saliva-soluble or insoluble forms of each macronutrient, and water-based liquid diets. To mask differences in smell, a different odorant (such as benzaldehyde, dibutyryl, or maltol) should be added to each diet, with nutrient-odorant pairings constant for one animal but balanced across animals. The weights of intakes of the isocaloric diets should be analysed to compare patterns of choice when the form of the presented diets is changed. The requirement for any interpretation in terms of macronutrient intake or selection is identical patterns across two such choices.

This protocol[2] provides similarly preferred but sensorily contrasting isocaloric diets high in protein, carbohydrate, or fat, for choice between two diets or among three diets. The solid diets should be presented to rats and mice as granules (not fine powder), or pelleted if feasible, as these are the physical forms preferred by rodents. Liquid diets must use an emulsifier (such as lecithin) to keep fat dispersed in water, and a thickener (such as guar gum) to keep insoluble or immiscible ingredients in suspension while available to the animal. Some fiber is also important: pectin serves as both soluble fiber and a thickener; cellulose flour is insoluble fiber that requires suspending. Equating energy densities necessitates adding fiber to the fat-rich diet in a three-way choice. Protein diets based on casein require choline chloride and methionine. All diets must include vitamins and minerals. Every diet should be moderately preferred so that there are no animals which refuse one diet or eat almost exclusively from one diet.

An early version of this protocol[1] was used to examine the first claims that a drug affects macronutrient selection.[118] The results showed that the original claims were unfounded. The few claims that are supported by similarity in choice pattern across sensorily contrasting sets of diets, identified in Sections 3.2 to 3.4, should be tested further using this protocol. If they all fail to be confirmed, it would have to be concluded that we know nothing about the neural processes underlying protein-specific or carbohydrate-specific selection, nor fat-specific selection, if it exists. Even claims to have observed macronutrient selection differences between strains of animal, times of day, or differences in ages or hormonal status are unsupported unless they have survived examination by such a protocol. The only example that we have seen as yet compares liquid choices with powder choices and confirms greater selection of fat as such by AKR/J mice and of carbohydrate by SWR/J mice.[118]

REFERENCES

1. Baker, B.J. and Booth, D.A., Effects of dl-fenfluramine on dextrin and casein intakes influenced by textural preferences, *Behav. Neurosci.*, 104, 153–159, 1990.

2. Thibault, L. and Booth, D.A., Macronutrient-specific dietary selection in rodents and its neural bases, *Neurosci. Biobehav. Rev.*, 23, 457–528.

3. Booth, D.A., Caloric compensation in rats with continuous or intermittent access to food, *Physiol. Behav.*, 8, 891–899, 1972.

4. Booth, D.A., Approaches to feeding control, in *Appetite and Food Intake*, T. Silverstone, Ed., Abakon Verlagsgesellschaft/Dahlem Konferenzen, West Berlin, 417–478, 1976.

5. Booth, D.A., Toates, F.M., and Platt, S.V., Control system for hunger and its implications in animals and man, in *Hunger: Basic Mechanisms and Clinical Implications*, D. Novin, W. Wyrwicka, and G.A. Bray, Eds., Raven Press, New York, 127–142, 1976.

6. Booth, D.A., Appetite and satiety as metabolic expectancies, in *Food Intake and Chemical Senses*, Y. Katsuki, M. Sato, S.F. Takagi, and Y. Oomura, Eds., University of Tokyo Press, Tokyo, 317–330, 1977.

7. Booth, D.A., Prediction of feeding behaviour from energy flows in the rat, in *Hunger Models: Computable Theory of Feeding Control*, D.A. Booth, Ed., Academic Press, London, 227–228, 1978.

8. Booth, D.A. and Mather, P., Prototype model of human feeding, growth and obesity, in *Hunger Models: Computable Theory of Feeding Control*, D.A. Booth, Ed., Academic Press, London, 279–322, 1978.

9. Booth, D.A., Acquired behavior controlling energy intake and output, in *Obesity*, A.J. Stunkard, Ed., W.B. Saunders, Philadelphia, 101–143, 1980.

10. Booth, D.A., Conditioned reactions in motivation, in *Analysis of Motivational Processes*, F.M. Toates and T.R. Halliday, Eds., Academic Press, London, 77–102, 1980.

11. Booth, D.A., Food-conditioned eating preferences and aversions with interoceptive elements: learnt appetites and satieties, *Ann. N.Y. Acad. Sci.*, 443, 22–37, 1985.

12. Booth, D.A., Conner, M.T., and Gibson, E.L., Measurement of food perception, food preference, and nutrient selection, *Ann. N.Y. Acad. Sci.*, 561, 226–242, 1989.

13. Booth, D.A., Protein- and carbohydrate-specific cravings: neuroscience and sociology, in *Chemical Senses, Volume 4: Appetite and Nutrition*, M.I. Friedman, M.G. Tordoff, and M.R. Kare, Eds., Marcel Dekker, New York, 261–276, 1991.

14. Booth, D.A., Acquired ingestive motivation and the structure of food recognition, in *Behavioral Aspects of Feeding*, B.G. Galef, M. Mainardi, and P. Valsecchi, Eds., Harwood Academic Publishers, Langhorne, PA, 37–61, 1994.

15. Booth, D.A., Gibson, E.L., Toase, A.-M., and Freeman, R.P.J., Small objects of desire: the recognition of foods and drinks and its neural mechanisms, in *Appetite: Neural and Behavioural Bases*, C.R. Legg and D.A. Booth, Eds., Oxford University Press, Oxford, 98–126, 1994.

16. Epstein, A.N., Thirst and salt intake, in *Thirst*, D.J. Ramsay and D.A. Booth, Eds., Springer-Verlag, New York, 481–501, 1991.

17. Booth, D.A., Summary: concluding session. How do eating disorders work? *Ann. N.Y. Acad. Sci.*, 575, 466–471, 1989.

18. Booth, D.A., Learned ingestive motivation and the pleasures of the palate, in *The Hedonics of Taste*, R.C. Bolles, Ed., Erlbaum, Hillsdale, NJ, 29–58, 1991.

19. Pilcher, C.W.T., Jarman, S.P., and Booth, D.A., The route of glucose to the brain from food in the mouth of the rat, *J. Comp. Physiol. Psychol.*, 87, 56–61, 1974.

20. Dibsdall, L.A., Wainwright, C.J., Read, N.W., and Booth, D.A., How fats and carbohydrates in familiar foods contribute to everyday satiety by their sensory and physiological actions, *Nutr. Food Sci.*, No. 5, 37–43, 1996.

21. Rogers, P.J., and Schutz, H.G., Influence of palatability on subsequent hunger and food intake: a retrospective replication, *Appetite*, 19, 155–156, 1992.

22. Booth, D.A., Mather, P., and Fuller, J., Starch content of ordinary foods associatively conditions human appetite and satiation, indexed by intake and eating pleasantness of starch-paired flavours, *Appetite*, 3, 163–184, 1982.

23. Berridge, K.C., Food reward: brain substrates of wanting and liking, *Neurosci. Biobehav. Rev.*, 20, 1–25, 1996.

24. Craig, W., Appetites and aversions as constituents of instincts, *Biol. Bull.*, 34, 91–107, 1918.

25. Booth, D.A., Taste reactivity in satiated, ready to eat and starved rats, *Physiol. Behav.* 8, 901–908, 1972.

26. Booth, D.A., Influences on human drinking behaviour, in *Thirst: Physiological and Psychological Aspects*, D.J. Ramsay and D.A. Booth, Eds., Springer-Verlag, London, 52–72, 1991

27. Le Magnen, J., Effets sur la prise alimentaire du rat blanc des administrations postprandiales d'insuline et le mécanisme des appétits caloriques, *J. Physiol. Paris*, 48, 789–802, 1956.

28. Le Magnen, J., La facilitation différentielle des réflexes d'ingestion par l'odeur alimentaire, *C. R. Seances Soc. Biol. Paris* 154, 1355–1388, 1963.

29. Yeomans, M., Palatability and the micro-structure of feeding in humans: the appetizer effect, *Appetite*, 27, 119–133, 1996.

30. Treit, D. and Deutsch, J.A., Pleasantness of food does not decrease at the end of a meal, *Physiol. Behav.*, 50, 991–999, 1991.

31. Rogers, P.J., Interaction between palatability and satiation, *Appetite*, 32, 290, 1999.

32. Richter, C.P., Schmidt, E.C.H., and Malone, P.D., Further observations on the self-regulatory dietary selection of rats made diabetic by pancreatectomy, *Bull. Johns Hopkins Hosp.*, 76, 192–219, 1945.

33. Denton, D.A., *The Hunger for Salt*, Springer-Verlag, New York, 1984.

34. Dethier, V.G. and Bodenstein, D., Hunger in the blowfly, *Z. für Tierpsychol.*, 15, 129–138, 1957.

35. Booth, D.A. and Brookover, T., Hunger elicited in the rat by a single injection of crystalline bovine insulin, *Physiol. Behav.* 3, 439–446, 1968.

36. McCaughey, S.A. and Scott, T.R., The taste of sodium, *Chem. Senses*, 22, 663–676, 1998.

37. Nicolaidis, S., Galaverna, O., and Metzler, C.H., Extracellular dehydration during pregnancy increases salt appetite of offspring, *Am. J. Physiol.*, 258, R281-R283, 1990.

38. Crystal, S.R. and Bernstein, I.L., Infant salt preferences and mother's morning sickness, *Appetite*, 30, 297–307, 1998.

39. Leshem, M. and Rudoy, Y., Hemodialysis increases the preference for salt in soup, *Physiol. Behav.*, 61, 65–69, 1997.

40. Booth, D.A., Thompson, A.L., Shepherd, R., Land, D.G., and Griffiths, R.P., Salt intake and blood pressure: the triangular hypothesis, *Med. Hypotheses*, 24, 325–328, 1987.

41. Fregly, M.J., Specificity of the sodium chloride appetite of adrenalectomized rats. Substitution of lithium chloride for sodium chloride, *Am. J. Physiol.*, 195, 645–653, 1958.

42. Beauchamp, G.K., Bertino, M., Burke, D., and Engelman, K., Experimental sodium depletion and salt taste in normal human volunteers, *Am. J. Clin. Nutr.*, 51, 881–889, 1990.

43. Conner, M.T., Haddon, A.V., Pickering, E.S., and Booth, D.A., Sweet tooth demonstrated: individual differences in preference for both sweet foods and foods highly sweetened, *J. Appl. Psychol.*, 73, 275–280, 1988.

44. Richardson, N.J., Booth, D.A., and Stanley, N.L., Effect of homogenization and fat content on oral perception of low and high viscosity model creams, *J. Sensory Stud.*, 8, 133–143, 1993.

45. Richardson, N.J. and Booth, D.A., Multiple physical patterns in judgments of the creamy texture of milks and creams, *Acta Psychol.*, 84, 93–101, 1993.

46. Mela, D., Langley, K.R., and Martin, A., Sensory assessment of fat content: effect of emulsion and subject characteristics, *Appetite,* 22, 67–82, 1994.
47. Hyde, R.J. and Witherly, S.A., Dynamic contrast: a sensory contribution to palatability, *Appetite,* 21, 1–16, 1993.
48. Booth, D.A., Conditioned satiety in the rat, *J. Comp. Physiol. Psychol.,* 81, 457–471, 1972.
49. Booth, D.A. and Davis, J.D., Gastrointestinal factors in the acquisition of oral sensory control of satiation, *Physiol. Behav.* 11, 23–29, 1973.
50. Booth, D.A., Lee, M., and McAleavey, C., Acquired sensory control of satiation in man, *Br. J. Psychol.,* 67, 137–147, 1976.
51. Booth, D.A., Lovett, D., and McSherry, G.M., Postingestive modulation of the sweetness preference gradient in the rat, *J. Comp. Physiol. Psychol.,* 78, 485–512, 1972.
52. Booth, D.A., Objective measurement of influences on food choice, *Appetite,* 7, 236–237, 1986.
53. Davis, J.D. and Smith, G.P., Learning to sham feed: behavioral adjustments to the absence of negative feedback signals, *Am. J. Physiol.,* 259, R1228–1235, 1990.
54. Sclafani, A., Cardieri, C., Tucker, K., Drucker, D.B., and Ackroff, K., Intragastric glucose but not fructose conditions robust flavor preferences in rats, *Am. J. Physiol.,* 265, R320–R325, 1993.
55. Davis, J.D., Smith, G.P., and Meisner, J., Learning to sham feed: behavioral adjustments to loss of postpyloric stimulation, *Am. J. Physiol.,* 264, R888–R895, 1993.
56. Schilstra, A., A stochastic model of food intake in the rat, in *Hunger Models,* D.A. Booth, Ed., Academic Press, London, 1978.
57. Rozin, P., Are carbohydrate and protein intakes separately regulated? *J. Comp. Physiol. Psychol.,* 65, 23–29, 1968.
58. Booth, D.A., Food intake compensation for increase or decrease in the protein content of the diet, *Behav. Biol.,* 12, 31–40, 1974.
59. Booth, D.A., Acquired sensory preferences for protein in diabetic and normal rats, *Physiol. Psychol.,* 2, 344–348, 1974.
60. Gibson, E.L. and Booth, D.A., Acquired protein appetite in rats: dependence on a protein-specific need state, *Experientia,* 42, 1003–1004, 1986.
61. Baker, B.J., Booth, D.A., Duggan, J.P., and Gibson, E.L., Protein appetite demonstrated: learned specificity of protein-cue preference to protein need in adult rats, *Nutr. Res.,* 7, 481–487, 1987.
62. Le Magnen, J., Le mécanisme d'etablissement d'un appétit différentiel pour les régimes de diverses densités caloriques, *J. Physiol.,* 49, 1105–1117, 1957.
63. Le Magnen, J., Control of food intake, in *Handbook of Physiology: Alimentary Canal,* C.F. Code, Ed., American Physiological Society, Washington, D.C., 1967.
64. Le Magnen, J., Peripheral and systemic actions of food in the caloric regulation of intake, *Ann. N.Y. Acad. Sci.,* 157, 1126–1156, 1969.
65. Rzoska, J., Bait shyness: a study in rat behaviour, *Br. J. Anim. Behav.,* 1, 128–135, 1953.
66. Sclafani, A. and Nissenbaum, J.W., Is gastric sham-feeding really sham-feeding? *Am. J. Physiol.,* 248, R387–R390, 1985.
67. Toates, F.M. and Booth, D.A., Control of food intake by energy supply, *Nature,* 251, 710–711, 1974.
68. Booth, D.A., Metabolism and the control of feeding in man and animals, in *Chemical Influences on Behaviour,* K. Brown and S.J. Cooper, Eds., Academic Press, London, 1979, 79–134.
69. McCleary, R.A., Taste and post-ingestion factors in specific-hunger behavior. *J. Comp. Physiol. Psychol.,* 46, 411–421, 1953.
70. D'Mello, G.D., Stolerman, I.P., Booth, D.A., and Pilcher, C.W.T., Factors influencing flavour aversions conditioned with amphetamine in rats, *Pharmacol. Biochem. Behav.,* 7, 185–190, 1977.
71. Lovett, D. and Booth, D.A., Four effects of exogenous insulin on food intake, *Q. J. Exp. Psychol.,* 22, 406–419, 1970.
72. Garcia, J., Kimeldorf, D.J., and Koelling, R.A., Conditioned aversion to saccharin resulting from exposure to gamma radiation, *Science,* 122, 157–158, 1955.
73. Rozin, P., Specific aversions as a component of specific hungers, *J. Comp. Physiol. Psychol.,* 64, 237–242, 1967.
74. Garcia, J., Ervin, F.R., Yorke, C., and Koelling, R.A., Conditioning with delayed vitamin injections, *Science,* 155, 716–718, 1965.

75. Simson, P.C. and Booth, D.A., Effect of CS-US interval on the conditioning of odour preferences by amino acid loads, *Physiol. Behav.* 11, 801–808, 1973.

76. Booth, D.A. and Simson, P.C., Food preferences acquired by association with variations in amino acid nutrition, *Q. J. Exp. Psychol.*, 23, 135–145, 1971.

77. Valenstein, E.S., Cox, V.C., and Kakolewski, J., Polydipsia elicited by the synergistic action of a saccharin and glucose solution, *Science,* 157, 552–554, 1967.

78. Bellows, R.T., Time factors in water drinking in dogs, *Am. J. Physiol.,* 125, 87–97, 1939.

79. Mook, D.G., Satiety, specifications and stop rules: feeding as voluntary action, in *Progress in Psychobiology and Physiological Psychology, Volume 14*, A.N. Epstein and A.R. Morrison, Eds., Academic Press, New York, 1991.

80. Le Magnen, J., La facilitation différentielle des réflexes d'ingestion par l'odeur alimentaire, *C. R. Seances Soc. Biol. Paris,* 157, 1165–1170, 1963.

81. Davidson, T.L., The nature and function of interoceptive signals to feed: toward integration of physiological and learning perspectives, *Psychol. Rev.,* 100, 640–657, 1993.

82. Gibson, E.L. and Booth, D.A., Dependence of carbohydrate-conditioned flavor preference on internal state in rats, *Learning and Motivation* 20, 36–47, 1989.

83. Davis, J.D. and Campbell, C.S., Peripheral control of meal size in the rat: effect of sham feeding on meal size and drinking rate, *J. Comp. Physiol. Psychol.,* 83, 379–387, 1974.

84. Mook, D.G., Culberson, R., Gelbart, R.G., and McDonald, K., Oropharyngeal control of ingestion in the rat: acquisition of sham drinking patterns, *Behav. Neurosci.,* 97, 574–584, 1983.

85. Warwick, Z.S. and Weingarten, H.P., Flavor-postingestive consequence associations incorporate the behaviorally opposing effects of positive reinforcement and anticipated satiety: implications for interpreting two-bottle tests, *Physiol. Behav.,* 60, 711–715, 1996.

86. Booth, D.A. and Toase, A.M., Conditioning of hunger/satiety signals as well as flavour cues in dieters, *Appetite,* 4, 235–236, 1983.

87. Birch, L.L., Children's food acceptance patterns, *Ann. Nestlé,* 56, 11–18, 1998.

88. Kern, D.L., McPhee, L., Fisher, J., Johnson, S., and Birch, L.L., The postingestive consequences of fat condition preferences for flavors associated with high dietary fat, *Physiol. Behav.,* 54, 71–76, 1993.

89. Bolles, R.C., Hayward, L., and Crandall, C., Conditioned taste preferences based on caloric density, *J. Exp. Psychol. Anim. Behav. Process.,* 7, 59–69, 1981.

90. Pérez, C., Ackroff, K., and Sclafani, A., Carbohydrate- and protein-conditioned flavor preferences: effects of nutrient preloads, *Physiol. Behav.,* 59, 467–474, 1996.

91. Gibson, E.L., Wainwright, C.J., and Booth, D.A., Disguised protein in lunch after low-protein breakfast conditions food-flavor preferences dependent on recent lack of protein intake, *Physiol. Behav.,* 58, 363–371, 1995.

92. Peck, J.W., Situational determinants of the body weights defended by normal rats and rats with hypothalamic lesions, in *Hunger,* D. Novin, W. Wyrwicka and G.A. Bray, Eds., Raven Press, New York, 297–311, 1976.

93. Baker, B.J. and Booth, D.A., Effects of dl-fenfluramine on dextrin and casein intakes influenced by textural preferences, *Behav. Neurosci.,* 104, 153–159, 1990.

94. Richardson, N.J., Booth, D.A., and Stanley, N.L., Effect of homogenization and fat content on oral perception of low and high viscosity model creams, *J. Sensory Stud.,* 8, 133–143, 1993.

95. Booth, D.A., Gibson, E.L., and Baker, B.J., Behavioral dissection of the intake and dietary selection effects of injection of fenfluramine, amphetamine or PVN norepinephrine, *Soc. Neurosci. Abstr.,* 15, 593, 1986.

96. Gibson, E.L., and Booth, D.A., Fenfluramine and amphetamine suppress dietary intake without affecting learned preferences for protein or carbohydrate cues, *Behav. Brain Res.,* 30, 25–29, 1988.

97. Booth, D.A., Central dietary "feedback onto nutrient selection": not even a scientific hypothesis, *Appetite,* 8, 195–201, 1987.

98. Evans, K.R. and Vaccarino, F.J., Amphetamine- and morphine-induced feeding: evidence for involvement of reward mechanisms, *Neurosci. Biobehav. Rev.,* 14, 9–22, 1990.

99. Bhakthavatsalam, P. and Leibowitz, S.F., Morphine elicited feeding: diurnal rhythm, circulating corticosterone and macronutrient selection, *Pharmacol. Biochem. Behav.,* 24, 911–917, 1986.

100. Shor-Posner, G., Azar, A.P., Filart, R., Tempel, D., and Leibowitz, S.F., Morphine-stimulated feeding: analysis of macronutrient selection and PVN lesions, *Pharmacol. Biochem. Behav.,* 24, 931–939, 1986.

101. Ottaviani, R. and Riley, A.L., Effect of chronic morphine administration on the self-selection of macronutrients in the rat, *Nutr. Behav.*, 2, 27–36, 1984.

102. Rowland, N.E. and Marques, D.M., Stress-induced eating: misrepresentation? *Appetite*, 1, 225–228, 1980.

103. Antelman, S.M. and Rowland, N., Endogenous opiates and stress-induced eating, *Science*, 214, 1149–1150, 1981.

104. Luo, S. and Li, E.T.S., Food intake and selection patterns of rats treated with dexfenfluramine, fluoxetine and RU24969, *Brain Res. Bull.*, 24, 729–733, 1990.

105. Luo, S. and Li, E.T.S., Effects of repeated administration of serotonergic agonists on diet selection and body weight in rats, *Pharmacol. Biochem. Behav.*, 38, 495–500, 1991.

106. Weiss, G.F., Rogacki, N., Fueg, A., Buchen, D., and Leibowitz, S.F., Impact of hypothalamic d-norfenfluramine and peripheral d-fenfluramine injection on macronutrient intake in the rat, *Brain Res. Bull.*, 25, 849–859, 1990.

107. Leibowitz, S.F., Alexander, J.T., Cheung, W.K., and Weiss, G.E., Effects of serotonin and the serotonin blocker metergoline on meal patterns and macronutrient selection, *Pharmacol. Biochem. Behav.*, 45, 185–194, 1993.

108. Leibowitz, S.F., Weiss, G.F., Walsh, U.A., and Viswanath, D., Medial hypothalamic serotonin: role in circadian patterns of feeding and macronutrient selection, *Brain Res.*, 503, 132–140, 1989.

109. Shor-Posner, G., Grinker, J.A., Marinescu C., Brown, O., and Leibowitz, S.F., Hypothalamic serotonin in the control of meal patterns and macronutrient selection, *Brain Res. Bull.*, 17, 663–671, 1986.

110. Newman, J.C. and Booth, D.A., Gastrointestinal and metabolic consequences of a rat's meal on maintenance diet ad libitum, *Physiol. Behav.* 27, 929–939, 1981.

111. Le Magnen, J. and Tallon, S., La périodicité spontanée de la prise d'aliments ad libitum du rat blanc, *J. Physiol.*, 58, 323–349, 1965.

112. Tempel, D.L., Shor-Posner, G., Dwyer, D., and Leibowitz, S.F., Nocturnal patterns of macronutrient intake in freely feeding and food-deprived rats, *Am. J. Physiol.*, 256, R541–R548, 1989.

113. Booth, D.A. and Baker, B.J., dl-Fenfluramine challenge to nutrient-specific textural preference conditioned by concurrent presentation of two diets, *Behav. Neurosci.* 104, 226–229, 1990.

114. Welch, C.C., Grace, M.K., Billington, C.J., and Levine, A.S., Preference and diet type affect macronutrient selection after morphine, NPY, norepinephrine, and deprivation, *Am. J. Physiol.*, 266, R426–R433, 1994.

115. Chandler, S.H., Goldbers, L.J., and Alba, B., Effects of serotonin agonists on cortically induced rhythmical jaw movements in the anesthetized guinea pig, *Brain Res.*, 334, 202–206, 1985.

116. Orthen-Gambill, N. and Kanarek, R.B., Differential effects of amphetamine and fenfluramine on dietary self-selection in rats, *Pharmacol. Biochem. Behav.*, 16, 303–309, 1982.

117. Gosnell, B.A., Krahn, D.D., and Majchrzak, M.J., The effects of morphine on diet selection are dependent upon baseline diet preferences, *Pharmacol. Biochem. Behav.*, 37, 207–212, 1990.

118. Wurtman, J.T. and Wurtman, R.T., Fenfluramine and fluoxetine spare protein consumption while suppressing caloric intake in rats, *Science*, 198, 1178–1180, 1977.

119. Smith, B.K., Andrews, P.K., York, D.A., and West, D.B., Divergent fat preference in AKR/J and SWR/J mice endures across diet paradigms, *Am. J. Physiol.*, in press.

7 Macronutrient-Conditioned Flavor Preferences

Anthony Sclafani

CONTENTS

1   INTRODUCTION

The ability of animals, including humans, to select a nutritionally balanced diet has been the subject of considerable discussion and research since the beginning of the 1900s.[1] For much of the century, the primary concern was the selection of an appropriate mix of micro- and macronutrients to avoid nutritional deficits and to maintain adequate growth. Although malnutrition remains a significant problem for much of the world, overnutrition and obesity have become a major health concerns in affluent societies. This has led to a renewed interest in diet selection, in particular, the selection of foods rich in macronutrients that may promote overeating and obesity, i.e., fat and simple carbohydrates.

One fundamental issue in diet selection is how animals recognize food and distinguish between foods of different nutritional compositions. The flavor of food, that is, its taste, smell, and texture, is one important source of information. Many animal species show an increased attraction to salty-tasting foods and fluids when sodium deficient, and there is compelling evidence for the innate basis of this salt hunger.[2] Animals are also innately attracted to other flavors, such as the sweet taste of sugar, but there is little evidence that they have an innate recognition of the nutritional significance of these flavors. Rather, flavor–nutrient associations appear to be learned through experience. Parental and other social interactions are one way that animals learn which flavors signify nutritionally adequate foods. Once a food is consumed, animals learn to associate the flavor of the food with its postingestive consequences. If the food contains toxic elements or poorly digested nutrients, animals readily learn to avoid the food by developing an aversion to its flavor.

If the food has positive postingestive effects, i.e., provides nutrition, the animal acquires a preference for its flavor. This chapter reviews the experimental literature on flavor preferences conditioned by the postingestive actions of macronutrients in laboratory rats. Data from ruminants[3] and humans[4] are discussed elsewhere.

2   CONDITIONED FLAVOR PREFERENCE PARADIGMS

Nutrient-conditioned changes in food preferences are often inferred from observations of altered food choices in animals presented with new foods that differ in their nutrient composition. In one recent study, for example, DiBattista and Holder[5] gave rats, in addition to their maintenance diet, 2 h/day access to two novel foods that were high in protein and carbohydrate, respectively. Rats fed a protein-free maintenance diet selectively increased their intake of the protein-rich novel food, while rats fed a nutritionally complete maintenance diet showed a small and nonselective increase in both novel foods. The authors concluded that the protein-restricted rats learned a preference for the flavor of protein-rich food based on its beneficial postingestive consequences. This is a plausible interpretation, but not the only one. Deutsch et al.[6] hypothesized that rats can detect the orosensory properties of proteins and that their attraction to protein flavor increases with protein deprivation. This innate protein appetite hypothesis has been challenged by other investigators.[7] Nevertheless, our understanding of the orosensory detection abilities of rats remains incomplete.

More definitive evidence that animals learn about the nutritional consequences of foods comes from studies which use a conditioned flavor preference (CFP) paradigm. With this paradigm, one arbitrary cue flavor (e.g., grape flavor) is paired with a nutrient source and a second cue flavor (e.g., cherry flavor) is paired with another nutrient source or a non-nutrient source during training trials. The specific flavor–nutrient parings are counterbalanced across subjects. Following training, the animals are given the choice between the two cue flavors. If the animals consume more of the flavor previously paired with the target nutrient, this demonstrates a learned preference. Conditioned flavor preferences are typically viewed as a form of classical (Pavlovian) conditioning with the nutrient considered to be the unconditioned stimulus (US) and its paired flavor the conditioned stimulus (CS+). The flavor paired with the non-nutrient source is labeled the CS–.

The CFP paradigm has many variations. The conditioned stimuli can be complex flavors that combine odor and tastes (e.g., artificial fruit flavors), specific tastes (e.g., sour and bitter), or specific odors (e.g., almond and vanilla odors) and can be presented in solution form or mixed into a solid food. The US can be a nutritionally complete diet, a specific macronutrient, or even ethanol. Animals can be trained to associate different flavors with different nutrients (e.g., carbohydrate and protein) or different nutrient densities (e.g., high and low energy). The nutrient US can be consumed by mouth or delivered by an intragastric (IG), intraduodenal (ID), or intravenous (IV) infusion. The simplest training procedure is to mix the CS+ flavor into the nutrient US which the animal consumes by mouth. There are many advantages to this method, but a major disadvantage is that animals may learn to associate the CS+ flavor with the flavor of the nutrient as well as its postingestive actions so that the nature of the US reinforcing the CS+ preference can be ambiguous. For this reason, the present chapter emphasizes conditioning studies in which the nutrient is delivered by a postoral route, in most cases by IG infusion, which eliminates the flavor of the nutrient as a possible US.

In some IG conditioning studies, animals are trained to drink a CS-flavored solution during a short session and are given a fixed infusion of nutrient just prior to, during, or just after the session. In other studies, as the animals drink the CS-flavored solution infusion pumps are automatically operated, delivering nutrient or water to their stomachs at a rate approximating their oral ingestion rate. In this case animals can be trained 23 h/day during which they can eat chow normally and can control the number and size of CS drinks and IG infusions they receive each day.

3   MACRONUTRIENT-CONDITIONED FLAVOR PREFERENCES

3.1   COMPLETE DIETS

As indicated in Table 1, intragastric infusions of a variety of nutrient sources are effective in conditioning flavor preferences. The first study to report a CFP also used the most complex nutrient source, a milk–egg–sucrose mixture (i.e., eggnog).[8] This particular diet was used because of prior reports that rats learned to bar press for IG infusions of eggnog in the absence of oral cues.[9] However, Holman[8] reported that rats failed to bar-press for an IG diet if all oral cues (including temperature cues) were eliminated. They did bar-press when the IG infusion was paired with oral intake of a dilute saccharin solution (0.01%) which, by itself, did not support bar-pressing. Holman concluded, therefore, that IG diet infusions were ineffective in reinforcing operant behavior directly, but rather "enhance the reward value" of oral stimuli. To test this idea, he trained food-deprived rats on alternate days to consume a flavored solution (the CS+) paired with an IG infusion of eggnog diet (7 to 10 ml), and a different flavored solution (the CS–) paired with IG water infusion. After three training sessions (5 min/day), the rats showed a reliable preference (66%) for the CS+ over the CS– during the first 20 min of a two-choice test, although the preference disappeared during the remainder of the 2-h test.

Deutsch and co-workers[10-12] reported conditioned preferences for a CS+ flavor paired with IG infusions of evaporated whole milk. A notable aspect of this series of studies is that preferences were obtained only with "predigested" milk, i.e., milk that had been consumed by donor rats and collected from their stomachs. IG infusions of regular evaporated milk failed to condition flavor preferences. However, the training conditions were stringent in that the rats had access to both the CS+ and CS– flavors during 10-min sessions. The fact that the rats acquired the preference with both flavors available during training was taken as evidence that nutrient infusions generate a rapid feedback signal from the stomach that serves to reinforce the flavor preference. Subsequent studies, however, question the rapidity of the learning in these experiments and indicate that postgastric rather than gastric feedback mediates preference conditioning.[13]

Recent reports indicate that IG infusions of milk-based diets are effective in conditioning flavor preferences in food-deprived rats trained 30 min/day[14] and in nondeprived rats trained 23 h/day.[15] As discussed further below, the milk diets were modified with the addition of fat and/or carbohydrate to produce isocaloric high-fat and high-carbohydrate diets. Predigestion of the milk was not required to obtain these preferences, but the training procedure was less demanding in that the rats had access to the CS+ and CS– during separate one-bottle training sessions.

3.2   CARBOHYDRATES

Nutrient-conditioned flavor preferences have been most extensively documented using carbohydrates, particularly glucose and maltodextrins (such as Polycose®). In the case of glucose, flavor preferences have been conditioned by IG, ID, hepatic portal, and IV infusions as well as by oral administration.[16-19] IG carbohydrate infusions reinforce flavor preferences in food-restricted animals trained 30 min/day as well as in animals given 23 h/day access to food and a CS+ solution paired with IG carbohydrate infusions. Furthermore, preference conditioning has been obtained in nondeprived rats infused with calorically dilute solutions (1 to 4% Polycose) as well as concentrated solutions (32% Polycose), and in animals given simultaneous access to chow and CS+ and CS– flavors paired with IG Polycose and water, respectively.[20-23] The latter findings indicate that animals are remarkably sensitive to the postingestive actions of carbohydrates, and need not be in an energy-deprived state to form flavor–nutrient associations.

Various findings indicate that IG carbohydrate infusions can produce substantial and long-lasting changes in the how animals evaluate flavors. First, rats can develop near-total (97 to 99%)

TABLE 1
Macronutrient-Conditioned Flavor Preference and Acceptance

Macronutrient	Route	Response	Ref.
Complete Diet			
Milk–egg diet	IG	Preference	8
Milk–predigested	IG	Preference	10–12
Milk–HF	IG	Preference	14, 15
Milk–HC	IG	Preference	14, 15
Carbohydrate			
Glucose	PO	Preference	17, 28, 52
	IG	Preference	27, 30
		Acceptance	26, 27
	ID	Preference	16
	HP	Preference	18
	IV	Preference	19
Fructose	PO	Preference	17, 28, 31
	IG	Preference	31–32
		Acceptance	32
Sucrose	PO	Preference	35
	IG	Preference	34
		Acceptance	34
Maltose	IG	Preference	34
		Acceptance	34
Maltodextrins	PO	Preference	35, 78
	IG	Preference	13, 20, 22, 30
		Acceptance	26, 55, 61
	ID	Preference	62, 65
Protein			
Amino acid mixtures	PO	Preference	48, 49
	IG	Preference	45–47, 79
Casein	PO	Preference	53, 80, 81
	IG	Preference	13, 50, 51
Fat			
Corn oil	PO	Preference	35, 36, 52
	IG	Preference	38, 39, 41–43
		Acceptance	55
	ID	Preference	62
Beef tallow	IG	Preference	43
Vegetable shortening	IG	Preference	43
MCT	IG	Preference	43

Note: PO = oral; IG = intragastric; ID = intraduodenal; HP = hepatic portal; IV = intravenous; Preference = increased intake in two-bottle tests; Acceptance = increased intake in one-bottle tests.

preferences for the CS+ flavor over the CS– flavor with repeated testing.[20,23] Second, IG carbohydrate infusions can convert a normally avoided flavor into a preferred flavor. For example, rats trained with bitter (sucrose octaacetate) or sour (citric acid) solutions paired with IG Polycose infusions subsequently strongly preferred these bitter and sour solutions to plain water.[24,25] Third, rats continue to prefer the CS+ to the CS- for several weeks or more when the CS+ is no longer reinforced with IG nutrient infusions and after both flavors were not available for a month.[20,25] That is, once conditioned, CS+ preferences show little or no extinction or forgetting over the time periods tested. Fourth, rats treat the CS+ as a food-relevant flavor even when it is no longer paired with IG nutrient infusions. Drucker et al.[25] reported that in two-choice extinction tests, food deprivation selectively increased the rats' intake of the CS+ flavor, whereas water deprivation increased their intake of both the CS+ and CS- flavors.

In addition to conditioning a preference for a CS+ flavor over a CS- flavor in two-bottle choice tests, IG carbohydrate infusions can dramatically increase the absolute intake (*acceptance*) of a flavored solution in one-bottle tests. Ramirez[26] gave rats *ad libitum* access to chow and a saccharin solution 23 h/day. Rats that had the solution paired with IG infusions of 6% carbohydrate consumed 70% more saccharin than did rats that had the solution paired with IG water infusions. Using a within-subject design, Pérez et al.[27] gave rats daily 2-h access to chow and water and, on alternate days, 20-h access to CS+ and CS- solutions (grape or cherry water) which were paired with concurrent IG infusions of 16% glucose and water, respectively. The animals greatly increased their CS+ intake over days and by the end of training they were drinking almost four times more CS+ than CS- (71 vs. 18 g/20 h). This increased acceptance was due to both increased CS+ bout size and number. The rats continued to drink more CS+ than CS- when both solutions were paired with IG water infusions (extinction test) which demonstrates that the elevated intake of the CS+-flavored solution was a learned response rather than a direct effect of the glucose infusions. These results demonstrate that flavor–nutrient learning not only influences food choice but can also dramatically influence how much food is consumed.

In contrast to the robust conditioning effects obtained with glucose and maltodextrins (which are digested to glucose), Ackroff, Sclafani, and co-workers[7,28-30] have found that fructose is much less effective in producing flavor preferences, but see Reference 31. In particular, food-deprived rats trained 2 h/day or nondeprived rats trained 23 h/day acquired weak or no preferences for a CS+ flavor paired with IG infusions of 16% fructose.[30] In addition, the rats significantly preferred a flavor (CS+Glu) paired with IG glucose over a flavor (CS+Fru) paired with IG fructose infusions.[30] More recently, Ackroff et al.[32] observed an 86% preference for a CS+Fru over a CS- in food-restricted rats trained 20 h/day with the CS+Fru paired with IG 16% fructose. Nevertheless, the same rats strongly preferred a CS+Glu (paired with IG 16% glucose) over the CS+Fru during two-choice tests (20 h/day). Furthermore, when retrained with new flavors paired with IG glucose, fructose, and water during 30 min/day sessions, the rats developed a strong CS+Glu preference but failed to prefer the CS+Fru over the CS-. Taken together, these findings indicate that fructose generates a postingestive US that is available (or effective) only in food-restricted animals given long-term training sessions, and is much less potent than the postingestive US generated by glucose.

One reason fructose may be relatively ineffective in conditioning flavor preferences is that, compared with glucose, fructose is slowly absorbed from the intestine which can result in malabsorption and discomfort at high concentrations.[33] Fructose absorption is facilitated when it is mixed with glucose.[33] Azzara and Sclafani[34] therefore compared the flavor conditioning effect of sucrose (a glucose-fructose disaccharide) with that of maltose (a glucose-glucose disaccharide). Following one-bottle training with CS+Suc, CS+Mal, and CS- flavors paired with IG sucrose, maltose, and water, respectively, the rats showed strong preferences (85 to 93%) for both CS+ flavors compared with the CS-. However, in a choice test with the two CS+ flavors, they preferred the CS+Mal to the CS+Suc by 78%, which is similar to the preferences for CS+Glu over CS+Fru observed in other studies.[29,30] These data suggest that even when mixed with glucose, fructose is less reinforcing than glucose.

3.3   FATS

In the first reports of fat-conditioned flavor preferences Tordoff et al.[52] and Mehiel and Bolles[35] trained rats to consume corn oil emulsions that contained the CS+ flavor or were paired with a chow containing the CS+ flavor. Elizalde and Sclafani[36] confirmed that rats learned to prefer a CS+ flavor mixed in a corn oil emulsion over a CS− flavor presented in a noncaloric solution. However, they also reported that rats developed a preference for a flavor mixed in a non-nutritive mineral oil emulsion, which indicates that the orosensory qualities of oils are sufficient to reinforce flavor preferences. Nevertheless, the mineral oil-conditioned preference was weaker than that produced by corn oil. In addition, a second experiment revealed that corn oil, but not mineral oil produced a CS+ preference when there was a 10-min delay between the intakes of the CS+ flavor and of the oil.[36] The latter findings support the idea that the postingestive actions of fat are required for some types of preference conditioning. Prior work indicates that a postingestive nutrient US, but not a flavor US is effective using a CS–US delay paradigm.[37]

More direct evidence for postingestive fat conditioning was provided by Lucas and Sclafani.[38] They trained food-deprived rats to drink a CS+ solution paired with an IG infusion of 7.1% corn oil, and a CS− solution paired with IG water during short daily sessions. In the choice test the rats showed a significant, although modest (~60%) preference for the CS+. In a second experiment,[38] nondeprived rats trained 23 h/day failed to develop a reliable preference for a CS+ paired with IG corn oil infusions. However, when food restricted and given additional training (20 h/day), the rats showed strong preferences (up to 95%) for the CS+ over the CS−.[38]

Lucas and Sclafani[39] attempted to enhance fat conditioning by feeding animals a high-fat maintenance diet based on prior reports that such diets enhance fat appetite and metabolism.[40] The animals were trained 30 min/day with a CS+ paired with IG infusions of 7.1% corn oil, and a CS− paired with IG water. In the choice test the rats fed a high-fat chow (48% fat energy) displayed a 90% preference for the CS+, whereas rats fed a standard low-fat chow (12% fat) showed only a 62% CS+ preference. In a subsequent experiment the high-fat chow produced a smaller, and not reliable, improvement in fat conditioning.[41] Other recent studies[42,43] obtained rather strong CS+ preferences (78 to 89%) in rats fed low-fat chow. The reason for this variability is not certain, although in the more recent studies the rats were trained with multiple CS+ solutions paired with different nutrients which may have facilitated fat conditioning.

The fat-conditioning data cited above all involved corn oil emulsions as the fat source. Ackroff et al.[43] recently compared the effectiveness of different fats in producing flavor preferences. The composition of the fat sources varied in their ratio of polyunsaturated/saturated fatty acids (corn oil vs. beef tallow or vegetable shortening) and in their fatty acid chain lengths (corn oil vs. medium-chain triglyceride oil, MCT). Rats were fed restricted rations of standard chow and were trained (30 min/day) with one flavor (CS+CO) paired with IG corn oil and a second flavor (CS+BT) paired with IG beef tallow; the fats were prepared as isocaloric emulsions. A third flavor (CS−) was paired with IG water. In subsequent choice tests the rats showed strong preferences for both the CS+CO (89%) and the CS+BT (82%) relative to the CS−. When given the choice of the two CS+ flavors, the rats preferred the CS+CO to the CS+BT by 67%. The same rats were subsequently trained with three new flavors that were paired with IG infusions of corn oil (CS+CO), vegetable shortening (CS+VS), and water (CS−). The rats displayed strong preferences for both the CS+CO (91%) and the CS+VS (86%) over the CS−, and they preferred the CS+CO to the CS+VS by 64%. Other rats trained with corn oil and MCT oil showed a strong preference for the CS+CO (85%) and a weaker preference for the CS+MCT (65%) relative to the CS−. In a direct choice test, the CS+CO was preferred to the CS+MCT by 75%.

These findings demonstrate that flavor preferences can be conditioned by a variety of fat sources: by vegetable and animal fats, by solid and liquid fats, and by fats containing long-chain and short-chain fatty acids. Corn oil was the most effective US of the fats tested. This may be because the fatty acids in corn oil are the most effective in generating postingestive reinforcing signals. However, nonspecific factors may also affect preference conditioning. MCT oil may have aversive consequences that counteract its positive feedback effects. In the case of corn oil, beef tallow, and vegetable shortening, it is possible that there are differences in the gastric emptying and/or digestion rates of the emulsions that affect preference conditioning.

3.4   PROTEINS

In contrast to carbohydrate and fat, protein provides not only energy but also essential nutrients, i.e., amino acids, needed for growth and thus is required in substantial amounts in the diet.[44] It is especially important, therefore, that animals select foods that contain sufficient amounts of protein. As previously noted, some investigators have proposed that rats have an innate recognition of the protein,[6] but other investigators have argued that protein appetite is largely learned.[7] There is, in fact, considerable evidence that rats learn to prefer flavors paired with high-quality protein or balanced amino acid mixtures and learn to avoid flavors associated with imbalanced amino acid mixtures.

In the first report of an amino acid–conditioned preference, Simson and Booth[45] trained rats to eat flavored, protein-free diets. One flavored diet was consumed following an IG infusion of a balanced amino acid mixture; the other flavored diet was consumed following an IG saline infusion. In a subsequent two-choice test, the rats showed a mild (~67%), but significant preference for the flavored diet that had been paired with the amino acid infusion. Other rats that were infused with an imbalanced amino acid mixture (histidine devoid) showed a strong avoidance of the paired diet in the choice test. Booth and Simson[45-47] have published other reports of preferences and aversions conditioned by balanced and imbalanced amino acid mixtures.[45-47] In related studies, Gietzen et al.[48,49] trained rats with a flavored solution paired with an imbalanced diet (threonine or isoleucine devoid) and another flavor with a balanced diet. In choice tests the rats preferred the flavor paired with the balanced diet over the flavor paired with the deficient diet or a novel flavor.

In the above studies, the rats were placed in a protein-deficient state by feeding them a protein-free or imbalanced diet, and the flavor preference conditioned by the balanced amino acid mixtures were attributed to a repletion effect. Flavor preferences have also been conditioned by protein infusions in rats maintained on chow diets containing abundant protein. In a study by Baker et al.[50] rats were deprived of chow for 4 h each day and given one-bottle training with a CS+-flavored solution paired with IG protein (10% calcium caseinate) and a CS– solution with an IG infusion of a noncaloric solution. In a subsequent choice test, the rats displayed a 70% preference for the CS+ over the CS–. In a more recent study, Pérez et al.[51] fed rats limited chow rations and gave them one-bottle training with one flavor (CS+Pro) paired with IG protein infusions (10% calcium caseinate), a second flavor (CS+Cho) paired with IG carbohydrate infusions (10% Polycose), and a third flavor (CS–) paired with IG water infusions. In subsequent two-choice tests, the rats preferred both the CS+Pro (70 to 89%) and CS+Cho (85 to 91%) over the CS–. As discussed further below, the CS+Pro and CS+Cho flavors were equally preferred. Although the rats in these studies were fed a balanced diet, they were mildly to moderately food restricted, and Baker et al.[50] argue that even a 4-h deprivation creates a mild protein "need." Whether protein infusions can condition flavor preferences in nondeprived animals, as carbohydrate infusions can, remains to be determined.

4   MACRONUTRIENT-SPECIFIC PREFERENCES

Several studies report that isocaloric solutions of different nutrients condition similar flavor preferences when compared with a CS– flavor.[35,52,53] The "equipotentiality" of different nutrients has suggested that the flavor preferences are reinforced by a common signal related to the energy value of the nutrients, for example, by signals generated by fuel oxidation in the liver.[54] However, other experiments report substantial differences in the preference-reinforcing effects of some isocaloric nutrient sources. As reviewed above, fructose is less effective than glucose,[30] and MCT oil is less effective than corn oil in conditioning a CS+ preference over a CS–. Furthermore, even when isocaloric infusions produce comparable preferences, relative to a CS– flavor, animals may still discriminate between the nutrients. For example, rats preferred a CS+ flavor paired with IG maltose over a CS+ flavor paired with IG sucrose, and a CS+ flavor paired with IG corn oil over a CS+ flavor paired with IG beef tallow.[34,43] Of particular interest to the present discussion are comparisons between different macronutrient classes.

4.1   PROTEIN AND CARBOHYDRATE

As noted above, Baker et al.[50] trained rats to prefer a CS+ flavor paired with IG protein over a CS– flavor. This preference was blocked when the rats were given an IG preload of protein prior to the CS+ vs. CS– test, but not when they were given an isocaloric preload of carbohydrate. Baker et al. concluded, therefore, that the conditioned preference for the CS+ flavor represented a specific appetite for protein.

Pérez et al.[51] further examined this issue by training rats to associate different flavors, a CS+Pro and a CS+Cho, with IG protein and IG carbohydrate infusions, respectively. When given the choice between the CS+Pro vs. CS+Cho following no preloads, the rats consumed comparable amounts of the two CSs. In one experiment, an IG protein preload prior to the choice test shifted the rats' preference to the CS+Cho, while an IG carbohydrate preload shifted their preference to the CS+Pro. However, in two other experiments IG protein and carbohydrate preloads did not produce comparable preference shifts. More consistent results were obtained when the rats were given an oral + IG preload prior to the choice tests. That is, in two experiments rats given a fixed amount of CS+Pro to drink paired with IG protein shifted their preference to CS+Cho in the subsequent choice test. Similarly, rats given CS+Cho to drink paired with IG carbohydrate shifted their preference to CS+Pro in the choice test. Preference shifts were not obtained when the rats were given oral-only preloads (i.e., CS+Pro or CS+Cho in the absence of an IG nutrient infusion). Taken together, these findings indicate rats can discriminate the postingestive actions of protein and carbohydrate but that the combined actions of flavor and postingestive cues are more effective than either type of cue alone in modifying subsequent food choices.

4.2   CARBOHYDRATE AND FAT

Several studies have compared the preference conditioning effects of fat vs. carbohydrate. This is of particular interest given the tendency of animals to prefer high-fat to high-carbohydrate diets in choice tests, and to overconsume and gain excess weight on high-fat diets, relative to high-carbohydrate diets. The preference for high-fat foods is often attributed to their "palatability" but the source of this palatability is unclear. Clearly, postingestive nutrient actions affect flavor preferences, and animals may learn to prefer high-fat to high-carbohydrate foods.

Lucas and Sclafani[41] trained rats to drink different flavored solutions (CS+Fat, CS+Cho) paired with IG infusions of fat (corn oil) and carbohydrate (Polycose) at isocaloric concentrations; a third flavor (CS–) was paired with IG water infusions. The rats were food restricted and fed either low-fat or high-fat rations to determine if maintenance diet affected flavor preference. In two-choice tests, both the CS+Cho and CS+Fat were preferred to the CS–. However, in direct choice tests the CS+Cho was preferred to the CS+Fat by ~70% and this preference was observed in both low-fat and high-fat fed groups. Pérez et al.[42] also investigated the effects of maintenance diet on IG fat and carbohydrate conditioning. In this case the rats were fed either a standard chow diet or a cafeteria diet containing 20 "supermarket" foods which differed in taste and nutrient composition. The two diet groups showed similar preferences for the CS+Fat and CS+Cho over the CS–, and both preferred the CS+Cho to the CS+Fat by about 77%. Using a different training paradigm, Ramirez[55] compared the ability of isocaloric fat and carbohydrate IG infusions to condition increased flavor acceptance. Different groups of rats were trained 24 h/day to drink a flavored solution paired with IG infusions of a corn oil emulsion or isocaloric carbohydrate solution. The corn oil infusions produced smaller and less consistent increases in the intake of the flavored solution than did infusions of starch or maltodextrin. Ramirez[55] concluded that IG fat is a less potent reinforcer of flavor conditioning than is IG carbohydrate.

In contrast to these IG results, two studies observed that rats equally preferred flavors paired with orally consumed corn oil and carbohydrate (glucose or Polycose). In one study, the rats drank noncaloric CS solutions, followed after a 10-min delay, by access to corn oil or Polycose.[51] In the other study, the CS flavors were presented in chow that was available along with a corn oil emulsion or glucose solution.[52] Another more recent experiment trained rats with the CS flavors mixed into the a corn oil emulsion and an isocaloric Polycose solution.[56] In a choice test with the CS flavors only, the Polycose-paired flavor was preferred to the corn oil flavor by 77%, which is comparable to the preferences observed in the IG studies. The reason for the discrepant results obtained in the different oral conditioning studies is not clear, but all studies are consistent in failing to obtain a preference for the fat-paired flavor over the carbohydrate-paired flavor.

In the above studies, rats were trained during short-term daily sessions using pure carbohydrate and fat sources. Different results were obtained by Lucas et al.[15] using a long-term testing paradigm (22 h/day) and IG infusions of milk-based diets that contained added fat (corn oil) or carbohydrate (sucrose or maltodextrin). The high-fat (HF) diet was 60% fat (in kcal) and 34% carbohydrate, whereas the high-carbohydrate (HC) diet was 79% carbohydrate and 15% fat; otherwise, the diets were identical in nutrient composition and energy density. Over 9 training days, the rats were given *ad libitum* access to chow, water, and one of three CS-flavored solutions. One flavor (CS+HF) was paired with IG infusions of the HF diet, a second flavor (CS+HC) with IG infusions of the HC diet, and a third flavor (CS–) with IG water infusions. A series of two-bottle choice tests were then conducted with the various CS solutions remaining paired with their respective infusions. In two separate experiments, the rats displayed strong preferences (84 to 92%) for the CS+HF and CS+HC over the CS–. In the critical choice test between the two CS+ flavors, the rats reliably preferred the CS+HF to the CS+HC by 64 to 72%. This is the first report of an acquired preference for an HF-paired flavor over an HC-paired flavor.

Not only did the rats consume more CS+HF than CS+HC in the choice tests, but they also consumed more on training days, and consequently self-infused themselves with more HF diet than HC diet, which confirms prior results.[57] The rats overconsumed the CS+HF, relative to the CS+HC, by taking more "meals" per day, indicating that the HF diet infusions produced less postmeal satiety than the HC diet infusions. Lucas et al.[15] investigated whether the preference for CS+HF over CS+HC was due to the rats being infused with more diet on CS+HF than CS+HC training days. This was done by training new rats with limited access to the CS solutions so that HF and HC diet infusions were matched. Despite this limitation, the rats still reliably preferred the CS+HF to the CS+HC in two-choice tests. Lucas et al.[15] then determined if the reduced satiety effect of the HF diet was responsible for the CS+HF preference by training new rats with the standard HF diet (2.1 kcal/ml) and a diluted version of the HC diet (1.4 ml/kcal). On training days, the rats infused themselves with equivalent volumes of the HF and diluted HC diets, and CS+HF and CS+HC meal sizes and numbers were comparable. Thus, on a volume basis, the HF and diluted HC diets had equivalent satiety effects. In the two-choice test the rats did not reliably prefer the CS+HF to the CS+HC.

The Lucas et al.[15] data revealed that the postingestive reinforcing effects of the HF milk diet were more potent than that of the isocaloric HC milk diet. This appeared to have occurred not because fat is inherently more reinforcing than carbohydrate, but because it reduced the satiety effects of the liquid diet. Although it might seem that foods that produce less satiety should also be less reinforcing, recent data indicate that just the opposite may be true: too much satiety may limit postingestive reinforcement.[58,59] Thus, the HF and HC results are not inconsistent with the conditioning data obtained with corn oil and carbohydrate infusions;[41,42] see Lucas and Sclafani[14] for further data relevant to this point. The generality of the HF and HC diet results remains to be established given that only liquid diets, one type of fat (corn oil), and a limited range of caloric density were investigated. Nevertheless, the data suggest that preferences for HF foods may be mediated, at least in part, by the postingestive actions of the foods. The results further indicate that preference conditioning effects of carbohydrates and fats depend on whether the nutrients are presented alone or as components of a mixed diet.

5   PHYSIOLOGICAL MECHANISMS

While flavor–nutrient conditioning is well documented, relatively little is known about the physiological mechanisms that mediate this learning. In particular, where and how the nutrient US is detected remain to be determined.[54] Deutsch and Wang[12] proposed that the US is generated by nutrient receptors in the stomach, but other investigators question this proposal.[13] A postgastric site of action is indicated by the findings that (1) IG glucose infusions failed to condition a flavor preference when the glucose was prevented from entering the duodenum by a pyloric cuff and (2) ID glucose infusions were as effective as IG infusions in conditioning a flavor preference.[16] Tordoff and Friedman[18] observed that hepatic-portal glucose infusions were more effective that jugular infusions in conditioning flavor preferences, indicating that detectors in the liver may mediate flavor–nutrient learning. The reinforcing effectiveness of duodenal and hepatic-portal nutrient infusions have not been directly compared and it is possible that both preabsorptive (intestinal) and postabsorptive (hepatic) detectors contribute to nutrient conditioning. Brain nutrient detectors have also not been ruled out as participating in flavor preference learning.[54]

How information from a nutrient US acting at a peripheral site reaches the brain is another unresolved issue. Selective gastric vagotomy was reported to block preference conditioning by IG carbohydrate infusions,[60] but a subsequent study reported flavor conditioning in rats with a complete abdominal vagotomy.[61] Also, visceral afferent deafferentation produced by systemic capsaicin treatment failed to block preference conditioning by ID carbohydrate or fat infusions.[62] Hormonal involvement in nutrient conditioning was suggested by reports that injections of cholecystokinin or insulin can reinforce flavor preferences in some situations.[63,64] However, blocking CCK_A receptors with devazepide failed to prevent carbohydrate-conditioned flavor preferences in rats.[65] Experimentally induced diabetes was reported to block glucose-conditioned preferences in one study[52] but not in another study,[29] and did not prevent flavor preference conditioning by dietary fat.[52]

To date, few studies have investigated the central neural and neurochemical systems that mediate flavor–nutrient learning. One brain site of particular interest is the parabrachial nucleus (PBN) in the hindbrain since it receives both gustatory and viscerosensory inputs and PBN lesions disrupt flavor aversion learning.[66] Initial findings indicate that PBN lesions do not block flavor preference learning,[66,67] but current research suggests that some forms of preference conditioning may be impaired by the lesions.[68] The involvement of other brain areas, particularly hypothalamic and amygdala sites implicated in feeding regulation and flavor learning, remain to be studied. Some drug studies suggest the involvement of serotonin and opioid systems in flavor–nutrient learning,[69-71] but conflicting results have been obtained[72] and the pharmacology of preference conditioning needs further study.

6   FLAVOR-NUTRIENT INTERACTIONS

The present chapter has focused on the postingestive conditioning actions of nutrients evaluated in studies using arbitrary flavor cues. Macronutrients, however, typically add their own flavor to foods and thus flavor–nutrient associations are usually not arbitrary. Consequently, macronutrient selection may be influenced by unlearned flavor preferences and aversions. Rats appear to have innate preferences for the taste of sugar and maltodextrins, and perhaps for the flavors of some starch, fats, and proteins, which may influence flavor–nutrient learning.[6,73,74] Adding a sweet and/or maltodextrin taste to a CS+, for example, will increase CS+ intake and the amount of nutrient self-infused.[20,22,75] Some findings indicate that preference and acceptance conditioning by IG nutrients varies as a function of the specific flavors or odors used.[26,76] Whether pairing nutrients with nutrient-specific flavors (e.g., sugar with sweet taste or fat with oily flavor) influences conditioning has not been investigated in detail and thus remains a possibility (but see Reference 39). The role of flavor-stimulated cephalic phase digestive responses in nutrient conditioning is another topic that requires further investigation.[77]

7   CONCLUSIONS

There is extensive evidence documenting the ability of animals to learn flavor preferences based on the postingestive actions of nutrients. This has been obtained with nutritionally complete diets as well as specific macronutrients. Not all nutrients are equally effective in conditioning preferences, however. Furthermore, preference conditioning effects may depend on whether nutrients are provided in pure form or as components of a mixed diet. Similarly, flavors may differ in their ability to serve as a CS although this issue has not received much attention. A major issue that requires further investigation is the nutrient specificity of these learned preferences. To what degree do animals learn to associate flavors with the energy value of nutrients and to what degree do they learn specific flavor–nutrient associations? The available data indicate that rats discriminate between the postingestive actions of at least some proteins and carbohydrates. How they use this information remains to be fully explored. One interesting question is whether rats that previously acquired preferences for flavors paired with IG protein and carbohydrate infusions while fed a protein-replete diet would increase their CS+Pro preference when placed on a protein-deficient diet. Specifically, would the rats require additional flavor–nutrient training while in the protein-deficient state, or would they show an enhanced CS+Pro preference without such training? Similarly, do pharmacological and metabolic treatments reported to alter fat and carbohydrate selection shift preferences for flavors previously associated with these nutrients? The impact of maintenance diet composition on fat- and carbohydrate-conditioned preferences requires clarification in light of the inconsistent results obtained to date. Another interesting issue is how individual and strain differences in macronutrient selection patterns relate to flavor–nutrient learning. For example, do animals that prefer HF food in self-selection paradigms show elevated preferences for fat-paired flavors in conditioning paradigms compared with animals preferring HC foods?

The peripheral and central neural mechanisms that mediate nutrient learning remain largely unknown. How and where a nutrient US is detected and how this information is communicated to the brain requires further investigation. Nutrient oxidation in the liver may be one but not necessarily the only source of feedback signals that reinforce flavor preferences; intestinal site of action remains a possibility. Available data indicate that the viscerosensory afferents damaged by vagotomy or systemic capsaicin treatment do not carry these signals to the brain. Only recently have researchers begun to investigate what brain sites and neurochemical systems are critical to the development of flavor–nutrient associations.

ACKNOWLEDGMENTS

Preparation of this chapter was supported by a National Institute of Mental Health Research Scientist Award (MH-00983). The research from the author's laboratory was supported by research grants from the National Institute of Diabetes and Digestive and Kidney Diseases (DK-31135) and the PSC-CUNY Award Program. The author thanks Dr. Karen Ackroff for her helpful comments on this chapter.

REFERENCES

1. Galef, B. G., Jr., A contrarian view of the wisdom of the body as it relates to dietary self-selection, *Psychol. Rev.,* 98, 218, 1991.
2. Schulkin, J., Innate, learned, and evolutionary factors in hunger for salt, in *Chemical Senses,* Vol. 4: *Appetite and Nutrition*, Friedman, M. I., Tordoff, M. G., and Kare, M. R., Eds., Marcel Dekker, New York, 1991, 211.
3. Provenza, F. D., Postingestive feedback as an elementary determinant of food preference and intake in ruminants, *J. Range Manage.,* 48, 2, 1995.

4. Capaldi, E. D., Ed., *Why We Eat What We Eat,* American Psychological Association, Washington, D.C., 1996.

5. DiBattista, D. and Holder, M. D., Enhanced preference for a protein-containing diet in response to dietary protein restriction, *Appetite,* 30, 237, 1998.

6. Deutsch, J. A., Moore, B. O., and Heinrichs, S. C., Unlearned specific appetite for protein, *Physiol. Behav.,* 46, 619, 1989.

7. Booth, D. A., Protein- and carbohydrate-specific cravings: neuroscience and sociology, in *Chemical Senses,* Vol. 4: *Appetite and Nutrition,* Friedman, M. I., Tordoff, M. G., and Kare, M. R., Eds., Marcel Dekker, New York, 1991, 260.

8. Holman, G. L., Intragastric reinforcement effect, *J. Comp. Physiol. Psychol.,* 69, 432, 1968.

9. Epstein, A. N. and Teitelbaum, P., Regulation of food intake in the absence of taste, smell, and other oropharyngeal sensations, *J. Comp. Physiol. Psychol.,* 55, 753, 1962.

10. Puerto, A., Deutsch, J. A., Molina, F., and Roll, P. L., Rapid discrimination of rewarding nutrient by the upper gastrointestinal tract, *Science,* 192, 485, 1976.

11. Puerto, A., Deutsch, J. A., Molina, F., and Roll, P., Rapid rewarding effects of intragastric injections, *Behav. Biol.,* 18, 123, 1976.

12. Deutsch, J. A. and Wang, M.-L., The stomach as a site for rapid nutrient reinforcement sensors, *Science,* 195, 89, 1977.

13. Baker, B. J. and Booth, D. A., Preference conditioning by concurrent diets with delayed proportional reinforcement, *Physiol. Behav.,* 46, 585, 1989.

14. Lucas, F. and Sclafani, A., Flavor preferences conditioned by high-fat vs. high-carbohydrate diets vary as a function of session length, *Physiol. Behav.,* 66, 389–395, 1999.

15. Lucas, F., Ackroff, K., and Sclafani, A., High-fat diet preference and overeating mediated by postingestive factors in rats, *Am. J. Physiol.,* 275, R1511, 1998.

16. Drucker, D. B. and Sclafani, A., The role of gastric and postgastric sites in glucose-conditioned flavor preferences in rats, *Physiol. Behav.,* 61, 351, 1997.

17. Sclafani, A. and Ackroff, K., Glucose- and fructose-conditioned flavor preferences in rats: taste vs. postingestive conditioning, *Physiol. Behav.,* 56, 399, 1994.

18. Tordoff, M. G. and Friedman, M. I., Hepatic-portal glucose infusions decrease food intake and increase food preference, *Am. J. Physiol.,* 251, R192, 1986.

19. Mather, P., Nicolaidis, S., and Booth, D. A., Compensatory and conditioned feeding responses to scheduled glucose infusions in the rat, *Nature,* 273, 461, 1978.

20. Elizalde, G. and Sclafani, A., Flavor preferences conditioned by intragastric Polycose: a detailed analysis using an electronic esophagus preparation, *Physiol. Behav.,* 47, 63, 1990.

21. Sclafani, A. and Nissenbaum, J. W., Robust conditioned flavor preference produced by intragastric starch infusions in rats, *Am. J. Physiol.,* 255, R672, 1988.

22. Ackroff, K. and Sclafani, A., Flavor preferences conditioned by intragastric infusions of dilute Polycose solutions, *Physiol. Behav.,* 55, 957, 1994.

23. Drucker, D. B., Ackroff, K., and Sclafani, A., Flavor preference produced by intragastric Polycose infusions in rats using a concurrent conditioning procedure, *Physiol. Behav.,* 54, 351, 1993.

24. Sclafani, A., Conditioned food preferences, *Bull.Psychon.Soc.,* 29, 256, 1991.

25. Drucker, D. B., Ackroff, K., and Sclafani, A., Nutrient-conditioned flavor preference and acceptance in rats: effects of deprivation state and nonreinforcement, *Physiol. Behav.,* 56, 701, 1994.

26. Ramirez, I., Stimulation of fluid intake by carbohydrates: interaction between taste and calories, *Am. J. Physiol.,* 266, R682, 1994.

27. Pérez, C., Lucas, F., and Sclafani, A., Flavor preference and acceptance conditioned by intragastric glucose infusions, *Physiol. Behav.,* 64, 483, 1998.

28. Ackroff, K. and Sclafani, A., Flavor preferences conditioned by sugars: rats learn to prefer glucose over fructose, *Physiol. Behav.,* 50, 815, 1991.

29. Ackroff, K., Sclafani, A., and Axen, K. V., Diabetic rats prefer glucose-paired flavors over fructose-paired flavors, *Appetite,* 28, 73, 1997.

30. Sclafani, A., Cardieri, C., Tucker, K., Blusk, D., and Ackroff, K., Intragastric glucose but not fructose conditions robust flavor preferences in rats, *Am. J. Physiol.,* 265, R320, 1993.

31. Tordoff, M. G., Ulrich, P. M., and Sandler, F., Flavor preferences and fructose: evidence that the liver detects the unconditioned stimulus for calorie-based learning, *Appetite,* 14, 29, 1990.

32. Ackroff, K., Williams, D., and Sclafani, A., Fructose-conditioned flavor preference and acceptance: effect of training paradigm, *Physiol. Behav.,* in preparation.

33. Riby, J. E., Fujisawa, T., and Kretchner, N., Fructose absorption, *Am. J. Clin. Nutr.,* 58 (Suppl.), 748S, 1993.

34. Azzara, A. and Sclafani, A., Flavor preferences conditioned by intragastric sugar infusions in rats: maltose is more reinforcing than sucrose, *Physiol. Behav.,* 64, 535, 1998.

35. Mehiel, R. and Bolles, R. C., Learned flavor preferences based on calories are independent of initial hedonic value, *Anim. Learn. Behav.,* 16, 383, 1988.

36. Elizalde, G. and Sclafani, A., Fat appetite in rats: flavor preferences conditioned by nutritive and non-nutritive oil emulsions, *Appetite,* 15, 189, 1990.

37. Holman, E. W., Immediate and delayed reinforcers for flavor preferences in the rat, *Learn. Motiv.,* 6, 91, 1975.

38. Lucas, F. and Sclafani, A., Flavor preferences conditioned by intragastric fat infusions in rats, *Physiol. Behav.,* 46, 403, 1989.

39. Lucas, F. and Sclafani, A., The composition of the maintenance diet alters flavor preferences conditioned by intragastric fat infusions in rats, *Physiol. Behav.,* 60, 1151, 1996.

40. Reed, D. R., Tordoff, M. G., and Friedman, M. I., Enhanced acceptance and metabolism of fats by rats fed a high-fat diet, *Am. J. Physiol.,* 261, R1084, 1991.

41. Lucas, F. and Sclafani, A., Differential reinforcing and satiating effects of intragastric fat and carbohydrate infusions in rats, *Physiol. Behav.,* 66, 381–388, 1999.

42. Pérez, C., Fanizza, L. J., and Sclafani, A., Flavor preferences conditioned by intragastric nutrients in rats fed chow or a cafeteria diet, *Appetite,* 32, 155–170, 1999.

43. Ackroff, K., Lucas, F., and Sclafani, A., Flavor preferences conditioned by different types of fat, in preparation.

44. Anderson, G. H., Black, R. M., and Li, E. T. S., Physiological determinants of food selection: association with protein and carbohydrate, in *The Biology of Feast and Famine: Relevance to Eating Disorders,* Anderson, G. H. and Kennedy, S. H., Eds., Academic Press, San Diego, 1992, 73.

45. Simson, P. C. and Booth, D. A., Olfactory conditioning by association with histidine-free or balanced amino acid loads in rats, *Q. J. Exp. Psychol.,* 25, 354, 1973.

46. Booth, D. A. and Simson, P. C., Food preferences acquired by association with variations in amino acid nutrition, *Q. J. Exp. Psychol.,* 23, 135, 1971.

47. Simson, P. C. and Booth, D. A., Effect of CS-US interval on the conditioning of odour preferences by amino acid loads, *Physiol. Behav.,* 11, 801, 1973.

48. Gietzen, D. W., McArthur, L. H., Theisen, J. C., and Rogers, Q. R., Learned preference for the limiting amino acid in rats fed a threonine-deficient diet, *Physiol. Behav.,* 51, 909, 1992.

49. Naito-Hoopes, M., McArthur, L. H., Gietzen, D. W., and Rogers, Q. R., Learned preference and aversion for complete and isoleucine-devoid diets in rats, *Physiol. Behav.,* 53, 485, 1993.

50. Baker, B. J., Booth, D. A., Duggan, J. P., and Gibson, E. L., Protein appetite demonstrated: learned specificity of protein-cue preference to protein need in adult rats, *Nutr. Res.,* 7, 481, 1987.

51. Pérez, C., Ackroff, K., and Sclafani, A., Carbohydrate- and protein-conditioned flavor preferences: effects of nutrient preloads, *Physiol. Behav.,* 59, 467, 1996.

52. Tordoff, M. G., Tepper, B. J., and Friedman, M. I., Food flavor preferences produced by drinking glucose and oil in normal and diabetic rats: evidence for conditioning based on fuel oxidation, *Physiol. Behav.,* 41, 481, 1987.

53. Pérez, C., Lucas, F., and Sclafani, A., Carbohydrate, fat and protein condition similar flavor preferences in rats using an oral-delay conditioning procedure, *Physiol. Behav.,* 57, 549, 1995.

54. Tordoff, M. G., Metabolic basis of learned food preferences, in *Appetite and Nutrition,* Friedman, M. I., Tordoff, M. G., and Kare, M. R., Eds., Marcel Dekker, New York, 1991, 239.

55. Ramirez, I., Stimulation of fluid intake by nutrients: oil is less effective than carbohydrate, *Am. J. Physiol.,* 272, R289, 1997.

56. Sclafani, A. and Rosenrauch, Y., Unpublished data, 1998.

57. Warwick, Z. S. and Weingarten, H. P., Determinants of high-fat diet hyperphagia: experimental dissection of orosensory and postingestive effects, *Am. J. Physiol.,* 269, R30, 1995.

58. Warwick, Z. S. and Weingarten, H. P., Flavor-postingestive consequence associations incorporate the behaviorally opposing effects of positive reinforcement and anticipated satiety: implications for interpreting two-bottle tests, *Physiol. Behav.,* 60, 711, 1996.

59. Lucas, F., Azzara, A. V., and Sclafani, A., Flavor preferences conditioned by intragastric Polycose in rats: more concentrated Polycose is not always more reinforcing, *Physiol. Behav.*, 63, 7, 1997.

60. Horn, C. C. and Mitchell, J. C., Does selective vagotomy affect conditioned flavor–nutrient preferences in rats? *Physiol. Behav.*, 59, 33, 1996.

61. Sclafani, A. and Lucas, F., Abdominal vagotomy does not block carbohydrate-conditioned flavor preferences in rats, *Physiol. Behav.*, 60, 447, 1996.

62. Lucas, F. and Sclafani, A., Capsaicin attenuates feeding suppression but not reinforcement by intestinal nutrients, *Am. J. Physiol.*, 270, R1059, 1996.

63. Pérez, C. and Sclafani, A., Cholecystokinin conditions flavor preferences in rats, *Am. J. Physiol.*, 260, R179, 1991.

64. Vanderweele, D. A., Deems, R. O., and Kanarek, R. B., Insulin modifies flavor aversions and preferences in real-feeding and sham-feeding rats, *Am. J. Physiol.*, 259, R823, 1990.

65. Pérez, C., Lucas, F., and Sclafani, A., Devazepide, a CCK_A antagonist, does not block flavor preference conditioning by intestinal carbohydrate or fat infusions in rats, *Pharmacol. Biochem. Behav.*, 59, 451, 1998.

66. Reilly, S., Grigson, P. S., and Norgren, R., Parabrachial nucleus lesions and conditioned taste aversion: evidence supporting an associative deficit, *Behav. Neurosci.*, 107, 1005, 1993.

67. Azzara, A., Sclafani, A., Grigson, P. S., and Norgren, R., Parabrachial nucleus (PBN) lesions block flavor aversion but not flavor preference conditioning, *Appetite*, 31, 286, 1998.

68. Azzara, A., Touzani, K., Sclafani, A., Grigson, P. S., and Norgren, R., Unpublished data, 1998.

69. Mehiel, R., The effects of naloxone on flavor-calorie preference learning indicate involvement of opioid reward systems, *Psychol. Rec.*, 46, 435, 1996.

70. Ramirez, I., Intragastric carbohydrate exerts both intake-stimulating and intake-suppressing effects, *Behav. Neurosci.*, 111, 612, 1997.

71. Yildiz, O., Bolu, E., Deniz, G., Senoz, S., Simsek, A., Ozata, M., Gundogan, N., and Gundogan, M. A., Effects of dexfenfluramine on glucose drinking and glucose-conditioned flavour preferences in rats: taste vs. postingestive conditioning, *Pharmacol. Res.*, 33, 41, 1996.

72. Sclafani, A., Bodnar, R. J., and Delamater, A. R., Pharmacology of food conditioned preferences, *Appetite*, 31, 406, 1998.

73. Sclafani, A., Starch and sugar tastes in rodents: an update, *Brain Res. Bull.*, 27, 383, 1991.

74. Ackroff, K., Vigorito, M., and Sclafani, A., Fat appetite in rats: the response of infant and adult rats to nutritive and nonnutritive oil emulsions, *Appetite*, 15, 171, 1990.

75. Sclafani, A., Lucas, F., and Ackroff, K., The importance of taste and palatability in carbohydrate-induced overeating in rats, *Am. J. Physiol.*, 270, R1197, 1996.

76. Lucas, F. and Sclafani, A., Carbohydrate-conditioned odor preferences in rats, *Behav. Neurosci.*, 109, 446, 1995.

77. Tordoff, M. G. and Friedman, M. I., Drinking saccharin increases food intake and preference — IV. Cephalic phase and metabolic factors, *Appetite*, 12, 37, 1989.

78. Elizalde, G. and Sclafani, A., Starch-based conditioned flavor preferences in rats: influence of taste, calories, and CS-US delay, *Appetite*, 11, 179, 1988.

79. Simson, P. C. and Booth, D. A., Dietary aversion established by a deficient load: specificity to the amino acid omitted from a balanced mixture, *Pharmacol. Biochem. Behav.*, 2, 481, 1974.

80. Gibson, E. L. and Booth, D. A., Acquired protein appetite in rats: dependence on a protein-specific state, *Experientia*, 42, 1003, 1986.

81. Baker, B. J. and Booth, D. A., Genuinely olfactory preferences conditioned by protein repletion, *Appetite*, 13, 223, 1989.

8 Evidence for Caloric, but Not Macronutrient, Compensation to Preloads Varying in Fat and Carbohydrate Content in Human Subjects

Barbara J. Rolls and Timothy H. Moran

CONTENTS

1   INTRODUCTION

Data from animal studies suggest that intake of individual macronutrients is regulated.[1] Central or peripheral infusions of various neuropeptides can influence macronutrient intake; for example, exogenous administration of neuropeptide Y (NPY) stimulates carbohydrate intake,[2] while administration of galanin stimulates the intake of fat.[3,4] Furthermore, levels of NPY within the paraventricular nucleus (PVN) are high during the time of day when carbohydrate intake is high,[5] and elevated levels of galanin within the PVN have been demonstrated in animals that naturally consume large amounts of fat.[6] On the other hand, serotonergic agents have been suggested to reduce carbohydrate intake preferentially,[1] while enterostatin appears to inhibit the intake of fat specifically.[7]

In humans the intake of dietary fat has been suggested to be important in both the etiology and maintenance of obesity.[8] Obese individuals have an elevated preference for fatty foods[9] and often consume a higher proportion of fat in their diets than normal-weight individuals.[10] As well as being associated with obesity, high levels of fat intake have been associated with coronary disease and cancer.[11] Based on these associations, public health authorities have advised that the percentage of dietary fat in the diet be reduced.[12] The degree to which dietary fat intake can be reduced is unclear. If, as suggested by some of the animal studies noted above, the brain contains neural systems that specifically control macronutrient intake, efforts to reduce overall fat intake may result in activation of these neural systems in opposition to such efforts.

The degree to which changes in macronutrient content of the diet at individual meals or over short periods of time results in compensatory behavior to maintain usual levels of macronutrient

0-8493-2752-0/00/$0.00+$.50
© 2000 by CRC Press LLC

intake has not been extensively investigated. We have, however, undertaken a series of experiments examining whether fat and carbohydrate have different effects on satiety, i.e., intake in a subsequent meal or meals. In these studies participants were served preloads varying in macronutrient content followed by a self-selected meal. Because these meals included a selection of foods varying in macronutrient composition, we could determine whether or not compensation for changes in macronutrient content of the preloads occurred.

2   DESIGN ISSUES IN PRELOAD EXPERIMENTS

Before presenting the experimental data, a discussion of some important aspects of experimental design is appropriate. Preloading, or administering a fixed amount of a given food or nutrient and measuring its effects on subsequent intake, is a widely accepted technique for studying satiety.[13] For example, experimenters may serve a preload, varying across one dimension, followed by a self-selected meal. Covert alterations in the composition of the preload provide an effective method for assessing the physiological effects of the preload.[14] It is important that the manipulation be covert since overt manipulations can affect intake not only through physiological mechanisms but also through a subject's beliefs about a food or nutrient. For example, if a subject believes that a high-fat food is very filling or high in calories, this can interact with or even override physiological satiety.[15] Studies that have provided subjects with false information about the fat content of preloads have clearly made this point.[16]

The choice of the preload must also be considered. If the preload is for oral consumption, the amount should be appropriate for the time of day and for the particular subject population. When comparing the effects of different macronutrients on subsequent intake, it is important that preloads be matched for energy content, weight, and sensory properties. This allows for the effects of macronutrient content, independent of other properties of the preload, to be assessed. It is also important to balance the order in which preloads are administered and use a "no load" condition to establish baseline energy and macronutrient intakes in a subsequent test meal. Finally, the manipulated variable in the preload should be of sufficient magnitude to allow detection by physiological mechanisms.

In order to determine the effects of macronutrients on food selection, subsequent test meals should provide a variety of well-liked foods that vary in both their macronutrient and caloric composition to allow subjects to adjust their intake to the preload manipulation. The test meals should also be offered in portion sizes that allow subjects to vary their intake appropriately without simply cleaning the plate. Finally, subjects need to be screened to ensure that they like and will eat the offered foods.

We have conducted a number of preloading experiments using these principles of experimental design. Although the aim was to assess whether there are differential effects of fat and carbohydrate on satiety, the use of these design principles has also allowed us to examine whether or not, in situations where caloric control was evident, any evidence for macronutrient control could be found. Three basic types of experiments were performed: responses to oral preloads in a residential laboratory over a period of several days, responses to oral preloads in subsequent lunch meals, and responses to intragastric or intravenous preloads in subsequent lunch meals. All of these studies provided the opportunity to address whether or not stable levels of macronutrient intake were defended.

3   RESIDENTIAL LABORATORY STUDIES

The first type of experiment took place in a residential laboratory setting in normal-weight unrestrained men. In the first experiment,[17] subjects' food intake was monitored continuously for 13 days and the carbohydrate or fat content of a required lunch was varied by 400 kcal on different days.

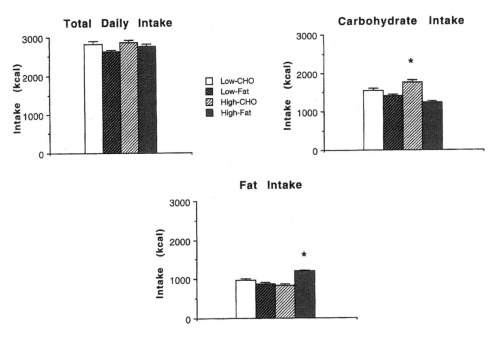

FIGURE 1   Total daily caloric (top, left panel), carbohydrate (top, right panel), and fat intake (bottom panel) including the lunch preload. Subjects compensated for the energy content of the preloads but not the macronutrient content. (From Foltin, R. W. et al., *Am. J. Clin. Nutr.,* 52, 969, 1990. With permission.)

An hour following consumption of the lunch meal, a food box containing a wide variety of foods, including meal items and conventional snack foods and drinks, was given to the subjects. Subjects also received a book containing the pictures of the fronts of the boxes of a wide variety of frozen meals that subjects could request. All items could be consumed at any time during the experimental day (from 1330 to 1200 h on the following day). Consumption was closely monitored by video- and audio-monitoring systems. In addition, subjects were instructed to inform the researchers via a computerized communication system whenever they ate or drank, specifying substance and portion size. Wrappers of all foods were color-coded and trash was removed and measured daily to validate the accuracy of the subject's reports.

Each of the four lunch conditions contained either ~431 or 844 kcal and consisted of sandwiches, salad, gelatin dessert, and a carbonated drink. The caloric differential was accomplished by varying the proportion of calories derived from fat or carbohydrate in commercially available products such as soft drinks, salad dressings, and processed meats and cheeses. The low-carbohydrate and low-fat lunches (431 kcal) were similar to the high-fat and high-carbohydrate lunches (844 kcal) in taste and appearance. The effect of each lunch condition was examined for 3 consecutive days. As shown in Figure 1, total daily caloric intake was unaffected by either the carbohydrate or fat manipulation. Subjects compensated for the difference in the energy content of the lunches later in the day so that their daily caloric intake did not significantly differ across conditions. Despite this caloric compensation, there was no evidence for macronutrient compensation in response to the carbohydrate or fat content of the lunches. As shown in Figure 1, total daily carbohydrate intake was elevated significantly in the high-carbohydrate condition and total daily fat intake was elevated significantly in the high-fat condition.

In a second residential study,[18] the effects of greater manipulations of fat or carbohydrate content of meals were assessed. In this experiment, subjects were given and required to consume breakfast, lunch, and an afternoon snack. Overall energy content of the required meals varied from ~700 to ~1700 kcal (representing 22 to 53% of spontaneous intake) with the majority of the differential derived from fat or carbohydrate. Because of the large caloric manipulation, it was not possible in this study to control tightly for the weight or sensory properties of the preloads. Each of six conditions (high-, medium-, and low-fat, and high-, medium-, and low-carbohydrate) was in effect for 2 days and caloric and macronutrient intakes were compared with intakes on control days when there were no required meals or snacks. Again, subjects were allowed to choose from a wide variety of foods that were available throughout the remainder of the day. Subjects compensated for the energy consumed in the required eating occasions except in the low-carbohydrate condition. In this condition the energy density (kcal/g) was the lowest, which meant that the weight or volume of the compulsory foods was the highest. We have found in several recent studies[19,20] that the energy density of foods can affect energy intake and caloric compensation. Thus, although in general the results again provided evidence for caloric compensation, this can be disrupted when large volumes of food are consumed before or as part of the meal. As in the previous study, there was no indication that macronutrient intake was regulated. When the macronutrient content of the required meals was included in the analysis, the results reflected the macronutrient content of the preloads. Subjects derived a higher overall proportion of their daily energy intake from carbohydrate in the medium- and high-carbohydrate conditions and a larger proportion of their energy intake was derived from fat in the high-fat condition. When the macronutrient content of the required eating occasions was excluded from the analysis, the proportion of energy derived from each macronutrient was not affected by the macronutrient content of the preloads. Manipulations of the macronutrient content of the required eating occasions that represented up to 800 to 990 kcal did not affect the subsequent macronutrient choices. Thus, while this experiment again provided evidence for overall caloric compensation, no evidence for macronutrient compensation was found.

4   ORAL PRELOADS

Experiments from a number of laboratories have compared the effects of fat or carbohydrate in preloads on hunger and satiety ratings and/or food intake in subsequent single meals (see Rolls and Hammer[14] for a complete discussion). We have carried out a series of experiments in which preloads consisted of yogurts formulated to have similar energy densities and sensory properties but large differences in their overall macronutrient content. The first study[21] was aimed at determining whether there were differences in the time course of the effects of preloads high in fat or carbohydrate content on subsequent food intake in a single self-selection meal. The self-selection meal contained a wide variety of foods differing in their caloric density and their carbohydrate and fat content. Subjects consumed either 350 (for females) or 500 (for males) g (~1.0 kcal/g) of one of two yogurt preloads in which the majority of the calories was derived from either fat or carbohydrate (Table 1). Yogurts were consumed either 30, 90, or 180 min before the self-selection

TABLE 1
Macronutrient Composition of the High-Carbohydrate and High-Fat Yogurts (percent total calories shown in parentheses)

(in 350 g)	Yogurt Type	
	Carbohydrate	Fat
Fat (kcal)	18 (5%)	232 (65%)
Carbohydrate (kcal)	289 (81%)	71 (20%)
Protein (kcal)	50 (14%)	54 (15%)
Total Energy (kcal)	357	357

Source: Rolls, B.J. et al., *Am. J. Physiol.,* 260, R756, 1991. With permission.

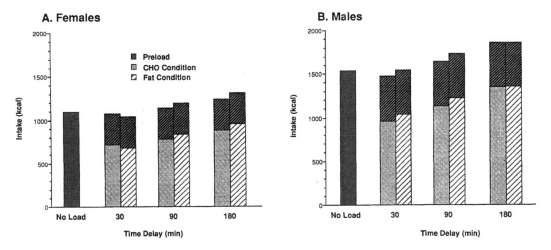

FIGURE 2 Mean total caloric intake (lunch + yogurt preload) for the no-load condition and for the high-carbohydrate and high-fat yogurt preload conditions in females (A) and males (B). Caloric compensation became less precise as the delay between the preload and the test lunch lengthened. (From Rolls, B. J. et al., *Am. J. Physiol.,* 260, R756, 1991. With permission.)

lunch which was offered at a standard time. Importantly, in these studies, the high-fat and high-carbohydrate yogurts had similar sensory ratings, so differences in subsequent intake could be attributed to the physiological effects of the yogurt preloads rather than to cognitive responses to sensory differences. The results of these experiments again provided support for caloric compensation, and there were no differences in the ability of subjects to compensate for calories derived from carbohydrate or fat. As shown in Figure 2, both male and female subjects accurately compensated for the energy in the preloads following a 30-min delay, but as the time interval between the preload and the test meal increased, energy compensation was less complete. The proportion of calories consumed at lunch that was derived from fat or carbohydrate was unaffected by the macronutrient content of the preload (Table 2). When the macronutrients in the preload were included in the analyses, fat intake was significantly higher in the high-fat preload condition and carbohydrate intake was significantly higher in the high-carbohydrate preload condition.

TABLE 2
Mean Macronutrient Intake in Each Condition as a Percent of Total Calories Consumed at Lunch

Condition	Carbohydrate Intake		Fat Intake	
	Males	Females	Males	Females
No load	34.7 ± 2.7	36.8 ± 2.0	46.4 ± 2.5	47.5 ± 2.2
30 min CHO	32.1 ± 2.8	33.1 ± 2.8	47.7 ± 2.4	48.2 ± 2.1
30 min Fat	33.5 ± 3.1	35.8 ± 3.1	44.3 ± 3.4	45.4 ± 3.8
90 min CHO	36.3 ± 3.1	36.3 ± 2.4	44.6 ± 3.2	46.7 ± 2.7
90 min Fat	32.7 ± 2.6	34.4 ± 1.7	44.8 ± 2.0	49.3 ± 2.2
180 min CHO	33.0 ± 2.8	39.4 ± 2.4	47.5 ± 2.8	45.2 ± 2.7
180 min Fat	35.3 ± 2.5	37.7 ± 2.7	45.4 ± 2.7	44.5 ± 2.8

Source: Rolls, B.J. et al., *Am. J. Physiol.,* 260, R756, 1991. With permission.

In the next experiment,[22] we examined responses to yogurt preloads containing varying amounts of carbohydrate and fat in lean and obese men and women to determine whether there were differences in caloric or macronutrient compensation that depended on, sex, body weight, and degree of restraint over eating. Responses to five different formulations of yogurt (control, medium-, and high-carbohydrate, and medium- and high-fat) were assessed in normal-weight restrained and unrestrained males and in normal-weight and obese, restrained and unrestrained females. The manipulations were covert and the yogurts were rated as equally palatable. The caloric density of yogurts differed (control yogurt: 0.46 kcal/g; medium-carbohydrate and medium-fat yogurt: 0.75 kcal/g; high-carbohydrate and high-fat yogurt: 1.02 kcal/g). Yogurts were consumed 30 min prior to a lunch meal. The degree of energy compensation depended upon the test population. Normal-weight, unrestrained males accurately compensated for the energy content of the yogurt regardless of the macronutrient composition. The other groups, however, did not show such precise compensation in that calorie-for-calorie the high-carbohydrate preloads suppressed intake to a slightly greater degree than did the high-fat preloads. Despite this difference in the degree of energy compensation, there were no differences in the pattern of macronutrient intake in the test meal. The percent of energy consumed at lunch derived from fat and carbohydrate did not differ across the six conditions for any group. Again, this experiment did not provide evidence for regulation of macronutrient intake in human subjects.

5   PARENTERAL AND ENTERAL PRELOADS

The third type of preloading study that was performed involved a comparison of the effects of intravenous and intragastric infusions of macronutrients on the amount and macronutrient pattern of subsequent intake. A significant advantage of infusion studies over the preload experiments discussed above was that any effects that the preloads might have had because of a subject's response to the sensory properties or previous experiences with the food were avoided. Also, pure macronutrients could be administered, and, thus, the amount of the macronutrient manipulation could be large and unaffected by the presence of even small amounts of the other macronutrients. In one study,[23] fat or carbohydrate was infused into either the circulatory system via an intravenous cannula or directly into the stomach via a nasogastric tube. The subjects were normal-weight males with low dietary restraint scores. Infusions were equal in volume and caloric content and infusion rates were controlled. The infusions were 500 ml of isotonic saline, or 500 cal of either fat in the form of Intralipid or carbohydrate as a dextrose solution. Parenteral preloads were administered as a "slow" infusion over 3½ h. Enteral preloads were administered either at the "slow" infusion rate over 3½ h or as "rapid" infusions over 15 min. Subjects were given a self-selection lunch 15 min after the termination of the 15-min infusion or 30 min after the end of the h infusion. Lunch and dinner for the test days were individual buffet-style, self-selection meals. As in the experiments discussed above, the foods offered varied in energy and macronutrient content such that the individuals could vary their caloric and macronutrient consumption.

Parenteral infusion of fat and carbohydrate over a 3½-h period 30 min prior to access to a self-selection lunch had no significant effect on either the total energy consumed or the macronutrient selection during the subsequent lunch and dinner. In contrast, intragastric infusions affected caloric intake depending upon both the macronutrient infused and the rate of infusion. As shown in Figure 3, the rapid infusion of either fat or carbohydrate significantly reduced subsequent lunch intake such that total caloric intake (preload plus lunch intake) was the same following intragastric fat or carbohydrate as following the intragastric saline infusions. This occurred despite different effects of the intragastric macronutrients on plasma glucose, insulin, and cholecystokinin levels. After slow infusion, fat was more effective at reducing subsequent intake than was carbohydrate (Figure 3). In contrast to these effects on overall energy intake, and in agreement with the results from the other preloading studies, there were no significant effects of intragastric fat or carbohydrate on

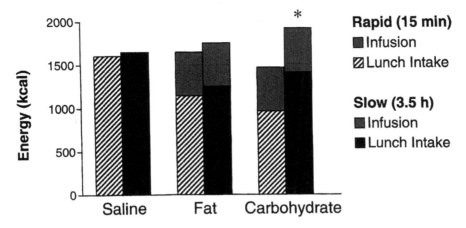

FIGURE 3 Mean caloric intake (lunch + preload) after rapid and slow intragastric infusions of isotonic saline, or 500 kcal of fat (Lyposyn) or carbohydrate (dextrose). (Adapted from Shide, D. J. et al., *Am. J. Clin. Nutr.*, 61, 754, 1995. With permission.)

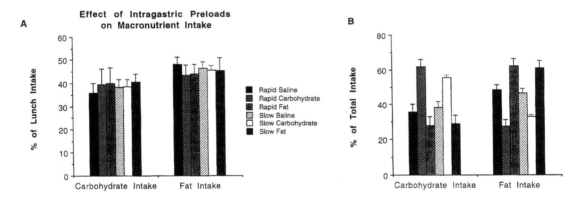

FIGURE 4 Effect of rapid and slow intragastric infusions of saline, fat, or carbohydrate on macronutrient choices at lunch following the infusions. (A) Carbohydrate and fat intake as a percent of calories consumed at lunch. (B) Carbohydrate and fat intake as a percent of total caloric intake (preload + lunch intake). (Adapted from Shide, D. J. et al., *Am. J. Clin. Nutr.*, 61, 754, 1995. With permission.)

macronutrient intake as a percent of overall energy intake (Figure 4A). When the amount of fat or carbohydrate provided in the infusions was included in the analysis, there was a significant effect such that total fat intake was greater in the fat infusion condition and total carbohydrate intake was greater in the carbohydrate infusion condition (Figure 4B). Thus, while slow or rapid intragastric fat infusions reduced overall energy intake, this reduction was not due to any selective reduction in fat intake. Similarly, while rapid intragastric carbohydrate infusions diminished overall energy intake, there was no indication that this reduction of energy intake was due to a preferential reduction in the amount of carbohydrate consumed. The conclusions of this study are supported by a recent report that even larger (1000 kcal) intragastric infusions of Intralipid or glucose over 15 min did not differentially affect hunger or caloric compensation at a test meal 1½ h later. Furthermore, adjustment of macronutrient intake at lunch in relation to the type of macronutrient in the infusions was not observed.[24]

6   DISCUSSION

These data demonstrate that, in a variety of human test populations, caloric compensation occurs in response to preloads which vary across a wide range of macronutrient levels. Overall, fat and carbohydrate provided in the preloads had similar effects on caloric compensation. This was evident in single test meals following oral or intragastric preloads as well as across days when subjects were required to consume multiple meals where the fat or carbohydrate manipulations were as large as 1000 kcal. While caloric compensation occurred in these tests, there was no evidence for macronutrient compensation. Even in situations where subjects were presented with a wide range of foods which varied greatly in their fat and carbohydrate content, preloads that had macronutrient contents that were markedly different from subjects' normal diets had no effect on macronutrient selection. A clear conclusion from these experiments was that macronutrient intake was not regulated in human subjects in response to large variations in the macronutrient content of preloads over periods of several days.

This conclusion is consistent with the results of studies that have examined the effects of fat substitutes on overall caloric intake and macronutrient choice. For example, when olestra, a non-absorbable fat substitute, replaced 20 or 36 g of fat at breakfast, daily fat intake, but not energy intake, was reduced. There was no evidence of increased fat intake later in the day.[25] Thus, this study, along with the others reviewed, indicates that dietary manipulations aimed at reducing fat consumption do not activate mechanisms that produce compensatory fat intake in subsequent meals.

In contrast to these ideas, De Castro[26] has recently reported data suggesting macronutrient-specific negative feedback occurring with a delay of 2 days. Conducting an autoregression analysis of intake data from dietary records from over 700 adults produced significant negative correlations between intake on 1 day and intake 2 days later. Significant effects were found for overall energy intake but also for each individual macronutrient. Noteworthy was the finding that each macronutrient had its greatest negative feedback effect on itself and little effect on the subsequent intake of other macronutrients. While these data suggest that intake of individual macronutrients is regulated, it must be stressed that the effects identified in the analysis, while significant, were small. The largest amount of variance accounted for was 16% for any total day's macronutrient intake. In these studies, macronutrient effects were noted at the same time as energy effects. In the preload experiments discussed above, precise energy compensation was evident at times when there was no macronutrient compensation. Finally, it must be noted that these findings of De Castro do not indicate that the compensation is occurring through a physiological mechanism. By definition, these effects were not the outcome of a covert manipulation. Subjects kept their own dietary records and decisions about intake on any day may have been influenced by their knowledge of energy and macronutrient consumption on prior days.

In conclusion, fat and carbohydrate manipulations have been shown to have similar effects on subsequent food intake. In situations where accurate caloric compensation was demonstrated, there was no evidence that individual macronutrient intake was under tight physiological control. While fat is often overconsumed in obese individuals, this is likely to be because fatty foods are often highly palatable[27] and high in energy density.[28] Ingestion of low-fat foods and fat substitutes is unlikely to produce a compensatory fat appetite and will likely be a useful dietary strategy for reducing the proportion of calories from fat in the diet.

REFERENCES

1. Leibowitz, S. F., Neurochemical-neuroendocrine systems in the brain controlling macronutrient intake and metabolism, *Trends Neurosci.*, 15, 491, 1992.
2. Stanley, B. G., Daniel, D. R., Chin, A. S., and Leibowitz, S. F., Paraventricular nucleus injection of peptide YY and neuropeptide Y preferentially enhance carbohydrate ingestion, *Peptides*, 6, 1205, 1985.
3. Kyrkouli, S., Stanley, B. G., Seirafi, R. D., and Leibowitz, S. F., Stimulation of feeding by galanin: anatomical localization and behavioral specificity of this peptides effect in the brain, *Peptides*, 11, 995, 1990.

4. Tempel, D. L., Leibowitz, K. L., and Leibowitz, S. F., Effects of PVN galanin on macronutrient selection, *Peptides,* 9, 309, 1988.
5. Akabayashi, A., Levin, N., Paez, X., Alexander, J. T., and Leibowitz, S. F., Hypothalamic neuropeptide Y and its gene expression: relation to light dark cycle and circulating corticosterone, *Mol. Cell. Neurosci.,* 5, 210, 1994.
6. Akabayashi, A., Koenig, J. I., Watanabe, Y., Alexander, J. T., and Leibowitz, S. F., Galanin containing neurons in the paraventricular nucleus: a neurochemical marker for fat ingestion and body weight gain, *Proc. Natl. Acad. Sci. U.S.A.,* 91, 10375, 1994.
7. Erlanson-Albertson, C., Mei, J., Okada, S., York, D., and Bray, G. A., Pancreatic procolipase propeptide, Enterostatin specifically inhibits fat intake, *Physiol. Behav.,* 49, 1191, 1991.
8. Klesges, R. C., Klesges, L. M., Haddock, C. K., and Peters, J. C., A longitudinal analysis of the impact of dietary intake and physical activity on weight change in adults, *Am. J. Clin. Nutr.,* 55, 818, 1992.
9. Drewnowski, A., Kurth, C., Holden-Wiltse, L., and Saari, J., Food preferences in human obesity; carbohydrates vs. fats, *Appetite,* 18, 207, 1992.
10. Miller, W. C., Lindeman, A. K., Wallace, J., and Niederpruem, M., Diet composition, energy intake and exercise in relation to body fat in men and women, *Am. J. Clin. Nutr.,* 52, 818, 1992.
11. Pi-Sunyer, X., Medical hazards of obesity, *Ann. Intern. Med.,* 119, 655, 1993.
12. National Research Council, *Diet and Health,* National Academy Press, Washington, D.C., 1989.
13. Kissileff, H., Satiating efficiency and a strategy for conducting preloading experiments, *Neurosci. Biobehav. Rev.,* 8, 129, 1984.
14. Rolls, B. J. and Hammer, V. A., Fat, carbohydrate and the regulation of energy intake, *Am. J. Clin. Nutr.,* 62S, 1086, 1995.
15. Shide, D. J. and Rolls, B. J., Information about the fat content of preloads influences energy intake in healthy women, *J. Am. Diet. Assoc.,* 95, 993, 1995.
16. Caputo, F. A. and Mattes, R. D., Human dietary responses to perceived manipulation of fat content in a midday meal, *Int. J. Obesity,* 17, 237, 1993.
17. Foltin, R. W., Fischman, M. W., Moran, T. H., Rolls, B. J., and Kelly, T. H., Caloric compensation for lunches varying in fat and carbohydrate content by humans in a residential laboratory, *Am. J. Clin. Nutr.,* 52, 969, 1990.
18. Foltin, R. W., Rolls, B. J., Moran, T. H., Kelly, T. H., McNelis, A. L., and Fischman, M. W., Caloric, but not macronutrient compensation by humans for required-eating occasions with meals and snacks varying in fat and carbohydrate, *Am. J. Clin. Nutr.,* 55, 331, 1992.
19. Bell, E. A., Castellanos, V. H., Pelkman, C. L., Thorwart, M. L., and Rolls, B. J., Energy density of foods affects energy intake in normal-weight women, *Am. J. Clin. Nutr.,* 67, 412, 1998.
20. Rolls, B. J., Castellanos, V. H., Halford, J. C., Kilara, A., Panyam, D., Pelkman, C. L., Smith, G. P., and Thorwart, M. L., Volume of food consumed affects satiety in men, *Am. J. Clin. Nutr.,* 67, 1170, 1998.
21. Rolls, B. J., Kim, S., McNelis, A. L., Fischman, M. W., Foltin, R. W., and Moran, T. H., Time course of effects of preloads high in fat or carbohydrate on food intake and hunger ratings in humans, *Am. J. Physiol.,* 260, R756, 1991.
22. Rolls, B. J., Kim-Harris, S., Fischman, M. W., Foltin, R. W., Moran, T. H., and Stoner, S. A., Satiety after preloads with different amounts of fat and carbohydrate: implications for obesity, *Am. J. Clin. Nutr.,* 60, 476, 1994.
23. Shide, D. J., Caballero, B., Reidelberger, R., and Rolls, B. J., Accurate energy compensation for intragastric nutrients in lean males, *Am. J. Clin. Nutr.,* 61, 754, 1995.
24. Cecil, J. E., Castiglione, K., French, S., Francis, J., and Read, N. W., Effects of intragastric infusions of fat and carbohydrate on appetite ratings and food intake from a test meal, *Appetite,* 30, 65, 1998.
25. Rolls, B. J., Pirraglia, P. A., Jones, M. B., and Peters, J. C., Effects of olestra, a noncaloric fat substitute, on daily energy and fat intake in lean men, *Am. J. Clin. Nutr.,* 56, 84, 1992.
26. De Castro, J. M., Prior day's intake has macronutrient-specific delayed negative feedback effects on the spontaneous food intake of free-living humans, *J. Nutr.,* 128, 61, 1998.
27. Rolls, B. J., Carbohydrates, fats and satiety, *Am. J. Clin. Nutr.,* 61(Suppl.), 960S, 1995.
28. Rolls, B. J. and Bell, E. A., Energy intake: effects of fat content and energy density of foods, in *Nutrition, Genetics and Obesity,* Vol. IX, G. A. Bray and D. H. Ryan, Eds., Louisiana State University Press, Baton Rouge, 1999, 172.

9 The Effects of Nutrient Preloads on Subsequent Macronutrient Selection in Animal Models

L. A. Foster

CONTENTS

1 INTRODUCTION

During the course of a typical meal, animals and people select a variety of foods. Controversy exists over what mechanisms underlie this preference for variety. Part of the mechanism is most likely a nutrient–receptor interaction. Due to the dynamics of a meal, i.e., food choices and consumption are completed before the entire meal has even emptied from the stomach, there would not be time for negative feedback from a repleted nutrient store. In the classic negative feedback model, one would predict that a nutrient consumed in the beginning of a meal would interact with a receptor. That interaction would lead to negative feedback on further consumption of that nutrient but not other nutrients. Different aspects of this model can be expanded into several more questions: How many different types of receptors are there? What do they interact with (what types of foods will have similar effects)? Where are these receptors located (orosensory, gastric, intestinal, hepatic)? The activation of which receptor type will stop the intake of what type of foods?

A nutrient preload is one way of addressing some of these questions. A nutrient preload will be loosely defined as a nutritive substance that is administered to an animal prior to access to test meals. Often, rather than allowing the animal to ingest the preload, it is infused into the stomach, intestine, portal, or peripheral vein in order to isolate the location of action. Different types of nutrients can be used for the preload and different types of foods can be used as the test meal to determine what types of nutrients affect what types of food.

One possible obstacle in elucidating the mechanisms involved with food selection during a meal is the likelihood that the regulation of calories can override the mechanisms for macronutrient selection. From an evolutionary standpoint, it might be advantageous for an animal to meet its caloric requirement from the "safest" food first, and then seek out small amounts of nutrients that were lacking in the "safe" food. An upper limit, also set by calories, would stop intake regardless of how appealing a food might be.

Thus, there would only be this "window of finicky eating" where differential selection would actively occur. Although this "window" would be small relative to an animal's caloric requirement, it might be very important as an explanation for why there is such a large percentage of obese people in countries where a wide variety of palatable foods are readily available.

One conclusion that can be drawn from the literature is that, in spite of all of these unknowns, there are instances where a nutrient preload has had a significant differential effect on subsequent food choices. This suggests that there is some type of nutrient recognition in the body, and that this information can be used to differentially effect further consumption of different nutrients. Next, the impact of these different methods used for preload experiments will be examined.

2   METHODOLOGICAL DIFFERENCES AND ISSUES

2.1   CHARACTERISTICS OF THE PRELOAD

2.1.1   Composition

One aspect that may be hampering progress in defining how a particular preload will affect subsequent nutrient selection is the nomenclature used. Investigators frequently describe the composition of preloads and test meals in terms of the type of macronutrient that was used. *Macronutrient* is a broad term. Within a macronutrient class are contained both essential and nonessential nutrients, nutrients that cross membranes via different transporters and nutrients that are metabolized differently. It is not surprising that there is evidence to suggest that not all representatives within a macronutrient class elicit similar responses. Glucose and sucrose are both CHOs (carbohydrates); yet, a preload of glucose suppressed intake of glucose as a test meal but not sucrose.[1] A preload of four different concentrations of fat (Intralipid) affected the sham intake of two CHOs (sucrose and Polycose) very differently.[2]

2.1.2   Dose (Amount) and Rate of Administration

Another variable that could cause differences in the results of preload experiments is the amount of preload given. This is important on both a caloric and volume basis. One can predict that only preloads of an intermediate size would elicit a differential nutrient selection; very small preloads would have no effect and large preloads (in either volume or calories) would stop intake altogether. It is not easy to predict what will be of an intermediate size for a given nutrient preload in a particular experiment. An infusion of 2.5 kcal of lipid suppressed intake when given while rats were sham feeding, whereas the same infusion had no effect when given just prior to the rats sham ingesting the food.[3]

Volume of a preload can have an effect separate from composition; this is particularly true for preloads given into the stomach. Infusions restricted to the stomach have been reported to affect total food intake based on volume but not composition.[4,5] It has also been suggested that increasing the volume of the stomach contents (i.e., gastric distension) stops intake via nausea (reviewed by Read[6]). This does not mean that volume changes could not have an effect on macronutrient selection; nor for that matter, that nausea could not have an effect on macronutrient selection. In fact, when cows become ill with clinical ketosis, they will reverse their normal preference for grain over hay to hay over grain.[7] So there is support that illness (or a preload that induced illness) could affect macronutrient selection. This has simply not been tested specifically using preloads.

2.2   CHARACTERISTICS OF THE TEST DIET (OR INGESTED PRELOAD)

Basically the same variables that affect preloads are important for test diets, except that, whereas preloads are often infused, test diets are ingested (although there are preparations where a flavored non-caloric test diet is yoked to a nutrient gastric infusion, as will be discussed later).

One consideration is whether to employ a diet (mixture of nutrients) or single form of macro-nutrient. Each has advantages and disadvantages. A diet formulation is more physiological, and can be identical to the type of food an animal would normally see and readily eat. Disadvantages would include the fact that a diet contains other nutrients, each with a potentially separate effect. If the macronutrient composition was changed but caloric density held constant, then one of the other macronutrients would have to be changed also. This confounds the results regarding whether they are due to the increase of one nutrient or a decrease in the other (Figure 1 and Table 1).[8]

There is evidence to suggest that a diet enriched with a particular macronutrient and a single nutrient source within that macronutrient class are not treated the same by the body. A preload of glucose suppressed glucose intake but not the intake of sweetened condensed milk.[9]

While using a single nutrient type can produce much cleaner results, there are numerous challenges. One problem is that there are few naturally occurring foodstuffs that consist of only one macronutrient; this is particularly true for protein. Another technical difficulty for protein is that most amino acid solutions are unpalatable and, if made to contain a similar caloric concentration as a typical food, will have a very high osmolarity.

Making the texture the same is another challenge in developing three test meals that are identical except for their macronutrient type. This is especially true for fat. Thus, there are several technical problems to overcome in developing three single-nutrient source test meals (or preloads), each representing one macronutrient class, that are equivalent in volume, texture, calories, and osmolarity.

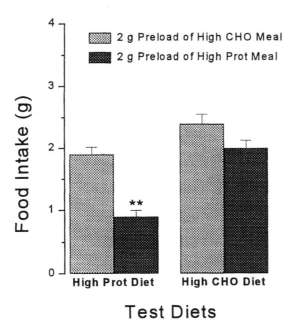

FIGURE 1   Bars represent the food intake of either a high-CHO diet or a high-protein diet after the rats had ingested either a high-protein preload or a high-CHO preload (**$p < 0.01$). (Adapted from Li, E.T.S. and Anderson, G.H., *Physiol. Behav.,* 29, 780, 1982. With permission.)

TABLE 1
Summary of Experimental Parameters Used in the Experiment Depicted in Figure 1

How preload was administered	Ingested
Amount of preload	2 g
Time interval between preload and test meal	30 min
Preload composition	**CHO:** Cornstarch 83.5%, corn oil 10%, minerals 4.5%, vitamins 2.5%
	Prot: casein 45%, cornstarch 38.5%, corn oil 10%, minerals 4%, vitamins 2.5%
Test meal composition	**High-Prot diet:** Casein 60%, cornstarch 23.5%, corn oil 10%, minerals 4.5%, vitamins 2.5%
	High-CHO diet: casein 10%, cornstarch 73.5%, corn oil 10%, minerals 4.5%, vitamins 2.5%

Adapted from Li and Anderson.[8] With permission.

There is evidence to suggest that intake and macronutrient selection are both critically affected by the interaction between the orosensory stimuli and postingestional stimuli of a food. That is, the same nutrient infused postgastrically can have a different effect on intake depending on the type of solution being ingested. When palatability was estimated by the amount that the rats would sham feed, it was found that intakes of the solutions with the highest palatability were suppressed by the smallest amount of fat infused into the intestine (Figure 2).[2] For the least palatable solution, the suppression induced by infusions of different concentrations of lipid (Intralipid) was similar to the "all-or-none" phenomenon as was reported when gastric infusions occurred without concurrent orosensory stimuli.[10]

2.3   NUMBER OF TEST MEALS TO OFFER AT ONCE

After administering the preload, the investigator can offer one, two, or more of the test meals at the same time to the animal. The rationale for using these different paradigms has been discussed previously by Sclafani.[11] The data would suggest that offering the animals a choice is a more sensitive test than offering only one test meal at a time. In an experiment where test meals were offered one at a time, duodenal infusions of either sucrose or lipid suppressed sham intake of sucrose to the same degree (Figure 3).[12] When animals were given a choice of sucrose or lipid (two-bottle test), a duodenal infusion of sucrose suppressed sucrose intake more than lipid intake.[3] This suggests the possibility that the mechanisms regulating caloric intake and those regulating nutrient selection might be different.

2.4   NUTRITIONAL STATUS AND DIET HISTORY

Another area that might prove insightful would be to alter the state or type of the test animal and to keep the preloads and test meals the same. These include (1) deprivation state; (2) maintaining animals on different diets that were high in different macronutrients; (3) using some animals that were naive and others that had previous experience to the test meals; or (4) genetically different animals (obese, diabetic, etc.). There is work to suggest that these different conditions can affect an animal's preference for different macronutrients (see Chapter 23 by Castonguay and Brown). Preloads of different nutrients in diabetic rats elicited similar results as normal rats on subsequent nutrient selection.[13] Obese Zucker rats were not as responsive as nonobese rats in making a selective response to different nutrient preloads.[14]

It has been reported that animals that have been food deprived will select more fat.[15] It would not be surprising if rats in different states of deprivation would respond differently to a preload, but it is difficult to predict how that difference might appear.

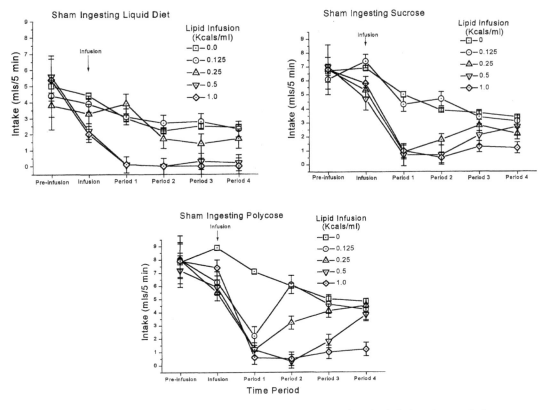

FIGURE 2   Sham intakes of a liquid diet, sucrose, or Polycose. During the infusion time period rats were infused with 10 ml of one of four concentrations of lipid solution (in kcal/ml) or saline (kcal/ml = 0). (From Foster, L.A., Nakamura, K., Greenberg, D., and Norgren, R., *Am. J. Physiol.,* 270, R1123, 1996. With permission.)

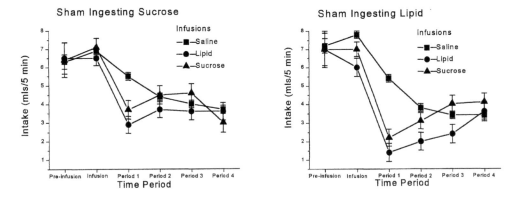

FIGURE 3   The lines represent the sham intakes of either sucrose or lipid. During the infusion time period rats were infused with lipid, sucrose, or saline. Each point represents mean intake ±SEM. (From Foster, L.A., Boeshore, K., and Norgren, R., *Physiol. Behav.,* 64, 451, 1998. With permission.)

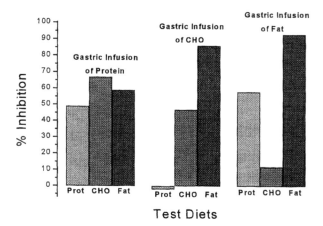

FIGURE 4   Bars represent the percentage inhibition of intake of test meals composed of one type of macronutrient, i.e., Prot, CHO, or Fat, induced by a gastric infusion of either Prot, CHO, or Fat. The percentage inhibition was calculated by (subtracting the intake after the vehicle preload from intake after the nutrient preload, divided by intake after control preload) × 100. (Adapted from Bartness, T.J. and Rowland, N., *Physiol. Behav.,* 31, 546, 1983. With permission.)

TABLE 2
Summary of Parameters Used in the Experiment Depicted in Figure 4

How preload was administered	Gastric infusion
Amount of preload	**Prot:** 20 ml 10 kcal
CHO:	**CHO:** 10 ml 10 kcal
Fat:	**Fat:** 10 ml 10 kcal
Time interval between preload and test meal	1 h
Preload composition	**Prot:** Sodium caseinate and water
	CHO: Sucrose and water
	Fat: Crisco vegetable oil, emulsifier, and water
	Control: Emulsifier and water
Test meal composition	**CHO diet:** cornstarch, sucrose, vitamins, minerals, and fiber
	Prot diet: Casein, methionine, vitamins, minerals, and fiber
	Fat diet: Vegetable oil

Adapted from Bartness et al.[13] With permission.

The type of diet the animal is being maintained on may also play a role in affecting how the animal will react to a preload. There is evidence to suggest that the type of diet an animal is being maintained on will affect that animal's preference in macronutrient.[16] It is logical that the maintenance diet would also affect an animal's preferences in macronutrients after a preload.

3   EFFECTS OF CHO PRELOADS

A gastric infusion of CHO (sucrose) decreased subsequent intake of CHO (cornstarch and sucrose) and fat (vegetable oil), but not protein (casein and methionine) (Figure 4 and Table 2).[13] A preload of CHO (cornstarch) given as a gastric infusion decreased the consumption of a CHO-enriched diet (73.5% cornstarch) more than a diet high in protein (60% casein) (Figure 5 and Table 3).[8] These experiments support the hypothesis that a CHO preload will have the greatest effect on suppressing intake of a CHO test meal. Neverthless, there are experiments where this result was not found. When a preload of CHO- or protein-enriched diets was ingested, there were no differences

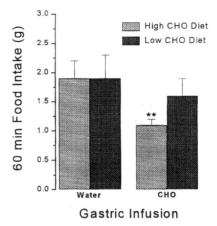

FIGURE 5 Bars represent intake of either a high-CHO diet or a high-protein diet after the rats had received either a gastric infusion of CHO or vehicle (** $p < 0.05$). (Adapted from Van Zeggeren, A. and Li, E.T.S., *J. Nutr.,* 120, 309, 1990.)

TABLE 3
Summary of Experimental Parameters Used in the Experiment Depicted in Figure 5

How preload was administered	Gastric infusion
Amount of preload	2 g/kg bodyweight
Time interval between preload and test meal	30 min
Preload compositions	**CHO:** Cornstarch and water
	Control: Water
Test meal composition	**High CHO:** Cornstarch 73.5%, casein 10%, corn oil 10%, minerals 4.5%, vitamins 2.5%
	High Prot: Cornstarch 23.5%, casein 60%, corn oil 10%, minerals 4.5%, vitamins 2.5%

Adapted from Van Zeggeren and Li.[14]

in intake of a test meal enriched in CHO (73.5% cornstarch) (see Figure 1).[14] In a third experiment both an intestinal infusion of CHO (sucrose) and an infusion of fat (Intralipid) suppressed the sham intake of CHO (sucrose) the same (see Figure 3).[12] In a similar experiment, except that the rats were given a choice of CHO (sucrose) or fat (Intralipid) to ingest at the same time, an intestinal infusion of CHO (sucrose) suppressed intake of sucrose but not fat.[3]

4   EFFECTS OF FAT PRELOADS

A gastric infusion of fat (vegetable oil) suppressed intake of fat (vegetable oil) and protein (casein) but not CHO (cornstarch and sucrose) (see Figure 4).[13] An intestinal infusion of fat in sham-feeding rats decreased the intake of fat (Intralipid) quicker and longer than an infusion of CHO (sucrose) (see Figure 3).[12] Again the suggestion is that a fat preload has the greatest effects on suppressing the intake of a fat meal.

5   EFFECTS OF PROTEIN PRELOADS

A gastric infusion of protein (casein) suppressed intake of protein (casein), CHO (cornstarch and sucrose), and fat (vegetable oil) (see Figure 4).[13] When a preload of a high protein (casein 45%)

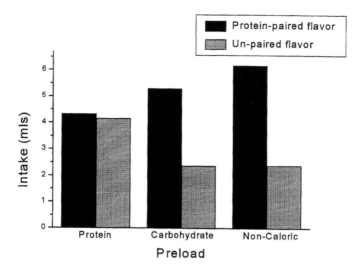

FIGURE 6 To determine whether there is a learned hunger specifically for protein, rats were deprived for 4 h and then allowed to drink a distinctly flavored, nonnutritive solution while a protein solution was infused into their stomachs. Rats showed a conditioned preference for the flavor. When protein was infused during the deprivation period (protein preload) the conditioned preference for protein was lost. Noncaloric and carbohydrate preloads maintained a preference for the protein-paired flavor. (Adapted from Baker, B.J., Booth, D.A., Duggan, J.P., and Gibson, E.L., *Nutr. Res.,* 7, 481, 1987.)

TABLE 4
Summary of Parameters Used in the Experiment Depicted in Figure 6

How preload was administered	Gastric infusion
Amount of preload	Three 2.5-ml doses
Time interval between preload and test meal	2 h, 0.5 h, and 1 min prior to preference testing
Preload composition	**Prot:** 10% casein hydrolysate
	CHO: 10% maltodextrin
Test meal composition	Two nonnutritive fluids, one flavored with a distinct odorant previously associated with 10% casein hydrolysate infusion and the other flavored with an odorant not associated with infusion

Adapted from Baker et al.[17]

or CHO (83.5% cornstarch) diet was ingested, intake of a high-protein diet (casein 60%) was suppressed more than a high-CHO diet (cornstarch 73.5%) (see Figure 1).[8]

In another experiment animals were trained so that whenever they ingested one non-nutritive flavor they received a gastric infusion of protein (casein) and when they ingested a different nonnutritive flavor they received a saline infusion. When the rats received a gastric infusion of protein, they decreased their consumption of the flavor associated with protein (Figure 6 and Table 4).[17]

Wurtman and co-workers have suggested that the reason that a preload of protein decreases subsequent consumption of protein is because the preload increases the concentration of amino acids in the blood. These amino acids compete with tryptophan to enter the brain, thus, decreasing the concentration of tryptophan (the precursor for serotonin) in the brain. They suggest that decreased serotonin in the brain causes rats to prefer a CHO meal over protein (as reviewed in References 18 through 21).

6   SITE OF ACTION FOR A PRELOAD

6.1   OROSENSORY

Another important difference in experiments is whether the preload is ingested or infused. If the preload is ingested then at least part of the effect may be orosensory in origin. Different nutrients within a macronutrient class tend to share more orosensory characteristics than between macronutrient classes; i.e., safflower oil tastes more like corn oil than either taste like sugar. Thus, an orosensory-specific satiety could account for part of the explanation of how ingesting a preload could lead to a differential nutrient selection along macronutrient lines. Moreover, there is evidence that animals can learn to associate the orosensory characteristics of a nutrient with its postingestive effects.[17,22]

On the other hand, in sham feeding, when orosensory effects have been isolated by allowing ingesta to drain from an esophageal or gastric fistula, the animal will continue to consume the same food, for an extended period.[23] This suggests that, although animals will stop ingesting eventually when only experiencing orosensory stimuli, orosensory stimuli alone are not responsible for termination of eating in the typical meal. However, this does not address the effects that orosensory stimulation alone could have on the selection of a variety of foods during a meal; i.e., whether preexposure to an orosensory stimulus would affect an animal's subsequent preference between two foods. To test this we gave sham-feeding rats access to either sucrose or lipid (Intralipid) for 3 or 20 min. Then the solution was removed, and 5 min later the rats were given access to both lipid and sucrose solutions simultaneously for 20 min. In this preliminary experiment, a 3-min exposure did not affect subsequent intake preferences differently than no prior exposure. However, 20 min of sham ingestion decreased preference for ingested nutrient, but did not affect (nonsignificant increase) the consumption of the other nutrient (Figure 7). Further work will need to be done to determine how intermediate times would affect preferences and if, as was the case for termination of eating, a small nutrient infusion will reduce the duration of preload ingestion necessary to affect subsequent food preferences.

6.2   POSTINGESTIVE

The locations with the highest potential for these putative nutrient-responsive receptors that might regulate nutrient selection during a meal would include the mouth and nose (orosensory); intestine; and liver. By infusing the preload downstream of these different locations, one can see what changes in nutrient selection occur with the absence of nutrient stimulation to one or more of these areas.

One finding with these experiments is that a small infusion that suppresses intake when occurring concurrently with orosensory stimulation (ingestion) may have no effect when given prior to orosensory stimulation.[3,24] This suggests that there is an interaction between the orosensory stimulation and postgastric (probably intestinal) stimulation. In fact, when infusions coincide with orosensory stimulation, there is a dose–response effect observed in the suppression of intake due to the caloric amount of the infusion into the duodenum.[2,25] However, when a gastric infusion occurred without any coincident orosensory stimulation, the response was described as all or none.[10]

Not surprisingly, similar findings have been reported for the effects of a duodenal infusion on nutrient selection. When lipid (Intralipid) was infused into the intestine, while rats sham fed either lipid or sucrose (two-bottle test), they decreased consumption of lipid more than that of sucrose. When the same lipid infusion was given just prior to spout access, there was no suppression of intake of either solution.[3]

These results suggest that concurrent orosensory and postingestional stimuli of a preload are critical for the maximal expression of a differential nutrient selection. However, there is evidence that the orosensory stimuli do not have to come from the same nutrient as the postgastric stimuli.

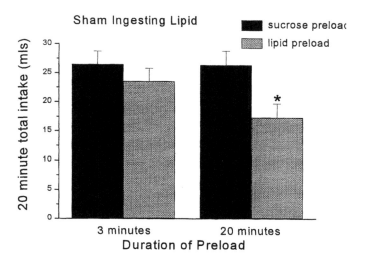

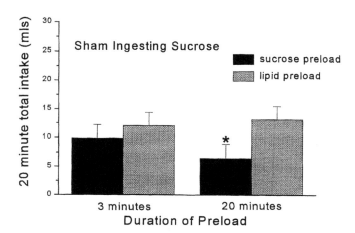

FIGURE 7 Sham intakes of either sucrose or lipid (two-bottle presentation) after pre-feeding with either sucrose or lipid for a duration of either 3 min or 20 min. Bars represent the mean sham intakes, ±SEM, for sucrose or lipid after ingesting either sucrose or lipid. There were no significant differences in lipid intake when rats previously ingested sucrose (either 3 or 20 min) or lipid for 3 min. There were no significant differences in sucrose intake when rats previously ingested lipid (either 3 or 20 min) or sucrose for 3 min. (From Foster and Hobrock, unpublished.)

Animals can learn to associate a particular flavor with a postgastric response.[17,26] In a study by Perez et al.,[26] rats were trained to associate one nonnutritive flavor with a gastric infusion of casein and another nonnutritive flavor with Polycose (see Table 5). A gastric preload of CHO or protein had no consistent effect on the subsequent ingestion of either flavor associated with CHO or protein. However, if the animals were allowed to taste the associated flavor simultaneously with the gastric preload, a preload of protein suppressed subsequent intake of the flavor associated with protein and a preload of CHO suppressed intake of the flavor associated with CHO. Along similar lines, a preload of sucrose, but not fat, suppressed intake of saccharin.[27]

TABLE 5
Experimental Parameters Used in Perez et al., 1996[26]

How preload was administered	Sham oral ingestion paired with intragastric infusion
Amount of preload	7.5 g infused at same rate as ingested
Time interval between preload and test meal	45 min
Preload composition	**Ingested:** Nonnutritive, flavor matched (see test meal)
	Infused Prot: 10% calcium caseinate (0.4 kcal/g)
	CHO: 10% Polycose (0.4 kcal/g)
Test meal composition	0.2% saccharin and 0.05% unsweetened cherry, grape, and orange Kool-Aid flavors in water previously associated with protein or CHO infusion

Data from Perez, C. et al., *Physiol. Behav.*, 59, 467, 1996.

7   CONCLUSIONS

One generalization that can be made based on the results of different preload experiments is that "the nutrient type of the preload tends to suppress the same type of nutrient as a test meal." This is not exclusive, in that sometimes the preload does not suppress intake of any of the test meals,[28] and sometimes the preload indiscriminately suppresses intake of all foods offered.[13] Nevertheless, when a preload differentially suppressed intake of test meals, the test meals most similar in composition to the preload tend to be suppressed first.

Based on the evidence presented in this chapter one can surmise that this differential selection does not seem to be based on a strictly macronutrient basis. One particular nutrient in the body does not seem to represent an entire macronutrient class or affect food selection on a strict macronutrient basis. At this time, it cannot be stated what aspect of the preload determines the specificity of the preference for a novel food or inhibits further consumption of a food similar to the preload. Future research may demonstrate that these components form subdivisions within the macronutrient classes. There are physiological processes where there are subdivisions within the general headings of macronutrients, such as method of absorption or metabolism. Fat is absorbed differently from CHO or protein, but, within the category of fat, medium-chain fatty acids are absorbed differently from long-chain fatty acids. Another possibility is that even as nutrients leave the stomach, they maintain similar categories as in the mouth, i.e., taste. It has been reported that certain receptors in the intestine exist that share characteristics with taste receptors.[29]

In conclusion, there is substantial evidence to suggest that nutrients recently introduced into the body can affect an animal's food choices, but the lines of demarcation seem to be narrower than the broad categories of fat, carbohydrate, and protein.

REFERENCES

1. Mook, D. G., Dreifuss, S., and Keats, P. H., Satiety for glucose solution in rat: the specificity is postingestive, *Physiol. Behav.*, 36, 897, 1986.
2. Foster, L. A., Nakamura, K., Greenberg, D., and Norgren, R., Intestinal fat differentially suppresses sham feeding of different gustatory stimuli, *Am. J. Physiol.*, 270, R1122, 1996.
3. Foster, L. A., Sham intake of lipid and sucrose (2 bottle presentation), when a duodenal infusion (lipid, sucrose, or saline) was initiated before, simultaneously with, or after, drink presentation, *Appetite*, 29, 395A, 1997.

4. Phillips, R. J. and Powley, T. L., Gastric volume rather than nutrient content inhibits food intake, *Am. J. Physiol.,* 271, R766, 1996.

5. Mathis, C., Moran, T. H., and Schwartz, G. J., Load-sensitive rat gastric vagal afferents encode volume but not gastric nutrients, *Am. J. Physiol.,* 274, R280, 1998.

6. Read, N. W., Role of gastrointestinal factors in hunger and satiety in man, *Proc. Nutr. Soc.,* 51, 7, 1992.

7. Foster, L. A., Clinical ketosis, *Vet. Clin. North Am. Food Anim. Pract.,* 4, 253, 1988.

8. Li, E. T. S. and Anderson, G. H., Meal composition influences subsequent food selection in the young rat, *Physiol. Behav.,* 29, 779, 1982.

9. Campbell, C. S. and Davis, J. D., Peripheral control of food intake: interaction between test diet and postingestive chemoreception, *Physiol. Behav.,* 12, 377, 1974.

10. Mook, D. G., Wagner, S., and Hartline, D. F., All-or-none suppression of glucose sham feeding by an intragastric mixed meal in rats, *Behav. Neurosci.,* 105, 712, 1991.

11. Sclafani, A., Carbohydrate taste, appetite, and obesity: an overview, *Neurosci. Biobehav. Rev.,* 11, 131, 1987.

12. Foster, L. A., Boeshore, K., and Norgren, R., Intestinal fat suppressed intake of fat longer than intestinal sucrose, *Physiol. Behav.,* 64, 451, 1998.

13. Bartness, T. J. and Rowland, N., Dietary self-selection in normal and diabetic rats after gastric loads of pure macronutrients, *Physiol. Behav.,* 31, 546, 1983.

14. Van Zeggeren, A. and Li, E. T. S., Food intake and choice in lean and obese Zucker rats after intragastric carbohydrate preloads, *J. Nutr.,* 120, 309, 1990.

15. Bernardini, J., Kamara, K., and Castonguay, T. W., Macronutrient choice following food deprivation: effect of dietary fat dilution, *Brain Res. Bull.,* 32, 543, 1993.

16. Lucas, F. and Sclafani, A., The composition of the maintenance diet alters flavor-preference conditioning by intragastric fat infusions in rats+, *Physiol. Behav.,* 60, 1151, 1996.

17. Baker, B. J., Booth, D. A., Duggan, J. P., and Gibson, E. L., Protein appetite demonstrated: learned specificity of protein-cue preference to protein need in adult rats, *Nutr. Res.,* 7, 481, 1987.

18. White, P. J., Cybulski, K. A., Primus, R., Johnson, D. F., Collier, G. H., and Wagner, G. C., Changes in macronutrient selection as a function of dietary tryptophan, *Physiol. Behav.,* 43, 73, 1988.

19. Wurtman, R. J. and Wurtman, J. J., Carbohydrate craving, obesity and brain serotonin, *Appetite,* 7 (Suppl.), 99, 1986.

20. Wurtman, J. J., Moses, P. L., and Wurtman, R. J., Prior carbohydrate consumption affects the amount of carbohydrate that rats choose to eat, *J. Nutr.,* 113, 70, 1983.

21. Wurtman, R. J. and Wurtman, J. J., Brain serotonin, carbohydrate-craving, obesity and depression, *Obesity Res.,* 3 (Suppl. 4), 477S, 1995.

22. Mehiel, R. and Bolles, R. C., Learned flavor preferences based on caloric outcome, *Anim. Learn. Behav.,* 12, 421, 1984.

23. Young, R. C., Gibbs, J., Antin, J., Holt, J., and Smith, G. P., Absence of satiety during sham feeding in the rat, *J. Comp. Physiol. Psychol.,* 87, 795, 1974.

24. Antin, J., Gibbs, J., and Smith, G. P., Intestinal satiety requires pregastric food stimulation, *Physiol. Behav.,* 18, 421, 1977.

25. Greenberg, D., Smith, G. P., and Gibbs, J., Intraduodenal infusions of fats elicit satiety in sham-feeding rats, *Am. J. Physiol.,* 259, R110, 1990.

26. Perez, C., Ackroff, K., and Sclafani, A., Carbohydrate- and protein-conditioned flavor preferences: effects of nutrient preloads, *Physiol. Behav.,* 59, 467, 1996.

27. Warwick, Z. S., Probing the causes of high-fat diet hyperphagia: a mechanistic and behavioral dissection, *Neurosci. Biobehav. Rev.,* 20, 155, 1996.

28. Burggraf, K. R., Willing, A. E., and Koopmans, H. S., The effects of glucose or lipid infused intravenously or intragastrically on voluntary food intake in the rat, *Physiol. Behav.,* 61, 787, 1997.

29. Höfer, D., Püschel, B., and Drenckhahn, D., Taste receptor-like cells in the rat gut identified by expression of α-gustducin, *Proc. Natl. Acad. Sci. U.S.A.,* 93, 6631, 1996.

Section II

Effects of Metabolic Processing on Energy Intake and Macronutrient Selection

10 Intermediary Metabolism of Macronutrients

Malcolm Watford

CONTENTS

1 INTRODUCTION

The macronutrients, those that provide energy to the body, are carbohydrates, fats, and proteins. In addition, alcohol can contribute a significant portion of the daily energy intake in some individuals and populations. Although we consume these nutrients in a variety of forms, by the end of digestion they enter the body as a relatively limited number of compounds. Most carbohydrate will have been digested to glucose with varying amounts of fructose and galactose. Triglycerides, the principal component of dietary lipid, will have been hydrolyzed to fatty acids and monoacylglycerols, absorbed and reesterified to pass into the body as triglycerides via the lymphatic system, and dietary protein will have been hydrolyzed to small peptides and then to amino acids before absorption.

The bulk of the daily intake of macronutrients is destined to be used for energy production. The immediate source of energy within the cell is ATP, and during the course of a day we synthesize and degrade over 40 kg of ATP but this is all intracellular and there is no interorganal flux of ATP. Therefore, tissues obtain fuels, glucose, free fatty acids, ketone bodies, lactate, amino acids, and ethanol, from the circulation. Glycogen is a storage form of glucose, and hence not a fuel by this definition; although it could be argued that triglycerides are a fuel for some cells, they are really a storage and transport form of fuel (fatty acids). One unusual fuel is butyrate produced by the colonic bacteria from the fermentation of fiber and resistant starch and is the major fuel for the colonocytes (epithelial cells lining the colon lumen).

Not all fuels are available at all times and not all cells are capable of using all fuels. As shown in Table 1 different tissues show distinct patterns in their utilization of fuels, the pathways employed to metabolize those fuels, and the products released back into the circulation. Table 1 should be

TABLE 1
Tissue-Specific Fuel Metabolism

Tissue	Fuel Used	Released	Comments
Brain	Glucose Ketone bodies Lactate	Lactate	Glucose is usually completely oxidized to CO_2 but during prolonged starvation some may be released as lactate; ketone bodies and lactate are only used when their levels are very high
Heart	Free fatty acids Ketone bodies Glucose Lactate		Heart is highly aerobic and will completely oxidize these fuels Glucose is only used when it is very abundant. Lactate is used in the absorptive phase
Skeletal Muscle	Glucose Free fatty acids Ketone bodies	Lactate	White (type IIb) muscle relies on anaerobic metabolism (mainly from glycogen) to lactate
	Branched-chain amino acids	Glutamine Alanine	Red (type I & IIa) muscle can completely oxidize fuels
Kidney	Glucose Free fatty acids Ketone bodies Lactate Glutamine	Glucose	Renal gluconeogenesis is linked to the use of glutamine to produce ammonia plus bicarbonate and is consequently of greater importance during metabolic acidosis
Red Blood Cells	Glucose	Lactate	These cells lack mitochondria and are therefore restricted to anaerobic metabolism
Intestine	Glucose Glutamate Glutamine	Lactate Alanine	Dietary glutamate and glutamine and circulating glutamine are used; the intestine also releases dietary glucose and amino acids into the bloodstream and lipids into the lymph
Adipose Tissue	Glucose	Lactate Glycerol Free fatty acids	Most of the glucose used is incorporated into the glycerol phosphate moeity of triglycerides; glycerol and free fatty acids are derived from the lipolysis of stored triglycerides
Liver	Amino acids Free fatty acids Lactate Glycerol Glucose Alcohol Fructose Galactose	Glucose Ketone bodies Lactate Triglycerides Acetate	The liver derives most of its energy from the partial oxidation of amino acids during the absorptive phase and from the partial oxidation of fatty acids during starvation; when available alcohol will be metabolized before other fuels, with a resultant release of acetate

used as reference throughout this chapter when mention is made of metabolism of a specific nutrient or tissue. Full details of the metabolic pathways are available in any standard biochemistry textbook.

In order to maintain homeostasis the body coordinates the utilization of the macronutrients. A striking feature is the maintenance of circulating glucose levels. After an overnight fast, plasma glucose will be about 5 mM (90 mg/dl) and even if we consume a high glucose meal the rise in peripheral glucose levels will seldom exceed 8 mM (140 mg/dl) and will return to around 5.5 mM relatively quickly (within 1 to 1.5 h) after the meal. Perhaps even more striking is the fact that even during prolonged starvation (up to 40 days) plasma glucose levels are maintained at about 70 mg/dl 3.5 mM.[1] This is related to the limited fuel utilization of the brain, the brain is only able to utilize glucose or ketone bodies, and as we will see, ketone bodies are seldom available and essentially never available in a well-fed healthy subject. However, the brain is only able to obtain sufficient glucose at plasma levels of >3 mM (60 mg/dl) below which brain function is severely compromised and so the body works to maintain this minimal level at all times. This chapter will consider the metabolic fates of the macronutrients as ingested in a standard meal and the changes in tissue-specific metabolism and regulation which occur during the absorptive phase and as the subject begins to starve.

2   FATE OF DIETARY MACRONUTRIENTS

On ingesting a standard meal of approximately 90 g carbohydrate, 30 g protein, and 30 g fat, the products of digestion will be dealt with in a reproducible and organized manner. Assuming that all of the carbohydrate enters the body as glucose (see later for fructose and galactose), that glucose is absorbed into the enterocyte up a concentration gradient by the sodium-linked active glucose transporter SGLT 1. A little glucose will be metabolized by the enterocyte, with a resultant production of lactate, but this is minor in terms of the total amount of glucose absorbed. The bulk of the glucose then passes into the portal circulation where the concentration may rise to 15 mM (270 mg/dl). This rise in glucose level (and the presence of food in the gastrointestinal tract) stimulates the β cells of the islets of Langerhans in the pancreas to secrete insulin. The presence of higher glucose in the portal vein than in the hepatic artery is sensed by the liver and in the presence of high levels of insulin the liver is able to take up,[2,3] and store as glycogen about one third of the glucose load.

The glucose that passes through the liver will cause a rise in circulating glucose levels to about 8 mM (140 mg/dl). A further third of the dietary glucose load will be taken up and oxidized by the brain, and most of the remainder will be taken up in an insulin-dependent manner by skeletal muscle. Within muscle, about half of the glucose will be deposited as glycogen while the remainder will be utilized to produce energy. About 2 g of glucose will be taken up by adipose tissue, where a little may be used for energy production, but the majority will be incorporated into the glycerol phosphate moiety of triglycerides. A small amount (2 g) will be utilized by the obligatory glycolytic tissues such as red blood cells, renal medulla, and the retina. These tissues and cells have little, or no, mitochondrial metabolism and so are limited in their fuel metabolism to glycolysis (glucose to lactate). Thus, by the time the carbohydrates from the meal have been absorbed the circulating glucose is back to 5 to 5.5 mM (90 to 100 mg/dl) and dietary glucose has been roughly equally distributed between energy production and glycogen storage.

Dietary proteins may arrive at the brush border of the enterocyte as free amino acids or as small peptides but the latter are hydrolyzed during absorption and therefore only free amino acids enter the cell. Most of the amino acid load will enter the portal blood unmodified, but any glutamate, aspartate, and glutamine will be utilized by the enterocyte. Since these three amino acids comprise about 20% of dietary protein, a considerable amount (6 g from the meal) of the dietary protein load is metabolized within the enterocyte. Glutamate and glutamine are the primary fuels utilized by these cells, and the products released into the portal blood are lactate, alanine, citrulline and proline, and ammonia, together with the remaining dietary amino acids.

The large surge of dietary amino acids results in a burst of protein synthesis in the liver and other tissues, but this will only utilize about one third of the dietary load and the remainder must be metabolized. Leucine, isoleucine, and valine, the branched-chain amino acids (about 25% of dietary protein) are not metabolized to any great degree by the liver and they pass out to the periphery, where they are taken up by skeletal muscle and other tissues and transaminated to their keto acids. Some of the branched-chain ketoacids will be oxidized within the muscle, but some may also be released into the circulation to be taken up by tissues such as liver and kidney. The nitrogen from branched-chain amino acid catabolism is transported back to the liver as alanine and glutamine. The liver is also the site of metabolism of the remaining amino acids through the expression of tissue-specific enzymes such as phenylalanine hydroxylase, tyrosine aminotransferase, tryptophan oxygenase, etc. Although textbooks often state that the liver "oxidizes" amino acids, this is not strictly accurate. Jungas and colleagues[4] calculated that to oxidize the daily load

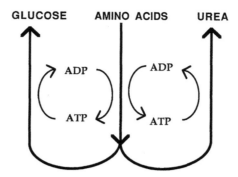

FIGURE 1   Amino acid catabolism is linked to glucose and urea synthesis.

of amino acids completely in the liver would require most, if not all, of the daily oxygen consumption of the liver. In fact, the liver only partially oxidizes amino acids and the carbon skeletons are used in gluconeogenesis (plus ketone body synthesis for leucine, lysine, and some other amino acids). Jungas and colleagues[4] proposed that the synthesis of glucose and urea, both ATP-consuming processes, allows the regeneration of ADP and so enables the liver to continue to dispose of excess amino acids (Figure 1). This means that the amino acid carbon skeletons are conserved as glucose (or, more accurately, glycogen in the case of dietary protein) and that the principal energy-producing process in the liver during the absorptive phase is the partial oxidation of amino acids.

Dietary lipid is more slowly absorbed. The end products of digestion are fatty acids and monoacylglycerols which are absorbed and then reesterified within the enterocyte to enter the lymphatic system as triglycerides in chylomicra. These are delivered into the blood via the subclavian vein and after extracellular hydrolysis by lipoprotein lipase in adipose tissue the fatty acids enter the cell to be reesterified using glycerol phosphate derived from glucose while the glycerol returns to the liver for further metabolism. Alcohol will be metabolized, in preference to other fuels, by the liver (via alcohol and acetaldehyde dehydrogenases) to yield acetate. Alcohol metabolism also results in a high $NADH/NAD^+$ ratio which may inhibit gluconeogenesis, while excessive alcohol consumption leads to the accumulation of large amounts of acetyl CoA and a fatty liver.

Thus, after the meal has been digested and absorbed, about half of the carbohydrate will have been oxidized and half stored as glycogen in liver and muscle; perhaps a third of the amino acids will have been used for protein synthesis, with the remainder metabolized primarily yielding glycogen and urea; and most of the lipid will have been stored as triglyceride in adipose tissue.

3   REGULATION AND CONTROL OF METABOLISM

Throughout the remainder of this chapter the tissue-specific utilization of fuels is considered from a regulatory viewpoint, and in considering regulation it is useful to define a few terms. Regulation is a system property in that we can say that blood glucose concentration is regulated by insulin, but this does not tell us anything about the mechanism. Thus, we introduce the concept of control, a local property; blood glucose is regulated by insulin which controls the activities of key enzymes and transporters in metabolism. For many years, regulation has been addressed by the concept of the "rate-limiting step." This implied that changing the activity of one step in a pathway would change flux throughout the pathway. However, apart from the fact that pathways do not exist in isolation *in vivo*, this concept is patently untenable. Control of flux through a metabolic process is shared by a number of steps. For example, insulin does stimulate the movement of glucose transporters to the cell surface in skeletal muscle, but without a stimulation of glucose metabolism within that cell there would be no change in glucose utilization. We can also describe two types of regulatory enzymes, those that act to stabilize metabolite concentrations (homeostasis) and those that actually control flux through the pathway. Modern approaches to the study of metabolic control ask the question, how much does flux change when the activity of a specific enzyme is changed by a known degree? Such approaches applied to whole-body physiological processes, rather than isolated pathways, have consistently shown that regulation is achieved by control of a number of key steps.[5]

3.1   REGULATION OF CARBOHYDRATE METABOLISM

Glucose can be used for glycolysis, the pentose phosphate pathway, or glycogen synthesis. Glycolysis functions primarily to provide energy (ATP, NADH), and the pyruvate formed can be converted to lactate (or alanine) or further oxidized via pyruvate dehydrogenase and the citric acid cycle. The pentose phosphate pathway functions to produce ribose phosphate (for nucleic acid synthesis) and NADPH, and flux is determined by need for these end products and will not be considered further here. Excess dietary carbohydrate is stored as glycogen, primarily in liver, where it functions as a store of glucose for the peripheral tissues, and skeletal muscle, where it is a local energy store for the muscle cell. Glycogen is synthesized on a protein backbone called glycogenin from glucose 6-phosphate via glucose 1-phosphate and uridyl diphosphate glucose which is added to the glycogen chain via the enzyme glycogen synthase. Glycogen is degraded by an enzyme glycogen phosphorylase which hydrolyzes the terminal glucose and transfers a phosphate directly using inorganic phosphate (Pi) to yield glucose 1-phosphate and then glucose 6-phosphate.

3.1.1   Carbohydrate Metabolism in Skeletal Muscle

Muscle glycogen is a local store of fuel and the concentration of glycogen is highest in white (type IIb) muscles which are fast acting and rely on rapid, anaerobic metabolism of glycogen to lactate to produce energy. This is a very fast process and allows very high intensity exercise, but the energy yield is low and the buildup of lactic acid results in rapid fatigue. However, even in very low intensity, long-duration exercise some muscle cells continuously utilize glycogen and glucose and there is a good correlation of initial muscle glycogen levels with endurance. This has led to the practice of glycogen loading involving the careful control of dietary and exercise patterns in the days before a sports event to achieve maximal glycogen stores.

If the muscle cell is not contracting, there is little need to utilize the glucose for energy and so most is stored as glycogen. The high levels of glucose and insulin after a meal result in increased glucose transport into skeletal muscle via the transporter GLUT 4. However, this transporter is facilitative in nature, meaning that glucose will only flow down a concentration gradient; therefore, to continue to bring glucose into the cell the glucose must be metabolized. This is achieved by the action of hexokinase which produces glucose 6-phosphate, and since this cannot cross the cell membrane the glucose is effectively trapped in the cell. The high levels of insulin also stimulate the utilization of glucose 6-phosphate for glycogen synthesis through activation of glycogen synthase and inhibition of glycogen phosphorylase.

Muscle contraction stimulates both glucose transport into the cell and the mobilization of muscle glycogen. The release of calcium into the muscle cytosol (sarcoplasm) is the event initiating contraction. Calcium binds to troponin C which stimulates the actinomyosin–ATPase resulting in ATP hydrolysis and muscle contraction. This is matched by the immediate resynthesis of ATP from ADP achieved by the calcium activation of glycogen phosphorylase kinase to phosphorylate glycogen phosphorylase. This activates the enzyme and results in breakdown of glycogen to provide substrate for glycolysis and so yields ATP. The catabolism of glucose to lactate usually involves the input of 2 ATP with the resultant production of 4 ATP, for a net yield of 2 ATP. However, when intracellular glycogen is the starting material, the initial ATP is not needed (Pi is used) and so the net yield is 3 ATP.

This system illustrates how the entire metabolic process is important, not just a single pathway or enzyme, since a buildup of glucose 6 phosphate in the cell from increased transport of glucose or hydrolysis of glycogen will not, in itself, stimulate flux in glycolysis. In fact, glucose 6 phosphate is a strong inhibitor of hexokinase and for flux to continue steps further down glycolysis must also be stimulated. A key glycolytic enzyme phosphofructokinase 1 (PFK 1) is very sensitive to inhibition by ATP and should be completely inhibited at physiological ATP concentrations. Since ATP levels do not change very much (when they do the cell dies) an important regulatory mechanism in muscle involves the relief of the ATP inhibition of PFK 1 by AMP. Through the action of adenylate kinase (myokinase), a small change in ATP concentration results in a change in AMP concentration of sufficient magnitude to relieve the ATP inhibition of PFK I and allow flux through glycolysis. Another important regulator of PFK 1 is inhibition by acidic pH (i.e., high levels of lactic acid arising from anaerobic contraction) which limits the time that such high rates of glycolysis can be maintained. Thus, the stimulation of glucose transport, glycogen breakdown, and glycolysis all contribute to resynthesis of ATP used in contraction. At rest, inhibition of PFK 1 results in an accumulation of fructose 6-phosphate which is in equilibrium with glucose 6-phosphate and glucose 6-phosphate builds up to inhibit hexokinase thus preventing the further utilization of glucose.

Skeletal muscle, and other tissues, can also oxidize glucose and the energy yield is considerably greater (38 ATP vs. 2 or 3 ATP) if the pyruvate produced in glycolysis enters the mitochondria to be further metabolized through pyruvate dehydrogenase and the citric acid cycle. Pyruvate dehydrogenase is tightly regulated since it has the potential to remove glucose-synthesizing units from the pool. The conversion of pyruvate (three carbons) to acetyl CoA (two carbons) is irreversible and no bypass reaction is present in the mammalian body. Therefore, we are unable to make glucose from two-carbon compounds (acetyl CoA and most fatty acids) which means that once pyruvate has been decarboxylated by pyruvate dehydrogenase it can never give rise to glucose. Thus, in times of glucose shortage (starvation) it becomes very important to limit flux through pyruvate dehydrogenase as much as possible. This is achieved by end product inhibition through high ratios of acetyl CoA:CoASH and NADH:NAD$^+$, and by phosphorylation and inactivation of the enzyme. At times of glucose abundance the dephosphorylation of pyruvate dehydrogenase is stimulated by insulin, and the complex is activated resulting in flux to acetyl CoA. Although of central importance in energy metabolism the rate of the TCA cycle is essentially driven by the demand for ATP synthesis and will not be considered further here. It is should be noted that the TCA cycle is not a just a pathway of carbohydrate metabolism, it is the central pathway of energy metabolism and will also oxidize carbon arising from carbohydrates, lipids (fatty acids, ketone bodies), ethanol, and amino acids.

3.1.2   Carbohydrate Metabolism in Liver

The liver stores glucose as glycogen after a meal and then releases that glucose in the postabsorptive state and synthesizes new glucose (gluconeogenesis) later in starvation. This is possible since the liver expresses glucose 6-phosphatase which liberates free glucose from glucose 6-phosphate.

3.1.2.1   *Glycolysis and gluconeogenesis*

The liver expresses a unique form of hexokinase, known as glucokinase (hexokinase IV or D), which has a relatively low affinity for glucose (half maximal activity at 5 to 10 mM glucose) and other hexoses, and is not subject to inhibition by physiological levels of glucose 6-phosphate. This means that the liver will only take up and metabolize glucose when it is very abundant, and will continue to use glucose even if glucose 6-phosphate levels rise considerably. This allows the liver to use whatever glucose it requires for glycolysis, but will still store glucose as glycogen even when glycolytic flux is very low and glucose 6-phosphate levels are high. Hepatic glycolysis (glucose to lactate) may be relatively more important in perivenous cells, but glycolytic flux probably occurs in most liver parenchymal cells if only to maintain high levels of intermediates such as malonyl CoA (see below).

Although many textbooks state that excess dietary glucose will be converted to fat, there is little evidence for this in humans consuming a typical Western diet of >30% fat. In subjects consuming a 3% fat diet with carbohydrate overfeeding, glycogen stores can expand considerably (up to 1 kg of glucosyl units) before any net fatty acid synthesis is detectable by calorimetry.[6] Tracer studies have confirmed such findings in that *de novo* fatty acid synthesis is undetectable in humans consuming a 30% or greater fat diet.[7] Only in subjects consuming a 10% eucaloric diet is *de novo* fatty acid synthesis detectable, but this also results in hypertriglyceridemia and the trigylcerides contain high levels of saturated fatty acids.[8] An additional, teleological, piece of evidence is that the fatty acid composition of human adipose tissue closely resembles the fatty acid composition of the diet; i.e., dietary fat is stored without extensive modification. This is in accord with studies showing that dietary carbohydrate and amino acids are oxidized before dietary lipid.[9]

Glycolysis in liver is subject to control at pyruvate kinase, PFK 1, and glucokinase. However, since the liver can also carry out gluconeogenesis, it expresses other enzymes to bypass these three irreversible enzymes. Pyruvate carboxylase and phosphoenolpyruvate carboxykinase (PEPCK) reverse pyruvate kinase, fructose 1,6-bisphosphatase reverses PFK 1, and glucose 6-phosphatase reverses glucokinase. Therefore, regulation of flux involves the control of enzymes operating in two directions.[10] At times when glycolytic flux is predominant the inhibition of liver PFK 1 by ATP is relieved, not by AMP, but by fructose 2,6-bisphosphate. This molecule, present in micromolar amounts, is synthesized by a bifunctional enzyme phosphofructokinase 2/fructose 2,6-bisphosphatase. In the fed state this enzyme exists in the nonphosphorylated state and acts as the kinase resulting in high levels of fructose 2,6-bisphosphate which relieves the ATP inhibition of PFK 1 and allows glycolytic flux. In the starved state, high levels of glucagon stimulate the production of cAMP within the liver cell and this activates protein kinase A which phosphorylates PFK 2 and so converts it to fructose 2,6-bisphosphatase. The subsequent decrease in fructose 2,6-bisphosphate levels results in a slowing of glycolysis. Fructose 2,6-bisphosphate is also an inhibitor of fructose 1,6-bisphosphatase and so when fructose 2,6-bisphosphate levels are low gluconeogenic flux through fructose 1,6-bisphosphatase is high. Protein kinase A also phosphorylates liver pyruvate kinase rendering it less active and more susceptible to inhibition by gluconeogenic substrates such as alanine, again promoting gluconeogenesis over glycolysis. At the same time there is an increase in the rate of transcription of the PEPCK gene, which increases the amount of the enzyme and potential for gluconeogenic flux. Conversely, transcription of PEPCK gene is inhibited by insulin, a signal of glucose abundance, which acts to decrease the gluconeogenic capacity.

3.1.2.2   Fructose and galactose metabolism

Dietary fructose comes from sucrose, fruit, and high-fructose corn syrup. Fructose will be metabolized by the liver by a unique enzyme, fructokinase, which phosphorylates fructose in the 1 position to yield fructose 1-phosphate. Fructose 1-phosphate is split by adolase B, another liver-specific enzyme, to triose phosphates which are part of the glycolytic/gluconeogenic pathway. Under normal conditions the rate of absorption of fructose from the diet will limit the rate of fructose use by the liver and the majority of fructose is probably used for gluconeogenesis (glycogen synthesis). Galactose is similar to fructose in that it is exclusively metabolized in the liver. Galactose is only found in milk and milk products and is a relatively minor part of our daily carbohydrate intake. Within the liver cell galactose is phosphorylated at the 1 position by galactokinase and then galactosyl-uridyl transferase yields glucose 1-phosphate which isomerases to glucose 6-phosphate to enter glycolysis, be released as free glucose, or, more likely, be used for glycogen synthesis.

3.1.2.3   Liver glycogen metabolism

The pathways of glycogen metabolism in liver are similar to that in muscle but are regulated differently in keeping with the physiological role of the liver.[11] At times of glucose abundance (high insulin, low glucagon), glycogen phosphorylase is phosphorylated and inactive which glycogen synthase is active (dephosphorylated). As glucose levels drop and glucagon levels rise, the stimulation of protein kinase A by cAMP results in phosphorylation and activation of glycogen phosphorylase kinase, which phosphorylates glycogen phosphorylase and results in the degradation of glycogen to ultimately yield glucose 6-phosphate. Concomitantly, glycogen synthase is phosphorylated and inactivated. Protein phosphatase 2 is responsible for the removal of the phosphate from proteins phosphorylated by protein kinase A, and thus the pattern can be reversed. Dephosphorylation of glycogen synthase activates the enzyme stimulating glycogen synthesis, while dephosphorylation of glycogen phosphorylase inactivates and stops glycogen breakdown. There has been considerable debate over the source of the glucose 6-phosphate for hepatic glycogen synthesis. Immediately prior to the meal the liver had probably been in a gluconeogenic state, and many studies *in vitro* raised doubts about the ability of the liver to take up free glucose. This became known as the "glucose paradox" and it was proposed that hepatic glycogen synthesis occurred, not from glucose, but from lactate and other gluconeogenic precursors. The existence of this indirect pathway was confirmed and it appears to account for 30 to 40% of the daily hepatic glycogen synthesis in humans consuming a typical Western diet.[12]

3.2   TISSUE-SPECIFIC AMINO ACID METABOLISM

The individual pathways of amino acid metabolism are beyond the scope of this book and this section will briefly outline interorganal aspects of amino acid metabolism which impact on the availability and utilization of these and other fuels. As described above, the intestinal mucosa will metabolize most of the dietary glutamine, glutamate, and aspartate during the absorptive process. In postabsorptive conditions this tissue will take up considerable amounts of glutamine from the circulation, again using this as a fuel. The metabolism of dietary branched-chain amino acids in muscle and the catabolism of excess other amino acids in the liver have also been described earlier. Amino acids also arise from the continuous degradation of proteins within the body, and these are essentially metabolized in a manner similar to that seen for dietary amino acids.

The major deposit of protein in the body is skeletal muscle, but the pattern of amino acids leaving muscle shows relatively low levels of the branched-chain amino acids when compared with the amino acid pattern of muscle protein, indicating extensive metabolism of the branched-chain amino acids within the muscle cell. In contrast, alanine and glutamine, which together with glutamate, comprise about 11 to 15% of muscle protein, make up well over 50% of the amino acids released by muscle, indicating their synthesis within the muscle cell. The synthesis of alanine is straightforward, branched-chain amino acids transaminated to form branched-chain ketoacids result in the formation of glutamate which then transaminates with pyruvate (derived from glucose metabolism) to form alanine via the alanine transaminase reaction. The alanine then leaves the muscle and is taken up by the liver where it is transaminated to yield pyruvate and glutamate. The pyruvate is used for gluconeogenesis and the glutamate passes the nitrogen to urea synthesis. Glutamine metabolism is more complex; within the muscle cell glutamate and ammonia are combined in the glutamine synthetase reaction to yield glutamine. However, although these intermediates are presumed to be derived from the catabolism of other amino acids (e.g., branched-chain amino acids) the exact pathway is not definitively established. Other sites of net glutamine synthesis in the body include the lungs, adipose tissue, and, under certain conditions, the liver. The major sites of glutamine catabolism are the mucosa of the small intestine and the cells of the immune system, where it serves as a major fuel; the liver especially in the absorptive phase and during diabetes, where glutamine is a precursor of glucose and urea synthesis; and the kidneys during metabolic acidosis where glutamine is the primary substrate for ammonia synthesis with the carbon skeleton yielding bicarbonate and glucose (renal gluconeogenesis). At times of stress, e.g., burns, sepsis, surgical trauma, there is a large release of glutamine from skeletal muscle which provides substrate for the liver, kidneys, and immune system but results in a fall in the intramuscular glutamine levels. This fall in muscle glutamine seems to signal increased net protein catabolism,

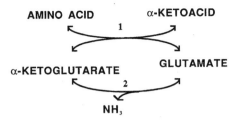

FIGURE 2 Glutamate plays a central role in amino acid metabolism. The enzymes indicated are (1) transaminases and (2) glutamate dehydrogenase. Since both reactions are freely reversible, glutamate can either transfer amino groups between amino acids, produce ammonia from amino acids, or incorporate ammonia into amino acids.

which provides more substrate for glutamine synthesis. Thus, glutamine and alanine, and to a limited degree serine and glycine, play major roles in transporting amino acid nitrogen between tissues; in addition these compounds also transport carbon (for gluconeogenesis), and in the case of glutamine can be used directly as a fuel in certain cell types.

The liver is the major site of amino acid metabolism,[13] and most will undergo transamination to yield glutamate and the respective ketoacid, or are deaminated to yield ammonia, while glutamine is deamidated to ammonia and glutamate. Glutamate plays a central role in all amino acid metabolism (Figure 2) since via transamination it is able to pass the amino group to other ketoacids to form different amino acids. It can be deaminated via glutamate dehydrogenase to form ammonia and α ketoglutarate (TCA cycle intermediate and gluconeogenic precursor), or, conversely, glutamate dehydrogenase can fix ammonia into glutamate and so put that ammonia into the general amino acid pool. The end products of amino acid metabolism in the liver are urea, glucose (glycogen), and ketone bodies.

3.3 LIPID METABOLISM

The fate of dietary lipid has been described above. The pathway of *de novo* fatty acid synthesis is of limited importance in humans in terms of storing calories, but at least part of the pathway must exist in liver to yield malonyl CoA which is an important regulator of hepatic fatty acid metabolism. In the fed state when glucose is abundant some acetyl CoA in liver will be converted to malonyl CoA through the action of acetyl CoA carboxylase, an enzyme that is very sensitive to activation by insulin. The malonyl CoA so formed could then be used for fatty synthesis via the fatty acid synthase complex to yield palmitate. The liver can thus receive long-chain fatty acids both from the circulation or via *de novo* synthesis. Within the liver cell, long-chain fatty acids can either be esterified to triglycerides or undergo partial oxidation via β oxidation with the resultant production of ketone bodies (acetoacetate and β hydroxybutyrate). In order to be oxidized the fatty acids must be transported into the mitochondrial matrix as carnitine derivatives. One of the enzymes responsible for this transport is carnitine palmitoyl transferase which is very sensitive to inhibition by malonyl CoA. Thus, when hepatic malonyl CoA levels are high, fatty acid oxidation is low since they are unavailable to the enzymes of β oxidation within the mitochondrial matrix. The fatty acids in the cytosol will be esterified to triglycerides using glycerol phosphate derived either from glucose metabolism or the phosphorylation of glycerol through a liver-specific enzyme glycerol kinase. Conversely, in the starved state, when glucose is no longer abundant and insulin levels are lower, malonyl CoA levels fall and carnitine palmitoyl acyltransferase is able to transport the fatty acids into the mitochondria for oxidation. Thus, the principal energy-yielding process in the liver in the postabsorptive state is the partial oxidation of fatty acids.

During the absorptive phase the bulk of dietary fatty acids had been stored in adipose tissue as triglycerides. In part this was due to the high level of insulin stimulating the action of adipose tissue lipoprotein lipase releasing free fatty acids to be taken up by the cell. Insulin also stimulated the transport of glucose into the adipose cell, and this was used to make the glycerol phosphate moiety of triglycerides. Within the adipose cell tryglycerides are hydrolyzed by hormone-sensitive lipase, but at times of high insulin lipolysis is very low since this enzyme is extremely sensitive to inhibition in response to insulin. Conversely, if insulin levels drop and catecholamines (epinephrine or norepinephrine) rise, there is a rise in cAMP in the adipose cell which stimulates protein kinase A to phosphorylate and activate hormone-sensitive lipase with a resultant increase in lipolysis.

Most tissues show a hierarchy of fuel use in that free fatty acids (or ketone bodies) will be used before glucose. Therefore, in the absorptive state fatty acid metabolism is low as free fatty acids are not available since they are being transported as triglycerides. However, during the postabsorptive phase the drop in insulin levels, together with a rise in catecholamines, results in an increase in adipose tissue lipolysis to yield free fatty acids and glycerol. Since there is little glucose available, reesterification rates are low and the free fatty acids are released into the circulation where they are transported bound to albumin. This results in a rise in the concentration of free fatty acids in the circulation, which increases their use as a fuel. Similarly, the liver will take up free fatty acids and produce ketone bodies which are released into circulation and will also be used, as their concentration rises, in preference to other fuels.

3.4   Integrated Metabolism during Starvation

Within 3 to 4 h after a meal the macronutrients have been absorbed and either metabolized or stored, and as the person enters the postabsorptive phase the body must use stored fuels. Liver glycogen is first to be mobilized but will only last about 20 to 24 h and then the body uses gluconeogenesis to supply the glucose required by the brain. Since we are only able to make glucose from three-carbon compounds, the possible substrates, lactate/pyruvate, alanine, glycerol, and amino acids, are available in a limited supply. However, not all gluconeogenesis contributes new glucose to the body pool. The degradation of glucose to lactate in red blood cells and other glycolytic tissues, and the synthesis of alanine from glucose-derived pyruvate, in tissues such as skeletal muscle result in the cycling of glucose carbon from the liver to peripheral tissues and back to the liver as three-carbon compounds. Although these processes, the Cori cycle for lactate and the glucose–alanine cycle for alanine, do transfer energy to peripheral tissues, they do not contribute any new glucose that would be available for the brain to oxidize. Therefore, the only substrates contributing new glucose during starvation are glycerol and amino acids. Glycerol is derived from the hydrolysis of triglycerides in adipose tissue and is released into the circulation to be used by the liver for gluconeogenesis since the liver is the only tissue expressing glycerol kinase. The contribution of glycerol to gluconeogenesis is about 18 g of glucose per day and remains relatively constant during even a prolonged fast. Therefore, amino acids are the major gluconeogenic substrate but to synthesize 1 g of glucose requires about 1.6 g of amino acids. Initially, in early starvation, most of these amino acids will be derived from the breakdown of liver protein, but as starvation progresses the major site of proteolysis will be skeletal muscle. Since the brain requires 100 to 120 g glucose per day this would require the breakdown of over 200 g of muscle protein or close to a kilogram of muscle tissue. Clearly, this cannot be sustained for prolonged periods and the body adapts by utilizing other fuels and so sparing glucose use and the rate of gluconeogenesis can decrease to perhaps as low as 35 to 40 g per day from glycerol and amino acids plus a further 30 to 40 g from the Cori and glucose/alanine cycles (Figure 3).

The increase in circulating free fatty acids has increased their rate of metabolism, and this has allowed them to be used in place of glucose in many tissues. However, this was of little direct value to the brain since the transport of long-chain free fatty acids across the blood–brain barrier is very slow and insufficient for the oxidative needs of the brain. Within the liver, now in a

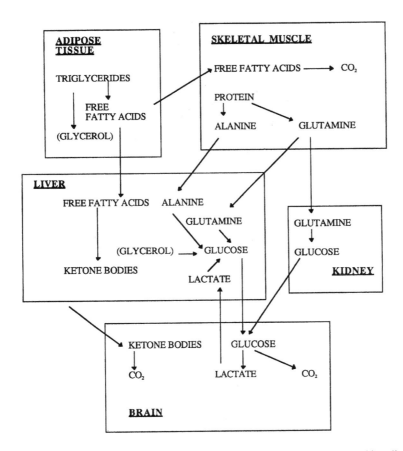

FIGURE 3 Interorganal metabolism during prolonged starvation. Glycerol is produced in adipose tissue and metabolized in liver but the arrows have been omitted for clarity. Similarly, the production of lactate from glucose in glycolytic tissues and the resynthesis of glucose in the liver (Cori cycle) is not shown.

gluconeogenic state, glycolysis is inactive and malonyl CoA levels are low so that fatty acids enter the mitochondria to be oxidized with the resultant production of ketone bodies. Ketone bodies are small (four carbon), water-soluble molecules that can readily cross the blood–brain barrier to be oxidized. In the healthy, fed individual the level of ketone bodies in the circulation is very low (<0.2 mM), but as the concentration rises during starvation the brain is able to use them and so reduce its glucose use to about 30 g per day. In addition, during prolonged starvation the brain will decrease its glucose utilization even further by limiting some glucose metabolism to glycolysis with a resultant release of lactate (i.e., conservation of glucose-synthesizing potential).

The ketone bodies β-hydroxybutyrate and acetoacetate are strong acids and a rise in their concentration produces metabolic acidosis. Ketone bodies also spill out into the urine; they are excreted as ammonium salts, conserving other cations (sodium and potassium). This ammonia is produced in the kidney from the metabolism of glutamine (derived from skeletal muscle proteolysis), and the glutamine carbon skeleton is used for the production of glucose and bicarbonate, thus helping to control acid–base balance. Renal gluconeogenesis is usually a very small proportion of total gluconeogenesis but may rise to 50% of total gluconeogenesis as a result of the ketoacidosis of prolonged starvation. However, it should be remembered that by this time the rate of hepatic gluconeogenesis has dropped considerably.

The utilization of free fatty acids and ketone bodies spares glucose so that total body glucose utilization is limited to the obligatory glycolytic tissues and those parts of the brain that absolutely require glucose. The utilization of fatty acids or ketone bodies results in a buildup of acyl-CoA/CoASH, NADH/NAD$^+$, and other metabolites related to oxidative metabolism. These compounds signal a decreased need for glucose oxidation and, together with the low levels of insulin, result in an almost complete suppression of glucose oxidation in most cells. Together with the liver and pancreas, this constitutes the glucose/fatty acid or Randle cycle whereby low circulating glucose results in low insulin secretion allowing the mobilization of fatty acids (and ketone bodies) which are then used to spare glucose oxidation.[14] Thus, the body is able to maintain circulating glucose levels for considerable lengths of time (over 40 days in the case of very obese men) during starvation through the utilization of alternative fuels which spares the need to maintain high rates of proteolysis for glucose synthesis.

REFERENCES

1. Cahill, G. F., Jr., Starvation in man, *New Engl. J. Med.,* 282, 668, 1975
2. Pagliassotti, M. J. and Horton, T. J., Hormonal and neural regulation of hepatic glucose uptake, in *The Role of the Liver in Maintaining Glucose Homeostasis*, Pagliassoti, M.J., Davis, S.N., and Cherrington, A.D., Eds., R.G. Landes Company, Austin, TX, 1994, Chap. 4, 45–70.
3. Stumpel, F. and Jungerman, K., Sensing by intrahepatic muscarinic nerves of a portal-arterial glucose concentration gradient as a signal for insulin-dependent glucose uptake in the perfused rat liver, *FEBS Lett.,* 406, 119, 1997.
4. Jungas, R. L., Halperin, M. L., and Brosnan, J. T., Quantitative analysis of amino acid oxidation and related gluconeogenesis in humans, *Physiol. Rev.,* 72, 419, 1992.
5. Fell, D., *Understanding the Control of Metabolism*, Portland Press, London, U.K., 1996, 301 pp.
6. Acheson, K., Schutz, Y., Bessard, T., Anantharaman, K., Flatt, J. P., and Jequier, E., Glycogen storage capacity and *de novo* lipogenesis during massive carbohydrate overfeeding in man, *Am. J. Clin. Nutr.,* 48, 240, 1988.
7. Hellerstein, M. K., Schwarz, J.-M., and Neese, R. A., Regulation of hepatic *de novo* lipogenesis in humans, *Annu. Rev. Nutr.,* 16, 523, 1996.
8. Hudgins, L. C., Hellerstein, M. K., Seidman, C., Neese, R., Diakun, J., and Hirsch, J., Human fatty acid synthesis is stimulated by a eucaloric low fat, high carbohydrate diet, *J. Clin. Invest.,* 97, 2081, 1996.
9. Horton, T. J., Drougas, H., Brachy, A., Reed, G. W., Peters, J. C., and Hill, J. O., Fat and carbohydrate overfeeding in humans: different effects on energy storage, *Am. J. Clin. Nutr.,* 62, 19, 1995.
10. Pilkis, S. J., Claus, T. H., Kurland, I. J., and Lange, A. J., 6-Phosphofructo-2-kinase/fructose-2,6-bisphosphatase: a metabolic signalling enzyme, *Annu. Rev. Biochem.,* 64, 799, 1996.
11. Bollen, M., Keppens, S., and Stalmans, W., Specific features of glycogen metabolism in liver, *Biochem. J.,* 336, 19, 1998.
12. Shulman, G. I. and Landau, B. R., Pathways of glycogen repletion, *Physiol. Rev.,* 72, 1019, 1992.
13. Meijer, A. J., Lamers, W. H., and Chamuleau, R. A. F. M., Nitrogen metabolism and ornithine cycle function, *Physiol. Rev.,* 70, 701, 1990.
14. Randle, P. J., Garland, P. B., Hales, C. N., and Newsholme, E. A., The glucose-fatty acid cycle. Its role in insulin sensitivity and the metabolic disturbances of diabetes mellitus, *Lancet,* 1, 1875, 1963.

11 Physiological and Metabolic Control of Macronutrient Balance

J. P. Flatt

CONTENTS

KEY WORDS: *Substrate balance, body fat, metabolic regulation, mass effects, insulin, respiratory quotient*

ABSTRACT

When lifestyle and food supply are stable, glucose and amino acid oxidation over a few days are essentially equal to carbohydrate and protein intake. When considered over periods of months or years, fat oxidation, too, rarely differs by more than a few percent from fat intake. Thus, one needs to understand why and how adjustment of substrate oxidation to nutrient intake occurs with such remarkable accuracy, thereby bringing about energy balance. Increases in the body's protein, glycogen, and fat contents promote, respectively, the turnover and oxidation of amino acids, glucose, or fatty acids. Gains or losses of protein, glycogen, and fat thus elicit "mass action effects." Evolution has led to the development of enzymatic and endocrine regulatory phenomena which allow some of these mass effects to manifest themselves promptly and more powerfully than others. The interactions between these effects is ultimately decisive in bringing about situations where influxes and outflows are in balance, over time, in all the body's compartments. Body weight stability can thereby be explained without the need to invoke the existence of a body weight "set point." This is in contrast to other regulated systems, such as the control of blood pressure or of body temperature, where products do not accumulate or become depleted and where mass action effects cannot come into play. In such cases, additional features and/or set points are needed to provide a reference relative to which stability can be sought. Obesity is a situation in which the phenomena involved in the regulation of substrate balance operate in the usual manner, except that an excessive amount of adipose tissue is needed to bring about the oxidation of a fuel mix containing, on average, as much fat as is consumed.

1   INTRODUCTION: GOALS OF METABOLIC REGULATION

In striving to understand substrate utilization and its regulation, it is important at the onset to consider what are the main goals of metabolic regulation, as this will help recognition and put into perspective the mechanisms that have evolved to meet these goals. These include the following:

1. Securing the high adenosine triphosphate/adenosine diphosphate (ATP/ADP) ratio necessary to drive the life-sustaining reactions. The body's ATP content is not much greater than the amount of ATP used in 1 min (some 25 g in an adult under resting conditions). ATP utilization can readily increase five to ten times during physical exertion, so that very fast responses are needed to ensure that ATP regeneration can keep pace. Foremost, these responses include powerful circulatory responses to meet the challenge of providing adequate oxygen distribution under all circumstances. The delivery of substrates requires somewhat less acute responses, since most tissues contain glycogen or some other substrates that can be used to drive ATP regeneration. However, mobilization and distribution of metabolic fuels must be sufficient over time to cover the substrate needs of the tissues fully, so as to prevent cells from burning up some of their important constituents.
2. Avoidance of undue accumulation or depletion of metabolic intermediates. This could alter the intracellular environment and impair cell functions.
3. Maintenance of a stable and favorable extracellular environment, i.e., homeostasis. The concept of homeostasis applies to the stability of the internal environment, not to the stability of body composition. Homeostasis is thus entirely compatible with changes in body size, such as may be brought about by excessive accumulation of body fat. The concentration ranges within which the components of the extracellular fluids may vary is quite narrow for some, notably for substances likely to affect membrane potentials, such as potassium and calcium. Among metabolic fuels, glucose levels are particularly critical. On the one hand, blood glucose levels need to remain above 3 mM, so that the rate of diffusion of glucose into the central nervous system (CNS) remains sufficient to cover the needs of the CNS, which is primarily dependent on glucose (about 120 g/day) for energy production. On the other hand, glucose concentration must remain below 10 mM to avoid glucosuria, even though meals commonly supply 100 g of glucose or more, which is four to five times as much as all of the free glucose present in the body, or 20 times the amount of glucose in the blood. The other main major metabolic fuel is free fatty acids (FFA), also known by the more appropriate name of "non-esterified fatty acids" (NEFA). Circulating FFA levels usually vary between 0.1 and 1.5 mM. They are almost entirely bound to albumin. Albumin, whose plasma concentration is 0.5 mM, has four to eight high-affinity binding sites per molecule,[1] thus preventing FFA from exerting undesirable detergent effects. On account of the small size of the FFA pool (0.1 to 1.5 g in adults, as compared with 15 to 25 g for glucose), its turnover is very rapid, corresponding to half-lives of 1 to 2.3 min.[2]

4. Maintenance of desirable substrate levels, in spite of the fact that food consumption is intermittent. This requires the ability to remove substrates from the circulation rapidly during the rapid postprandial influx of nutrients from the gut, as well as means to mobilize substrates from endogenous reserves between meals.

5. Endowing the organism with functional reserve capacities. This is another important goal to enable the organism to cope with changing conditions, be they due to excessive nutrient loads, to food deprivation, stress, or to unusual physical activity demands.

6. Economic use of substrates. Given the challenge of finding enough food to survive, it is important that substrates be used economically. Such economy is demonstrated by the very rapid decline in substrate oxidation when physical activity stops. Indeed, the excess (i.e., above resting) postexercise oxygen consumption (EPOC) amounts to only about 15% of the increment in energy expenditure caused by physical activity, being spread over many hours.[3]

7. Restraint of the use of amino acids. Another critical aspect of fuel economy is the need to restrain the use of amino acids (AA) for energy production. This is mandated by the fact that only 10 to 18% of the energy consumed is typically in the form of protein, with some essential AAs present in only very small proportions. The amount of protein degraded and resynthesized in 1 day — about 5 g/kg body weight (BWt)/day in adults[4] — is some four to six times greater than the amount of protein consumed, typically 0.7 to 1.2 g/kg BWt/day. (The RDA is currently set at 0.8 g/kg BWt/day.[5]) Thus, most of the AAs formed by protein degradation must be preserved, to be reutilized for protein synthesis. Control over AA degradation must be achieved in spite of their relatively high intracellular concentrations, needed to sustain protein synthesis. (By contrast, glucose concentrations are lower in most cells than in the extracellular fluid.) The availability of other fuels, i.e., mainly glucose and fatty acids, as well as the ability to produce ketone bodies (KB) (some 120 g/day) during starvation[6] is essential, as well as effective in preserving AAs from excessive rates of oxidation. Less often appreciated, but very important as well, is the fact that the regulation of AA metabolism also needs features to increase their oxidation when protein intakes are high. It would indeed be a disadvantage to build up an excessively large lean body mass (LBM), which is costly to maintain on account of increases in energy expenditure for protein turnover and for moving a heavier body.

8. The ability to shift quickly and effectively from parsimonious to nearly exclusive use of glucose as a fuel. This is crucial, because carbohydrate (CHO) often provides the bulk of dietary energy, while the ability to store glycogen is limited by the fact that glycogen is associated with three times its weight of water. Glycogen reserves are indeed not much greater than the amount of CHO usually consumed in 1 day (Figure 1).[7,8] There are, in addition, substantial amounts of glucogenic precursors present in the body, in the form of glucogenic AAs and as glycerol in proteins and triglycerides. However, they become available only at the rates at which protein and fat are oxidized, so that they rarely generate more than 100 g of glucose per day. Switching from high to low rates of glucose oxidation, and vice versa, also implies an effective ability to control fatty acid oxidation. In the face of these challenging needs, powerful regulatory mechanisms have evolved, which shape the interplay between nutrient intake and metabolic fuel utilization.[9-11]

9. Finally, regulation of metabolism and of food intake to bring about a tendency to build up and maintain a fuel reserve sufficient to survive periods of food deprivation. Because fat yields more than twice as many calories per gram as CHO and protein, and because fat is hydrophobic and can be stored free of water (an even greater advantage!), energy reserves are primarily accumulated in the form of body fat. Triglycerides typically make up 80 to 90% of adipose tissue, which allows storage of some 8000 kcal/kg (vs. 1000 kcal/kg of lean tissue or per kilogram of hydrated glycogen). Fat reserves in adult

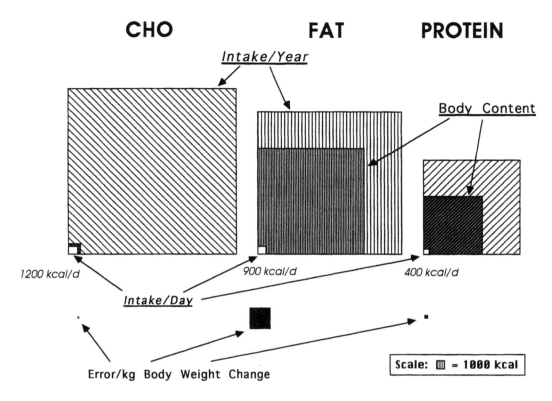

FIGURE 1   Energy reserves and daily and yearly turnover. (From Flatt, J. P., *Am. J. Clin. Nutr.,* 62, 820, 1995. With permission.)

men and women generally reach some 100,000 kcal (see Figure 1), but they are commonly much greater. The very size of this fuel reserve implies that the goal of precisely regulating the body's fat stores is neither necessary, nor achievable, as daily fluctuations are too small to be detectable or to alter fuel availability and energy expenditure. Short-term errors in the fat balance, and hence in the energy balances, are thus readily committed, and progressive fat accumulation and obesity well tolerated.

Whereas excessive adiposity exerts an adverse impact on health only after several decades, when the reproductive age is past, the ability to accumulate fat reserves had survival value in the biological evolution of humans. This ability is not so much due to a higher metabolic efficiency, as suggested by the "thrifty gene" concept,[12] but it rather reflects a tendency to eat in excess of requirements when food is available.[13] This evolutionary history, combined with the lack of mechanisms serving to adjust fat oxidation to fat intake,[14] surely has much to do with the rising prevalence of obesity now that food scarcity is replaced by a constant offer of desirable foods.

2   MEANS BY WHICH THE GOALS OF METABOLIC REGULATION ARE ACHIEVED

The simple fact that the same coenzymes (NAD, FAD, ATP, ADP, Coenzyme A) are required for the oxidation of glucose, fats, and AA forces total substrate oxidation to be stoichiometrically integrated and adjusted to ATP dissipation in each cell. A further "coordination" is brought about by the fact that the metabolic pathways for substrate oxidation converge toward the reactions of the citric acid cycles, where the ultimate conversion of the carbons derived from CHO, protein,

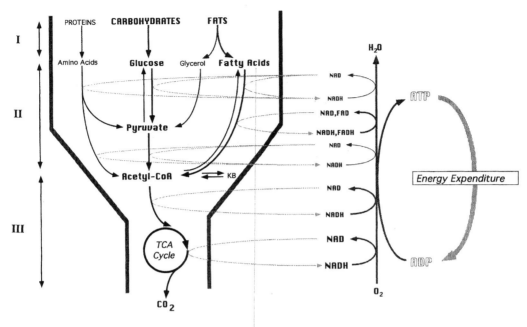

FIGURE 2   The metabolic funnel. The pathways for protein, CHO, and fat oxidation converge toward the tricarboxylic acid (TCA) cycle where most of the ultimate oxidation to CO_2 occurs. Overall substrate oxidation is limited and integration of substrate oxidation forced by the fact that the same coenzymes are needed in all pathways and that reoxidation of NADH to NAD (nicotinamide-adenine-dinucleotide, carrying or devoid of transferable hydrogen atoms) and of FADH to FAD (flavin-adenine-dinucleotide, with or without H) is coupled to the resynthesis of ATP from ADP. (From Flatt, J. P., *Diabetes Rev.,* 4, 443, 1996. Copyright John Wiley & Sons Limited. Reproduced with permission.)

and fat to CO_2 takes place (Figure 2).[15] Complete oxidation is assured and undue accumulation of intermediates avoided because Coenzyme A (CoA) is needed at key steps in the degradation of glucose, fatty acids, and AAs, with CoA availability depending on acetyl-CoA being oxidized to CO_2. The oxidation of acetyl-CoA and of other intermediates is in effect limited by the rate at which ATP is used and ADP and NAD become available, so that total substrate oxidation cannot be greater than needed to drive oxidative phosphorylation. Because CoA is a reactant required for the oxidation of pyruvate, the branched-chain amino acids (BCAA), fatty acids, and KBs, one can also understand why pyruvate and/or of fatty acids plus KBs availability is important in restraining the oxidation of BCAA, for which the first irreversible degradation step is the reaction of their α-keto-derivatives with CoA. Indeed, if fat mobilization is inhibited when pyruvate availability is restricted by food deprivation, rapid "protein wasting" occurs.[16]

The ability to adjust to different substrate mixtures and to live on foods providing nutrients in various combinations is made possible by the fact that most cells can exchangeably use substrates derived from CHO, fat, or proteins. The most significant exception is the CNS. Being unable to use FFA as a fuel, it is therefore dependent on an adequate glucose supply, unless KBs are present in the circulation in substantial amounts (>2 to 3 mM) to provide an alternate fuel, as is the case during prolonged starvation[6,17] and in the neonatal period.[18] These features make it possible for animals to be omnivores, to thrive on some peculiar food source, or, as in the case of humans, to live and function well on diets of very different composition.

The ability to shift from one type of substrate to another, in a manner benefiting the whole organism, depends on the presence in the body's various tissues of enough enzymes to handle even unusual physical exertions and nutrient intakes, without being hampered by metabolic bottlenecks. The enzymes involved in intermediary metabolism are thus present in amounts greatly exceeding

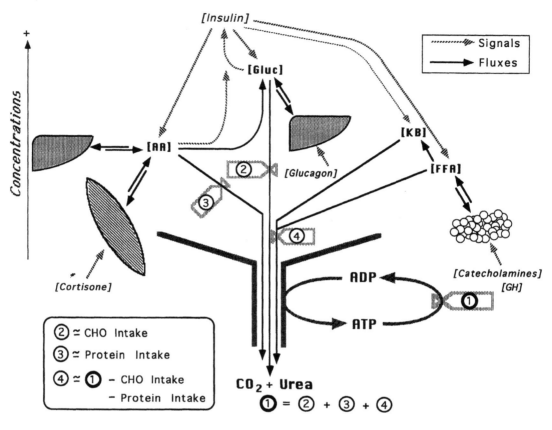

FIGURE 3 Metabolic fuel regulatory system, energy expenditure, and integration of substrate oxidation rates. ATP utilization determines total substrate oxidation. Since glucose (Gluc) oxidation and AA oxidation adjust themselves to CHO and protein intake, fat oxidation is determined by the gap between energy expenditure and the energy consumed as CHO and Protein. (KB = ketone bodies; FFA = free fatty acids; GH = growth hormone). (From Flatt, J. P., *Am. J. Clin. Nutr.*, 62, 820, 1995. With permission.)

the activity needed under resting or inactive conditions. However, a few enzymes are susceptible to be reversibly inhibited or activated, by allosteric effects and/or by phosphorylation/dephosphorylation mechanisms. These key enzymes typically carry out the first irreversible reaction in a metabolic sequence, thereby controlling substrate influx into particular metabolic pathways. The process of enzyme induction, combined with the universal phenomenon of protein degradation and turnover, allows the organism to adapt to prevailing conditions by maintaining a sufficient, but not excessive "enzyme machinery."

To permit acquisition of fat reserves when the foods consumed do not provide appreciable amounts of fat, fatty acids can be synthesized from glucose. In humans, this process of *de novo* lipogenesis becomes quantitatively significant only when glycogen levels have risen to unusually high levels, as a result of deliberate consumption of excessive amounts of CHOs during several consecutive days.[19] In individuals consuming mixed diets, only a few grams of glucose are converted into fat per day,[20] a fact that still remains to be incorporated into many people's thinking.

Regulation of nutrient storage and of fuel mobilization is achieved through an elaborate set of endocrine regulatory phenomena.[10] Insulin promotes the storage of all fuels, as illustrated in Figure 3 by arrows depicting the action of insulin in "pushing down" circulating substrate levels.[8] The universal storage-promoting effect of insulin can be offset by several *counterregulatory hormones* whose action opposes that of insulin in specific tissues. Thus glucagon acts primarily in the liver, to accelerate gluconeogenesis and hepatic glucose release to maintain adequate blood glucose levels. Glucagon is also important in inducing ketogenesis during starvation or glucose deprivation[21] (an action not explicitly shown in Figure 3). Catecholamines, by activating glycogen phosphorylase, can quickly stimulate intramuscular glycogen breakdown in response to physical exertion. Catecholamines also activate a "hormone-sensitive lipase" in adipose tissue, thereby stimulating fat mobilization.[10] Growth hormone also promotes FFA release and utilization, but in a more sustained, rather than acute manner.[22,23] As there is no special form of "storage-protein," the role of maintaining an adequate supply of circulating AAs falls to to the large skeletal muscle compartment. It captures AAs after meals by incorporating them into muscle proteins, to release them later under fasting conditions.[10] In fact, muscle proteins constitute a protein reserve from which AAs can be withdrawn when protein intake is insufficient to sustain protein synthesis in organs and visceral tissues. This transfer is activated by glucocorticoids, under whose influence liver protein content can increase even during periods of negative overall protein balance.[24]

Overall integration of substrate utilization is helped by the fact that increases and decreases in the concentrations of substrates and metabolic intermediates can greatly influence the rates of enzyme-catalyzed reactions by keeping enzymes more or less loaded with substrate. These self-adjusting influences are often most manifest in the range in which these intermediates are typically present, thanks to the fact that the k_m of the enzymes evolved to be in that same range. However, in highly complex and therefore "well-buffered" systems, such as the intracellular environment, or the organism as a whole, these effects tend to be quite sluggish. There were therefore compelling reasons for the development of regulatory mechanisms capable of greatly enhancing and speeding up such mass effects. Thus many regulatory effects reinforce and amplify the mass effects, and nutrient availability, the size of the body's protein, glycogen, and fat reserves, as well as circulating fuel levels consequently influence metabolic fluxes and substrate utilization.[15]

3   MASS EFFECTS

A basic notion in chemistry is that the concentrations of the reactants in a chemical reaction influence reaction rates, as described by the "mass action law." In view of the tremendous progress made in the ability to characterize metabolic regulatory phenomena, the importance which the amounts and/or the concentrations of available substrates can exert on metabolic fluxes is now often disregarded. The impact of substrate concentrations and availability need to be duly considered, however, as these *mass effects are ultimately decisive in bringing about steady-state situations and stability of body composition.* Indeed, weight maintenance is not so much due to the attainment of a preset body composition as it is the result of the balance achieved between various mass effects, as they are bound to bring about a steady-state where influxes and outflows are equal over time in all the body's compartments. However, whereas this phenomenon is universal, *the body composition for which this steady-state occurs can vary considerably among individuals.*

3.1   CARBOHYDRATE AVAILABILITY AND FUEL METABOLISM

The impact of carbohydrate availability on glucose oxidation provides a good illustration of a hormone-enhanced mass effect. For instance, consumption of CHO-containing foods elicits a prompt increase in glucose oxidation. This can be readily assessed, by measuring the production of CO_2 and the consumption of oxygen, or the ratio of CO_2 produced/O_2 consumed, known as the *respiratory quotient,* or RQ. When fat is oxidized, 0.71 mol of CO_2 are produced per mol of O_2 consumed, whereas 1 mol of CO_2 is produced per mol of O_2 consumed in the oxidation of CHO. RQ measurements therefore provide information about the relative proportions of glucose and triglycerides being oxidized. The postprandial rise in the RQ, as well as its duration, depends on the amount and nature of the CHOs consumed, and to a lesser extent on the presence of other nutrients in meals, through their effects on gastric transit and intestinal absorption.[25] After ingestion of a 500 g load of carbohydrate, the RQ rises toward 1.0 and remains at this value for many hours. This nearly complete shift toward exclusive use of glucose helps to dissipate such a large accumulation of glycogen,[26] a response which is enhanced (or attenuated) by the presence of large (or small) glycogen reserves at the time of CHO consumption.[27]

The main effector of this response is insulin, a hormone which facilitates glucose transport, stimulates glycogen synthesis, and activates pyruvate dehydrogenase,[28] the enzyme catalyzing the irreversible step in the oxidation of CHO. Increased release of insulin when blood glucose levels rise thus allows an effective shift toward the predominant use of glucose as a fuel when plenty of glucose is available after meals. This shift is further enhanced by the decrease in FFA availability brought about by the antilipolytic of insulin action on adipose tissue.

A further component of mass effect–related, hormone-enhanced regulation of glucose metabolism relates to the fact that insulin inhibits hepatic glucose production. It is noteworthy (although often overlooked) that higher insulin levels are needed to restrain hepatic glucose release when liver glycogen stores are high than when they are low.[29] This reflects the fact that the mobilization of metabolic fuels is determined not only by the intensity of the signal promoting or inhibiting their release, but also by the size of the fuel stores. High insulin levels need therefore not necessarily be taken to imply a decrease in insulin sensitivity.[29] In fact, maintenance of relatively high insulin levels when glycogen reserves are abundant has the effect of restricting FFA release from adipose tissue, allowing glucose to remain the principal fuel oxidized for ATP regeneration once the intestinal absorption of a meal has been completed. This feature of metabolic regulation is particularly important when CHOs provide the bulk of the food energy consumed.

Although insulin promotes the storage and restrains the mobilization of all fuels, i.e., glucose, fatty acids, KBs, and AAs, its secretion is most directly affected by the conditions prevailing in the CHO sector of metabolism (see Figure 3). Indeed, given the relatively small storage capacity for glycogen, the fact that CHOs often provide most of the dietary energy but are consumed intermittently, and the importance of maintaining adequate blood glucose levels, evolution was compelled to develop a system of metabolic regulation that gives higher priority to the control of CHO metabolism than to the control of AA and lipid metabolism (see Figure 3).

3.2   PROTEIN STORES AND AMINO ACID OXIDATION

The body of an adult contains some 12 kg of protein. Of this, about half is extracellular, mainly as collagen. The remainder is mostly intracellular. Intracellular proteins, as well as plasma proteins and enzymes secreted into the intestine, are degraded and resynthesized more rapidly than collagen, some having half-lives as short as a few hours. Nitrogen (N) balance measurements have revealed long ago that over a few days the organism effectively adjusts AA oxidation to protein intake. When changing from situations of low to high protein intake, or vice versa, a few days are required before N balance is again achieved. The small gains or losses of proteins which are thereby incurred appear to be prerequisites for the reestablishment of the balance between oxidation and intake. This demonstrates that relatively small changes in the body's protein content can significantly influence the rate of AA oxidation. Mass effects therefore appear to contribute substantially to the corrective responses needed to offset deviations from even N balance caused by daily variations in food consumption and physical activity.

3.3   ADIPOSITY AND THE ADJUSTMENT OF RQ TO FQ

The effect of the size of the fat mass on fuel utilization can be understood by considering changes in body weight, or more appropriately changes in body fat content as a function of time (Figure 4). When living conditions are stable, adults tend to approach a plateau of weight maintenance. This can only occur when the amounts of protein, CHO, and fat oxidized are equal, over time, to the amounts of protein, CHO, and fat consumed.[30] The RQ is then equal, on average, to the *food*

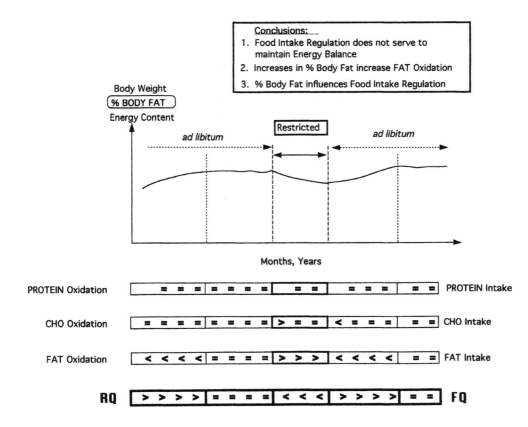

FIGURE 4 Effects of food restriction and weight regain on the RQ and on protein (PRO), CHO, and fat balances (ox = oxidation; in = intake). (From Flatt, J. P., *Am. J. Clin. Nutr.,* 62, 820, 1995. With permission.)

quotient or FQ, which describes the ratio of CO_2 produced to oxygen consumed during the oxidation of a representative sample of the diet.[31]

The role that the adipose tissue mass plays in the establishment of the steady state of weight maintenance becomes apparent when one considers what happens when *ad libitum* food consumption is reinstituted after a period of dietary deficit. Except during the first days of dietary restriction and later during the first days of restored ad libitum food consumption, protein and glycogen balances are equilibrated or nearly so, whereas the fat balance is not. Due to the contribution made by endogenous fat during energy deprivation to the substrate mix provided by food, the RQ is lower than the FQ during such periods. During weight regain, on the other hand, part of the fat consumed is stored, resulting in the oxidation of a fuel mix containing a higher proportion of CHO than the diet. The average RQ is thus greater than the FQ during periods of weight gain. If circumstantial factors have not changed, food intake becomes once more commensurate with energy expenditure when the RQ again matches the FQ, which happens when the adipose mass has been restored to its original size and exerts again the same leverage on fat oxidation as before diet restriction. Expansion of the adipose tissue mass is thus a factor inhibiting food consumption. Regain of the fat lost by dieting is indeed a well-known and all-too-common phenomenon. The opposite response, weight loss after overeating, though less frequently observed, has now also been well documented.[32-35] The seasonal weight changes which occur in many individuals further illustrate that spontaneous weight loss after a period of weight gain is a common experience. Taken together, these responses imply that expansion of the fat mass restrains food consumption, and vice versa.

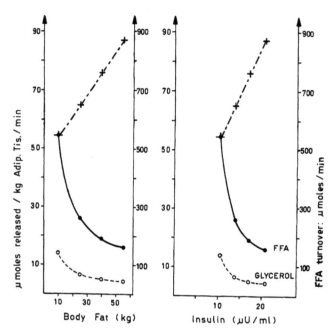

FIGURE 5 Effect of fat mass on FFA and glycerol turnover in the postabsorptive state. The left panel shows rates of FFA (●—●) and glycerol (○---○) turnover in terms μmol/kg adipose tissue/min (left ordinate axes) and total FFA turnover (×---×) in terms of μmol/min (right ordinate axis), as a function of adipose tissue mass. The right panel shows the same values as a function of the insulin concentrations (μUnits/ml) which prevail for fat masses of 10, 25, 40, and 55 kg. (Based on data of Björntorp et al.[38])

Furthermore, one comes to understand that the percent body fat maintained by a given individual is the result of current interactions between body composition, genetic makeup, and environmental and circumstantial factors. This provides a very different perspective than commonly held views, which attribute obesity to the fact that at some time energy intake must have exceeded energy expenditure. Although factually true, this explanation confuses past and present. It stands in the way of a much more pertinent concept, namely, that lean and obese individuals tend to remain that way interpretation.[36] This is so because the particular degree of adiposity which they have reached brings about an average rate of fat oxidation commensurate with fat intake, that food intake under these conditions tends to adjust itself to energy expenditure, and that this level of adiposity therefore tends to maintain itself.[8]

Substrate fluxes are determined by the intensity of regulatory signals, as well as of the size of the target tissues to which these signals are conveyed.[37] FFA release from adipose tissue, as well as rates of glucose and fat uptake vary when the amount of body fat is different (Figure 5).[38] The impact of visceral adipose tissue depots tends to be greatest, as fatty acid turnover is more rapid in visceral than in peripheral sites.[39] An increase in FFA release proportional to increments in the fat mass is avoided by maintaining higher insulin levels, although this rise in insulin levels is not sufficient to prevent totally an increase in the overall release of FFA into the circulation.[40] Because expansion of the adipose tissue mass does not cause a great increase in lean body mass (LBM) and resting energy expenditure (REE), FFA release exceeds their use as fuel in obese subjects, to an even greater extent than in lean.[40] Much of the increment in FFA release must therefore be reesterified by the liver and reexported to adipose tissue, explaining why lipoprotein fluxes are raised in obesity.[41,42]

Enlargement of the adipose tissue mass and hyperinsulinemia that stimulates lipoprotein lipase (LPL) increase the capacity of adipose tissue to capture fatty acids from the chylomicrons transporting the fats absorbed from the intestine. However, the impact of this increased capacity for picking up fatty acids is limited by dietary fat intake, so that the mouth, rather than adipose tissue,

LPL activity acts as "gatekeeper" for fat storage. Expansion of the fat mass thus inherently has the effect of promoting FFA release and thereby fat oxidation.

3.4   FAT MASS AND INSULIN RESISTANCE

The apparent paradox of the coexistence of elevated FFA levels, essentially normal blood glucose levels, and higher insulin levels has long been noted, giving rise to the concept of "insulin resistance." As one can judge by considering FFA release relative to the size of the adipose tissue mass, the antilipolytic action of insulin remains clearly manifest (see Figure 5).[38,40,41] It seems more puzzling that insulin levels in obesity should be elevated out of proportion to the minimal increments in fasting blood glucose levels. This again should be attributed in part to a simple mass effect. Consider how insulin levels, raised by the evening meal, decline progressively during the night, as nutrient uptake wanes and metabolic fuel fluxes evolve toward a condition where rates of glucose and FFA release match their rates of uptake. In the presence of an enlarged adipose tissue mass, particularly when visceral fat depots are expanded, the approximate steady state which prevails after the overnight fast is approached while insulin levels are still relatively high, as otherwise FFA levels would rise and alter glucose oxidation. Because sustained high hormone levels tend to induce some degree of "desensitization," this would be expected to contribute to the decline in the responsiveness of the tissues to insulin generally observed in obesity.[43] In regard to body weight regulation, the insulin resistance induced by excessive enlargement of the fat stores can be expected to reduce glucose, but to promote fatty acid oxidation. By helping to bring about rates of fat oxidation commensurate with fat intake, this serves to limit further expansion of the adipose tissue mass.[13,44-46]

4   WEIGHT MAINTENANCE

Many factors are known to come into play in determining body weights and energy balance, some inherited,[47] others related to lifestyle, and socioeconomic and dietary conditions.[8,48,49] In the face of the complexities of these interactions, it is helpful to realize that the problem of weight maintenance and obesity essentially boils down to two issues: (1) The phenomena eliciting adjustments in food intake, in such a way that energy intake remains commensurate with energy expenditure, once a particular body composition has been reached, and (2) the factors that determine the size of the adipose tissue mass (or the percent body fat) for which weight maintenance tends to become established, in a given individual, under a particular set of circumstances (Figure 6). While differences in adiposity is the issue of practical importance, the mechanisms which one may propose to explain them must accommodate and be consistent with the phenomena explaining weight maintenance.

Body weight regulation ought to be studied under conditions in which food intake is self-determined, rather than imposed. To this effect, variations in daily food intakes were measured in *ad libitum* fed mice. Concomitant measurements of their 24-h respiratory exchanges allowed calculation of the amounts of CHO and fat oxidized over 24 h, as well as to establish daily CHO, fat, and energy balances during many consecutive days.[50] In comparing C57BLK and ob/ob mice, energy turnover was found to be essentially identical, as greater spontaneous physical activity among the control mice made their overall energy expenditure similar to that of the much bigger, but less active ob/ob mice (Figure 7). Variations in intake and expenditure, and the size of the errors committed in maintaining energy and nutrient balances were also similar, as shown by the size of the standard deviations affecting the means in Figure 7. Deviations from equilibrated balance are substantially smaller for CHO than for fat, but comparable in size in ob/ob and control mice. Furthermore, food consumption was found to adjust itself to energy expenditure in a similar manner in the ob/ob and the control mice, with variations in daily food intake more responsive to errors in the CHO than to errors in the fat balances committed on the preceding day.[51] The dominant role of these corrective effects in stabilizing body weights needs to be taken into account in trying to explain the action of various factors on adiposity, for instance, the effect of leptin.[52]

The Problem of Body Weight Regulation boils down to TWO ISSUES !

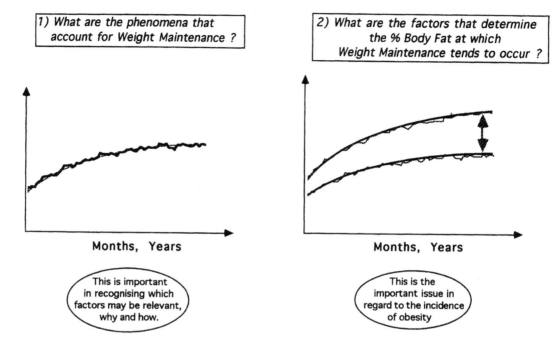

FIGURE 6   The two main issues in body weight maintenance and obesity. (Adapted from Flatt.[8])

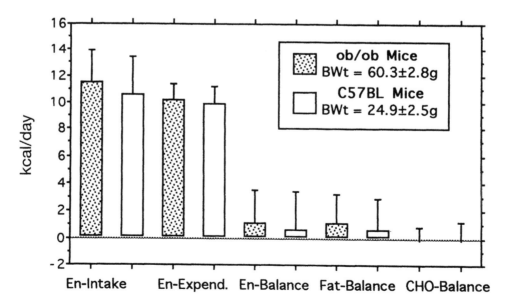

FIGURE 7   Variability of daily food intake and energy expenditure, and of 24-h energy, CHO, and fat balances in five female ob/ob and five female C57BLK mice maintained *ad libitum* on a diet providing 19% of its energy as casein, 27% as fat, and the balance as cornstarch plus sucrose. Substrate oxidation was determined by indirect calorimetry as described in Reference 50. (From Flatt, J. P., *Diabetes Rev.,* 4, 443, 1996. Copyright John Wiley & Sons Limited. Reproduced with permission.)

4.1   Energy Turnover and Energy Balance

Whether weight maintenance and energy balance come about by adjustment of energy expenditure to energy intake, or vice versa, has long been a central preoccupation in obesity research. In humans, a key consideration is that energy expenditure is only modestly altered by energy intake. The corollary is that changes in intake affect the energy balance very powerfully. Because changes in metabolic rates can only slightly attenuate (by 10 to 20%), but not reverse, energy imbalances created by routine variations in daily intakes and physical activities, weight maintenance under *ad libitum* food intake conditions depends primarily on appropriate adjustments in food intake.

When mice are given free access to food after a period of dietary restriction, a compensatory increase in food intake quickly leads to regain of the lost weight. The animals' unusually high food consumption abates when the fat mass had returned to its previous size. As shown in Figure 4, the balance between energy expenditure and food intake is brought about by a decrease in food consumption, not by an increase in energy expenditure as weight is regained. The linkage between a particular degree of expansion of the adipose tissue mass and the fact that food intake becomes commensurate with energy expenditure is of particular interest. The fact that this occurs when the RQ is again equal to the FQ supports the view that adjustment of the composition of the fuel mixed oxidized to the nutrient distribution in the diet is a critical factor for weight maintenance.[44,53]

4.2   Metabolic Efficiency and Body Weight Regulation

There has long been a fascination about the possible role of even small differences in energy expenditure on long-term body weight regulation.[54] When energy intake is imposed, the gap between intake and expenditure progressively decreases as changes in body size alter the costs for maintaining and moving it. When food is insufficient, physical activity is spontaneously curtailed and energy expenditure per kilogram of LBM also declines somewhat.[55,56] However, the effect of adaptive changes in resting energy expenditure on body weight regulation cannot be assessed from such observations, because the most critical phenomenon in body weight regulation, regulation of food intake, cannot manifest itself under conditions of imposed food intake.

To assess how differences in REE may affect adiposity under free-living conditions, one can examine the impact of stature on adiposity, given that stature influences the size of the fat-free mass (FFM), the main predictor of REE.[57] In the NHANES3 data pool on some 6000 women and 6000 men, height accounts for 12 and 16% of the variance FFM, respectively. In spite of this, height has essentially no impact on indexes of obesity (i.e., R^2 values ≤ 0.01).[58] Differences in REE and, for that matter, differences in metabolic efficiency, do not therefore appear to play an important role in the development or prevention of obesity.

4.3   Modulation of Food Intake during Weight Maintenance

Large short-term errors in the energy balance occur even during periods of weight maintenance. Indeed, food intake varies greatly from day to day, such that coefficients of variation (CV) for intraindividual food intake average ±23%.[59] Furthermore, variations in food consumption are not closely correlated with variations in physical activity,[60] resulting in substantial short-term deviations from energy balance. How is it that body weights nevertheless remain stable during prolonged periods of most individuals' lives?

In *ad libitum* fed mice, studies of food consumption and substrate balances during many consecutive days have shown that food intake on a given day is negatively correlated with the CHO and the fat balances on the previous day. As evident from the numerical value of the regression coefficients (Table 1), the CHO imbalances exert a substantially greater influence. Furthermore, their effect is more consistent. Thus, the negative correlation between energy intake and the CHO balance during the preceding day manifesting itself during periods of weight gain or weight loss, whereas this is not the case with deviations from even fat balances.[61] Stubbs et al.[62] established

TABLE 1
Factors Influencing Energy Intake in *ad Libitum* Fed Mice[a]

	Previous day's				
	Energy Intake	CHO Balance	FAT Balance	Revolutions/day[b]	D Revolutions/day
	Female CD1 Mice on Mixed Diet (kJ/g BWt/day)				
Energy Intake = 0.04[c]	+0.84	−1.16	−0.50	-1.1×10^{-6}	-1×10^{-6}
P = 0.0001[a]	(P ≤ 0.0001)	(P ≤ 0.0001)	(P ≤ 0.0001)	(P = 0.75)	(P = 0.1)
$R^2 = 0.32$[d]; N = 703	{19%	7%	5%	0%	0%}[e]
	Male CD1 Mice on Various Diets (kJ/g BWt/day)				
Energy Intake = 0.04[c]	+0.95	−1.21	−0.50	$+0.09 \times 10^{-6}$	1.4×10^{-6}
P = 0.0001[d]	(P ≤ 0.0001)	(P ≤ 0.0001)	(P ≤ 0.0001)	(P = 0.7)	(P = 0.0007)
$R^2 = .56$[d]; N = 2704[d]	{36%	15%	4%	0%	0.3%}[e]

[a] *Ad libitum* food intake, spontaneous use of a running wheel, and daily CHO and fat oxidation and balances were established for each animal in a group of 10 female and a group of 10 male CD1 mice during consecutive days. (Modified from Flatt.[8])
[b] Average number of revolutions/day (±S.D.): 7550 ± 9480 in females, 7000 ± 7240 in males.
[c] Multiple regression equation which shows the intercept and the relationship of a given day's food intake with the animal's running activity and with its CHO and fat balances during the preceding day.
[d] P, R^2 values and number of days studied.
[e] {% of variance}.

substrate balances and monitored food consumption in volunteers who stayed in a respiratory chamber during 7 consecutive days. They could serve themselves freely from foods presented to them in large portions. A negative correlation between food intake and the previous days' CHO balances was also observed (accounting for 5 to 10% of the variance in food intake). By contrast, the correlation with the previous days' fat balances was positive. In assessing the biological significance of such small effects, one has to consider that variations in food intake from day to day are considerable)[59] and that energy and substrate imbalances are tolerated with remarkable ease. Evidently, control over food intake is rather loose in the short run, so that it is difficult to recognize with statistical significance the physiological mechanisms involved in down- and upregulating food intake. If the main causes of variations in food intake occur essentially at random, relatively weak phenomena such as those exerted by deviations from CHO balance in "modulating" food intake (see, e.g., Table 1 and Reference 62), could nevertheless exert "steering effects" capable of significantly affecting body composition in the long run.

5   FACTORS INFLUENCING THE BODY FAT CONTENT NEEDED FOR WEIGHT MAINTENANCE

Nitrogen balance is spontaneously maintained under all kinds of conditions; furthermore, protein contributes only a minor part of the food energy consumed. The regulation of body weight and of its fat content is thus primarily the result of the interactions that occur in the metabolism of CHO and fat. Many features of the interactions between CHO and fat metabolism can be illustrated with a two-compartment model, designed to illustrate the competition for oxidation between glucose and FFA (Figure 8), as proposed by Randle et al.[9] The small and the large reservoir represent the human body's limited capacity for storing glycogen and its large capacity for fat storage. The numbers describe the size of these reserves in terms of kilocalories, as typically present in adults. The exclusive use of glucose by the brain is illustrated by a small turbine deriving all of its fuel

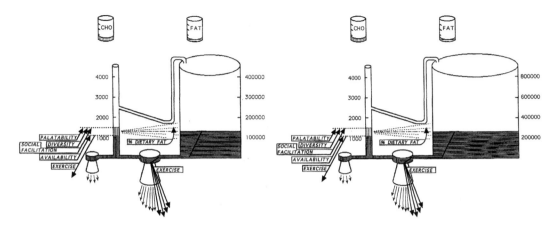

FIGURE 8   Two-compartment model illustrating the impact of circumstantial, lifestyle, and genetic factors on adiposity (for explanations, see text). (From Flatt, J. P., *Am. J. Clin. Nutr.*, 62, 820, 1995. With permission.)

from the small reservoir. The relative proportions of glucose and of fatty acids used by the rest of the body (i.e., the flux through the large turbine) is influenced by the availability of glucose and FFAs, which can be thought to be proportional to the levels to which the two reservoirs are filled at a given time.

Replenishment of the body's glycogen and fat stores occurs from time to time by consumption of meals (illustrated by the contents of the small containers shown above the reservoirs), of which the fraction corresponding to the diet's fat content is delivered into the large reservoir. Food consumption is determined by the habitual pattern of meal consumption, complemented when necessary by physiological regulatory mechanisms to assure that glycogen reserves are sufficient to avoid hypoglycemia, while also preventing their buildup to levels at which *de novo* lipogenesis would be induced. (i.e., through the conduit from the small to the large reservoir). Additions of fuel to the large reservoir cause only insignificant changes in its level, whereas they markedly affect the content of the small reservoir. This causes the outflow from the small reservoir to increase temporarily, just as glucose oxidation increases after meals and adjusts itself to glycogen availability. The interplay between these changes in glucose oxidation and the factors controlling food intake causes glycogen levels to oscillate up and down within a particular operating range. When outflow from the large reservoir (= fat oxidation) is not commensurate with inflow (= fat intake), its content (= adipose tissue mass) will change slowly over time, until it is filled to the level that causes its contribution to the flow through the large reservoir to be equal on average to the amount of fuel added to it. A steady-state is then reached.

Increases in the proportion of fuel falling into the large reservoir causes its content to increase, until its level is high enough to make its outflow commensurate again with the inflow. Increases in dietary fat content thus tend to cause increases in adiposity in animal,[63] as well as in human populations,[64] although not in all subjects. A rise in the level of the large reservoir can indeed be avoided if its outflow is facilitated by an appropriate decline in the content of the small reservoir. To explain the high prevalence of obesity in the world, other factors than dietary fat content are obviously important as well. For instance, the availability of appetizing foods and their diversity as well as good company tend to increase the amounts of food consumed at a meal.[65,66] In the model this corresponds to raising the upper limit of the operating range in the small reservoir. The ubiquitous availability of foods tends to promote food consumption between meals, an effect illustrated in the model by the arrow which opposes the decline in the content of the small reservoir between meals. Together, these circumstantial variables raise the range in which glycogen levels are habitually maintained, which has the effect of curtailing fat oxidation. A higher fat mass is then required (i.e., the level in the large reservoir must rise) for fat to make a contribution to the fuel mix oxidized commensurate with dietary fat intake.

Regular physical activity limits excessive accumulation of body fat. In the model, this leverage can be visualized by two effects: (1) exercise increases flux through the large turbine (i.e., substrate oxidation in skeletal muscle which uses both glucose and FFAs), thereby reducing the impact of the small turbine (i.e., the impact of the tissues preferentially using glucose) on the composition of the fuel mix is diminished, and (2) exercise leads to greater glycogen depletion between meals, thereby extending the periods during which the body operates at a relatively low RQ. In effect, if one considers calories spent, rather hours elapsed, exercise lengthens the interval between meals and this can readily be seen to promote fat oxidation, relative to glucose oxidation, accounting for the overall impact of exercise.[67]

The influence of inheritance on body composition and on the susceptibility to develop obesity can be extremely powerful.[68,69] Genetic differences may act by altering the balance in the competition between glucose and fat for oxidation.[63] This effect can be mimicked in the model by shrinking or expanding the diameter of the large reservoir, which alters the amount of fuel that must be present in the large reservoir to reach the level that brings about a given rate of outflow (i.e., in the case illustrated in Figure 8, the amount needed in the case at right is twice that in the left). Such a change does not prevent the influence of circumstantial factors, nor does it prevent food intake regulation from bringing about weight maintenance.

Differences in the relative use of glucose and fat are most conveniently evaluated in terms of the RQ, which provides a direct assessment of the relative contributions made by glucose and fatty acids. Indeed, it has been found that individuals with relatively high RQs tend to gain more weight during subsequent years than others.[70-72] These considerations illustrate that fat will tend to accumulate when the combined impact of metabolic regulation and expansion of the adipose tissue mass falls short in promoting fat, relative to glucose oxidation. Conversely, the presence of an enlarged adipose tissue mass indicates that the impact of the adipose tissue is too weak in lowering the RQ. This may be a manifestation of different inherited properties at the level of enzyme or hormone regulation, or it may be due to the habit of maintaining relatively high glycogen levels. In turn, the latter may be imputable to inherited differences in the systems regulating food intake, or in individual food preferences and responsiveness to an environment in which foods are freely available.

6   OBESITY

Obesity is commonly characterized by anthropometric parameters, among which the body mass index (BMI) is the most frequently used. Actual assessment of body fat content provides a more precise characterization and allows description of the relative proportions of the lean body and the fat mass, as well as percent body fat, which has the advantage of being independent of stature. Monitoring of body fat content over extended periods provides an integrated measure of the balance between energy intake and expenditure. Because of the insurmountable difficulties in cumulating data on intake and expenditure in natural settings over extended periods, such an integrated measure is extremely valuable, and in fact more reliable than other parameters that one may attempt to measure. Percent body fat also offers a useful perspective to think about weight maintenance, obesity, and its possible causes (see Figures 4 and 6). With this parameter in mind, one realizes that obesity is a situation where excessive expansion of the adipose tissue mass is needed to bring about rates of substrate oxidation that result in the oxidation of a fuel mix matching the nutrient distribution in the diet. In effect, the degree of adiposity depends on how fat one has to be, to burn as much fat as one eats.[31]

When weight maintenance is approached only once the adipose tissue mass is enlarged, this merely shows that the organism fails to burn as much fat as consumed while the body fat content is in the desirable range. The corollary to this is that obesity implies that CHO oxidation is somewhat too high, while adiposity is normal. This view is of greater interest, considering that metabolic regulation is dominated by the conditions related to the metabolism of glucose (see Figure 3). Because pyruvate can readily be reconverted into glucose by gluconeogenesis, the irreversible step in the degradation of CHO is the conversion of pyruvate to acetyl-CoA by pyruvate dehydrogenase (PDH). Inhibition of PDH is of crucial importance in minimizing glucose oxidation during starvation, when glucose oxidation can be reduced to some 40 g/day, thanks to the concomitant induction of ketogenesis in the liver[21] and the use of KBs by the brain instead of glucose.[6] Inhibition of PDH is brought about by phosphorylation, as well as by allosteric inhibitors, including fatty acids, fatty acyl-CoA, acetyl-CoA, and NAD.[28,73,74] On the other hand, it is important as well that PDH be promptly activated when the CHO supply is ample, as after meals. This activation is brought about by insulin, via stimulation of protein phosphatases that dephosphorylate and activate PDH, as well as by the direct effect of high pyruvate levels on the enzyme.[28]

When the adipose tissue mass increases, insulin and FFA levels both rise, in a manner susceptible to be influenced by dietary and lifestyle factors, as well as by genetic predispositions.[47] The balance between their opposing effects on pyruvate oxidation may be quite variable, a fact that could explain the wide variability of adiposity in human populations. The high prevalence of obesity in the U.S. as well as its increase from 25 to 33% during the last decade[75] imply that environmental factors can affect the conditions determinig the relative rates of pyruvate and fat oxidation, *and that this is offset by the impact of an enlarged adipose tissue mass*. The need for greater fat mass expansion to approach weight stability, in populations whose gene pool is stable, suggests that higher glycogen levels are maintained and/or that fat consumption increased. In recent years, the latter does not appear to be the case in the U.S., but it seems plausible that an increasing variety and availability of desirable foods could cause people habitually to maintain higher glycogen levels.[8]

7 CONCLUSIONS

It is useful to take stock of the phenomena about which we have substantial data, but about which perceptions still differ, and to identify issues that remain poorly resolved (Figure 9). There is still much debate about the role of differences in metabolic rates and/or "metabolic efficiency," but since differences in stature have essentially no influence on adiposity,[58] it appears that differences in resting metabolic rates exert little leverage on the prevalence of obesity. Another frequently implicated notion relates to *substrate partitioning*. While relevant during rapid growth and in dealing with meat production, this concept is hardly applicable to situations where body composition is stable and when essentially all the substrates consumed are oxidized. One would have to consider that the nutrients absorbed are handled in different ways in lean and overweight adults during the postprandial period, although this would be compensated during the rest of the day. The fact that fat oxidation in the postabsorptive state is generally higher in obese than in lean subjects[76] suggests that this may be the case, although it is difficult to rule out that this is not due to differences in prior food consumption or in physical activity. If lower RQs in the postabsorptive state are indeed a feature of metabolism in weight-maintaining adults, as compared with weight-maintaining lean subjects consuming a diet of the same composition, one has to expect the RQ to be higher in the obese subjects during other periods of the day. This could be related to the fact that the postprandial rises in blood glucose and insulin levels tend to last longer in obese subjects.

The exploration of the role of possible inherited variations on the relative use of glucose vs. fatty acids has been undertaken, but the recognized effects at the enzyme or intermediate level are more suggestive than established.[49] Lack of information on liver and muscle glycogen levels, on their variations during the day, and on their impact on the RQ is a particularly important limitation in trying to understand why adiposity can differ so much between individuals.

As emphasized in this review, substrate utilization over 24 h is essentially the same in weight-maintaining lean and obese subjects, thanks in part to the metabolic impact of cumulative changes in the adipose tissue mass. This has led to attempts at recognizing possible metabolic differences between lean subjects and individuals who are in a sustained phase of weight maintenance at a

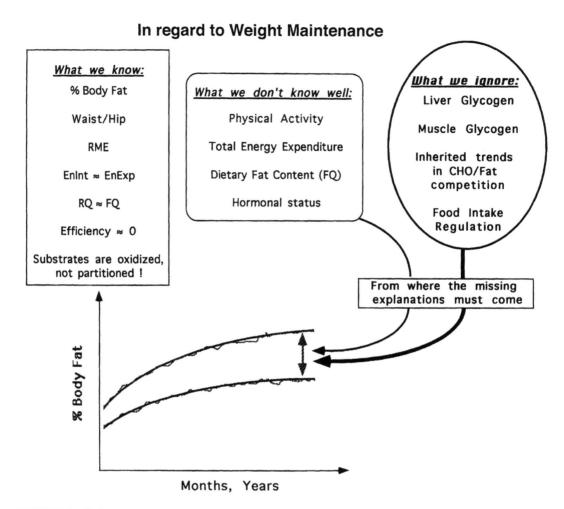

FIGURE 9 Well-known and less well-known parameters affecting adiposity and their potential role in explaining weight maintenance at different degrees of adiposity. (RME = resting metabolic rate; RQ = respiratory quotient; FQ = food quotient; CHO = carbohydrate.) (From Flatt, J. P., *Am. J. Clin. Nutr.*, 62, 820, 1995. With permission.)

reduced body weight.[77] Among such reduced-obese subjects, fat oxidation tends to be lower and to increase more sluggishly when the fat content of the diet is raised.[78,79] These observations complement the insights offered by the two-compartment model, which show that the mass of adipose tissue is bound to increase until it is large enough to cause rates of fat oxidation to become commensurate with fat intake. Limiting fat intake and promoting fat oxidation by physical activities[67] can thus provide leverage in limiting the amount of body fat needed to achieve weight maintenance. Indeed, these are the main measures adopted by individuals who have successfully maintained substantial weight losses for extended periods.[80]

In dealing with the obesity problem, one will ultimately have to understand why the impact of the adipose tissue mass fails to restrict pyruvate oxidation appropriately in subjects prone to developing obesity. In particular, one will need to recognize when this is due to inherited metabolic characteristics, when it is a consequence of a food intake regulation that leads to the maintenance of inordinately high glycogen and insulin levels, whose effects override the signals usually restraining pyruvate oxidation, and when it is due to environmental pressures which cause habitual glycogen levels to be too high for diets with a substantial fat content.

ACKNOWLEDGMENTS

This publication was made possible by Grant DK 33214 from the National Institutes of Health. Its contents are solely the responsibility of the author and do not necessarily represent the official views of the National Institutes of Health.

REFERENCES

1. Spector, A. A., Fatty acid binding to plasma albumin, *J. Lipid Res.*, 16, 165, 1975.
2. Fredrickson, D. S. and Gordon, R. S., The metabolism of albumin-bound C14-labeled unesterified fatty acids in normal human subjects, *J. Clin. Invest.*, 37, 1504, 1958.
3. Bahr, R., Ingnes, I., Vaage, O., Sejersted, O. M., and Newsholme, E. A., Effect of duration of exercise on excess postexercise O_2 consumption, *J. Appl. Physiol.*, 62, 485, 1987.
4. Garlick, P. J., Clugston, G. A., Swick, R. W., and Waterlow, J. C., Diurnal pattern of protein and energy metabolism in man, *Am. J. Clin. Nutr.*, 33, 1983, 1980.
5. *Recommended Dietary Allowances,* National Academy Press, Washington, D.C., 1989.
6. Cahill, G. F., Starvation in man, *N. Engl. J. Med.*, 282, 668, 1970.
7. Björntorp, P. and Sjöström, L., Carbohydrate storage in man: speculations and some quantitative considerations, *Metabolism*, 27, 1853, 1978.
8. Flatt, J. P., McCollum Award Lecture, 1995: Diet, lifestyle and weight maintenance, *Am. J. Clin. Nutr.*, 62, 820, 1995.
9. Randle, P. J., Hales, C. N., Garland, P. B., and Newsholme, E. A., The glucose fatty-acid cycle: its role in insulin sensitivity and the metabolic disturbances of diabetes mellitus, *Lancet*, 1, 785, 1963.
10. Cahill, G. F., Jr., Physiology of insulin in man, *Diabetes*, 20, 785, 1971.
11. Flatt, J. P. and Blackburn, G. L., The metabolic fuel regulatory system: implications for protein sharing therapies during caloric deprivation and disease, *Am. J. Clin. Nutr.*, 27, 175, 1974.
12. Neel, J. V., Diabetes mellitus: a "thrifty" genotype rendered detrimental by "progress," *Am. J. Hum. Genet.*, 14, 352, 1962.
13. Ravussin, E. and Swinburn, B. A., Pathophysiology of obesity, *Lancet*, 340, 404, 1992.
14. Flatt, J. P., Ravussin, E., Acheson, K. J., and Jéquier, E., Effects of dietary fat on postprandial substrate oxidation and on carbohydrate and fat balances, *J. Clin. Invest.*, 76, 1019, 1985.
15. Flatt, J. P., Sustrate utilization and obesity, *Diabetes Rev.*, 4, 443, 1996.
16. Talke, H., Maier, K. P., Kersten, M., and Gerok, W., Effect of nicatinamide on carbohydrate metabolism in the rat liver during starvation, *Eur. J. Clin. Invest.*, 3, 467, 1973.
17. Hasselbalch, S. G., Madsen, P. L., Hageman, L. P., Olsen, K. S., Justesen, N., Holm, S., and Paulson, O. B., Changes in cerebral blood flow and carbohydrate metabolism during acute hyperketonemia, *Am. J. Physiol.*, 270, E746, 1996.
18. Mitchell, G. A., Kassavoska-Bratinova, S., Boukaftane, Y., Robert, M.-F., Wang, S. P., Ashmarina, L., Lambert, M., Lapierre, P., and Potier, E., Medical aspects of ketone body metabolism, *Clin. Invest. Med.*, 18, 193, 1995.
19. Acheson, K. J., Schutz, Y., Bessard, T., Anantharaman, K., Flatt, J. P., and Jéquier, E., Glycogen storage capacity and *de novo* lipogenesis during massive carbohydrate overfeeding in man, *Am. J. Clin. Nutr.*, 48, 240, 1988.
20. Hellerstein, M. K., Synthesis of fat in response to alterations in diet: insights from new stable isotope methodologies, *Lipids*, 31, S117, 1996.
21. Keller, U., Lustenberger, J., Müller-Brand, J., Gerber, P. P. G., and Stauffacher, W., Human ketone body production and utilization studied using tracer techniques: regulation by free fatty acids, insulin, catecholamines, and thyroid hormones, *Diabetes Metab. Rev.*, 5, 285, 1989.
22. Davidson, M. B., Effect of growth hormone on carbohydrate and lipid metabolism, *Endocr. Rev.*, 8, 115, 1987.
23. Salomon, F., Cuneo, R. C., Hesp, R., and Sönksen, P. H., The effects of treatment with recombinant human growth hormone on body composition and metabolism in adults with growth hormone deficiency, *N. Engl. J. Med.*, 321, 797, 1989.

24. Silber, R. H. and Porter, C. C., Nitrogen balance, liver protein repletion and body composition of cortisone treated rats, *Endocrinology*, 52, 518, 1953.

25. Jenkins, D. J. A., Jenkins, A. L., Wolever, T. M. S., Vuksan, V., Rao, A. V., Thompson, L. U., and Josse, R. G., Low glycemic index: lente carbohydrates and physiological effects of altered food frequency, *Am. J. Clin. Nutr.*, 59, 706S, 1994.

26. Acheson, K. J., Flatt, J. P., and Jéquier, E., Glycogen synthesis vs. lipogenesis after a 500 gram carbohydrate meal in man, *Metabolism*, 31, 1234, 1982.

27. Acheson, K. J., Schutz, Y., Bessard, T., Ravussin, E., Jéquier, E., and Flatt, J. P., Nutritional influences on lipogenesis and thermogenesis after a carbohydrate meal, *Am. J. Physiol.*, 246, E62, 1984.

28. Mandarino, L. J., Regulation of skeletal muscle pyruvate dehydrogenase and glycogen synthase in man, *Diabetes Metab. Rev.*, 5, 475, 1989.

29. Clore, J. N., Helm, S. T., and Blackard, W. G., Loss of hepatic autoregulation after carbohydrate overfeeding in normal man, *J. Clin. Invest.*, 96, 1967, 1995.

30. Flatt, J. P., Dietary fat, carbohydrate balance, and weight maintenance: effects of exercise, *Am. J. Clin. Nutr.*, 45, 296, 1987.

31. Flatt, J. P., The biochemistry of energy expenditure, *Rec. Adv. Obesity Res.*, 2, 211, 1978.

32. Roberts, S. B., Young, V. R., Fuss, P., Fiatarone, M. A., Richard, B., Rasmussen, H., Wagner, D., Joseph, L., Holehouse, E., and Evans, W. J., Energy expenditure and subsequent nutrient intakes in overfed young men, *Am. J. Physiol.*, 259, R461, 1990.

33. Diaz, E. O., Prentice, A. M., Goldberg, G. R., Murgatroyd, P. R., and Coward, W. A., Metabolic response to experimental overfeeding in lean and overweight healthy volunteers, *Am. J. Clin. Nutr.*, 56, 641, 1992.

34. Tremblay, A., Després, J. P., Thériault, G., Fournier, G., and Bouchard, C., Overfeeding and energy expenditure in humans, *Am. J. Clin. Nutr.*, 56, 857, 1992.

35. Pasquet, P. and Apfelbaum, M., Recovery of initial body weight and composition after long-term massive overfeeding in men, *Am. J. Clin. Nutr.*, 60, 861, 1994.

36. Flatt, J. P., How NOT to approach the obesity problem, *Obesity Res.*, 5, 632, 1997.

37. Flatt, J. P., Role of the increased adipose tissue mass in the apparent insulin insensitivity of obesity, *Am. J. Clin. Nutr.*, 25, 1189, 1972.

38. Björntorp, P., Bergman, H., Varnauskas, E., and Lindholm, B., Lipid mobilization in relation to body composition in man, *Metabolism*, 18, 840, 1969.

39. Zamboni, M., Armellini, F., Turcato, E., de Pergola, G., Todesco, T., Bissoli, L., Bergamo Andreis, I. A., and Bosello, O., Relationship between visceral fat, steroid hormones and insulin sensitivity in premenopausal obese women, *J. Intern. Med.*, 236, 521, 1994.

40. Campbell, P. J., Carlson, M. G., and Nurjhan, N., Fat metabolism in human obesity, *Am. J. Physiol.*, 266, E600, 1994.

41. Coppack, S. W., Jensen, M. D., and Miles, J. M., *In vivo* regulation of lipolysis in humans, *J. Lipid Res.*, 35, 177, 1994.

42. Frayn, K. N., Williams, C. M., and Arner, P., Are increased plasma non-esterified fatty acid concentrations a risk marker for coronary heart disease and other chronic diseases? *Clin. Sci.*, 90, 243, 1996.

43. Felber, J. P., Haesler, E., and Jéquier, E., Metabolic origin of insulin resistance in obesity with and without Type 2 (non-insulin-dependent) diabetes mellitus, *Diabetologia*, 36, 1221, 1993.

44. Flatt, J. P., Importance of nutrient balance in body weight regulation, *Diabetes Metab. Rev.*, 4, 571, 1988.

45. Swinburn, B. A., Nyomba, B. L., Saad, M. F., Zurlo, F., Raz, I., Knowler, W. C., Lillioya, S., Bogardus, C., and Ravussin, E., Insulin resistance associated with lower rates of weight gain in Pima Indians, *J. Clin. Invest.*, 88, 168, 1991.

46. Eckel, R. H., Insulin resistance: an adaptation for weight maintenance, *Lancet*, 340, 1452, 1992.

47. Bouchard, C. and Pérusse, L., Current status of the human obesity gene map, *Obesity Res.*, 4, 81, 1996.

48. Sobal, J. and Stunkard, A. J., Socioeconomic status and obesity: a review of the literature, *Psychol. Bull.*, 105, 260, 1989.

49. Bouchard, C., Can obesity be prevented? *Nutr. Rev.*, 54, S125, 1996.

50. Flatt, J. P., Assessment of daily and cumulative carbohydrate and fat balances in mice, *J. Nutr. Biochem.*, 2, 193, 1991.

51. Flatt, J. P., Weight maintenance in ob/ob and C57BL mice, *Obesity Res.*, 4, 16S, 1996.

52. Halaas, J., Gajiwala, K., Maffei, M., Cohen, S., Chait, B., Rabinowitz, D., Lallone, R., Burley, S., and Friedman, J., Weight reducing effects of the plasma protein encoded by the obese gene, *Science*, 269, 543, 1995.

53. Flatt, J. P., The RQ/FQ concept and weight maintenance, in *Progress in Obesity Research*, Vol. 7, Angel, A., Anderson, H., Bouchard, C., Lau, D., Leiter, L., and Mendelson, R., Eds., Libbey, London, 1996, 49.

54. Leibel, R. L., Rosenbaum, M., and Hirsch, J., Changes in energy expenditure resulting from altered body weight, *N. Engl. J. Med.*, 332, 621, 1995.

55. Keys, A., Brozek, J., Henschel, A., Mickelsen, O., and Taylor, H. L., *The Biology of Human Starvation*, University of Minnesota Press, Minneapolis, 1950.

56. Waterlow, J. C., Metabolic adaptation to low intakes of energy and protein, *Annu. Rev. Nutr.*, 6, 495, 1986.

57. Ravussin, E., Lillioja, S., Anderson, T. E., Christin, L., and Bogardus, C., Determinants of 24-h energy expenditure in man, *J. Clin. Invest.*, 78, 1568, 1986.

58. Flatt, J. P. and Gupte, S., Metabolic efficiency, in *Pennington Center Nutrition Series*, Vol. 9, Bray, G.A. and Ryan, D.H., Eds., Louisiana State University Press, Baton Rouge, 1999, 73.

59. Bingham, S. A., Gill, C., Welch, A., Day, K., Cassidy, A., Khaw, K. T., Sneyd, M. J., Key, T. J. A., Roe, L., and Day, N. E., Comparison of dietary assessment methods in nutritional epidemiology: weighed records v. 24h recalls, food-frequency questionnaires and estimated-diet records, *Br. J. Nutr.*, 72, 619, 1994.

60. Edholm, O. G., Adam, J. M., Healey, M. J., Wolff, H. S., Goldsmith, R., and Best, T. W., Food intake and energy expenditure of army recruits, *Br. J. Nutr.*, 24, 1091, 1970.

61. Flatt, J. P., Carbohydrate balance and body-weight regulation, *Proc. Nutr. Soc.*, 55, 449, 1996.

62. Stubbs, J. R., Ritz, P., Coward, W. A., and Prentice, A. M., Covert manipulation of the ratio of dietary fat to carbohydrate and energy density: effect on food intake and energy balance in free-living men eating ad libitum, *Am. J. Clin. Nutr.*, 62, 330, 1995.

63. Flatt, J. P., The difference in the storage capacity for carbohydrate and for fat, and its implications in the regulation of body weight, *Ann. N.Y. Acad. Sci.*, 499, 104, 1987.

64. Lissner, L. and Heitmann, B. L., Dietary fat and obesity: evidence from epidemiology, *Eur. J. Clin. Nutr.*, 49, 79, 1995.

65. Rolls, B. J., Rowe, E. A., Rolls, E. T., Kingston, B., Megson, A., and Gunary, R., Variety in a meal enhances food intake in man, *Physiol. Behav.*, 26, 215, 1981.

66. de Castro, J. M. and Brewer, E. M., The amount eaten in meals by humans is a power function of the number of people present, *Physiol. Behav.*, 51, 121, 1992.

67. Flatt, J. P., Integration of the overall effects of exercise, *Int. J. Obesity*, 19, S31, 1995.

68. Sorensen, T. I. A., Holst, C., and Stunkard, A. J., Childhood body mass index — genetic and familial enviromental influences assessed in a longitudinal adoption study, *Int. J. Obesity*, 16, 705, 1992.

69. Bouchard, C., Déspres, J.-P., and Mauriège, P., Genetic and nongenetic determinants of regional fat distribution, *Endocr. Rev.*, 14, 72, 1993.

70. Zurlo, F., Lillioja, S., Esposito-Del Puente, A., Nyomba, B. L., Raz, I., Saad, M. F., Swinburn, B. A., Knowler, W. C., Bogardus, C., and Ravussin, E., Low ratio of fat to carbohydrate oxidation as predictor of weight gain: study of 24h RQ, *Am. J. Physiol.*, 259, E650, 1990.

71. Seidell, J. C., Muller, D. C., Sorkin, J. D., and Andres, R., Fasting respiratory exchange ratio and resting metabolic rate as predictors of weight gain: the Baltimore Longitudinal Study on Aging, *Int. J. Obesity*, 16, 667, 1992.

72. Larson, D. E., Ferraro, R. T., Robertson, D. S., and Ravussin, E., Energy metabolism in weight-stable, postobese individuals, *Am. J. Clin. Nutr.*, 62, 735, 1995.

73. Boden, G., Jadali, F., White, J., Liang, Y., Mozzoli, M., Chen, X., Coleman, E., and Smith, C., Effects of fat on insulin stimulated carbohydrate metabolism in normal men, *J. Clin. Invest.*, 88, 960, 1991.

74. Randle, P. J., Priestman, D. A., Mistry, S. C., and Halsall, A., Glucose fatty acid interactions and the regulation of glucose disposal, *J. Cell Biochem.*, 55S, 1, 1994.

75. Kuczmarski, R. J., Flegal, K. M., Campbell, S. M., and Johnson, C. L., Increasing prevalence of overweight among US adults, *J. Am. Med. Assoc.*, 272, 205, 1994.

76. Schutz, Y., Tremblay, A., Weinsier, R. L., and Nelson, K. M., Role of fat oxidation in the long term stabilization of body weight in obese women, *Am. J. Clin. Nutr.*, 55, 670, 1992.

77. Eckel, R., Ailhaud, G., Astrup, A., Flatt, J. P., Hauner, H. L. A. S., Prentice, A. M., Ricquier, D., Steffens, A. B., and Woods, S. C., What are the metabolic and physiological mechanisms associated with the regulation of body weight? in *Regulation of Body Weight: Biological and Behavioral Mechanisms,* Bouchard, C., and Bray, G.A., Eds., John Wiley & Sons, Chichester, 1996, 225.

78. Astrup, A., Buemann, B., Christensen, N. J., and Toubro, S., Failure to increase lipid oxidation in response to increasing dietary fat content in formerly obese women, *Am. J. Physiol.,* 266, E592, 1994.

79. Horton, T. J., Drougas, H., Brachey, A., Reed, G. W., Peters, J. C., and Hill, J. O., Fat and carbohydrate overfeeding in humans: different effects on energy storage, *Am. J. Clin. Nutr.,* 62, 19, 1995.

80. Klem, M. L., Wing, R. R., Mcguire, M. T., Seagle, H. M., and Hill, J. O., A descriptive study of individuals successful at long-term maintenance of substantial weight loss, *Am. J. Clin. Nutr.,* 66, 239, 1997.

12 Carbohydrate and Fat Metabolism, Appetite, and Feeding Behavior in Humans

R. James Stubbs and Leona M. O'Reilly

CONTENTS

0-8493-2752-0/00/$0.00+$.50
© 2000 by CRC Press LLC

1   INTRODUCTION

1.1   OVERVIEW

This review critically discusses the evidence (and lack of it) for the role of metabolic processing (metabolism and stores) of fat and carbohydrate (CHO) as components of postingestive satiety which can affect subsequent energy intake and nutrient intake. With this purpose in mind the review will consider (1) the nutrient balance concept and the constraints under which macronutrient balance is physiologically regulated and the implications this may have for the control of body weight and (2) macronutrient balance/oxidation as determinants of energy intake and macronutrient selection relative to other influences on feeding. The chapter will then discuss (3) the evidence that postabsorptive nutrient utilization affects feeding behavior in relation to fat and CHO, and (4) the importance of considering the combined effects of macronutrients and their metabolism/stores on appetite and nutrient intake in relation to integrative models for the physiological control of feeding. (5) Finally, the (sparse) evidence for the effects of nutrient metabolism/stores on nutrient selection will briefly be discussed. The work primarily focuses on human studies, drawing on animal models or referring to other chapters were appropriate. Although protein does not fall within the scope of this discussion, the profound effects that its ingestion, metabolism, and storage are likely to exert on energy and nutrient intake should not be forgotten.[1]

1.2   LIMITATIONS TO HUMAN EXPERIMENTATION

Examination of the relationship between macronutrient metabolism and nutrient intake in humans necessarily entails indirect, largely noninvasive, experiments. Much of the evidence for or against specific hypotheses is fragmentary or controversial. Experimental dissections of the cause–effect relationships between macronutrient metabolism and feeding behavior are inevitably contaminated by a number of extraneous variables.[2] These include other physiological signals arising from the sequence of nutrient digestion and processing. The idea that some aspect of postabsorptive nutrient processing or macronutrient storage compartment plays a key role in determining feeding behavior is not new.[3-6] Difficulties in direct experimental tests of hypotheses have led to over half a century of debate concerning various models of appetite control based on macronutrient metabolism. With the increase in investigative capabilities at the cellular and molecular level an increasing number of such molecular signals have been identified. It has become particularly important to evaluate the quantitative importance of these putative mechanisms in the feeding behavior of the whole organism. Only then can their contribution to appetite control be fully evaluated. In order to understand the likely influence of macronutrient oxidation/balance on feeding behavior the physiological constraints under which nutrient balance is regulated should be appreciated.

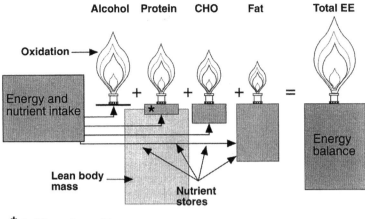

FIGURE 1 A schematic diagram illustrating the constraints under which nutrient balance is regulated. The energy ingested, stored, and expended in the oxidation of the metabolic fuels, alcohol, protein, CHO, and fat, summate to determine energy intake, storage, and expenditure. There is a constraint on the ability to up- or down-regulate resting energy expenditure. The net interconvertibility of the macronutrients is limited under Western diet conditions. There is a hierarchy in the capacity to store the macronutrients. These constraints determine a hierarchy in the immediacy with which recently ingested macronutrients are oxidized or stored. (From Stubbs, R. J., *Macronutrients, Appetite and Energy Balance in Humans,* British Nutrition Foundation, London, 1994, 64. With permission.)

2   THE NUTRIENT BALANCE CONCEPT

Macronutrients in foods summate to determine total energy intake. Energy expenditure can be divided into the energy expended in the oxidation of each of the macronutrients. Body energy stores also comprise proteins (structural, functional, and labile amino acid pool), glycogen (stored in liver and muscles), and fat (adipose tissue). Figure 1 illustrates the physiological and thermodynamic constraints under which macronutrient balance is regulated.

The profile of metabolic fuels being utilized by the body changes with the composition of the diet and the level of energy intake.[7-11] There is a hierarchy in the immediacy with which the stores of the macronutrients (alcohol, protein, CHO, and fat) are regulated by increases in their own oxidative disposal.[12] This hierarchy is determined by the following constraints:

1. Total energy expenditure is largely determined by metabolic body size and level of physical activity (see Reference 12 for a review). There appears to be very little scope to up- or downregulate background metabolic turnover in relation to changes in energy intake. Thus, there is a ceiling on total energy expenditure for a given level of physical activity, which limits the rate at which recently ingested energy can be disposed of by oxidation. Energy ingested in excess of total energy expenditure is therefore stored. Within these thermodynamic constraints, all nutrients are not equal, when considering the tendency to oxidize or store them.
2. For a subject in approximate energy balance, or the fed state, there is a hierarchy in the extent to which the macronutrients can be stored and this has implications for their metabolic fate, once ingested. Alcohol is a toxic drug which cannot be stored. The storage capacity for protein and CHO is limited and converting these nutrients to a more readily stored form is energetically expensive. The storage capacity for fat is potentially very large.

3. It appears that under Western dietary conditions at least, the *net* interconvertibility of protein, CHO, and fat is fairly limited, although this is clearly not the case under extreme conditions such as massive CHO overfeeding.[8,13]

These three constraints and the subject's energy balance status determine the metabolic fate of protein, CHO, and fat once they are ingested. A positive balance of protein will lead to rapid metabolism of a high proportion of the amount ingested, the percentage depending on the body's requirements for specific amino acids. Similarly, as CHO intake increases, more of it is disposed of by oxidation. Because there is a "ceiling" on adaptive changes in energy expenditure, an increased oxidation of protein and CHO will lead to a decreased oxidation of fat. Conversely, fat intake (long-chain triglyceride) does not promote fat oxidation and leads to fat storage.[10,11] Indeed, the contribution of fat oxidation to total energy expenditure appears to be primarily determined by the amount of protein and CHO being oxidized. Alcohol cannot be stored and is disposed of by obligatory oxidation, ultimately at the expense of fat oxidation, so this too promotes fat storage.[11] There is therefore a specific order in which recently ingested macronutrients are metabolized, which is inversely related to the body's capacity to store them (alcohol < protein < CHO < long-chain triglycerides).[12]

How are the physiological mechanisms involved in the regulation of nutrient balance likely to affect body weight? If the total level of energy expenditure is not greatly influenced by the nutrient composition of the diet (except under extreme conditions of overfeeding), then any diet-induced changes in nutrient balance can only significantly influence energy balance if fat, protein, CHO, and alcohol have different effects on appetite control, or different satiating efficiencies (i.e., the extent to which each megajoule of net positive nutrient balance suppresses subsequent energy intake compared with other macronutrients). What are the theoretical reasons to suppose that macronutrient metabolism or storage exerts feedback effects on energy and nutrient intake?

3   MACRONUTRIENT BALANCE/OXIDATION AS DETERMINANTS OF ENERGY INTAKE AND MACRONUTRIENT SELECTION RELATIVE TO OTHER INFLUENCES ON FEEDING

3.1   WHY SHOULD MACRONUTRIENT METABOLISM ACT AS A POTENT SATIETY SIGNAL?

The reason perhaps most frequently cited for supposing that macronutrient metabolism or stores exert important feedback effects on energy and nutrients intake is that a sufficient environmental supply of macronutrients is required for normal physiological functions. Any decrease or cessation in the supply of macronutrients should therefore invoke physiological signals promoting the intake of these nutrients — the strength of the signal presumably being proportional to the physiological requirement. However, the requirements for proteins, CHOs, and lipids in adult humans are largely met by all but except perhaps the most extreme diets. Part of human adaptation to variations in the environmental supply of fats and CHOs (and proteins) consists of a huge degree of physiological plasticity in meeting our requirements for specific metabolic fuels.[14] This does not mean that altered physiological levels or utilization of fat and CHO do not act as cues which drive feeding behavior toward altered energy and nutrient intakes. However, in well-fed Western subjects, these cues may not be heavily deterministic. This view is supported by the following arguments:

1. Over half of the adult population of several Western countries are collectively overweight and obese, suggesting poor physiological defense against increments in energy balance.[15]
2. Models of appetite control based on feedback from physiological processing of individual nutrients do not appear to predict a large proportion of the variance in human feeding behavior.[12]

3. Certain conditions such as transient periods of intense physical activity or the use of covertly manipulated, unfamiliar foods appear to result in a very poor coupling of energy intake with energy expenditure.[16]

4. Although a large number of studies of human feeding show varying degrees of compensation for a number of experimental interventions, few have recorded clear evidence of energy balance regulation.

There are a number of reasons against models that claim peripheral events related to nutrient metabolism/storage exert tight physiological control over appetite and body weight regulation. Sociocultural and psychological factors appear to exert large influences on behavior in Western humans.[17] Gut-borne signals have an important role to play in meal-to-meal feeding behavior.[18] Nutrient ingestion and absorption stimulates the release of numerous primary and secondary messengers which may not relate to nutrient oxidation per se.[19] Changes in nutrient oxidation and metabolism may constitute an integral component of the physiology of appetite control, but are unlikely to be the primary determining factor. Thus, while CHO ingestion and oxidation appear to correlate well with subsequent satiety, it is still possible to overeat and gain weight on high CHO diets.[20]

3.2 FROM SINGLE NUTRIENT FEEDBACK MODELS TO INTEGRATIVE MODELS OF FEEDING

Earlier models viewed macronutrient metabolism or stores as exerting direct deterministic feedback on energy and nutrients intake as one or more single-feedback loops (see, for example, References 3 through 6). More recently, research has focused on the multiple signaling systems involved in the maintenance of nutrient and energy balance. Thus, the concept of nutrient metabolism affecting intake through some direct feedback leverage on intake has been replaced by hierarchical feedback models in which macronutrients can influence satiety at nearly every anatomical level of nutrient ingestion, digestion, processing, storage, and metabolism.[1] Figure 2 illustrates the primary ways in which food and nutrient ingestion influences the main compartments of the human appetite system. Macronutrient metabolism is embedded at a level of organization in which many putative satiety signals precede and follow it. The signaling systems involved in appetite control are redundant and complex, acting as cues for, rather than being determinants of, behavior. Many satiety signals have other primary functions involved in nutrient processing. The potency of these cues is likely to be affected by a number of exogenous and endogenous factors, including the energy balance status of the subject.

3.3 METABOLIC FEEDBACK FROM NUTRIENT BALANCE/OXIDATION WILL BE INFLUENCED BY ENERGY BALANCE STATUS

It is highly likely that normal feeding under modern conditions of dietary superabundance is controlled by learned behavior in response to a range of influences, which include physiological processes associated with ingestion, absorption, storage, and metabolism of nutrients. Other influences include salient social and cultural variables. It can be appreciated that as energy imbalances accrue, the influence of additional physiological signals (e.g., depletion of fat mass or lean tissue mass) on feeding behavior will escalate. It therefore appears that there is a margin of energy balance (that typifies the majority of Western adults) within which a large number of factors influence appetite and energy intake. Within this zone changes in energy and nutrient balance do not induce large detectable sensations that lead to corrective changes in feeding behavior. Outside of this range (which may vary among different individuals) physiological signals may exert more potent restorative influences on appetite and energy balance. These influences appear to be greater in response to energy deficits and so appetite control in humans appears skewed toward a positive energy

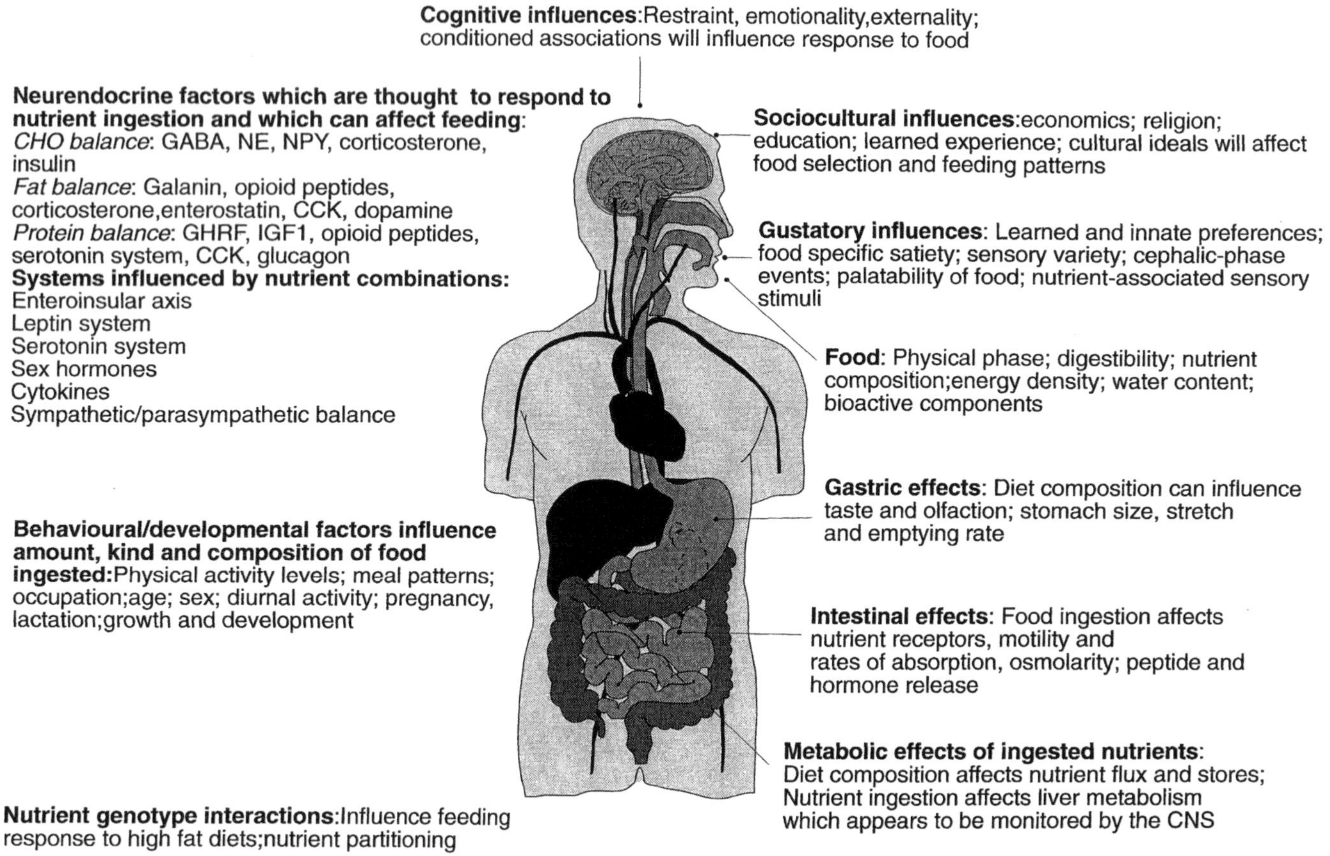
Cognitive influences:Restraint, emotionality,externality; conditioned associations will influence response to food
Neurendocrine factors which are thought to respond to nutrient ingestion and which can affect feeding:
CHO balance: GABA, NE, NPY, corticosterone, insulin
Fat balance: Galanin, opioid peptides, corticosterone,enterostatin, CCK, dopamine
Protein balance: GHRF, IGF1, opioid peptides, serotonin system, CCK, glucagon
Systems influenced by nutrient combinations:
Enteroinsular axis
Leptin system
Serotonin system
Sex hormones
Cytokines
Sympathetic/parasympathetic balance
Sociocultural influences:economics; religion; education; learned experience; cultural ideals will affect food selection and feeding patterns
Gustatory influences: Learned and innate preferences; food specific satiety; sensory variety; cephalic-phase events; palatability of food; nutrient-associated sensory stimuli
Food: Physical phase; digestibility; nutrient composition;energy density; water content; bioactive components
Gastric effects: Diet composition can influence taste and olfaction; stomach size, stretch and emptying rate
Behavioural/developmental factors influence amount, kind and composition of food ingested:Physical activity levels; meal patterns; occupation;age; sex; diurnal activity; pregnancy, lactation;growth and development
Intestinal effects: Food ingestion affects nutrient receptors, motility and rates of absorption, osmolarity; peptide and hormone release
Metabolic effects of ingested nutrients: Diet composition affects nutrient flux and stores; Nutrient ingestion affects liver metabolism which appears to be monitored by the CNS
Nutrient genotype interactions:Influence feeding response to high fat diets;nutrient partitioning

FIGURE 2 The primary ways in which food and nutrient ingestion influences the main compartments of the human appetite system. Food influences this system through multiple feedbacks at multiple levels which can be traced through the processes of food location, ingestion, digestion, absorption, and metabolism. Satiety is therefore maintained by a functional sequence or cascade of sequential physiological events which reinforce each other. Removing parts of the effects of a food or nutrient on this sequence will therefore diminish its impact on satiety. (Reprinted from Blundell, J.E. and Stubbs, R.J., *Handbook of Obesity,* Bray, G.A., Bouchard, C., and James, W.P.T., Eds., Marcel Dekker, New York, 1997, Chap. 13. With permission.)

balance and defense against negative energy balances, since positive energy balances appear more comfortably tolerated than energy deficits.[21] Physiological signals may affect feeding behavior either directly or through perceptible changes in functional integrity (e.g., feeling weak and tired during energy restriction), which can act as a learning cues for feeding.

These considerations raise a number of important questions, which are not necessarily easy to answer:

1. What is the evidence that macronutrient metabolism plays *any* role in appetite control of normal human subjects?
2. How potent are signals relating to postabsorptive metabolism in affecting energy and nutrients intake?
3. Are some macronutrients more important than others in their metabolic contribution to satiety?
4. Is there any evidence that nutrient metabolism directionally affects nutrient intake?

4 EVIDENCE THAT POSTABSORPTIVE PERIPHERAL NUTRIENT UTILIZATION AFFECTS FEEDING BEHAVIOR

Several sources of evidence suggest that nutrient oxidation is an important component of satiety (see below). Individually, each piece of evidence is relatively weak. However, taken together there is a reasonable body of data which suggests that increases in nutrient oxidation, subsequent to nutrient ingestion, correlate with and are probably causatively involved in satiety.

4.1 Carbohydrate Metabolism and Stores: Their Effect on Subsequent Intake

4.1.1 Feeding Studies

In assessing role of macronutrient metabolism/stores in appetite control, it is pertinent to examine the impact that macronutrients exert on feeding behavior (1) under naturalistic conditions where fats contribute disproportionately to dietary energy density (ED), (2) in the laboratory where differences in the energy density of different macronutrients can be controlled. Under naturalistic conditions, there appears to be a hierarchy in the satiating efficiency of the dietary macronutrient such that per megajoule of energy ingested, protein ingestion suppresses intake to a greater extent than CHO, which suppresses subsequent energy intake to a greater extent than fat intake. Protein appears to produce hypercaloric compensation, CHO induces approximately caloric compensation and fat produces hypocaloric compensation.[22-24] These effects are extant in the laboratory.[1,25,26]

When the energy density of fat- and CHO-rich, systematically manipulated diets is equalized, they appear to have similar effects on intake over periods of 2 weeks.[27,28] When the short-term effects of dietary macronutrients are compared at the same level of energy density, protein-rich foods are contrasted to fat- and CHO-rich foods by their high satiety value. Under these conditions CHO appears to exert a more acute effect on satiety than fat,[29] and three other studies have found this relatively subtle effect to be independent of energy density.[30-32] In one of these studies the satiating effect of fat was delayed rather than diminished per se, relative to CHO, so that average 24-h values were the same.[31] These effects are consistent with the fact that CHOs are more rapidly absorbed into the portal circulation and metabolized than fats which are absorbed more slowly through the lymphatics. The more rapid effect of CHOs on satiety (within the first hour after meal initiation) may play a role in meal termination.

4.1.2   Relationships between Inferred Carbohydrate (CHO) Metabolism and Appetite

Some studies have reported that CHO oxidation, estimated by indirect calorimetry, correlates with subjectively expressed satiety.[33,34] Raben et al.[33] examined the relationship between net CHO oxidation and subjectively expressed satiety across three different studies and found a close relationship between the two ($r = 0.86$, $p < 0.05$).[33] We have also found a significant inverse correlation between CHO oxidation and hunger in the intermeal interval when six subjects fed *ad libitum* on low-, medium-, and high-fat diets for 7 days in a calorimeter chamber (Figure 3).[34]

Ostensibly, these correlations suggest that glucose oxidation is an important process underlying metabolic satiety signals.[35-37] However, CHO is usually the main substrate being oxidized in the intermeal interval. Other signals such as gastric emptying and nutrient absorption from the gut may also be involved and a conclusion that CHO oxidation was the main signal underlying satiety in the intermeal period would be premature. Some studies have shown a correlation between arteriovenous differences in plasma glucose concentration (>15 mg/dl) and subjectively expressed satiety and a correlation between minimal areteriovenous differences in glucose concentration and subjective hunger.[35] While arteriovenous differences in plasma glucose are indicative of tissue uptake of glucose, they are not a direct measure of CHO oxidation.[3] Other studies have failed to demonstrate this relationship.[38,39]

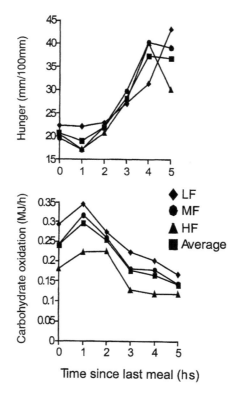

FIGURE 3   The relationship between CHO oxidation after a meal and subjective hunger in six men feeding *ad libitum* on low-, medium-, and high-fat diets while continually resident in a whole-body indirect calorimeter for 7 days. Each plot represents the average values for all subjects and all meals per dietary treatment. There is a clear inverse relationship between CHO oxidation and hunger, which was highly significant ($p < 0.001$). However, a number of other physiological processes also occur during this time period, and the correlation may not necessarily amount to causation.

The glucostatic theory has also been supported by positive correlations between postprandial satiety and changes in plasma glucose. In their analysis of three studies using test meals, Raben et al.[33] demonstrated negative correlations between hunger and CHO oxidation, AUC for glucose, insulin, GIP, GLP-1, noradrenaline, and adrenalin. The authors concluded that the amount of food consumed, its energy density, the CHO content, and a number of physiological responses involved in the net uptake and utilization of CHOs were involved in the changes of postprandial hunger and satiety sensations after a meal. However, they also noted that "due to the covariation between the single variables, it is not possible … to distinguish between the different factors involved."[33] A number of authors have found results which suggest that CHO-induced increases of satiating hormones (insulin, norepinephrine, GIP, GLP-1) may also play an important role in satiety.[40-42] However, the different mechanisms mentioned above make it difficult to separate these from each other.

4.1.3   Blood Glucose Dynamics and Satiety

Studies in rats using continuous online monitoring of blood glucose have shown that meal initiation is preceded by a decrease in blood glucose concentration.[43-46] Reports that a microinfusion of glucose that can delay meal initiation for up to 3 h strengthens the evidence that peripheral CHO status may affect feeding in rodents.[45] More recently, Campfield et al.[46] have studied 18 human subjects in approximate time isolation, while continuously monitoring blood glucose in relation to subjective hunger and meal requests. The authors reported that in 83% of subjects both changes in hunger ratings and spoken meal requests were preceded by, and significantly correlated with, transient declines in plasma glucose, as previously seen in rats.[46] In addition, infusion of insulin, which produced a decline in blood glucose of a similar magnitude and duration to those occurring spontaneously in the 18 men, led to an elevated desire to eat. Other work has not provided evidence in favor of this concept.[47] The correlation between decline in plasma glucose and meal initiation implies but does not demonstrate causation. If the relationship is causal, what mechanism(s) are in operation?

1. Plasma glucose could be directly affecting hypothalamic or other glucoreceptors.[42]
2. Transient declines in plasma glucose could have been caused by a cephalic phase insulin release in relation to the anticipation of food.[48] This in itself may be affected by temporal cues out of the experimenter's control. Gastric emptying could act as such a temporal cue.
3. The transient decline in plasma glucose could reflect a decreased rate of glucose absorption from the gut, as the gut empties of nutrients.
4. Conversely, a change from mobilization of exogenous (gut) to endogenous (glycogen) CHO stores could be involved.
5. The falloff in nutrient absorption from the small intestine could decrease the impact of small intestinal glucoreceptors on satiety.[49]
6. The falloff in plasma glucose could reflect a decreased rate of peripheral glucose utilization, which itself would elevate hunger.

Some, all, or none of these mechanisms could be involved. Herein lies a major dilemma to the investigator — the more specific and mechanistic the investigation, the less generalizable the results.

4.1.4   Carbohydrate Stores and Subsequent Intake

Recently, the notion that glycogen stores exert negative feedback on subsequent energy intake[5] has received renewed attention as a possible explanation for the influence of high-fat diets on energy intake and body weight.[10] This model was initially supported by experiments in *ad libitum* feeding mice which have shown that there was a negative correlation between changes in CHO stores and the subsequent day's *ad libitum* energy intake ($r = -0.2$; $p < 0.025$). The association between fat balance on one day and energy balance on the subsequent day was positive, stronger, and more significant, ($r = 0.35$; $p < 0.001$). There was no apparent feedback from energy balance on one day and energy intake on the subsequent day. This evidence appeared to be supported by studies showing that high-fat, energy-dense diets promote higher levels of energy intake than lower-fat, less-energy-dense diets in rodents and humans (see Reference 34 for a review). Is there evidence that glucostatic or glycogenostatic mechanisms exert a large enough influence on human feeding behavior to be quantitatively important in the laboratory and in real life?

4.1.4.1   Do people eat to maintain a constant range of CHO stores?

Cross-sectional and epidemiological studies suggest that people who consume high-fat diets appear to be at greater risk of having a higher BMI.[50,51] Laboratory studies have repeatedly demonstrated that increasing the energy density of the diet by covertly adding fat has led to increasing energy intake but not to increasing the amount of food ingested.[34] Conversely, decreasing dietary energy density by removing the dietary fat has led to lower levels of energy intake. These data are consistent with the predictions of CHO-based models of feeding but do not test those predictions directly. This is because CHO balance was not measured in any of those studies.

One study used indirect calorimetry to monitor nutrient balance (and nutrient oxidation) continually as predictors of subsequent feeding behavior while subjects were given access to low-, medium-, and high-fat diets (energy density increased with percent fat). Six men were each studied three times in experiments conducted over a week per treatment (randomized, within-subject design). During this time subjects were continually resident in a whole-body indirect calorimeter.[52] As the fat content of the diet increased, the total weight of food eaten stayed the same, so energy intake increased, leading to a progressively positive energy balance. The relationship between the changes in actual, cumulative nutrient balance over time and the subsequent day's energy intake was examined to assess possible feedback from nutrient balance on energy intake and so assess whether CHO stores exerted negative feedback on energy intake. The previous day's cumulative balance of CHO and protein was negatively related to the subsequent day's energy intake (and balance), but there was no apparent suppression of intake in relation to the previous day's cumulative fat balance. The effect of a positive protein balance was greater than that for CHO storage. The effect for CHO alone accounted for only 5.5% of the variance in subsequent energy intake. The model relating to all three macronutrients accounted for 27.8% of the variance in subsequent energy intake. The oxidation of all of the macronutrients predicted a reduction in the subsequent day's energy balance, but this effect was again hierarchical, with protein oxidation exerting a stronger predicted fall in intake than CHO which, in turn, exerted a marginally stronger effect than fat.[52] Thus, the previous day's cumulative stores *and* the oxidation of protein and CHO were negatively related to subsequent energy intake. Importantly, while fat oxidation seemed to predict, to a modest extent, the subsequent day's energy intake, alterations in fat stores did not show any effect on the subsequent day's energy intake. These data suggest (1) that at the level of nutrient balance, macronutrients appear to exert hierarchical effects in their potentially suppressive effects on subsequent energy intake and (2) that it is the component of obligatory oxidative disposal which underlies the potentially suppressive effects that protein and CHO, but not fat balance, exert on subsequent energy intake.

4.1.4.2   Is excess energy intake on high-fat diets due to a drive to maintain CHO balance?

CHO-based models of feeding predict that excess energy intakes occur on high-fat diets, not because they are energy dense, but because the negative feedback from CHO oxidation or stores is diminished and subjects are actively increasing energy intake to optimize CHO status. Two studies have examined the effects of allowing women[27] and men[28] to feed *ad libitum* on high-fat and low-fat isoenergetically dense diets, for 2 weeks per treatment. Despite large differences in CHO intake (and, hence, CHO status) there were no differences in energy intake. These data cast doubt on the notion that increases in CHO oxidation and/or storage per se are the major factors which exert powerful, unconditioned, negative feedback on food or energy intake.

4.1.4.3   The effect of manipulating whole-body CHO status on subsequent energy intake

CHO-based models of feeding also predict that manipulation of CHO status (oxidation or stores) will reciprocally influence energy intake. Three studies have examined this issue in detail. In each of these studies CHO balance (but not energy balance) was depleted using diet over 1,[53] 2,[54] or 3 days, using diet and exercise.[55] The impact of these different degrees of CHO (but not energy) depletion on intake over the day(s) subsequent to the depletion protocols was assessed. These manipulations did not affect the subsequent day's *ad libitum* energy intake, compared with the control in any of the three studies. Instead, CHO balance was reestablished by directing more dietary CHO toward storage and maintaining high rates of fat oxidation throughout the *ad libitum* day. Thus, plasticity in fuel utilization, not appetite, was the primary mechanism for reestablishing nutrient balance. Taken together, this series of studies does suggest that in the short- to medium-term dietary-induced changes in CHO stores per se do not exert powerful unconditioned negative feedback on energy intake.

4.1.4.4   Is it difficult to consume excess energy intakes on high-carbohydrate diets?

CHO-based models of feeding predict that excess energy intakes on high CHO diets are unlikely to occur because of the strength of negative feedback arising from CHO ingestion. We tested this prediction in six normal-weight men who were each studied twice during 14 days.[20] During this time they had *ad libitum* access to one of two covertly manipulated diets of fixed composition. The fat, CHO, and protein in each diet, expressed in terms of energy, were in the proportions of 22:65:13. The energy density of the diets was 348 and 617 kJ/100 g. *Ad libitum* energy intakes were 8.56 and 14.56 MJ/day, respectively. Intake was not influenced by perceived pleasantness of the diets. Subjects were able to detect this difference in the energy density of high-CHO diets and felt significantly more hungry on the low-energy density diet. This was not apparent in a previous study where the energy density of the diet was altered using fat.[52] Thus, excess energy intakes are possible on high-carbohydrate, high-energy diets in *ad libitum* feeding subjects, where conditions preclude diet selection and hunger rose above normal levels when CHO depletion and a negative energy balance coincided.

4.1.5   Glycemic Index and Satiety

The glycemic index of CHO-rich foods indicates the rapidity with which CHOs are digested and absorbed. This may influence appetite and energy intake. However, indications from the literature are contradictory. Foods with low glycemic indexes (GI) have been found to be more satiating than foods with high indexes,[56,57] or to promote higher intake than low-GI foods.[58] These differences probably relate to differences in the time window of measurement, the foods eaten, and perhaps the physiological status of the subjects. Foods with a higher GI may have a more acute but transient effect on satiety than those with lower GIs. In addition, the effect of lower-GI foods may also depend on other materials present in the small intestine, with which they may interact.

Different starches have different physiological effects which relate to the type, bioavailability, amylose/amylopectin content, as well as preparation method of the starch. Bioavailability is greatly influenced by the physical state of the starch. Cooking greatly improves the digestibility of poorly digestible starches due to increased susceptibility to the amylase enzymes.[59] For example, Raben et al.[60] have found that pregelatinized potato starch increased plasma concentrations of glucose, insulin, and gastrointestinal hormones, CHO oxidation, and satiety relative to native (resistant) potato starch.[60] However, when added to a normal diet over a week, the effects from the acute

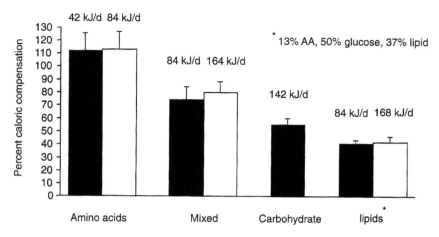

FIGURE 4   Data from rodent studies in which amino acids, glucose, intralipid, or a mixture of these nutrients was parenterally infused in rodents.[62] These data demonstrate that when the mouth and gastrointestinal tract are bypassed by parenteral infusion that there is still a hierarchy in their satiating efficiencies. The rank order is protein or amino acids (greatest), CHO (glucose), and lipid (weakest). This in turn suggests that the hierarchy in the satiating efficiency of administered macronutrients has a strong postabsorptive or metabolic component. (From Walls, E. K. and Koopmans, H. S., *Am. J. Physiol.,* 262, R225, 1992. With permission.)

substitution studies could not be reproduced.[61] The acute suppression of appetite seen in short-term studies may be counterbalanced by a more-delayed elevation of satiety arising from less bioavailable starches, over longer periods. Different types of starch also exert other effects on appetite, through mechanisms independent of their effects on CHO oxidation.[1]

4.1.6   Parenteral Infusion Studies

Parenteral infusions of amino acids, glucose, or lipid in rats lead to compensatory decreases in oral energy intake. However, the degree of caloric compensation was only complete when amino acids were infused, whereas around 70% compensation occurred with glucose and less than 50% when fat was infused (Figure 4).[62] Similar effects have been found in one study in humans.[26] These data demonstrate that when the mouth and gastrointestinal tract are bypassed by parenteral infusion, there is still a hierarchy in the satiating efficiency of the macronutrients, the rank order being protein or amino acids (greatest), CHO (glucose), and lipid (weakest). This in turn suggests that the hierarchy in the satiating efficiency of administered macronutrients has a strong postabsorptive or metabolic component. Langhans and Scharrer[42] note that in general the effect of glucose infusions is greater than that of fat in suppressing oral intake in nonhuman monogastric species. However, glucose is not always effective in this regard. This may relate to the timing and rate of nutrient delivery in relation to oral intake is since the greatest suppression of oral intake "at a time when concomitant oral food intake stimulates other satiety signals, which interact with and strengthen the glucose effect."[42] This is an important point because parenteral infusions of glucose and/or fats are often used as a means of inducing a positive energy balance during nutritional rehabilitation of some subjects whose oral intakes are precluded or decreased by some disease states.[63] Further-more, parenteral glucose infusions have been found to be less effective at suppressing oral intake than glucose plus insulin, which promotes uptake and utilization of the infused glucose.[64]

4.1.7   Use of Metabolic Inhibitors

If relatively large changes in the pattern of oxidation of metabolic fuels are important in influencing feeding behavior, then one might expect that the inhibition of nutrient oxidation should stimulate

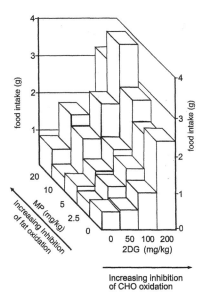

FIGURE 5   Data from studies in rodents[65] have shown that pharmacological inhibition of glucose oxidation (using 2-DG) increases food intake in a dose-dependent manner. Similarly, pharmacological inhibition of beta oxidation (using methyl palmoxyrate, which prevents transport of fatty acid derivates into the mitochondrion) also increases food intake in a dose-dependent manner. However, use of both inhibitors tends to have an additive effect. At maximal doses the fat and CHO oxidation inhibitors lead to a massive increase in food intake. Since endogenous fuel supplies are critically diminished, the animal attempts to obtain those fuels from the environment by eating more. (From Friedman, M. I. and Tordoff, M. G., *Am. J. Physiol.,* 251 (5 pt 2): R840–R845, 1986. With permission.)

feeding. In a remarkable study, rats were given graded doses of 2-deoxy-glucose (2DG) which inhibits glucose oxidation by competitive inhibition of phosphohexoisomerase activity. There was a dose-dependent increase in food intake as glucose oxidation was inhibited — a glucoprivic feeding response. 2DG has also been found to increase hunger and food intake in humans when given at doses of 50 mg/kg body weight[36,37] (Figure 5). There is also anecdotal evidence that 2,4-dinitrophenol has been administered to human subjects in an attempt to develop obesity treatments by uncoupling oxidation from phosphorylation. 2,4-Dinitrophenol does this by collapsing proton gradients across the inner mitochondrial membrane and so decreases the availability of ATP for cellular metabolism. Apparently, a net effect of these treatments was to elevate appetite in human subjects (Livesey, personal communication). If these reports are accurate and the compound is not affecting appetite through other mechanisms, then this is potentially powerful evidence to suggest that the metabolism of macronutrients in the provision of reducing equivalents is an important factor underlying satiety in humans. This effect has been demonstrated in a number of studies using a number of metabolic inhibitors in rodent models.[42,65,66]

4.2   FAT METABOLISM AND STORES: THEIR EFFECT ON SUBSEQUENT INTAKE

4.2.1   Feeding Studies

As discussed above, the combined data relating to the physiological regulation of nutrient balance and to the effect of macronutrients on feeding behavior suggest that dietary fats are less quickly metabolized than CHOs and exert a correspondingly weaker action in suppressing subsequent energy intake. Some feeding studies have attempted to use more readily oxidized forms of lipid in an attempt to assess whether they differentially affect appetite and energy balance. A short-term[67] and longer-term study[68] of the effects of MCTs on food and energy intake both suggested that

MCTs are more satiating than the less readily metabolized LCTs. In the longer of these studies, energy intakes were suppressed, by 9% on a high-fat diet that was high in MCT relative to high-fat diets with medium and low levels of MCT.[68] This study supported the notion that a readily oxidized form of fat (MCT) was more effective at suppressing energy intake than the less readily oxidized LCT. It should be noted that fat oxidation was not actually measured in these studies and its role in suppressing energy intake on high MCTs diets remains inferential.

4.2.2   The Relationship between Inferred Fat Metabolism and Subsequent Intake in Humans

The short-term studies that have shown a correlation between CHO oxidation and satiety in the intermeal interval have tended to suggest that fat oxidation bears either a positive relationship ($r = 0.85$; $p < 0.005$)[33] or no relationship[34] to satiety. However, in these short-term assessments, where substrate oxidation is inferred from respiratory exchange data, fat oxidation bears a reciprocal relationship to CHO oxidation.

It is frequently inferred or suggested that the poor response of fat oxidation to ingestion is a factor favoring weight gain when people consume high-fat diets. This implies there is poor feedback from fat oxidation which would limit weight gain either through acute elevations of metabolic turnover (which are not apparent) or through suppression of appetite. A necessary condition that needs to be met for this to be true is that increased levels of fat oxidation should increase satiety, limit energy intake, and thus prevent subsequent weight gain. There is relatively little evidence of this because in a positive energy balance fat tends to be stored and in a negative energy balance fat tends to be mobilized. The tendency to oxidize fat is therefore a secondary response to the balance status of protein and CHO and energy. Thus, higher rates of fat oxidation on high-fat diets are usually because those diets are low in CHO.

It has been suggested that elevated levels of fat storage and increased rates of CHO oxidation in the postobese constitute a defect in fat oxidation which predisposes subjects to subsequent weight gain. Restrained eaters and the postobese exhibit "impaired" fat oxidation under standardized conditions.[69,70] This reduction in fat oxidation may reflect the preferential storage of fat subsequent to periods of prior energy deficit that characterize restraint and weight loss rather than an inherent tendency to store preferentially rather than oxidize fat. These examples illustrate further difficulties in unraveling cause and effect in the relationship between physiology and behavior.

A constitutional tendency to store preferentially rather than oxidize fat has been inferred in the Pima Indians in Arizona.[71] In over 100 Pima Indians studied over 3 years, high respiratory quotient (RQ) values were predictive of weight gain relative to those with low RQ values. Similarly, the Baltimore longitudinal study[72] suggested that high RQ is predictive of subsequent weight gain. Bouchard et al.[73] have shown that RQ shows a reasonably high heritability, under controlled dietary conditions. However, given that it takes a number of days for RQ to track food quotient (FQ), these relationships between short-term measurements showing than high RQ or high tendency to store fat predicts subsequent weight gain may be reflecting the behaviors involved in the process of weight gain (see Reference 69). High RQ values in some twins and Western subjects may coincide with rather than predict the process of weight gain. Despite great care to control experimental conditions, many experimental claims of physiological differences in nutrient balance as predictors of weight gain may be contaminated by the very behavior they are supposed to predict. Studies in which different types of subject categories are maintained at, above, and below energy balance on high- and low-fat diets over periods far in excess of 1 to 2 days are urgently required to assess inherent tendency to store or oxidize fat, uncontaminated by prior dietary selection, or energy balance status.

Higher rates of fat oxidation are apparent (1) in obese subjects on a 24-h basis (part of this may reflect the usual diet rather than the circulating NEFA arising from adipose tissue[69]), (2) during energy restriction, as the body draws on energy reserves from adipose tissue, (3) in some cases of impaired glucose metabolism such as diabetes.[14] These are three situations more likely to result in subsequent weight gain than weight loss. Furthermore, the above discussion suggests that low rates of fat oxidation are also predictive of subsequent weight gain, in some cases. The inability to control entirely for the behavioral influences possibly underlying some of these physiological traits may have produced a paradox — namely, increases and decreases in fat oxidation can be predictive of subsequent weight gain. A further paradox is apparent. If high rates of CHO oxidation are believed to be an important factor underlying satiety, why do low rates of fat oxidation and high rates of CHO oxidation predict subsequent weight gain? Perhaps these differences reflect short-term glucostatic and longer-term lipostatic factors.

Thus, the literature relating fat oxidation to satiety and or to subsequent weight change appears at the present time to be somewhat confused. Part of the confusion may arise from measuring physiological traits in an experiment that may be more reflective of habitual behavior outside the experiment. It is reasonable to suppose that factors favoring net uptake and storage of fat will diminish the metabolic feedback from fats and favor higher levels of energy intake. The converse should also be true. High rates of fat oxidation should elevate satiety but this may be more difficult to demonstrate physiologically.

More impressive evidence that fat oxidation affects satiety comes from animal models. These are discussed in detail elsewhere in this book and so will only be referred to here. There is a large and growing body of evidence in nonhuman monogastric species (mainly rodents) which suggests that fatty acid oxidation in the liver of fed animals does constitute a component of satiety. There is evidence that pharmacological inhibition of hepatic fatty acid oxidation leads to increases in food intake.[42,65,66] In selection paradigms this increase of intake is not nutrient specific.[42] Friedman and Tordoff[65] have shown that in rats, pharmacological inhibition of beta oxidation can elevate intake in a dose-dependent manner.

The fact that fat appears at the bottom of an oxidative hierarchy in *ad libitum* feeding humans suggests that its contribution to satiety through metabolic feedback will be less immediate than for protein or CHO. These suggestions are consistent with the physiological regulation of nutrient balance on the one hand and the known effects of macronutrients on appetite and feeding behavior on the other. This raises other issues. Why should the enhanced fat oxidation during starvation not enhance satiety? How is the body able to detect differences in fat oxidation arising from earlier meals in a fed animal, as opposed to fat stores being mobilized during a progressive negative energy balance? The first question may be answered by considering the nutrient mixture being oxidized, which reflects fed vs. fasted state and its effects on appetite. A potential answer to the second question has been suggested by Le Magnen[74] who has shown in the rat, by ^{14}C labeling of substrates, that those metabolites eaten and stored at night (the rat's active period) are mobilized and oxidized during the subsequent day. He argues that the rat's feeding follows a diurnal periodicity characterized by high rates of ingestion, tissue uptake, and storage of CHOs and lipids in the dark period, and the opposite in the light period.[74]

The purported mechanism underlying this effect is a circadian rhythm in glucose tolerance. He argues that both basal and glucose stimulated insulin release are higher during the lipogenic nocturnal phase in rats, producing higher glucose tolerance.[75] This produces a greater rate of storage of substrates in the night when active and also accelerates the frequency of meal initiations. The reverse occurs in the day giving rise to a relative glucose intolerance, greater release of substrates, greater satiety, and low rates of meal initiation.[74] Humans also appear to exhibit high daytime glucose and food stimulated insulin release, relative to the night. They also show lower tissue sensitivity to exogenous insulin at night promoting release of substrates (see Reference 74 for a discussion). There may therefore be a liporegulatory mechanism "which modulates (both within the diurnal cycle, and between days)," the glucosensitive short-term control system of food intake.[74]

4.2.3   Changes in Plasma Fat Metabolites and Satiety

While correlations have been reported for changes in plasma glucose dynamics and subsequent intake in humans, the same is not necessarily true for fat. However, there are some indications that elevations in plasma fat metabolites that are rapidly oxidized may correlate with satiety.[42] Most of these indications come from animal studies and are discussed elsewhere in this book. Short-chain fatty acids are rapidly oxidized relative to longer-chain fats.[14] The administration of ketones have been implicated as anorectic agents, when administered to rats either orally[76] or intravenously.[77] In general the shorter the chain length of a fat derivative the greater the tendency to oxidize it. It remains to be determined whether the chain length of administered fatty acids correlates inversely with satiety.

Langhans and Scharrer[42] have also suggested that while fat oxidation may not decrease whole-body RQ subsequent to a meal, there may be a postprandial elevation of hepatic fat oxidation which contributes to satiety and additionally may exert an impact on satiety which is distinct from the effect of increased whole-body fat oxidation during a period of negative energy balance. Evidence derived from a number of sources suggest that fat and CHO oxidation interact to affect feeding, at least in relation to glucoprivic and lipoprivic feeding models.[42,65,66]

4.2.4   Changes in Fat Stores and Subsequent Nutrient Intake

It is almost axiomatic than elevations in fat mass do not exert profound feedback that leads to rapid suppression of subsequent intake and the maintenance of a stable body weight. At least this is the case for the majority of adults in the U.S. and U.K. who are either overweight and obese (e.g., Reference 15). However, decreases in fat mass, due to prior weight loss in humans appear to invoke subsequent, compensatory weight gain, presumably brought about through compensatory changes in energy and possibly nutrient intake.[21] Weight loss leads to increased lipoprotein lipase activity, which favors and increased tendency to store ingested fat.[78,79] According to some models this would favor higher rates of energy intake.[42,65] The initial rush of excitement in leptin as a simple circulating lipostatic factor that could readily be manipulated to enhance satiety signals arising from adipose tissue has largely evaporated as the full complexities of the leptin system begin to unravel.[80] Furthermore, weight loss does not constitute mere loss of fat tissue. Lean and adipose tissue tend to be lost in a 40:60% ratio for subjects who are normal weight or mildly overweight (corresponding to 25% lean body mass and 75% fat mass, respectively).[81] Loss of lean tissue compromises functional integrity which in itself would constitute a powerful cue driving energy intake up. There is indeed evidence that repletion of lean body mass subsequent to prolonged semistarvation may be a major factor involved in the cessation of the hyperphagia which attends rapid weight regain.[82] Interesting data could be gleaned from the large body of literature on cosmetic lipectomy. However, Kral[83] notes, "There are no long-term clinical follow-up studies of liposuction or lipectomy to resolve the controversy over re-growth or compensation after surgical removal of fat. There is no evidence that extensive lipectomy influences energy balance in humans. Anecdotally, patients having reduced weight by dieting with subsequent lipectomy are not spared from weight (re)gain."

4.2.5   Does Type of Fat Influence Nutrient Utilization, Appetite, and Energy Balance?

The majority of studies that have examined the effects of dietary fat on appetite and energy intake in humans do not discriminate between the types of dietary fat used, and these studies tend to use mixed fats in their dietary preparations. Fats can vary in structure in terms of (1) chain length, (2) degree of saturation, and (3) degree of esterification of the glycerol backbone.

There is fragmentary evidence that the chain length of fat may influence energy intake, short-chain fat metabolites may elevate satiety (see above). Isoenergetic substitution of MCT for LCT at high levels in high-fat energy-dense diets limits the excess energy intakes and weight gains produced by such diets in rodents[84] and humans.[68]

The effects of large doses of 1-monoglyceride on appetite and energy intake have recently been examined both within day[85] and between day.[86] Together these studies found that large isoenergetic doses of dietary monoglyceride or triglyceride did not exert differential effects on subjective motivation to eat, or nutrient oxidation, or food intake (FI). Thus, the degree of saturation[87] and chain length of orally ingested dietary fat[68] can have significant effects on appetite and energy intake, but orally ingested 1-monacyl glycerols did not differentially influence appetite and energy intake relative to triacylglycerols. It is possible that 2-monacyl glycerols may exert a greater appetite-suppressing effect since these molecules are more readily absorbed across the gut wall.[88] In summary, it appears that fats and fat derivatives which tend to be more readily absorbed and/or metabolized exert a greater suppressive effect on appetite and energy intake than less readily utilized fats. The magnitude of these effects under experimental conditions appears to be significant but modest (not exceeding 1 MJ/day differentials in energy balance). Their significance in the population at large is currently unclear.

4.2.6   Parenteral Infusion Studies

In can be inferred from two studies that parenteral infusions of amino acids appear to suppress oral intake more effectively than glucose, which is more effective than fat at suppressing oral intake in rodents[62] and humans.[26] Similar experiments carried out in rats found intraduodenal infusions inhibited FI while intravenous infusions did not.[89] These findings suggest that after the ingestion of fat, potent fat-induced satiety signals are generated by preabsorptive (orosensory, stomach, and small intestine receptors) rather than postabsorptive physiological responses. The postabsorptive action (nutrient absorption, substrate utilization, and oxidation) of fats is weaker than other macronutrients.[1]

5   THE IMPORTANCE OF CONSIDERING THE COMBINED EFFECTS OF MACRONUTRIENT BALANCE/OXIDATION ON APPETITE AND NUTRIENT INTAKE

A major development in the field of ingestive behavior in recent years has been the general acceptance that macronutrients exert multiple feedbacks on appetite and feeding behavior at the different levels of nutrient ingestion, digestion, absorption, assimilation, and metabolism. Furthermore, there is growing evidence that the major macronutrients protein, CHO, and fat simultaneously exert differential effects on satiety at these different levels of organization. A major challenge for the future is to develop integrative models which would account for the combined effects of changes in nutrient intake on physiological signaling and subsequent intake. This is true of nutrient oxidation and balance as for other aspects of physiological signaling in the control of food intake. Thus, the effects of a given nutrient needs to be considered, *inter alia*, in the context of (1) the presence or absence of other nutrients, (2) the levels of physiological organization it operates at, (3) the metabolic state and energy balance profile of the subject concerned. What then are the combined effects of nutrient oxidation on appetite and subsequent feeding likely to be? As regards diet composition, nutrient processing, and peripheral signals related to feeding, there is now sufficient evidence to link provisionally changes in diet composition to changes in the signaling systems concerned with the control of macronutrient balance. This notion is schematically illustrated in Figure 6:

1. Changes in diet composition can influence peripheral fuel selection[7-14]
2. Peripheral changes in fuel selection are determined by physiological and thermodynamic constraints which determine a hierarchy in the immediacy with which the balance of recently ingested macronutrients are autoregulated by increases in their own oxidative disposal.[7-14]

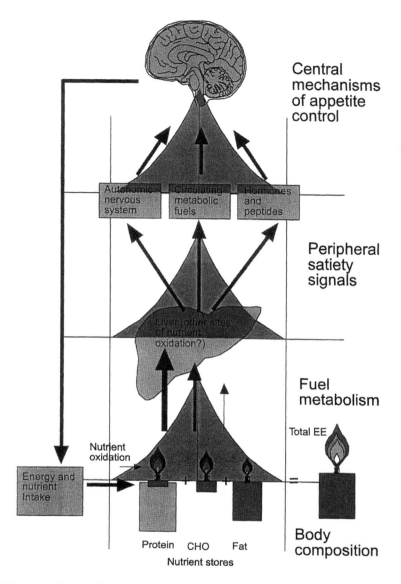

FIGURE 6   Schematic diagram illustrating putative connection between diet composition, fuel metabolism, peripheral satiety signals, and central control of feeding. Diet composition affects satiety. Protein is more satiating than CHO which is more satiating than fat. Diet composition also affects postingestive fuel metabolism, since increases in protein and carbohydrate, but not fat balance, are tightly modulated by autoregulatory increases in their oxidative disposal. It is known that nutrient oxidation in peripheral tissues appears to be associated with satiety. The hierarchical regulation of macronutrient balance by oxidative disposal may partially underlie the hierarchy in the satiating efficiency of the macronutrients. Indeed, high levels of protein and CHO (but not fat) oxidation are indicative of the fed state. High rates of fat oxidation are usually synonymous with energy deficits. It has also been suggested that changes in peripheral fuel metabolism may act as an additional trigger for peripheral satiety signals which are relayed to feeding centers of the brain believed to be concerned with the control of macronutrient balance.

3. This hierarchy appears to parallel a hierarchy in the satiating efficiency of the macronutrients.[12]

4. These two hierarchies may be causatively related since a growing body of literature suggests
   that nutrient oxidation in the periphery is monitored by the central nervous system (CNS)

as a component of satiety.[42,65,66] In particular, Ritter and Clingasan[66] have provided important evidence which suggests that neural pathways monitor fat oxidation in the periphery and CHO oxidation (perhaps more precisely) in both the periphery and the CNS.[66]

5. Langhans[90] notes that the midbrain centers that are concerned with monitoring peripheral fuel utilization, are connected via extensive neural relays to the areas of the forebrain (especially the hypothalamus and the paraventricular nucleus) that are concerned with the control of protein, CHO, and fat balance.

Thus, an integrative model is beginning to emerge which may account for the manner in which the CNS is capable of monitoring physiological signals concerned with overall macronutrient balance and fuel flux, and which accounts for the manner in which feeding behavior responds to changes in peripheral physiology. In essence, feeding responses are coupled to physiological changes rather than being directly determined by them. The evolution of a flexible and adaptive system concerned with the control of feeding behavior is likely to have bestowed a far greater survival advantage on an opportunistically foraging species such as humans (or for that matter rodents) than a system in which behavioral outcomes are an inevitable and rigid outcome of physiological signaling.

6   NUTRIENT STORAGE/UTILIZATION AND APPETITE FOR SPECIFIC NUTRIENTS

The title of this book is *Neural and Metabolic Control of Macronutrient Selection*. It is ironic in this respect that it is not currently possible to make statements about the effect of fat or CHO oxidation or storage on nutrient selection in humans. The effects of metabolic blockers, food deprivation, starvation, exercise, and nutrient preloads on nutrient selection in animals is extensively discussed in Chapters 22 through 24. There is very little evidence at the present time that changes in nutrient selection have been detected in response to changes in macronutrient metabolism. In this context it is of interest that the macronutrient whose ingestion is most likely to affect subsequent nutrient selection is protein. It has been well demonstrated that a number of species can learn to optimize the protein/energy ratio of their diet in order to match their genetically determined requirements for growth.[91] There are a number of hypotheses in relation to nutrient utilization and subsequent selection. There is not enough evidence to build a case in favor or against them at the present time. DeCastro[92] has reported remarkable autocorrelations for nutrient intake across days in free-living subjects. It is currently unclear whether this reflects some physiologically guided process or a repeated pattern of behavior (i.e., habit).

It has been suggested that intense exercise which depletes CHO stores will stimulate CHO intake. While there is evidence for this in some rodent studies, others do not support this hypothesis. In a recent review King et al.[93] conclude that at present there is no clear evidence that in the short term exercise induces changes in food or nutrient preferences. Similarly, it is reasonable to suppose that depletion of key body compartments (e.g., lean body mass) should stimulate changes in feeding behavior that will alleviate such an acute stress.[82] Again, the data are limited. Subjects receiving nutritional support are usually prescribed a set regimen and so cannot alter the foods or nutrients they select.

A major problem in detecting changes in nutrient intake in humans is that changes in nutrient selection can be derived from self-reported intakes of subjects eating their normal diets (these measures can be imprecise) or in the laboratory using a dietary design to detect changes in energy and nutrient intake. While a number of studies provided a "range of food items" which vary in composition, rarely is evidence provided to demonstrate statistically or empirically that the model in use is sufficiently sensitive to detect changes in macronutrient selection. There is currently a need for protocols that are capable of detecting changes in nutrient selection over periods of days and even weeks.

7   CONCLUSION

The above discussion of evidence relating to the postabsorptive metabolic satiety signal arising from ingestion of CHO and fat suggests that

1. All sources of evidence are individually inconclusive but collectively suggest that post-absorptive processing of CHO and fat are involved in satiety.
2. In most of the experiments concerned there is more evidence pertaining to CHO than fat.
3. The same evidence suggests a stronger role for metabolic feedback from CHO than from fat.
4. Increments and decrements in CHO utilization appear to correlate with satiety and hunger, respectively. Decreases in the oxidation of CHO and protein (attended by increases in fat oxidation) should indicate a decrease in nutrient delivery from the small intestine and could well act as a physiological cue for hunger. Increases in the oxidation of these macronutrients are physiologically indicative of the fed state and could act a component of satiety.
5. There is more evidence that decrements in fat oxidation stimulate intake than evidence that increments decrease intake. The role of fat oxidation as a signal for hunger is theoretically less certain, because while beta oxidation generates reducing equivalents, elevated fat oxidation is often indicative of fasting.
6. The effects of fat and CHO metabolism/stores are by no means deterministic. These effects are enhanced if linked to physiological signals preceding (e.g., small intestinal signals) or attending (e.g., hormone release) delivery of the nutrient to the circulation and tissue, under normal physiological conditions.

Combining the evidence from a number of disparate sources suggests that macronutrient ingestion and processing induces a number of physiological changes, which affect neuroendocrine signaling systems that connect peripheral signals to the major feeding centers of the brain. Changes in nutrient metabolism and storage appear to be an integral component of these peripheral signals. Understanding (1) the conditions under which these signals are most potent, (2) their importance relative to other influences on appetite, and (3) the potential to target these pathways for therapeutic purposes remains a challenge for the future.

REFERENCES

1. Blundell, J. E. and Stubbs, R. J., Diet and food intake in humans, in *International Handbook of Obesity*, G. A. Bray, C. Bouchard, and W.P.T. James, Eds., Marcel Dekker, New York, 1998.
2. Stubbs, R. J., Johnstone, A. M., O' Reilly, L. M., and Poppitt, S. D., Methodological issues relating to the measurement of food, energy and nutrient intake in human laboratory-based studies, *Proc. Nutr. Soc.*, 57(3), 357, 1998.
3. Mayer, J., Regulation of energy intake and the body weight, *Ann. N.Y. Acad. Sci.*, 63, 15, 1955.
4. Kennedy, G. C., The role of depot fat in the hypothalamic control of food intake in the rat, *Proc. Nutr. Soc.*, 140, 578, 1953.
5. Russek, M., An hypothesis on the participation of hepatic glucoreceptors in the control of food intake, *Nature*, 197, 79, 1963.
6. Mellinkoff, S. M., Franklin, M., Boyle, D., and Geipell, M., Relationship between serum amino acid concentration and fluctuation in appetite, *J. Appl. Physiol.*, 8, 535, 1956.
7. Abbot, W. G. H., Howard, B. V., Christin, L., Freymond, D., Lillioja, S., Boyce, V. L., Anderson, T. E., Bogardus, C., and Ravussin, E., Short-term energy balance: relationship with protein, carbohydrate and fat balances, *Am. J. Physiol.*, 255, E332, 1988.
8. Acheson, K. J. and Jequier, E., Glycogen synthesis vs. lipogenesis after a 500 gram carbohydrate meal in man, *Metabolism*, 31, 1234, 1982.

9. Acheson, K. J., Schutz, Y., Bessard, T., Anantharaman, K., Flatt, J. P., and Jequier, E., Glycogen storage capacity and *de novo* lipogenesis during massive carbohydrate overfeeding in man, *Am. J. Clin. Nutr.,* 48, 240, 1988.

10. Flatt, J. P., The difference in storage capacities for carbohydrate and for fat, and its implications for the regulation of body weight, *Ann. N.Y. Acad. Sci.,* 499, 104, 1987.

11. Jequier, E., Calorie balance vs. nutrient balance, in *Energy Metabolism: Tissue Determinants and Cellular Corollaries,* J. M. Kinney, Ed., Raven Press, New York, 1992, 123.

12. Stubbs, R. J., Appetite, feeding behaviour and energy balance in human subjects, *Proc. Nutr. Soc.,* 57, 1, 1998.

13. Hellerstein, M. K., Christiansen, M., and Kaempfer, S., Measurement of *de novo* hepatic lipogenesis in human beings using stable isotopes, *J. Clin. Invest.,* 87, 1841, 1991.

14. Newsholme, E. A. and Leech, A. R., *Biochemistry for the Medical Sciences,* John Wiley & Sons, New York, 1983.

15. Seidell, J. C., Obesity in Europe: scaling an epidemic, *J. Obesity,* 19, 51, 1955.

16. King, N. A., The relationship between physical activity and food intake, *Proc. Nutr. Soc.,* 57, 77, 1998.

17. Spitzer, L. and Rodin, J., Human eating behaviour: a critical review of studies in normal weight and overweight individuals, *Appetite,* 2, 293, 1981.

18. Morely, J. E., Appetite regulation by gut peptides, *Annu. Rev. Nutr.,* 10, 383, 1990.

19. Koopmans, H. S., Endogenous gut signals and metabolites control daily food intake, *Int. J. Obesity,* 14, 93, 1990.

20. Stubbs, R. J., Harbron, C. G., and Johnstone, A. M., The effect of covertly manipulating the energy density of high carbohydrate diet on ad libitum food intake in "pseudo free-living" humans, *Proc. Nutr. Soc.,* 56, 133a, 1997.

21. Garrow, J. S., *Obesity and Related Diseases,* Churchill Livingstone, Edinburgh, 1988.

22. DeCastro, J. M., Macronutrient relationships with meal patterns and mood in the spontaneous feeding behaviour of humans, *Physiol. Behav.,* 39, 561, 1987.

23. DeCastro, J. M. and Elmore, D. K., Subjective hunger relationships with meal patterns in the spontaneous feeding behaviour of humans: evidence for a causal connection, *Physiol. Behav.,* 43, 159, 1988.

24. Bingham, S. A., Gill, C., Welch, A., Day, K., Cassidy, A., Khaw, K. T., Sneyd, M. J., Key, T. J. A., Roe, L., and Day, N. E., Comparison of dietary assessment methods in nutritional epidemiology: weighed records v. 24 h recalls, food-frequency questionnaires and estimated-diet records, *Br. J. Nutr.,* 72, 619, 1994.

25. Westrate, J. A., Effects of nutrients on the regulation of food intake, in Unilever Research, The Netherlands, *Vlaardin Nutr. Food Sci.,* 5, 267, 1992.

26. Gil, K., Skeie, B., Kvetan, V., Askanazi, J., and Friedman, M. I., Parenteral nutrition and oral intake: effect of glucose and fat infusion, *J. Parenteral Enteral Nutr.,* 15, 426, 1991.

27. Van Stratum, P., Lussenburg, R., Van Wezel, L., Vergroesen, A., and Cremer, H., The effect of dietary carbohydrate: fat ratio on energy intake by adult women, *Am. J. Clin. Nutr.,* 31, 206, 1978.

28. Stubbs, R. J., Harbron, C. G., and Prentice, A. M., The effect of covertly manipulating the dietary fat to carbohydrate ratio of isoenergetically dense diets on ad libitum food intake in free-living humans, *Int. J. Obesity,* 20, 651, 1996.

29. Cotton, J. R., Burley, V. J., Westrate, J. A., and Blundell J. E., Dietary fat and appetite: similarities and differences in the satiating effect of meals supplemented with either fat or carbohydrate, *Br. J. Hum. Nutr. Diet.,* 7, 11, 1994.

30. Rolls, B., Kim-Harris, S., Fischman, M. W., Foltin, R., Moran, T. and Stoner, S., Satiety after preloads with different amounts of fat and carbohydrate: implications for obesity, *Am. J. Clin. Nutr.,* 60, 476, 1994.

31. Johnstone, A. M., Harbron, C. G. and Stubbs, R. J., Macronutrients, appetite and day-to-day food intake in humans, *Eur. J. Clin. Nutr.,* 50, 418, 1996.

32. Stubbs, R. J., Van Wyk, M. C. W., Johnstone, A. M., and Harbron, C., Breakfasts high in protein, fat or carbohydrate: effect on within-day appetite and energy balance, *Eur. J. Clin. Nutr.,* 50, 409, 1996.

33. Raben, A., Holst, J. J., Christensen, N. J., and Astrup, A., Determinants of postprandial appetite sensations: macronutrient intake and glucose metabolism, *Int. J. Obesity Relat. Metab. Disorders,* 20, 161, 1996.

34. Stubbs, R. J., Dietary macronutrients and glucostatic control of feeding, *Proc. Nutr. Soc.,* 55, 467, 1996.

35. Van Itallie, T. B., Beaudoin, R., and Mayer, J., Ateriovenous glucose differences, metabolic hypoglyceamia and food intake in man, *J. Clin. Nutr.*, 1, 208, 1953.
36. Thompson, D. A. and Campbell, R. G., Hunger in humans induced by 2 deoxy-D-glucose glucoprivic control of taste preference and food intake, *Science*, 198, 1065, 1977.
37. Thompson, D. A. and Campbell, R. G., Experimental hunger in man: behavioural and metabolic correlates of intracellular glucopenia, in *Central Mechanisms of Anorectic Drugs*, S. Garrattini and R. Samanin, Eds., Raven Press, New York, 1978, 437.
38. Bernstein, L. M. and Grossman, M. I., An experimental test of the glucostatic theory of the regulation of food intake, *J. Clin. Invest.*, 35, 627, 1956.
39. Van Itallie, T. B. and Hashim, S. A., Biochemical concomitants of hunger and satiety in man, *Am. J. Clin. Nutr.*, 8, 587, 1960.
40. Smith, G. P. and Gibbs, J., The effect of gut peptides on hunger, satiety, and food intake in humans, *Ann. N.Y. Acad. Sci.*, 499, 132, 1987.
41. Lebowitz, S. F., Neurochemical-neurendocrine systems in the brain controlling macronutrient intake and metabolism, *Trends Neurosci.*, 15, 491, 1992.
42. Langhans, W. and Scharrer, E., The metabolic control of food intake, *World Rev. Nutr. Diet.*, 70, 1, 1992.
43. Louis-Sylvestre, J. and Le Magnen, J., A fall in blood glucose level precedes meal onset in free feeding rats, *Neurosci. Biobehav. Rev.*, 4, 13, 1980.
44. Campfield, L. A. and Smith, A., Functional coupling between transient declines in blood glucose and feeding behaviour: temporal relationships, *Brain Res. Bull.*, 17, 427, 1986.
45. Campfield, L. A. and Smith, A., Transient declines in blood glucose signal meal initiation, *Int. J. Obesity*, 14, S15, 1990.
46. Campfield, L. A., Smith, A., Rosenbaum, M., and Hirsch, J., Human eating: evidence for a physiological basis using a modified paradigm, *Neurosci. Biobehav. Rev.*, 20, 133, 1996.
47. Pollack, C. P., Green, J., and Smith, G. P., Blood glucose prior to meal request in humans isolated from all temporal cues, *Physiol. Behav.*, 46, 529, 1989.
48. Mattes, R. D., Sensory influences on food intake and utilisation in humans, *Hum. Nutr. Appl. Nutr.*, 41A, 77, 1987.
49. Lavin, J. H., Wittert, G., Sun, W. M., Horowitz, M., Morley, J. E., and Read, N. W., Appetite regulation by carbohydrate: role of blood glucose and gastrointestinal hormones, *Am. J. Physiol.*, 271, E209, 1996.
50. Lissner, L. and Heitmann, B. L., Dietary fat and obesity: evidence from epidemiology, *Eur. J. Clin. Nutr.*, 49, 79, 1995.
51. Macdiarmid, J. I., Cade, J. E., and Blundell, J. E., High and low fat consumers, their macronutrient intake and body mass index: further analysis of the National Diet and Nutrition Survey of British Adults, *Eur. J. Clin. Nutr.*, 50, 505, 1996.
52. Stubbs, R. J., Harbron, C., Murgatroyd, P., and Prentice, A., Covert manipulation of dietary fat and energy density: effect substrate flux and food intake in men eating ad libtum, *Am. J. Clin. Nutr.*, 62, 1, 1995.
53. Stubbs, R. J., Goldberg, G. R., Murgatroyd, P. R., and Prentice, A. M., Carbohydrate balance and day-to-day food intake in man, *Am. J. Clin. Nutr.*, 57, 897, 1993.
54. Shetty, P. S., Prentice, A. M., Goldberg, G. R., Murgatroyd, P. R., McKenna, A. P. M., Stubbs, R. J., and Volschenk, P. A., Alterations in fuel selection and voluntary food intake in response to iso-energetic manipulation of glycogen stores in man, *Am. J. Clin. Nutr.*, 60, 534, 1994.
55. Snitker, S., Larson, D. E., Tataranni, P. A., and Ravussin, E., Ad libitum food intake in humans after manipulation of glycogen stores, *Am. J. Clin. Nutr.*, 65, 941, 1997.
56. Holt, S., Brand-Miller, J., Soveny, C., and Hansky, J., Relationship of satiety to postprandial glycemic, insulin and cholecystokinin responses, *Appetite*, 18, 129, 1992.
57. Jenkins, D. J. A., Wolever, T. M. S., Buckley, G., Lam, K. Y., Giudici, S., Kalmusky, J., Jenkins, A. L., Patten, R. L., Bird, J., Wong, G. S., and Josse, R. G., Low-glycemic-index starchy foods in the diabetic diet, *Am. J. Clin. Nutr.*, 48, 248, 1988.
58. Ludwig, D. S., Majzoub, J. A., Dallal, G., and Roberts, S. B., High glycemic index foods, overeating, and obesity, *Obesity Res.*, 5, 25S, 1997.
59. Chiang, Y. and Johnson, J. A., Measurement of total and gelatinized starch by amyloglucosidase and toluidine reagent, *Cereal Chem.*, 54, 429, 1977.

60. Raben, A., Tagliabue, A., Christensen, N. J., Madsen, J., Holst, J. J., and Astrup, A., Resistant starch: the effect on postprandial glycemia, hormonal response and satiety, *Am. J. Clin. Nutr.,* 60, 544, 1994.

61. de Roos, N., Heijnen, M. L., de Graf, C., Woestenenk, G., and Hobbel, E., Resistant starch has little effect on appetite, food intake and insulin secretion of healthy young men, *Eur. J. Clin. Nutr.,* 49, 532, 1995.

62. Walls, E. K. and Koopmans, H. S., Differential effects of intravenous glucose, amino acids and lipid on daily food intake in rats, *Am. J. Physiol.,* 262, R225, 1992.

63. Pullicino, E. and Elia, M., Intravenous carbohydrate overfeeding — a method for rapid nutritional repletion, *Clin. Nutr.,* 10, 146, 1991.

64. Wanderweele, D. A., Hyperinsulinism and feeding; not all sequences lead to the same behavioural outcomes or conclusions, *Appetite,* 6, 47, 1985.

65. Friedman, M. I. and Tordoff, M. G., Fatty acid oxidation and glucose utilisation interact to control food intake in rats, *Am. J. Physiol.,* 251 (5 pt 2), R840, 1986.

66. Ritter, S. and Clingasan, N. Y., Neural substrates for metabolic controls of feeding, in *Appetite and Body Weight Regulation: Sugar, Fat and Macronutrient Substitutes,* J. D. Fernstrom and G. D. Miller, Eds., CRC Press, Boca Raton, FL, 1994.

67. Rolls, B. J., Gnaizak, N., Summerfelt, A., and Laster, L. J., Food intake in dieters and no dieters after a liquid meal containing medium chain triglycerides, *Am. J. Clin. Nutr.,* 48, 66, 1988.

68. Stubbs, R. J. and Harbron, C. G., Covert manipulation of the ratio of medium- to long-chain triglycerides in isoenergetically dense diets: effect on food intake in ad libitum feeding men, *Int. J. Obesity,* 20, 435, 1996.

69. Astrup, A., Buemann, B., Toubro, S., and Raben, A., Defects in substrate oxidation involved in predisposition to obesity, *Proc. Nutr. Soc.,* 55, 817, 1996.

70. Verboeket-van de Venne, W. P. H. G., Westerterp, H. R., and Hoor, F., Substrate utilization in man: effects of dietary fat and carbohydrate, *Metabolism,* 43, 152, 1994.

71. Zurlo, F., Lillioja, S., Esposito-Del Puente, A., Nyomba, B. L., Raz, I., Saad, M. F., Swinburn, B. A., Knowler, W. C., Bogardus, C., and Ravussin, E., Low ratio of fat to carbohydrate oxidation as predictor of weight gain: study of 24 hr RQ, *Am. J. Physiol.,* 259, E650, 1990.

72. Seidell, J. C., Muller, D. C., Sorkin, J. D., and Andres, R., Fasting respiratory exchange ratio and resting metabolic rate as predictors of weight gain: the Baltimore Longitudinal Study on Ageing, *Int. J. Obesity,* 16, 667, 1992.

73. Bouchard, C., Tremblay, A., Nadeau, A., Depres, J. P., Theriault, G., Dussault, J., Moorjani, S., Pinault, S., and Fourinier, G., The response to long term overfeeding in identical twins, *N. Engl. J. Med.,* 322, 1477, 1990.

74. Le Magnen, J., *Neurobiology of Feeding and Nutrition,* Academic Press, San Diego, 1992, Chap. 2.

75. Larue-Achagiotis, C. and Le Magnen, J., The different effects of continuous night and daytime insulin infusion on the meal pattern of normal rats: comparison with the meal pattern of hyperphagic hypothalamic rats, *Physiol. Behav.,* 22, 435, 1979.

76. Rich, A. J., Chambers, P., and Johnston, I. D. A., Are ketones an appetite suppressant? *J. Parenteral Enteral Nutr.,* 13, 7S, 1988.

77. Carpenter, R. G. and Grossman, S. P., Plasma fat metabolites and hunger, *Physiol. Behav.,* 30, 57, 1982.

78. Kern, P. A., Ong, J. M., Saffari, B., and Carty, J., The effects of weight loss on the activity and expression of adipose-tissue lipoprotein lipase in very obese humans, *N. Engl. J. Med.,* 322, 1053, 1990.

79. Eckel, R. H. and Yost, T. J., Weight reduction increases adipose tissue lipoprotein lipase responsiveness in obese women, *J. Clin. Invest.,* 80, 992, 1987.

80. Campfield, L. A., Smith, L. A., and Burn, P., OB protein: a hormonal controller of central neural networks mediating behavioural, metabolic and neuroendocrine responses, *Endocrinol. Metab.,* 4, 81, 1997.

81. Forbes, G. B., Brown, M. R., Welle, S. L., and Lipinski, B. A., Deliberate overfeeding in women and men: energy cost and composition of the weight gain, *Br. J. Nutr.,* 56, 1, 1986.

82. Dulloo, A. G., Human pattern of food intake and fuel-partitioning during weight recovery after starvation: a theory of autoregulation of body composition, *Proc. Nutr. Soc.,* 56, 25, 1997.

83. Kral, J. G., Surgical treatment of obesity, in *International Handbook of Obesity,* G. A. Bray, C. Bouchard, and W. P. T. James, Eds., Marcel Dekker, New York, 1998.

84. Furuse, M., Choi, Y.-H., Mabayo, R. T., and Okumura, J.-I., Feeding behaviour in rats fed diets containing medium chain triglyceride, *Physiol. Behav.,* 52, 815, 1992.

85. Johnstone, A. M., Ryan, L. M., Reid, C. A., and Stubbs, R. J., Breakfasts high in monoglyceride or triglyceride: no differential effect on appetite or energy balance, *Eur. J. Clin. Nutr.,* 52(8), 603, 1998.

86. Johnstone, A. M., Ryan, L. M., Reid, C. A., and Stubbs, R. J., Overfeeding fat as monoglyceride or triglyceride: effect on appetite, nutrient balance and the subsequent day's energy intake, *Eur. J. Clin. Nutr.,* 52(8), 610, 1998.

87. Lawton, C., Delargy, H., Smith, F., and Blundell, J. E., Does the degree of saturation of fatty acids affect postingestive satiety? *Int. J. Obesity,* 21, S35, 1997.

88. Bistrian, B. R., Novel lipid sources in parenteral and enteral nutrition, *Proc. Nutr. Soc.,* 56, 471, 1997.

89. Greenburg, D., Gibbs, J., and Smith, G. P., Infusions of lipid into the duodenum elicit satiety in rats, while similar infusions into the vena cava do not, *Appetite,* 12, 213, 1989.

90. Langhans, W., Metabolic and glucostatic control of feeding, *Proc. Nutr. Soc.,* 55, 497, 1996.

91. Forbes, J. M., Voluntary Food Intake and Selection in Farm Animals, Centre for Agriculture and Biosciences International, Wallington, Oxon, U.K., 1995.

92. De Castro, J. M., How can energy balance be achieved by free-living human subjects? *Proc. Nutr. Soc.,* 56, 1, 1997.

93. King, N. A., Tremblay, A., and Blundell, J. E., Effects of exercise on appetite control: implications for energy balance, *Med. Sci. Sports Exercise,* 29, 1076, 1997.

13 Effects of Metabolic Blockade on Macronutrient Selection

Sue Ritter, Frank H. Koegler, and Michael Wiater

CONTENTS

1   METABOLIC CONTROL OF FOOD INTAKE: DIFFERENTIAL ACTIVATION OF CENTRAL AND PERIPHERAL MECHANISMS BY 2-DEOXY-D-GLUCOSE AND MERCAPTOACETATE

Experimental and clinical findings in a variety of animal species, including humans, have established that feeding behavior is controlled in part by internal signals arising from the metabolic availability of glucose and fatty acids (for reviews, see References 1 and 2). Smith and Epstein[3] reported the first definitive evidence for a metabolic control of feeding. They showed that pharmacological blockade of intracellular glucose utilization stimulates food intake. In their experiments, they used the antimetabolic glucose analogue, 2-deoxy-D-glucose (2DG), which competitively blocks phosphohexose isomerase and thus reduces the further metabolism of glucose.[4,5] Subsequently, a number of related substances that reduce glucose availability or metabolism, including 5-thioglucose,[6] 3-O-methylglucose,[7] 2,5-anhydro-D-mannitol (2,5-AM),[8] and phlorizin[9] have been shown to stimulate food intake.

Food intake is also increased by pharmacological blockade of fatty acid oxidation. The contribution of fatty acid oxidation in controlling food intake was first demonstrated by Scharrer and Langhans[10] using mercaptoacetate (MA), a substance that blocks mitochondrial beta oxidation of fatty acids.[11] In the same year, Friedman and colleagues[12] reported a similar finding using methyl palmoxirate, a carnitine palmitoyl transferase I inhibitor that blocks transport of long-chain fatty acids into mitochondria.[13]

1.1 MA AND 2DG ACTIVATE DIFFERENT METABOLICALLY SENSITIVE RECEPTOR CELLS

Several lines of evidence indicate that glucose and fatty acids are monitored by different receptor cells, although in neither case have the specific receptor cells been identified. Fatty acids are important substrates for peripheral energy metabolism, and their availability is monitored peripherally by vagal sensory neurons or by vagally innervated receptor cells. Subdiaphragmatic vagotomy[14] and selective lesion of fine-diameter unmyelinated vagal sensory neurons by capsaicin administration[15] abolishes MA-induced feeding. In contrast, glucose is the essential metabolic fuel for the central nervous system and glucose utilization is monitored in the brain. Administration of 2DG or 5-thioglucose directly into the brain stimulates feeding at doses that do not cause peripheral glucoprivation.[16-18] In addition, the vagus nerve is not required for stimulation of feeding by systemic 2DG.[14,15]

1.2 MA AND 2DG ACTIVATE DIFFERENT CENTRAL NEURAL PATHWAYS

2DG and MA also appear to stimulate feeding by engaging different central neural pathways. MA induces the expression of the c-*fos* gene in neurons of the nucleus of the solitary tract, the lateral parabrachial nucleus and the central nucleus of the amygdala,[19] indicating that MA activates these neurons.[20] The induction of c-*fos* expression in these areas by MA, like MA-induced feeding, is entirely dependent on the vagus nerve and does not occur in vagotomized rats.[19] These areas are essential components of the central neural pathway for MA-induced feeding, as revealed by results showing that electrolytic or chemical lesion of any one of these nuclear areas abolishes feeding in response to MA.[14,21,22]

Although 2DG induces c-*fos* in some of the same brain sites as MA, as well as in other sites, the neural circuitry underlying 2DG-induced feeding differs in several ways from the neural pathway thought to be involved in MA-induced feeding. First, the vagus nerve is not required for induction of c-*fos* in the brain by systemic 2DG.[19] Second, area postrema/nucleus of the solitary tract lesions impair both MA and 2DG-induced feeding,[14,23] but for apparently different reasons in each case. This brain region receives the central terminals of vagal sensory neurons crucial for MA-induced feeding and has been suggested to contain receptor cells mediating glucoprivic feeding. Third, the lateral parabrachial nucleus and the central nucleus of the amygdala, both essential for MA-induced feeding, are not essential for 2DG-induced feeding.[14,21,22] However, lesions of the central nucleus of the amygdala raise the threshold dose required for elicitation of feeding by 2DG,[22,24] suggesting that some confluence of the neural circuitry for both controls may occur in this nucleus.

1.3 THE BLOOD–BRAIN BARRIER DEFINES THE FUNCTIONAL DIFFERENCES BETWEEN CENTRAL AND PERIPHERAL METABOLIC RECEPTOR CELLS

The localization of the putative metabolic receptors responsible for 2DG-induced feeding to sites within the brain itself suggests that the metabolic fuels they monitor are restricted to those that enter the brain. Under most circumstances, glucose is the only substrate that crosses the blood–brain barrier in quantities sufficient to meet the energy requirements of the brain.[25] Hence, it is probably fair to call these receptors glucoreceptors. It should not be assumed, however, that glucose per se is the transduction signal. During 2DG-induced glucoprivation, intravascular and intracellular glucose levels are elevated and only products of metabolism subsequent to the point of 2DG-induced blockade (phosphohexose isomerase) are diminished intracellularly. Therefore, some product of glucose metabolism such as a phosphorylated intermediate or ATP is more likely to provide the transduction signal. This is consistent with the finding that the ketone body, beta hydroxybutyrate, which enters the glycolytic pathway beyond the phosphohexose isomerase step, is capable of reducing 2DG-induced feeding.[26] In states of ketosis during fat catabolism, ketone bodies are capable of entering the brain and can supply part, but not all, of the energy requirement of the brain.[27] It is still not certain, however, whether ketone bodies block 2DG-induced feeding by a metabolic action in glucoreceptive cells, in different metabolically sensitive cells, or by a nonspecific mechanism such as the production of malaise, although the first of these mechanisms seems most likely. Intravenous infusions of fructose and lipids — metabolic fuels that are excluded from the brain, but utilized by peripheral tissues — do not reduce feeding in response to systemic 2DG.[28] In addition, central administration of glucose directly into the brain ventricles is sufficient to reverse the feeding induced by systemic 2DG-induced glucoprivation.[29] The isolation of central glucoreceptors by the blood–brain barrier potentially means that they function entirely in response to brain energy supply without direct modulation by deficits or surfeits in peripherally available nutrients.

On the other hand, the localization outside the central nervous system of the metabolic receptors responsible for MA-induced feeding may enable these receptors to respond more broadly to the variety of metabolic fuels available to peripheral tissues from systemic blood. In support of this view, it has been shown that intravenous infusion of glucose, fructose, and lipids all reduce MA-induced feeding and that, to be effective, glucose must be administered systemically.[28,29] Administration of glucose into the brain does not reduce MA-induced feeding. Thus, peripheral metabolic receptors must detect availability of multiple energy-yielding substances. Decreased fatty acid oxidation may activate these receptors under the appropriate dietary conditions, but decreased availability of other metabolic fuels may also be effective.

The hypothesis that carbohydrate as well as fatty acid utilization is monitored peripherally is supported by a number of findings. As just noted, glucose attenuates the effects of energy deficit induced by blockade of fatty acid oxidation, providing that the glucose is available to peripheral tissues. Results from experiments utilizing the antimetabolic fructose analogue, 2,5-AM, are also important in this regard. 2,5-AM interferes with carbohydrate metabolism[30,31] and stimulates food intake.[32] However, this analogue, like fructose, is excluded from the brain by the blood–brain barrier. Thus, the stimulation of feeding presumably arises from the peripheral antagonism of carbohydrate metabolism of 2,5-AM. This peripheral action is confirmed by the finding that 2,5-AM-induced feeding is dependent on an intact vagus nerve.[33,34] Finally, we and others have found that MA is less effective in stimulating feeding in animals maintained on a high-carbohydrate, low-fat diet than in those maintained on a medium-fat diet,[10,35] indicating that the increased availability of glucose compensates metabolically to some extent for the MA-induced decrease in fatty acid oxidation. Together, these results suggest that peripheral receptor cells are sensitive to overall energy availability but incapable of discriminating between alternative metabolic fuels. Thus, the peripheral system is complementary to the glucose-specific central system in serving the diverse nutrient-specific and metabolic requirements of body tissues and brain.

2   MACRONUTRIENT SELECTION INDUCED BY METABOLIC BLOCKADE

At the outset of our work on macronutrient selection in response to acute metabolic challenges, we suspected that 2DG and MA would affect macronutrient selection differently. Our thinking was influenced in particular by the finding that 2DG and MA appear to activate afferent neural pathways that maintained their distinctness through several synaptic relays within the brain. We speculated that separate information channels would only be maintained if information arriving at integrative sites from each of these channels were crucial in its own right for regulation of appetite and metabolism. According to this line of reasoning, it seemed possible that this distinctness would be reflected in the nutrients selected during activation of each metabolic control. Our intent was to utilize a paradigm in which the macronutrient selection in response to MA and 2DG would be examined in the same rats. Results reported in the literature led us to expect that 2DG would selectively increase carbohydrate intake.[36] At the time this work was begun, macronutrient selection during MA-induced lipoprivation had not been assessed. However, since MA blocks oxidation of fat, we expected initially that MA would selectively increase the intake of fat. Our results were not entirely consistent with these initial expectations.[37,38]

In our first set of experiments, we examined the hypothesis that 2DG and MA would induce different macronutrient preferences. Adult male rats were placed on a maintenance diet in which carbohydrate, protein, and fat were consumed *ad libitum* from separate food cups containing cornstarch, casein, and corn oil. In this and all of the experiments described below, the carbohydrate and protein were supplemented with vitamins, minerals, and fiber. Rats were adapted to this diet for 3 weeks until their daily intakes stabilized. This same diet was utilized during the drug tests. Fresh food was presented at the end of the light cycle, but not at the beginning of feeding tests. Food available to each rat during the night was weighed and returned to that same rat at the beginning of the feeding tests. The position in the rats' cages of the food cups containing each macronutrient was varied when fresh food was provided to avoid biases in intake due to place preferences. All macronutrients were available at all times, except in one experiment noted below. Intake of each macronutrient was measured in 24-h tests and during the 4 h immediately following injection of 2DG, MA, or saline. The drug and saline tests were conducted in random order during the middle of the illuminated phase of the light/dark cycle, unless noted below. One drug test and one saline test were conducted each week with at least 3 days rest following each drug test.

We found that under basal conditions rats offered cornstarch, casein and oil consistently selected nearly equal amounts of their total daily caloric intake from each macronutrient source (Figure 1, insert). In the 4-h tests with metabolic inhibitors, food intake was significantly increased above control levels by both 2DG and MA (Figure 1). However, the pattern of macronutrient selection differed for 2DG and MA. 2DG increased intake of all three macronutrients to a similar degree.

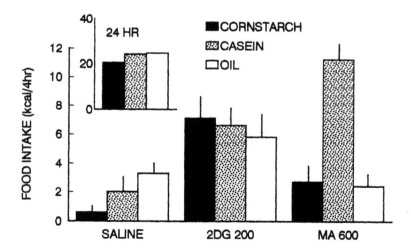

FIGURE 1 Macronutrients (expressed as kcal) consumed by male Sprague-Dawley rats ($n = 10$) in 4-h tests following systemic injections of 0.9% saline (1 ml/kg), 2DG (200 mg/kg, s.c.), and MA (600 μmol/kg, i.p.). Separate sources of carbohydrate (cornstarch), protein (casein), and fat (vegetable oil) were simultaneously available during the test. Rats were maintained on the three-macronutrient self-selection diet for at least 3 weeks prior to initiation of testing. Insert shows spontaneous 24-h intake (kcal) of the three macronutrients in the same rats. After 2DG injection, rats increased their intake of all three macronutrients. The pattern of intake was similar to the pattern observed in spontaneous feeding. MA induced a different pattern of intake in which protein intake was selectively enhanced. Means and S.E.M.s (line above bars) are shown.

In contrast, in these same rats, MA induced a strong preference for casein. After 2DG, rats ingested the macronutrients in the same relative proportions as in their 24-h spontaneous feeding, whereas after MA their macronutrient selection was distinctly different from either their spontaneous 24-h or their 2DG-induced macronutrient intake. In additional experiments with the same macronutrient choices, similar macronutrient preference profiles were expressed in repeated tests, across drug doses and in different groups of rats.

The fact that the same rats expressed different preferences in response to the two metabolic inhibitors rules out palatability or pre-existing preferences for particular macronutrients as the sole determinant of the preferences expressed during metabolic blockade. Clearly, the different metabolic states induced by 2DG and MA alter appetite in distinctly different ways. This finding is consistent with other data cited above, suggesting that 2DG and MA stimulate food intake by activation of different neural systems.

2.1   MA DOES NOT STIMULATE FAT INTAKE

We found several additional aspects of these data to be particularly intriguing and unexpected. First, fat intake was not increased 4 h after MA injection, as shown in the figure, or at other measurement times (1, 2, or 24 h following the injection). To determine whether the lack of fat appetite was due to specific parameters employed in our experimental design, we examined MA-induced macronutrient selection under a variety of conditions. For example, we tested rats at different times during the circadian cycle, including dark onset and middle and late dark cycle. However, MA did not increase fat intake above baseline levels at any of our testing times. Furthermore, offering albumin, an apparently unpalatable protein, instead of casein as the protein source, decreased MA-induced protein intake and increased carbohydrate intake, but did not increase MA-induced fat intake (Figure 2). We also varied the fat source. Instead of corn oil, rats were adapted to and tested with solid fat (hydrogenated vegetable fat). Even though rats given hydrogenated fat as their fat source consumed significantly more fat during spontaneous feeding than rats given oil (Figure 3, insert), they did not increase their intake of fat in response to MA (Figure 3). In contrast, the relative proportion of fat ingested in response to 2DG was increased greatly in rats consuming

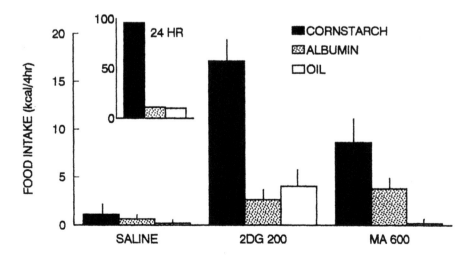

FIGURE 2   Macronutrients (expressed as kcal) consumed by male Sprague-Dawley rats ($n = 6$) in 4-h tests when albumin was presented as the protein source. Rats were injected with 0.9% saline (1 ml/kg), 2DG (200 mg/kg, s.c.) and MA (600 umol/kg, i.p.) immediately before the test. Rats were maintained on the 3-macronutrient diet consisting of cornstarch, albumin, and corn oil for at least three weeks prior to initiation of testing. All three macronutrients and water were available during the tests. Insert shows spontaneous 24-h intake (kcal) of the three macronutrients in the same rats. Means and S.E.M.s (line above bars) are shown.

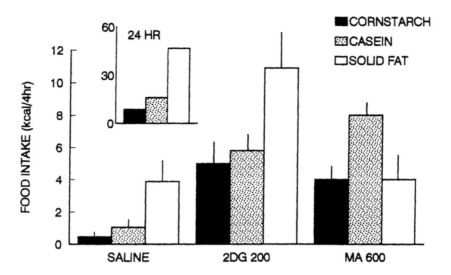

FIGURE 3   Pattern of macronutrient consumption in 4-h tests when hydrogenated fat (unflavored Crisco) was provided as the fat source. Rats (n = 10) were injected with 0.9% saline, 2DG (200 mg/kg, s.c.), and MA (600 μmol/kg, i.p.) immediately prior to the test. Rats were maintained on the three-macronutrient test diet (cornstarch, casein, and unflavored hydrogenated vegetable fat) for at least three weeks prior to initiation of testing. All three macronutrients and water were available during the tests. Insert shows spontaneous 24-h intake (kcal) of the three macronutrients in the same rats. Means and S.E.M.s (line above bars) are shown.

hydrogenated fat, in comparison with those consuming oil. Finally, in another experiment, rats (n = 10) adapted to the macronutrient self-selection diet were presented with only one macronutrient during the MA test. After saline injection, rats consumed 0.2 ± 0.2 kcal of cornstarch, 0.43 ± 0.15 kcal of casein, and 2.3 ± 0.6 kcal of fat in the three separate tests. In response to MA, they consumed 1.6 ± 0.6 kcal of cornstarch, 2.4 ± 0.7 of casein, and 2.6 ± 1.2 kcal of fat. In response to 2DG, they consumed 9.9 ± 1.9, 9.3 ± 1.8, and 13.2 ± 2.3 kcal of cornstarch, casein, and fat, respectively. Thus, in response to MA, rats increased their intake when cornstarch or casein were presented, but not when oil was presented. These same rats increased their intake significantly in response to 2DG, regardless of which macronutrient was present during the test.

   One potential explanation of the failure of rats to increase their fat intake in response to MA is that MA blocks the metabolism of fat such that consumption of fat does not positively reinforce consummatory behavior. However, intravenous infusion of fat (Intralipid) effectively attenuates or blocks MA-induced feeding.[28] This blockade appears to be mediated by metabolic effects of the lipid and not nonspecific behavioral suppression since 2DG-induced feeding is not blocked by infusion of lipid. More importantly, it has been shown that MA stimulates sham feeding. In the sham feeding paradigm, no nutrients are absorbed since ingested food is drained from the stomach.[39] Therefore, while we cannot rule out the reinforcement hypothesis completely, it seems unlikely that it will yield the correct explanation of the data. Another possibility is that an interaction of the self-selection diet composition with the test conditions could have modified the response of the rats to fat. For example, it has been shown that animals adapted to low-fat diets are more sensitive to the satiating effect of fat than rats adapted to high-fat diets.[40,41] However, our rats on self-selection diets including corn oil as the fat source had relatively low daily intake of fat, while those on solid fat had very high daily fat intake. Neither group increased fat intake in response to MA. Moreover, rats on both fat sources significantly increased fat intake in response to 2DG. At this point, we cannot offer a reasonable explanation for this very clear aspect of our results. It is interesting to note, however, that the failure to increase fat intake in response to MA has also been observed in Osborne-Mendel rats.[42]

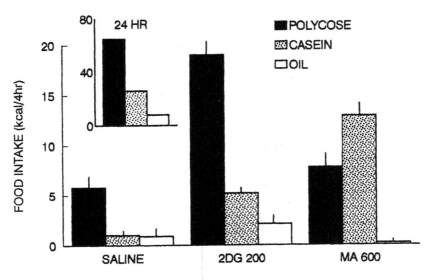

FIGURE 4   Pattern of macronutrient consumption in 4-h tests when powdered Polycose was provided as the carbohydrate source. Rats ($n = 6$) were injected with 0.9% saline, 2DG (200 mg/kg, s.c.), and MA (600 μmol/kg, i.p.) immediately prior to the test. Rats were maintained on the three-macronutrient test diet (Polycose, casein, and corn oil) for at least 3 weeks prior to initiation of testing. All three macronutrients and water were available during the tests. Insert shows spontaneous 24-h intake (kcal) of the three macronutrients in the same rats. Means and S.E.M.s (line above bars) are shown.

2.2   MA STIMULATES PROTEIN INTAKE SELECTIVELY

The macronutrient selection profiles of rats treated with MA raise a second important question. Why does blockade of fatty acid oxidation stimulate protein intake? One possibility that immediately suggests itself is that events signaling peripheral energy deficit may stimulate gluconeogenesis. If so, the increased intake of protein may be compensatory for the increased demand for glucogenic amino acids. Indeed, a number of studies have shown that MA-induced blockade of fatty acid oxidation leads to an enhancement of glucose utilization.[43] This enhanced glucose utilization can be unmasked by food deprivation, which reduces glucose (or glycogen) supply, and by exercise-induced enhancement of energy metabolism. Since MA-induced decrease in fatty acid oxidation does not stimulate adrenal medullary secretion directly,[43-45] the glucose needed to replace the deficient fuels must arise at least in part from gluconeogenesis rather than hepatic glycogenolysis. Unfortunately, neither the amino acid profiles nor the secretion of gluconeogenic hormones has been thoroughly investigated in MA-treated rats.

Again, we cannot provide a physiological explanation for the protein appetite, but additional experiments provide convincing evidence for the strength of this phenomenon. In one series of experiments, we varied the macronutrient sources making up the rats' diets, substituting more highly palatable carbohydrates for cornstarch. When given powdered Polycose, a highly palatable carbohydrate,[46] along with casein and corn oil the 24-h spontaneous intake of carbohydrate was increased to 300% of the amount consumed by rats given pure cornstarch as their carbohydrate source (Figure 4, insert). Results were similar when carbohydrate mixtures comprising cornstarch plus either 15 or 40% sucrose were substituted for the pure cornstarch (data not shown). When rats were offered these highly palatable carbohydrates along with casein and oil, relatively more carbohydrate was consumed in response to MA than when pure cornstarch was used as the carbohydrate source. Yet, MA-treated rats still revealed a preference for casein, consuming significantly more protein than carbohydrate during the test (Figure 4). In contrast, under these same dietary conditions, food ingested in response to 2DG consisted almost entirely of carbohydrate, as reported previously in macronutrient self-selection paradigms utilizing highly palatable carbohydrate sources.[36] As noted above (see Figure 2), only when rats were given an unpalatable protein source (albumin) in place of the casein, did they consume more carbohydrate than protein during MA tests. And even then, protein intake was significantly elevated above baseline.

These results indicate that stimulation of protein appetite by MA is a robust effect, but that the protein appetite is not absolute. MA will increase carbohydrate intake under certain conditions. Increase in intake of carbohydrates in response to MA when protein is unpalatable (e.g., albumin) or not available (e.g., in the single macronutrient test) may be related to their protein-sparing effect.[47] The protein-sparing effect of carbohydrates may also account for the reduction of protein intake when a highly palatable carbohydrate is available during the MA test. Although these explanations may fit with the data, additional work will be required to identify the actual mechanisms underlying the induction of protein appetite by metabolic blockers.

2.3   THE VAGUS NERVE IS INVOLVED IN PROTEIN PREFERENCE INDUCED BY METABOLIC BLOCKADE

In addition to the fact that 2DG and MA impair utilization of different metabolic fuels, an important difference between the mechanisms of action of 2DG and MA is that MA induces feeding by a vagally mediated action,[14,15,48] while 2DG induces feeding predominantly by its actions within the brain.[14-16] Therefore, it is reasonable to assume that information arriving in the brain from the vagus nerve may be responsible for the macronutrient appetite induced by MA, and possibly also some of the characteristics of feeding induced by systemically administered 2DG. To assess the role of the vagus in macronutrient selection in response to metabolic inhibitors, rats were adapted to the macronutrient self-selection diet. When intakes had stabilized, the vagus nerve was transected bilaterally just distal to the diaphragm. Vagotomized rats recovered quickly and remained healthy on the macronutrient diet.

As reported previously for mixed-nutrient diets, vagotomized rats did not increase their intake of the macronutrient diets in response to MA. This result is consistent with the hypothesis that the vagus nerve accounts for the entire effect of MA on food intake, including the stimulation of protein intake. Although total subdiaphragmatic vagotomy did not block the stimulation of macronutrient intake in response to 2DG, the macronutrient selection of the vagotomized rats was strikingly different from that in normal rats (Figure 5). After 2DG treatment, the increased food intake in vagotomized rats consisted almost entirely of cornstarch, while the controls ingested significant amounts of all three macronutrients as observed in the experiments discussed above. The change in macronutrient selection in response to 2DG and MA was correlated with a dramatic decrease in the vagotomized rats' intake of protein during spontaneous feeding. Measurements of spontaneous feeding in 24-h tests revealed that vagotomized rats consumed 58% less casein than controls (Figure 5, insert), without reduction of total daily caloric intake. Total 24-h caloric intake averaged 75.1 kcal for sham-operated rats and 77.9 kcal for vagotomized rats. The reduction in protein intake in response to MA, 2DG, and during spontaneous feeding in vagotomized rats may reflect the loss of a crucial mechanism controlling protein intake in all three situations. However, before this important conclusion can be fully accepted, it will be necessary to consider more thoroughly the potential contribution of other, possibly nonspecific, interactions of vagotomy and protein intake.

It is interesting that vagotomy blocked the increase in fat intake in response to 2DG, but not in spontaneously feeding rats (Figure 5). Intake of fat during spontaneous feeding was actually elevated in the vagotomized rats in comparison with the intake of sham-operated rats. This may suggest that fat intake is controlled by mechanisms in addition to those requiring the vagus nerve.

One question raised by studies with MA and 2DG is whether the particular metabolic fuel interfered with by the antimetabolic drugs or the specific neural substrate each activates is the crucial factor in determining the macronutrient selection. We hoped that the antimetabolic fructose analogue

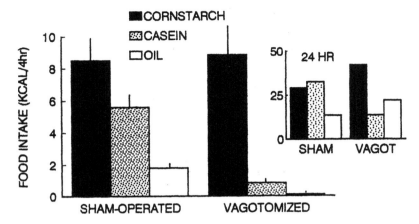

FIGURE 5 Pattern of macronutrient consumption in vagotomized and sham-operated rats in 4-h tests following systemic injection of 2DG (200 mg/kg, s.c.). Intake after 0.9% saline solution was not greater than 1 kcal for any of the macronutrients. In the same vagotomized rats, MA did not stimulate intake of any macronutrient (data not shown). Insert shows spontaneous 24-h intake (kcal) of the three macronutrients in the same rats. Total subdiaphragmatic vagotomy was accomplished by bilateral transection of the vagal trunks just distal to the diaphragm ($n = 6$). Controls were sham vagotomized ($n = 6$). Rats were adapted to the three-macronutrient diet prior to surgery. Testing was done after recovery from surgery. Means and S.E.M.s (line above bars) are shown.

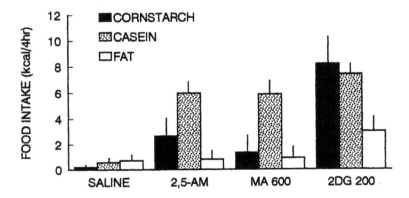

FIGURE 6 Pattern of macronutrient consumption in 4-h tests following systemic injections of 0.9% saline (1 ml/kg), 2DG (200 mg/kg, s.c.), MA (600 μmol/kg, i.p.), and 2,5-AM (300 mg/kg, i.p.). Rats ($n = 10$) were maintained on the 3-macronutrient test diet for at least 3 weeks prior to initiation of testing. All three macronutrients and water were available during the tests. Means and S.E.M.s (line above bars) are shown.

2,5-AM might shed some light on this question. As discussed above, the direct antimetabolic effects of 2,5-AM on food initake, like those of MA, are limited to the periphery and are vagally mediated.[32,33] However, like 2DG, 2,5-AM impairs carbohydrate, not fatty acid, utilization. To examine the effects of 2,5-AM on macronutrient selection, rats adapted to a three-macronutrient diet in which cornstarch, casein, and solid vegetable fat were provided as the macronutrient sources. They were then tested for feeding in response to 2,5-AM, MA, and 2DG. Under these conditions, MA and 2,5-AM induced macronutrient appetites that were nearly identical (Figure 6). Intake of casein was significantly and preferentially enhanced. Unlike MA and 2,5-AM, 2DG stimulated intake of all macronutrients, as shown in the experiments discussed previously. Although it has been shown that feeding induced by both 2,5-AM and MA requires vagal neurons, these results are consistent with the suggestion that the same vagal neurons and the same central neural pathway may be stimulated by both drugs. From these results, it appears that activation of metabolically sensitive vagal neurons results in a selective enhancement of protein appetite. Results also strongly suggest that protein appetite stimulated by systemic 2DG is a vagally mediated component of glucoprivic feeding.

An additional facet of 2DG-induced macronutrient selection is that the intake of carbohydrate in response to 2DG was virtually identical in vagotomized and sham-operated rats, in spite of the fact that the sham-operated rats also consumed significant amounts of protein and fat. This observation suggests that carbohydrate appetite is induced by a mechanism that is independent of the mechanism inducing protein and fat intake in response to 2DG. If so, the contribution of the vagus nerve to 2DG-induced feeding in normal rats might not be detectable under testing conditions in which a high carbohydrate diet is used. In addition, the seemingly selective involvement of the vagus in protein and fat intake leads to the prediction that central as opposed to systemic administration of a glucoprivic agent would stimulate carbohydrate appetite selectively. In preliminary studies we have found this to be the case.[37] However, additional work is required to substantiate these initial findings under a wider range of dietary conditions.

A role for the vagus nerve in protein appetite has been suggested previously. Li and Anderson[49] have reported that vagotomized rats chronically reduce their *ad libitum* protein intake to 50% of their presurgical protein intake, without reduction of total caloric intake. The vagus nerve has also been implicated in detection of amino acid imbalance by data showing that the hypophagic effect of an amino acid–imbalanced diet is attenuated by vagotomy.[50] Regulation of protein intake is an important but poorly understood aspect of physiology. Our results indicating that MA, and possibly also 2,5-AM, elicit selective, reproducible, and vagally dependent increases in protein appetite suggest that metabolic inhibitors may be useful tools with which to probe the underlying mechanisms of protein intake.

2.4   Stimulation of Carbohydrate Preference Is the Most Consistent Effect of 2DG across a Variety of Dietary Conditions

Our finding that 2DG elicits a broad-spectrum appetite, including all three macronutrients, would appear to be in conflict with earlier reports that 2DG induces a specific carbohydrate appetite. Therefore, it is important to note that these earlier experiments (e.g., Reference 36) utilized a highly palatable carbohydrate in their macronutrient self-selection protocol. When we utilized palatable carbohydrate sources (Polycose or sweetened cornstarch), we also observed a selective increase in carbohydrate appetite in 2DG-treated rats. In fact, across all of our dietary conditions, we found no cases in which carbohydrate intake was not increased by 2DG. In contrast, the intake of protein and fat was dependent on the specific choices available. We therefore conclude that the most powerful effect of glucoprivation on macronutrient selection is the stimulation of carbohydrate appetite. To date, we have not conducted macronutrient selection studies using an unpalatable carbohydrate source, although this would be an important experiment.

2.5   2DG May Activate the Same Neural Circuits That Control Spontaneous Feeding

The potential involvement of the glucoprivic control in spontaneous feeding has been debated. In this regard, it is interesting that Benoit and Davidson[51] have shown that interoceptive cues conditioned in response to 24-h food deprivation generalize to those produced by 2DG administration, but not those produced by MA or by a combination of 2DG and MA. Our macronutrient selection data also indicate a similarity between 2DG-induced and spontaneous feeding. The data show that the ratio of carbohydrate, protein, and fat intake ingested in response to 2DG was very similar to the ratio of macronutrients ingested under the same dietary conditions during spontaneous feeding, suggesting that glucoprivation and spontaneous feeding activate the same neural circuits.

3 NEUROCHEMISTY OF MACRONUTRIENT PREFERENCES INDUCED BY 2DG AND MA

The neurochemical mechanisms involved in 2DG- and MA-induced feeding are the subject of continuing investigation. However, a number of studies suggest that norepinephrine (NE) and neuropeptide Y (NPY) may have special importance for the mediation of carbohydrate appetite induced by 2DG. Pharmacological and neurochemical studies indicate that NE- and NPY-containing neurons are activated by glucoprivation and contribute to the mediation of glucoprivically induced responses.[52-57] In addition, both NE and NPY are particularly effective in stimulating carbohydrate appetite.[58,59] Glucocorticoids may be involved since they are released in response to glucoprivation, increase hypothalamic NPY levels, and stimulate carbohydrate appetite when injected into the paraventricular hypothalamus (for review, see Reference 60). Opiate peptides[61,62] and gamma aminobutyric acid (GABA)[63] have been suggested to be involved in glucoprivic feeding and continuing work with these substances with regard to 2DG-induced macronutrient appetites should yield interesting results.

The neuropeptide galanin increases food intake when administered centrally into specific brain sites, including sites crucial for MA-induced feeding.[64-68] In addition, the galanin antagonist, M40, attenuates MA- but not 2DG-induced feeding when injected into the nucleus of the solitary tract or fourth ventricle.[69] This evidence makes galanin a potential mediator of MA-induced feeding. However, the role of galanin in macronutrient selection during MA-induced hyperphagia is not clear. Galanin has been shown to stimulate either fat or carbohydrate intake, depending upon the initial macronutrient preference of the animal and other factors,[64,67,70] but apparently does not stimulate intake of protein, the macronutrient preferred by MA-treated rats.

4 SUMMARY AND CONCLUSIONS

Data presented show that MA and 2DG elicit distinctly different macronutrient preferences. 2DG elicits intake of all three macronutrients in the same relative proportion consumed during spontaneous feeding across a number of dietary conditions, suggesting that glucoprivation activates some of the same interoceptive signals and some of the same neural pathways as normal hunger. 2DG-induced intake of fat and protein, but not carbohydrate, appears to be controlled by peripheral signals, since intake of fat and protein is abolished by vagotomy. Stimulation of carbohydrate appetite, the most prominent effect of 2DG, appears to be mediated centrally. MA elicits a selective vagally mediated intake of protein. Conditions in which carbohydrate palatability is enhanced or protein palatability is diminished lead to a relative increase in carbohydrate intake in response to MA. However, MA did not increase the intake of fat. In conclusion, the fact that 2DG and MA elicit different patterns of macronutrient selection indicates that the metabolic challenges induced by these antimetabolic substances activate different central neural pathways. These findings provide strong evidence that intake of each macronutrient is subject to separate control. Finally, we suggest that MA may be a particularly useful tool for studying mechanisms controlling protein intake, since it produces an acute and selective increase in protein intake without significant toxicity.

REFERENCES

1. Ritter, S. and Calingasan, N. Y., Neural substrates for metabolic controls of feeding, in *Nutrition and Central Nervous System Function*, Fernstrom, J. and Miller, G., Eds., CRC Press, Boca Raton, FL, 1994, 77.
2. Ritter, S., Calingasan, N. Y., Hutton, B., and Dinh, T. T., Cooperation of vagal and central neural systems in monitoring metabolic events controlling feeding behavior, in *Vagal Afferents: Anatomy, Physiol. Behav.*, Ritter, S., Ritter, R. C., and Barnes, C. D., Eds., CRC Press, Boca Raton, FL, 1992, 249.

3. Smith, G. P. and Epstein, A. N., Increased feeding in response to decreased glucose utilization in the rat and monkey, *Am. J. Physiol.*, 217, 1083, 1969.

4. Wick, A. N., Dury, D. R., Nakada, H. I., and Wolfe, J. B., Localization of the primary metabolic block produced by 2-deoxyglucose, *J. Biol. Chem.*, 224, 963, 1957.

5. Brown, J., Effects of 2-deoxy-D-glucose on carbohydrate metabolism: review of the literature and studies in the rat, *Metabolism*, 11, 1098, 1962.

6. Ritter, R. C. and Slusser, P. J., 5-Thio-D-glucose causes increased feeding and hyperglycemia in the rat, *Am. J. Physiol.*, 238, E141, 1980.

7. Booth, D. A., Modulation of the feeding response to peripheral insulin, 2-deoxyglucose or 3-*O*-methyl glucose injection, *Physiol. Behav.*, 8, 1069, 1972.

8. Tordoff, M. G., Rafka, R., DiNovi, M. J., and Friedman, M. I., 2,5-Anhydro-D-mannitol: a fructose analogue that increases food intake in rats, *Am. J. Physiol.*, 254, R150, 1988.

9. Flynn, F. W. and Grill, H. J., Fourth ventricular phlorizin dissociates feeding from hyperglycemia in rats, *Brain Res.*, 341, 331, 1985.

10. Scharrer, E. and Langhans, W., Control of food intake by fatty acid oxidation, *Am. J. Physiol.*, 250, R1003, 1986.

11. Bauche, F., Sabourault, D., Giudicelli, Y., Nordmann, J., and Nordmann, R., Inhibition *in vitro* of acyl-CoA-dehydrogenases by 2-mercaptoacetate in rat liver mitochondria, *Biochem. J.*, 215, 457, 1983.

12. Friedman, M. I., Tordoff, M. G., and Ramirez, I., Integrated metabolic control of food intake, *Brain Res. Bull.*, 17, 855, 1986.

13. Tutwiler, G. F., Brentzel, H. J., and Kiorpes, T. C., Inhibition of mitochondrial carnitine palmitoyl transferase A *in vivo* with methyl 2-tetradecylglycidate (methyl palmoxirate) and its relationship to ketonemia and glycemia, *Proc. Soc. Exp. Biol. Med.*, 178, 288, 1985.

14. Ritter, S. and Taylor, J. S., Vagal sensory neurons are required for lipoprivic but not glucoprivic feeding in rats, *Am. J. Physiol.*, 258, R1395, 1990.

15. Ritter, S. and Taylor, J. S., Capsaicin abolishes lipoprivic but not glucoprivic feeding in rats, *Am. J. Physiol.*, 256, R1232, 1989.

16. Miselis, R. R. and Epstein, A. N., Feeding induced by intracerebroventricular 2-deoxy-D-glucose in the rat, *Am. J. Physiol.*, 229, 1438, 1975.

17. Berthoud, H. R. and Mogenson, G. J., Ingestive behavior after intracerebral and intracerebroventricular infusions of glucose and 2-deoxy-D-glucose, *Am. J. Physiol.*, 233, R127, 1977.

18. Ritter, R. C., Slusser, P. J., and Stone, S., Glucoreceptors controlling feeding and blood glucose: location in hindbrain, *Science*, 213, 451, 1981.

19. Ritter, S. and Dinh, T. T., 2-Mercaptoacetate and 2-deoxy-D-glucose induce Fos-like immunoreactivity (Fos-li) in rat brain, *Brain Res.*, 641, 111, 1994.

20. Morgan, J. I. and Curran, T., Stimulus-transcription coupling in neurons: role of cellular immediate-early genes, *Trends Neurosci.*, 12, 459, 1989.

21. Calingasan, N. Y. and Ritter, S., Lesions of specific lateral parabrachial subnuclei abolish feeding induced by mercaptoacetate but not by 2-deoxy-D-glucose, *Am. J. Physiol.*, 265, R1168, 1993.

22. Ritter, S. and Hutton, B., Mercaptoacetate-induced feeding is impaired by central nucleus of the amygdala lesions, *Physiol. Behav.*, 58, 1215, 1995.

23. Contreras, R. J., Fox, E., and Drugovich, M. L., Area postrema lesions produce feeding deficits in in the rat: effects of preoperative dieting and 2-deoxy-D-glucose, *Physiol. Behav.*, 29, 875, 1982.

24. Tordoff, M. G., Geiselman, P. J., Grijalva, C. V., Keifer, S. W., and Novin, D., Amygdaloid lesions impair ingestive responses to 2-deoxy-D-glucose but not insulin, *Am. J. Physiol.*, 242, R129, 1982.

25. Clarke, D. D. and Sokoloff, L., Circulation and energy metabolism of the brain, in *Basic Neurochemistry: Molecular, Cellular, and Medical Aspects*, 5th ed., Siegel, G. J., Agranoff, B. W., Albers, R. W., and Molinoff, P. B., Eds., Raven Press, New York, 1994, 645.

26. Stricker, E. M. and Rowland, N., Hepatic vs. cerebral origin of stimulus for feeding induced by 2-deoxy-D-glucose in rats. *J. Comp. Physiol. Psychol.*, 92, 126, 1978.

27. Owen, O. E., Morgan, A. P., Kemp, H. G., Sullivan, J. M., Herrera, M. G., and Cahill, G. F., Jr., Brain metabolism during fasting, *J. Clin. Invest.*, 46, 1589, 1967.

28. Singer, L. K. and Ritter, S., Differential effects of infused nutrients on 2DG- and MA-induced feeding, *Physiol. Behav.*, 56, 193, 1994.

29. Singer L. K. and Ritter, S., Intraventricular glucose blocks feeding induced by 2-deoxy-D-glucose but not mercaptoacetate, *Physiol. Behav.,* 59, 921, 1996.

30. Riquelme, P. T., Wernette-Hammond, M. E., Kneer, N. M., and Lardy, H. A., Regulation of carbohydrate metabolism by 2,5-anhydro-D-mannitol, *Proc. Natl. Acad. Sci. U.S.A.,* 80, 4301, 1983.

31. Riquelme, P. T., Wernette-Hammond, M. E., Kneer, N. M., and Lardy, H. A., Mechanism of action of 2,5-anhydro-D-mannitol in hepatocytes: effects of phosphorylated metabolites on enzymes of carbohydrate metabolism, *J. Biol. Chem.,* 259, 5115, 1984.

32. Tordoff, M. G., Rafka, R., DiNovi, M. J., and Friedman, M. I., 2,5-Anhydro-D-mannitol: a fructose analogue that increases food intake in rats, *Am. J. Physiol.,* 254, R150, 1988.

33. Tordoff, M. G., Rawson, N., and Friedman, M. I., 2,5-Anhydro-D-mannitol acts in the liver to initiate feeding, *Am. J. Physiol.,* 261, R283, 1991.

34. Ritter, S., Dinh, T. T., and Friedman, M. I., Induction of Fos-like immunoreactivity (Fos-li) and stimulation of feeding by 2,5-anhydro-D-mannitol (2,5-AM) require the vagus nerve, *Brain Res.,* 646, 53, 1994.

35. Singer, L. K., Magluyan, P., and Ritter, S., The effects of low, medium and high fat diets on 2-deoxy-D-glucose (2DG)- and mercaptoacetate (MA)-induced feeding, *Physiol. Behav.,* 60, 321, 1996.

36. Kanarek, R. B., Marks-Kaufman, R., Ruthazer, R., and Gualtieri, L., Increased carbohydrate consumption by rats as a function of 2-deoxy-D-glucose administration, *Pharmacol. Biochem. Behav.,* 18, 47, 1983.

37. Cromer, L., Koegler, F. H., and Ritter, S., 2-Mercaptoacetate (MA) and 2-deoxy-D-glucose (2DG) induce qualitatively different macronutrient appetites, *Soc. Neurosci. Abstr.,* 19, 1695, 1993.

38. Ritter, S., Ritter, J. B., and Cromer, L., 2-Deoxy-D-glucose and mercaptoacetate induce different patterns of macronutrient ingestion, *Physiol. Behav.,* 66, 709, 1999.

39. Stockinger, Z. and Geary, N., 2-Mercaptoacetate stimulates sham feeding in rats, *Physiol. Behav.,* 47, 1283, 1990.

40. Covasa, M. and Ritter, R. C., Rats maintained on high fat diets exhibit reduced satiety in response to CCK and bombesin, *Peptides,* 19, 1407, 1998.

41. Brenner, L. A., Covasa, M., and Ritter, R. C., Dietary adaptation increases digestive capacity and decreases satiety response to macronutrients, *Soc. Neurosci. Abstr.,* 23, 253, 1997.

42. Singer, L. K., York, D. A., and Bray, G. A., Feeding response to mercaptoacetate in Osborne-Mendel and S5B/PL rats, *Obesity Res.,* 5, 587, 1997.

43. Van Dijk, G., Scheurink, A., Ritter, S., and Steffens, A., Glucose hemostasis and sympathoadrenal activity in mercaptoacetate-treated rats, *Physiol. Behav.,* 57(4), 759, 1995.

44. Scheurink, A. and Ritter, S., Sympathoadrenal responses to glucoprivation and lipoprivation in rats, *Physiol. Behav.,* 53, 995, 1993.

45. Ritter, S., Singer, L. K., and Scheurink, A., 2-Deoxy-D-glucose but not mercaptoacetate increases Fos-like immunoreactivity in adrenal medulla and sympathetic preganglionic neurons, *Obesity Res.,* 3, 729S, 1995.

46. Sclafani, A., Nissenbaum, J. W., and Vigorito, M., Starch preference in rats, *Neurosci. Biobehav. Rev.,* 11, 253, 1987.

47. Aoki, T. T., Muller, W. A., Brennan, M. F., and Cahill, G. F., Metabolic effects of glucose in brief and prolonged fast in man, *Am. J. Clin. Nutr.,* 28, 507, 1975.

48. Langhans, W. and Scharrer, E., Evidence for a vagally mediated satiety signal derived from hepatic fatty acid oxidation, *J. Auton. Nerv. Syst.,* 18, 13, 1987.

49. Li, E. T. S. and Anderson, G. H., A role for vagus nerve in regulation of protein and carbohydrate intake, *Am J. Physiol.,* 247, E815, 1984.

50. Washburn, B. S., Jiang, J. C., Cummings, S. L., Dixon, K., and Gietzen, D. S., Anorectic responses to dietary amino acid imbalance: effects of vagotomy and tropisetron, *Am. J. Physiol.,* 266, R1922, 1994.

51. Benoit, S. D. and Davidson, T. L., Interoceptive sensory signals produced by 24-h food deprivation, pharmacological glucoprivation and lipoprivation, *Behav. Neurosci.,* 110, 168, 1996.

52. Bellin, S. I., and Ritter. S., Insulin-induced elevation of hypothalamic norepinephrine turnover persists after glucorestoration unless feeding occurs, *Brain Res.,* 217, 327, 1981.

53. McCaleb, M. L., Myers, R. D., Singe, C., and Willis, G., Hypothalamic norepinephrine in the rat during feeding and push-pull perfusion with glucose, 2DG, or insulin, *Am. J. Physiol.,* 236, R312, 1979.

54. McCaleb, M. L. and Myers, R. D., 2-Deoxy-D-glucose and insulin modify release of norepinephrine from rat hypothalamus, *Am. J. Physiol.*, 242, R596, 1982.

55. Thompson, D. A., Penicaud, L., and Welle, S. L., Alpha-2 adrenoreceptor stimulation inhibits thermogenesis and food intake during glucoprivation in humans, *Am. J. Physiol.*, 247, R560, 1984.

56. Booth, D. A., Modulation of the feeding response to peripheral insulin, 2-deoxyglucose or 3-*O*-methyl glucose injection, *Physiol. Behav.*, 8, 1069, 1972.

57. He, B., White, D., Edwards, G. L., and Martin, R. J., Neuropeptide Y antibody attenuates 2-deoxy-D-glucose induced feeding in rats, *Brain Res.*, 781, 348, 1998.

58. Stanley, B. G., Daniel, D. R., Chin, A. S., and Leibowitz, S. F., Paraventricular nucleus injections of peptide YY and neuropeptide Y preferentially enhance carbohydrate ingestion, *Peptides*, 6, 1205, 1985.

59. Leibowitz, S. F., Weiss, G. F., Yee, F., and Tretter, J. B., Noradrenergic innervation of the paraventricular nucleus, specific role in control of carbohydrate ingestion, *Brain Res. Bull.*, 14, 561, 1985.

60. Stanley, B. G., Neuropeptide Y in multiple hypothalamic sites controls eating behavior, endocrine, and autonomic systems for body energy balance, in *The Biology of Neuropeptide Y and Related Peptides*, Colmers, W. F. and Wahlestedt, C., Eds., Humana Press, Totowa, NJ, 1993, 457.

61. Koch, J. E. and Bodnar, R. J., Selective alterations in macronutrient intake of food-deprived or glucoprivic rats by centrally-administered opioid receptor subtype antagonists in rats, *Brain Res.*, 657, 191, 1994.

62. Arjune, D. and Bodnar, R. J., Suppression of nocturnal, palatable and glucoprivic intake in rats by the κ-opioid antagonist, nor-binaltorphamine, *Brain Res.*, 534, 313, 1990.

63. Beverly, J. L., Beverly, M. F., and Meguid, M. M., Alterations in extracellular GABA in the ventral hypothalamus of rats in response to acute glucoprivation, *Am. J. Physiol.*, 269, R1174, 1995.

64. Corwin, R. L., Robinson, J. K., and Crawley, J. N., Galanin antagonists block galainin-induced feeding in the hypothalamus and amygdala of the rat, *Eur. J. Neurosci.*, 5, 1528, 1993.

65. Crawley, J. N., Austin, M. C., Fiske, S. M., Martin, B., Sonsolo, S., Berhold, M., Langel, U., Fisone, G., and Bartfai, T., Activity of centrally administered galanin fragments on stimulation of feeding behavior and on galanin receptor binding in the rat hypothalamus, *J. Neurosci.*, 10, 3695, 1990.

66. Kyrkouli, S. E., Stanley, B. G., Seifafi, R. D., and Leibowitz, S. F., Stimulation of feeding by galanin: anatomical localization and behavioral specificity of this peptide's effects in the brain, *Peptides*, 11, 995, 1990.

67. Smith, B. K., York, D. A., and Bray, G. A., Effects of dietary preference and galanin administration in the paraventricular or amygdaloid nucleus on diet self-selection, *Brain Res. Bull.*, 39, 149, 1996.

68. Koegler, F. H., and Ritter, S., Aqueduct occlusion does not impair feeding induced by either third or fourth ventricle galanin injection, *Obesity Res.*, 5(3), 262, 1997.

69. Koegler, F. H., and Ritter, S., Feeding induced by pharmacological blockade of fatty acid metabolism is selectively attenuated by hindbrain injections of the galanin receptor antagonist, M40, *Obesity Res.*, 4, 329, 1996.

70. Temple, D. L. and Leibowitz, S. F., Diurnal variations in the feeding responses to norephinephrine, neuropeptide Y and galanin in the PVN, *Brain Res. Bull.*, 25, 821, 1988.

14 Memory and Macronutrient Regulation

T. L. Davidson, Javier R. Morell, and Stephen C. Benoit

CONTENTS

1 INTRODUCTION

The survival of an animal depends, in part, on how efficiently it can regulate its supply and use of metabolic fuels and essential nutrients. Traditionally, the question of how animals regulate their bodily energy supplies has been addressed within homeostatic models. According to these views, animals eat when departures from energy balance give rise to internal deficit or hunger signals. Food intake restores energy balance and removes the energy deficit signals that served to stimulate intake.[1,2] The regulation of macronutrient intake has also been described within a homeostatic framework. For example, Bray[3] made the case that the intake of fats, carbohydrates, and proteins is "regulated separately and then integrated into total energy needs" (p. 19). According to Bray, different stores exist for each of these metabolic fuels and feedback specific to each store signals departures from macronutrient homeostasis.

The efficient utilization of energy resources and macronutrients also depends critically on the operation of behavioral mechanisms that are involved with obtaining and consuming food. Modern views often link the operation of these mechanisms to the encoding, storage, and retrieval of sensory information.[4] It is well established that animals can encode and remember detailed information about the events that they encounter.[5] Animals not only learn about the events themselves, but also store information about predictive relationships between different events and about specific conditions under which those predictive relationships might be modified.[6] There is little doubt that food and cues related to food are among the most salient stimuli that are encountered by animals. Therefore, understanding the control of feeding behavior will involve identifying what food-related stimuli are encoded by animals, specifying what animals "know" about the relationships among those cues, and describing how this information is utilized to yield specific behavioral outcomes.

2   A ROLE FOR PAVLOVIAN CONDITIONING

How can the contents of animal memory be determined? How do memories determine behavior? Pavlovian conditioning provides a methodological and conceptual framework for dealing with these questions. An idea common to many models of Pavlovian conditioning is that once a stimulus has been detected by a sensory system it can become part of a memory network that is made up of many distinct nodes or units.[4,7] The sum of what an animal "knows" about anything, including food, is defined with respect to the events that are represented in this network. Predictive relationships between different events can also become part of an animal's knowledge structure. For example, if a conditioned stimulus (CS) consistently predicts the occurrence of an unconditioned stimulus (US), the detection of the CS not only activates its own memory unit, but also the representational unit of the US. Because the US is capable of evoking a behavioral response (known as the unconditioned response) when it is presented on its own, activation of the memory of the US following presentation of the CS will also be capable of evoking a response (known as a conditioned response, CR). The ability of CS to evoke a CR that anticipates of the occurrence of a US is usually viewed as evidence that the representations of the CS and US are connected or associated in memory.

2.1   SIMPLE PAVLOVIAN CONDITIONING

This type of association formation is considered to be the basic outcome of simple Pavlovian conditioning. This outcome is recognizable in several situations that involve the control of feeding behavior. Most notably, the ability of animals to avoid toxins and to select nutritious foods rather than non-nutritious substances is commonly attributed to the formation of associations between tastes or other orosensory cues and the postingestive consequences of intake.[8] Within this framework, orosensory stimulation serves as a CS that directly activates or retrieves the memory of a postingestive US with which it is associated. If the postingestive US is aversive (e.g., intragastric malaise) the orosensory CS will evoke defensive CRs. If the postingestive US is appetitive (e.g., nutrient or metabolic repletion) the CS will evoke approach and consummatory behaviors.

2.2   PAVLOVIAN OCCASION SETTING

A fact that may not be widely appreciated by investigators of the controls of food intake, is that Pavlovian conditioning also involves a learning process distinct from the simple conditioning process described thus far. We will refer to this process as *occasion setting*[9] although it is also known as conditioned facilitation.[10] A stimulus can be established as an occasion setter by embedding it within a discrimination problem of the general form $X \rightarrow A+$, $X-$, $A-$. In this problem, presentation of an occasion setting cue (X) signals when a target CS (A) will be followed by delivery of a US (on $X \rightarrow A+$ trials). The US is not delivered when the occasion setter ($X-$ trials) or the target CS ($A-$ trials) is presented alone. Animals show that they have solved this problem by exhibiting more CRs on $X \rightarrow A+$ trials than $X-$ or $A-$ trials. In contrast to a stimulus trained as simple CS, a stimulus trained as an occasion setter (i.e., stimulus X) usually has little capacity to evoke CRs. Rather, occasion setters are thought to promote discrimination performance by augmenting the capacity of the target CS (stimulus A) to evoke CRs.

According to a number of formulations, an occasion setter accomplishes this by lowering the threshold for activation of the memorial representation of any US with which it has been trained.[10-12] By lowering the threshold for memory activation, the occasion setter makes it easier for the target CS, as well as other associates of that US, to excite the US representation and thereby evoke CRs.

Occasion Setting as a Mechanism
for Memory Modulation

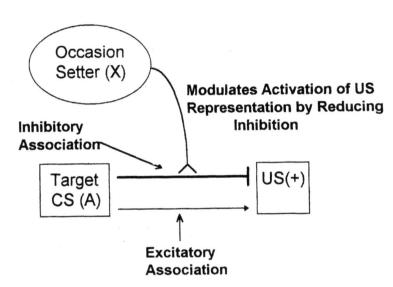

FIGURE 1 Hypothetical mechanism by which Pavlovian occasion-setting cues augment the capacity of a target CS to excite the memorial representation of a US. The strength of conditioned feeding behavior is assumed to be a positive function of activation of this US representation.

In other words, occasion setters are distinct from CSs, in that they have little ability to excite directly the memory of the US with which they were trained. Rather, occasion setters promote conditioned responding by increasing the likelihood that an animal will remember the US when a CS associated with that US is encountered.

Figure 1 outlines a mechanism that would enable occasion setters to perform their modulatory function. According to the figure, a CS is an appropriate target for occasion setting only if it has concurrent excitatory and inhibitory associations with the US. That is, a target CS must have been subjected to excitatory training that enables presentation of that CS to activate or retrieve the memory of the US. Also, the target must have been exposed to extinction or some other procedure that would result in the formation of an inhibitory association that would reduce the capacity of the CS to excite the US representation. As the diagram shows, an occasion setter lowers the threshold for activation of the US representation by removing or reducing the strength of the inhibitory target CS–US association. Reducing inhibition makes it more likely the opposing excitatory association with the CS will be able to activate the memory of the US. The results of several studies confirm that subjecting CSs to simple training followed by extinction makes them especially effective targets for occasion setting.[12,13]

This analysis also predicts that an occasion setter will make it easier for the US memory to be activated by *any* CS as long as the CS has concurrent excitatory and inhibitory associations with the same US that was trained with that occasion setter. In contrast, an occasion setter will be unable to promote responding to a transfer CS that is associated with a US different from the US that was trained with the occasion setter. In other words, the capacity of an occasion setter to reduce or remove inhibition is US specific.

Data confirming this prediction were obtained recently in our laboratory. Javier Morell trained rats on a X → A+, X–, A– problem in which a visual (i.e., a light) occasion setter (X) signaled when an auditory (i.e., a tone) target CS (A) would be followed by delivery of a US(+). For half the rats in this study the US was peanut oil and for the remaining rats it was sucrose pellets. After all rats solved the discrimination problem, they received simple excitatory training then partial extinction with a novel auditory (i.e., a clicker) CS (B). For half of the rats, this transfer target was trained with the same US that was used in original training, and for half of the rats the other US was used to train the transfer CS. The capacity of the occasion setter to promote responding to the transfer target (B) was then in extinction assessed as a function of the US (same or different from the US used in training) that was associated with that transfer CS.

When training was completed, presentation of the occasion setter strongly augmented responding to the originally trained target, regardless of whether that target had signaled peanut oil or sucrose pellets. The left portion of Figure 2 shows the results of the transfer test when the transfer target had been trained with the same US as was used in original training. Under this condition, presenting the occasion setter in conjunction with the transfer target (X → B– trials) promoted robust conditioned responding (i.e., approach to the food cup) compared to when either the occasion setter (X– trials) or the transfer target (B– trials) was presented alone. In contrast, the right portion of Figure 2 shows that the occasion setter was unable to promote responding to the transfer target when the target had undergone training with a US different from that trained originally with the occasion setter.

This outcome is consistent with the previously stated hypothesis that an occasion setter will make it easier for the US memory to be activated by *any* CS that has concurrent excitatory and inhibitory associations with that specific US memory. Also important, these results confirm our earlier[14] findings that the peanut oil and sucrose pellets produce distinct US memories. Neither of these USs was presented during the test phase so the differences in performance shown in Figure 2 must be based on what the rats remembered about those USs. In addition, if rats did not differentiate

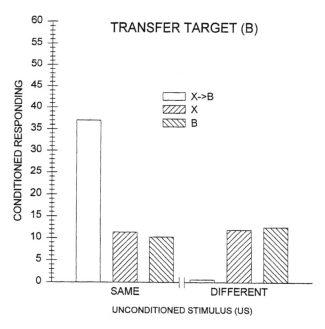

FIGURE 2   Mean percent appetitive behavior (i.e., percent of time that an infrared photobeam near the food cup was interrupted) during transfer testing on trials with the occasion setter presented in conjunction with the transfer target (X → B), and on trials where the occasion setter (X) and the transfer target (B) were presented separately. Performance when the transfer target had been trained with the same US employed during original training is shown on the left side of the figure, whereas performance when the transfer target had been trained with a different US than was employed during original training is shown on the right.

between the memories of the peanut oil and sucrose USs it is difficult to explain why test performance depended on which US (the same or different) had been associated with the transfer target prior to testing.

3   OCCASION SETTING AS A MODEL FOR INTAKE REGULATION

The purpose of this chapter is to examine the implications of occasion setting for understanding macronutrient regulation. As a point of departure for this examination, we begin with the assumption that adaptive feeding behavior requires animals to solve a set of discrimination problems that have the same general form as the $X \rightarrow A+$, $X-$, $A-$ problem discussed in the preceding section. As will be illustrated momentarily, when animals solve this problem as it relates to intake regulation, the stimuli that serve as occasion setters, target CSs, and USs are different from the cues that have these roles in laboratory studies of discrimination learning. Nonetheless, in both situations, solving this problem enables animals to predict more accurately the occurrence of appetitive USs. Furthermore, the principles that determine how animals solve these problems are also assumed to be the same both inside and outside of the laboratory.

3.1   OCCASION SETTING AND ENERGY REGULATION

Within the present theoretical framework, the problem of macronutrient regulation can be viewed as a special case of the more general problem of energy regulation. That is, from the perspective of an animal, maintaining energy homeostasis is essentially a problem of determining when to eat and when to refrain from eating. On the other hand, to solve the problem of maintaining macronutrient balance, animals must not only decide when to eat but also what foods to eat and to avoid eating. Because our ideas about macronutrient regulation are derived primarily from an earlier model that describes a role for occasion setting mechanisms in maintaining energy homeostasis, it makes sense to summarize briefly the main points of that model here. For a more complete discussion of occasion setting and energy regulation see Davidson.[15]

We proposed that animals solve the problem of when to eat and when to refrain from eating by using interoceptive stimuli arising from their condition of energy balance to signal when stimuli related to food (exteroceptive or orosensory cues) will be followed by appetitive postingestive consequences. For convenience, these appetitive postingestive consequences can be viewed as the stimulus concomitants of a return to energy homeostasis (the possibility that nonhomeostatic mechanisms may be important contributors of postingestive stimuli is not ruled out by our model). This problem has the same general form as the $X \rightarrow A+$, $X-$, $A-$ problem described above. That is, animals must learn that food cues (A) will be followed by the occurrence of an appetitive postingestive US(+) only when energy deficit or hunger cues (X) are also present (i.e., on $X \rightarrow A+$ trials). Appetitive postingestive USs (produced by restoration of energy balance) do not occur when food cues are present but hunger cues are not (i.e., on $A-$ trials) or when hunger cues are present but food cues are not (i.e., on $X-$ trials)

Within this arrangement, hunger cues influence feeding behavior by lowering the threshold for activation of the memorial representation of an appetitive postingestive US. This enhances the capacity of conditioned food cues to evoke that memory and to elicit conditioned responses. Thus, unlike simple excitatory cues, behavioral control by hunger cues does not depend on their involvement in a direct association with a US or a response. Rather, hunger cues are seen as modulating or setting the occasion for the evocation of feeding behaviors by conditioned food cues.

One feature of the model that deserves special mention is that it emphasizes the CS properties of orosensory stimuli while neglecting the US properties of these cues. That is, there is little doubt that in addition to serving as signals for important events, orosensory cues (e.g., sweet tastes) can be potent USs that are capable of evoking unconditioned appetitive behaviors. However, according to the present analysis, hunger stimuli do not set the occasion (i.e., lower the threshold) for the activation of the memories of orosensory USs. This means that conditioned or unconditioned responses controlled *only* by orosensory USs will not be potentiated by hunger cues. Thus although the presence of hunger cues make it easier to remember the appetitive postingestive USs that are associated with tastes, hunger does not make it easier to remember the tastes themselves.

This is not an arbitrary feature of the model. As a general rule, the capacity of any stimulus to serve as an occasion setter appears to depend on the amount of information it provides about the reinforcement of other stimuli.[11] According to the model, hunger cues are informative about whether or not a food-related cue will be followed by a postingestive US that accompanies a movement toward homeostasis. In contrast, whether a particular orosensory cue gives rise to the sensation of sweetness does not depend on whether or not the animal is hungry. Thus, the performance of appetitive responses based on learning about orosensory USs will not be modulated by hunger cues, whereas hunger stimuli will modulate the strength of responses that are based on learning about appetitive postingestive USs.

3.2   MACRONUTRIENT REGULATION: FATS AND CARBOHYDRATES

Table 1 reviews the conceptual linkage between conventional Pavlovian occasion-setting training with ordinary lights and tones as stimuli and occasion setting that involves learning about hunger cues, food CSs, and appetitive postingestive USs. The bottom portion of Table 1 extends these ideas to describe the stimuli that animals would use when solving the problem of when to eat and when to refrain from eating fats and carbohydrates, respectively. To solve this problem, animals must be able to detect distinct sensory stimuli that signal the occurrence of deficiencies in their fat and carbohydrate supplies, respectively, or in the ability to utilize these supplies. These lipoprivic and glucoprivic cues would function as occasion setters within the present Pavlovian framework.

In addition, animals must be able to learn which foods contain which of these two macronutrients. Thus, to provide distinct target stimuli, the foods that replenish fat supplies must be different from foods that replenish supplies of carbohydrate. Orosensory cues (sweet taste, oily texture) are likely to provide one basis for differentiation between fat and carbohydrate target cues. Finally, separate regulation of fat and carbohydrate also requires each macronutrient to produce distinct postingestive feedback stimuli that would be capable of serving the role of Pavlovian USs. In other words, when the animal is in negative fat balance, consuming fat would produce a pattern of postingestive feedback unlike that produced when carbohydrate is consumed or when fat is consumed under conditions of fat homeostasis. Thus, fat regulation could be accomplished, to the extent that lipoprivic cues signaled selectively that oily tastes would be followed by at least partial restoration of fat homeostasis. Likewise, carbohydrate intake could be regulated, at least in part, to the extent that glucoprivic cues signaled selectively that sweet tastes would be followed by movement toward carbohydrate homeostasis.

Within our theoretical framework, lipoprivic and glucoprivic stimuli serve as occasion-setting cues that selectively lower the threshold for excitation of the memories of fats and carbohydrates,

TABLE 1
A Comparison of Three Potential Occasion-Setting Paradigms

Paradigm	Occasion Setter	Target	US
Conventional	Light	Tone	Food
Energy regulation	Hunger cues	Food	Restoration of energy balance
Macronutrient regulation	Lipoprivic cues	Oily foods	Restoration of lipid balance
	Glucoprivic cues	Sweet foods	Restoration of carbohyrate balance

respectively. Lowering the threshold for activation of a particular US representation makes it easier for the target CSs (e.g., sweet or oily orosensory cues) associated with that memory to evoke conditioned approach and consummatory responses. To the extent that different target CSs are associated with the memories of different macronutrients, the capacity for macronutrient deficit cues to promote selectively the activation of the memories of specific macronutrients provides a learned basis for adaptive food selection.

4   EVALUATION OF THE MODEL

Whether or not animals actually do behaviorally regulate their intake of fat and carbohydrate over time or in response to dietary deficiencies remains an open question. Although some early findings showed that rats self-select a relatively constant proportion of their calories from carbohydrate, fat, and protein when given the opportunity to choose freely among separate sources of these metabolites,[16,17] this pattern has not always been found.[18] Part of the difficulty in establishing that animals regulate behaviorally their bodily supplies of fat and carbohydrates may have to do with the use of ingestion as the sole index of regulation.

Previous research has often relied on intake of pure or relatively pure sources of different metabolic fuels to assess whether or not animals regulate their bodily supplies of these fuels in response to signals of specific needs or deficiencies. Yet, as Rozin[18] pointed out, differences in the palatability of the test foods could give rise to strong preferences that might mask or reduce the effectiveness of regulatory processes in the face of dietary deficiencies. Conversely, Galef[19] cautioned that a given set of flavor preferences could also yield a pattern of intake consistent with the operation of a homeostatic regulatory process even when such processes are absent.

In terms of the present learning analysis, intake tests can pit the capacity of orosensory cues to evoke unconditioned ingestive responses against their capacity to evoke conditioned or learned feeding behaviors. These unconditioned responses have the apparent advantage of being evoked by *direct physical contact* with a potentially potent orosensory US (e.g., sweet taste). In comparison, according to the present model, orosensory cues evoke conditioned responses not by direct contact with a US, but by exciting the *memory* of a US. Thus, although an animal may recall that it needs a particular macronutrient, activation of that knowledge in memory may not always be enough to overcome direct stimulus control by a highly palatable taste. Unfortunately, it is usually difficult to estimate which response, the unconditioned or the conditioned, will be evoked most strongly by a given orosensory cue. Presumably, the outcome of such a competition is determined by many factors, including the intensity of the orosensory US and the degree of activation of the memory of the postingestive US.

Our way of dealing with this problem is to assess the capacity of animals to regulate macronutrient intake under conditions that avoid this type of response competition. In the experiments that follow, we accomplished this by eliminating intake measures as the index of self-regulation. Instead, we examined the capacity of conventional Pavlovian CSs to activate the memories of fat and carbohydrate USs, following pharmacological treatments designed to produce interoceptive glucoprivic or lipoprivic cues.

4.1   METABOLIC MANIPULATIONS CAN SELECTIVELY ACTIVATE THE MEMORIES OF PEANUT OIL AND SUCROSE USs

A Pavlovian occasion setter promotes the performance of CRs is by making it easier for a Pavlovian CS to activate the memory of the US with which it is associated. This analysis, applied to the behavioral regulation of bodily supplies of fat and carbohydrate, proposes that separate regulation of fat and carbohydrates stores depends, in part, on the formation of separate memorial representations of fat and carbohydrate USs.

As we reported earlier in this chapter, a stimulus trained as an occasion setter promoted responding to a transfer target only if that transfer target was associated with the same US that was subject to occasion setting in original training. It is difficult to account for this finding without assuming that the rats remembered that the US associated with the training CS was different from the US associated with the transfer CS. Because no USs were presented in the test phase, judgment whether the USs were the same or different must have been based on a comparison of the memories of each US. The USs employed were also peanut oil and sucrose pellets. Therefore, this study also supports the idea that rats form distinct memories of fats and carbohydrates.

For fat and carbohydrate regulation to be based on the occasion-setting mechanism that we propose, lipoprivic and glucoprivic stimuli must set the occasion for activating the memories of fat and carbohydrate USs, respectively. Assuming that the representations of peanut oil and sucrose pellets contain at least some information about their respective macronutrient compositions, lipoprivic cues should selectively activate the memory for peanut oil and glucoprivic cues should selectively activate the memory for sucrose pellets. If the representations of peanut oil and sucrose only contain information that is irrelevant or unrelated to their distinct macronutrient contents, there would seem to be little basis to expect that the activity of these representations would be selectively modulated by lipoprivic or glucoprivic signals.

To test these implications, Davidson et al.[14] gave food-deprived rats simple Pavlovian training in which one CS signaled peanut oil and another signaled sucrose pellets. We trained rats to approach a food cup to obtain either 0.3 ml of peanut oil (100% fat) or two 0.45-g sucrose pellets (100% carbohydrate). The delivery of each of these USs was signaled by the presentation of different punctate CSs (auditory, visual stimuli). The identities of the CSs were counterbalanced for each US. The index of conditioned responding was the percent of time during each 10-s CS that the rats interrupted a photobeam that was located directly in front of the food cup, minus the percent of time that photobeam was interrupted during the 10-s pre-CS period that ended immediately prior to presentation of each CS. Throughout the experiment, a computerized system recorded photobeam disruptions during pre-CS periods and during presentation of each CS. At the end of training the rats showed robust conditioned responding to both CSs.

After completing this training, all of the rats were food sated. Conditioned responding to each CS was then tested following treatments with antimetabolites that were designed to produce either lipoprivic or glucoprivic cues. In one experiment, one group of rats received systemic Na-2-mercaptoacetate (400 μmol/kg) prior to testing. This dose of Na-2-mercaptoacetate (MA) gives rise to lipoprivation by blocking fatty acid oxidation. Another group received an equivalent volume of isotonic saline (SAL) injected interperitoneally (ip). A third group received a 350 mg/kg injection ip of 2-deoxy-D-glucose (2DG). 2DG is a glucose antimetabolite that is thought to be capable of inducing glucoprivic signals. All injections occurred approximately 1 h before testing began. No USs were delivered during testing.

The results of the test phase, expressed as the percent of appetitive behavior recorded during the each 10-s CS minus the percent of appetitive behavior during the 10-s baseline period that preceded the delivery of each CS, are presented in Figure 3. The figure shows that rats treated with MA responded significantly more to the CS for oil than to the CS for sucrose pellets. Responding for rats treated with saline did not differ dependent on the CS. Neither did responding depend on the CS for the rats that received 2DG. However, this dose of 2DG appeared to produce a nonspecific reduction in behavioral activity that rendered the results uninterpretable for rats treated with this compound.

To address this difficulty, we conducted additional experiments that assessed the effects of intracerebroventricular (ICV) infusion of the glucose antimetabolite 5-thio-d-glucose (5-TG). We used a dose (150 μg) that produced little evidence of nonspecific behavioral deactivation. The results of one such study[20] are shown in Figure 4 (also see Reference 14). Here, ICV infusion of 5-TG augmented responding to the CS for sucrose more than to the CS for oil. This difference was

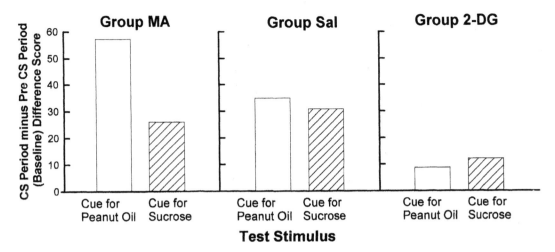

FIGURE 3 Mean percent appetitive behavior (food cup approach) during the 10-s CS period minus that observed during the immediately preceding 10-s pre-CS baseline period. The data are shown for the CS that previously signaled peanut oil (open bars) and the CS that previously signaled sucrose pellets (diagonal bars), following treatment with 400 mmol/kg na-2-MA (Group MA), isotonic saline (Group Sal), or 350 mg/kg 2DG (Group 2-DG).

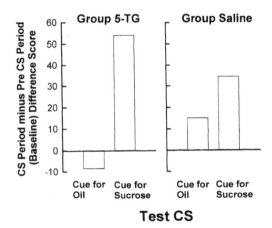

FIGURE 4 Mean percent appetitive behavior (food cup approach) during the 10-s CS period minus that observed during the immediately preceding 10-s pre-CS baseline period. The data are presented for the CS that previously signaled peanut oil and the CS that previously signaled sucrose pellets, following ICV infusion of 150 mg 5-TG (Group 5-TG) or isotonic saline (Group SAL).

much larger for rats treated with 5-TG than for rats that had been administered ICV saline prior to testing.

These experiments agree with the hypothesis that lipoprivic signals induced by administration of MA selectively promote conditioned responding based on the memory of the peanut oil US. In contrast, glucoprivic cues, at least those produced by ICV infusion of 5-TG, selectively promoted responding based on the memory of sucrose pellets. Moreover, approach to the food cup was not augmented as a direct effect of administration of MA or of ICV 5-TG. These treatments selectively promoted behavior only in the presence of cues that were associated with delivery of the macronutrient that was most relevant to the metabolic challenge induced by treatment. The results suggest that what rats "remember" about peanut oil and sucrose pellet USs includes at least some information about their respective macronutrient contents. There would seem to be little basis for differential performance in the face of lipoprivic and glucoprivic challenges, if rats were unable to encode this type of information.

4.2   INHIBITORY ASSOCIATIONS AND MACRONUTRIENT REPRESENTATIONS

The basic occasion-setting model also has implications with respect to the effects of restoring energy and macronutrient balance. Recall that the ability of an occasion setter to modulate the response-eliciting capacity of a target CS depends on the existence of an inhibitory association between the target CS and the memory of the US. Specifically, occasion setters are thought to remove or reduce the strength of these inhibitory associations, thereby enabling a previously established excitatory association between the same CS and US representation to be more strongly expressed in performance. One implication of this analysis is that occasion setters will have little capacity to influence memory activation when inhibitory associations are absent.

Davidson[15] proposed that eating food when food sated results in the formation of inhibitory associations between the CS properties of the food and the memory of its appetitive postingestive consequences. Based on conventional learning principles, an inhibitory association forms when an excitatory CS activates the memory of its US but the actual US does not occur. Accordingly, Davidson[15] proposed that eating food when food sated involves activation of the memory of the appetitive postingestive US by a food CS, under conditions where the actual appetitive postingestive US (e.g., return to homeostasis) does not occur. If animals have not eaten food when sated, an inhibitory association between food cues and the memory of the appetitive postingestive US is not likely to develop.

This theoretical framework yields the prediction that the performance of appetitive CRs will be largely unaffected by changes in hunger level for animals that have not been exposed previously to the CS when food sated. To evaluate this prediction, we modified a research design like that described by Dickinson and Baleine.[20] We[21] placed food-deprived rats in an apparatus where sucrose pellets were delivered about once every 2 min during each of eight 30-min training sessions. As training proceeded, the rats spent an increasing amount of time in front of the food cup as indexed by interruption of the photobeam. After training was completed, the rats were placed in a different apparatus. Half the rats were exposed to the apparatus when food sated and half remained food deprived. Within each of these groups, half the rats were given the opportunity to consume 30 sucrose pellets and half were not given sucrose pellets. Both the food-deprived and the food-sated rats consumed all of the pellets. Finally, all of the rats were returned to the original training apparatus and tested under the same deprivation conditions that were in effect during the preceding US exposure phase of the experiment. Testing was conducted in the same manner as original training except that no sucrose pellets were delivered. Incidence of photobeam interruptions was recorded for all rats.

The results of test phase are presented in Figure 5. The test phase data are expressed as a percentage of baseline performance recorded on the last day of the original training. Satiation markedly suppressed conditioned responding during extinction testing for rats that had been previously been exposed to sucrose pellets when sated. That is, rats that were both exposed to sucrose and tested under satiation (Group SE) showed a much greater drop from baseline than did rats that were exposed to sucrose and tested food deprived (Group DE). In contrast, rats that were tested under satiation but that were not exposed to sucrose when they were sated (Group SN) did not show this large drop from baseline performance levels. Test performance for these rats was quite like controls that had not been exposed to sucrose and remained food deprived during testing (Group DN). In fact, the test performance of Group SN did not differ significantly from either group (DE or DN) that was tested when food deprived. In summary, a strong suppressive effect

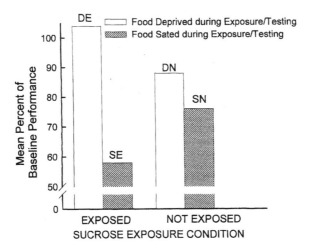

FIGURE 5 Mean percent of baseline appetitive behavior during testing. Rats exposed to sucrose pellets when food sated prior to testing are designated SE. Rats food sated but not exposed to pellets prior to testing are designated SN. Rats exposed to pellets but not food sated prior to testing are designated DE. Rats not food sated or exposed to pellets prior to testing are designated DN. Baseline refers to percent of time photobeam nearest the food cup was interrupted at the end of initial training.

of satiation was observed only for Group SE, which had been previously exposed to sucrose pellets when food-sated.

The results are consistent with the idea that consuming sucrose pellets when food sated provided the opportunity to learn that the orosensory CS produced by sucrose was no longer followed by an appetitive postingestive US. This opportunity for inhibitory learning reduced the capacity of cues in the training apparatus to activate the memory the appetitive US when the rats were sated during the test phase. The fact that satiation appeared to have little suppressive effect on the performance of rats that did not have the opportunity to learn this inhibitory relationship is also consistent with this interpretation. The results indicate that animals must learn that the consequences of intake are different under satiation, compared with food deprivation, before satiation will have a strong suppressive effect on the performance of food approach CRs. This was also the basic conclusion of Dickinson and Balleine[20] based on their studies of food satiation and bar pressing.

If supplies of fats and carbohydrates are regulated separately at homeostatic levels, animals might learn this relationship separately for each of these macronutrients. As part of his Ph.D. dissertation research, Benoit[21] used a variation of the US exposure technique to examine if rats could be selectively satiated for sucrose pellets, without being satiated for peanut oil. In one experiment, Benoit trained one group of hungry rats with a tone CS that signaled the delivery of peanut oil. For another group, the tone signaled the delivery of sucrose pellets. After training was completed, all the rats were given *ad libitum* food for several days and were then placed in another apparatus where they were allowed to eat the US that they received during initial training. However, 1 h prior to US exposure, half the rats in each US condition were administered MA (400 μmol/kg) and half received isotonic SAL. All the rats ate all of the peanut oil (9.0 ml) or sucrose (30 45-g pellets) presented during the exposure session. The next day, the rats were returned to the training apparatus, where the capacity of the CS to evoke CRs was tested. All rats remained food sated during the test.

Based on simple associative principles, initial training under hunger with peanut oil would enable the tone CS to excite the memory of the taste of the peanut oil. This memory would itself be associated with the memory of the postingestive US produced by peanut oil. Similarly, following training with sucrose pellets, the tone would excite the memory of the taste of sucrose and the taste would excite the memory of the appetitive postingestive US produced by sucrose. Upon consuming either peanut oil or sucrose during the exposure phase, the saline control rats would be expected to learn that the taste peanut oil or sucrose pellets are no longer followed by an appetitive postingestive US. That learning would be expressed in the test phase. The presentation of the tone would be less able to activate the memory of the appetitive postingestive US, because the rats had the opportunity to learn that this US does not occur when they are food sated.

The main question of interest was how would the administration of MA affect what the rats learned during the satiated exposure phase. If fat and carbohydrate supplies were regulated separately, rats injected with MA could be simultaneously "hungry" for peanut oil and satiated for sucrose pellets. Under these conditions, rats exposed to peanut oil following MA would be expected to learn little if anything about the effects of satiation on the occurrence of the appetitive postingestive US produced by peanut oil. On the other hand, learning about the postingestive US produced by sucrose pellets would be similar to that for saline controls. Behavioral evidence for this state of affairs would take the form of more conditioned responding during subsequent satiated testing for rats whose CS signaled peanut oil than for rats whose CS signaled sucrose pellets. An interesting feature of this design is that no rats received MA at the time of testing. Thus, differences between the MA and saline treatments must reflect differences between these conditions in learning that occurred during the prior exposure phase.

The results of the test phase are presented in Table 2. As the table shows, responding to the CS for sucrose did not depend on whether the rats had been treated with MA or saline when they consumed sucrose in the prior exposure phase. In contrast, responding to the CS for peanut oil was substantially greater for rats that received MA prior to consuming peanut oil during the exposure phase than for rats that received saline prior to ingesting peanut oil. This pattern of results is consistent with the hypothesis that lipodeprived rats were selectively hungry for oil but were sated for sucrose.

The results suggest that lipodeprived rats learned different things about peanut oil and sucrose pellet USs during the exposure phase, and these differences in learning influenced the activation of the memories of those USs during CS testing. The findings support the hypothesis that rats can be satiated for one macronutrient without being satiated for another. Moreover, it appears that rats modify their behavior toward different macronutrients in ways that help them to compensate for depletions in supplies of specific macronutrients. This ability would enable rats to regulate their intakes of fat and carbohydrates separately.

TABLE 2

The Effects Satiation on Conditioned Responding[a] Evoked by a CS for Sucrose Pellets and a CS for Peanut Oil, for Rats That Had Received MA or Saline during Prior Satiated Exposure to Peanut Oil

		Test CS	
		Cue for Sucrose	Cue for Oil
Treatment during	MA	18.5	41.5
satiated exposure	SAL	20.8	14.9

[a] The data reflect responding during the CS period minus responding during the immediately preceding baseline period.

---

5   CONCLUSIONS AND IMPLICATIONS

Adaptive feeding behavior, including that involved with macronutrient regulation, depends on the integration of biological and behavioral control mechanisms. The central problem addressed by the present chapter has been to describe how changes in the biological mechanisms that underlie regulation might be linked to the occurrence of specific behavioral outcomes.

We approached this problem within the framework provided by contemporary research and theory on animal learning and memory. A basic assumption of this approach is that the ability to encode, remember, and retrieve information enables animals to behave in ways that anticipate the occurrence of important events. We suggested that some of the same mechanisms that enable animals to anticipate events in laboratory experiments on learning and memory also operate extraexperimentally to enable animals to anticipate important events that relate to normal feeding. Specifically, we attempted to use an occasion-setting model to describe how animals might regulate their bodily supplies of fat and carbohydrate.

The involvement of occasion-setting mechanisms in macronutrient regulation would require animals to possess several different types of information. First, animals must "know" which foods contain which macronutrient. We suggested that fats and carbohydrates, respectively, are normally associated with distinct types of orosensory cues. That is, oily tastes are normaly distinctly associated with postingestive stimulation that accompanies repletion of fat stores, whereas sweet tastes are normally associated with postingestive cues that accompany a return to carbohydrate balance. Thus, animals know which foods contain which macronutrients to the extent that they have formed associations between the unique stimulus properties of foods (e.g., sweet or oily tastes) and distinct patterns of postingestive stimulation that usually follow contact with food stimuli.

Of course, everything that an animal knows is assumed to be stored in memory. What determines which memories will be activated at any given moment? More specifically, what determines when the memories for foods will be activated relative to the memories for other things? From the perspective of the present occasion-setting model, the activation of the memories of peanut oil and sucrose depends on the ability of animals to detect sensory stimuli that signal the occurrence of specific macronutrient deficiencies. Detection of these interoceptive stimuli selectively lower the threshold for excitation of memory of foods that supply the deficient macronutrient. Lowering the threshold for activation of a particular US representation makes it easier for the food CSs (e.g., visual, orosensory, or other types of cues) associated with that memory to evoke conditioned approach and consummatory responses. To the extent that different food CSs are associated with the memories of different macronutrients, the capacity for distinct macronutrient deficit cues to promote the activation of the memories of specific macronutrients selectively provides a learned basis for adaptive food selection.

In this chapter, we provided evidence that rats encode and retrieve much, if not all, of the information that is needed for them to regulate separately their bodily supplies of fats and carbohydrates. The data show that rats form distinct representations of peanut oil (100% fat) and carbohydrate (100% sucrose) USs. The activation of these memorial representations is selectively promoted by pharmacologically induced lipoprivic and glucoprivic signals, respectively. This is not a direct effect of lipoprivation or glucoprivation. That is, lipoprivic and glucoprivic signals have little capacity to evoke appetitive behavior on their own. Rather, as specified by the occasion-setting model, these cues appear to augment selectively the capacity of CSs that are associated with peanut oil and sucrose, respectively, to evoke conditioned responses. Finally, previous findings showed that the ability of an occasion setter to promote conditioned responding is limited to CSs that have a history of both excitatory training followed by extinction (i.e., one form of inhibitory training). Data presented in this chapter indicate that the occasion-setting capacities of food deprivation cues and pharmacologically induced lipoprivic signals also appear to be subject to similar constraints.

The present data fall short of providing definitive evidence that the intake of fats and carbo-hydrates is regulated separately in the manner proposed by Bray.[3] We did not provide evidence of any dietary requirement for carbohydrate or fat. Our findings did not show that carbohydrate or fat intake is regulated over time. Neither have we provided evidence that intake of either fat or carbohydrate is defended in response to environmental scarcity. These types of evidence have sometimes been described as fundamental to establishing the existence of a system for controlling the intake of any macronutrient.[22]

However, our findings do indicate that mechanisms capable of serving this type of regulatory function are present in rats. Our data show that rats distinguish between at least one source of pure fat (peanut oil) and one source of pure carbohydrate (sucrose pellets). Animals can associate these sources of different macronutrients with different conditioned stimuli. Thus, animals have a way of identifying different food-related stimuli with sources of different macronutrients. The behavior-promoting capacity of stimuli that are differentially associated with those sources of fat and carbohydrate, respectively, can be augmented by the presence of specific metabolic depletions. Thus, rats seem to be capable of responding to signals of metabolic challenges in ways that would enable them to reduce or eliminate those specific challenges. The existence of separate regulatory systems is also supported by data indicating that it is possible to deprive a rat of one macronutrient (i.e., fat) pharmacologically while leaving it sated for another (i.e., carbohydrate). Finally, all of these findings seem to be integrated within the theoretical framework offered by the present occasion-setting model. The implications of this model for describing the regulation of macronu-trient intake under normal dietary conditions and with different sources of fat and carbohydrates remain to be investigated.

REFERENCES

1. Kissileff, H. R. and Van Itallie, T. B., Physiology of the control of food intake, *Am. Rev. Nutr.*, 2, 371, 1982.
2. Stricker, E. M., Homeostatic origins of ingestive behavior, in *Handbook of Behavioral Neurobiology*, Stricker, E. M., Ed., Plenum Press, New York, 1990, 45.
3. Bray, G. A., Appetite control in adults, in *Appetite and Body Weight Regulation*, Fernstrom, J. D. and Miller, G. D., Eds., CRC Press, Boca Raton, FL, 1994.
4. Bouton, M. E., Context, ambiguity and classical conditioning, *Curr. Directions Psychol. Sci.*, 3, 49, 1994.
5. Holland, P. C., Event representation in Pavlovian conditioning: image and action, *Cognition*, 37, 105, 1990.
6. Rescorla, R. A., Probability of shock in the presence and absence of CS in fear conditioning, *J. Comp. Physiol. Psychol.*, 66, 1, 1968.
7. Wagner, A. R. and Brandon, S. E., Evolution of a structured connectionist model of Pavlovian conditioning (AESOP), in *Contemporary Learning Theories: Pavlovian Conditioning and the Status of Traditional Learning Theory*, Klein, S. B. and Mowrer R. R., Eds., Erlbaum, Hillsdale, NJ, 1989.
8. Capaldi, E. D., Conditioned food preferences, in *The Psychology of Learning and Motivation*, Medin, D., Ed., Academic Press, New York, 1992.
9. Holland, P. C., Occasion-setting in Pavlovian feature positive discrimination, in *Quantitative Analyses of Behavior: Discrimination Processes*, Commons, M. L., Herrnstein, R. J., and Wagner, A. R., Eds., Ballinger, New York, 1983.
10. Rescorla, R. A., Facilitation and inhibition, in *Information Processing in Animals: Conditioned Inhibition*, Miller, R. R. and Spear, N. E., Eds., Erlbaum, Hillsdale, NJ, 1985.
11. Davidson, T. L. and Rescorla, R. A., Transfer of facilitation in the rat, *Anim. Learn. Behav.*, 4, 380, 1986.
12. Swartzentruber, D. and Rescorla, R. A., Modulation of trained and extinguished stimuli by facilitators and inhibitors, *Anim. Learn. Behav.*, 22, 309, 1994.

13. Davidson, T. L. and Benoit, S. C., The learned function of food deprivation cues: a role for conditioned modulation, *Anim. Learn. Behav.*, 24, 46, 1996.
14. Davidson, T. L., Altizer, A. M., Benoit, S. C., Walls, E. K., and Powley, T. L., Encoding and selective activation of "metabolic memories" in the rat, *Behav. Neurosci.*, 111, 1014, 1997.
15. Davidson, T. L., Hunger cues as modulatory stimuli, in *Occasion Setting: Theory and Data,* Schmajuk, N. and Holland, P. C., Eds., American Psychological Association, Washington, D.C., 1998.
16. Richter, C. P., Total self-regulatory functions in animals and human beings, *Harvey Lect.*, 38, 63, 1943.
17. Richter, C. P., Holt, E. L., and Barelare, B., Nutritional requirements for normal growth and reproduction in rats, studied by the self-selection method, *Am. J. Physiol.,* 122, 734, 1938.
18. Rozin, P., Are carbohydrates and protein intakes separately regulated? *J. Comp. Physiol. Psychol.*, 65, 23, 1968.
19. Galef, B. G., Jr., A contrarian view of the wisdom of the body as it relates to dietary self-selection, *Psychol. Rev.*, 98, 218, 1991.
20. Dickinson, A. and Balleine, B. W., Motivational control of goal-directed action, *Anim. Learn. Behav.,* 22, 1, 1994.
21. Benoit, S. C., The Role of the Central Nucleus of the Amygdala in the Modulation of Metabolic Memories, Ph.D. dissertation, Purdue University, West Lafayette, IN, 1998.
22. Fernstrom, J. D., The effects of dietary macronutrients on brain serotonin formation, in *Appetite and Body Weight Regulation*, Fernstrom, J. D. and Miller, G. D., Eds., CRC Press Boca Raton, FL, 1994.
23. Altizer, A. A. and Davidson, T. L., The effects of NPY and 5-TG on responding to cues for fats and carbohydrates, *Physiol. Behav.,* 65, 685, 1999.

15 Effects of Food Deprivation, Starvation, and Exercise on Dietary Selection in the Rat

T. W. Castonguay and Lynda M. Brown

CONTENTS

1   OVERVIEW

This chapter attempts to outline and summarize what is known about the effects of food deprivation, starvation, and exercise on macronutrient selection patterns in laboratory rodents. Like most reviews, the work summarized here is not an exhaustive canvassing of the literature, but rather an attempt at highlighting what are arguably the most important contributions to our knowledge of these influences on food choice. Table 1 provides a summary. The chapter ends with a presentation of some of our work in this area and of how our findings support the working hypothesis of our laboratory that dietary selection patterns reflect existing patterns of substrate utilization.

Richter [1-8] is often credited with having developed the paradigm that is most frequently used in investigations of dietary selection, although selection studies were performed as early as 1915 (see Evvard).[9] Starting with the early demonstrations that rats would increase salt intake when made sodium deficient, Richter went on to demonstrate that several endocrine secretions had effects on the choice of foods that tended to replace or make up for a nutrient deficiencies. For example, Richter found that calcium intake was decreased in parathyroidectomized rats.[5] Further, he showed that pancreatectomized rats increased their intake of carbohydrate,[8] and that adrenalectomized (ADX) rats altered their intake of carbohydrate and sodium.[6] Many years later, Kanarek et al.[10]

TABLE 1
Summary of Selection Studies

	Deprivation	
Piquard et al.[17]	Three macronutrient selection and fast	Increase carbohydrate and fat during refeed
		Increase protein if prefed low protein
Bligh et al.,[18]	Three macronutrient selection and fast	Increase carbohydrate and fat
Welch et al.[19]	Three macronutrient selection and fast	Increase fat intake if prefed control diet
		Increase carbohydrate if prefed high-carbohydrate or high-fat diet
Lucas and Sclafani[26]	Three macronutrient selection and fast	Increase fat intake
DiBattista[27]	Restrict access to protein or carbohydrate	Increase protein if restrict protein
		No increase in carbohydrate if restrict carbohydrate
Tempel and Leibowitz[29]	Three macronutrient selection and fast	Increase carbohydrate and fat
Bernardini et al.[66]	Three macronutrient selection and fast	Increase fat

	Starvation	
White et al.[34]	Three macronutrient selection and starvation	Increase protein and decrease carbohydrate
Thouzeau et al.[38]	Three macronutrient selection and starvation	Phase 2 — decrease in fat
		Phase 3 — increase fat and protein
Piquard et al.[39]	Three macronutrient selection and starvation	Decrease fat intake

	Exercise	
Oudot et al.[42]	Three macronutrient selection and activity wheels	Increase carbohydrate and total calories
Collier et al.[44]	High-carbohydrate vs. high-protein choice and activity wheels	Increase carbohydrate
Rieth and LaRue-Achagiotis[45]	Three macronutrient selection and treadmill	Increase protein at night; increase fat in day and night
Miller et al.[46]	Three macronutrient selection and treadmill	Decrease fat
LaRue-Achagiotis et al.[47,48]	Three macronutrient selection and treadmill	Initially increase fat and protein, decrease carbohydrate; long-term decrease fat increase carbohydrate
Castonguay et al.[53,56]	Three macronutrient selection and treadmill	No change in obese or lean Zucker rats
Andik et al.[69]	Varied food sources	Decrease "starchy" food

showed that a rat's preference for sugar solutions could be altered by insulin, an observation that once again supported Richter's initial speculation that dietary preferences are reflections of ongoing metabolic processes. Similarly, when glucose utilization was blocked by injection of 2-deoxy-D-glucose, sugar solution preference was affected. Adding to these demonstrations has been the work of Ritter and Taylor[11] who demonstrated that rats readily make the switch in dietary patterns to take advantage of the availability of energy sources. If for some reason fatty acid utilization is retarded or blunted, then rats increase their intake of carbohydrate. Conversely, if glucose utilization is blocked, fat intake is preferentially increased.

All of these observations lend support to the hypothesis that dietary selection patterns reflect ongoing metabolic processes. This thesis is further supported by the reports of changes in dietary preferences and selection patterns following deprivation, starvation, and exercise.

1.1 BACKGROUND

The observations that have been used to develop hypotheses about food choice are very limited. Conclusions drawn about nutrient selection are generalizations that sometimes go beyond appropriate bounds, the results of food choice experiments are often "paradigm specific." It was initially thought that the study of macronutrient choice might simplify the task of developing models for the control of food intake. Since intake is controlled by factors such as nutrients, energy, sensory properties and post ingestive events, it might be simpler to study its controls if they were somehow reduced in at least one dimension, that of macronutrient content. It was argued that since rats must pick and choose their food sources for both energy and nutrient needs simultaneously, the study of selection patterns might be advanced by separating "food" into separate macronutrient sources. In theory, it would then be easier to attribute changes in intake by an experimental intervention to alterations in the mechanism(s) controlling the intake of protein, carbohydrate, or fat. Although several important findings have been borne out of this proposition, more confusion than light has been shed on the true nature of the controls of intake. The food selection literature is replete with contradictions and overgeneralizations. The reader is cautioned to keep in mind that all of the factors that contribute to determining food choice have not yet been controlled for in any single experiment. As a consequence, the results from each experiment must be considered as only the reflection of a process that has multiple controls, any one of which can alter the final outcome.

1.2   DEFINITIONS

Since there is a long history of research dealing with how and why alterations in metabolism influence food selection, it is important to make some distinctions in the terms that will be used. *Food deprivation* in this chapter will refer to either partial or complete withholding of caloric sources for some short period of time. The distinction between deprivation and starvation is not always straightforward. Calorie restriction leads to attenuation of body weight gain in some models, and in weight loss in others. Thus, making the distinction between deprivation and starvation a function of rate of weight loss was initially considered. However, since species differences as well as age, sex, and initial body composition all influence the rate of weight loss with food restriction or starvation, we decided not to use rate of weight loss as a defining characteristic. Rather, for purposes of this review, *starvation* will be defined operationally as the complete withholding of energy sources for a period of at least 2 days. Chronic caloric restriction (such as used in weight loss diets), individual macronutrient source restriction, and the withholding of food for less than 3 days will be defined as *deprivation*.

Another issue that needs to be clarified is the distinction between exercise and activity. These terms have been used interchangeably in the food intake literature. *Activity* will be referred to as movement in the absence of pronounced oxygen debt. For example, in most humans, a leisurely 1-mile walk would constitute activity, whereas the same distance traveled in under 7 or 8 min would constitute exercise in all but the fittest among us. Similarly, the behavioral and metabolic consequences of giving rats access to running wheels will be distinguished from those of exercising rats on a treadmill or forced to swim for some period of time. Despite the fact that large muscle groups will be used to maintain both running wheel activity and treadmill running, there is no evidence that activity places the animal in oxygen debt, whereas exercise does. Thus, *exercise* produces sustained oxygen debt.

2   LITERATURE REVIEW

2.1   FOOD DEPRIVATION

Deprivation and starvation are procedurally simple, but have complex, dynamic physiological consequences that in turn have numerous influences on the controls of food intake and the mechanisms that influence food choice. For example, deprivation is characterized by dissociation between the sympathetic nervous system (SNS) and the adrenal medulla. Although the SNS is suppressed, plasma epinephrine rises, resulting in lipolysis and increased substrate mobilization. Deprivation and starvation result not only in body weight loss, but also in hypothermogenesis, and decreased metabolic rate, especially at night.[12] There are several other important behavioral and neuroendocrine adaptations that take place with deprivation and starvation. These make it all the more difficult for the experimenter to identify any one of these changes as the critical element influencing food intake or food selection.

Investigators have used a number of ways to induce deprivation. The lack of a standard protocol or definition of deprivation has led to a myriad of different conclusions about its effect on the regulation of food intake and food item choice. For purposes of this chapter, deprivation will be defined as the withholding of food or a macronutrient for 24 to 48 h. By comparison, food restriction has been achieved using several procedures, including permitting an animal access to food or macronutrient source for only a limited time each day, or to only a fraction of its *ad libitum* daily intake, either by weight or by calories. Starvation, on the other hand, will be defined in this chapter as food deprivation of periods longer than 48 h.

Another characteristic of the deprivation literature is the inconsistent nutritional backgrounds that have been used prior to studying the effects of deprivation on intake. For example, Duggan and Smart[13] reported that rats that had been underfed prior to weaning preferred the odors associated with carbohydrate test solutions than the odors of protein test solutions. However, well-nourished controls preferred the odors associated with the protein test solutions. No differences in dietary selection patterns were observed when undernourished or controls were allowed to compose their own diets from macronutrient sources. By comparison, Welch and associates[14] have convincingly demonstrated that prior experience with diets varying in fat or carbohydrate can play a major role in determining the selection patterns of self-selecting rats. Rats raised on a high-fat or high-carbohydrate diet that were then subjected to a 24-h fast preferred carbohydrate during the refeeding period, whereas rats raised on control diet exposed to the same deprivation consumed more fat during refeeding. Clearly, prior nutritional experience (or nutritive plane) can have an impact upon the response to deprivation, at least as measured by dietary selection patterns.

2.2 FEEDING BEHAVIOR

Contrary to expectation, food deprivation does not always lead to increased intake. Bousfield and Elliott,[15] using rats that were maintained on a fixed-time eating regimen, found a decrease in food intake following deprivation. Their rats ate at a slower rate following deprivation that correlated with the length of the fast. Results were explained by a reduced tonicity of the stomach. Using an *ad libitum* refeeding procedure, Levitsky and Collier[16] reported that following a 24-h fast rats recovered by increasing the duration of meals, but the number of meals remained unchanged.

2.3 MACRONUTRIENT SELECTION

In general, deprivation results in an increase in dietary fat selection, although the generality of this observation is limited. These findings support the hypothesis that dietary selection reflects ongoing metabolic processes, as deprivation results in the depletion of glycogen and the mobilization of fat stores to meet energy needs. For example, Piquard et al.[17] allowed rats to self-select macronutrients and found that fasted rats increased carbohydrate and fat intake during refeeding. An increase in protein intake was only observed when animals were fed a low-protein diet. Several years later Bligh et al.[18] and Castonguay[19] also examined self-selection patterns following deprivation. Both fat and carbohydrate intakes were increased during refeeding (see Figure 1).

Kim et al.[14] measured macronutrient selection after either deprivation and norepinephrine administration or neuropeptide Y (NPY) administration. NPY increased carbohydrate intake, while norepinephrine and deprivation increased fat intake. Whether or not this preference was expressed depended on whether rats were allowed to select macronutrients or were fed high-fat or high-carbohydrate regimens. After 24 or 48 h deprivation self-selecting rats consumed more fat, but animals fed high-fat or high-carbohydrate diets consumed more carbohydrate during refeeding. The

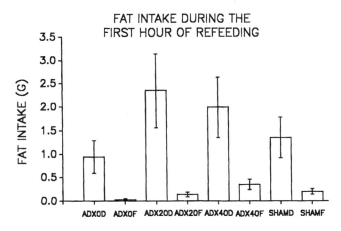

FIGURE 1 Fat intake during the first hour of refeeding following 0-h (F) or 24-h food deprivation (D). Rats were either ADX or sham-operated (SHAM). ADX rats were treated with 0, 20, or 40 µg/ml corticosterone in saline. For example, the group designated as ADX0D was adrenalectomized, with 0 µg/ml corticosterone replacement, deprived for 24 h. See text for further details. (Adapted from Bligh et al.[18])

Welch et al. findings once again reinforce the importance of previous plane of nutrition and type of diet choice presented when interpreting metabolic responses during refeeding (see also Duggan and Smart[13]).

As noted above, the generality of the conclusion that deprivation promotes fat intake is limited. This limitation is apparent when considering the results of other ways of inducing deprivation. For example, caloric dilution has been used to study deprivation. Ramirez[20] investigated the motivational factors responsible for rats' preferences for dilute suspensions of vegetable oils and starch. He hypothesized that rats ingest these suspensions for their nutritive value. Rats would therefore be expected to increase intake of starch and oil when their normal food is withheld. This applies to an array of nutrients including sugars, starches, and vegetable oils.[20] Contrary to his hypothesis, Ramirez found that rats exposed to oil suspensions while they are food deprived reduced (rather than increased) their preference for the oil suspension. This did not occur when other nutritive substances were substituted for the oil. He concluded that rats readily learned to associate oil flavor, but not other flavors, with the aversive effects of food deprivation. Although some investigators have examined the effects of food deprivation on intake of low-calorie fluids[21,22] and others have examined ingestion of fats by food-deprived rats,[23-25] no one had described anything like this specific taste aversion.

Although Ramirez showed that rats could form an aversion to the taste of fat following deprivation, Lucas and Sclafani[26] found that deprivation increases preference for fat under choice conditions. Since high- and low-fat foods have different postingestive consequences and flavors, they concluded that both factors may be important to long-term energy changes.

Several other investigators have reported that macronutrient intake is controlled by several mechanisms, and that specific nutrients are not all influenced by deprivation similarly. For example, DiBattista[27] showed that time-restricted access to protein and carbohydrate sources results in different patterns of intake. He restricted access to either a protein or carbohydrate source to 1 h/day, and then measured the animal's intake of that source. He found that rats compensate for the restricted access by increasing their intake of the protein source, but fail to compensate for the restriction when carbohydrate was withheld. These studies are similar to Simson and Booth's[28] earlier report that rats will compensate for individual amino acid deficiencies by increasing intake of solutions containing the limiting amino acid. These studies further support the hypothesis that dietary selection or dietary preferences are at least in part formed as a reflection of ongoing metabolic processes.

They demonstrate that when either body stores needed for protein synthesis or for energy balance are reduced, intake of nutrients capable of restoring those levels is favored.

Since rats are nocturnal, analyzing intake by time of day produces some interesting patterns after deprivation. Tempel and Leibowitz[29] noted a rhythm of meal-to-meal shifts during nocturnal feeding. Carbohydrate ingestion was favored at dark onset. Protein and fat ingestion was favored at the end of the dark cycle. Deprivation produced compensatory feeding only when it occurred at the beginning of the dark cycle. Mild, 2 h deprivation produced a shift to fat and carbohydrate intake. The authors suggest that these feeding patterns may relate to the activity the hypothalamic norepinephrine and serotonin, which are known to be important in modulating temporal feeding patterns.

2.4   NEURAL AND METABOLIC CONTROLS

As already noted, it is very difficult to specify which mechanism(s) is controlling intake of individual macronutrients, given the lack of uniformity in experimental paradigms. Despite that limitation, several investigators have advanced theories of how selection is achieved. For example, Wurtman et al.[30,31] have proposed that increases and decreases in brain serotonin play an important role in influencing protein and carbohydrate selection. More recently, Stanley et al.[32] qualified the importance of the role of serotonin in the hypothalamic (paraventricular nucleus, PVN) regulation of intake. They demonstrated that the usual peak in concentration of the serotonin metabolite 5-hydroxyindolacetic acid (5-HIAA) in the PVN found at the onset of dark was abolished by deprivation. However, the endogenous rhythm in 5HIAA (without the characteristic peak) was preserved in the deprived state. They concluded that the peak was in all likelihood an artifact of eating. PVN serotonin metabolism may increase in association with feeding specifically in the early portion of the nocturnal eating period, when it may play a role in controlling food intake and macronutrient selection.

The biochemical response to carbohydrate restriction can be observed at both neurochemical and neuroanatomical levels. Thibault and Roberge[33] used an 8-day restriction of carbohydrate in rats and found a reduction in the serum tyrosine and tryptophan compared to the other neutral amino acids. In the brain, the decreased tyrosine corresponded to a high dopamine/norepinephrine ratio that was observed in the hypothalamus, thalamus, and raphe nuclei. The authors suggested that this observation implied low dopamine usage. Brain tryptophan content was increased in the hypothalamus and neostratium with a decrease in the serotonin/5-HIAA ratio.

White et al.[34] found that low-protein diets increased NPY gene expression in the hypothalamus. While energy-restricted and protein-restricted animals elevate NPY mRNA in the PVN and basomedial hypothalamus, no effect was observed during carbohydrate or fat restriction. They suggested that dietary protein may regulate NPY expression.

The previous plane of nutrition has a major impact on the observed response to deprivation. Kim and Freake[35] explored tissue-specific regulation of lipogenesis. The synthesis of fatty acids is required for the storage of energy and provision of cellular components. This pathway is under hormonal and nutritional control. High-carbohydrate, fat-free diets induce lipogenesis in the liver, whereas deprivation leads to decreased lipogenesis. Animals on a high-carbohydrate, fat-free diet responded to deprivation by reducing cholesterol synthesis in the liver while fatty acids synthesis was not changed. Similarly, Erskine et al.[36] found that following a 24-h fast, only rats on a high-fat diet increased lipoprotein lipase (LPL), while those on a high-carbohydrate diet showed no change.

2.5   STARVATION

As already noted, we have made the distinction between deprivation and starvation based on the duration of withheld food access. The studies included in this section have all incorporated more than a 24-h fast to study selection.

2.6   Feeding Behavior/Macronutrient Selection

Starvation generally leads to an increase in protein intake during the initial phases of refeeding. Hunsicker et al.[37] found starved rats increased protein and decreased carbohydrate when refed. Fat intake was unchanged. Thouzeau et al.[38] reported that, when permitted to select macronutrients, moderately starved rats (those that primarily rely on lipid fuels) progressively decreased fat intake. Thouzeau and colleagues refer to these animals as having entered into Phase 2 of starvation. Severely starved rats (that are undergoing protein breakdown — or Phase 3) progressively increased fat and protein intake. Both groups were hyperphagic during refeeding. The results suggest each group selected macronutrients to optimize the restoration of body fuels according to the metabolic state reached during starvation.

Piquard et al.[39] found a decrease in fat intake in starved-refed rats. They described an overshoot phenomenon in refeeding self-selecting rats that involved a number of metabolic changes. Food intake in rats fed a composite diet increased significantly during refeeding subsequent to a fast. By contrast, in self-selecting rats, total food intake remained the same, but lipid and carbohydrate levels varied reciprocally. By the third day of refeeding, lipid intake decreased while carbohydrate intake increased. These findings suggest that carbohydrate selection is favored over fat selection when lean body mass has been reduced by starvation. One of the best-established consequences of starvation is a drop in metabolic rate. This reduction can be thought to decrease energy requirements. However, by mobilizing lean tissue and its glycogen during starvation, the deficit in lean body mass cannot be reversed expeditiously without carbohydrate intake. Thus, starved rats increase carbohydrate intake during refeeding, all the while maintaining total intake within the bounds established by their caloric needs.

2.7   Exercise

The large variability in experimental protocols in type of exercise, intensity, duration, frequency, time of day, in subjects (male vs. female), age, environmental conditions have contributed to the inconsistencies in the effect of exercise on dietary selection patterns. For purposes of this chapter, we are separating the studies giving animals access to activity with those using forced exercise. The activity studies reviewed below all make use of running wheels, whereas the exercise studies have used treadmill running to promote increased energy expenditure. Like deprivation and starvation, the distinction between activity and exercise is one of degree. Richter[1] and later Collier et al.[40] have noted that rats run in bursts of activity when allowed access to running wheels. Rats have been reported to run for up to 18 h/day when given access to wheels. By comparison, most forced exercise treadmill sessions last under 2 h/day. These sessions are characterized by continuous running at a given rate and slope, often with aversive consequences to the animal (such as tail shock) should it not comply with the day's protocol. As a result, the study of forced exercise often, but not always, includes the study of stressed animals. More recent investigators have developed protocols that habituate and train animals gradually to run on treadmills, and in that way minimize the stress to the animal.

When given access to a running wheel and placed on limited access to food, rats develop gastric lesions. These ulcerations are possibly the reason food intake decreases under some exercise conditions, but the relationship of food intake and exercise to this pathology is not known. Rats develop gastric lesions following various types of stress, like conflict, activity, or electric shock. The extreme stress presumed to lead to such profound ulceration has been linked behaviorally to interaction between the food deprivation and the activity schedules but the mechanism is unknown. This chapter will not cover the studies that produced gastric lesions.

2.8   ACTIVITY VS. EXERCISE BEHAVIOR

Access to a running wheel has temporary effects on food intake. When first given access to the wheel, rats eat less. When access to the wheel is removed, rats are hyperphagic. Body weight also initially decreases with wheel access. Differences in food intake and body weight reach a plateau after 2 weeks.[41]

Oudot et al.[42] explored the effects of exercise, dietary self-selection, and body weight on total caloric intake in female rats. Voluntary wheel running did not reduce body weight; however, it increased both carbohydrate intake and total caloric intake. Forced exercise decreased body weight, but did not change total calories. It decreased fat intake while increasing protein intake. Chiel[43] also reported that changing the ratio of carbohydrate to protein altered activity patterns. As the ratio increased, activity during the nocturnal period increased. No correlation between calories consumed and activity was found. The amount of fat in the diet varied from 15 to 45%.[43]

2.9   EXERCISE

Exercised rats have different nutritional requirements than sedentary rats. Collier et al.[44] found that active rats choose a higher proportion of carbohydrate to protein than controls. When compared with controls, exercise-trained rats reduce body weight, mostly by decreasing white adipose tissue. When allowed to self-select, food intake was slightly increased due to an increase in protein intake during the nocturnal period. Fat intake increased throughout the day and night, and plasma insulin levels were reduced. Chow-fed rats decreased food intake during the daytime period. Rats placed on a self-selection regimen increased fat intake while decreasing white adipose tissue mass. This could be the result of increased lipolytic capacity of the adipocytes. The results also point out that the carbohydrate-to-fat ratio in the diet is an important factor in understanding the interaction among exercise, body weight, and body composition.[45] Miller et al.[46] also used female rats and found that exercise reduced fat selection. Again, exercise training did not change total energy intake, but reduced dietary fat.[46]

Would delayed access to food change the selection of macronutrients? LaRue-Achagiotis et al.[47] examined this hypothesis by delaying access to food 0, 30 or 90 min after forced exercise. Food intake after deprivation was reduced for the first 3 h with restored food access, but total intake over 24 h remained unchanged. Carbohydrate and protein intakes were reduced just after exercise whereas a decrease in fat intake appeared later. Deprivation after exercise significantly reduced carbohydrate intake.[47]

When first exposed to exercise, rats increase protein and fat intake and lower carbohydrate intake when allowed to select macronutrients. After adaptation to exercise, total food intake of trained rats is lower than that of controls because fat intake is reduced while carbohydrate is increased. When compared with sedentary controls, exercise-trained rats have a lower body weight.[48]

2.10   NEURAL AND METABOLIC CONTROLS

Little is known about the specific neural dietary selection mechanisms affected by exercise. A marked increase in epinephrine appears in fasted exercising rats.[49] Epinephrine-dependent glycogenolysis in types I and II muscle fibers provide essential quantities of lactate for hepatic gluconeogenesis in fasted exercised rats. During prolonged exercise, liver glycogen is depleted and gluconeogenesis becomes the only source of blood glucose. As this shift occurs in rats the epinephrine level markedly increases. Chronic exercise results in decreased food intake and increases both epinephrine and norepinephrine levels. These neurotransmitters seem to contribute to the postexercise decrease in body weight and inhibition of food intake.[50]

Meal feeding (2 h access/22 h deprivation) causes several adaptations that lead to storage of energetic resources while exercise is a behavior that requires mobilization of energetic resources. When exercise and deprivation are combined, rats generally do not lose weight. Rather, they mobilize less free fatty acids, and maintain a slight hyperglycemia and insulinemia.[51]

Applegate and Stern[52] found exercise depressed food intake in the light cycle while food intake in the dark cycle and total 24 h food intake were not affected. Effects of exercise termination were measured 24, 48, 60, and 84 h after exercise termination. Exercised rats decreased plasma insulin as well as epididymal and retroperitoneal depot weight and adipocyte cell size, and decreased adipose tissue LPL activity. In 48 h, plasma insulin increased to normal levels. By 60 h, dark-cycle food intake was increased. At 84 h, dark-cycle food intake, plasma triglycerides, and epididymal LPL activity per cell were significantly greater than controls. The metabolic response to exercise termination resulted in rapid lipid deposition, increased food intake, plasma insulin, and LPL activity.

3   OUR CONTRIBUTIONS TO THE LITERATURE

Finally, our own early work in dietary selection is consistent with the thesis that nutrient selection patterns are at least in part a reflection of ongoing utilization patterns. We showed that genetically obese Zucker rats, that are characterized by both hyperinsulinemia and insulin resistance, will select a diet that is high in fat.[53]

It has been argued that an obese rat is a starving rat. Certainly both conditions have several characteristics in common. For example, obese rats are hyperphagic.[54] Similarly, starving rats that gain access to food eat significantly more food/day than do *ad libitum* fed controls. Neither the starving animal nor the insulin-resistant obese rat is using much glucose to fuel metabolism. Starving rats are using bodily stores to maintain existing metabolic processes. Thus, both obese and starving animals are metabolizing primarily fat and protein to meet energy demands.

It had been openly speculated that the hyperphagia of these animals might be due to an inborn deficit in this animal's ability to develop lean body mass.[55] Hyperphagia was thought to be a consequence of the animal's attempt to make up for a protein utilization deficit. Radcliffe and Webster[55] believed that the obese animal was attempting to eat that additional protein by increasing total diet intake, and as a consequence became obese.

Our data refuted this hypothesis by showing that when obese rats had free access to separate sources of protein, fat, and carbohydrate, they composed a diet that was higher in fat and lower in protein than the diets selected by lean littermate controls. Further, the obese animals that were allowed to self-select continued to eat significantly more calories per day than did controls.

Not long after we reported that Zucker obese rats selected a high-fat diet, we reported that the preference for an increased level of fat was preserved even in the face of serial water dilution.[56] Obese rats that were given access to 50% fat emulsions nevertheless maintained their relatively high intake of fat. When the fat was diluted with water so as to make a 25%, then a 12.5%, then a 6.25% fat emulsion, obese rats maintained increased intake of the diluted source up until the final dilution, when intake equaled that observed in lean controls. Clearly, we were tapping into a very motivated behavior.

Just as early learning theorists had demonstrated that varying magnitude of reinforcement could control operant behavior that preceded food delivery,[57,58] we showed that varying the delivery of fat via dilution resulted in far greater appetitive behavior directed at that source. However, like Adolph,[59] who demonstrated that dilution supports caloric compensation only so far, we showed that both obese and lean rats would compensate for the dilution of the fat source only so much. What made this experiment interesting is that the obese animals tolerated the dilution to a much greater extent than did the lean rats.

We also demonstrated that obese rats have an unusual diurnal corticosterone rhythm that makes them hypercorticoid during a large portion of the dark cycle.[60] Honma and colleagues[61,62] speculated that increases in corticosterone promote intake. They hypothesized that the pronounced peak in circulating corticosterone in rats that is typically observed at the onset of the dark period is part of the signal used to stimulate feeding activity observed at that time of day. Not only does the quiescent daytime promote increased circulating corticosterone, but food deprivation can also lead to hypercorticoidism. Young rats fasted for 24 h have significantly higher basal corticosterone in circulation.[18,63] Similarly, adult male rats that have been food deprived for 48 h have significantly higher circulating corticosterone than *ad libitum* fed controls.[63] Rats that are fasted for 24 to 48 h that are then given *ad libitum* access to food eat significantly larger meals than do *ad libitum* fed controls. However, ADX rats fail to respond to deprivation by increasing intake when access to food is restored.[64]

These parallels led to an obvious experiment in which the pattern of food choice would be measured subsequent to food deprivation. Bligh and colleagues[63] studied the dietary selection patterns of rats after they had been completely food deprived for 24 or 48 h. Rats consumed most of their calories as fat upon initial restored access to food. During the first hour of restored access, rats that had been deprived of food for 24 h ate 23.7 kcals, of which 55% were from fat. Undeprived controls ate only 8 kcals and composed a diet that was only 21% fat during the same test hour. Rats that had been food deprived for 48 h refed by composing a high-fat diet not only during the first hour of refeeding but also throughout the first 6 h of refeeding. What made this pattern of intake even more interesting was that the deprivation promoted changes in basal corticosterone in circulation at the time of restored access. Rats that had been deprived for 24 h had significantly higher basal corticosterone titers than undeprived controls. Deprivation for 48 h led to higher basal corticosterone in circulation. It should be noted, however, that these increases in hormone titers were short-lived, and that restored access to food restored normal basal corticosterone within 1 h.

As a sequel to these observations we repeated the basic protocol of fasting for 24 or 48 h and then measured intake of separate macronutrient sources every hour. However, this time we examined the role of the adrenal in influencing the pattern of selection. Berdanier and her colleagues[65] had already shown that ADX rats do not refeed subsequent to fast-like sham-operated controls. In particular, they noted that the characteristic overeating that is observed during the initial period of restored food access is missing in the ADX rat. Wurdeman et al.[66] demonstrated that corticosterone replacement was sufficient to restore the ADX rat to normal. We anticipated that corticosterone would also reverse the effects of starvation on dietary selection patterns. We used two groups of surgically prepared rats: those that had been bilaterally ADX and those that were sham operated. We then divided the ADX groups into three treatment subgroups: those that had no corticosterone replacement (ADX-0), those that had a low level of corticosterone (20 µg/ml) added to their drinking saline (ADX-20), and those that had a higher level of corticosterone (40 µg/ml) added to their drinking saline. These rats were allowed to recover from surgery and stabilize both weight and intake for 6 days prior to the start of a deprivation period. Rats from each surgical group were assigned to one of two deprivation periods: 0 h or 24 h. Intake of three separate macronutrient sources following deprivation was measured every hour for 12 h, and then every 12 h for 2 days. Food deprivation resulted in greater body weight loss in sham-operated rats and in corticosterone-supplemented groups when compared with the unsupplemented (ADX-0) group. During the first hour of refeeding, corticosterone-supplemented ADX rats ate significantly more fat than did unsupplemented ADX rats. This attenuation of fat intake continued throughout the 24-h refeeding period. Both high and low corticosterone replacement doses were sufficient to restore the refeeding selection patterns of ADX rats. Refer to Figure 1.

Finally, Bernardini et al.[67] in our laboratory set out to address a lingering hypothesis that developed while conducting the studies described above. We had convincingly demonstrated that rats eat fat subsequent to deprivation, especially early during refeeding. The question was "Why?" Do rats prefer fat because of its association with more energy, or do they prefer fat because they are already metabolizing fat (their endogenous body stores)? To begin to answer these questions, Bernardini adapted rats to the standard three macronutrient sources and then deprived some of them for 48 h. Not surprisingly, during the first hour of refeeding, the deprived rats ate a significant quantity of fat. A second group of rats was treated similarly, with the exception that they had been given access to a fat source that had been diluted with celluflour in sufficient quantity to make the fat source equicaloric with the protein and carbohydrate sources. During refeeding, these animals also preferentially ate fat. When compared with the fat intake of the first group given the concentrated fat source, rats fed the diluted fat ate as much fat (as measured in calories) during refeeding. In order to achieve this level of intake, they ate more than twice as much of their fat source (in grams). Compensation in response to dilution is one way of measuring the specificity of the animal's preference for that nutrient. It seems clear that when given the opportunity, rats will compose a high-fat diet during the initial stages of recovery from food deprivation. At least under these conditions, we concluded that rats were responding to the caloric content of the fat source, and not simply to its taste.

As a test of the hypothesis that altered metabolic state will be reflected in dietary choice patterns, Castonguay, Upton, and Stern (unpublished manuscript) examined the selection patterns of obese and lean female Zucker rats that were subjected to an exercise protocol that is known to promote shifts in carcass composition.[68] Exercise is known to alter several metabolic parameters in obese Zucker rats, including decreased plasma insulin and increased adipose tissue LPL activity. Obese and lean female adult Zucker rats were gradually trained to run on a treadmill (20 m/min, 60 min/day, 7 days/week) for 31 days. Both phenotypes were given *ad libitum* access to three separate macronutrient sources and allowed to compose their own diets. No differences were found between exercised and sedentary rats in body weight, total caloric intake, or in diet selection. Both exercised and sedentary obese rats selected diets that were higher in fat and lower in carbohydrate and protein than the diets composed by their lean littermates. Refer to Figure 2. Note that both

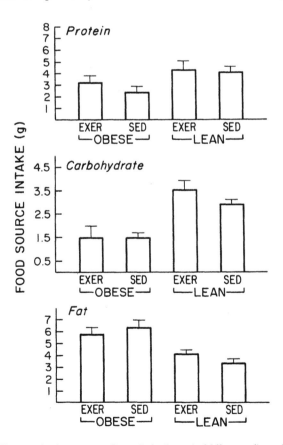

FIGURE 2 Average daily protein (upper panel), carbohydrate (middle panel), and fat (lower panel) intakes of obese and lean, exercised and sedentary female Zucker rats. Note that obese rats composed a diet that was higher in fat and lower in carbohydrate than the diets composed by lean littermates. The effects of treadmill exercise in both obese and lean rats were statistically insignificant.

exercised and sedentary obese rats composed a diet that provided more than 70% of total daily calories as fat. By comparison, both sedentary and exercised lean rats composed a diet that provided approximately 55% of total calories as fat. Carcass composition analyses revealed that exercise significantly increased body protein and decreased body fat and body water.

These data were in marked contrast to several reports of the efficacy of exercise in changing dietary selection as well as total caloric intake. For example, Mayer et al.[69] noted that rodents compensate for exercise by increasing total caloric intake. By comparison, the Zucker rats tested in the present study failed to alter either total caloric intake or dietary selection. Perhaps dietary selection patterns change only secondary to an adjustment to total daily calories. Such a hypothesis is consistent with the findings reported by Andik et al.,[70] who observed a decrease in the intake of "starch" food sources and an increase in protein and fat sources when rats were placed in a revolving drum for up to 8 h/day.

REFERENCES

1. Richter, C. P., Animal behavior and internal drives., *Q. Rev. Biol.*, 2, 307, 1927.
2. Richter, C. P., The role played by the thyroid gland in the production of gross body activity, *Endocrinology*, 17, 73, 1933.
3. Richter, C. P. and Eckert, J. F., Increased calcium appetite of parathyroidectomized rats, *Endocrinology*, 21, 50, 1937.
4. Richter, C. P., Holt, L. E. J., and Barelare, B. J., Vitamin B1 craving in rats, *Science*, 86, 354, 1937.
5. Richter, C. P. and Eckert, J. F., Mineral appetite of parathyroidectomized rats, *Am. J. Med. Sci.*, 198, 9, 1939.
6. Richter, C. P., Sodium chloride and dextrose appetite of untreated and treated adrenalectomized rats, *Endocrinology*, 29, 115, 1941.
7. Richter, C. P., Sodium chloride and dextrose appetite of untreated and treated adrenalectomized rats, *Endocrinology*, 29, 115, 1941.
8. Richter, C. P. and Schmidt, E. C. H., Increased fat and decreased carbohydrate appetite of pancreatectomized rats, *Endocrinology*, 28, 179, 1941.
9. Evvard, J. M., Is the appetite of swine a reliable indication of physiological needs? *Proc. Iowa Acad. Sci.*, 22, 375, 1915.
10. Kanarek, R. B., Marks-Kaufman, R., and Lipeles, B. J., Increased carbohydrate intake as a function of insulin administration in rats, *Physiol. Behav.*, 25, 779, 1980.
11. Ritter, S. and Taylor, J. S., Vagal sensory neurons are required for lipoprivic but not glucoprivic feeding in rats, *Am. J. Physiol.*, 258, R1395, 1990.
12. Siyamak, A. Y. and Macdonald, I. A., Sub-acute underfeeding but not acute starvation reduces noradrenaline induced thermogenesis in the conscious rat, *Int. J. Obesity*, 16, 113, 1992.
13. Duggan, J. P. and Smart, J. L., Effects of early under-nutrition on dietary self-selection and relative carbohydrate- and protein-conditioned preferences, *Appetite*, 10, 227, 1988.
14. Welch, C., Grace, M., Billington, C., and Levine, A., Preference and diet type affect macronutrient selection after morphine, NPY, norepinepherine, and deprivation, *Am. J. Physiol.*, 266, R426, 1994.
15. Bousfield, W. A. and Elliott, M. H., The effect of fasting on the eating behavior of rats, *J. Genet. Psychol.*, 45, 227, 1934.
16. Levitsky, D. A. and Collier, G., Effects of diet and deprivation on meal eating behavior in rats, *Physiol. Behav.*, 3, 137, 1968.
17. Piquard, F., Schaeffer, A., and Habery, P., Influence of fasting and protein deprivation on food self-selection in the rat, *Physiol. Behav.*, 20, 771, 1978.
18. Bligh, M. E., DeStefano, M. B., Kramlik, S. K., Douglass, L. W., Dubuc, P., and Castonguay, T. W., Adrenal modulation of enhanced fat intake subsequent to fasting, *Physiol. Behav.*, 48(3), 373, 1990.
19. Castonguay, T. W., Glucocorticoids as modulators in the control of feeding, *Brain Res. Bull.*, 27, 423, 1991.
20. Ramirez, I., Food deprivation reduces rats' oil preference, *Appetite*, 21, 53, 1993.

21. Campbell, D., Capaldi, E., and Myers, D., Conditioned flavor preferences as a function of deprivation level: preferences or aversions? *Anim. Learn. Behav.*, 15, 193, 1987.

22. Revusky, S. H., Hunger level during food consumption: effects on subsequent preference, *Psychonomic Sci.*, 4, 109, 1967.

23. Ackroff, K., Fat appetite in rats: the response of infant and adult rats to nutritive and non-nutritive oil emulsions, *Appetite*, 15, 171, 1990.

24. Deutsch, J. and Gonzales, M., Gastric fat content and satiety, *Physiol. Behav.*, 26, 676, 1981.

25. Mindell, S., Smith, G., and Greenberg, D., Corn oil and mineral oil stimulate sham feeding in rats, *Physiol. Behav.*, 48, 283, 1990.

26. Lucas, F. and Sclafani, A., Food deprivation increases the rat's preference for a fatty flavor over a sweet taste, *Chem. Senses*, 21, 169, 1996.

27. DiBattista, D., Efffects of time-restricted access to protein and to carbohydrate in adult mice and rats, *Physiol. Behav.*, 49, 263, 1991.

28. Simson, P. C. and Booth, D. A., Dietary aversion established by a deficient load: specificity to the amino acid omitted from a balanced mixture, *Pharmacol. Biochem. Behav.*, 2, 481, 1974.

29. Tempel, D. L. and Leibowitz, S. F., PVN steroid implants: effects on feeding patterns and macronutrient selection, *Brain Res. Bull.*, 23, 553, 1989.

30. Wurtman, J., Wurtman, R., Mark, S., Tsay, R., Gilbert, W., and Growdon, J., d-Fenfluramine selectively suppresses carbohydrate snacking by obese subjects, *Int. J. Eating Disorders*, 4(1), 89, 1985.

31. Wurtman, R. J. and Wurtman, J. J., Carbohydrates and Depression, *Sci. Am.*, 50, 1989.

32. Stanley, B. G., Schwartz, D. H., Hernandez, L., Hoebel, B. G., and Leibowitz, S. F., Patterns of extracellular norepinephrine in the paraventricular hypothalamus: relationship to circadian rhythm and deprivation-induced eating behavior, *Life Sci.*, 45, 275, 1989.

33. Thibault, L. and Roberge, A. G., Dietary protein-induced changes in serum dopamine-hydroxylase activity, glucose, insulin, cortisol and pyridoxal-5 phosphate levels in normal and immobilized cats, *J. Clin. Biochem. Nutr.*, 3, 55, 1987.

34. White, B., He, B., Dean, R., and Martin, R., Low protein diets increase neuropeptide Y gene expression in the basomedial hypothalamus of rats, *J. Nutr.*, 124, 1152, 1994.

35. Kim, T.-S. and Freake, H. C., Tissue specific regulation of lipogenesis by carbohydrate feeding and twenty four hour starvation in the rat, *Nutr. Res.*, 13, 297, 1993.

36. Erskine, J., Jensen, D., and Eckel, R., Macronutrient regulation of lipoprotein lipase is posttranslational, *J. Nutr.*, 124, 500, 1994.

37. Hunsicker, K. D., Mullen, B. J., and Martin, R. J., Effect of starvation or restriction on self-selection of macronutrients in rats, *Physiol. Behav.*, 51, 325, 1992.

38. Thouzeau, C., Le Maho, Y., and LaRue-Achagiotis, C., Refeeding in fasted rats: dietary self-selection according to metabolic status, *Physiol. Behav.*, 58, 1051, 1995.

39. Piquard, F., Schaefer, A., Haberey, P., Chanez, M., and Peret, J., The effects of dietary self-selection upon the overshoot phenomenon in starved-refed rats, *J. Nutr.*, 109, 1035, 1979.

40. Collier, G., Johnson, D. F., CyBulski, K. A., and McHale, C. A., Activity patterns in rats (*Rattus norvegicus*) as a function of the cost of access to four resources, *J. Comp. Psychol.*, 104(1), 53, 1990.

41. Looy, H. and Eikelboom, R., Wheel running, food intake, and body weight in male rats, *Physiol. Behav.*, 45, 403, 1989.

42. Oudot, F., LaRue-Achagiotis, C., Anton, G., and Verger, P., Modifications in dietary self-selection specifically attributable to voluntary wheel running and exercise training in the rat, *Physiol. Behav.*, 59, 1123, 1996.

43. Chiel, H. J., Short-term variations in diet composition change the pattern of spontaneous motor activity in rats, *Science*, 213, 676, 1981.

44. Collier, G., Leshner, A. I., and Squibb, R. L., Dietary self-selection in active and non-active rats, *Physiol. Behav.*, 4, 79, 1969.

45. Rieth, N. and LaRue-Achagiotis, C., Exercise training decreases body fat more in self-selecting than in chow-fed rats, *Physiol. Behav.*, 62, 1291, 1997.

46. Miller, G., Dimond, A., and Stern, J., Exercise reduces fat selection in female Sprague-Dawley rats, *Med. Sci. Sports Exercise*, 26, 1466, 1994.

47. LaRue-Achagiotis, C., Martin, C., Verger, P., Chabert, M., and Louis-Sylvestre, J., Effects of acute treadmill exercise and delayed access to food on food selection in rats, *Physiol. Behav.*, 53, 403, 1993.

48. LaRue-Achagiotis, C., Rieth, N., and Louis-Sylvestre, J., Exercise training modifies nutrient self-selection in rats, *Physiol. Behav.*, 56, 367, 1994.

49. Winder, W. W., Terry, M. L., and Mitchell, V. M., Role of plasma epinephrine in fasted exercising rats, *Am. J. Physiol.*, 248, R302, 1985.

50. Guilland, J. C., Moreau, D., Genet, J. M., and Klepping, J., Role of catecholamines in regulation by feeding of energy balance following chronic exercise in rats, *Physiol. Behav.*, 42(4), 365, 1988.

51. Curi, R., Hell, S., and Timo-Iara, C., Meal-feeding and physical effort. 1. Metabolic changes induced by exercise training, *Physiol. Behav.*, 47, 869, 1990.

52. Applegate, E. A. and Stern, J. S., Exercise termination effects on food intake, plasma insulin, and adipose lipoprotein lipase activity in the osborne-mendel rat, *Metabolism*, 36(8), 709, 1987.

53. Castonguay, T. W., Hartman, W. J., Fitzpatrick, E. A., and Stern, J. S., Dietary self-selection and the Zucker rat, *J. Nutr.*, 112(4), 796, 1982.

54. Zucker, L. M. and Antoniades, H. N., Insulin and obesity in the Zucker genetically obese rat "fatty," *Endocrinology*, 90, 1320, 1972.

55. Radcliffe, J. D. and Webster, A. J. F., Regulation of food intake during growth in fatty and lean female Zucker rats given diets of different protein content, *Br. J. Nutr.*, 36, 457, 1976.

56. Castonguay, T. W., Burdick, S. L., Guzman, M. A., Collier, G. H., and Stern, J. S., Self-selection and the obese Zucker rat: the effect of dietary fat dilution, *Physiol. Behav.*, 33, 119, 1984.

57. Collier, G., Hirsch, E., and Hamlin, P. H., The ecological determinants of reinforcement in the rat, *Physiol. Behav.*, 9, 705, 1972.

58. Hirsch, E. and Collier, G., The ecological determinants of reinforcement in the guinea pig, *Physiol. Behav.*, 12, 239, 1974.

59. Adolph, E.F., Urges to eat and drink in rats, *Am. J. Physiol.*, 151, 110, 1947.

60. Gibson, M. J., Liotta, A. S., and Krieger, D. T., The Zucker fa/fa rat: absent circadian corticosterone periodicity and elevated b-endorphin concentrations in brain and neurointermediate pituitary, *Neuropeptides*, 1, 349, 1981.

61. Honma, K., Honma, S., and Hiroshige, T., Feeding-associated corticosterone peak in rats under various feeding cycles, *Am. J. Physiol.*, 246, R721, 1984.

62. Shiraishi, I., Honma, K.-I., Honma, S., and Hiroshige, T., Ethosecretogram: relation of behavior to plasma corticosterone in freely moving rats, *Am. J. Physiol.*, 16, R40, 1984.

63. Bligh, M. E., Douglass, L. W., and Castonguay, T. W., Corticosterone modulation of dietary selection patterns, *Physiol. Behav.*, 53, 975, 1993.

64. Kaul, L. and Berdanier, C. D., Effect of pancreatectomy or adrenalectomy on the responses of rats to meal-feeding, *J. Nutr.*, 105, 1176, 1975.

65. Berdanier, C. D., Wurdeman, R., and Tobin, R. B., Further studies on the role of the adrenal hormones in the responses of rats to meal-feeding, *J. Nutr.*, 106, 1791, 1976.

66. Wurdeman, R., Berdanier, C. D., and Tobin, R. B., Enzyme overshoot in starved-refed rats: role of the adrenal glucocorticoid, *J. Nutr.*, 108, 1457, 1978.

67. Bernardini, J., Kamara, K., and Castonguay, T. W., Macronutrient choice following food deprivation: effect of dietary fat dilution, *Brain Res. Bull.*, 32, 543, 1993.

68. Applegate, E. A., Upton, D. E., and Stern, J. S., Food intake, body composition and blood lipids following treadmill exercise in male and female rats, *Physiol. Behav.*, 28(5), 917, 1982.

69. Mayer, J., Marchall, N. B., Vitale, J. J., Christensen, J. H., Mashayekhi, M. B., and Stare, F. J., Exercise, food intake and body weight in normal rats and genetically obese adult mice, *Am. J. Physiol.*, 177, 544, 1954.

70. Andik, I., Bank, J., Moring, I., and Szegari, G., The effect of exercise on the intake and selection of food in the rat, *Acta Physiol.*, 5, 457, 1954.

16 Effects of Fat Substitutes and Inhibitors of Absorption on Macronutrient Intake

Karen Ackroff

CONTENTS

ABSTRACT

Fat substitutes and inhibitors of fat absorption have been developed as therapeutic agents to reduce the contribution of dietary fat energy to total energy intake. Substitution of materials which mimic the orosensory properties of fat reduces the amount of fat in foods while sparing palatability. The absorption inhibitor orlistat reduces the net energy of consumed fat by partial inactivation of lipase secreted in response to meals. Orlistat and the nonabsorbable fat-substitute olestra have not yet been utilized in studies that specifically address neural controls of macronutrient intake. Existing studies of food intake with controlled use of these agents offer no support for a fat-specific appetite, but often were not designed as specific tests of this possibility. When future studies take advantage of these agents as experimental tools, any effects on macronutrient selection are likely to be indirect results of the reduction in products of triglyceride hydrolysis.

1   INTRODUCTION

In the application of research on the controls of macronutrient intake to practical dietary concerns, the nutrient that probably receives the most attention is fat. Many analyses suggest that overconsumption of dietary fat is a primary cause of overweight and obesity, in humans and in animal models. Accordingly, much research has attempted to discover ways of reducing the contribution of fat in the diet, while retaining some of the characteristics that make high-fat foods attractive.

This chapter will focus on two of the major techniques for reducing fat energy in food: the use of fat substitutes to replace dietary fat and the inhibition of dietary fat absorption. In both cases, the desirable orosensory characteristics of fat are retained, but the net contribution of fat to total energy intake is reduced. Both techniques represent only partial removal of fat, but the reductions have yielded significant alterations in the macronutrient composition and total energy intakes of self-selected diets.

Fortunately for the consumer but unfortunately for attempts to understand the control of nutrient selection, much of the research in these areas has focused on issues of safety and efficacy, in the service of preparing products for approval in the marketplace. The majority of published research has been conducted on the fat substitute olestra and the lipase inhibitor orlistat. The process of approving the nonabsorbable and nonenergy fat substitute olestra for U.S. consumption took years; the lipase inhibitor orlistat is nearing approval in the U.S. after many years of study to demonstrate safety. Many of the studies required fixed diets and other procedures that precluded the collection of data relevant to the controls of nutrient intake. However, the interest in possible compensation for the energy replaced by olestra has increased the number of studies that measured the composition of the diet selected after olestra consumption in target foods. Orlistat studies have almost always used fixed diets, so little is known about the possible impact of the drug on food selection. Because they have received the most experimental attention, this chapter will focus on these two agents.

Fat substitutes and absorption inhibitors have in common a reduction in net fat energy without any known direct effects on neural control systems. Because their actions are peripheral and because they have no demonstrable direct effects on the digestion and absorption of dietary protein and carbohydrate, any effects they might have on macronutrient selection must be indirect.

The basic findings with these two technologies can be summarized easily: while there may be substantial compensation for the energy reduction of a fat manipulation, there is no evidence at present for any fat-specific compensation. An important caveat is that many of the appropriate studies to test for macronutrient specificity have not been done, especially for absorption inhibition.

2 BACKGROUND: WORK WITH ALTERED FAT IN MACRONUTRIENT SELECTION

Prior to the development of specific therapeutic agents, there were many studies of the control of energy and nutrient intake that manipulated the fat content and orosensory characteristics of diets. A brief review of these data will provide some context for evaluation of current substitute and inhibitor studies.

2.1 ANIMAL STUDIES

Early examination of altered dietary fat in rats focused on the regulation of energy intake and appetite for fatty substances. While rats are initially attracted to foods containing nonenergy fat mimics (petrolatum, mineral oil), they rapidly learn to prefer foods containing real fat.[1-3] Although rats eat more chow adulterated with mineral oil, energy intake is less than that of unadulterated chow.[2] Diluting a fat source with water sometimes increases intake of the fat source in rats. Total fat intake was greater when an oil emulsion rather than pure oil was available as an option to chow, but energy intakes of pure vegetable shortening and a 35% shortening gel did not differ.[4] In contrast, total fat intake fell and carbohydrate intake increased as the oil emulsion was increasingly diluted in a macronutrient choice paradigm.[5] Dilution of the fat source with cellulose often yields fat selection lower than that with undiluted fat. For example, rats familiar with separate macronutrient sources responded to a 25% dilution of the fat by cellulose with increased intake of the fat source to maintain (but not exceed) undiluted levels; rats given a 50% dilution ate only a third the fat energy of the undiluted group and compensated by increasing carbohydrate intake.[6] An exception, in keeping with a greater need for fat as fuel, was the response of diabetic rats, which compensated for as much as 50% dilution with greater intake of the fat source. The degree of compensation may also depend on experience with the undiluted fat source: selection-naive rats offered a cellulose-diluted fat that is isoenergetic with the protein and carbohydrate sources consume only 65% of the fat energy taken by rats with undiluted fat.[7]

Another avenue that has received some attention is the use of less digestible fats in place of more readily digested ones. Hydrogenated soybean oil (HSO) has only 2.79 kcal/g of digestible energy, about 30% of the typical 9 kcal/g of common dietary fats.[8] Lean but not obese Zucker rats compensated for 20% HSO in a mixed diet by eating more food than rats given a 20% corn oil diet.[9,10] Similarly, rats fed 40% HSO diets consumed more food than rats fed highly digestible 40% soybean oil, hydrogenated coconut oil, or medium-chain triglyceride diets,[8] thereby maintaining the same absorbable energy intake. These responses resemble those to dilution of fat with bulking agents: within limits, rats compensate for bulk dilution by increasing the weight of food eaten, so that total energy intake is conserved. Although not yet studied for its effect on macronutrient selection, HSO is analogous in effect to the substitutes and inhibitors, reducing the net calories from food with the orosensory properties of fats.

2.2   HUMAN STUDIES

In humans, the general finding is a lack of macronutrient-specific compensation for foods with reduced fat content (e.g., References 11 and 12). The degree to which energy is compensated varies, which may be due to subject characteristics, test duration, and the nature of the available foods. Intervention studies involving a wide variety of reduced-fat products produced significant reductions in total and proportional energy from fat, accompanied by increased protein and/or carbohydrate intake so that total energy intake was conserved.[13,14] This conclusion has also been drawn from a review of the controlled, short-term laboratory studies of reduced-fat foods,[15] which also noted that the impact of these foods on overall fat intake has not been assessed in long-term monitoring of free-living consumers.

3   FAT SUBSTITUTES

Because fat is both the source of flavor and textural aspects of food palatability and of the most concentrated energy, efforts to replace the energy contribution of fat without the loss of its sensory characteristics have been widespread. The many fat replacers that have been developed are the subject of numerous reviews (e.g., References 15 through 21). Accordingly, their properties will not be covered in detail here. There are two basic types: energy and nonenergy. The former represent mostly carbohydrate- or protein-based products modified to carry some of the sensory characteristics of fat. For example, Simplesse® (Simplesse Company, Deerfield, IL) is a microparticulated protein used in selected foods to give the mouth feel of fat while reducing the proportion of energy from fat, often to near zero. One of many carbohydrate-based substitutes is Paselli® (Avebe America, Princeton, NJ), a suspension of maltodextrin in water. Naturally, foods using these replacers can have a markedly altered nutrient composition, higher in carbohydrate and/or protein than their normal-fat counterparts. They usually reduce the energy density of the food and thus can lead to reduced energy intake if portion sizes do not increase. However, because they are effectively reduced-fat foods rather than fat-substituted foods, they are not as instructive as the nonenergy substitutes, the most prominent of these being olestra.

Olestra is a formulation of sucrose polyester (SPE), a nondigestible, nonabsorbable group of esters of sucrose and fatty acids, which has physical and organoleptic properties indistinguishable from ordinary dietary fats. It can be included in a variety of prepared foods, withstanding heating and other forms of processing. Ingested olestra does not alter the absorption of fat[22] or other macronutrients,[23,24] and has minimal effects on gastrointestinal transit.[25] Unlike dietary fat, olestra does not stimulate gastric acid secretion or cholecystokinin (CCK) release.[26]

3.1 Animal Work with Fat Substitutes

The animal studies of SPE have largely focused on micronutrient absorption and other safety issues that had to be addressed prior to its approval for human consumption. There is no information on possible shifts in selection, because the rats in these studies were fed fixed diets rather than options among several food sources. However, their total intake of altered diets provides a measure of responsiveness to fat substitutes. In several published feeding trials,[27-29] rats compensated for the reduction in energy density of fixed diets with SPE by eating more food and thereby maintaining total energy. This response to SPE resembles that to nonenergy bulk such as cellulose.[30]

Some studies of other fat replacers have found unchanged or elevated intake in rats. Rats overate a diet that was matched for macronutrient composition with a control diet (25% fat) but differed in orosensory qualities by the addition of a fat mimetic (Paselli, 5% maltodextrin solids in water suspension). Like a group given a high-fat diet (45% of energy), the fat mimetic group overate and gained weight relative to rats eating the control diet.[31] However, the incorporation of water, rather than any successful mimicry of fat, may have been largely responsible for this result, given the overeating that has been produced by simple hydration of a diet.[32] Additionally, rats are attracted to maltodextrin flavor, particularly in hydrated form.[33] In another study, a protein/carbohydrate/water mimetic was substituted for portions of the fat in a high-fat (63% of energy) mixed diet. When the proportion of fat was halved, overeating was still apparent relative to intake of the 21% fat control diet. With near-total substitution that reduced fat to 2% of energy, the rats compensated with increased food intake to maintain energy intake at the 21% fat control level.[34] Nonenergy mimetics (cellulose/xanthan/water) may also improve intake of low-fat diets to levels of higher-fat energy intake[35] but reduced the intake of a diet with 30% fat energy relative to the 40% fat basal diet.[36]

When rats are offered full-fat or reduced-fat foods as options to a basal mixed diet, their food selection may depend on the nature of the basal diet. When rats were given chow (12% fat energy) and either full-fat (41% fat energy) or fat-free cake, they took more than 80% of energy as cake, so that the macronutrient composition of their diets was dictated largely by cake composition (36.8 and 2.3% fat energy in the full-fat and fat-free groups).[37] But rats maintained on a 29% fat basal diet and offered a rotating menu of high-fat foods or their fat-free counterparts as options selected only 30 to 46% of energy from the optional foods, composing diets that were 43% fat energy with high-fat foods and 24% with fat-free foods.[38] The many differences between the studies make it difficult to determine the critical factors; a better understanding of rats' responses to these foods could yield a useful animal model to predict human usage.

3.2 Human Studies of Olestra

There have been two general methods of studying the effects of olestra on nutrient and energy intake: short-term (days) and long-term (weeks) studies. In studies lasting up to 2 days, intake following an olestra meal period is followed and compared with a control period. These studies have found varying levels of energy compensation which is not fat specific. In the longer studies, which used a variety of protocols, subjects are required to consume olestra-containing foods but other intake is unrestricted. These too have found no fat-specific compensation for the energy replaced by olestra, and have sometimes found reductions in total energy intake. There have been some suggestions that the degree of "compensation" may reflect the subjects' habitual intake rather than that imposed during control periods. That is, in some cases the full-fat diet period represented an increase in subjects' fat and energy intake compared with their normal consumption patterns.

The short-term studies of olestra-containing foods have taken a variety of approaches to compare intake and food choices with normal-fat foods. Most used an olestra intervention in one or more meals eaten in the laboratory, followed by monitoring in the laboratory or via food diaries for 1 or 2 days. Two studies collaborated to examine the effect of an olestra breakfast on subsequent intake in two populations of lean men.[39,40] Others evaluated the effect of breakfast, snack, and lunch items with olestra in women[41] and in young children.[42] Several studies required the midday consumption of a preload with fat or olestra prior to the next meal.[43-45] Additional studies of lean men manipulated the fat/olestra content of lunch or dinner[46] or an entire day's meals.[47,48] All studies used crossover designs in which each subject received all conditions; they differed in the locus of the olestra-containing meal, and in the gender, age, and body weight characteristics of the subjects.

The Leeds/Penn collaboration[39,40] was among the earliest to evaluate possible compensation for olestra. In a crossover design, lean men ate breakfasts with normal fat and with fat replacement by 20 and 30 g of olestra, reducing energy contribution of fat by 23 and 41%. In both studies lunch intake was the same but the energy was made up within 24 h, primarily by increased carbohydrate intake at dinner. This is suggestive of a somewhat delayed response to the missing energy, without any fat-specific compensation. Fat intake was reduced from ~41 to ~36%.

Studies that prolonged the period of olestra substitution from breakfast through the midday meal found partial energy compensation (22 to 44%) in women and total energy compensation in children. The women did not show nutrient-specific changes in intake[41]; however, the variety of foods available to affect nutrient selection was limited to snacks and dessert, because the nutrient composition of the dinner was fixed. Fat intake was reduced from more than 35% to 31 to 32% on the olestra day. Children (2 to 5 years old) did not compensate for a 10% energy reduction on the day of the olestra manipulation but gradually made up the difference by the end of the second day.[42] They did not increase their selection of fat, which fell from 39% in the full-fat condition to 36%, with a reciprocal increase in the proportion from carbohydrate.

One strategy that differs somewhat from the standard meal-based replacement is the use of full-fat and olestra-substituted preloads, which can test for very short-term as well as day-long compensation for energy and fat. Two studies required consumption of croissants[45] or hot meals[44] with widely varying energy and fat contents but similar sensory characteristics, followed 15 min to 4.75 h later by the next meal. Men compensated more than women for the reduced-energy hot meal preload when they ate from a test buffet 2 h later, and neither sex altered its nutrient selection when olestra replaced fat. The fat/olestra content of the croissant preload did not affect nutrient selection in the subsequent meal at any delay, nor did it alter the day's energy intake. A third preload study examined the role of the sensory properties of olestra in determining subsequent intake.[43] Following a soup preload (high-fat, fat-free, or fat-free with olestra), lunch intake was reduced by the fat-free loads but not the high-fat load, so that only the high-fat total intake was greater than the no-preload control intake. This suggests that the mechanism relating intake to fat content of food was not dependent on learned sensory associations with postingestive consequences. There was no effect of preload on macronutrient selection at lunch; subjects took 30% of energy from fat regardless of preload fat content and sensory characteristics. The lack of specific compensation was seen not only when fat was reduced, but when it was increased by the high-fat load.

In contrast to most of the short-term studies, which included olestra in foods eaten early in the day, a few experiments presented these foods at other times of day. In all cases lean males were studied in all conditions. Possible effects of the timing of a single olestra meal were evaluated by comparing lunch and dinner manipulation.[46] Again, the full-fat meal was rather high in fat, at 53%, and was reduced to 27% by partial fat replacement with olestra. Subsequent intakes did not differ; although there were some increases in fat intake following the fat-replaced dinner, there was no clear nutrient compensation, suggesting insensitivity to the fat reduction. A similar result was obtained when olestra was used to replace fat across an entire day's menu of meals or snacks[48]: neither energy intake nor nutrient selection on the free-intake day was altered when the preceding fixed-intake day's nutrient composition was reduced from 40 to 30% energy from fat. However, in a similar design that reduced the proportion of fat from 30 to 20%, lean men compensated for 74% of the energy by the end of the next day by increasing both carbohydrate and fat consumption.[47] This result points to the recurring baseline problem in these experiments: judgments of compensation must be made relative to some standard, and when the standard is imposed rather than subject-defined, it is not clear that all subjects should be expected to compensate.

In the long-term studies, consumption of olestra-containing foods was required as part of daily intake, with selection of nonolestra foods making up the remainder of intake. The earliest study[49] had only 10 subjects, obese inpatients whose normal food preferences were used to create the foods given to them. They were required to consume 60% of their baseline intake in three meals, with free selection of evening snacks from an array of foods comprising another 60% of baseline intake, so that they were free to exceed their normal intakes. After a baseline period, they were given olestra-containing foods which replaced 40 g of fat per 1200 required energy (33% reduction of energy) for 20 days and normal-fat foods for 20 days, in counterbalanced order. Snack choice and intake were similar in the two periods, so that the overall proportion of intake from fat was lower and that of carbohydrate and protein higher during the olestra period. There was no compensation for the missing energy, and subjects lost weight in both periods. A similar result was obtained in a 2-week study of lean men[50] when ⅓ of the fat in a 40% fat diet was replaced. No compensation for the missing fat occurred, and proportional intake of fat fell while carbohydrate and protein increased.

Recent long-term studies have extended the analysis to include both men and women and lean and obese subjects. In one, normal-weight subjects ate full-fat or SPE-reduced evening meals in the laboratory and reported all other intake during counterbalanced 12-day periods.[51] There was partial compensation for the energy reduction, which varied with gender and information about the SPE manipulation, but no specific fat compensation was detected and no differences in macronutrient selection occurred. A potential problem with this study was the use of fixed portions in the manipulated meals, which may have increased some subjects' fat intake above their habitual consumption. In another study with 2-week counterbalanced periods, the olestra substitution was covert and the foods were tailored to the estimated energy requirements of the subjects.[52] Lean and obese men and women ate breakfast and dinner in the laboratory, with required items plus freely selected food available in both meals; at other times they ate from a wide variety of provided foods, which allowed macronutrient and energy compensation. The olestra substitution reduced core-food energy by about 10% of each subject's requirements, but they only compensated for 20% of the missing energy and 15% of the missing fat. There were no differences in carbohydrate and protein intake or in responses as a function of gender or body weight status. Therefore, even when the olestra substitution is spread over the day and designed to fit subject characteristics, there is no evidence that people seek to replace the missing fat and not much support for sensitivity to olestra-reduced energy intake.

3.3   SUMMARY

The many studies of olestra using a variety of protocols are the best data available on the effects of a fat substitute on macronutrient selection. However, there are potential problems with the information gathered so far. Procedural variables which could compromise the study of human macronutrient selection during manipulation of dietary fat content include:

- Fat content of the diet during the control period (if higher than habitual intakes, responses may not reflect subjects' normal regulation of intake);
- Laboratory-provided food as opposed to a freely chosen diet (the constraint may provoke choices for extraneous reasons: initial novelty, insufficient variety, overeating from buffet);
- Unnatural meal times (changes in normal schedule might compromise subjects' ability to adjust to challenges).

However, in spite of the variety of approaches and their potential problems, the consensus is that olestra does not produce compensatory changes in macronutrient selection, with the result that the proportion of dietary fat goes down. Compensation for energy varies, but when it occurs, it is generally more from carbohydrates than from protein (probably reflecting the greater proportion of intake from carbohydrates overall). There is no support for a fat-specific compensatory mechanism, and only partial support in humans for energy compensation. The latter is a function of subject and study variables.

4   INHIBITORS OF FAT ABSORPTION

A second way to reduce net fat intake is to allow fat consumption but reduce its postingestive availability to the body. Fat absorption in the intestine can be inhibited by inactivation of lipase, so that the large triglyceride molecules are not reduced to their component fatty acids and glycerol. Fat that is not digested in the small intestine is not absorbed, reducing the net energy contribution of ingested fat. The only significant agent in this category is orlistat, or tetrahydrolipstatin. It is already available in many countries and has been through most phases of U.S. testing prior to approval as the therapeutic drug Xenical. While there are a few discussions of other inhibitors (e.g., the surfactant pluronic L-101,[53,54] ebelactone B,[55] and an assortment of other agents[56]) there has been little work on these to date and the existing studies do not assess macronutrient choice.

Orlistat binds essentially irreversibly with lipase, but does not interact with other hydrolases in the intestine, resulting in a specific interference with fat digestion and absorption. However, secretion of lipase is much greater than that needed for hydrolysis of ingested fat, and the nonabsorbed drug–lipase complex moves along the intestine, limiting the inhibitory action of the drug to the directly contacted fraction of the secreted enzyme. Orlistat itself does not disturb gastrointestinal processes such as gastric emptying and bile and pancreatic secretions.[57] However, a recent study showed that intraduodenal delivery of a 120-mg dose of orlistat completely inhibited duodenal lipase activity during a mixed-nutrient test meal, which in turn prevented triglyceride hydrolysis and reduced plasma CCK by 77%.[58] In addition, amylase, trypsin, and bilirubin were significantly depressed, showing that triglyceride digestion is necessary for normal CCK, gallbladder, and pancreatic secretion. A similar study showed that these effects of orlistat were specific to fatty meals.[59] An orlistat-induced reduction in CCK, which is known to participate in the process of satiation, would also be expected to reduce the satiating effect of fat. These indirect effects of orlistat could lead to alterations in the digestion of other nutrients and ultimately to changes in diet selection, but the appropriate studies have not been done.

In humans, orlistat prevents absorption of 25 to 35% of dietary fat, as measured in fixed diets with 60 to 80 g fat/day, without much dose–response over the range 80 to 400 mg/day.[60-62] The incomplete effect is thought to reflect protection of the remaining fat by the normal fat-induced slowing of gastric emptying, because the drug and its sequestered lipase have moved farther along the digestive tract.[60] However, orlistat dosing after meals (rather than during meals) did not improve the inhibition.[61] Recent long-term studies found that 120-mg doses with each meal yielded significantly greater weight loss than placebo treatment in obese subjects on mildly hypoenergetic diets for 12 weeks,[63] 6 months,[64] or 1 year.[65] No conclusions about nutrient selection can be made since the dietary regimen in all studies was controlled at 30% of energy from fat.

Although the demonstrations of efficacy and safety as a drug have predominated in the animal literature, two published studies have examined food intake. In dietary-obese rats, 27 mg/kg/day orlistat prevented the absorption of over 80% of dietary fat.[66] The rats increased their intake of the fixed-composition diet (32% of energy from fat) but not compensatorily, losing weight relative to controls.

Thus far, there is one study[67] addressing the effect of orlistat on macronutrient selection in rats. After a 20-day period of adaptation to macronutrient selection from fat (vegetable shortening), carbohydrate (cornstarch), and protein (casein) sources, separate groups of female rats were given 0, 1, or 4 mg orlistat per gram of the fat diet. The drug was incorporated in the fat source to ensure that it accompanied all fat intake. After 28 days, the drug was removed for a 24-day recovery period.

The drug reduced fat intake dose dependently, with gradual increases in protein and carbohydrate intakes during the drug phase. The drug groups were nominally hyperphagic relative to controls, but their net energy intakes were approximately compensatory. For example, the 4-mg group was getting the dose that inhibited 80% of fat absorption in rats,[66] and they ate enough extra protein and carbohydrate to obtain control levels of energy intake. Although total fat intake decreased, the fat source was still consumed, suggesting that the animals did not form an aversion to it. This was supported in subsequent acceptance and flavor conditioning tests, showing that rats could detect the drug in the fat, reducing preference relative to unadulterated fat but not avoiding the drugged fat completely. While the control group continued to gain weight, body weights were stable in the drug groups, with growth resuming and fat intake recovering when the drug was removed from the food.

Several features of the study limit the generalizations that can be made. One is the lack of measures of fat absorption, which prevent a complete assessment of rats' compensation. Another is the high baseline rate of fat intake (70% of energy); the rats may not have been able to compensate for lost fat energy by eating even more fat. However, they were clearly responsive to the reduced energy content of their selected diet in two ways: shifting to greater consumption of the other food sources and eating more total energy. Finally, the fact that growth stopped relative to control rats is in part due to the ongoing fat deposition as the controls ate unadulterated high-fat diets. It would be interesting to compare this situation with one in which the animals had stable body weights prior to the introduction of the drug; the probable outcome would be weight loss as a function of drug dose, but this study has not been done.

There are a number of ways that this work could be extended to obtain more information on the effects of orlistat. One method, which would remove the important potential confound of drug flavor as a cue, would be to administer the drug postorally whenever the animal ate any of the available foods. This could be done with existing automated systems that infuse material into the gastrointestinal tract when oral ingestion is detected by sensors attached to food containers.[68] Any alterations in food selection would then be attributable to the physiological effect of the drug and not its flavor in the food. When the drug was presented in orally consumed fat, it reduced fat selection (and drug intake) as well as reducing fat absorption. Another valuable technique for assessing more immediate effects of fat inhibition on macronutrient choice is the use of a preload, followed at various delays by access to a varied meal. Comparison of high-fat and low-fat preloads with and without the drug would provide basic information on the possible interdependence of nutrient intakes over time. That is, if a high-fat meal tends to shift intake to other nutrients or if a low-fat meal later yields increased fat intake, repeated experience with drug effects on fat availability may alter animals' nutrient selection.

The net result of orlistat is a reduction in the energy available from the fat source, effectively diluting it as might be done with a physical agent. When physical dilution has been done in other studies of macronutrient selection, the animals took less fat and more carbohydrate; protein intake was not altered.[5,6] However, other characteristics of the nutrient sources also differed among the studies, so that it is not possible to conclude that orlistat differs because protein intake was affected. A useful study would compare orlistat directly with dilution in separate groups fed otherwise similar macronutrient sources.

How could macronutrient selection information be obtained in future work with orlistat in human subjects? The challenge is to provide an opportunity for macronutrient selection to change, which means overcoming well-established food preferences. The best design would be a straight-forward baseline-drug-recovery trial with cafeteria feeding: provide a wide enough assortment of foods, adjusted to the subjects' reported food preferences, varying over days but keeping the availability of high-fat and low-fat foods constant. Ideally, some foods that have reduced fat content but full-fat orosensory characteristics would also be included covertly to assess the extent of choices based on nutritional beliefs rather than actual fat content. From the onset, a placebo is taken with every meal, until the subject is well-adapted. Then the drug is introduced covertly or overtly at one of several doses while monitoring of intake continues for at least 2 weeks. Finally, the drug is replaced by the placebo to allow recovery.

The ingestion of fat with normal orosensory properties but reduced net energy, if it continues in a consistent manner, is expected to lead to an increase in food intake to the extent that the lost energy is detected. This appears to occur in rats, but in humans it is not clear from the long-term data that the modest reductions in energy are compensated. The lack of experimental and clinical data on macronutrient intake during orlistat administration limits the conclusions that can be drawn. However, the alteration of gastrointestinal handling of fat may in turn affect responses to long-term use of the drug.

5   CONCLUSIONS

Studies employing these therapeutic agents offer no support for a concept of specific regulation of fat intake. When the net intake of fat is artificially reduced, subsequent intake of fat is not elevated to make up the "deficit," although compensation for the missing energy is observed under some conditions. In these cases, the orosensory features of fat are preserved, so that responses both learned and unlearned can be sustained. To the extent that a fatty taste signals high energy content to the consumer, such food may be eaten in smaller amounts than a food that signals low fat. In contrast to this idea are increasing reports that high-fat diets are overeaten, possibly as a result of tendencies to eat a particular volume of food rather than a energetically determined amount (e.g., References 69 through 71). In this scenario high-fat diets are overeaten because they are more energetically dense (and no learning occurs to change the evaluation, perhaps because of the established palatability of such food).

The practical constraints on total reduction of fat energy for human use (~35% maximal reduction by orlistat; the current restriction of olestra to savory snack foods in the U.S.) limit the potential impact of these agents on nutrient selection. Significant absorbable fat remains in the diet. Both fat substitutes and inhibitors of fat absorption leave intact the orosensory properties of fat in foods, and this would sustain the intake of and preference for fat. There may be a chemosensory basis for physiological responsiveness to fat, as shown by more pronounced metabolism of a fat meal in people exposed to oral fat but not to an indistinguishable nonfat stimulus.[72] This mechanism could be disrupted with the use of fat substitutes that do not possess the required (and as yet uncharacterized) properties, but exposure to the real fat remaining in meals would be expected to suffice.

The most likely mechanism for potential alteration in macronutrient selection via these agents appears to lie in the reduced presence of the products of fat digestion. Reduced triglyceride hydrolysis has a variety of effects on gastrointestinal responses to meals which could ultimately alter nutrient selection. Both agents have considerable potential as research tools, due to their selective removal of dietary fat features. For example, as noted above, when lipase activity is strongly inhibited and the triglycerides of a fatty meal are not digested, CCK, gallbladder, and pancreatic secretion are suppressed;[58,59] these changes, if sustained by treatment, might lead to altered food choices. A recent review indicates that for both fat substitutes and absorption inhibitors, altered enzyme activity and intestinal transport capability could occur, which could lead to more efficient absorption of the remaining nutrients.[73] Future work should take advantage of the many fat substitutes and the ability to inhibit fat absorption in assessing the contributions of fat intake on macronutrient selection.

REFERENCES

1. Ackroff, K., Vigorito, M., and Sclafani, A., Fat appetite in rats: the response of infant and adult rats to nutritive and non-nutritive oil emulsions, *Appetite*, 15, 171, 1990.
2. Carlisle, H. J. and Stellar, E., Caloric regulation and food preference in normal, hyperphagic, and aphagic rats, *J. Comp. Physiol. Psychol.*, 69, 107, 1969.
3. Hamilton, C. L., Rats' preference for high-fat diets, *J. Comp. Physiol. Psychol.*, 58, 459, 1964.
4. Lucas, F., Ackroff, K., and Sclafani, A., Dietary fat-induced hyperphagia in rats as a function of fat type and physical form, *Physiol. Behav.*, 46, 937, 1989.

5. Castonguay, T. W., Burdick, S. L., Guzman, M. A., Collier, G. H., and Stern, J. S., Self-selection and the obese Zucker rat: the effect of dietary fat dilution, *Physiol. Behav.*, 33, 119, 1984.

6. Tepper, B. J. and Kanarek, R. B., Selection of protein and fat by diabetic rats following separate dilution of the dietary sources, *Physiol. Behav.*, 45, 49, 1989.

7. Bernardini, J., Kamara, K., and Castonguay, T. W., Macronutrient choice following food deprivation: effect of dietary fat dilution, *Brain. Res. Bull.*, 32, 543, 1993.

8. Kaplan, R. J. and Greenwood, C. E., Poor digestibility of fully hydrogenated soybean oil in rat: a potential benefit of hydrogenated fats and oils, *J. Nutr.*, 128, 875, 1998.

9. Comai, K., Triscari, J., and Sullivan, A. C., Differences between lean and obese Zucker rats: the effect of poorly absorbed dietary lipid on energy intake and body weight gain, *J. Nutr.*, 108, 826, 1978.

10. Sullivan, A. C., Triscari, J., and Comai, K., Caloric compensatory responses to diets containing either nonabsorbable carbohydrate or lipid by obese and lean Zucker rats, *Am. J. Clin. Nutr.*, 31, S261, 1978.

11. Foltin, R. W., Rolls, B. J., Moran, T. H., Kelly, T. H., McNelis, A. L., and Fischman, M. W., Caloric, but not macronutrient, compensation by humans for required-eating occasions with meals and snack varying in fat and carbohydrate, *Am. J. Clin. Nutr.*, 55, 331, 1992.

12. de Graaf, C., Drijvers, J. J., Zimmermanns, N. J., van het Hof, K., Weststrate, J. A., van den Berg, H., Velthuis-te Wierik, E. J., Westerterp, K. R., Verboeket-van de Venne, W. P., and Westerterp-Plantenga, M. S., Energy and fat compensation during long-term consumption of reduced fat products, *Appetite*, 29, 305, 1997.

13. Gatenby, S. J., Aaron, J. I., Morton, G. M., and Mela, D. J., Nutritional implications of reduced-fat food use by free-living consumers, *Appetite*, 25, 241, 1995.

14. Gatenby, S. J., Aaron, J. I., Jack, V. A., and Mela, D. J., Extended use of foods modified in fat and sugar content: nutritional implications in a free-living female population, *Am. J. Clin. Nutr.*, 65, 1867, 1997.

15. Miller, G. D. and Groziak, S. M., Impact of fat substitutes on fat intake, *Lipids*, 31 (Suppl.), S293, 1996.

16. Akoh, C. C., Lipid-based fat substitutes, *Crit. Rev. Food Sci. Nutr.*, 35, 405, 1995.

17. Gershoff, S. N., Nutrition evaluation of dietary fat substitutes, *Nutr. Rev.*, 53, 305, 1995.

18. Glueck, C. J., Streicher, P. A., Illig, E. K., and Weber, K. D., Dietary fat substitutes, *Nutr. Res.*, 14, 1605, 1994.

19. Mela, D. J., Impact of macronutrient-substituted foods on food choice and dietary choice, *Ann. N.Y. Acad. Sci.*, 819, 96, 1997.

20. Peters, J. C., Fat substitutes and energy balance, *Ann. N.Y. Acad. Sci.*, 827, 461, 1997.

21. Peters, J. C., Nutritional aspects of macronutrient-substitute intake, *Ann. N.Y. Acad. Sci.*, 819, 169, 1997.

22. Daher, G. C., Cooper, D. A., Zorich, N. L., King, D., Riccardi, K. A., and Peters, J. C., Olestra ingestion and dietary fat absorption in humans, *J. Nutr.*, 127, 1694S, 1997.

23. Peters, J. C., Lawson, K. D., Middleton, S. J., and Triebwasser, K. C., Assessment of the nutritional effects of olestra, a nonabsorbed fat replacement: summary, *J. Nutr.*, 127, 1719S, 1997.

24. Lawson, K. D., Middleton, S. J., and Hassall, C. D., Olestra, a nonabsorbed, noncaloric replacement for dietary fat: a review, *Drug Metab. Rev.*, 29, 651, 1997.

25. Aggarwal, A. M., Camilleri, M., Phillips, S. F., Schlagheck, T. G., Brown, M. L., and Thomforde, G. M., Olestra, a nondigestible, nonabsorbable fat: effects on gastrointestinal and colonic transit, *Dig. Dis. Sci.*, 38, 1009, 1993.

26. Maas, M. I., Hopman, W. P., van der Wijk, T., Katan, M. B., and Jansen, J. B., Sucrose polyester does not inhibit gastric acid secretion or stimulate cholecystokinin release in men, *Am. J. Clin. Nutr.*, 65, 761, 1997.

27. Miller, K. W. and Jones, P. H., A 91-day feeding study in rats with heated olestra/vegetable oil blends, *Food Chem. Toxicol.*, 28, 307, 1990.

28. Mattson, F. H., Hollenbach, E. J., and Kuehlthau, C., The effect of a non-absorbable fat, sucrose polyester, on the metabolism of vitamin A by the rat, *J. Nutr.*, 109, 1688, 1979.

29. Nolen, G. A., Wood, F. E. J., and Dierckman, T. A., A two-generation reproductive and developmental toxicity study of sucrose polyester, *Food Chem. Toxicol.*, 25, 1, 1987.

30. Adolph, E. F., Urges to eat and drink in rats, *Am. J. Physiol.*, 151, 110, 1947.

31. Harris, R. B., Factors influencing energy intake of rats fed either a high-fat or a fat mimetic diet, *Int. J. Obesity*, 18, 632, 1994.

32. Ramirez, I., Feeding a liquid diet increases caloric intake, weight gain and body fat in rats, *J. Nutr.*, 117, 2127, 1987.

33. Sclafani, A., Carbohydrate-induced hyperphagia and obesity in the rat: effects of saccharide type, form, and taste, *Neurosci. Biobehav. Rev.*, 11, 155, 1987.

34. Harris, R. and Jones, W., Physiological response of mature rats to replacement of dietary fat with a fat substitute, *J. Nutr.*, 121, 1109, 1991.

35. Harris, R. B. S., Growth measurements in Sprague Dawley rats fed diets of very low fat concentration, *J. Nutr.*, 121, 1075, 1991.

36. Harris, R. and Kor, H., Insulin insensitivity is rapidly reversed in rats by reducing dietary fat from 40 to 30% of energy, *J. Nutr.*, 122, 1811, 1992.

37. Sclafani, A., Weiss, K., Cardieri, C., and Ackroff, K., Feeding response of rats to no-fat and high-fat cakes, *Obesity Res.*, 1, 173, 1993.

38. Harris, R. B., The impact of high- or low-fat cafeteria foods on nutrient intake and growth of rats consuming a diet containing 30% energy as fat, *Int. J. Obesity*, 17, 307, 1993.

39. Blundell, J. E., Burley, V. J., and Peters, J. C., Dietary fat and human appetite: effects of non-absorbable fat on energy and nutrient intakes, in *Obesity: Dietary Factors and Control*, Romsos, D. R., Himms-Hagen, J., and Suzuki, M., Eds., Japan Scientific Society Press, Tokyo, 1991.

40. Rolls, B. J., Pirraglia, P. A., Jones, M. B., and Peters, J. C., Effects of olestra, a noncaloric fat substitute, on daily energy and fat intakes in lean men, *Am. J. Clin. Nutr.*, 56, 84, 1992.

41. Westerterp-Plantenga, M. S., Wijckmans-Duijsens, N. E., ten Hoor, F., and Weststrate, J. A., Effect of replacement of fat by nonabsorbable fat (sucrose polyester) in meals or snacks as a function of dietary restraint, *Physiol. Behav.*, 61, 939, 1997.

42. Birch, L. L., Johnson, S. L., Jones, M. B., and Peters, J. C., Effects of a nonenergy fat substitute on children's energy and macronutrient intake, *Am. J. Clin. Nutr.*, 58, 326, 1993.

43. Rolls, B. J., Castellanos, V. H., Shide, D. J., Miller, D. L., Pelkman, C. L., Thorwart, M. L., and Peters, J. C., Sensory properties of a nonabsorbable fat substitute did not affect regulation of energy intake, *Am. J. Clin. Nutr.*, 65, 1375, 1997.

44. Hulshof, T., de Graaf, C., and Westrate, J. A., Short-term satiating effect of the fat replacer sucrose polyester (SPE) in man, *Br. J. Nutr.*, 74, 569, 1995.

45. Hulshof, T., de Graaf, C., and Westrate, J. A., Short-term effects of high-fat and low-fat/high-SPE croissants on appetite and energy intake at three deprivation levels, *Physiol. Behav.*, 57, 377, 1995.

46. Cotton, J. R., Burley, V. J., Weststrate, J. A., and Blundell, J. E., Fat substitution and food intake: effect of replacing fat with sucrose polyester at lunch or evening meals, *Br. J. Nutr.*, 75, 545, 1996.

47. Burley, V. J., Cotton, J. R., Westrate, J. A., and Blundell, J. E., Effect on appetite of replacing natural fat with sucrose polyester in meals or snacks across one whole day, in *Obesity in Europe 1993*, Ditschuneit, H., Gries, F. A., Hauner, H., Schusdziarra, V., and Wechsler, J. G., Eds., John Libbey & Company, London, 1994.

48. Cotton, J. R., Weststrate, J. A., and Blundell, J. E., Replacement of dietary fat with sucrose polyester: effects on energy intake and appetite control in nonobese males, *Am. J. Clin. Nutr.*, 63, 891, 1996.

49. Glueck, C. J., Hastings, M. M., Allen, C., Hogg, E., Baehler, L., Gartside, P. S., Phillips, D., Jones, M., Hollenbach, E. J., Braun, B., and Anastasia, J. V., Sucrose polyester and covert caloric dilution, *Am. J. Clin. Nutr.*, 35, 1352, 1982.

50. Sparti, A., Windhauser, M., Lovejoy, J., and Bray, G., Subjects eat for carbohydrate not calories after dietary fat replacement with olestra, *Am. J. Clin. Nutr.*, 61, 902, 1995.

51. de Graaf, C., Hulshof, T., Weststrate, J. A., and Hautvast, J. G., Nonabsorbable fat (sucrose polyester) and the regulation of energy intake and body weight, *Am. J. Physiol.*, 270, R1386, 1996.

52. Hill, J. O., Seagle, H. M., Johnson, S. L., Smith, S., Reed, G. W., Tran, Z. V., Cooper, D., Stone, M., and Peters, J. C., Effects of 14 d of covert substitution of olestra for conventional fat on spontaneous food intake, *Am. J. Clin. Nutr.*, 67, 1178, 1998.

53. Comai, K. and Sullivan, A. C., Antiobesity activity of pluronic L-101, *Int. J. Obesity*, 4, 33, 1980.

54. Puls, W., Krause, H. P., Müller, L., Schutt, H., Sitt, R., and Thomas, G., Inhibitors of the rate of carbohydrate and lipid absorption by the intestine, *Int. J. Obesity*, 8 (Suppl. 1), 181, 1984.

55. Nonaka, Y., Ohtaki, H., Ohtsuka, E., Kocha, T., Fukuda, T., Takeuchi, T., and Aoyagi, T., Effects of ebelactone B, a lipase inhibitor, on intestinal fat absorption in the rat, *J. Enzyme Inhib.*, 10, 57, 1996.

56. Ransac, S., Gargouri, Y., Marguet, F., Buono, G., Beglinger, C., Hildebrand, P., Lengsfeld, H., Hadvary, P., and Verger, R., Covalent inhibition of lipases, *Methods Enzymol.*, 286, 190, 1997.

57. Guerciolini, R., Mode of action of orlistat, *Int. J. Obesity*, 21 (Suppl. 3), S12, 1997.

58. Hildebrand, P., Petrig, C., Burckhardt, B., Ketterer, S., Lengsfeld, H., Fleury, A., Hadvary, P., and Beglinger, C., Hydrolysis of dietary fat by pancreatic lipase stimulates cholecystokinin release, *Gastroenterology*, 114, 123, 1998.

59. Schwizer, W., Asal, K., Kreiss, C., Mettraux, C., Borovicka, J., Remy, B., Guzelhan, C., Hartmann, D., and Fried, M., Role of lipase in the regulation of upper gastrointestinal function in humans, *Am. J. Physiol.*, 273, G612, 1997.

60. Hauptman, J. B., Jeunet, F. S., and Hartmann, D., Initial studies in humans with the novel gastrointestinal lipase inhibitor Ro 18-0647 (tetrahydrolipstatin), *Am. J. Clin. Nutr.*, 55, 309S, 1992.

61. Hartmann, D., Hussain, Y., Guzelhan, C., and Odink, J., Effect on dietary fat absorption of orlistat, administered at different times relative to meal intake, *Br. J. Clin. Pharmacol.*, 36, 266, 1993.

62. Guzelhan, C., Odink, J., Niestijl Jansen-Zuidema, J. J., and Hartmann, D., Influence of dietary composition on the inhibition of fat absorption by orlistat, *J. Int. Med. Res.*, 22, 255, 1994.

63. Drent, M. L., Larsson, I., William-Olsson, T., Quaade, F., Czubayko, F., von Bergmann, K., Strobel, W., Sjostrom, L., and van der Veen, E. A., Orlistat (Ro 18-0647), a lipase inhibitor, in the treatment of human obesity: a multiple dose study, *Int. J. Obesity*, 19, 221, 1995.

64. Van Gaal, L. F., Broom, J. I., Enzi, G., and Toplak, H., Efficacy and tolerability of orlistat in the treatment of obesity: a 6-month dose-ranging study, *Eur. J. Clin. Pharmacol.*, 54, 125, 1998.

65. James, W. P., Avenell, A., Broom, J., and Whitehead, J., A one-year trial to assess the value of orlistat in the management of obesity, *Int. J. Obesity*, 21 (Suppl. 3), S24, 1997.

66. Hogan, S., Fleury, A., Hadvary, P., Lengsfeld, H., Meier, M. K., Triscari, J., and Sullivan, A. C., Studies on the antiobesity activity of tetrahydrolipstatin, a potent and selective inhibitor of pancreatic lipase, *Int. J. Obesity*, 11 (Suppl. 3), 35, 1987.

67. Ackroff, K. and Sclafani, A., Effects of the lipase inhibitor orlistat on intake and preference for dietary fat in rats, *Am. J. Physiol.*, 271, R48, 1996.

68. Ackroff, K. and Sclafani, A., Conditioned flavor preferences: evaluating the postingestive reinforcing effects of nutrients, in *Current Protocols in Neuroscience*, Crawley, J., Gerfen, C., McKay, R., Rogawski, M., Sibley, D., and Skolnick, P., Eds., Wiley, New York, 1999.

69. Lissner, L., Levitsky, D. A., Strupp, B. J., Kalkwarf, H. J., and Roe, D. A., Dietary fat and the regulation of energy intake in human subjects, *Am. J. Clin. Nutr.*, 46, 886, 1987.

70. Mattes, R. D., Pierce, C. B., and Friedman, M. I., Daily caloric intake of normal-weight adults: response to changes in dietary energy density of a luncheon meal, *Am. J. Clin. Nutr.*, 48, 214, 1988.

71. Poppitt, S. D. and Prentice, A. M., Energy density and its role in the control of food intake: evidence from metabolic and community studies, *Appetite*, 26, 153, 1996.

72. Mattes, R. D., Oral fat exposure alters postprandial lipid metabolism in humans, *Am. J. Clin. Nutr.*, 63, 911, 1996.

73. Thomson, A. B., De Pover, A., Keelan, M., Jarocka-Cyrta, E., and Clandinin, M. T., Inhibition of lipid absorption as an approach to the treatment of obesity, *Methods Enzymol.*, 286, 3, 1997.

Section III

Detection of Macronutrients and Their Metabolites

17 Taste, Olfactory, Visual, and Somatosensory Representations of the Sensory Properties of Foods in the Brain, and Their Relation to the Control of Food Intake

Edmund T. Rolls

CONTENTS

1   INTRODUCTION

This chapter describes how the sensory properties of the main macronutrients in food are represented in the cortical areas of primates and how these representations are affected by feeding to satiety. In the primary taste cortex, there is a representation of sweet, salt, bitter, sour. There is also a representation of umami or protein taste, as exemplified by monosodium glutamate (MSG) and inosine 5'-monophosphate (IMP), and this represents the taste of protein. The representation in the primate primary taste cortex is of the identity of taste, in that the representation is not affected by feeding to satiety. In contrast, in the secondary taste cortex in the orbitofrontal area, single neurons respond to the taste of a food only if hunger is present. The neurons here reflect the reward value of taste. They show sensory-specific satiety. For example, feeding to satiety with glucose decreases the responses of orbitofrontal cortex neurons to sweet taste to zero, and feeding to satiety with MSG decreases the responses of these neurons to protein taste to zero. In addition, there are separate representations produced through the somatosensory system in the orbitofrontal cortex of the mouth feel produced by fat, and by astringency (e.g., tannic acid). These orosensory inputs converge in some cases onto neurons with taste responses. In the orbitofrontal cortex, olfactory inputs converge onto neurons with taste inputs, forming representations of flavour. Neurons in this region may also respond to the sight of food, and to its texture. The olfactory representation for some neurons reflects the taste association of odours; and olfactory sensory-specific satiety is represented in this part of the brain. Rapid learning of visual to taste associations is also a feature of the neural processing which occurs in the orbitofrontal cortex. Thus, the primate orbitofrontal cortex contains representations of the orosensory inputs produced by different macronutrients (including sweet indicating carbohydrate, protein, fat, and also astringency). In addition, visual and olfactory inputs converge onto these orosensory representations, form representations that are established by associative learning, and enable a vast range of different foods to be represented. These different (taste, olfactory, visual, and somatosensory) representations of food each show a sensory-specific effect of feeding to satiety. Although a major modulation by ingestion of the responsiveness of these orbitofrontal cortex representations of food is sensory-specific, it will be of interest to determine to what extent the modulation may be contributed to also by postingestive consequences of ingesting different macronutrients.

The aims of this chapter are to describe the orosensory (taste and somatosensory) representation of different foods in the cerebral cortex, how olfactory and visual inputs combine with these representations, and how hunger affects the representations in different cortical areas. The neural representations in the cortex of the sensory inputs produced by different macronutrients are thus described, together with ways in which the acceptability of different foods is affected by sensory-specific satiety. Particular attention is paid to investigations in nonhuman primates, macaques, because of their relevance for understanding the information processing in the same areas in humans.

2   THE TASTE PATHWAYS IN PRIMATES

A diagram of the taste and related pathways in primates is shown in Figure 1. The three taste nerves, the facial (seventh, chorda tympani and greater superficial petrosal branches), glossopharyngeal (ninth, lingual branch), and vagus (tenth, superior laryngeal branch) terminate in the rostral part of the nucleus of the solitary tract (NTS) in the rostrolateral medulla.[1] Second-order taste neurons in this first central relay in the taste system project monosynaptically to the thalamic taste nucleus, the parvicellular division of the ventralposteromedial thalamic nucleus (VPMpc).[1,2] A remarkable difference from the taste system of rodents is this direct projection from the NTS to the taste thalamus. In rodents, there is an obligatory relay from the NTS to the pontine parabrachial taste nuclei (the "pontine taste area"), which in turn project to the thalamus.[1] The pontine taste nuclei also project to the hypothalamus and amygdala in rodents,[3] providing direct subcortical access to these subcortical structures important in motivational behaviour (e.g., feeding) and learning.[4] In contrast, in primates there may well be no such direct pathway from the brain stem taste areas to the hypothalamus and amygdala.[1] This remarkable difference in the anatomy of the taste pathways between rodents and primates shows that even in such an apparently phylogenetically

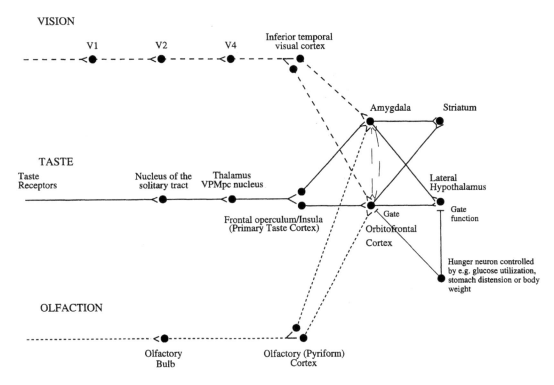

FIGURE 1   Schematic diagram of the taste and olfactory pathways in primates showing how they converge with each other and with visual pathways. The gate functions shown refer to the finding that the responses of taste neurons in the orbitofrontal cortex and the lateral hypothalamus are modulated by hunger. VPMpc = ventralposteromedial thalamic nucleus; V1, V2, V4 = visual cortical areas.

old system as the taste system, the way in which the system functions and processes information may be very different in primates. This difference may be due to the great development of the cerebral cortex in primates, and the advantage of using extensive and similar cortical analysis of inputs from every sensory modality before the analysed representations from each modality are brought together in multimodal regions, as is documented below.

The thalamic taste area, VPMpc, then projects to the cortex which in primates forms the rostral part of the frontal operculum and adjoining insula (Figure 2), so that this is by definition the primary taste cortex.[5] In macaques, this is at the anterior end of the Sylvian fissure. This region of cortex was implicated in gustatory function by Bornstein[6,7] who observed ageusias (inability to taste) in a dozen patients with bullet wounds in this area.

A secondary cortical taste area has been discovered by Rolls et al.[8] in the caudolateral orbitofrontal cortex, extending several millimeters in front of the primary taste cortex (see Figure 2). Injections of wheat germ agglutinin-horseradish peroxidase (WGA-HRP) for retrograde neuronal tracing were made into this region in five monkeys in which the exact location of this cortical taste region had been identified by recordings of the activity of single taste neurons. Labelled cell bodies were found in the frontal opercular taste cortex and in the insular taste cortex,[9] showing that the caudolateral orbitofrontal taste cortex is a secondary taste cortical area. Afferents were also shown to reach the caudolateral orbitofrontal taste cortex from the more ventral part of the rostral insular cortex, the amygdala, the substantia innominata, the rhinal sulcus, and from the surrounding orbitofrontal cortex. Through some of these pathways visceral information may reach the caudolateral orbitofrontal taste cortex.

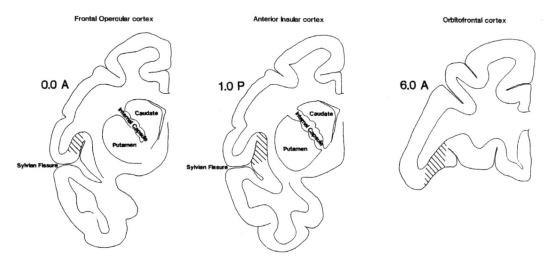

FIGURE 2   Coronal sections to show the locations of the primary taste cortices in the macaque in the frontal operculum and rostral insula, and of the secondary taste cortex in the caudolateral orbitofrontal cortex. The coordinates are in mm anterior (A) or posterior (P) to sphenoid.

3   TASTE PROCESSING IN THE NUCLEUS OF THE SOLITARY TRACT

Taste neurons have been found and their responses analysed in the rostral part of the nucleus of the solitary tract of macaque monkeys.[10] Different neurons were found which responded best to glucose, NaCl, HCl (sour), and quinine HCl (bitter), but the tuning of the neurons was in most cases broad, in that, for example, 84% of the neurons had at least some response to three or more of these four prototypical taste stimuli.

4   GUSTATORY NEURONAL RESPONSES IN THE TASTE THALAMUS

Pritchard et al.[5] were able to confirm that the parvicellular division of the ventroposteromedial (VPMpc) nucleus of the thalamus is the thalamic taste relay nucleus in primates by showing that single neurons in it responded to taste stimuli in macaque monkeys. The neurons were relatively broadly tuned, with a mean breadth of tuning of 0.73 (measured across responses to sucrose, NaCl, HCl, and quinine HCl). Responses to sweet and salt were most common, with 56% responding best to sucrose and 24% responding best to NaCl. Relatively few neurons had best responses to HCl or QHCl (14%), and most of the responses to HCl and QHCl were described as sideband responses in NaCl-best neurons. In addition to gustatory neurons, some neurons in the nucleus responded to tactile stimuli.

5   GUSTATORY RESPONSES IN THE PRIMARY TASTE CORTEX

In an extensive series of investigations of the neurophysiology of the taste system of the primate, Rolls and his colleagues have analysed the responses of single neurons in several cortical areas concerned with taste in the cynomolgus macaque monkey, Macaca fascicularis. In a first area studied, the responses of 165 single neurons in the frontal operculum with gustatory responses to stimuli which included NaCl, glucose, HCl, quinine (QHCl), water, and a complex taste stimulus, black currant juice, were analysed.[11] The taste region was found to be located in the dorsal part of the frontal operculum (see Figure 2). The neurons were found to be more specifically tuned to the prototypical stimuli glucose, NaCl, HCl, and QHCl than were neurons recorded in the same monkeys in the nucleus of the solitary tract, with a mean breadth of tuning for the opercular neurons of 0.67.[10,11] Neurons with gustatory responses were also found localised in the adjoining rostral and dorsal part of the insula[12] (see Figure 2). These neurons were also found to be a little more specifically tuned to the gustatory stimuli than were neurons recorded in the same monkeys in the nucleus of the solitary tract. By using glucose and NaCl, a corresponding area has been demonstrated in fMRI investigations in humans.[13]

6   GUSTATORY RESPONSES IN THE SECONDARY CORTICAL TASTE AREA, THE CAUDOLATERAL ORBITOFRONTAL CORTEX

In a study of the role of the orbitofrontal cortex in learning, Thorpe et al.[14] found a small proportion of neurons (7.9%) with gustatory responses in the main part of the orbitofrontal cortex. In some cases these neurons were very selective for particular gustatory stimuli. Therefore, when they set out to search for a secondary taste cortical area, Rolls et al.,[8] started recording at the anterior boundary of the opercular and insular cortical taste areas, and worked forward toward the orbito-frontal area investigated by Thorpe et al.[14] In recordings made from 3120 single neurons, Rolls et al.[8] found a secondary cortical taste area in the caudolateral part of the orbitofrontal cortex of the cynomolgus macaque monkey, *Macaca fascicularis*. The area is part of the dysgranular field of the orbitofrontal cortex, OFdg (see Figure 2), and is situated anterior to the primary taste cortical area in the frontal opercular and adjoining insular cortices. The responses of 49 single neurons with gustatory responses in the caudolateral orbitofrontal taste cortex were analysed using the taste stimuli glucose, NaCl, HCl, quinine HCl, water, and blackcurrant juice. Examples of the responses of orbitofrontal cortex neurons to the stimuli are shown by Rolls et al.,[8] and quite sharp tuning was evident in many cases, in that some of the neurons responded primarily to the taste of one tastant. This tuning is much finer than that of neurons in the nucleus of the solitary tract of the monkey, and finer than that of neurons in the primary frontal opercular and the insular taste cortices. By using glucose and NaCl, a corresponding area in the medial orbitofrontal cortex close to the subgenual cingulate area has been demonstrated in fMRI investigations in humans.[13]

7   PROTEIN TASTE

An important food taste which appears to be different from that produced by sweet, salt, bitter, or sour is the taste of protein. At least part of this taste is captured by the Japanese word *umami*, which is a taste common to a diversity of food sources including fish, meats, mushrooms, cheese, and some vegetables. Within these food sources, it is glutamates and 5′-nucleotides, sometimes in a synergistic combination, that create the umami taste.[15-18] MSG, guanosine 5′-monophosphate (GMP), and IMP are examples of umami stimuli.

These findings raise the question of whether umami taste operates through channels in the primate taste system which are separable from those for the "prototypical" tastes sweet, salt, bitter, and sour. To investigate the neural encoding of glutamate in the primate, Baylis and Rolls[19] made recordings from 190 taste-responsive neurons in the primary taste cortex and adjoining orbitofrontal cortex taste area in macaques. Single neurons were found that were tuned to respond best to MSG (umami taste), just as other cells were found with best responses to glucose (sweet), sodium chloride (salty), HCl (sour), and quinine HCl (bitter). Across the population of neurons, the responsiveness to glutamate was poorly correlated with the responsiveness to NaCl, so that the representation of glutamate was clearly different from that of NaCl. Further, the representation of glutamate was shown to be approximately as different from each of the other four tastants as they are from each other, as shown by multidimensional scaling and cluster analysis. It was concluded that in primate taste cortical areas, glutamate, which produces umami taste in humans, is approximately as well represented as are the tastes produced by glucose (sweet), NaCl (salty), HCl (sour), and quinine HCl (sour).[19]

In a recent investigation, these findings have been extended beyond the sodium salt of glutamate to other umami tastants, which have the glutamate ion but which do not introduce sodium ion into the experiment, and to a nucleotide umami tastant.[20] In recordings made mainly from neurons in the orbitofrontal cortex taste area, it was shown that single neurons that had their best responses to sodium glutamate also had good responses to glutamic acid. It was also shown that the responses of these neurons to the nucleotide umami tastant IMP were more correlated with their responses to MSG than to any prototypical tastant.[21]

Thus, neurophysiological evidence in primates does indicate that there is a representation of umami flavour in the cortical areas which is separable from that to the prototypical tastants sweet, salt, bitter, and sour. This representation is probably important in the taste produced by proteins, and to complement these findings, recent evidence has started to accumulate that there may be taste receptors on the tongue specialized for umami taste.[22,23]

8   THE MOUTH FEEL OF FAT

Texture in the mouth is also an important indicator of whether *fat* is present in a food, which is important not only as a high-value energy source, but also as a potential source of essential fatty acids. In the orbitofrontal cortex, Rolls et al.[76] have found a population of neurons that responds when fat is in the mouth. An example of such a neuron is shown in Figure 3. The neuron illustrates that information about fat as well as about taste can converge onto the same neuron in this region.

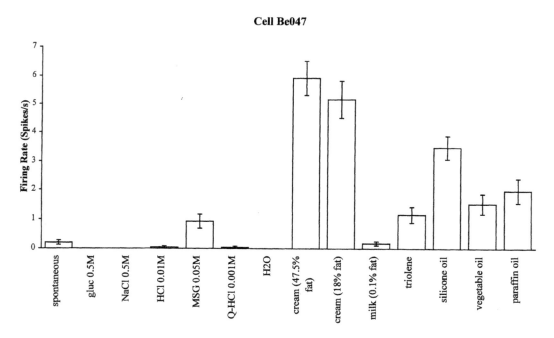

FIGURE 3   A neuron in the primate orbitofrontal cortex responding to the texture of fat in the mouth. The cell (Be047) increased its firing rate to cream (double and single cream), and responded to texture rather than the chemical structure of the fat in that it also responded to 0.5 ml of silicone oil (SiO$_2$) or paraffin oil (hydrocarbon). The cell has a taste input too, in that it had a consistent but small response to umami taste (MSG). Gluc, glucose; NaCl, salt; HCl, sour; Q-HCl, quinine, bitter. The spontaneous firing rate of the cell is also shown. (After Rolls et al.[76])

The neuron responded to taste in that its firing rate was significantly different within the group of tastants sweet, salt, bitter, and sour. However, its response to fat in the mouth was larger. The fat-related responses of these neurons are produced at least in part by the texture of the food rather than by chemical receptors sensitive to certain chemicals, in that such neurons typically respond not only to foods such as cream and milk containing fat, but also to paraffin oil, which is a pure hydrocarbon, and to silicone oil, which contains $(SiO)n$. Some of the fat-related neurons do, however, have convergent inputs from the chemical senses, in that in addition to taste inputs, some of these neurons respond to the odour associated with a fat, such as the odour of cream.[24] Feeding to satiety with fat (e.g., cream) decreases the responses of these neurons to zero of the food eaten to satiety, but if the neuron receives a taste input from, for example, glucose taste, it is not decreased by feeding to satiety with cream. Thus, there is a representation of the macronutrient fat in this brain area, and the activation produced by fat is reduced by eating fat to satiety.

9 EFFECTS OF HUNGER AND SATIETY ON TASTE PROCESSING AT DIFFERENT STAGES OF THE TASTE PATHWAY AND THEIR RELEVANCE TO THE CONTROL OF EATING AND SENSORY-SPECIFIC SATIETY

In order to analyse the neural control of feeding, the activity of single neurons is being recorded during feeding in brain regions implicated in feeding in the monkey.[4,25-29] It has been found that a population of neurons in the lateral hypothalamus and adjoining substantia innominata of the monkey respond to the sight and/or taste of food.[30] Part of the evidence that these neurons are involved in the control of the responses which are made to food when hungry is that these neurons only respond to food when the monkey is hungry.[31,32] Indeed, the modulation of the reward or incentive value of a motivationally relevant sensory stimulus such as the taste of food by motivational state, for example, hunger, is one important way in which motivational behaviour is controlled.[4,33] The subjective correlate of this modulation is that food tastes pleasant when hungry and tastes hedonically neutral when it has been eaten to satiety.

These findings raise the question of the stage in sensory processing at which satiety modulates responsiveness. It could be that the effects of satiety are manifest far peripherally in the taste pathways, for example, in the nucleus of the solitary tract. Or it could be that peripheral processing is concerned primarily with stimulus analysis and representation and, in order to optimise this, is independent of hunger. In order to provide evidence on where hunger controls taste processing in primates, the responses of single neurons at different stages of the taste system have been analysed while macaque monkeys are fed to satiety, usually with 20% glucose solution. To ensure that the results were relevant to the normal control of feeding (and were not due, for example, to abnormally high levels of artificially administered putative satiety signals such as gastric distension or plasma glucose), we allowed the monkeys to feed until they were satiated, and determined whether this normal and physiological induction of satiety influenced the responsiveness of neurons in the taste system, which were recorded throughout the feeding until satiety was reached. The recordings were made in the monkey in order to make the results as relevant as possible to our understanding of sensory processing and the control of feeding, and its disorders, in humans.

We have found that this modulation of taste-evoked signals by motivation is not a property found in early stages of the primate gustatory system. The responsiveness of taste neurons in the nucleus of the solitary tract is not attenuated by feeding to satiety.[34] Thus taste processing at this early stage of the taste system does not appear to be modulated by satiety, the signals for which include gastric distension[35] as well as other signals.[36] We have also found that in the primary taste cortex, both in the frontal opercular part[37] and in the insular part,[38] hunger does not modulate the responsiveness of single neurons to gustatory stimuli.

In contrast, in the secondary taste cortex, in the caudolateral part of the orbitofrontal cortex, it was found that the responses of the neurons to the taste of the glucose decreased to zero while the monkey ate it to satiety, during the course of which its behaviour turned from avid acceptance to active rejection.[39] This modulation of responsiveness of the gustatory responses of the orbitofrontal cortex neurons by satiety could not have been due to peripheral adaptation in the gustatory system or to altered efficacy of gustatory stimulation after satiety was reached, because modulation of neuronal responsiveness by satiety was not seen at the earlier stages of the gustatory system, including the nucleus of the solitary tract, the frontal opercular taste cortex, and the insular taste cortex. Evidence was obtained that gustatory processing involved in thirst also becomes interfaced to motivation in the caudolateral orbitofrontal cortex taste projection area, in that neuronal responses here to water were decreased to zero while water was drunk until satiety was produced.[39]

In the secondary taste cortex, it was also found that the decreases in the responsiveness of the neurons were relatively specific to the food with which the monkey had been fed to satiety. For example, in seven experiments in which the monkey was fed glucose solution, neuronal responsiveness decreased to the taste of the glucose but not to the taste of black currant juice (see example in Figure 4). Conversely, in two experiments in which the monkey was fed to satiety with fruit juice, the responses of the neurons decreased to fruit juice but not to glucose.[39]

This evidence shows that the reduced acceptance of food which occurs when food is eaten to satiety, and the reduction in the pleasantness of its taste,[25,29,36,40-44] are not produced by a reduction in the responses of neurons in the nucleus of the solitary tract or frontal opercular or insular gustatory cortices to gustatory stimuli. Indeed, after feeding to satiety, humans reported that the taste of the food on which they had been satiated tasted almost as intense as when they were hungry, although much less pleasant.[45] This comparison is consistent with the possibility that activity in the frontal opercular and insular taste cortices as well as the nucleus of the solitary tract does not reflect the pleasantness of the taste of a food, but rather its sensory qualities independently of motivational state. On the other hand, the responses of the neurons in the caudolateral orbitofrontal cortex taste area and in the lateral hypothalamus[32] are modulated by satiety, and it is presumably in areas such as these that neuronal activity may be related to whether a food tastes pleasant, and to whether the food should be eaten.[46-48]

10   THE REPRESENTATION OF FLAVOUR: CONVERGENCE OF OLFACTORY AND TASTE INPUTS

At some stage in taste processing, it is likely that taste representations are brought together with inputs from different modalities, for example, with olfactory inputs to form a representation of flavour. Takagi and colleagues[49,50] have found an olfactory area in the medial orbitofrontal cortex. In a mid-mediolateral part of the caudal orbitofrontal cortex is the area investigated by Thorpe et al.[14] in which are found many neurons with visual and some with gustatory responses. We therefore subsequently investigated whether there are neurons in the secondary taste cortex and adjoining more medial orbitofrontal cortex which respond to stimuli in other modalities, including the olfactory and visual modalities, and whether single neurons in this cortical region in some cases respond to stimuli from more than one modality. We found[51] that in the orbitofrontal cortex taste areas, of 112 single neurons which responded to any of these modalities, many were unimodal (taste 34%, olfactory 13%, visual 21%), but were found in close proximity to each other. Some single neurons showed convergence, responding, for example, to taste and visual inputs (13%), taste and olfactory inputs (13%), and olfactory and visual inputs (5%). Some of these multimodal single neurons had corresponding sensitivities in the two modalities, in that they responded best to sweet tastes (e.g., 1 M glucose), and responded more in a visual discrimination task to the visual stimulus which signified sweet fruit juice than to that which signified saline; or responded to sweet taste, and in an olfactory discrimination task to fruit odor. The different types of neurons (unimodal in different modalities, and multimodal) were frequently found close to one another in tracks made into this region, consistent with the hypothesis that the multimodal representations are actually being formed from unimodal inputs to this region.

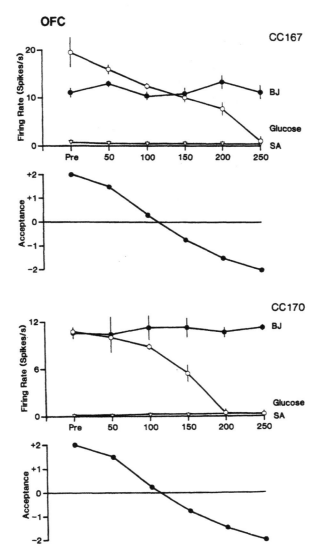

FIGURE 4   The effect of feeding to satiety with glucose solution on the responses of two neurons in the secondary taste cortex to the taste of glucose and of black currant juice (BJ). The spontaneous firing rate is also indicated (SA). Below the neuronal response data for each experiment, the behavioural measure of the acceptance or rejection of the solution on a scale from +2 to –2 (see text) is shown. The solution used to feed to satiety was 20% glucose. The monkey was fed 50 ml of the solution at each stage of the experiment as indicated along the abscissa, until it was satiated as shown by whether it accepted or rejected the solution. Pre = the firing rate of the neuron before the satiety experiment started. The values shown are the mean firing rate and its standard error (s.e.) (From Rolls et al.[39])

These results show that there are regions in the orbitofrontal cortex of primates where the sensory modalities of taste, vision, and olfaction converge; and that in many cases the neurons have corresponding sensitivities across modalities. It appears to be in these areas that flavour representations are built, where flavour is taken to mean a representation which is evoked best by a combination of gustatory and olfactory input. This orbitofrontal region does appear to be an important region for convergence, for there is only a low proportion of bimodal taste and olfactory neurons in the primary taste cortex.[51]

11   THE RULES UNDERLYING THE FORMATION OF OLFACTORY REPRESENTATIONS IN THE PRIMATE CORTEX

Critchley and Rolls[52] showed that 35% of orbitofrontal cortex olfactory neurons categorised odours based on their association in an olfactory-taste discrimination task on the basis of the taste reward association of the odorants. Rolls et al.[20] found that 68% of orbitofrontal cortex odour-responsive neurons modified their responses in some way following changes in the taste reward associations of the odorants during olfactory-taste discrimination reversals. (In an olfactory discrimination experiment, if the monkey makes a lick response when one odour, the S+, is delivered it obtains a drop of glucose reward; if the monkey incorrectly makes a lick response to another odour, the S–, it obtains a drop of aversive saline. At some time in the experiment, the contingency between the odour and the taste is reversed. The monkey relearns the discrimination, showing reversal. It is of interest to investigate in which parts of the olfactory system the neurons show reversal, for, where they do, it can be concluded that the neuronal response to the odour depends on the taste with which it is associated.) Full reversal of the neuronal responses was seen in 25% of the neurons analysed. (In full reversal, the odour to which the neuron responded reversed when the taste with which it was associated reversed.) Extinction of the differential neuronal responses after task reversal was seen in 43% of these neurons. (These neurons simply stopped discriminating between the two odours after the reversal.) These findings demonstrate directly a coding principle in primate olfaction whereby the responses of some orbitofrontal cortex olfactory neurons are modified by and depend upon the taste with which the odour is associated. This modification is likely to be important for setting the motivational or reward value of olfactory stimuli for feeding and other rewarded behaviour. It was of interest, however, that this modification was less complete, and much slower, than the modifications found for orbitofrontal visual neurons during visual–taste reversal.[20] This relative inflexibility of olfactory responses is consistent with the need for some stability in odour–taste associations to facilitate the formation and perception of flavours.

12   OLFACTORY AND VISUAL SENSORY-SPECIFIC SATIETY AND THEIR REPRESENTATION IN THE PRIMATE ORBITOFRONTAL CORTEX

It has also been possible to investigate whether the olfactory representation in the orbitofrontal cortex is affected by hunger. In satiety experiments, Critchley and Rolls[46] have been able to show that the responses of some olfactory neurons to a food odour are decreased when the monkey is fed to satiety with a food (e.g., fruit juice) with that odour. In particular, seven of nine olfactory neurons that were responsive to the odours of foods, such as black currant juice, were found to decrease their responses to the odour of the satiating food. The decrease was typically at least partly specific to the odour of the food that had been eaten to satiety, potentially providing part of the basis for sensory-specific satiety. (It was also found for eight of nine neurons that had selective responses to the sight of food that they demonstrated a sensory-specific reduction in their visual responses to foods following satiation.) These findings show that the olfactory and visual representations of food, as well as the taste representation of food, in the primate orbitofrontal cortex are modulated by hunger. Usually, a component related to sensory-specific satiety can be demonstrated. The findings link at least part of the processing of olfactory and visual information in this brain region to the control of feeding-related behaviour. This is further evidence that part of the olfactory representation in this region is related to the hedonic value of the olfactory stimulus, and in particular that at this level of the olfactory system in primates, the pleasure elicited by the food odour is at least part of what is represented.

As a result of the neurophysiological and behavioural observations showing the specificity of satiety in the monkey, experiments were performed to determine whether satiety was specific to foods eaten in humans. It was found that the pleasantness of the taste of food eaten to satiety decreased more than for foods that had not been eaten.[41] One implication of this finding is that if one food is eaten to satiety, appetite reduction for other foods is often incomplete, and this should mean that in humans also at least some of the other foods will be eaten. This has been confirmed in an experiment in which either sausages or cheese with crackers were eaten for lunch. The liking for the food eaten decreased more than for the food not eaten and, when an unexpected second course was offered, more was eaten if a subject had not been given that food in the first course than if the subject had been given that food in the first course (98 vs. 40% of the first course intake eaten in the second courses, $p < 0.01$[41]). A further implication of these findings is that if a variety of foods is available, the total amount consumed will be more than when only one food is offered repeatedly. This prediction has been confirmed in a study in which humans ate more when offered a variety of sandwich fillings than one filling or a variety of types of yoghurt which differed in taste, texture, and color.[42] It has also been confirmed in a study in which humans were offered a relatively normal meal of four courses, and it was found that the change of food at each course significantly enhanced intake.[53] Because sensory factors such as similarity of colour, shape, flavour, and texture are usually more important than metabolic equivalence in terms of protein, carbohydrate, and fat content in influencing how foods interact in this type of satiety, it has been termed "sensory-specific satiety."[29,36,41-43,54] It should be noted that this effect is distinct from alliesthesia, in that alliesthesia is a change in the pleasantness of sensory inputs produced by internal signals (such as glucose in the gut),[40,55-56] whereas sensory-specific satiety is a change in the pleasantness of sensory inputs which is accounted for at least partly by the external sensory stimulation received (such as the taste of a particular food), in that as shown above it is at least partly specific to the external sensory stimulation received.

The parallel between these studies of feeding in humans and of the neurophysiology of hypothalamic neurons in the monkey has been extended by the observations that in humans, sensory-specific satiety occurs for the sight as well as for the taste of food.[43] Further, to complement the finding that in the hypothalamus neurons are found which respond differently to food and to water,[4] and that satiety with water can decrease the responsiveness of hypothalamic neurons which respond to water, it has been shown that in humans motivation-specific satiety can also be detected. For example, satiety with water decreases the pleasantness of the sight and taste of water but not of food.[44]

To investigate whether the sensory-specific reduction in the responsiveness of the orbitofrontal olfactory neurons might be related to a sensory-specific reduction in the pleasure produced by the odour of a food when it is eaten to satiety, Rolls and Rolls[57] measured humans' responses to the smell of a food which was eaten to satiety. It was found that the pleasantness of the odour of a food, but much less significantly its intensity, was decreased when the subjects ate it to satiety (Figure 5). It was also found that the pleasantness of the smell of other foods (i.e., foods not eaten in the meal) showed much less decrease. This finding has clear implications for the control of food intake; for ways to keep foods presented in a meal appetitive; and for effects on odour pleasantness ratings that could occur following meals. In an investigation of the mechanisms of this odour-specific sensory-specific satiety, Rolls and Rolls[57] allowed humans to chew a food without swallowing, for approximately as long as the food is normally in the mouth during eating. They demonstrated a sensory-specific satiety with this procedure, showing that the sensory-specific satiety does not depend on food reaching the stomach. Thus, at least part of the mechanism is likely to be produced by a change in processing in the olfactory pathways. It is not yet known which is the earliest stage of olfactory processing at which this modulation occurs. It is unlikely to be in the receptors, because the change in pleasantness found was much more significant than the change in the intensity.[57]

The enhanced eating when a variety of foods is available, as a result of the operation of sensory-specific satiety, may have been advantageous in evolution in ensuring that different foods with important different nutrients were consumed, but today in humans, when a wide variety of foods is readily available, it may be a factor which can lead to overeating and obesity. In a test of this in the rat, it has been found that variety itself can lead to obesity.[58,59]

In addition to the sensory-specific satiety described above which operates primarily within and in the postmeal period,[53] there is now evidence for a long-term form of sensory-specific satiety.[60]

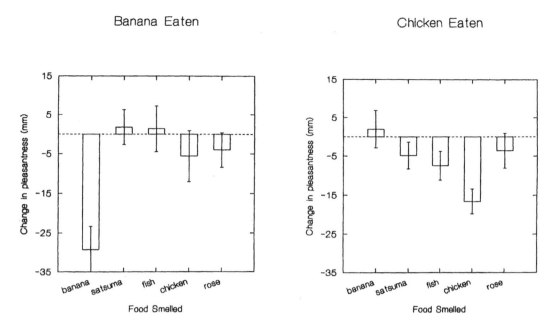

FIGURE 5 Olfactory sensory-specific satiety in humans. The pleasantness of the smell of a food became less when the humans ate that food (banana or chicken) to satiety. A similar reduction was not found for other foods not eaten in the meal. The changes in pleasantness were measured on a 100 mm visual analogue rating scale. The number of subjects was 12, and the results (as shown by the interaction term in a two-way within subjects ANOVA) were very significant ($p < 0.001$). (From Rolls and Rolls.[57])

This was shown in a study in an Ethiopian refugee camp, in which it was found that refugees who had been in the camp for 6 months found the taste of their three regular foods less pleasant than that of three comparable foods which they had not been eating. The effect was a long-term form of sensory-specific satiety in that it was not found in refugees who had been in the camp and eaten the regular foods for 2 days.[60] It is suggested that it is important to recognise the operation of long-term sensory-specific satiety in conditions such as these, for it may enhance malnutrition if the regular foods become less acceptable and so are rejected, exchanged for other less nutritionally effective foods or goods, or inadequately prepared. It may be advantageous under these circumstances to attempt to minimise the operation of long-term sensory-specific satiety by providing some variety, perhaps even with spices.[60]

In addition to this modulation of neuronal responses to the taste and smell of foods eaten, there will be effects of the energy ingested on taste and smell responses to food. These are likely to depend on factors such as gastric distension, and the concentration of glucose and other indicators of hunger/satiety in the systemic circulation.[4,26,27,61-65] Further investigation is needed of whether there is a postingestive macronutrient-specific modulation of the representation of the sensory properties of food described in this chapter. The major effect of hunger investigated so far is the sensory-specific reduction in the pleasantness of a particular food just eaten to satiety, and the findings reviewed in this chapter show that an important component of this modulation is partly specific to the sensory properties of a food which has been ingested.

13 THE RESPONSES OF ORBITOFRONTAL TASTE AND OLFACTORY NEURONS TO THE SIGHT AND TEXTURE OF FOOD

Many of the neurons with visual responses in this region also show olfactory or taste responses,[51] reverse rapidly in visual discrimination reversal,[21] and only respond to the sight of food if hunger is present.[46] This part of the orbitofrontal cortex thus seems to implement a mechanism which can flexibly alter the responses to visual stimuli depending on the reinforcement (e.g., the taste) associated with the visual stimulus.[14,47] This enables prediction of the taste associated with ingestion of what is seen, and thus in the visual selection of foods.[4,25-27]

The orbitofrontal cortex of primates is also important as an area of convergence for somatosensory inputs, related, for example, to the texture of food and in addition to that produced by the texture of fat in the mouth. We have shown, for example, in recent recordings that single neurons influenced by taste in this region can in some cases have their responses modulated by the texture of the food. This was shown in experiments in which the texture of food was manipulated by the addition of methyl cellulose or gelatine, or by puréeing a semisolid food.[4,28]

14   CONCLUSIONS

The primate orbitofrontal cortex is an important site for the convergence of representations of the taste, smell, sight, and mouth feel of food, and this convergence allows the sensory properties of each food to be defined in detail. The primate orbitofrontal cortex is also the region where a short-term, sensory-specific control of appetite and eating is implemented, in the form of sensory-specific satiety. Moreover, it is likely that visceral and other satiety-related signals reach the orbitofrontal cortex and there modulate the representation of food, resulting in an output that reflects the reward (or appetitive) value of each food. It will be of interest to determine to what extent this modulation is macronutrient selective. Part of the evidence that the reward value, and pleasantness of food in humans, is represented in the orbitofrontal cortex is that macaques will work to obtain electrical stimulation of this brain region if they are hungry, but much less if they are satiated.[4] Further, monkeys or humans with damage to this brain region show altered, often less selective, food preferences.[4,66] Some other brain areas which receive inputs from the orbitofrontal cortex, such as the amygdala[67] and ventral striatum[68] also contain neurons whose responses reflect visual, taste, and olfactory stimuli produced by food, and in the amygdala, there may be modulation of neuronal responsiveness by hunger.[69,70]

ACKNOWLEDGMENTS

The author has worked on some of the experiments described here with Drs. A. Browning, H. Critchley, T. R. Scott, Z. J. Sienkiewicz, E. A. Wakeman, L. L. Wiggins (L. L. Baylis), and S. Yaxley, and their collaboration is sincerely acknowledged. The research from the author's laboratory was supported by the Medical Research Council, PG8513790.

REFERENCES

1. Norgren, R., Central neural mechanisms of taste, in *Handbook of Physiology — The Nervous System III, Sensory Processes 1,* Darien-Smith, I., Brookhart, J., and Mountcastle, V. B., Eds., American Physiological Society, Washington, D.C., 1984, 1087.
2. Pritchard, T. C., Hamilton, R. B., and Norgren, R., Neural coding of gustatory information in the thalamus of Macaca mulatta, *J. Neurophysiol.,* 61, 1, 1989.
3. Norgren, R., Taste pathways to hypothalamus and amygdala, *J. Comp. Neurol.,* 166, 17, 1976.
4. Rolls, E. T., *The Brain and Emotion,* Oxford University Press, Oxford, 1999.
5. Pritchard, T. C., Hamilton, R. B., Morse, J. R., and Norgren, R., Projections of thalamic gustatory and lingual areas in the monkey, *Macaca fascicularis, J. Comp. Neurol.,* 244, 213, 1986.
6. Bornstein, W. S., Cortical representation of taste in man and monkey. I. Functional and anatomical relations of taste, olfaction and somatic sensibility, *Yale J. Biol. Med.,* 12, 719, 1940.
7. Bornstein, W. S., Cortical representation of taste in man and monkey. II. The localization of the cortical taste area in man, a method of measuring impairment of taste in man, *Yale J. Biol. Med.,* 13, 133, 1940.

8. Rolls, E. T., Yaxley, S., and Sienkiewicz, Z. J., Gustatory responses of single neurons in the orbito-frontal cortex of the macaque monkey, *J. Neurophysiol.,* 64, 1055, 1990.

9. Baylis, L. L., Rolls, E. T., and Baylis, G. C., Afferent connections of the orbitofrontal cortex taste area of the primate, *Neuroscience,* 64, 801, 1994.

10. Scott, T. R., Yaxley, S., Sienkiewicz, Z. J., and Rolls, E. T., Taste responses in the nucleus tractus solitarius of the behaving monkey, *J. Neurophysiol.,* 55,182, 1986.

11. Scott, T. R., Yaxley, S., Sienkiewicz, Z. J., and Rolls, E. T., Gustatory responses in the frontal opercular cortex of the alert cynomolgus monkey, *J. Neurophysiol.,* 56, 876, 1986.

12. Yaxley, S., Rolls, E. T, and Sienkiewicz, Z. J., Gustatory responses of single neurons in the insula of the macaque monkey, *J. Neurophysiol.,* 63, 689, 1990.

13. Francis, S., Rolls, E. T., Bowtell, R., McGlone, F., O'Doherty, J., Browning, A., Clare, S., and Smith, E., The representation of pleasant touch in the brain and its relationship with taste and olfactory areas, *NeuroReport,* 10, 453, 1999.

14. Thorpe, S. J., Rolls, E. T., and Maddison, S., Neuronal activity in the orbitofrontal cortex of the behaving monkey, *Exp. Brain Res.,* 49, 93, 1983.

15. Ikeda, K., On a new seasoning, *J. Tokyo Chem. Soc.,* 30, 820, 1909.

16. Yamaguchi, S., The synergistic taste effect of monosodium glutamate and disodium 5'-inosinate, *J. Food Sci.,* 32, 473, 1967.

17. Yamaguchi, S. and Kimizuka, A., Psychometric studies on the taste of monosodium glutamate, in *Glutamic Acid: Advances in Biochemistry and Physiology,* Filer, L. J., Garattini, S., Kare, M. R., Reynolds, A. R., and Wurtman, R. J., Eds., Raven Press, New York, 1979, 35–54.

18. Kawamura, Y. and Kare, M. R., Eds., *Umami: A Basic Taste,* Marcel Dekker, New York, 1987.

19. Baylis, L. L. and Rolls, E. T., Responses of neurons in the primate taste cortex to glutamate, *Physiol. Behav.,* 49, 973, 1991.

20. Rolls, E. T., Critchley, H., Wakeman, E. A., and Mason, R., Responses of neurons in the primate taste cortex to the glutamate ion and to inosine 5'-monophosphate, *Physiol. Behav.,* 59, 991, 1996.

21. Rolls, E. T., Critchley, H., Mason, R., and Wakeman, E. A., Orbitofrontal cortex neurons: role in olfactory and visual association learning, *J. Neurophysiol.,* 75, 1970, 1996.

22. Chaudhari, N., Yang, H., Lamp, C., Delay, E., Cartford, C., Than, T., and Roper, S., The taste of monosodium glutamate: membrane receptors in taste buds, *J. Neurosci.* 16, 3817, 1996.

23. Chaudhari, N. and Roper, S. D., Molecular and physiological evidence for glutamate (umami) taste transduction via a G protein-coupled receptor, *Ann. N.Y. Acad. Sci.,* 855, 398, 1998.

24. Rolls, E. T., Critchley, H. D., Browning, A., and Hernadi, I., The neurophysiology of taste and olfaction in primates, and umami flavor, *Ann. N.Y. Acad. Sci.,* 855, 426, 1998.

25. Rolls, E. T., Neuronal activity related to the control of feeding, in *Feeding Behavior: Neural and Humoral Controls,* Ritter, R. C., Ritter, S., and Barnes, C. D., Eds., Academic Press, New York, 1986, chap. 6.

26. Rolls, E. T., The neural control of feeding in primates, in *Neurophysiology of Ingestion,* Booth, D. A., Ed., Pergamon, Oxford, 1993, chap. 9.

27. Rolls, E. T., Neural processing related to feeding in primates, in *Appetite: Neural and Behavioural Bases,* Legg, C. R. and Booth, D. A., Eds., Oxford University Press, Oxford, 1994, chap. 2.

28. Rolls, E. T., Taste and olfactory processing in the brain and its relation to the control of eating, *Crit. Rev. Neurobiol.,* 11, 263, 1997.

29. Rolls, E. T., and Rolls, B. J., Brain mechanisms involved in feeding, in *Psychobiology of Human Food Selection,* Barker, L. M., Ed., AVI Publishing Company, Westport, CT, 1982, chap. 3.

30. Rolls, E. T., Burton, M. J., and Mora, F., Hypothalamic neuronal responses associated with the sight of food, *Brain Res.,* 111, 53, 1976.

31. Burton, M. J., Rolls, E. T., and Mora, F., Effects of hunger on the responses of neurones in the lateral hypothalamus to the sight and taste of food, *Exp. Neurol.,* 51, 668, 1976.

32. Rolls, E. T., Murzi, E., Yaxley, S., Thorpe, S. J., and Simpson, S. J., Sensory-specific satiety: food-specific reduction in responsiveness of ventral forebrain neurons after feeding in the monkey, *Brain Res.,* 368,79, 1986.

33. Rolls, E. T., *The Brain and Reward,* Pergamon, Oxford, 1975.

34. Yaxley, S., Rolls, E. T., Sienkiewicz, Z. J., and Scott, T. R., Satiety does not affect gustatory activity in the nucleus of the solitary tract of the alert monkey, *Brain Res.,* 347, 85, 1985.

35. Gibbs, J., Maddison, S. P., and Rolls, E. T., Satiety role of the small intestine examined in sham-feeding rhesus monkeys, *J. Comp. Physiol. Psychol.*, 95, 1003, 1981.

36. Rolls, E. T. and Rolls, B. J., Activity of neurons in sensory, hypothalamic and motor areas during feeding in the monkey, in *Food Intake and Chemical Senses*, Katsuki, Y., Sato, M., Takagi, S., and Oomura, Y., Eds., University of Tokyo Press, Tokyo, 1977, 525–549.

37. Rolls, E. T., Scott, T. R., Sienkiewicz, Z. J., and Yaxley, S., The responsiveness of neurones in the frontal opercular gustatory cortex of the macaque monkey is independent of hunger, *J. Physiol.*, 397, 1, 1988.

38. Yaxley, S., Rolls, E. T., and Sienkiewicz, Z. J., The responsiveness of neurones in the insular gustatory cortex of the macaque monkey is independent of hunger, *Physiol. Behav.*, 42, 223, 1988.

39. Rolls, E. T., Sienkiewicz, Z. J., and Yaxley, S., Hunger modulates the responses to gustatory stimuli of single neurons in the orbitofrontal cortex, *Eur J. Neurosci.*, 1, 53, 1989.

40. Cabanac, M., Physiological role of pleasure, *Science*, 173, 1103, 1971.

41. Rolls, B. J., Rolls, E. T., Rowe, E. A., and Sweeney, K., Sensory specific satiety in man, *Physiol. Behav.*, 27, 137, 1981.

42. Rolls, B. J., Rowe, E. A., Rolls, E. T., Kingston, B., and Megson, A., Variety in a meal enhances food intake in man, *Physiol. Behav.*, 26, 215, 1981.

43. Rolls, B. J., Rowe, E. A., and Rolls, E. T., How sensory properties of foods affect human feeding behavior, *Physiol. Behav.*, 29, 409, 1982.

44. Rolls, B. J., Rolls, E. T., and Rowe, E. A., Body fat control and obesity, *Behav. Brain Sci.*, 4, 744, 1983.

45. Rolls, E. T., Rolls, B. J, and Rowe, E. A., Sensory-specific and motivation-specific satiety for the sight and taste of food and water in man, *Physiol. Behav.*, 30, 185, 1983.

46. Critchley, H. D. and Rolls, E. T., Hunger and satiety modify the responses of olfactory and visual neurons in the primate orbitofrontal cortex, *J. Neurophysiol.*, 75, 1673, 1996.

47. Rolls, E. T., The orbitofrontal cortex, *Philos. Trans. R. Soc. B*, 351, 1996.

48. Scott, T. R., Yan, J., and Rolls, E. T., Brain mechanisms of satiety and taste in macaques, *Neurobiology*, 3, 281, 1995.

49. Tanabe, T., Yarita, H., Iino, M., Ooshima, Y., and Takagi, S. F., An olfactory projection area in orbitofrontal cortex of the monkey, *J. Neurophysiol.*, 38, 1269, 1975.

50. Tanabe, T., Iino, M., and Takagi, S. F., Discrimination of odors in olfactory bulb, pyriform-amygdaloid areas, and orbitofrontal cortex of the monkey, *J. Neurophysiol.*, 38, 1284, 1975.

51. Rolls, E. T. and Baylis, L. L., Gustatory, olfactory and visual convergence within the primate orbito-frontal cortex, *J. Neurosci.*, 14, 5437, 1994.

52. Critchley, H. D. and Rolls, E. T., Olfactory neuronal responses in the primate orbitofrontal cortex: analysis in an olfactory discrimination task, *J. Neurophysiol.*, 75, 1659, 1996.

53. Rolls, B. J., Van Duijenvoorde, P. M., and Rolls, E. T., Pleasantness changes and food intake in a varied four course meal, *Appetite*, 5, 337, 1984.

54. Rolls, B. J., The role of sensory-specific satiety in food intake and food selection, in *Taste, Experience, and Feeding*, Capaldi, E. D. and Powley, T. L., Eds., American Psychological Association, Washington, D.C., 1990, chap. 14.

55. Cabanac, M. and Fantino, M., Origin of olfacto-gustatory alliesthesia: intestinal sensitivity to carbo-hydrate concentration? *Physiol. Behav.*, 10, 1039, 1977.

56. Cabanac, M. and Duclaux, R., Specificity of internal signals in producing satiety for taste stimuli, *Nature*, 227, 966, 1970.

57. Rolls, E. T. and Rolls, J. H., Olfactory sensory-specific satiety in humans, *Physiol. Behav.*, 61, 461, 1997.

58. Rolls, B. J., Van Duijenvoorde, P. M., and Rowe, E. A., Variety in the diet enhances intake in a meal and contributes to the development of obesity in the rat, *Physiol. Behav.*, 31, 21, 1983.

59. Rolls, B. J. and Hetherington, M., The role of variety in eating and body weight regulation, in *Handbook of the Psychophysiology of Human Eating*, Shepherd, R., Ed., Wiley, Chichester, 1989, chap. 3.

60. Rolls, E. T. and de Waal, A. W. L., Long-term sensory-specific satiety: evidence from an Ethiopian refugee camp, *Physiol. Behav.*, 34, 1017, 1985.

61. Oomura, Y., Nishino, H., Karadi, Z., Aou, S., and Scott, T. R., Taste and olfactory modulation of feeding related neurons in the behaving monkey, *Physiol. Behav.*, 49, 943, 1991.

62. Karadi, Z., Oomura, Y., Nishino, H., Scott, T. R., Lenard, L., and Aou, S., Complex attributes of lateral hypothalamic neurons in the regulation of feeding of alert monkeys, *Brain Res. Bull.*, 25, 933, 1990.

63. Karadi, Z., Oomura, Y., Nishino, H., Scott, T. R., Lenard, L., and Aou, S., Responses of lateral hypothalamic glucose-sensitive and glucose-insensitive neurons to chemical stimuli in behaving rhesus monkeys, *J. Neurophysiol.*, 67, 389, 1992.

64. LeMagnen, J., *Neurobiology of Feeding and Nutrition,* Academic Press, San Diego, 1992.

65. Campfield, L. A., Smith, F. J., Guisez, Y., Devos, R., and Burn, P., Recombinant mouse OB protein: evidence for a peripheral signal linking adiposity and central neural networks, *Science*, 269, 475, 1995.

66. Baylis, L. L. and Gaffan, D., Amygdalectomy and ventromedial prefrontal ablation produce similar deficits in food choice and in simple object discrimination learning for an unseen reward, *Exp. Brain Res.*, 86, 617, 1991

67. Sanghera, M. K., Rolls, E. T., and Roper-Hall, A., Visual responses of neurons in the dorsolateral amygdala of the alert monkey, *Exp. Neurol.*, 63, 610, 1979.

68. Williams, G. V., Rolls, E. T., Leonard, C. M., and Stern, C., Neuronal responses in the ventral striatum of the behaving macaque, *Behav. Brain Res.*, 55, 243, 1993.

69. Scott, T. R., Karadi, Z., Oomura, Y., Nishino, H., Plata-Salaman, C. R., Lenard, L., Giza, B. K., and Aou, S., Gustatory neural coding in the amygdala of the alert macaque monkey, *J-Neurophysiol*, 69, 1810, 1993.

70. Yan, J. and Scott, T. R., The effect of satiety on responses of gustatory neurons in the amygdala of alert cynomolgus macaques, *Brain Res.*, 740, 193, 1996.

71. Critchley, H. D. and Rolls, E. T., Responses of primate taste cortex neurons to the astringent tastant tannic acid, *Chem. Senses*, 21, 135, 1996.

72. Graham, H. N., Green tea composition, consumption and polyphenol chemistry, *Prev. Med.*, 21, 334, 1992.

73. Hladik, C. M., Adaptive strategies of primates in relation to leaf-eating, in *The Ecology of Arboreal Folivores*, Montgomery, G. G., Ed., Smithsonian Institute Press, Washington, D.C., 1978, 373.

74. Johns, T. and Duquette, M., Detoxification and mineral supplementation as functions of geophagy, *Am. J. Clin. Nutr.*, 53, 448, 1991.

75. Uma-Pradeep, K., Geervani, P., and Eggum, B. O., Common Indian spices: nutrient composition, consumption and contribution to dietary value, *Plant Foods Hum. Nutr.*, 44, 138, 1993.

76. Rolls, E. T., Critchley, H. D., Browning, A. S., Hernadi, A., and Lenard, L., Responses to the sensory properties of fat of neurons in the primate orbitofrontal cortex, *J. Neurosci.*, 19, 1532, 1999.

18 Satiation in Response to Macronutrient Signals from the Intestine: Mechanisms and Implications for Macronutrient Selection

Mihai Covasa and Robert C. Ritter

CONTENTS

1 INTRODUCTION

Satiation is the reduction of food intake that occurs in response to ingestion of food and ultimately results in meal termination (satiety). The process of satiation is mediated by feedback signals from the gastrointestinal tract. Experimental data indicate that mechanical signals from the stomach and chemical signals from the intestine each are sufficient to produce satiety.[1] However, it is likely that both gastric and intestinal signals participate in satiation during normal meals. Reduction of food intake by gastrointestinal feedback is rapid, and satiation is evident prior to significant absorption of nutrients.[2,3] Although, there has been considerable progress in our understanding of how gastrointestinal feedback signals participate in satiation, very little experimental attention has been paid to the possibility that gastrointestinal feedback may contribute to food choice or macronutrient selection. Yet, by rapidly and selectively terminating intake of foods rich in a particular nutrient, or by selectively suppressing intake of one nutrient for longer than intake of another, gastrointestinal satiety signals could conceivably contribute to macronutrient-selective ingestion. Nonetheless, only very limited data are available to support the effects of gastrointestinal stimulation on macronutrient choice. Therefore, this chapter will be primarily prescriptive in nature. We will begin with discussion of the anatomy, physiology, and pharmacology of gastrointestinal feedback signals that participate in satiation and that could serve as macronutrient-discriminative cues, and hence contribute to macronutrient selection. Second, we will discuss recent results, which suggest that intestinal feedback mechanisms are plastic and speculate on how diet-induced changes in these mechanisms could potentially contribute to altered ingestion of individual macronutrients, specifically of fat. Finally, we will briefly review some of the existing experimental evidence that suggests that dietary components passing through the alimentary tract could influence macronutrient selection.

2   ROLE OF INTESTINAL VS. GASTRIC MECHANISMS IN MACRONUTRIENT FEEDBACK SIGNALS CONTROLLING FOOD INTAKE

The results of sham-feeding experiments provide compelling evidence that gastrointestinal feedback signals are important for meal termination.[2,4-6] Specifically, when food eaten by rats or monkeys is allowed to drain from open gastric cannulas, meal termination is delayed and animals eat continuously.[2,7,8] In contrast, if the same animals are offered the same diet while the gastric cannulas are closed, the rate of ingestion begins to diminish shortly after the start of the meal and eating has all but stopped within 15 min. Results such as these indicate that postoral consequences of ingestion provide signals that result in meal termination, but they cannot specify the site(s) at which satiation signals are generated. Experiments in which the stomach is distended[9,10] or in which gastric emptying is prevented by occluding the pylorus[11-13] indicate that the stomach provides mechano-responsive information and that some of that information contributes to meal termination. The possibility that the stomach also is directly sensitive to chemical stimulation by nutrients also has been proposed.[13-15] However, recent experimental results strongly support the conclusion that gastric feedback signals result from mechanical stimuli only and that the stomach itself does not provide behaviorally relevant signals concerning specific nutrients.[11,12,16] For example, in recent experiments Phillips and Powley[12] demonstrated that when gastric emptying is prevented, by inflating a pyloric cuff, intragastric liquid loads reduced food intake in direct proportion to the infused volume, and independent of the caloric content of the infusate. On the other hand, if the pylorus is not occluded, then the efficacy of an intragastric load for reducing food intake depends on caloric content, as well as volume. These results indicate that gastric feedback signals are purely mechanical in nature. In addition, the results support the existence of sensory feedback signals related to nutrient content (calories) and the fact that these signals arise postgastrically, probably from the intestine.

A variety of experimental and clinical observations provide evidence that the small intestine provides feedback signals that participate in satiation. Perhaps the earliest evidence for intestinal participation in satiation comes from the 1858 report of a human patient with a fistula of the upper small intestine, through which much ingested food drained.[17] This patient reported sensations of gastric fullness, but not satiation. While this report is tantalizing, it is difficult to dissociate possible malabsorption of calories, with resultant caloric deficiency, from reduced intestinal feedback as causes of diminished satiation.

More direct evidence of the participation of intestinal feedback in satiation comes from experiments in which nutrient solutions are directly infused into the small intestine. Such experiments have been described for a variety of species, including rats,[4,8,18] monkeys,[2] and humans,[19] with fundamentally similar results. The principal finding in all of these studies is that infusion of liquid diets into the upper small intestine results in reduced food intake and, in the case of humans, reduced hunger ratings.[20] Furthermore, in experiments where intestinal infusates of differing caloric concentrations have been compared, reduction of food intake varied with caloric content of the infusate for individual macronutrients.[21,22]

Experiments dealing with satiation effects of intestinal infusions have been performed using both real-feeding and sham-feeding preparations. Each preparation has distinct advantages and disadvantages. Intestinal infusions that are made into the intestine of people and animals engaged in real feeding trigger enterogastric reflexes, which alter gastric motility and reduce the rate of gastric emptying, and thereby are likely to increase mechanical feedback signals from the stomach.[23,24] Such gastric feedback signals may play an important role in feelings of satiation induced by intragastric nutrients. For example, Castiglione et al.[25] reported that intraduodenal infusion of fat has little effect on desire to eat, prior to meal onset. However, after eating began, subjects reported greater feelings of fullness and earlier meal termination when they were infused with fat than when they got control infusions. This observation suggests that interaction of intestinal signals with gastric or other preintestinal signals are important for maximizing the satiation effects of intraintestinal fat, and perhaps other nutrients as well.

Although an enterogastric servomechanism may participate in physiological satiation, results from real feeding experiments beg the question of whether intestinal chemical signals can directly contribute to satiation, or whether they are only capable of reducing food intake via a gastric servomechanism. Experiments utilizing intraintestinal infusions during sham feeding obviate the contribution of altered gastric emptying to satiation, and thereby more assuredly assess the potential for postgastric satiation signals for reducing food intake. Reduction of food intake by intestinal infusions in sham-feeding animals has been clearly demonstrated in a variety of mammals, including rats[4,18,26] and monkeys,[2] which indicates that postgastric signals are sufficient to reduce food intake, even in the absence of enterogastric inhibition of gastric emptying. Furthermore, the fact that sham-feeding rates are reduced within 5 min of the start of intestinal infusion suggests that reduction of intake occurs prior to significant absorption and therefore probably is a response to intestinal chemosensory signals. Finally, it should be apparent that adequacy of intestinal infusions to reduce food intake in sham-feeding animals in no way diminishes the likely importance of gastric signals in the reduction of food intake during real feeding. Indeed, intestinal control of gastric emptying might provide important discriminative stimuli, which could influence macronutrient choices during ongoing or subsequent meals. Thus, it is probable that under physiological conditions mechanical feedback signals from the stomach, perhaps enhanced by enterogastric amplification, together with chemical signals from the intestine, both contribute to satiation (Figure 1).

3 PATHWAYS AND MECHANISMS BY WHICH SPECIFIC MACRONUTRIENTS REDUCE FOOD INTAKE

It is now clear that the intestine is sensitive to more than one component of ingesta. Houpt and co-workers[27] have systematically examined the effect of intestinal osmotic stimulation on reduction of food intake in pigs. They found that hyperosmotic solutions reduced food intake when osmotic concentrations exceeded 480 mOsm. Furthermore, the same investigators demonstrated that the postprandial osmotic concentrations of intestinal ingesta were well within the range of concentrations that reduced food intake when they were infused into the intestine.

To determine whether individual macronutrients can reduce food intake in the absence of hyperosmotic activity, Yox and Ritter[18] and others[4,8,28] have made intraintestinal infusions of individual macronutrients in solutions that were isosmotic with plasma. They found dose-related reduction of sham feeding of sucrose by a long-chain fatty acid and oligosacharides, such as maltose or maltotriose (Figure 2). In addition, they replicated the previous finding of Gibbs and Smith[29] demonstrating that intraintestinal infusion of L-phenylalanine, also isosmotic with plasma, reduced sham feeding of sucrose. While the magnitude of reduction of intake appears to be related to the caloric content (concentration) of individual nutrient infusions, caloric content per se does not entirely account for reduction of food intake by intestinal nutrients. For example, infusions of

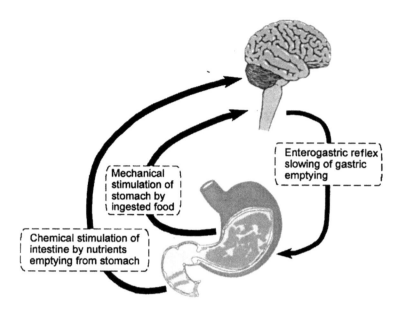

FIGURE 1 Feedback signals that reduce food intake arise from both the stomach and the small intestine. Chemical signals from the stomach and chemical signals from the intestine both are sufficient to reduce food intake. However, these two modalities appear to work in concert during real feeding. Furthermore, although intestinal signals can reduce food intake in the absence of enterogastric effects on gastric emptying, it remains possible, although unproved, that the enterogastric reflex could contribute to satiation.

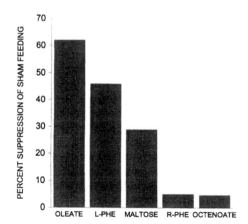

FIGURE 2 Suppression of sham feeding following intraintestinal infusion of three classes of nutrients (fatty acids, oligosaccharides, and amino acids). While long-chain fatty acids (oleate, C-18) are potent for suppressing feeding, medium-chain fatty acids (octanoic, C-8) is ineffective. Sham feeding is also reduced by intestinal infusion of L-phenylalanine (L-Phe), while R-phenylalanine (R-Phe) has no effect. Intestinal infusion of maltose significantly reduced sham feeding.

medium-chain fatty acid (octenoic) at the same caloric concentration as oleic acid do not significantly reduce intake (see Figure 2).[18] Likewise, experimental evidence indicates that intestinal infusion of triglyceride is less effective for reducing food intake than equicaloric infusion of long-chain fatty acid,[30,31] which is a product of fat digestion. In addition, infusion of hydrolyzed casein is less effective for reducing food intake than equicaloric infusions of L-phenylalanine, oleate, or maltose.[18] Finally, Greenberg and Smith[30] have reported that more highly unsaturated fatty acids reduce food intake more than equicaloric infusions of saturated fatty acids. These observations all indicate that the intestine is sensitive to specific chemical configurations of nutrient compounds not just caloric content or colligative properties of the infusate. Finally, the fact that behavioral responses differ not only between the three macronutrient classes, but even between chemically different members of the same macronutrient class, provides a prospective mechanism by which intestinal signals might be applied toward macronutrient selection.

Electrophysiological recordings provide support for the interpretation that the intestinal innervation can respond to chemically distinct macronutrient signals. For example, recordings from nerve fibers innervating the small intestine have revealed responses to infusions of fatty acids, amino acids, and glucose.[32-34] Electrophysiological recordings also suggest that intestinal signals converge with gastric mechanoreceptive signals in the hindbrain, supporting the notion that chemical and mechanical signals from the alimentary tract are likely to act in concert to control food intake.[35-37] Finally, there is some evidence that gastrointestinal signals may converge with gustatory signals in the hindbrain, providing an avenue by which such visceral signals may influence orosensory cues in selection of macronutrients, based on the nutrients that are already in the intestine and stomach.[38] Thus, available behavioral and electrophysiological data suggest that the intestine is capable of providing information on the macronutrient composition of a meal and that this information is used to adjust meal size.

The role of the vagal innervation in reduction of food intake by intestinal nutrients has been systematically investigated by several laboratories. Yox et al.[5] demonstrated that reduction of food intake by intraintestinal infusion of maltose and oleate was essentially abolished in rats with total subdiaphragmatic vagotomies (Figure 3A). Reduction of food intake by L-phenylalanine was attenuated by vagotomy but not abolished. These results indicated that the vagal innervation participates in reduction of food intake by nutrients, and may be the sole neural mediator for reduction of intake by oleate and maltose. In another report,[39] Yox et al. also found that systemic or fourth ventricular pretreatment with the neurotoxin capsaicin attenuates reduction of food intake by oleate and maltose, indicating that small, unmyelinated vagal sensory neurons participate in reduction of food intake by these nutrients (Figure 3B).

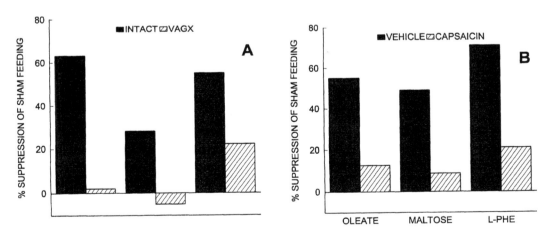

FIGURE 3 Suppression of sham feeding by intestinal nutrients: oleate, maltose, and L-phenylalanine (L-Phe) depends on intact vagus and capsaicin-sensitive neural substrate(s). (**A**) Suppression of sham feeding by oleate and maltose is eliminated by subdiaphragmatic vagotomy, whereas suppression of sham feeding by L-Phe is attenuated but not eliminated. (**B**) Suppression of sham feeding by all three nutrients is attenuated in rats pretreated with the neurotoxin capsaicin 1 month prior to the beginning of behavioral testing.

Walls et al.[40] have reported that reduction of real feeding by 3% glucose, 3% maltose, 3% L-phenylalanine, and 1.4% oleic acid are eliminated by surgical deafferentation of the coeliac vagal branch, which innervates the upper small intestine. While this study is in general agreement with the results from sham-feeding experiments by Yox et al.,[5,18,39] the complete elimination of phenylalanine-induced reduction of intake differs from the Yox et al. report for reasons that are not yet appreciated. One possible explanation for the apparent discrepancy is that Yox et al.[5] sectioned the vagal trunks near the diaphragmaic hiatus, while Walls et al.[40] unilaterally sectioned the vagal sensory rootlets at the brainstem, and selectively cut the coeliac vagal branch. If nonvagal neurons, headed for the coeliac branch, were to join the vagus distal to diaphragmatic hiatus, these nonvagal fibers would be spared by the Yox et al. procedure, but would be sectioned in the Walls et al. procedure. Thus, the Walls et al. results may actually provide circumstantial evidence that nonvagal sensory fibers contribute to reduction of food intake by L-phenylalanine.

The density of vagal sensory innervation to the intestine has been examined by Berthoud et al.[41] and by Jagger et al.[42] Both groups found that the duodenum was densely innervated by vagal sensory nerve endings. The jejunum also received heavy vagal sensory innervation. Sensory endings in the ileum were relatively scarce, compared with the upper small intestine. Furthermore, both the Berthoud et al. and Jagger et al. studies found that systemic treatment with capsaicin nearly eliminated the intestinal sensory innervation to the small intestine, while Berthoud et al. found that significant numbers of gastric vagal endings survived capsaicin treatment. Thus, reduction of food intake by intestinal nutrients most likely depends upon vagal sensory innervation of the duodenum and jejunum.

In an attempt to further localize neural substrates participating in reduction of food intake by intestinal nutrient, Tamura and Ritter[43] examined the effects of intraintestinal capsaicin infusion on the vagal and peptidergic innervation of the intestine and on reduction of food intake by intraintestinal oleate infusion. They found that, unlike systemic capsaicin treatment, low doses of intraintestinal capsaicin were not associated with degeneration of vagal sensory fibers in the nucleus of the solitary tract, but did deplete calcitonin gene-related peptide (CGRP) from the submucosal plexus of the duodenum and jejunum for a period of 24 h, after which CGRP immunoreactivity returned to preinfusion levels.[44] Tamura and Ritter[43] also found that reduction of sham feeding by intraintestinal oleate was attenuated for 24 h, following intraintestinal capsaicin infusion, but sensitivity to oleate-induced reduction of food intake returned to preinfusion values by 48 h postinfusion. These results, taken together with those reviewed in the previous paragraph, suggest that the intestinal innervation is sufficient to mediate reduction of food intake by intestinal fat infusion. The results of Tamura and Ritter also raise the possibility that enteric neurons might participate in reduction of food intake by intestinal nutrients. This intriguing possibility, however, remains to be experimentally substantiated.

Vagal sensory innervation is most dense for the duodenum and jejunum. Furthermore, the digestion and absorption of most macronutrients is complete within the orad two thirds of the small intestine (duodenum and jejunum). Consequently, termination of food intake by intestinal nutrients would likely be controlled by signals from the upper small intestine. In fact, Huff and Ritter (unpublished observations) have found that infusion of maltose, oleate, and L-phenylalanine are effective for reduction of real and sham feeding when they are infused into the duodenum or jejunum, but not when infused into the ileum (Figure 4). These results suggest that satiation by intestinal nutrients results from signals generated in the upper small intestine. However, a seemingly contradictory set of data has been published by Woltman and Reidelberger.[21] They found that infusion of glucose or oleate into the ileum produced reductions of food intake comparable with, and in some cases greater than, that produced by duodenal infusion. The apparent discrepancy between the results of Woltman and Reidelberger and those of the previously mentioned investigators may be explained by revealing differences in their experimental designs. Most importantly, Huff and Ritter examined reduction of food intake during a single liquid meal, lasting only 30 to 40 min, while Woltman and Reidelberger[21] examined the mean meal size and meal frequency for

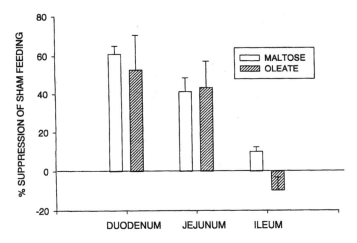

FIGURE 4 Intestinal infusion of maltose and oleate is more inhibitory in the proximal small intestine (duodenum and jejunum) vs. the distal small intestine (ileum). These results support the hypothesis that nutrient-sensitive mechanisms in the proximal but not distal small intestine can act to reduce food intake.

solid food intake over a 4-h period, with infusion of either glucose or oleate during the first 2 h of measurement. Therefore, the Woltman and Reidelberger results may reveal the presence of controls that operate over more than a single meal, whereas results from experiments where infusions are confined to a period of ingestion circumscribing a single meal may reflect only those controls that operate to terminate that meal. Because virtually all studies of neural substrates mediating nutrient-induced reduction of food intake have been done using the single liquid meal design it is not possible to infer whether the effects obtained by Woltman and Reidelberger depend upon similar neural substrates to those obtained by other investigators. Furthermore, the length of infusions (2 h) and measurements (4 h) in the Woltman and Reidelberger study increase the probability that the effects they report may be due in part to endocrine or postabsorptive signals in addition to or instead of those arising from the intestine itself.

We are still largely ignorant concerning the transductive mechanisms by which intestinal macronutrients reduce food intake. However, in the case of fat and carbohydrate, mucosal transport of the nutrient or some related process appears to be necessary for reduction of intake. Results from our own laboratory indicate that blockade of the sodium/glucose symport using phlorizin attenuates reduction of food intake by maltose and maltotriose.[36] These results suggest that entero-cyte transport of glucose is necessary for reduction of food intake by glucose-oligosaccharides. In addition, these results might be taken to indicate that reduction of food intake by intestinal carbohydrate infusion is postintestinal. However, intravenous infusion of glucose is far less effective for reducing food intake than intraintestinal infusion. Furthermore, when equimolar solutions of glucose, maltose, and maltotriose are infused into the intestine, the order of efficacy for elevation of plasma glucose during the first 15 min after the infusion, when satiation is occurring, is glucose > maltose > maltotriose. On the other hand, the order of efficacy for reduction of food intake is maltotriose > maltose ≫ glucose. In other words, even though intraintestinal maltotriose produces a lesser elevation of prandial plasma glucose concentrations than intraintestinal glucose, it is much more effective than glucose for reduction of food intake. Therefore, these data indicate that while transport is important for transduction of satiation by intestinal oligosaccharides, elevation of glucose in the blood is not responsible for reduction of food intake, following intraintestinal oligosaccharide infusion. Rather, the transductive signal originates in or near the intestine itself.

Greenberg et al.[3] have demonstrated that intravenous infusion of fat is ineffective for production of satiation during a meal, whereas intraintestinal infusion produces robust reductions of food intake. In addition, this group has demonstrated that satiation following intestinal fat infusion occurs prior to appearance of radiolabeled fat in the circulation.[45] Nonetheless, results recently published by Raybould et al.[46] indicate that transport of triglyceride across the basolateral enterocyte membrane is essential for inhibition of gastric emptying by intestinal fat infusion. Thus, if the signal for satiation by intestinal fat is similar to that which inhibits gastric emptying, it may originate somewhere beyond the enterocyte, but before fat enters the blood circulation. In this regard, it is interesting that secretion of cholecystokinin (CCK), which appears to be a signal for fat-induced satiety, also depends on absorption of lipolytic products. CCK secretion does not occur if chylomicron secretion from the enterocyte is blocked.[46]

4 CHOLECYSTOKININ PARTICIPATION IN REDUCTION OF FOOD INTAKE BY INTESTINAL MACRONUTRIENTS

Gibbs and co-workers[47] first proposed that CCK released from intestinal endocrine cells might serve as a satiety signal. They demonstrated that exogenous CCK reduced real and sham food intake in a dose-dependent manner. In addition, Smith et al.[48] were first to demonstrate that reduction of food intake by exogenous CCK is abolished by subdiaphragmatic vagotomy, and Smith et al.[49] subsequently demonstrated that surgical section of abdominal vagal sensory fibers attenuated CCK-induced reduction of food intake. Ladenheim and Ritter[50] and South and Ritter[51] demonstrated that capsaicin, applied systemically or to vagal trunks or terminals attenuated reduction of food intake by exogenous CCK. Two CCK receptor subtypes have been identified[52] and cloned[53-55] (CCK-A and CCK-B receptors). The vagal sensory neurons express CCK-A receptors,[52,56] as well as CCK-B receptors.[57] However, reduction of food intake by exogenous CCK is mediated by vagal CCK-A receptors, because reduction of food intake by CCK is abolished following systemic injections of CCK-A receptor antagonists.[58,59] Neither CCK-A nor CCK-B receptor antagonists abolish reduction of food intake by exogenous CCK when the antagonists are injected centrally.[58,59] CCK-A receptor antagonists also increase food intake, when they are injected in the absence of exogenous CCK. Hence, a working hypothesis has been that nutrients entering the intestine release CCK, which then reduce food intake via an action at CCK-A receptors, located on vagal sensory neurons, innervating abdominal viscera.

Involvement of CCK-A receptors in reduction of food intake by intestinal nutrient infusions is supported by experimental results from several laboratories (see, for example, Figure 5). Yox et al.,[6] Greenberg et al.,[60] and Woltman and Reidelberger[31,61,62] all have demonstrated that systemic injection of CCK-A receptor antagonists, but not CCK-B receptor antagonists, attenuate or abolish reduction of food intake by intraintestinal triglyceride or long-chain fatty acid. CCK-A antagonists also have been reported to attenuate reduction of food intake by intraintestinally infused protein.[63] Release of CCK by unhydrolyzed protein and peptones is well demonstrated in the rat. Furthermore, Brenner et al.[64] have demonstrated that intraintestinal oleate elevates plasma CCK concentrations in the rat, and others have demonstrated that triglycerides also elevate plasma CCK. Consequently, these results support the hypothesis that release of intestinal CCK mediates reduction of food intake by some intestinal nutrients. The hypothesis is further strengthened by the fact that CCK receptor antagonists do not attenuate reduction of food intake by L-phenylalanine (Figure 5B), which does not release CCK in the rat.[6,65]

Nevertheless, there are some problems with the simple hypothesis that CCK mediates reduction of food intake only by CCK-releasing nutrients. First, CCK is not released by intestinal infusions of maltose or maltotriose in the rat. However, reduction of food intake by these oligosaccharides is attenuated by CCK-A receptor antagonists.[64,65] In addition, soybean trypsin inhibitor, which releases CCK by interfering with negative feedback control of secretion, via CCK-releasing peptide, has generally been reported not to reduce food intake or to have weak effects on feeding in adult rats[66] (but, see also Weller et al.[67]). Thus, although the necessity of acute elevation of plasma CCK in reduction of food intake remains to be proved experimentally, the fact remains that there is compelling evidence that CCK-A receptors are necessary for reduction of feeding by some intestinal nutrients.

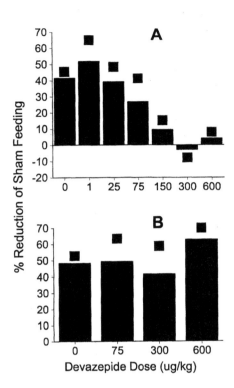

FIGURE 5 The CCK antagonist MK-329 (Devazepide) attenuates suppression of sham feeding induced by intraintestinal oleate but not L-Phenylalanine (L-Phe). (**A**) Percent suppression of sham feeding induced by intraintestinal oleate (0.065 kcal/ml) is significantly attenuated by the CCK antagonist MK-329 in a dose-dependent manner. The 300 and 600 μg/kg doses completely blocked suppression of sham feeding by intraintestinal oleic acid. (**B**) Suppression of sham feeding induced by L-phenylalanine is not attenuated by prior treatment with MK-329. These results suggest that endogenous CCK mediates suppression of feeding by some, but not all, nutrients.

5   DIET-INDUCED ALTERATIONS IN SENSITIVITY TO REDUCTION OF FOOD INTAKE BY INTESTINAL MACRONUTRIENT INFUSION

Satiety signals control meal size and thereby limit the amount of ingesta presented to the alimentary tract during a meal. However, the need for particular nutrients is not constant. Furthermore, the capacity of the alimentary tract to digest and absorb various macronutrients varies, according to their presence in the diet. Hence, there is no reason to assume that individual macronutrients have the same satiating effects under all conditions. In fact, Covasa and Ritter[75] recently have demonstrated that rats maintained on high-fat diets exhibit diminished sensitivity for reduction of food intake by intestinal oleate (Figure 6A). On the other hand, reduction of food intake by intestinal infusion of maltotriose does not differ between rats maintained on high- or low-fat diets. Thus, sensitivity to the satiating effects of fat appears to be diminished under dietary conditions that present large amounts of fat for digestion and absorption. While reduced sensitivity to satiation by fat seems counterintuitive, relative to the control of energy balance, it may be adaptive for animals living in an environment where fat constitutes a rich source of storable calories and the future is energetically uncertain.

Adaptation to high-fat diets also diminishes sensitivity to the satiation-producing effect of exogenous CCK (Figure 6B), further strengthening the relationship between CCK and satiation by

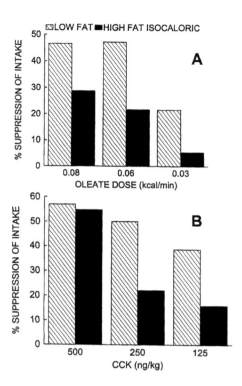

FIGURE 6   Percent reduction of 30-min food intake following intraintestinal infusion of oleate (**A**) or CCK administration (**B**) in rats adapted to low- and high-fat isocaloric diet. Low rates of intestinal infusion of oleate (0.48 kcal/ml) at three different caloric concentrations (0.08, 0.06, and 0.03 kcal/ml) reduced intake significantly higher in low-fat adapted rats than in high-fat adapted rats. Likewise, CCK caused significantly greater reduction of food intake at 250 and 125 ng/kg doses in low-fat adapted rats than in high-fat adapted rats. Thus, rats adapted to a high-fat diet exhibit reduced satiety in response to CCK and intestinal oleate.

intestinal fat.[68] In addition, Covasa and Ritter recently have reported that both CCK[69] and intestinal oleate[70] activate fewer neurons in the nucleus of the solitary tract and area postrema of rats maintained on high-fat diets than of rats maintained on low-fat diets, as indicated by expression of Fos-like immunoreactivity. Induction of Fos-like immunoreactivity by intestinal maltotriose does not differ between high- and low-fat fed rats. In support of this interpretation, Covasa and Ritter have demonstrated that inhibition of gastric emptying following exogenous CCK or intestinal oleate infusion is markedly diminished in rats maintained on high-fat diets.[76] Inhibition of gastric emptying by CCK and oleate is mediated by vagal sensory neurons. Therefore, these results suggest that high-fat diets may actually reduce vagal sensory neuron sensitivity to CCK released by fat in the intestine. In addition to our results from rats, there also is evidence that high-fat maintenance diets reduce satiation in humans. French et al.[71] have reported that humans maintained on a high-fat diet eat more of a test meal and report less satiation and more hunger than individuals maintained on low-fat rations. An implication of our results is that chronic exposure to high-fat diets reduces their satiating potency, thus enabling greater consumption of fat during each meal. Although no macronutrient-selection experiments have yet been performed in this experimental paradigm, one might predict that fat adaptation could lead to increased fat consumption in a situation where the animals had a choice between different macronutrients (Figure 7).

In Covasa and Ritter's experiments, diminished fat-induced satiation is not accompanied by any change in the satiating potency of carbohydrate (maltotriose). However, in an earlier report by Geary et al.,[72] rats maintained on high-fat diet reduced their 24-h food intake and lost weight when switched to a low-fat, high-carbohydrate diet. Although it is not clear what role alterations in

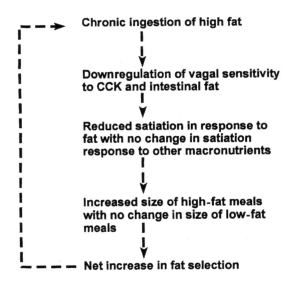

Chronic ingestion of high fat

Downregulation of vagal sensitivity to CCK and intestinal fat

Reduced satiation in response to fat with no change in satiation response to other macronutrients

Increased size of high-fat meals with no change in size of low-fat meals

Net increase in fat selection

FIGURE 7 Potential scheme by which plasticity of intestinal fat sensitivity could conceivably alter macronutrient selection, by increasing intake of fat.

gastrointestinal feedback signals may play in the reduction of intake observed by Geary et al., their results clearly demonstrate that adaptation to a particular macronutrient may have profound affects on the subsequent consumption of a different macronutrient. Experiments utilizing the designs developed by both Covasa and Ritter and Geary et al., which provide animals with macronutrient choices, may help to resolve the question of whether or not changes in sensitivity to macronutrient specific satiety cues can contribute to changes in macronutrient preference and consumption.

6   PARTICIPATION OF INTESTINAL MACRONUTRIENT SIGNALS IN CONTROL OF MEAL PARAMETERS AND MACRONUTRIENT SELECTION

Intestinal feedback signals potentially could contribute to the selective increase or decrease in consumption of specific macronutrients either by differentially reducing the size of meals containing a predominance of one nutrient, by suppressing intake of one macronutrient for a longer period than for another macronutrient or by providing sensory information that results in a change in effect regarding a particular macronutrient. The ability of intestinal nutrient infusions to reduce the size of deprivation-induced meals has been well documented, while there is less experimental information relative to the effect of infusions on intermeal interval (IMI).

Burton-Freeman et al.[73] demonstrated that duodenal infusion of either fat or carbohydrate reduced meal size. Moreover, both nutrient infusions diminished meal size by similar amounts. However, these investigators also found that fat infusion nearly doubled IMI, as compared with infusion of comparable amounts of carbohydrate. Inspection of the results of Woltman and Reidelberger[21] also suggests that duodenal oleate infusion reduced meal frequency over a 4-h measurement period more than duodenal glucose infusion. The disproportionate lengthening of the IMI by fat infusion was associated with a greater decrease in total food intake than was observed following duodenal carbohydrate infusion. The fact that IMI effects of fat infusion occurred over 4- or 6-h periods makes attribution of IMI effect to intestinal signals uncertain. However, in other work, Burton-Freeman et al. have demonstrated that fat-induced lengthening of the IMI can be attenuated by a CCK-A receptor antagonist, suggesting that IMI effects of fat infusions may be related to CCK secretion from the intestine. In further support of an association between increased

IMI and CCK, Burton-Freeman et al.[73] have reported that intestinal infusion of protein, which, like fat infusion, releases CCK, also induces greater increases in IMI than calorically comparable carbohydrate infusions.

Burton-Freeman and co-workers' work does not specifically address potential effects of intestinal feedback signals on macronutrient selection. However, their demonstration of differential effects of fat and carbohydrate infusions on length of IMI might have implications for macronutrient selection when more than one diet is available for ingestion. Following this line of reason, a recent report by Foster et al.[74] suggests that nutrient-specific intestinal feedback signals may in fact suppress intake of one macronutrient for a longer period of time than another. Foster et al. measured sham intake of either 6% sucrose or 20% Intralipid over 90 min, which included a 20 min period when the rats were given an infusion of either sucrose or fat (Intralipid). The results revealed that fat infusion reduced sham consumption of fat more and for a longer period of time than did infusion of sucrose. On the other hand, the amount of sucrose intake and the length of time that sucrose intake was reduced by intestinal infusion of fat or sucrose were not significantly different. Thus, intestinal feedback from fat may have greater efficacy for reducing fat consumption than carbohydrate consumption. While these sham-feeding results cannot be selectively related to reductions of meal size, as opposed to decreased meal frequency, a selective lengthening of the IMI between high-fat meals, as opposed to between carbohydrate meals might be inferred from the existing data.

In humans, Cecil et al.[20] have reported that a mixed intragastric infusion, in which 50% of the calories were supplied by fat and 50% by glucose, produced a reduction in the selection of fat from a test meal. However, in a second experiment these investigators also found that intragastric infusion of Intralipid alone did produce greater feelings of fullness and reduced hunger ratings, relative to glucose infusion. Intralipid infusions did not selectively reduce fat intake more than infusion of glucose. Both infusions did reduce total caloric intake of the test meal. While the effects of the Intralipid/glucose mixture were not compared with the effects of glucose alone in the same experiment, these results do not support differential reduction of fat intake by prior upper gastrointestinal fat loads, as might have been expected based on the Foster et al. results.

7 SUMMARY AND CONCLUSIONS

In conclusion, the results of the past 10 years have provided evidence that intestinal feedback signals contribute to satiation. Furthermore, there is evidence that different macronutrients may contribute to satiation by activating different mucosal, neuronal, and hormonal mechanisms. Sensitivity to the satiation-producing effects of some nutrients (fat) is not static and appears to diminish following a period of adaptation to high-fat diet. These aspects of intestinal feedback signals could provide avenues by which the intestinal contents may influence macronutrient selection. However, at the present time, there is little evidence that intestinal feedback signals influence macronutrient selection.

REFERENCES

1. Ritter, R. C., Brenner, L., and Yox, D. P., Participation of vagal sensory neurons in putative satiety signals from the upper gastrointestinal tract, in *Neuroanatomy and Physiology of Abdominal Vagal Afferents.*, Ritter, S., Ritter, R. C., Barnes, C. D., Eds., CRC Press, Boca Raton, FL, 1992, 221.
2. Gibbs, J., Maddison, S. P., and Rolls, E. T., Satiety role of the small intestine examined in sham-feeding rhesus monkeys, *J. Comp. Physiol. Psychol.*, 95, 1003, 1981.
3. Greenberg, D., Smith, G. P., and Gibbs, J., Intravenous triglycerides fail to elicit satiety in sham-feeding rats, *Am. J. Physiol.*, 264, R409, 1993.
4. Greenberg, D., Smith, G. P., and Gibbs, J., Intraduodenal infusions of fats elicit satiety in sham-feeding rats, *Am. J. Physiol.*, 259, R110, 1990.

5. Yox, D. P., Stokesberry, H., and Ritter, R. C., Vagotomy attenuates suppression of sham feeding induced by intestinal nutrients, *Am. J. Physiol.*, 260, R503, 1991.

6. Yox, D. P., Brenner, L., and Ritter, R. C., CCK-receptor antagonists attenuate suppression of sham feeding by intestinal nutrients, *Am. J. Physiol.*, 262, R554, 1992.

7. Young, R. C., Gibbs, J., Antin, J., Holt, J., and Smith, G. P., Absence of satiety during sham feeding in the rat, *J. Comp. Physiol. Psychol.*, 87, 795, 1974.

8. Reidelberger, R. D., Kalogeris, T. J., Leung, P. M., and Mende, E., Postgastric satiety in the sham-feeding rat, *Am. J. Physiol.*, 244, R872, 1983.

9. Schwartz, G. J., Tougas, G., and Moran, T. H., Integration of vagal afferent responses to duodenal loads and exogenous CCK in rats, *Peptides*, 16, 707, 1995.

10. Schwartz, G. J. and Moran, T. H., Sub-diaphragmatic vagal afferent integration of meal-related gastrointestinal signals, *Neurosci. Biobehav. Rev.*, 20, 47, 1996.

11. Seeley, R. J., Kaplan, J. M., and Grill, H. J., Effect of occluding the pylorus on intraoral intake: a test of the gastric hypothesis of meal termination, *Physiol. Behav.*, 58, 245, 1995.

12. Phillips, R. J. and Powley, T. L., Gastric volume rather than nutrient content inhibits food intake, *Am. J. Physiol.*, 40, R766, 1996.

13. Davis, J. D., Smith, G. P., and Sayler, J. L., Reduction of intake in the rat due to gastric filling, *Am. J. Physiol.*, 41, R1599, 1997.

14. Deutsch, J. A. and Gonzalez, M. F., Gastric nutrient content signals satiety, *Behav. Neural. Biol.*, 30, 113, 1980.

15. Davis, J. D. and Smith, G. P., Learning to sham feed: behavioral adjustments to loss of physiological postingestional stimuli, *Am. J. Physiol.*, 259, R1228, 1990.

16. Kaplan, J. M., Siemers, W., and Grill, H. J., Ingestion, gastric fill, and gastric emptying before and after withdrawal of gastric contents, *Am. J. Physiol.*, 267, R1257, 1994.

17. Busch, W., Contribution to the physiology of the digestive organs, *Arch. Pathol. Anat. Physiol. Kin. Med.*, 14, 140, 1858.

18. Yox, D. P. and Ritter, R. C., Capsaicin attenuates suppression of sham feeding induced by intestinal nutrients, *Am. J. Physiol.*, 255, R569, 1988.

19. Welch, I. M., Sepple, C. P., and Read, N. W., Comparisons of the effects on satiety and eating behaviour of infusion of lipid into the different regions of the small intestine, *Gut*, 29, 306, 1988.

20. Cecil, J. E., Castiglione, K., French, S., Francis, J., and Read, W., Effects of intragastric infusions of fat and carbohydrate on appetite ratings and food intake from a test meal, *Appetite*, 30, 65, 1998.

21. Woltman, T. and Reidelberger, R., Effects of duodenal and distal ileal infusions of glucose and oleic acid on meal patterns in rats, *Am. J. Physiol.*, 269, R7, 1995.

22. Geoghegan, J. G., Cheng, C. A., Lawson, C., and Pappas, T. N., The effect of caloric load and nutrient composition on induction of small intestinal satiety in dogs, *Physiol. Behav.*, 62, 39, 1997.

23. McHugh, P. R. and Moran, T. H., Calories and gastric emptying: a regulatory capacity with implications for feeding, *Am. J. Physiol.*, 236, R254, 1979.

24. Hunt, J. N. and Stubbs, D. F., The volume and energy content of meals as determinants of gastric emptying, *J. Physiol. London*, 245, 209, 1975.

25. Castiglione, K. E., Read, N. W., and French, S. J., Food intake responses to upper gastrointestinal lipid infusions in humans, *Physiol. Behav.*, 64, 141, 1998.

26. Brenner, L., Yox, D. P., and Ritter, R. C., Suppression of sham feeding by intraintestinal nutrients is not correlated with plasma cholecystokinin elevation, *Am. J. Physiol*, 264, R972, 1993.

27. Houpt, T. R., Houpt, K. A., and Swan, A. A., Duodenal osmoconcentration and food intake in pigs after ingestion of hypertonic nutrients, *Am. J. Physiol.*, 245, R181, 1983.

28. Foster, L. A., Nakamura, K., Greenberg, D., and Norgren, R., Intestinal fat differentially suppresses sham feeding of different gustatory stimuli, *Am. J. Physiol.*, 39, R1122, 1996.

29. Gibbs, J. and Smith, G. P., The neuroendocrinology of postprandial satiety, in *Frontiers in Neuroendocrinology*, Vol. 8, Martini, L. and Ganong, W. F., Eds., Raven Press, New York, 1984, 223.

30. Greenberg, D. and Smith, G. P., The controls of fat intake, *Psychosom. Med.*, 58, 559, 1996.

31. Woltman, T., Castellanos, D., and Reidelberger, R., Role of cholecystokinin in the anorexia produced by duodenal delivery of oleic acid in rats, *Am. J. Physiol.*, 269, R1420, 1995.

32. Jeanningros, R., Vagal unitary responses to intestinal amino acid infusions in the anesthetized cat: a putative signal for protein induced satiety, *Physiol. Behav.*, 28, 9, 1982.

33. Mei, N., Vagal glucoreceptors in the small intestine of the cat, *J. Physiol. London*, 282, 485, 1978.

34. Melone, J., Vagal receptors sensitive to lipids in the small intestine of the cat, *J. Auton. Nerv. Syst.*, 17, 231, 1986.

35. Ewart, W. R. and Wingate, D. L., Central representation of arrival of nutrient in the duodenum, *Am. J. Physiol.*, 246, G750, 1984.

36. Ritter, R. C., Ritter, S., Ewart, W. R., and Wingate, D. L., Capsaicin attenuates hindbrain neuron responses to circulating cholecystokinin, *Am. J. Physiol.*, 257, R1162, 1989.

37. Schwartz, G. J., McHugh, P. R., and Moran, T. H., Integration of vagal afferent responses to gastric loads and cholecystokinin in rats, *Am. J. Physiol.*, 261, R64, 1991.

38. Glenn, J. F. and Erickson, R. P., Gastric modulation of gustatory afferent activity, *Physiol. Behav.*, 16, 561, 1976.

39. Yox, D. P., Stokesberry, H., and Ritter, R. C., Fourth ventricular capsaicin attenuates suppression of sham feeding induced by intestinal nutrients, *Am. J. Physiol.*, 260, R681, 1991.

40. Walls, E. K., Phillips, R. J., Wang, F. B., Holst, M. C., and Powley, T. L., Suppression of meal size by intestinal nutrients is eliminated by celiac vagal deafferentation, *Am. J. Physiol.*, 38, R1410, 1995.

41. Berthoud, H. R., Patterson, L. M., Willing, A. E., Mueller, K., and Neuhuber, W. L., Capsaicin-resistant vagal afferent fibers in the rat gastrointestinal tract: anatomical identification and functional integrity, *Brain Res.*, 746, 195, 1997.

42. Jagger, A., Grahn, J., and Ritter, R. C., Reduced vagal sensory innervation of the intestinal myenteric plexus following capsaicin treatmen of adult rats, *Neurosci. Lett.*, 236, 103, 1997.

43. Tamura, C. S. and Ritter, R. C., Intestinal capsaicin transiently attenuates suppression of sham feeding by oleate, *Am. J. Physiol*, 267, R561, 1994.

44. Tamura, C. T. and Ritter, R. C., Intraintestinal capsaicin transiently reduces CGRP-like immunoreactivity in the rat submucosal plexus, *Brain Res.*, 770, 248, 1997.

45. Greenberg, D., Kava, R. A., Lewis, D. R., Greenwood, M. R., and Smith, G. P., Time course for entry of intestinally infused lipids into blood of rats, *Am. J. Physiol.*, 269, R432, 1995.

46. Raybould, H. E., Meyer, J. H., Tabrizi, Y., Liddle, R. A., and Tso, P., Inhibition of gastric emptying in response to intestinal lipid is dependent on chylomicron formation — rapid communication, *Am. J. Physiol.*, 43, R1834, 1998.

47. Gibbs, J., Young, R. C., and Smith, G. P., Cholecystokinin decreases food intake in rats, *J. Comp. Physiol. Psychol.*, 84, 488, 1973.

48. Smith, G. P., Jerome, C., Cushin, B. J., Eterno, R., and Simansky, K. J., Abdominal vagotomy blocks the satiety effect of cholecystokinin in the rat, *Science*, 213, 1036, 1981.

49. Smith, G. P., Jerome, C., and Norgren, R., Afferent axons in abdominal vagus mediate satiety effect of cholecystokinin in rats, *Am. J. Physiol.*, 249, R638, 1985.

50. Ladenheim, E. E. and Ritter, R. C., Capsaicin attenuates bombesin-induced suppression of food intake, *Am. J. Physiol.*, 260, R263, 1991.

51. South, E. H. and Ritter, R. C., Capsaicin application to central or peripheral vagal fibers attenuates CCK satiety, *Peptides*, 9, 601, 1988.

52. Moran, T. H., Robinson, P. H., Goldrich, M. S., and McHugh, P. R., Two brain cholecystokinin receptors: implications for behavioral actions, *Brain. Res.*, 362, 175, 1986.

53. de-Weerth, A., Pisegna, J. R., Huppi, K., and Wank, S. A., Molecular cloning, functional expression and chromosomal localization of the human cholecystokinin type A receptor, *Biochem. Biophys. Res. Commun.*, 194, 811, 1993.

54. Sykes, R. M., Spyer, K. M., and Izzo, P. N., Central distribution of substance P, calcitonin gene-related peptide and 5-hydroxytryptamine in vagal sensory afferents in the rat dorsal medulla, *Neuroscience*, 59, 195, 1994.

55. Wank, S. A., Pisegna, J. R., and de-Weerth, A., Cholecystokinin receptor family. Molecular cloning, structure, and functional expression in rat, guinea pig, and human, *Ann. N.Y. Acad. Sci.*, 713, 49, 1994.

56. Hill, D. R., Campbell, N. J., Shaw, T. M., and Woodruff, G. N., Autoradiographic localization and biochemical characterization of peripheral type CCK receptors in rat CNS using highly selective nonpeptide CCK antagonists, *J. Neurosci.*, 7, 2967, 1987.

57. Wank, S. A., Pisegna, J. R., and de-Weerth, A., Brain and gastrointestinal cholecystokinin receptor family: structure and functional expression, *Proc. Natl. Acad. Sci. U.S.A.*, 89, 8691, 1992.

58. Corp, E. S., Curcio, M., Gibbs, J., and Smith, G. P., The effect of centrally administered CCK-receptor antagonists on food intake in rats, *Physiol. Behav.*, 61, 823, 1997.
59. Brenner, L. A. and Ritter, R. C., Intracerebroventricular cholecystokinin A-receptor antagonist does not reduce satiation by endogenous CCK, *Physiol. Behav.*, 63, 711, 1998.
60. Greenberg, D., Torres, G. P., Smith, G. P., and Gibbs, J., The satiating effects of fats is attenuated by the cholecystokinin antagonists lorglumide, *Ann. N.Y. Acad. Sci.*, 575, 517, 1989.
61. Woltman, T. A. and Reidelberger, R. D., Effects of CCK-A antagonist devazepide on inhibition of feeding by duodenal infusion of oleic acid in rats, *Ann. N.Y. Acad. Sci.*, 713, 372, 1994.
62. Woltman, T. and Reidelberger, R., Role of cholecystokinin in the anorexia produced by duodenal delivery of glucose in rats, *Am. J. Physiol.*, 271, R1521, 1996.
63. Trigazis, L., Orttmann, A., and Anderson, G. H., Effect of a cholecystokinin-A receptor blocker on protein-induced food intake suppression in rats, *Am. J. Physiol.*, 272, R1826, 1997.
64. Brenner, L., Yox, D. P., and Ritter, R. C., Suppression of sham feeding by intraintestinal nutrients is not correlated with plasma cholecystokinin elevation, *Am. J. Physiol.*, 264, R972, 1993.
65. Brenner, L. A. and Ritter, R. C., Type A CCK receptors mediate satiety effects of intestinal nutrients, *Pharmacol. Biochem. Behav.*, 54, 625, 1996.
66. Smith, G. P., Greenberg, D., Falasco, J. D., Avilion, A. A., Gibbs, J., Liddle, R. A., and Williams, J. A., Endogenous cholecystokinin does not decrease food intake or gastric emptying in fasted rats, *Am. J. Physiol.*, 257, R1462, 1989.
67. Weller, A., Smith, G. P., and Gibbs, J., Endogenous cholecystokinin reduces feeding in young rats, *Science*, 247, 1589, 1990.
68. Covasa, M. and Ritter, R. C., Rats maintained on high-fat diets exhibit reduced satiety in response to CCK and bombesin, *Peptides*, 19, 1407, 1998.
69. Covasa, M. and Ritter, R. C., Attenuated cholecystokinin satiety and hindbrain fos expression following adaptation to high-fat diets in rats, *Soc. Neurosci. Abstr.*, 23, 1074, 1997.
70. Covasa, M. and Ritter, R. C., Reduced fos expression in enteric and hindbrain neurons of rats maintained on high-fat diet following intestinal oleate infusion, *Soc. Neurosci. Abstr.*, 24, 1440, 1998.
71. French, S. J., Murray, B., Rumsey, R. D., Fadzlin, R., and Read, N. W., Adaptation to high-fat diets: effects on eating behaviour and plasma cholecystokinin, *Br. J. Nutr.*, 73, 179, 1995.
72. Geary, N., Scharrer, E., Freudlsperger, R., and Raab, W., Adaptation to high-fat diet and carbohydrate-induced satiety in the rat, *Am. J. Physiol.*, 237, R139, 1979.
73. Burton-Freeman, B. G., Gietzen, D. W., and Schneeman, B. O., Meal pattern analysis to investigate the satiating potential of fat, carbohydrate, and protein in rats, *Am. J. Physiol.*, 42, R1916, 1997.
74. Foster, L. A., Boeshore, K., and Norgren, R., Intestinal fat suppressed intake of fat longer than intestinal sucrose, *Physiol. Behav.*, 64, 451, 1998.
75. Covasa, M. and Ritter, R. C., Reduced sensitivity to the satiating effect of intestinal oleate in rats adapted to high fat, *Am. J. Physiol.*, 277, R279, 1999.
76. Covasa, M. and Ritter, R. C., Adaptation to high fat diet reduces inhibition of gastric emptying by cholecystokinin and intestinal oleate, *Am. J. Physiol.*, in press.

19 Fat Absorption and the Role of Lymphatic Apolipoprotein A-IV in the Regulation of Food Intake

P. Tso and K. Fukagawa

CONTENTS

1 SOURCES OF LIPID IN THE LUMEN OF THE GASTROINTESTINAL TRACT

Dietary lipids provide as much as 40% of the daily caloric intake in the Western diet. The daily intake of lipids by humans in the Western world ranges between 60 and 100 g.[1] Triacylglycerol (TG) is the major dietary fat in humans. Major long-chain fatty acids present in the diet are palmitate (16:0), stearate (18:0), oleate (18:1), linoleate (18:2), and linolenate (18:3). In most infant diets, fat becomes an even more important source of calories. In human milk and in human formulas, as much as 50% of total calories are present as fat.[2] In human milk, there is also an abundance of medium-chain fatty acids. The human small intestine is presented daily with other lipids such as phospholipids (PL), cholesterol, and plant sterols. Both PL and cholesterol are major constituents of bile. In humans, the biliary PL is a major contributor of luminal PL and as much as 11 to 20 g of biliary PL enters the small intestinal lumen daily, whereas the dietary contribution is between 1 and 2 g.[3,4] The small intestinal epithelium undergoes rapid turnover and this also contributes to the luminal PL and cholesterol. The predominant sterol in the Western diet is cholesterol. However, plant sterols account for 20 to 25% of total dietary sterol.[5,6]

2   DIGESTION OF LIPIDS

Lipid digestion begins in the stomach. Lipase activity has been reported to be present in the human gastric juice.[7] In humans, the gastric lipase activity is mainly contributed by the stomach and the highest activity can be detected in the fundus of the stomach. Human gastric lipase has a pH optimum ranging from 3.0 to 6.0 and therefore has been called acid lipase. It hydrolyzes medium-chain TG (predominantly 8 to 10 carbon chain length) better than the long-chain TG.[8] The main hydrolytic products of gastric lipase are diacylglycerols (DG) and free fatty acids (FA).[9,10] Gastric lipase does not hydrolyze PL or cholesteryl esters. It is interesting that in rodents, the main gastric lipase activity is derived from the lipase secreted by the salivary glands and is therefore called lingual lipase.

In the stomach, the lipid is emulsified (broken into small oil droplets). Lipid emulsion enters the small intestinal lumen as fine lipid emulsion droplets less than 0.5 μm in diameter.[11,12] The combined action of the bile and the pancreatic juice brings about a marked change in the physical and chemical form of the luminal lipid emulsion. Pancreatic lipase is secreted into the duodenum and it hydrolyzes TG to form 2-monoacylglycerol (2-MG) and FAs. The most potent gastrointestinal hormone that stimulates the release of enzymes by the pancreas is cholecystokinin (CCK)[13] and CCK-A receptor has been demonstrated to be present in the pancreas.[14] The pancreatic lipase works at the oil and water interphase. Therefore, the rate of lipolysis is influenced by factors modifying the physicochemical properties of the interface as well as the surface area.[15-17]

In vitro studies using purified pancreatic lipase have demonstrated a potent inhibitory effect of bile salts on lipolysis of TG at concentration above the critical micellar concentration.[17] The inhibitory effect of bile salt is physiological since the concentration of bile salts in the duodenum is normally higher than the concentration of bile salt needed to observe the inhibitory effect. If this is the case, why then is pancreatic lipase so efficient in digesting TG? The explanation lies in the fact that the pancreas secretes another protein that counteracts this inhibition. The factor is called colipase. This factor was first isolated by Morgan et al.[18] from rat pancreatic juice. The structure and mechanism of action of colipase have been elucidated by the elegant work from the laboratories of Dr. Desnuelle[19] and Borgstrom.[20] Colipase acts by attaching to the ester bond region of the TG molecule. In turn, the lipase binds strongly to the colipase by electrostatic interactions, thereby allowing the hydrolysis of the TG by the lipase molecule.[21-23] Colipase is secreted as a procolipase and is converted to the active form through the removal of a five amino acid fragment by trypsin. The five amino acid fragment released is called the enterostatin. Enterostatin has been demonstrated to inhibit fat intake — an action independent of the route of administration. For instance, it has been reported that both the intraduodenal,[24] intraperitoneal,[25,26] intravenous,[27] and intracerebroventricular administration of enterostatin reduces fat intake, but both the dose and the time required for the administered enterostatin to exert its action are quite different. Investigators have questioned the physiological role of enterostatin in the inhibition of fat intake asking how it could get taken up by the small intestine and still gain access to the circulation. This criticism should be reevaluated particularly since a number of recent reports have documented the presence of both the mRNA as well as the enterostatin protein in the stomach and the small intestine of the rat.[28-30] It would be very interesting to see if enterostatin is secreted by the stomach or by the intestine.

3   APOLIPOPROTEIN A-IV AS A SATIETY FACTOR

Apolipoprotein A-IV (apo A-IV) was discovered almost 20 years ago but its physiological role has not been firmly established until recently. Elegant *in vitro* experiments have suggested roles for apo A-IV in certain aspects of lipoprotein metabolism[31-35]; however, there has been no direct evidence to date that apo A-IV plays such roles *in vivo*. Two concerns of the *in vitro* studies is that

TABLE 1
Food Intake in 24-h Fasted Rats after Infusion of 2 ml Test Solution through Indwelling Atrial Catheter

Test Solutions	Food Consumption After Refeeding, g	
	0–30 min	30–60 min
Control (physiological saline)	3.90 ± 0.40	1.31 ± 0.33
Mesenteric lymph samples		
Fasting Lymph	3.58 ± 0.33	1.75 ± 0.26
6–8 h after lipid infusion	0.60 ± 0.26[a]	1.20 ± 0.11
6–8 h after lipid + L-81 infusion	3.91 ± 0.21[b]	1.30 ± 0.25
5–7 h after cessation of L-81 infusion	0.40 ± 0.17[a]	1.60 ± 0.22
Intralipid (nutrition control)		
2% Intralipid	4.00 ± 0.29	1.65 ± 0.038
Mesenteric lymph samples 6–8 h after lipid infusion		
Immunoprecipitation with apo A-IV antiserum	3.41 ± 0.33[c]	1.22 ± 0.46
Control (treated by normal goat serum)	0.87 ± 0.24[a]	1.25 ± 0.48
Apolipoprotein A-IV, μg		
60	3.35 ± 0.46	1.13 ± 0.31
135	2.14 ± 0.16[a]	1.16 ± 0.26
200	0.90 ± 0.18[a,d]	1.10 ± 0.19
Apolipoprotein A-I, μg		
200	3.90 ± 0.48	1.20 ± 0.25

Values are means ± SE. Five rats were treated in each group.

[a] $P < 0.01$ compared with values for saline control and fasting lymph.

[b] $P < 0.01$ compared with values for lymph sample from rats infused with lipid only.

[c] $P < 0.01$ compared with its control.

[d] $P < 0.01$ compared with value for 135-μg apo A-IV.

apo A-I can perform similar physiological roles prescribed to apo A-IV and that apo A-I is present at higher concentration in blood than apo A-IV.

Recent *in vivo* studies[36-38] have provided evidence that apo A-IV may be involved in the inhibition of food intake following the ingestion of fat. Fujimoto et al.[36] demonstrated that intravenous infusion of physiological saline in rats with indwelling right atrial catheter ate 3.90 ± 0.40 g during the first 30 min after refeeding (Table 1). Infusion of fasting intestinal lymph collected from donor lymph-fistula rats had very little effect on food intake compared with saline infusion in the 24-h fasted rats (see Table 1). In contrast, the mesenteric lymph samples collected during the 6th to 8th hour of dietary lipid infusion (active lipid absorption) markedly suppressed food intake during the first 30 min ($P < 0.01$) when compared to the fasting lymph. This suppression of food intake was not observed in the following 30 min.

The fact that intestinal lymph collected from rats actively absorbing fat inhibits food intake indicates that it is one or more of the factors that change during active fat absorption that actually inhibits food intake. The lipid content of lymph increases as much 10 to 15 times during fat absorption. To test if the change in the lipid content of lymph is responsible for reducing food intake in 24-h fasted rats, rats were infused intravenously with a diluted Intralipid solution. When 2 ml of a 2% Intralipid in saline containing 42 μmol of triglyceride and 3.1 μmol of PL (composition comparable to the lymph collected during active lipid absorption) were infused intravenously, food intake was not suppressed (see Table 1). This result indicates that the anorectic effect of chylous lymph was not due to its lipid content.

If it is not the lipid component responsible for inhibiting food intake, we reasoned that it may be the apo A-IV in the lymph because apo A-IV is the only apolipoprotein secreted by the small intestine that is markedly stimulated by lipid feeding.[39,40] Hayashi et al.[39] also demonstrated that the stimulation of apo A-IV production by lipid feeding is associated with the formation and secretion of chylomicrons and the stimulation of apo A-IV production by fat absorption can be blocked by Pluronic L-81, which blocks the formation of chylomicrons. When the chylomicron block caused by Pluronic L-81 was reversed by stopping Pluronic L-81 infusion, chylomicron formation and secretion resumed and stimulation of apo A-IV secretion appeared. If the inhibition of food intake by intestinal lymph collected from animals actively absorbing lipid is caused by apo A-IV, what effect, then, will the lymph from an animal fed lipid plus Pluronic L-81 have on food intake? Table 1 shows that lymph from Pluronic L-81–treated animals does not inhibit food intake but that the lymph collected during the reversal of Pluronic L-81 inhibition is very potent in inhibiting food intake. This result strongly indicates that apo A-IV may be the factor in lymph that inhibits the intake of food.

To validate that apo A-IV inhibits food intake, we studied the effect of apo A-IV–deficient chylous lymph on feeding. As shown in Table 1, the chylous lymph treated with normal goat serum suppressed food intake significantly in the first 30 min. In contrast, chylous lymph that was treated with apo A-IV antiserum had no effect on food intake — the animal consumed an amount of food similar to that consumed by the saline controls. In contrast, lymph treated with apo A-I antiserum was just as effective as the untreated lymph on inhibiting food intake. Next, 200 μg of either apo A-IV or apo A-I dissolved in 2 ml physiological saline was infused intravenously in 24-h fasted rats. As shown in Table 1, 200 μg apo A-IV, an amount comparable to that present in 2 ml of lymph collected from a rat actively absorbing lipid, suppressed food intake significantly and to the same extent as the chylous lymph collected during 6 to 8 h of lipid infusion ($P < 0.01$, apo A-IV vs. fasting lymph). Feeding was significantly suppressed by 135 μg apo A-IV ($P < 0.01$), but the effect was less than 200 μg of apo A-IV ($P < 0.01$). A small dose of 60 μg apo A-IV equal to the amount in fasted lymph did not suppress feeding. In contrast, apo A-I had no effect on food intake. No nonphysiological reactions such as sedation, ataxia, or hyperthermia were observed after apo A-IV and chylous lymph infusion. This series of physiological studies finally led us to conclude that apo A-IV is a circulating signal released by the small intestine in response to fat feeding and it is likely the mediator for the anorectic effect of a lipid meal. This function is unique to apo A-IV and is not shared by apo A-I. This is an important point because all the functions ascribed to apo A-IV described in the *in vitro* studies are also functions that can be ascribed to apo A-I and plasma apo A-I concentration is higher than apo A-IV concentration.

4   SITE OF APO A-IV ACTION

Feeding behavior is influenced by many circulating chemical factors, and chemosensitive monitoring systems for these factors exist both in the peripheral organs as well as the central nervous system.[41,42] Fujimoto et al.[37] therefore investigated whether the inhibition of food intake by apo A-IV is mediated centrally. Third cerebroventricular administration of apo A-IV decreases food intake in a dose-dependent manner (Table 2) and with a potency that is about 50-fold higher than intravenous administration (Figure 1). In contrast, when apo A-I was infused into the third ventricle, it had no effect on food intake. The hypothesis that apo A-IV suppresses food intake via the central nervous system is further supported by the following experiment. When goat anti-rat apo A-IV serum was administered into the third ventricle in *ad libitum* fed rats at 11:00 h (light phase when rats usually do not eat), it resulted in feeding in all the animals tested (Table 3). One possible explanation for the observation made by Fujimoto et al. that administration of apo A-IV antiserum in the third ventricle elicits feeding is that apo A-IV antiserum probably removes any endogenous apo A-IV present.

TABLE 2

The Effect of Apo A-IV Infused into the Third Ventricle on Food Intake

	Food Consumption after Re-Feeding (g)	
	0–30 min	30–60 min
Apo A-IV		
0.5 µg	4.2 ± 0.3	1.0 ± 0.3
1.0 µg	2.8 ± 0.3[a]	1.2 ± 0.4
2.0 µg	1.3 ± 0.2[a]	1.0 ± 0.2
4.0 µg	0.6 ± 0.2[a]	1.1 ± 0.2
Apo A-I		
4.0 µg	4.0 ± 0.9	1.1 ± 0.4
Saline control	4.4 ± 0.4	1.4 ± 0.2

Rats were fasted for 24 h before the feeding study. Apo A-IV, apo A-I, or saline (total volume = 10 µl) was administered at 1 µl/min. for 10 min before refeeding. Values are expressed as mean ± SE. Five rats were tested in each group.

[a] $P < 0.01$ compared with saline controls.

TABLE 3

Feeding and Drinking Behavior during the Hour after Infusion of Goat Anti-Rat Apo A-IV or A-I Serum

Treatment	Incidence	Feeding Latency (min)	Duration (min)	Drinking Incidence
Goat anti-rat				
apo A-IV	5/5[a,b]	25.9 ± 2.3	2.8 ± 0.5	3/5
apo A-I	0/5	n.d.	n.d.	0/5
Saline	0/5	n.d.	n.d.	0/5

The feeding and drinking response was measured at 11:00 h. Five animals were studied in each group. n.d. = not detected.

[a] $p < 0.05$, compared with either apo A-I or saline;
[b] Number of rats with elicited feeding response/number of rats tested; both feeding latency and duration are expressed as mean ± SE.

Because available evidence suggests that *de novo* synthesis of apo A-IV in the brain is unlikely,[43,44] it has been proposed by Fujimoto et al.[37,38] that apo A-IV released by the small intestine (or perhaps a fragment thereof) may traverse the blood–brain barrier and act in the central nervous system. This hypothesis is supported by the following indirect evidence. First, Fujimoto et al.[37] have demonstrated that apo A-IV, or a fragment of apo A-IV, is present in the third ventricular cerebrospinal fluid.[37] Second, apo A-IV concentration in third ventricular cerebrospinal fluid increases as a result of lipid feeding. Third, Fukagawa et al.[45] using immunohistochemical technique demonstrated specific staining for apo A-IV in astrocytes and tanycytes throughout both white and gray matter. The granular nature and perinuclear distribution of apo A-IV immunoreactivity suggests that apo A-IV may be contained in perinuclear organelles or vesicles. The demonstration of

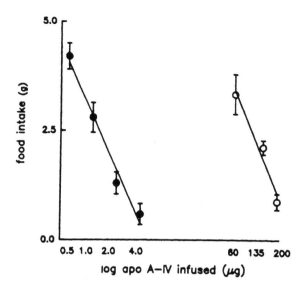

FIGURE 1 The relationship between the suppression of food intake during the first 30 min of refeeding in 24-h fasted rats and the logarithm of the dose in μg of apo A-IV infused either into the third ventricle (closed circles) or intravenously (open circles). (Data derived from reference 37.) Five animals were studied at each dose and the values are expressed as mean ± SE. Food intake diminished linearly with increasing amounts of apo A-IV infused into the third ventricle ($y = -1.78 \log X + 2.84$, $r = 0.98$, $P < 0.01$). (Reproduced from Fujimoto, K., Fukagawa, K., Sakata, T., and Tso, P., *J. Clin. Invest.*, 91, 1830–1833, 1993. By copyright permission of The American Society for Clinical Investigation.)

immunoreactive apo A-IV in tanycytes does not necessarily indicate a selective uptake mechanism for apo A-IV since tanycytes are known to take up a variety of neurotransmitters and nonmetabolizable amino acids. However, the presence of apo A-IV immunostaining in astrocytes may indicate the uptake of apo A-IV. Whether astrocytes are involved in satiety mechanisms associated with lipid feeding is unknown.

5   MECHANISM OF INHIBITION OF FOOD INTAKE BY APO A-IV

An important question regarding the inhibition of food intake by apo A-IV is how does it work? Most of the evidence regarding this aspect of apo A-IV action is derived from the elegant work of Okumura et al.[46-48] Okumura et al. demonstrated that intracisternal injections of purified apo A-IV inhibited gastric acid secretion[46,47] and gastric motility[48] in rats in a dose-dependent manner. As shown in Figure 2, intracisternal infusion of physiological doses of apo A-IV markedly inhibited gastric motility as well as gastric acid secretion in a dose-dependent manner. The doses of apo A-IV used in these studies are thought to reproduce the levels of apo A-IV measured in cerebrospinal fluid after lipid feeding.[37] Intravenous infusion of similar doses of apo A-IV failed to elicit any gastric response. As proposed by Okumura et al.,[46] apo A-IV acts like an enterogastrone, that is, a humoral mediator released by the distal intestine that mediates the humoral inhibition of gastric acid secretion as well as motility by the ingestion of fat. At present, it is not clear if there is a direct link between the effects of apo A-IV on food intake and its effects on gastric function. Apo A-IV could directly influence central feeding mechanisms; alternatively, it could affect feeding through its effects on gastric function, especially via inhibition of gastric emptying.[49] Further work will be needed to clarify this important issue.

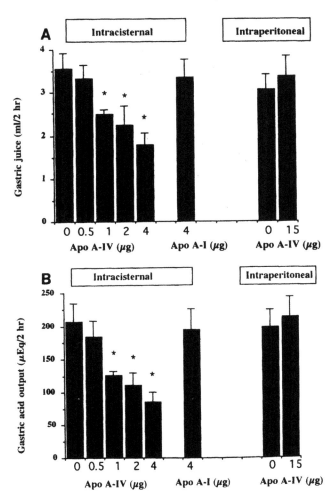

FIGURE 2   Effect of intracisternal injection of apo A-IV on gastric secretion in pylorus-ligated conscious rats. Under brief isoflurane anesthesia, rats received intracisternal injection of apo A-IV and the pylorus was ligated; 2 h after intracisternal injection, the animals were killed and the stomach was removed. (A) The volume was measured and (B) gastric acid output was determined. Each column represents the mean ± SE of four to nine animals. * $P < 0.05$ when compared with saline control. (From Okumura, T., Fukagawa, K., Tso, P., Taylor, I. L., and Pappas, T. N., *Gastroenterology,* 107, 1861–1864, 1994. With permission.)

6   IS APO A-IV A SHORT-TERM SATIETY SIGNAL?

Although the evidence presented thus far supports the idea that apo A-IV is a satiety signal released by the small intestine in response to lipid feeding, we have not yet considered the temporal relationship between intestinal synthesis and secretion of apo A-IV and satiety. Our question is whether the increases in plasma levels of apo A-IV in response to lipid feeding are of short enough time span and magnitude to elicit satiety. This point is important because hungry rats terminate ingestion within 30 min after meal initiation.[50] As shown in Figure 3, when a gastric bolus of 0.5 ml of a 20% Intralipid solution (0.1 g of TG) was fed, there was a rapid increase of the plasma level of apo A-IV within 15 min of ingestion of the lipid meal and the increment was statistically significant at 30 min after the meal.[51] Thus, Rodriguez et al.[51] concluded that there is a rapid increase in plasma apo A-IV level following the ingestion of lipid. Next, Rodriguez et al. showed the changes

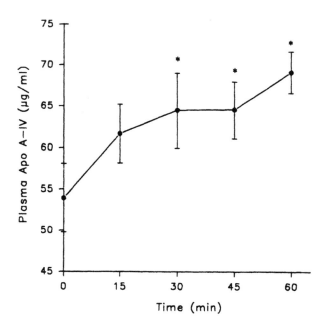

FIGURE 3 Plasma apo A-IV levels after a gastric bolus of 0.1 g of TG (0.5 ml of 20% Intralipid). Blood was collected every 15 min for 1 h after delivery of lipid. Plotted values are means ± SE concentration for six rats. *Significant change from basal value ($P < 0.05$). (From Rodriguez, M. D., Kalogeris, T. J., Wang, X. L., Wolf, R., and Tso, P., *Am. J. Physiol.*, 272, R1170–R1177, 1997. With permission.)

in plasma apo A-IV following a continuous intravenous infusion of different doses of apo A-IV. The fasting plasma apo A-IV levels were 51.2 ± 3.7 ($N = 6$) µg/ml. After the infusion of 60, 135, and 200 µg of purified apo A-IV, the increments of plasma apo A-IV levels averaged 2.9 ± 0.5, 8.6 ± 2.9, and 14 ± 0.6 µg/ml, respectively (Figure 4). Regression analysis showed a significant dose-dependent effect of exogenous dose of apo A-IV and the increment in plasma apo A-IV level ($t = 9.48, p < 0.001$). This increment in plasma apo A-IV level is similar to that reported by Fujimoto et al.,[36] showing that infusion of 135 and 200 µg of apo A-IV produced a significant dose-dependent inhibition of food intake. Rodriguez et al.[51] therefore concluded that the increase in plasma levels of apo A-IV produced in response to lipid meals are rapid and sufficient enough to produce satiety, therefore supporting a role for apo A-IV in the short-term control of food intake.

7   ROLE OF APO A-IV IN THE LONG-TERM CONTROL OF FOOD INTAKE

The experiments demonstrating the satiety effect of apo A-IV were performed in rats that had been deprived of food for relatively extended periods of time. This experimental paradigm ensures the absence of other food intake-inhibitory influences but may produce a situation of unusually high sensitivity in rats to the food intake-inhibitory effects of apo A-IV. This has been shown with other putative satiety factors. For example, severity of food deprivation increases the sensitivity to the food intake-inhibitory effects of a CCK analogue in dogs.[52] Although it has been demonstrated that intravenous administration of apo A-IV decreases food intake in *ad libitum* fed rats,[38] it is unknown to what extent endogenous apo A-IV may act to control intake under *ad libitum* conditions, especially during the dark phase when rats usually eat. In rats given free access to food, central administration of apo A-IV antiserum stimulates feeding during the light cycle.[37] Similar studies done during the dark phase may help clarify the role of apo A-IV in the control of food intake.

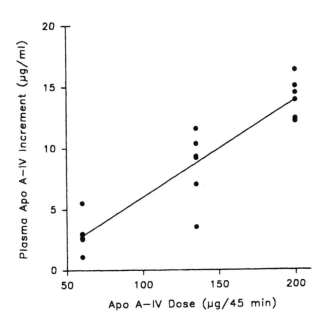

FIGURE 4   Effect of increasing dose of intravenously infused, purified apo A-IV on plasma apo A-IV levels. Values are increments in plasma concentration above levels measured in response to saline infusion; $n = 6$ rats. Linear regression analysis revealed a significant dose effect of apo A-IV on increment in plasma apo A-IV: $y = 0.079x - 1.889$; $r = 0.92$ (one tailed), $P < 0.001$. (From Rodriguez, M. D., Kalogeris, T. J., Wang, X. L., Wolf, R., and Tso, P., *Am. J. Physiol.*, 272, R1170–R1177, 1997. With permission.)

8   REGULATION OF APO A-IV SYNTHESIS AND SECRETION

It has been demonstrated that the formation of chylomicrons stimulates the synthesis and secretion of chylomicrons. However, the mechanism of how this occurs is still unknown. This would be a very interesting question for a molecular biologist to investigate. We have recently found that the formation and secretion of chylomicrons is not the only way to stimulate apo A-IV synthesis and secretion. Impetus for this finding came from studies in which we gave duodenal infusions containing graded doses of TG to rats and quantified both regional lipid distribution and mucosal synthesis of apo A-IV at various sites along the length of the intestine.[53] We found that despite significant amounts of lipid present only in the proximal half of the small intestine, apo A-IV synthesis was stimulated in the proximal three quarters of the gut, even in segments where there was negligible amount of lipid. This raises an interesting question of whether there may be factors other than lipid transport, but independent of the presence of lipid itself, that are capable of stimulating apo A-IV production by the gut. To address this interesting question, we have performed a series of experiments comparing the effects of proximal vs. distal intestinal infusion of lipid on the synthesis of apo A-IV in both the proximal and distal intestine. Kalogeris et al.[54] found that after duodenal lipid infusion, both apo A-IV synthesis and mRNA levels were elevated two- to threefold compared with control infusions (glucose-saline) in the jejunum, but ileal apo A-IV synthesis and mRNA levels were unaffected. Previous work from our laboratory demonstrated that under the conditions of our duodenal infusion, the amount of lipid reaching the ileum was negligible, suggesting that the lack of effect of duodenal lipid infusion on ileal apo A-IV expression was due to an insufficient exposure of the distal gut to lipid.

In contrast, delivery of lipid to the ileum stimulated both ileal as well as jejunal apo A-IV synthesis. Subsequent experiments in rats equipped with jejunal or ileal Thiry-Vella fistulas (segment of intestine isolated luminally from the rest of the gastrointestinal tract) demonstrated the following interesting findings: (1) ileally infused lipid elicits an increase in proximal jejunal apo A-IV synthesis independent of the presence of jejunal lipid and (2) both ileum and more distal sites may be involved in the stimulation. These results strongly suggest the existence of a signal arising from the distal gut capable of stimulating synthesis of apo A-IV in the proximal gut. These interesting findings have important physiological implications. The distal intestine is known to play an important role in the control of gastrointestinal function. Nutrient (especially lipid) delivered to the ileum results in the inhibition of gastric emptying,[55,56] decreased intestinal motility and transit,[56,57] and decreased pancreatic secretion.[58] Ileal nutrient also inhibits food intake.[59,60] The mechanism for these effects have been collectively termed the "ileal brake"[57] and appears to be related to the release of one or more peptide hormones from the distal intestine.[61-67] These effects have traditionally been considered operative only in the event of abnormal delivery of undigested nutrients to the distal gut, such as the malabsorptive state.[57] However, growing evidence supports the notion that, because of the rapid gastric emptying during the early phases of a meal, nutrient reaches the distal gut even under normal conditions.[51,55,60] We recently studied the intraluminal and mucosal distribution of a bolus of ^{3}H-triolein-labeled Intralipid (0.5 ml of a 20% emulsion) fed by gavage. By 15 to 30 min, radiolabeled lipid was spread evenly throughout the entire gut with 10 to 15% of the load recovered in the ileum and cecum combined. Presence of substantial amounts of lipids in these distal sites persisted for at least 4 h after the meal. When we examined apo A-IV synthesis by the small intestine, we found rapid stimulation (between 15 to 30 min) of apo A-IV synthesis throughout the intestine, including the ileum. This was associated with significant stimulation of lymphatic output and plasma levels of apo A-IV by 30 min after the gastric lipid load.[51] Consequently, it is becoming increasingly clear that even under normal conditions, a far greater length of the intestine could be involved both in the absorption of a lipid meal and the control of gastric and upper gut function than has been previously recognized. Thus, the ileal brake may play an important role in the normal control of gut function.

The most likely peptide to mediate the phenomenon of "ileal brake" is peptide tyrosine-tyrosine (PYY), which is a member of the peptide family including the pancreatic polypeptide (PP), neuropeptide Y (NPY), and fish pancreatic peptide Y (PY).[68] PYY is synthesized by the endocrine cells in the ileum and large intestine[61,69-72] and is released in response to intestinal nutrients, especially long-chain FAs.[70] However, PYY may not be the only mediator of the ileal brake. For example, perfusion of the intestine with fat produces a greater suppression of pentagastrin-stimulated acid secretion than does PYY[67] indicating that the enterogastrone effect of fat is mediated by more than one factor. We have now obtained evidence that PYY stimulates jejunal apo A-IV synthesis and secretion. Continuous intravenous infusion of physiological doses of PYY elicits significant increases in both synthesis and lymphatic transport of apo A-IV in rats.[73] We believe that this is the first demonstration indicating the involvement of a gastrointestinal hormone in the control of expression and secretion of an intestinal apolipoprotein thus bringing together two heretofore separate areas of research in gastrointestinal physiology.

9   CONCLUDING REMARKS

In summary, intestinal apo A-IV is a very interesting protein stimulated by dietary lipid with a potentially important physiological role in the integrated control of digestive function and ingestive behavior as well as a presumed role in cholesterol and lipoprotein metabolism. In terms of its role in the regulation of upper gut function as well as satiety, many issues still remain to be addressed. For instance, we do not have any information regarding what molecular form of apo A-IV is involved, i.e., free monomer, a homodimeric form (Weinberg and Spector, 1985), HDL (high-density lipoprotein)-bound apo A-IV, or perhaps apo A-IV–derived bioactive peptides. This important issue will have to be addressed before a comprehensive understanding of the physiology of apo A-IV can be achieved. Figure 5 integrates available evidence into a working model on the

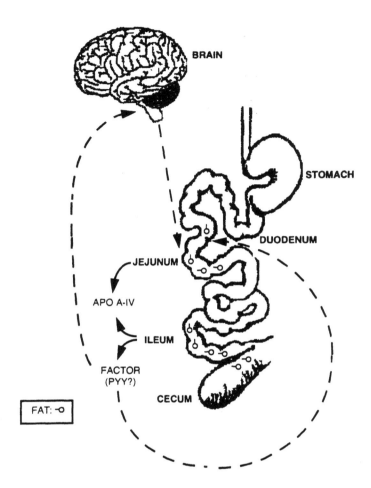

FIGURE 5 Proposed pathway for the control of apo A-IV synthesis by intestinal lipid. Lipid in the proximal intestine stimulates the synthesis and secretion of apo A-IV in a proximal-distal gradient in the intestine, depending upon the total lipid load. This effect is dependent on presence of lipid in the regions where apo A-IV is expressed. Lipid in the distal gut (ileum, cecum) also stimulates apo A-IV, both in the ileum and in the proximal jejunum. This latter effect is independent on the presence of jejunal lipid and is probably mediated by a signal released in response to the presence of lipid in the distal intestine. This signal is probably peptide YY (tyrosine-tyrosine) (PYY), although other gut hormones have not been unequivocally ruled out. (From Kalogeris, T. J., Rodriguez, M. D., and Tso, P., *Am. Soc. Nutr. Sci.*, p. 541S, 1997. With permission.)

overall control of apo A-IV synthesis and secretion following the ingestion of lipid. Direct exposure of both the jejunum and ileum to fat results in the stimulation of apo A-IV synthesis and secretion by the respective segments of intestine. Exposure of the ileum to lipid also results in the secretion of a factor (PYY) or factors, which in turn stimulate further synthesis and release of apo A-IV by the jejunum. Whether the stimulation of apo A-IV biosynthesis by lipid absorption and by PYY is mediated by the same molecular mechanism inside the enterocyte is being actively investigated in our laboratory.

ACKNOWLEDGMENTS

This work was supported by grants from the National Institutes of Health DK 32288 and DK 53444.

REFERENCES

1. Davenport, H. W., *Physiology of the Digestive Tract*, 3rd ed., Yearbook Medical Publishers, Chicago, 1971.
2. Hamosh, M., The role of lingual lipase in neonatal fat digestion, in *Development of Mammalian Absorptive Process,* Ciba Foundation Symposium 70 (new series), Elliot, K. and Whelan, J., Eds., Excerpta Medica, Amsterdam, 1979, 69–98.
3. Northfield, T. C. and Hofmann, A. F, Biliary lipid output during three meals and an overnight fat, *Gut*, 16, 1–11, 1975.
4. Borgström, B., Phospholipid absorption, in *Lipid Absorption: Biochemical and Clinical Aspects*, Rommel, K., Goebell, H., and Böhmer, R., Eds., MTP Press, London, 1976, 65–72.
5. Taylor, C. B. and Gould, R. G., A review of human cholesterol metabolism, *Arch. Pathol.*, 84, 2–14, 1967.
6. Gould, R. G., Jones, R. J., LeRoy, G. V., Wissler, R. W., and Taylor, C. B., Absorbability of β-sitosterol in humans, *Metabolism*, 18, 652–662, 1969.
7. Schonheyder, G. and Volquatz, K., Gastric lipase in man, *Acta Physiol. Scand.*, 11, 349–360, 1946.
8. Liao, T. H., Hamosh, P., and Hamosh, M., Fat digestion by lingual lipase: mechanism of lipolysis in the stomach and upper small intestine, *Pediatr. Res.*, 18, 402–409, 1984.
9. Paltauf, Esfandi, F., Holasek, A., Stereospecificity of lipases. Enzymatic hydrolysis of enantiomeric alkyl diacylglycerols by lipoprotein lipase, lingual lipase, and pancreatic lipase, *FEBS Lett.*, 40, 119–123, 1974.
10. Roberts, I. M., Montgomery, R. K., and Carey, M. C., Rat lingual lipase: partial purification, hydrolytic properties and comparison with pancreatic lipase, *Am. J. Physiol.*, 247, G385–393, 1984.
11. Senior, J. R., Intestinal absorption of fats, *J. Lipid Res.*, 5, 495–521, 1964.
12. Carey, M. C., Small, D. M., and Bliss, C. M., Lipid digestion and absorption, *Ann. Rev. Physiol.*, 45, 651–677, 1983.
13. Solomon, T. E., Control of exocrine pancreatic secretion, in *Physiology of the gastrointestinal tract,* 3rd ed., Johnson, L. R., Ed., Raven Press, New York, 1994, 1173–1207.
14. Rosenzweig, S. A., Miller, L. J., and Jamieson, J. D., Identification and localization of cholecystokinin-binding sites on rat pancreatic plasma membranes and acinar cells: a biochemical and autoradiographic study, *J. Cell Biol.*, 96, 1288–1297, 1983.
15. Brockerhoff, H., Substrate specificity of pancreatic lipase, *Biochim. Biophys. Acta*, 159, 296–303, 1968.
16. Mattson, F. H. and Beck, L. W., The specificity of pancreatic lipase or the primary hydroxyl groups of glycerides, *J. Biol. Chem.*, 219, 735–740, 1956.
17. Simmonds, W. J., Fat absorption and chylomicron formation, in *Blood Lipids and Lipoproteins: Quantitation, Composition and Metabolism*, Nelson, G. J., Ed., Wiley Interscience, New York, 1972, 705–743.
18. Morgan, R. G. H., Barrowman, J., and Borgström, B., The effect of sodium taurodeoxycholate and pH on the gel filtration behaviour of rat pancreatic protein and lipases, *Biochim. Biophys. Acta*, 175, 65–75, 1969.
19. Benzonana, G. and Desnuelle, P., Action of some effectors on the hydrolysis of long-chain triglycerides by pancreatic lipase, *Biochim. Biophys. Acta*, 164, 47–58, 1968.
20. Borgström, B. and Erlanson, C., Pancreatic lipase and colipase. Interactions and effects of bile salts and other detergents, *Eur. J. Biochem.*, 37, 60–69, 1973.
21. Maylie, M. F., Charles, M., Gache, C., and Desnuelle, P., Isolation and partial identification of a pancreatic colipase, *Biochim. Biophys. Acta*, 229, 286–2289, 1971.
22. Borgström, B., Erlanson-Albertson, C., and Wieloch, T., Pancreatic colipase — chemistry and physiology, *J. Lipid Res.*, 20, 805–816, 1979.
23. Erlanson-Albertsson, C., Pancreatic colipase. Structural and physiological aspects, *Biochim. Biophys. Acta*, 1125, 1–7, 1992.
24. Mei, J. and Erlanson-Albertsson, C., Role of intraduodenally administered enterostatin in rat: inhibition of food intake, *Obesity Res.*, 4, 161–165, 1996.

25. Okada, S., York, D. A., Bray, G. A., and Erlanson-Albertsson, C., Enterostatin (Val-Pro-Asp-Pro-Arg) the activation peptide of procolipase selectively reduces fat intake, *Physiol. Behav.*, 49, 1185–1189, 1996.

26. Lin, L., Okada, S., York, D. A., and Bray, G. A., Structural requirements for the biological activity of enterostatin, *Peptides*, 15, 849–854, 1994.

27. Mei, J. and Erlanson-Albertsson, C., Effect of enterostatin given intravenously and intracerebroventrically on high-fat feeding in rats, *Regul. Pept.*, 41, 209–218, 1992.

28. Sorhede, W., Erlanson-Albertsson, C., Mei, J., Nevalainen, T., Aho, A., and Sundler, F., Enterostatin in gut endocrine cells — immuocytochemical evidence, *Peptides*, 17, 609–614, 1996.

29. Sorhede, W., Mulder, H., Mei, J., Sundler, F., and Erlanson-Albertsson, C., Procolipase is produced in rat stomach — a novel source of enterostatin, *Biochim. Biophys. Acta*, 1301, 207–212, 1996.

30. Sorhede, W., Enterostatin in the Gastrointstinal Tract: Production and Possible Mechanism of Action, Ph.D. thesis, University of Lund, 1998.

31. Goldberg, I. J., Scheraldi, C. A., Yacoub, L. K., Saxena, U., and Bisgaier, C. L., Lipoprotein C-II activation of lipoprotein lipase. Modulation by apolipoprotein A-IV, *J. Biol. Chem.*, 265, 4266–4272, 1990.

32. Fielding, C. J., Shore, V. G., and Fielding, P. E., A protein cofactor of lecithin: cholesterol acyltransferase, *Biochem. Biophys. Res. Commun.*, 46, 1493–1498, 1972.

33. Bisgaier, C. L., Sachdev, O. P., Lee, E. S., Williams, K. J., Blum, C. B., and Glickman, R. M., Effect of lecithin: cholesterol acyl transferase on distribution of apolipoprotein A-IV among lipoproteins of human plasma, *J. Lipid Res.*, 28, 693–703, 1987.

34. Dvorin, E., Gorder, N. L., Benson, D. M., and Gotto, A. M., Apolipoprotein A-IV. A determinant for binding and uptake of high density lipoproteins by rat hepatocytes, *J. Biol. Chem.*, 261, 15714–15718, 1988.

35. Stein, O., Stein, Y., and Roheim, P., The role of apolipoprotein A-IV in reverse cholesterol transport studied with cultured cells and liposomes derived from an ether analog of phosphatidylcholine, *Biochim. Biophys. Acta*, 878, 7–13, 1986.

36. Fujimoto, K., Cardelli, J. A., and Tso, P., Increased apolipoprotein A-IV in rat mesenteric lymph after lipid meal acts as a physiological signal for satiation, *Am. J. Physiol.*, 262, G1002–G1006, 1992.

37. Fujimoto, K., Fukagawa, K., Sakata, T., and Tso, P., Suppression of food intake by apolipoprotein A-IV is mediated through the central nervous system in rats, *J. Clin. Invest.*, 91, 1830–1833, 1993.

38. Fujimoto, K., Machidori, H., Iwakiri, R., Yamamoto, K., Fujisaki, J., Sakata, T., and Tso, P., Effect of intravenous administration of apolipoprotein A-IV on patterns of feeding, drinking and ambulatory activity of rats, *Brain Res.*, 608, 233–237, 1993.

39. Hayashi, H., Nutting, D. F., Fujimoto, K., Cardelli, J. A., Black, D., and Tso, P., Transport of lipid and apolipoproteins A-I and A-IV in intestinal lymph of the rat, *J. Lipid Res.*, 31, 1613–1625, 1990.

40. Hayashi, H., Fujimoto, K., Cardelli, J. A., Nutting, D. F., Bergstedt, S., and Tso, P., Fat feeding increases size, but not number, of chylomicrons produced by small intestine, *Am. J. Physiol.*, 259, G709–G719, 1990.

41. Novin, D., Rogers, R. C., and Hermann, G., Visceral afferent and efferent connections in the brain, *Diabetologia*, 20, 331–336, 1981.

42. Niijima, A., Glucose-sensitive afferent nerve fibers in the hepatic branch of the vagus nerve in the guinea pig, *J. Physiol. (London)*, 332, 315–323, 1982.

43. Elsbourhagy, N. A., Walker, D. W., Paik, Y. K., Boguski, M. S., Freeman, M., Gordon, J. I., and Taylor, J. M., Structure and expression of the human apolipoprotein A-IV gene, *J. Biol. Chem.*, 262, 7973–7981, 1987.

44. Srivastava, R. A. K., Srivastava, N., and Schonfeld, G., Expression of low density lipoprotein receptor, apolipoprotein AI, AII and AIV in various rat organs utilizing an efficient and rapid method for RNA isolation, *Biochem. Int.*, 27, 85–95, 1992.

45. Fukagawa, K., Knight, D. S., Hamilton, K. A., and Tso, P., Immunoreactivity for apolipoprotein A-IV in tanycytes and astrocytes of rat brain, *Neurosci. Lett.*, 199, 17–20, 1995.

46. Okumura, T., Fukagawa, K., Tso, P., Taylor, I. L., and Pappas, T. N., Intracisternal injection of apolipoprotein A-IV inhibits gastric secretion in pylorus-ligated conscious rats, *Gastroenterology*, 107, 1861–1864, 1994.

47. Okumura, T., Fukagawa, K., Tso, P., Taylor, I. L., and Pappas, T. N., Mechanisms of action of intracisternal apolipoprotein A-IV in inhibiting gastric acid secretion in rats, *Gastroenterology*, 109, 1583–1588, 1995.

48. Okumura, T., Fukagawa, K., Tso, P., Taylor, I. L., and Pappas, T. N., Apolipoprotein A-IV acts in the brain to inhibit gastric emptying in the rat, *Am. J. Physiol.*, 270, G49–G53, 1996.

49. McHugh, P. R. and Moran, T. H., The stomach: a conception of its dynamic role in satiety, *Prog. Psychobiol. Physiol. Psychol.*, 11, 197–232, 1985.

50. Kalogeris, T. J., Reidelberger, R. D., and Mendel, V. E., Effect of nutrient density and composition of liquid meals on gastric emptying in feeding rats, *Am. J. Physiol.*, 270, R865–R871, 1983.

51. Rodriguez, M. D., Kalogeris, T. J., Wang, X. L., Wolf, R., and Tso, P., Rapid synthesis and secretion of intestinal apolipoprotein A-IV after gastric fat loading in rats, *Am. J. Physiol.*, 272, R1170–R1177, 1997.

52. Reidelberger, R. D., Kalogeris, T. J., and Solomon, T. E., Comparative effects of caerulein on food intake and pancreatic secretion in dogs, *Brain Res. Bull.*, 17, 445–449, 1986.

53. Kalogeris, T. J., Fukagawa, K., and Tso, P., Synthesis and lymphatic transport of intestinal apolipoprotein A-IV in response to graded doses of triglyceride, *J. Lipid Res.*, 35, 1141–1151, 1994.

54. Kalogeris, T. J., Tsuchiya, T., Fukagawa, K., Wolf, R., and Tso, P., Apolipoprotein A-IV synthesis in proximal jejunum is stimulated by ileal lipid infusion, *Am. J. Physiol.*, 270, G277–G286, 1996.

55. Lin, H. C., Doty, J. E., Reedy, T. J., and Meyer, J. H., Inhibition of gastric emptying of sodium oleate depends upon length of intestine exposed to the nutrient, *Am. J. Physiol.*, 259, G1031–G1036, 1990.

56. MacFarlane, A., Kinsman, R., and Read, N. W., The ileal brake: ileal fat slows small bowel transit and gastric emptying in man, *Gut*, 24, 471–472, 1983.

57. Spiller, R. C., Trotman, I. F., Higgens, B. E., Ghatel, M. A., Grimble, G. K, Lee, Y. C., Bloom, S. R., Misiewics, J. J., and Silk, D. B. A., The ileal brake — inhibition of jejunal motility after ileal fat perfusion in man, *Gut*, 25, 365–374, 1984.

58. Harper, A. A., Hood, J. C. A., and Mushens, J., Inhibition of external pancreatic secretion by intracolonic and intraileal infusions in the cat, *J. Physiol. (London)*, 292, 445–454, 1979.

59. Welch, I., Saunders, K., and Read, N. W., Effect of ileal and intravenous infusions of fat emulsions on feeding and satiety in human volunteers, *Gastroenterology*, 89, 1293–1297, 1985.

60. Meyer, J. H., Elashoff, J. D., Doty, J. E., and Gu, Y. G., Disproportionate ileal digestion on canine food consumption. A possible model for satiety in pancreatic insufficiency, *Dig. Dis. Sci.*, 39, 1014–1024, 1994.

61. Aponte, G. W., Fink, A. S., Meyer, J. H., Tatemoto, K., and Taylor, I. L., Regional distribution and release of peptide YY with fatty acids of different chain length, *Am. J. Physiol.*, 249, G745–G750, 1985.

62. Aponte, G. W., Park, K., Hess, R., Garcia, R., and Taylor, I. L., Meal-induced peptide tyrosine-tyrosine inhibition of pancreatic secretion in the rat, *FASEB J.*, 3, 1949–1955, 1989.

63. Jin, H., Gai, L., Lee, K., Chang, T. M., Li, P., Wagner, D., and Chey, W. Y., A physiological role of peptide YY on exocrine pancreatic secretion in rats, *Gastroenterology*, 105, 208–215, 1993.

64. Pappas, T. N., Debas, H. T., and Taylor, I. L., Peptide YY: metabolism and effect pancreatic secretion in dogs, *Gastroenterology*, 89, 1387–1392, 1985.

65. Pappas, T. N., Debas, H. T., Chang, A. M., and Taylor, I. L., Peptide YY release by fatty acids is sufficient to inhibit gastric emptying in dogs, *Gastroenterology*, 91, 1386–1389, 1986.

66. Pappas, T. N., Debas, H. T., and Taylor, I. L., The enterogastrone-like effect of peptide YY is vagally mediated in the dog, *J. Clin. Invest.*, 77, 49–53, 1986.

67. Savage, A. P., Adrian, T. E., Carolan, G., Chattarjee, V. K., and Bloom, S. R., Effects of peptide YY (PYY) on mouth to caecum intestinal transit time and on the rate of gastric emptying in healthy volunteers, *Gut*, 28, 166–170, 1987.

68. Larhammar, D., Söderberg, C., and Blomqvist, A. G., Evolution of the neuropeptide Y family of peptides, in *The Biology of Neuropeptide Y and Related Peptides*, Colmers, W. F. and Wahlestadt, C., Eds., Humana Press, Totowa, NJ, 1–41, 1993.

69. Adrian, T. E., Bacarese, H. A., Smith, H. A., Chohan, P., Manolas, K. J., and Bloom, S. R., Distribution and postprandial release of porcine peptide YY, *J. Endocrinol.*, 113, 11–14, 1987.

70. Hill, F. L. C., Zhang, T., Gomez, G., and Greeley, G. H., Jr., Peptide YY, a new gut hormone, *Steroids*, 56, 77–82, 1991.

71. Tatemoto, K., Isolation and characterization of peptide YY (PYY), a candidate gut hormone that inhibits pancreatic exocrine secretion, *Proc. Natl. Acad. Sci. U.S.A.*, 79, 2514–2518, 1982.
72. Taylor, I. L., Distribution and release of peptide YY in dog measured by specific radioimmunoassay, *Gastroenterology*, 88, 731–737, 1985.
73. Kalogeris, T. J., Sato, M., Wang, X., Monroe, F., and Tso, P., Intravenous infusion of peptide YY stimulates jejunal synthesis and lymphatic secretion of apolipoprotein A-IV in rats, *Gastroenterology*, 110, A809 (Abstr.).
74. Weinberg, R. B. and Spector, M. S., The self-association of human apolipoprotein A-IV. Evidence for an in vivo circulating dimeric form, *J. Biol. Chem.*, 260, 14279, 1985.

20 ENTEROSTATIN AS A REGULATOR OF FAT INTAKE

*David A. York, Ling Lin, Brenda K. Smith,
and Jian Chen*

CONTENTS

1   INTRODUCTION

A large number of peptides and neurotransmitters have been shown to affect food intake in rodents, but there is often little information on the component of feeding behavior that is affected to cause the observed response. One component of feeding behavior that is receiving increasing attention is the ability to select macronutrients and micronutrients. The ability to control the intake of specific nutrients is illustrated by the concept of sensory-specific satiety first proposed by Rolls and colleagues.[1] It is now recognized that overfeeding with one type of macronutrient may lead to a reduced intake of that particular macronutrient subsequently, but intake of other nutrients may not be altered. This is true for rat[2] as well as humans.[3]

In rodents, macronutrient intake is normally measured by giving animals either a choice of individual macronutrients (carbohydrate, protein, and fat) each supplemented with vitamins and minerals (a three-choice diet), or by providing selections from diets of different composition, e.g., high-fat (HF) and low-fat/high-carbohydrate (LF) diets which are equicaloric in protein. Using these types of experimental paradigm, we and others have shown that there is a wide variation in macro-nutrient choice between different strains of rodent and even within some strains.[4-7] Further, macro-nutrient selection may vary during the normal diurnal cycle of feeding,[8] it maybe altered in response to metabolic changes,[5,9] and it may be influenced by a range of hormones, neuropeptides, and aminergic transmitters.[8,10,11] The ability of salt[12] or amino acid–depleted rats[13] to select the missing dietary component is a further illustration of the existence of controls of the intake of individual

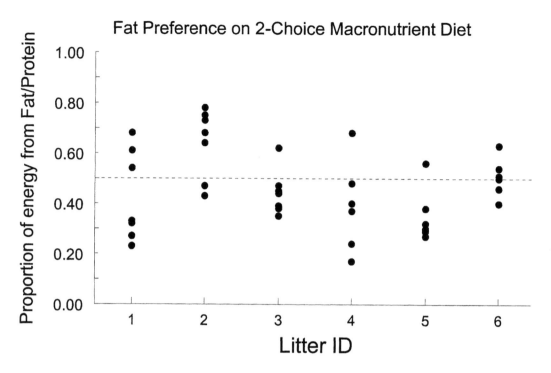

FIGURE 1   Within litter and between litter variation of diet selection by SD rats. Individual litters were reduced to six to seven males. At 8 weeks of age rats were given a two-choice diet of either fat/protein or carbohydrate/protein each supplemented with vitamins and minerals and equicaloric for protein. After a 7-day adaptation period, daily intake was measured over a 7-day period and the means for each individual rat are shown. (Data are the work of Smith and York.)

nutrients. Finally, we have recently shown that the intralitter variation in macronutrient selection is less than the interlitter variation in Sprague-Dawley (SD) rats (Figure 1). These and other data clearly suggest that both genetic and environmental factors influence macronutrient selection and that there must be a neurochemical basis for the regulation of the intake of each macronutrient.

Several of the neuropeptides/neurotransmitters that affect feeding behavior have selective effects on intake of specific macronutrients. One example of this is the peptide enterostatin which has a dose-dependent and selective effect to inhibit fat intake in a number of dietary paradigms.[14-19] Enterostatin is the aminoterminal pentapeptide of procolipase that is released by proteolytic activity when procolipase is converted into colipase.[20] There appears to be some polymorphism in this region so that two sequences, H_2N-Val-Pro-Asp-Pro-Arg-COOH and H_2N-Val-Pro-Gly-Pro-Arg-COOH have been identified in rats.[21,22] Similar polymorphism may be present in humans where the major form of the peptide is H_2N-Asp-Pro-Gly-Pro-Arg-COOH.

The ability to demonstrate selective macronutrient effects of a peptide in experimental situations does not prove that this is a physiological mechanism that is operative in the normal feeding cycle of the animal. In this chapter, we will suggest a number of experimental criteria that should be met to establish that a peptide has a physiological role in controlling the amount of a macronutrient that is ingested and review the status of enterostatin in meeting these criteria.

2   FEEDING RESPONSE TO ENTEROSTATIN

The first criterion for establishing the physiological role of a peptide on feeding behavior is that it will inhibit food or macronutrient intake in a home-cage environment. Enterostatin clearly meets

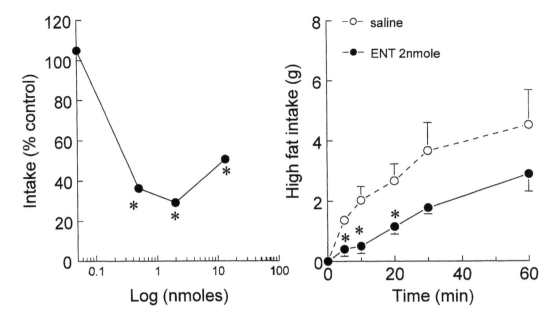

FIGURE 2 Dose dependence and time course of the response of food intake to near-celiac arterial infusions of enterostatin or saline vehicle. The dose–response data are for food intake 10 min after injection of overnight fasted rats. Immediately after injection of enterostatin or vehicle, the HF diet (56% energy fat) was provided. The time course illustrates the effects of enterostatin given at a dose of 2 nmol. *$p < 0.01$ compared to control value.

this criterion. In rats adapted to a three-choice macronutrient diet of fat, carbohydrate, and protein[4,17] enterostatin reduces intake of the fat macronutrient but has no effects on either carbohydrate or protein intakes. On a two-choice HF and LF diet paradigm, enterostatin only reduces intake of the HF diet not of the LF diet.[15,16] Similarly, enterostatin reduces intake of single diets when the diet is high, but not when it is low, in fat content.[23] The ability to inhibit fat intake selectively on a two- or three-choice feeding paradigm has been demonstrated after both intraperitoneal[4,14,17] and intracerebroventricular (icv)[14,23,24] injection of enterostatin. Enterostatin is also effective in reducing the ingestion of HF diets after intragastric,[25] intraduodenal,[26] intravenous,[24] and intracarotid[27] injection, but the ability to inhibit fat intake selectively in choice feeding studies has not been investigated for these routes of administration. We[27] have recently extended these observations to show that near celiac arterial injection of very low doses of enterostatin (10 to 1000 pmol) dose dependently suppresses intake of an HF diet in SD rats within minutes of administration providing strong evidence for a gastric/proximal duodenum site of action for the peptide (Figure 2). Further, we (Lin and York, unpublished observations) have recent evidence to show that near celiac arterial enterostatin has a selective fat effect; it reduces intake of HF diet but increases intake of LF diet in rats selecting from a two-choice HF/LF diet.

While the majority of the feeding studies with enterostatin have been performed in overnight fasted rats that have been previously adapted to the experimental diets, the selective effects toward dietary fat have been shown in free-feeding rats injected at the start of the dark cycle.[17] The potency of the action of enterostatin is reflected in its long duration of action, its effect on feeding lasting up to 6 h after a single injection in rats adapted to a 6-h feeding schedule[26,28] or lasting up to 24 h after a single injection in *ad libitum* fed rats.[17,18] Chronic icv administration of enterostatin from miniosmotic pumps also attenuates the daily intake of dietary fat in rats fed either a single-choice HF diet[29] or a two-choice HF/LF diet.[15] The decrease in daily food intake was accompanied by a reduction in fat deposition and body weight gain. However, in SD rats chronically treated with enterostatin and fed a LF diet for 7 days, there was no significant reduction in energy intake or change in body weight gain.[24] An intriguing characteristic of the response to enterostatin in both acute and chronic studies is that the reduction in intake of dietary fat is not compensated by any increase in the intake of other macronutrients when a choice is available. We have speculated that this may result from a concomitant increase in corticotropin releasing hormone (CRH) secretion since enterostatin is known to activate the hypothalamic–pituitary–adrenal axis.[29-31]

3   SPECIES DEPENDENCE OF THE FEEDING RESPONSE

We would anticipate that a peptide which had a significant role in the regulation of feeding behavior would show such an effect across a range of animal species. The majority of studies with enterostatin has been performed with rats where it is effective in the majority of strains tested, the exceptions appearing to be very thin strains which habitually have a very low preference for dietary fat and may have high levels of endogenous production of enterostatin. Enterostatin has also been shown to reduce food intake in rabbits,[18,31] sheep,[32] and baboons[33] but all of these studies were performed with single-choice diets. In humans, intravenous enterostatin had no effect on food intake.[34] However, the design of this study may have contributed to this negative result. Enterostatin was administered intravenously, a route that in rats has been shown to have a long delay (>2 h) in the response, yet only ingestion of an immediate meal was studied. Further, compared with studies in the baboon, the dose given was very high and enterostatin has a U-shaped dose–response curve in rodents.[14,35] Further, it is now clear that chronic exposure to high levels of fat intake is a prerequisite for the enterostatin inhibition of fat intake.[36] The need for good data on circulating levels of enterostatin is critical for the design of definitive studies in humans. We attempted to circumvent this problem by studying the response to oral enterostatin in male volunteers that were adapted to a 40% (by energy) fat diet for 2 weeks before study. While the majority of subjects studied that had stable baseline intakes of the test meal did show a reduction in intake of the test meal (7 of 11) in response to enterostatin there was still a large variation in response and 4 individuals showed an increase in food intake (York, Bray, Smeets, and Geiselman, unpublished observations).

4   DOSE DEPENDENCE OF THE RESPONSE TO ENTEROSTATIN

A physiological regulator of feeding behavior must be effective at dose levels that are present in the animal; i.e., it is not a pharmacological response. There is a lack of data to satisfy this criterion for enterostatin at this time. There are a number of reasons for this. It has proved difficult to raise antibodies that are selective to enterostatin and that yield believable values for circulating or tissue levels when used in an immunoassay. Thus, it has been suggested that plasma serum of enterostatin are 5 to 40 nM in humans and rats,[30,37,38] cerebral spinal fluid enterostatin is 18 to 92 ng/ml,[39] and brain enterostatin levels are ~2.5 nmol/g tissue,[39] all of which appear to be very high. A further complication is the suggestion that there are multiple forms of enterostatin in rats and in humans because of genetic polymorphisms in the enterostatin region of the procolipase parent molecule.[21,22] Despite these problems, enterostatin-like immunoreactivity in human serum and urine has been shown to increase after a meal in a biphasic manner and to be increased in lymph of cats[40] and serum of rats after feeding.[41]

It is inherently difficult to assess the concentration of a peptide at its site of action in proximity to its receptor. Thus, the ability of an antagonist to alter normal feeding behavior is often taken as a surrogate indicator of activity at normal physiological concentrations. β-Casomorphins 1–7 and 1–5 have bridged proline sequences similar to enterostatin, stimulate fat intake, and antagonize the response to enterostatin.[42] It is possible that these peptides are antagonists for enterostatin, although this will not be known until the enterostatin receptors that modulate fat intake are identified (see Section 8).

5   THE BEHAVIORAL SEQUENCES OF NORMAL FEEDING

If enterostatin is a physiological inhibitor of fat intake, then it will not induce nausea or an alternative behavior that would prevent feeding. To rule out these alternatives, the ability of enterostatin to induce a conditioned taste aversion (CTA) and the effects on the normal behavioral sequence of satiety have been studied. In overnight fasted rats, the inhibition of fat intake by peripheral enterostatin has been related to an early induction of satiety[43]; the time spent eating, grooming, and in physical activity were decreased, while the period of resting or sleeping was significantly increased. The behavior of the animals following injection of enterostatin was thus similar to natural satiety, suggesting that enterostatin was not mediating its effects through an alternative behavior.

The effect of central enterostatin injected either icv or locally into the amygdala and paraventricular nucleus (PVN) on the microstructure of feeding has been investigated recently by use of an automated feeding apparatus.[44] In normal feeding, at the beginning of the dark cycle, enterostatin delayed the onset of the first meal, shortened its duration, and reduced its size. Subsequent meals were unaffected, and there was no compensatory increase for the smaller initial meal. Similar effects were observed when enterostatin was given after an overnight starvation or at the beginning of the meal time in 6-h meal-fed rats, with one exception: the delayed onset of feeding was not observed in these situations.

It also appears unlikely that the enterostatin effects on food intake are related to induction of nausea. First, if this were the case, it would be anticipated that intake of all macronutrients, not just fat, would be reduced. Further, enterostatin does not induce a CTA after administration into either the brain ventricles[24] or directly onto the amygdala,[44] an area that previously has been associated with taste aversions.

6   FEEDBACK REGULATION OF ENTEROSTATIN SYNTHESIS AND SECRETION

A physiological regulator of fat intake must be responsive to the level of dietary fat. Thus, we would anticipate that enterostatin synthesis and secretion would be responsive to dietary fat. Enterostatin is the N-terminal pentapeptide of procolipase that is released during tryptic cleavage of the parent 106 amino acid peptide to colipase.[45] No other source of enterostatin is known. To date, the only tissue sites known to express the procolipase gene are exocrine pancreas, gastric mucosa, and possibly the duodenal mucosa.[45-47] We have been unable to detect procolipase mRNA in brain tissue. The role of colipase in the intestinal lumen is to activate pancreatic lipase, to which it binds forming a 1:1 molar complex[45] to facilitate fat digestion. The initial finding of procolipase mRNA in the rat antral stomach,[46] the subsequent identification of procolipase-like immunoreactivity in the chief cells[47] and in enterochromaffin cells of the gastrointestinal tract, and the measurement of procolipase and enterostatin in rat gastric juice[47] and the duodenum[37] suggest that the stomach is an alternative source of enterostatin and colipase. Immunohistochemical approaches have also revealed the presence of enterostatin in enterochromaffin cells of the gastrointestinal tract, being most abundant in the antral part of the stomach, less in the duodenum and jejunum, and with only a few cells in the ileum[48].

The ability to assay tissue colipase mRNA or colipase and enterostatin levels provide alternative methods for studying the effects of dietary fat on enterostatin synthesis and or secretion. Procolipase mRNA and protein are increased in the pancreas,[21,49] and procolipase mRNA levels in the gastric mucosa (Figure 3) are increased after feeding of fat. This increased procolipase synthesis occurs in proportion to the amount of fat ingested[37] and occurs within 24 h of presentation of the HF diet.[21] It is consistent with enterostatin acting as a long-term feedback signal to attenuate the levels of fat ingested. Conversely, short-term starvation reduced pancreatic procolipase levels significantly in SD rats.[50]

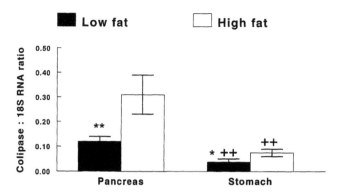

FIGURE 3 The effect of dietary fat on the expression of procolipase mRNA in the pancreas and gastric mucosa of OM rats adapted to either the HF (56% energy as fat) or LF (12% energy as fat) diet for 7 days. Food was available *ad libitum*. Values represent Mean ± SEM for four rats in each group. $*p < 0.05$; $**p < 0.01$ compared with HF group; $++p < 0.01$ compared with respective pancreas group. (Data are the work of Chen and York.)

The mechanism through which enterostatin/procolipase production is increased by dietary fat is not known. Gastric inhibitory polypeptide may be an important modulator of this response since it is released in the gastrointestinal tract during a fat-containing meal and has been shown to stimulate procolipase synthesis.[51] Two other key hormones in energy metabolism, insulin and corticosterone, inhibit procolipase synthesis,[52-54] whereas cyclic adenosine monophosphate (AMP) has a stimulatory effect.[55] The regulation of enterostatin synthesis by corticosterone was demonstrated further by the response to adrenalectomy which increased procolipase gene expression[30] as well as procolipase synthesis[56] concomitant with reduced intake of an HF diet and a reduction in body weight gain. However, since HF diets normally activate the hypothalamic–pituitary–adrenal axis,[11] corticosteroids would appear to counteract the fat-induced increase of procolipase production and promote fat ingestion.

7   RELATIONSHIP OF ENDOGENOUS SYNTHESIS OF ENTEROSTATIN TO HABITUAL FAT INTAKE

With the development of transgenic techniques it has become conventional to study the effects of peptides on feeding behavior either through overexpression or knockout of the relevant gene or receptor gene. To date, these genetic approaches have not been applied to the procolipase gene and enterostatin production. However, there are a number of studies in which either natural or induced differences in procolipase gene expression have been associated with differences in fat ingestion.

Various rat strains have a natural preference for dietary fat or dietary carbohydrate. The Osborne-Mendel (OM) rat has a strong dietary preference for fat in contrast to the S5B/Pl rat which prefers dietary carbohydrate and highly restricts its fat intake.[18,57] S5B/Pl rats have twofold to threefold higher levels of pancreatic procolipase than OM rats, and voluntary fat intake has been reciprocally related to pancreatic procolipase levels. This relationship holds both within and across rat strains when rats were allowed to choose their dietary macronutrients, higher endogenous pancreatic procolipase activities being associated with lower intakes of dietary fat, and vice versa. This relationship has been confirmed in SD rats.[50] Since dietary fat itself induces procolipase synthesis, these data strongly imply a feedback role of enterostatin release on fat intake. This was further supported by the observation that the response to exogenous enterostatin was also strain dependent; S5B/Pl rats which have a high pancreatic procolipase activity do not respond to exogenously administered enterostatin in contrast to OM rats which had a robust response and had low endogenous levels of enterostatin release.[18] This suggests that the tonic inhibitory effect of enterostatin is already maximal in S5B/Pl rats. A similar link between fat and carbohydrate preferences and response to exogenous enterostatin has been reported in outbred rats.[7]

The association between low levels of procolipase gene expression and the magnitude of the response to exogenous peptide has also been observed in obese fa/fa rats.[31] Pancreatic colipase mRNA levels of obese fa/fa rats are low compared with lean fa/? rats.[31] Adrenalectomy of the obese fa/fa rat increased the levels of procolipase mRNA concomitant with an inhibition of further obesity and abolished the feeding response to exogenous enterostatin. Conversely, corticosterone treatment reduces endogenous production of colipase but enhances the response to exogenous enterostatin.[58]

Further studies, using direct assay of circulating enterostatin, will be necessary to confirm the hypothesis of a feedback system through which endogenous levels of enterostatin secretion determine appetite for dietary fat.

8   ENTEROSTATIN RECEPTORS

For a physiological peptide signaling system, it is necessary to identify the presence of specific receptors that selectively bind the peptide and show that the characteristics of these receptor or receptors (e.g., Kd) and their location is consistent with the biological response at the concentration of peptide agonist that is present at the site of action. Further, we anticipate that other receptor agonists or antagonists will also affect the biological response in relation to their relative affinities. This aspect of research with enterostatin remains severely deficient. The absence of a tyrosine residue in enterostatin and lack of activity of some tyrosine-substituted analogues[14] has been a major limitation to the development of a radiolabeled ligand that could be used for binding studies and receptor isolation. To date, the enterostatin receptor has not been identified. However, we anticipate from behavioral and metabolic studies that the receptor(s) will be located in brain, endocrine pancreas, and the gastrointestinal tract and possibly also the pituitary because of the effects of enterostatin on feeding behavior, insulin secretion,[59,60] and activity of the hypothalamic–pituitary–adrenal axis.[29,30]

Enterostatin does not bind to either the galanin or neuropeptide Y (NPY) Y1 receptors[23] or compete for binding to kappa-opioid receptors.[42] We identified a low-affinity enterostatin site (Kd = 0.1 μM) on a brain membrane preparation that was competed by the β-casomorphins 1-7, 1-5, and 1-4.[42] This is of particular interest since the β-casomorphins 1-7 and 1-5 have opposing effects to enterostatin and stimulate the intake of dietary fat.[42] The dose–response curve to enterostatin is U-shaped exhibiting an inhibition of food intake at lower doses, but stimulation of food intake at higher doses.[26,35,61] This biphasic response may be explained in a number of ways. There could be two receptor subtypes with differing affinity binding sites for enterostatin, one high-affinity, suggested to be inhibitory to fat intake, and one low-affinity binding site, possibly the casomorphin-binding receptor that stimulates fat intake.[42] Alternatively, at higher doses enterostatin may become a partial antagonist. However, since enterostatin appears to be biologically active on food intake at extremely low doses compared with numerous other peptides[62] and since it will inhibit insulin secretion from isolated pancreatic islets at doses of 10^{-10} to 10^{-6} M[60] it is unlikely that this low-affinity casomorphin-binding site is the biologically important enterostatin receptor that inhibits fat intake and insulin secretion.

9   THE SIGNALING PATHWAY

9.1   PERIPHERAL SITE OF ACTION

Enterostatin, as a physiological regulator, must activate or inhibit a neurochemical pathway that is consistent with its effects on fat intake. Appropriate modulation of this signaling pathway should also alter fat intake. There appear to be two major sites of action for enterostatin, a peripheral site in the gastroduodenal area and a central site that may be located in the amygdala and/or PVN.

Food intake is reduced by oral procolipase,[61] and fat intake is reduced by both intragastric and intraperitoneal enterostatin in rats.[25,26] Further, it is clear that enterostatin is released peripherally either from the exocrine pancreas or the gastrointestinal tract in response to dietary fat.[44-46] The afferent vagus is essential for the response to peripheral enterostatin. Transection of the hepatic vagus and capsaicin treatment completely blocked the inhibitory response to intraperitoneal enterostatin on HF diet consumption in rats after overnight starvation.[63,64] The importance of neuronal transmission for the feeding response of enterostatin was also suggested by the attenuation of the feeding response to intraintestinal enterostatin after tetracain administration to block peripheral nerve endings.[26] These studies thus suggest neuronal transmission of the enterostatin response from the intestine to the brain. However, it is not clear how enterostatin activates the local vagal nerve terminals. It is possible that lumenal enterostatin is transported across the mucosal layers to activate the nerve terminals in submucosal areas directly. Alternatively, and probably more likely, enterostatin could activate a paracrine system to affect vagal activity indirectly.

Peripheral enterostatin induced c-*fos* immunoreactivity in specific brain sites including the nucleus tractus solitarius, parabrachial, paraventricular, and supraoptic nuclei, and this effect was absent in rats with selective hepatic vagotomies.[63] Serotonergic activity was also enhanced in a number of these brain regions by peripheral enterostatin.[64] This observation is particularly relevant since there is accumulating evidence to show that serotonin is an inhibitor of fat intake and that this response may be modulated at the level of the PVN (see Reference 65 and Chapter 27 in this book).

The enterostatin inhibition of food intake is only normally seen in rats adapted to an HF diet or allowed access to fat in a dietary choice paradigm. The fat signal that is perceived by the rat is unknown. The possibility that afferent vagal information was required for this "fat signal" was ruled out by the demonstration that capsaicin treatment, which destroys small nonmyelinated nerve fibers, prevented the response to peripheral enterostatin but not to intracerebroventricular enterostatin.[66]

Enterostatin also has effects on gastrointestinal motility and gastric emptying. The inhibition of gastric emptying is only observed after icv administration of enterostatin and not after either intraperitoneal or intragastric administration,[67] suggesting that enterostatin may also affect efferent vagal activity. However, the inhibitory effect of enterostatin on the consumption of an HF diet is not related to the slowdown of gastric emptying. Enterostatin also has direct effects on pig intestine to prolong the quiescent phase I period of peristalsis,[68] which would slow down the absorption of nutrients and prolong intestinal transit time. The outcome of this response might be a greater or more prolonged stimulation of other gastrointestinal satiety signals.

9.2 CENTRAL SITE OF ACTION

Initial mapping studies to identify the central site of action of enterostatin have shown that it is effective after local injection onto the central bed nucleus of the amygdala and the PVN, but not onto the ventromedial nucleus, lateral hypothalamus, or nucleus tractus solitarius.[28] Both κ-opioidergic and serotonergic components have been implicated in the pathway that responds to central enterostatin. κ-Opioid agonists stimulate feeding and induce a concomitant preference for fat intake[69] and reverse the enterostatin-induced inhibition of feeding at low doses that had no independent effects on food intake.[70] Conversely, the κ-opioid antagonist nor-BNI mimics the effects of enterostatin, and subthreshold doses of nor-BNI and enterostatin, which individually were without effect, reduced fat intake when they were given in combination.[19] Opiates are recognized to have an important role in preference acquisition.[71] Recent mapping studies (Lin and York, unpublished observations) indicate that this κ-opioidergic component of the enterostatin pathway is localized in the nucleus tractus solitarius region of the forebrain, a region that receives afferent information from both the oral cavity and the gastrointestinal tract (Figure 4).

The serotonin antagonist metergoline attenuates the inhibition of fat intake by enterostatin.[72] This effect may be modulated through 5-HT1 subtype receptors since the response to enterostatin

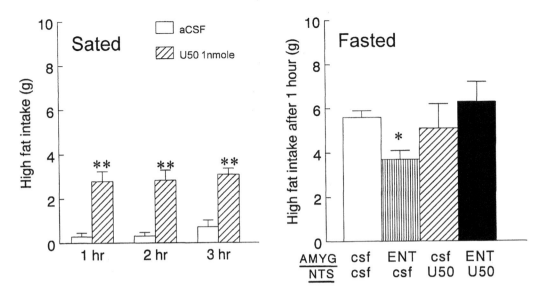

FIGURE 4 The effect of U50488 (1 nmol) administered into the nucleus tractus solitarius (NTS) on food intake of sated rats and on the feeding response to enterostatin (0.01 nmol) injected into the amygdala of overnight-fasted rats. All rats were adapted to an HF diet for 10 days before the experiment. The data show mean ± SEM ($n = 5$/group) for food intake. The right-hand panel shows food intake 1 h after simultaneous injections of enterostatin (0.01 nmol) or saline vehicle into the amygdala or U50488 (1 nmol) or saline vehicle into the NTS. *$p < 0.05$, **$p < 0.01$ compared with control group that received artificial cerebrospinal fluid (csf).

was not attenuated by the 5-HT2 receptor antagonist ketanserin (Figure 5). The stimulation of feeding by the κ-agonist U50488 was blocked by D-fenfluramine.[73] However, since the inhibition of fat intake by the kappa-opioid antagonist nor-binaltorphamine (nor-BNI) was not attenuated by metergoline,[73] it suggests that the basal activity of this serotonergic pathway may be relatively low and that it requires activation (enterostatin or D-fenfluramine treatment) for expression of its effect on fat intake.

10 BRAIN ENTEROSTATIN

If enterostatin is not synthesized in the brain, then circulating enterostatin must provide the signal to the central nervous system (CNS) systems. If this is so, its appearance into the circulation and its uptake into the CNS must be consistent with this from both temporal and quantitative viewpoints.

Immunoreactive enterostatin appears in the gastrointestinal tract of rats[37] and humans[74] after a meal containing fat or in the former case after cholecystokinin (CCK) stimulation. However, absorption across the intestine appears to be limited and slow.[75] Further, the major route of absorption is thought to be through the lymphatics.[40] This route would also only lead to a slow increase in plasma enterostatin after a meal. At this time there is no detailed information of the changes in plasma enterostatin or brain uptake of enterostatin after a meal that would allow a temporal comparison with the termination of feeding and the development of satiety. The data that are available suggest the rise in plasma immunoreactive-like enterostatin activity is slow and does not peak until at least 60 min after feeding.[38,41] These data are inconsistent with an increase in circulating enterostatin playing any role in the termination of the immediate meal.

We have proposed that the vagal afferent signaling pathway that is activated by enterostatin is important in terminating a meal containing high levels of fat. If this is the case, then what is the

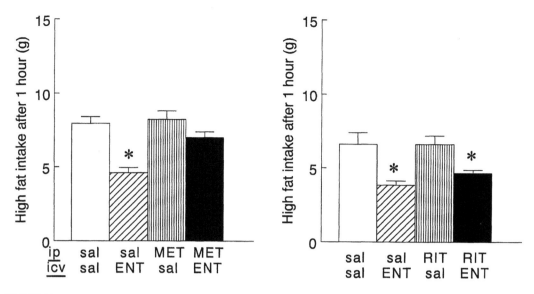

FIGURE 5   The effect of metergoline (1 mg/kg) or ketanserin (0.5 mg/kg) on the feeding response to lateral ventricular enterostatin (1 nmol). Overnight fasted rats that had been adapted to a HF diet for 10 days were injected with the serotonin antagonists 15 min before enterostatin. Food was made available at this time. Values represent means ± SEM for five to six rats/group. *$p < 0.05$ compared to vehicle (saline) control group.

function of the central system that is so sensitive to enterostatin? Using cDNA probes for procolipase, we have been unable to demonstrate the presence of procolipase mRNA in the CNS[48] although enterostatin-like immunoreactivity has been identified at high levels in the cerebrospinal fluid of rats.[39] This suggests that circulating enterostatin may gain access to the CNS either through a specific transport system or through the circumventricular organs outside the blood–brain barrier. We hypothesize that this central system is important in determining the appetite for dietary fat consistent with the evidence previously described that endogenous production of enterostatin appears to be reciprocally related to voluntary selection of fat across and within rat strains. Only future work and the development of good assays for enterostatin will enable us to evaluate this hypothesis.

Another caveat to add to this hypothesis is that we do not currently know what is the biologically active form of enterostatin *in situ*. First, there appears to be some polymorphism in the enterostatin region of the procolipase gene.[21,22] Second, a structure–function analysis of the feeding response to enterostatin analogues showed that the critical amino acid sequence was the aspartyl-proline (DP) structure at amino acids 3 to 4.[14] This structure, when cyclized into the diketopiperazine peptide, cyclo-Asp-Pro, has all the biological activities of enterostatin. The two peptides have similar potency, both are selective toward fat, both have U-shaped dose–response curves, and both are effective after peripheral and central administration.[14] The actual production of the DP peptide was observed in studies of the metabolism of enterostatin,[75] opening the possibility for a physiological role of the dipeptide in appetite regulation. However, it is not clear at this time if enterostatin must be converted into cyclo-aspartyl-proline for its biological activity or whether both molecules have a similar three-dimensional structure that allows them to interact individually with the receptor.

11   SUMMARY

The experimental evidence to date suggests that enterostatin has both a peripheral and central site of action (Figure 6). The peripheral signal is rapid and transmitted through the afferent vagus, via the nucleus tractus solitarius (NTS) to the medial hypothalamus and other regions. This system has

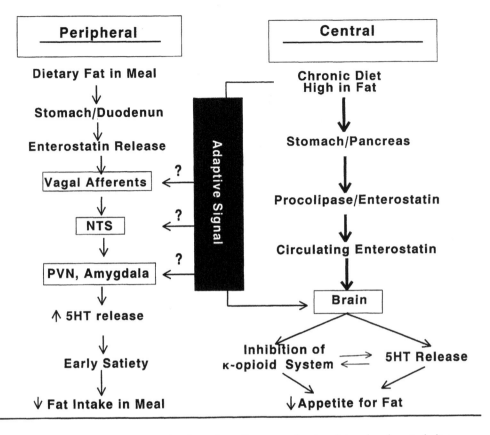

FIGURE 6   General scheme for the action of enterostatin at peripheral and central sites.

the capability of regulating the amount of fat eaten in the immediate meal. The central response is dependent upon a permissive signal that is associated with the chronic consumption of dietary fat.[36] The identify of this signal is not known Among the possibilities, there might be a complementary change in secretion of a second circulating factor or the induction of a specific receptor or signaling system associated with long-term ingestion of dietary fat. Leptin does not appear to be the signal since enterostatin is very effective in the obese Zucker fa/fa rat[31] which has an impaired leptin signaling system due to a mutation in the leptin receptor. This central mechanism may have the physiological role of regulating the appetite for fat since current evidence suggests that it is dependent upon the uptake of enterostatin from the circulation.

At this time there are large gaps in our knowledge of the biology of enterostatin. Undoubtedly, the identification of the receptor(s) and development of good assays will benefit this field and help to show if enterostatin has a role in the physiology of macronutrient selection.

REFERENCES

1. Rolls, B. J., Hetherington, M., and Burley, V. J., The specificity of satiety: the influence of foods of different macronutrient content on the development of satiety, *Physiol. Behav.,* 43, 145, 1988.
2. Leibowitz, S. F., Specificity of hypothalamic peptides in the control of behavioural and physiological processes, *Ann. N.Y. Acad. Sci.,* 739, 12, 1994.
3. Van Amelsvoort, J. M., Van Stratum, P., Kraal, J. H., Lussenberg, R. N., and Houtsmuller, V. M. T., Effects of varying the carbohydrate: fat ratio in a hot lunch on postprandial variables in male volunteers, *Br. J. Nutr.,* 61, 267, 1988.

4. Okada, S., York, D. A., Bray, G. A., Mei, J., and Erlanson-Albertsson, C., Differential inhibition of fat intake in two strains of rat by the peptide enterostatin, *Am. J. Physiol.,* 262, R1111, 1992.

5. Smith, B. K., Berthoud, H., York, D. A., and Bray, G. A., Differential effects of baseline macronutrient selection after galanin, NPY, and an overnight fast, *Peptides,* 18, 207, 1997.

6. Smith, B. K., West, D. B., and York, D. A., Carbohydrate vs. fat intake: differing patterns of macronutrient selection in two inbred mouse strains, *Am. J. Physiol.,* 272, R357, 1997.

7. Cook, C., Gatchair-Rose, A., Herminghuysen, D., Nair, R., Prasad, A., Mizuma, H., and Prasad, C., Individual differences in the macronutrient profile of outbred rats; implications for nutritional, metabolic and pharmacological studies, *Life Sci.,* 55, 1463, 1994.

8. Leibowitz, S. F., Specificity of hypothalamic peptides in the control of behavioural and physiological processes, *Ann. N.Y. Acad. Sci.,* 739, 12, 1994.

9. Singer, L. K., York, D. A., and Bray, G. A., Feeding response to 2-mercaptoacetate in Osborne-Mendel and S5B/Pl rats maintained on a macronutrient self-selection diet, *Appetite,* 7, 305, 1996 (Abstr.).

10. Bray, G. A., Peptides affect the intake of specific nutrients and the sympathetic nervous system, *Am. J. Clin. Nutr.,* 55, 2655, 1992.

11. Leibowitz, S. F., Hypothalamic paraventricular nucleus: Interaction between $\alpha2$-noradrenergic system and circulating hormones and nutrients in relation to energy balance, *Neurosci. Biobehav. Rev.,* 12, 101, 1998.

12. Fluharty, S. J. and Epstein, A. N., Sodium appetite elicited by intracerebroventricular infusion of angiotensin II in the rat: synergistic interaction with systemic mineralocorticoids, *Behav. Neurosci.,* 97, 746, 1983.

13. Gietzen, D. W., Neural mechanisms in the response to amino acid deficiency, *J. Nutr.,* 123, 610, 1993.

14. Lin, L., Okada, S., York, D. A., and Bray, G. A., Structural requirements for the biological activity of enterostatin, *Peptides,* 15, 849, 1994.

15. Lin, L., Chen, J., and York, D. A., Chronic icv enterostatin preferentially reduced fat intake and lowered body weight, *Peptides,* 18, 657, 1997.

16. Erlanson-Albertsson, C., Jie, M., Okada, S., York, D., and Bray, G. A., Pancreatic procolipase propeptide, enterostatin, specifically inhibits fat intake, *Physiol. Behav.,* 49, 1191, 1991.

17. Okada, S., York, D. A., Bray, G. A., and Erlanson-Albertsson, C., Enterostatin (Val-Pro-Asp-Pro-Arg) the activation peptide of procolipase selectively reduces fat intake, *Physiol. Behav.,* 49, 1185, 1991.

18. Okada, S., York, D. A., Bray, G. A., Mei, J., and Erlanson-Albertsson, C., Differential inhibition of fat intake in two strains of rat by the peptide enterostatin, *Am. J. Physiol.,* 262, R1111, 1992.

19. Ookuma, K. C., Barton, C., York, D. A., and Bray, G. A., Effect of enterostatin and kappa-opioids on macronutrient selection and consumption, *Peptides,* 18, 785, 1997.

20. Erlanson-Albertsson, C., Pancreatic colipase. Structural and physiological aspects, *Biochim. Biophys. Acta,* 1125, 1, 1992.

21. Wicker, C. and Puigserver, A., Effects of inverse changes in dietary lipid and carbohydrate on the synthesis of some pancreatic secretory proteins, *Eur. J. Biochem.,* 162, 25, 1987.

22. Sörhede, M., Rippe, C., Mulder, H., and Erlanson-Albertsson, C., Enterostatin is produced in three different forms in the rat intestine, *Int. J. Obesity,* 19, 115, 1995.

23. Lin, L., Gehlert, D. R., York, D. A., and Bray, G. A., Effect of enterostatin on the feeding responses to galanin and NPY, *Obesity Res.,* 1, 186, 1993.

24. Mei, J. and Erlanson-Albertsson, C., Effect of enterostatin given intravenously and intracerebroventricularly on high-fat feeding in rats, *Regul. Pept.,* 41, 209, 1992.

25. York, D. A., Enterostatin: new information, in *IBC Second International Symposium on Obesity: Advances in Understanding and Treatment,* Weston, L. A. and Savage, L. M. Eds., IBC Biomedical Library Series, Southborough, 1996, 2.3.1.

26. Mei, J. and Erlanson-Albertsson, C., Role of intraduodenally administered enterostatin in rat: inhibition of food intake, *Obesity Res.,* 4, 161, 1996.

27. Lin, L., York, D. A., and Bray, G. A., Enterostatin suppresses food intake in rats after near celiac arterial and carotid injection, *FASEB J.,* 12, A348, 1998 (Abstr).

28. Lin, L. and York, D. A., Enterostatin actions in the amygdala and PVN to suppress feeding in the rat, *Peptides,* 18, 1341, 1997.

29. Okada, S., Lin, L., York, D. A., and Bray, G. A., Chronic effects of intracerebral ventricular enterostatin in Osborne-Mendel rats fed a high-fat diet, *Physiol. Behav.,* 54, 325, 1993.

30. Mei, J. and Erlanson-Albertsson, C., Plasma insulin response to enterostatin and effect of adrenalectomy in rat, *Obesity Res.,* 4, 513, 1996.
31. Okada, S., Onai, T., Kilroy, G., York, D. A., and Bray, G. A., Adrenalectomy of the obese Zucker rat: effects on the feeding response to enterostatin and specific mRNA levels, *Am. J. Physiol.,* 265, R21, 1993.
32. Miner, J. L., Erlanson-Albertsson, C., Paterson, J. A., and Baile, C. A., Reduction of feed intake in sheep by enterostatin, the procolipase activation peptide, *J. Anim. Sci.,* 72, 1578, 1994.
33. Weatherford, S. C., Lattemann, D. F., Sipols, A. J., Chavez, M., Kermani, Z. R., York, D. A., Bray, G. A., Porte, D., Jr., and Woods, S. C., Intraventricular administration of enterostatin decreases food intake in baboons, *Appetite,* 19, 225, 1992.
34. Rössner, S., Barkeling, B., Erlanson-Albertsson, C., Larsson, P., and Wahlin-Boll, E., Intravenous enterostatin does not affect single meal food intake in man, *Appetite,* 34, 37, 1995.
35. Shargill, N. S., Tsujii, S., Bray, G. A., and Erlanson-Albertsson, C., Enterostatin suppresses food intake following injection into the third ventricle of rats, *Brain Res.,* 544, 137, 1991.
36. Lin, L. and York, D. A., Chronic ingestion of dietary fat is a prerequisite for inhibition of feeding by enterostatin. *Am. J. Physiol.,* 1998.
37. Mei, J., Bowyer, R. C., Jehanli, A. M. T., Patel, G., and Erlanson-Albertsson, C., Identification of enterostatin, the pancreatic procolipase activation peptide, in the intestine of rat: effect of CCK-8 and high-fat feeding, *Pancreas,* 8, 488, 1993.
38. Bowyer, R. C., Rowston, W. M., Jehanli, A. M. T., Lacey, J. H., and Hermon-Taylor, J., The effect of a satiating meal on the concentrations of procolipase activation peptide in the serum and urine of normal and morbidly obese individuals, *Gut,* 34, 1520, 1993.
39. Imamura, M., Sumar, N., Hermon-Taylor, J., Robertson, H. J. F., and Prasad, C., Distribution and characterization of enterostatin-like immunoreactivity in human cerebrospinal fluid, *Peptides,* 19, 1385, 1998.
40. Townsley, M. I., Erlanson-Albertsson, C., Ohlsson, A., Rippe, C., and Reed, R. K., Enterostatin efflux in cat intestinal lymph: relation to lymph flow, hyaluronan and fat absorption, *Am. J. Physiol.,* 271, G714, 1996.
41. Mei, J., Enterostatin. A Peptide Regulating Fat Intake. An Experimental Study in Rat, Thesis. University of Lund, Sweden, 1998.
42. Lin, L., Umahara, M., York, D. A., and Bray, G. A., β-Casomorphins stimulate and enterostatin inhibits the intake of dietary fat in rats, *Peptides,* 19, 325, 1998.
43. Lin, L., McClanahan, S., York, D. A., and Bray, G. A., The peptide enterostatin may produce early satiety, *Physiol. Behav.,* 53, 789, 1993.
44. Lin, L. and York, D. A., Changes in the microstructure of feeding after administration of enterostatin into the paraventricular nucleus and the amygdala, *Peptides,* 19, 557, 1998.
45. Erlanson-Albertsson, C., Pancreatic colipase. Structural and physiological aspects, *Biochem. Biophys. Acta,* 1125, 1, 1992.
46. Okada, S., York, D. A., and Bray, G. A., Procolipase mRNA: tissue localization and effects of diet and adrenalectomy, *Biochem. J.,* 292, 787, 1993.
47. Sörhede, M., Mulder, H., Mei, J., Sundler, F., and Erlanson-Albertsson, C., Procolipase is produced in the rat stomach — a novel source of enterostatin, *Biochim. Biophys. Acta,* 1301, 207, 1996.
48. Sörhede, M., Erlanson-Albertsson, C., Mei, J., Nevalainen, T., Aho, A., and Sundler, F., Enterostatin in gut endocrine cells — immunocytochemical evidence, *Peptides,* 17, 609, 1996.
49. Okada, S., York, D. A., and Bray, G. A., Procolipase mRNA: tissue localization and effects of diet and adrenalectomy, *Biochem. J.,* 292, 787, 1993.
50. Erlanson-Albertsson, C. and York, D., Enterostatin — a peptide regulating fat intake, *Obesity Res,,* 5, 360, 1997.
51. Duan, R.-D. and Erlanson-Albertsson, C., Gastric inhibitory polypeptide stimulates pancreatic lipase and colipase synthesis in rats, *Am. J. Physiol.,* 262, G779, 1992.
52. Duan, R. and Erlanson-Albertsson, C., Pancreatic lipase and colipase activity increase in pancreatic acinar tissue of diabetic rats, *Pancreas,* 4, 329, 1989.
53. Duan, R. and Erlanson-Albertsson, C., The anticoordinate changes of pancreatic lipase and colipase activity to amylase activity by adrenalectomy in normal and diabetic rats, *Int. J. Pancreatol.,* 6, 271, 1990.

54. Duan, R.-D., Wicker, C., and Erlanson-Albertsson, C., Effect of insulin administration on contents, secretion and synthesis of pancreatic lipase and colipase in rats, *Pancreas,* 6, 595, 1991.

55. Duan, R. and Erlanson-Albertsson, C., Evidence of a simulatory effect of cyclic AMP on pancreatic lipase and colipase synthesis in rats, *Scand. J. Gastroenterol.,* 27, 644, 1992.

56. Brindley, D. N., Cooling, J., Glenny, H. P., Burditt, S. L., and McKechnie, I. S., Effects of chronic modification of dietary fat and carbohydrate on the insulin, corticosterone and metabolic responses of rat fed acutely with glucose, fructose or ethanol, *Biochem. J.,* 200, 275, 1981.

57. Bray, G. A., Fisler, J. S., and York, D. A., Neuroendocrine control of the development of obesity: understanding gained from studies of experimental models of obesity, *Prog. Neuroendocrinol.,* 128, 1990.

58. Mizuma, H., Abadie, J., and Prasad, C., Corticosterone facilitation of inhibition of fat intake by enterostatin (Val-Pro-Asp-Pro-Arg), *Peptides,* 15, 447, 1994.

59. Mei, J., Cheng, Y., and Erlanson-Albertsson, C., Enterostatin — its ability to inhibit insulin secretion and to decrease high-fat food intake, *Int. J. Obesity,* 17, 701, 1993.

60. Ookuma, M. and York, D. A., Inhibition of insulin release by enterostatin, *Int. J. Obesity,* 22, 800, 1998.

61. Erlanson-Albertsson, C. and Larsson, A., A possible physiological function of procolipase activation peptide in appetite regulation, *Biochimie,* 70, 1245, 1988.

62. Leibowitz, S. and Hoebel, G. G., Behavioral neuroscience of obesity, in *Handbook of Obesity,* Bray, G. A., Bouchard, C., and James, W. P. T., Eds., Marcel-Dekker, New York, 1998, 313.

63. Tian, Q., Nagase, H., York, D. A., and Bray, G. A., Vagal-central nervous system interactions modulate the feeding response to peripheral enterostatin, *Obesity Res.,* 2, 527, 1994.

64. York, D. A, Waggener, J., and Bray, G. A., Brain amine responses to peripheral enterostatin, *Int. J. Obesity,* 18, 102, 1994 (Abstr.).

65. Smith, B. K., York, D. A., and Bray, G. A., Chronic d-fenfluramine treatment reduces fat intake independent of macronutrient preference, *Pharmacol. Biochem. Behav.,* 60, 105, 1998.

66. York, D. A. and Lin, L., Enterostatin and fat intake: current insights into possible mechanisms, in *Progress in Obesity Research,* Angel, A., Anderson, H., and Bouchard, C., Eds., John Libbey, London, 1996, 495.

67. Lin, L. and York, D. A., Comparisons of the effects of enterostatin on food intake and gastric emptying in rats, *Brain Res.,* 745, 205, 1997.

68. Pierzynowski, S. G., Erlanson-Albertsson, C., Podgurniak, P., Kiela, P., and Weström, B., Possible integration of the electrical activity of the duodenum and pancreas secretion through enterostatin, *Biomed. Res.,* 15, 257, 1994.

69. Cooper, S. J., Jackson, A., and Kirkham, T. C., Endorphins and food intake: kappa opioid receptor agonists and hyperphagia, *Pharmacol. Biochem. Behav.,* 23, 889, 1985.

70. Barton, C., Lin, L., York, D. A., and Bray, G. A., Differential effects of enterostatin, galanin and opioids on high-fat diet consumption, *Brain Res.,* 702, 55, 1995.

71. Lynch, W. C., Opiate blockade inhibits saccharin intake and blocks normal preference acquisition, *Pharmacol. Biochem. Behav.,* 24, 833, 1986.

72. Lin, L. and York, D. A., Metergoline blocks the feeding response to enterostatin, *Appetite,* 23, 313, 1994.

73. Lin, L. and York, D., Does serotonergic activity influence κ-opioid induced feeding, *FASEB J.,* 11, A172, 1997 (Abstr.).

74. Erlanson-Albertsson, C., Pancreatic lipase, colipase and enterostatin — a lipolytic triad, in *Esterases, Lipases and Phospholipases: From Structure to Clinical Significance,* Mackness, M. I. and Clerc, M., Eds., Plenum Press, New York, 1994, 159.

75. Bouras, M., Huneau, J. F., Luengo, C., Erlanson-Albertsson, C., and Tomé, D., Metabolism of enterostatin in rat intestine, brain membranes and serum: differential involvement of proline-specific peptidases, *Peptides,* 16, 399, 1995.

21 Portal-Hepatic Sensors for Glucose, Amino Acids, Fatty Acids, and Availability of Oxidative Products

Wolfgang Langhans

CONTENTS

1   INTRODUCTION

Physiological regulatory systems typically consist of a feedback loop that comprises peripheral sensors for specific stimuli, the relay of the resulting neural signal to the central nervous system, central signal processing, and a neural or endocrine efferent pathway that restores homeostasis or ensures proper organ function. The portal-hepatic area is an ideal location for metabolic sensors in the control of eating, gut function, and metabolism because it is exposed to the nutrient flow from absorption and because the liver plays a major role in metabolism. Prior to the postabsorptive activation of portal-hepatic sensors, food ingestion stimulates taste-specific receptors in the oral cavity and triggers the release of gastrointestinal hormones, which provide another set of feedback signals in the control of food intake as well as gut functions, and indirectly also metabolism. The nutrient composition of the ingested food largely determines the composition of the gastrointestinal peptide "cocktail" released. In addition, nutrient-specific neural signals from the intestine are involved in upper gut functions and presumably also in the control of eating, and some evidence suggests that postabsorptive signals from the portal-hepatic area, which contribute to food intake control, are at least in part substrate specific and may also influence nutrient selection. Thus, a general principle of some kind of specificity in various sensory functions appears to line the alimentary tract and may extend into the portal-hepatic area. This could provide the basis for coordinated ingestive, gut function, and metabolic responses to feeds, which are fine-tuned to ensure adequate energy intake and nutrient selection, to prepare the organism for the arrival of particular nutrients, and to facilitate their metabolic handling.

The portal-hepatic metabolic sensors and their physiological functions in this context are the focus of this chapter. I will first present the evidence for the existence of such sensors mainly derived from electrophysiological studies, in which a variety of different metabolites and antimetabolites have been shown to affect hepatic afferent nerve activity in isolated perfused rat or guinea pig liver preparations and in situ.[1] I will then discuss possible sensory or coding mechanisms, i.e., the question of how different substrates or metabolic events in the liver might stimulate afferent nerves. This issue is related to the questions of whether or not there is any kind of sensor specificity, and whether stimulus intensity is coupled to signal strength. Finally, I will deal with what is known about the physiological relevance of signals derived from these sensors in gut functions, metabolism, and, in particular, energy intake and nutrient selection.

2   PORTAL-HEPATIC SENSORS FOR GLUCOSE

2.1   EVIDENCE FOR THEIR EXISTENCE

About 30 years ago, Niijima[2] demonstrated that glucose infusion into the hepatic portal vein decreased the mean firing rate of hepatic vagal afferents in the in vitro perfused guinea pig liver. Similarly, intraduodenal infusion of a 5% glucose solution resulted in a gradual increase in portal vein blood glucose and a concomitant decrease in hepatic vagal afferent activity.[3] The afferent discharge rate in these experiments was inversely related to the hepatic portal vein glucose concentration. These results indicate that hepatic vagal afferent nerves respond to the glucose concentration in the portal-hepatic area. Glucose also increased hepatic splanchnic afferent activity in one study,[4] suggesting that splanchnic afferents from the liver are also sensitive to glucose.

2.2   CODING MECHANISM

Russek[5,6] proposed that hepatic glucose sensors are hepatocytes which are hyperpolarized by glucose or some metabolite of the glycolytic chain, related both to liver glycogen content and glucose uptake. Glucose has in fact a short-term hyperpolarizing effect on hepatocyte membranes.[7] This effect is independent of intracellular glucose utilization because the nonmetabolizable glucose antimetabolite 2-deoxy-D-glucose (2DG) also hyperpolarized hepatocytes.[7] In contrast, the effect of glucose on afferent nerve activity is presumably related to glucose utilization, because glucose decreased and 2DG increased the firing rate of hepatic vagal afferents.[8] Furthermore, only glucose decreased afferent nerve activity, but not fructose, mannose, or galactose, or the pentoses arabinose and xylose which are not utilized by peripheral nerves.[3] A direct effect of glucose on hepatic vagal afferent nerve activity is suggested by results from experiments in which glucose superfusion of an isolated preparation of the portal vein with the hepatic branch of the vagus resulted in a rapid (5 min) suppression of the afferent discharge rate.[1] Thus, hepatic parenchyma is not necessary for the effect of glucose on hepatic vagal afferent firing rate. Because glucose is readily utilized by peripheral nerves, all these data are consistent with and support the hypothesis that hepatic vagal afferents, which terminate in the hepatic portal wall, register their own rate of glucose utilization and thus function as hepatic glucose sensors as originally suggested by Niijima.[2] The effect of glucose on the spike frequency of hepatic vagal afferents was blocked by ouabain,[8] indicating that it is mediated by the sodium pump. Presumably, glucose provides adenosine triphosphate (ATP) which drives the sodium pump, thus increasing membrane potential and thereby decreasing spike frequency. Whether there is an additional indirect effect of glucose on afferent nerve activity in which hepatocytes are involved remains to be investigated. In addition to hepatic glucose sensors, central, i.e., hindbrain, and hypothalamic glucose-sensitive neurons exist, which appear to share the same coding mechanism with portal-hepatic glucose sensors.[9]

2.3   PHYSIOLOGICAL RELEVANCE

2.3.1   Energy Intake and Nutrient Selection

Russek's original report of the differential effect of intraportal and intrajugular infusion of glucose on food intake[10] has long been controversial (see Reference 11 for review). Yet, today it is widely accepted that physiological amounts of glucose reduce food intake more effectively after infusion into the hepatic portal vein than after infusion into the jugular vein.[12-14] In addition, rats given hepatic portal glucose infusions developed a preference for the flavor that had been paired with the glucose infusion.[12] The effect of the infusion was localized to the liver because rats given systemic glucose infusions did not develop a flavor preference.[12] In another study, drinking saccharin solution increased food intake and food preference in intact rats, but not rats with hepatic vagotomy.[15] In this setting, saccharin may increase food intake by temporarily shifting fuels toward storage and away from oxidation, and flavored food eaten after drinking saccharin may become preferred because it provides glucose to counteract this reduction.[15] Together these findings indicate that signals from portal-hepatic glucose sensors can influence taste signal processing and nutrient selection, but their exact physiological role in this context remains to be characterized.

Usually there is no clear relationship between the amounts of glucose or the concentration of the solution infused into the hepatic portal vein and the food intake reduction.[14,16] In addition, glucose infusions into the hepatic portal vein seem to affect hepatic vagal afferent nerve activity faster than food intake (see Reference 1 vs. References 12, 13, and 16). In one study, hepatic portal glucose delivery reduced oral glucose solution intake only when the infusions began 60 or 45 min prior to the intake test.[16] We could also not detect a reliable short-term satiating effect of hepatic portal glucose infusions during spontaneous meals, even at doses which appear to reduce food intake in longer-term tests (Grossmann and Langhans, unpublished results). Therefore, an additional stimulus may be required for the glucose-induced decrease in afferent nerve activity to inhibit eating. This additional stimulus could be a metabolic consequence of the glucose infusion. Candidates include hepatic glucose uptake and glycogen formation. Interestingly, transgenic mice expressing human glucokinase in liver display decreases in body weight.[17] Glukokinase is important for hepatic glucose uptake and, hence, glycogen formation. The amount of liver glycogen present at meal onset may influence meal size; i.e., the level of liver glycogen could act as a permissive factor for meal-related signals of satiation. Liver glycogen also decreases during spontaneous meals in rats (see Reference 12). Yet, it is unknown how liver glycogen content — or changes of it — might influence food intake.

Some studies with 2DG also suggest a role of a hepatic glucose-sensitive mechanism in eating control. In rabbits, the stimulation of eating in response to hepatic portal 2DG infusion was reduced after subdiaphragmatic vagotomy.[18] In rats, hepatic branch vagotomy did not reduce the stimulatory effect of 2DG on eating during the day,[19-21] but did so during the early dark phase of the lighting cycle and especially after consumption of a test meal.[19,20] Complete subdiaphragmatic and hepatic branch vagotomy also disrupted the otherwise reliable translation of the transient premeal decline in blood glucose levels into meal initiation in the rat.[22] Finally, given the putative role of sodium pump activity in portal-hepatic glucose sensor signal transduction, it is interesting that the sodium pump inhibitor ouabain stimulates eating after intraperitoneal injection in rats and that this effect appears to be mediated by hepatic vagal afferent nerves.[23] In summary, portal-hepatic glucose sensors appear to play a role in eating, but the exact mechanism(s) underlying the glucose effects on energy intake and nutrient selection remain(s) to be determined.

2.3.2   Gut Function

Several lines of evidence indicate that hepatic portal glucose modulates gastric motor and secretory functions. Glucose solution injected into the hepatic portal vein has been shown to influence the efferent activity of the vagal nerve innervating the stomach,[24] suggesting a functional link between hepatic vagal branch afferents and gastric vagal efferents in the brain. In line with this assumption, glucose injected into the hepatic portal vein decreased gastric acid output[24] and gastric pressure associated with insulin hypoglycemia.[25,26] Hepatic branch vagotomy abolished these responses. Electrical stimulation of hepatic vagal afferents also changes gastric pressure and gastric acid output.[24]

2.3.3   Metabolic Homeostasis

Hepatic portal glucose infusion causes a reflex inhibition of the efferent activity in the adrenal nerve and hepatic splanchnic nerve, and a facilitation of the pancreatic vagal activity.[27] These effects were absent after section of the hepatic branch of the vagus. Electrical stimulation of the central stump of the cut vagus yielded opposite effects in normal animals and in midpontine-transected animals.[27] The reflexive changes of efferent nerve activity presumably influences catecholamine secretion from the adrenal medulla, and insulin and glucagon secretion from the pancreas.

An increase in portal-hepatic glucose concentration facilitates hepatocyte glucose uptake in response to an increase in the portal-arterial glucose gradient.[28] As a result, net hepatic glucose uptake is much greater during oral or intraportal glucose loading than during peripheral intravenous glucose delivery, even when similar glucose loads and hormone levels reaching the liver are maintained. An intact nerve supply to the liver appears to be vital for the normal response of the liver to intraportal glucose delivery. The portal signal relieves the sympathetic inhibition of hepatic glucose uptake and enhances hepatic glucose uptake directly by stimulating the parasympathetic innervation to the liver and indirectly by enhancing insulin release.[29] The neurally mediated enhancement of pancreatic insulin secretion also favors glucose uptake by nonhepatic tissues. The importance of insulin for the augmented hepatic glucose uptake in response to intraportal glucose was confirmed in experiments using pancreatectomized dogs, and in dogs in which somatostatin release-inhibiting factor was used to induce acute insulin deficiency, and glucose was infused into the portal vein or a peripheral vein.[30] The low net hepatic glucose uptake with low insulin levels was dramatically increased by insulin substitution. The portal glucose-dependent effect of insulin on hepatic glucose uptake is blocked by atropine. In turn, acetylcholine administration increases glucose uptake without a portal arterial glucose gradient. Apparently, the gradient is sensed and transformed into a metabolic signal by intrahepatic nerves, releasing acetylcholine to muscarinic receptors.[31] In addition, hepatic portal insulin has been shown to stimulate intestinal glucose absorption through a neurally mediated signal that originates in the liver.[32]

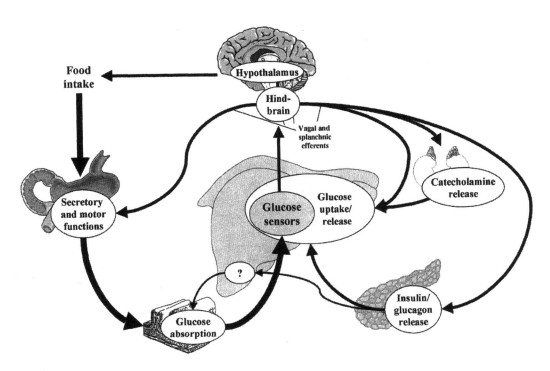

FIGURE 1 Possible homeostatic functions of portal-hepatic glucose sensors. See text for further details.

A decrease of the portal-hepatic glucose concentration stimulates hepatic splanchnic efferents and triggers glucose release from the liver. Likewise, electrical stimulation of the hepatic nerves increased norepinephrine outflow and glucose output from the liver. This effect was not abolished by bilateral adrenalectomy and pancreatectomy, and appeared to be mediated by an alpha-adrenergic mechanism and in part by eicosanoids.[33]

All in all, current neurophysiological and behavioral data indicate that portal-hepatic glucose sensors play a major role in a neural glucose-sensing system that also comprises glucose-sensitive neurons in the nucleus of the solitary tract (NST) and lateral hypothalamus. This system influences food intake and probably nutrient selection, gut functions, and hepatic glucose handling to maintain glucose homeostasis in response to changes in portal vein glucose concentration (Figure 1). The benefits of specific glucose sensors at the entrance of glucose into the circulation are obvious in such a system. In fact, the portal-hepatic glucose sensors are presumably part of an even more general mechanism of metabolic homeostasis because reflexive changes in hepatic efferent nerves induced by portal vein glucose can also be expected to have profound effects on hepatic fat and protein metabolism.[34] In addition, hepatic portal glucose also reduces plasma free fatty acid levels and lipolysis,[35,36] which may well be due to reflex inhibition of catecholamine secretion.

3 PORTAL-HEPATIC AMINO ACID SENSORS

3.1 EVIDENCE FOR THEIR EXISTENCE

Excitatory and inhibitory amino acid-sensitive afferent vagal neurons have been demonstrated in the portal-hepatic area. After hepatic portal infusion, arginine, alanine, histidine, leucine, lysine, serine, tryptophan, and valine enhanced hepatic vagal afferent activity, whereas cysteine, glycine, isoleucine, methionine, phenylalanine, proline, and threonine inhibited it.[1] More recently, monosodium glutamate has also been shown to activate hepatic vagal afferents,[36] but so far, no effects on hepatic vagal afferent nerve activity have been reported for glutamine, the most abundant amino acid in plasma.

3.2   Coding Mechanism

Little is known about the coding mechanism of portal-hepatic amino acid sensors. Second messengers probably mediate the action of amino acids on afferent nerves, given the (usually) long latency of the response(s), and the timing of their peak or nadir. Nerve activation by amino acids may be secondary to metabolic effects, provided that hepatocytes can communicate metabolic events to afferent nerves (see below). It is very unlikely that an osmotic effect is involved because osmosensors affect hepatic vagal afferent activity quite rapidly.[1] Moreover, as increases in perfusate osmolality usually increase afferent nerve activity, only the effects of the excitatory amino acids could be explained by osmotic effects.

3.3   Physiological Relevance

3.3.1.   Energy Intake and Nutrient Selection

Portal-hepatic amino acid sensors may be involved in the suppression of food intake observed after peripheral administration of amino acid solutions.[37-39] Yet, amino acids that increase and those that decrease hepatic vagal afferent activity in electrophysiological experiments, both reduce food intake after intraperitoneal injection.[37] Thus, either the hypophagic effects of various amino acids are mediated by different fibers, or they are independent from the induced changes in afferent nerve activity. Russek[37] suggested that deamination and ammonia production is responsible for the hypophagic effects of amino acids. He reported that ammonium chloride reduced food intake more efficiently after hepatic portal than after peripheral vein infusion in 22-h fasted dogs.[37] This differential effect suggests a hepatic origin, but so far, no effects of ammonium chloride on hepatic vagal afferent nerve activity have been reported. It is also unknown whether hepatic portal vein amino acid infusions affect nutrient selection. Thus, further studies are necessary to clarify the role of portal-hepatic amino acid or, perhaps, ammonia sensors in the control of energy intake and nutrient selection.

3.3.2   Gut Function

In one recent study, an intraportal injection of monosodium glutamate enhanced efferent activity in the gastric vagal branch,[36] suggesting that there is a reflex modulation of stomach function from portal-hepatic afferents sensitive to glutamate. Apart from that, no other effects of amino acid–derived hepatic vagal afferent signals on gut function have as yet been described.

3.3.3   Metabolic Homeostasis

Hepatic portal vein arginine delivery increased hepatic vagal afferent activity and caused a reflex stimulation of pancreatic splanchnic branch efferent activity as well as a reflex inhibition of pancreatic vagal efferent activity in electrophysiological experiments.[40] Moreover, intraperitoneally injected arginine stimulated insulin and glucagon secretion from the pancreas more in hepatic branch vagotomized rats than in sham-operated rats, indicating that the hepatic branch of the vagus mediates a signal that inhibits arginine-stimulated insulin and glucagon secretion.[41] One physiological role of portal-hepatic arginine sensors may therefore be to prevent an exaggerated pancreatic hormone secretion by arginine and, hence, to facilitate metabolic homeostasis in response to a protein-rich meal. Similarly, alanine, leucine, and glycine sensors in the portal-hepatic area appear to exert a reflex regulation of pancreatic hormone secretion through hepatic branch vagal afferents.[42,43] All in all, there is evidence for important homeostatic functions of portal-hepatic amino acid sensors. Yet, the physiological role of these sensors is not fully understood, and it is unknown whether signals from these sensors affect nutrient selection.

4 PORTAL-HEPATIC SENSORS FOR FATTY ACID OXIDATION

4.1 EVIDENCE FOR THEIR EXISTENCE

Orbach and Andrews[44] reported that perfusion of rabbit livers with sodium salts of several fatty acid solutions including palmitate-stimulated hepatic vagal afferent nerve activity. Yet, the physiological relevance of the observed effects is questionable because they disappeared when plasma was added to the perfusion medium. More recently, the fatty acid oxidation inhibitor mercaptoacetate (MA) was shown to increase multiunit hepatic vagal afferent nerve activity,[45] similar to 2DG. This suggests that portal-hepatic sensors respond to fatty acid oxidation.

4.2 CODING MECHANISM

The coding mechanism of portal-hepatic sensors for fatty acid oxidation is still unknown. Whereas afferent nerves readily utilize glucose, allowing glucose to be sensed by vagal afferents directly, it is very unlikely that afferent nerves oxidize fatty acids. Hepatocytes rely primarily on fatty acid oxidation[46] for ATP generation and may therefore be involved in the transduction of fatty acid oxidation into an afferent neural signal. Fatty acids stimulate respiration and ATP synthesis in isolated hepatocytes, and much of this ATP is utilized by the sodium pump because ouabain partially (50 to 60%) inhibited the stimulation of hepatocyte respiration by fatty acids.[47] Intracellular ATP, sodium pump activity, and hepatocyte membrane potential have been implicated in this sensory process, and according to Russek's "potentiostatic" theory,[5] increases in hepatocyte membrane potential should translate into decreases in afferent nerve activity. In liver perfusion studies, hepatocyte membranes were in fact hyperpolarized by palmitate.[48] This effect was antagonized by inhibitors of fatty acid oxidation, by ouabain, and by potassium-channel blockers,[48] indicating that sodium pump activation and an opening of potassium channels mediate the hyperpolarization of hepatocytes by palmitate oxidation.[48] These findings are compatible with a role of hepatocyte membrane potential in the transduction of hepatocellular fatty acid oxidation into changes in afferent nerve activity. However, discrepancies between the effects of other substances on hepatocyte membrane potential and hepatic vagal afferent activity (e.g., Reference 6; see Reference 49 for review) question a general role of hepatocyte membrane potential in the coding of metabolic information and, hence, in the effects of fatty acid oxidation on afferent nerve activity.

Using anterograde tracing of the hepatic vagal innervation, Berthoud et al.[50] failed to detect afferent nerve endings in rat liver parenchyma. This questions the presumed role of hepatocytes in signal transduction. Yet, fine nerve endings that are not detected by this technique may still be in close proximity to at least some hepatocytes. As hepatocytes are electrically coupled, innervation of some hepatocytes would be enough to ensure the transmission of metabolic signals to vagal afferents. Hepatocytes might also release some modulator of nerve activity. For fatty acid oxidation, ketone bodies have been suggested to serve such a function[51] because ketogenesis is closely related to fatty acid oxidation, and β-hydroxybutyrate (BHB) is readily taken up and oxidized by peripheral nerves. Whereas this possibility needs to be thoroughly tested, it also requires that hepatic parenchyma is innervated or that sensory nerve endings are accessible from the blood downstream of hepatocytes, which is also unproven. In general, a contribution of hepatocytes to the sensing of fatty acid oxidation by hepatic vagal afferents appears likely, but it is still unproven, and the coding mechanism for afferent nerve signals is unknown.

The neural signal in response to an inhibition of fatty acid oxidation has been traced from the periphery to the brain. Recently, *c-fos* expression in response to oral methyl palmoxirate administration has been observed in the NST, area postrema, lateral parabrachial nucleus, central lateral nucleus of the amygdala, dorsal lateral bed nucleus of the stria terminalis, and the paraventricular nucleus of the hypothalamus.[52] This suggests that peripheral inhibition of fatty acid oxidation activates an afferent pathway projecting from the hindbrain to the forebrain.

4.3   PHYSIOLOGICAL RELEVANCE

4.3.1   Energy Intake and Nutrient Selection

An inhibition of fatty acid oxidation is associated with enhanced eating (lipoprivic eating), especially when a fat-rich diet is consumed. This was first shown with MA, which increased food intake in rats fed a fat-rich diet (18% fat, 46% starch; w/w), but not in rats fed a low-fat diet (3.3% fat, 77% starch).[53] MA blocks fatty acid oxidation by inhibiting the acyl-coenzyme-A (CoA)-dehydrogenases located in the mitochondrial matrix.[54] Methyl palmoxirate, an inhibitor of the enzyme carnitine palmitoyltransferase-1 (CPT-1), also increased food intake in rats fed a diet high in long-chain fatty acids, but did not affect food intake in rats fed a diet high in medium-chain fatty acids, which do not require the CPT-1 for mitochondrial uptake and oxidation.[55] The hyperphagic effect of MA was more pronounced during the day, when lipolysis and fatty acid oxidation are accelerated, than during the night.[53] MA increased food intake primarily through an acute effect on the duration of the intermeal interval (IMI),[56] indicating that fatty acid oxidation contributes to the maintenance of postprandial satiety.

The hyperphagia of rats in response to MA is markedly attenuated by hepatic branch vagotomy,[57,58] which also eliminated the stimulating effect of MA on eating in rats receiving total parenteral nutrition.[57] Since MA retained its potency to increase food intake after peripheral blockade of cholinergic transmission,[58] fatty acid oxidation seems to affect eating mainly through vagal afferents and not efferents. Capsaicin pretreatment, subdiaphragmatic vagotomy, lesions of the vagal sensory terminal fields in the area-postrema NST, and lesions of the lateral parabrachial nucleus abolished lipoprivic eating.[59-61] In contrast, hypothalamic paraventricular lesions failed to affect lipoprivic eating,[61] and eating could not be elicited by intracerebroventricular injection of MA.[59] Finally, induction of *fos*-like immunoreactivity in the brain by peripheral administration of MA was abolished by subdiaphragmatic vagotomy.[62] All these results suggest that fatty acid oxidation affects eating through vagally mediated signals from portal-hepatic sensors.

Studies showing that ingestion or intragastric administration of medium-chain triglycerides can inhibit eating in animals[63,64] and humans[65,66] provide indirect evidence for a role of hepatic fatty acid oxidation in control of eating. After a fatty meal, at least some medium-chain fatty acids get to the liver through the portal vein and feed into the mitochondrial fatty acid oxidation pathway.

In rats maintained on a macronutrient self-selection diet, MA increased intake of carbohydrate and protein and decreased intake of fat;[67] i.e., the rats appear to avoid the nutrient they cannot oxidize in response to MA and overconsume the nutrients they can utilize. This is line with previous findings indicating that the greater acceptance of fat by rats adapted to a fat-rich diet is associated with an increased capacity of fat absorption and hepatic fatty acid oxidation.[68] The dependence of MA-induced stimulation of eating on intact hepatic vagal afferents together with its macronutrient specificity suggest that signals derived from portal-hepatic sensors for fatty acid oxidation can affect nutrient selection. In another study,[69] rats injected with MA responded more to a previously conditioned stimulus for oil than for a previously conditioned stimulus for sucrose, and this effect was abolished by subdiaphragmatic vagotomy. This suggests that an inhibition of fatty acid oxidation has the capacity to activate representations of fat-conditioned stimuli, and that this effect depends on vagal afferents. In this context it is also worth mentioning that high doses of MA have aversive effects and consistently reduce rather than stimulate food intake in rats (Langhans and Scharrer, unpublished data). This may reflect a modulation of taste signal processing by signals derived from portal-hepatic sensors for fatty acid oxidation. Thus, like portal-hepatic signals derived from glucosensors, signals derived from hepatic fatty acid oxidation apparently possess the potential to modulate nutrient selection, at least under certain conditions.

4.3.2   Gut Function

To the best of my knowledge, there is no direct evidence for an effect of hepatic fatty acid oxidation on gut function. Perhaps such a feedback loop is not to be expected because a major fuel source for hepatic oxidation of fatty acids is the adipose tissue, particularly when exogenous energy is not available and consequently gut function is less important. On the other hand, fatty acid oxidation and glucose utilization appear to affect hepatic vagal afferent nerve activity similarly. If the same hepatic vagal fibers respond to glucose utilization and fatty acid oxidation, which is possible, but not known, hepatic fatty acid oxidation may also affect motor and secretory functions of the stomach, as does glucose. Yet, the hepatic signal may of course be modulated by (or depend on) other factors which influence its CNS integration and, hence, the efferent response. These modulating factors may be different in situations in which the portal-hepatic signal is derived from glucose utilization and fatty acid oxidation.

4.3.3   Metabolic Homeostasis

It is well documented that MA triggers metabolic counterregulatory responses, which are mainly based on sympathetic activation.[70,71] As neurons do not readily oxidize fatty acids and the responses to MA appear to be primarily peripherally mediated, it is feasible that the metabolic responses to MA are also derived from portal hepatic sensors for fatty acid oxidation. So far, however, a contribution of hepatic vagal afferents to the metabolic responses to changes in hepatic fatty acid oxidation has not directly been shown. Portal-hepatic sensors for fatty acid oxidation are presumably involved in the control of food intake and nutrient selection, but their exact coding mechanism and their potential role in other regulatory systems are still unknown.

5   PORTAL-HEPATIC SENSORS FOR OTHER SUBSTANCES

5.1   GLUCOSE INTERMEDIATES

In one study, hepatic vagal afferents decreased their firing rate in response to hepatic portal vein administration of pyruvate.[2] A responsiveness of portal-hepatic sensors to glucose intermediates makes physiological sense. Hepatic portal and peripheral vein plasma lactate and pyruvate increase in response to carbohydrate ingestion,[72,73] and gluconeogenesis from extrahepatically produced lactate is a prominent feature of the role of the liver in glucose homeostasis. The coding mechanism of the effects of lactate on hepatic vagal afferent activity is unknown. Glucose intermediates have been shown to hyperpolarize hepatocyte membranes,[74] and this could somehow be involved. However, a direct effect of lactate and pyruvate on afferent nerves appears more likely because a general sensory function of hepatocyte membrane potential is questionable.[49] Independent of the coding mechanism for hepatic afferent nerve signaling, lactate and pyruvate presumably inhibit feeding through portal-hepatic sensors. Both metabolites potently decrease food intake after parenteral administration,[75-77] and their hypophagic effects depend on hepatic branch vagal afferents.[78,79] Moreover, remotely controlled, prandial hepatic portal vein and inferior vena cava infusions of sodium L-lactate acutely reduced meal size, but only hepatic portal lactate infusion also increased the satiety ratio — duration of subsequent IMI (min)/meal size (g) — of the target meal (Silberbauer et al., unpublished results). Hepatic portal vein lactate infusions did not affect the systemic glucose level and did not increase systemic lactate above the levels usually observed around meal termination. These results indicate that an increase in circulating lactate can prematurely end a meal and enhance its satiating effect. Together with the blockade of the hypophagic effect of lactate by hepatic branch vagotomy,[78,79] the findings suggest that at least the delayed effect of prandial lactate infusions on the satiating effect of a meal originates in the liver. Glucose intermediates may also contribute to a hepatic signal involved in metabolic homeostasis during exercise, because hepatic portal but not jugular vein infusion of pyruvate prevented the normally observed exercise-induced increase in the peripheral glucagon/insulin molar ratio.[80] Whether there is an effect of hepatic portal lactate or pyruvate infusions on nutrient selection remains to be investigated.

5.2 FRUCTOSE

Unlike glucose, hepatic portal infusion of fructose did not affect hepatic vagal afferent activity in the isolated guinea pig liver preparation.[2] On the other hand, like glucose, fructose potently reduces food intake[13] and increases flavor preference[81] after hepatic portal infusion. A preference for a flavor paired with hepatic portal vein fructose infusion could not be detected in rats with hepatic branch vagotomy.[81] Together these data suggest that fructose too can somehow trigger a hepatic vagal afferent signal that influences energy intake and nutrient selection. Hepatic fructose utilization appears to be crucial for the effects on energy intake and nutrient selection.[81] Yet, the coding mechanism for such a fructose-derived signal is unknown, and it remains to be investigated whether the signal is triggered by fructose itself or by one of its metabolites (lactate or pyruvate, for instance). It is also unclear whether hepatic portal vein fructose does perhaps affect hepatic vagal afferent nerve activity in rat and mice, i.e., in species in which it also hyperpolarizes hepatocytes.[7,74]

5.3 2,5-ANHYDRO-MANNITOL

The fructose analogue 2,5-anhydro-mannitol (2,5-AM) has also been shown to increase hepatic afferent nerve activity,[82] suggesting that sensory afferents somehow respond to 2,5-AM. 2,5-AM, which stimulates eating presumably through trapping of phosphate and depletion of ATP in the liver (see Reference 83), has recently also been shown to hyperpolarize hepatocytes.[84,85] The hyper-phagic effect of 2,5-AM is presumably mediated by portal-hepatic sensors because it is more pronounced after hepatic portal than after intrajugular infusion and because the effect of low 2,5-AM doses on feeding is blocked by hepatic branch vagotomy.[86] The same holds for 2,5-AM-induced *fos*-like immunoreactivity in the brain.[87] Thus, the increase in food intake after administration of 2,5-AM is consistent with a stimulatory effect of hepatic ATP depletion on eating, but this does not appear to require changes in hepatocyte membrane potential to be transformed into an afferent neural signal.[82] Perhaps 2,5-AM increases hepatic afferent nerve activity and stimulates food intake by decreasing ATP in nerve endings. An effect of 2,5-AM on neuronal metabolism appears possible because 2,5-AM stimulates eating also after intracerbroventricular administration.[88] High doses of 2,5-AM stimulate eating independent of the hepatic branch of the vagus.[87] Thus, fibers in other vagal branches may independently mediate *c-fos* induction and the stimulation of eating at higher doses of 2,5-AM. 2,5-AM induces brain *c-fos* and elicits eating behavior in rats maintained on a low-fat/high-carbohydrate diet, but not in rats on a high-fat/low-carbohydrate diet.[89] Yet, it is unknown whether or not 2,5-AM specifically influences nutrient selection.

6   PERSPECTIVES

Important questions concerning the morphological identity, the functioning, and the physiological relevance of the portal-hepatic metabolic sensors remain to be answered. Thus, with the exception of the glucose-sensitive sensory afferents in the wall of the portal vein,[1] the morphological substrate of metabolic sensors in the portal-hepatic area has not yet been identified. Whether the paraganglia, which have been found around the liver hilus and which are innervated by hepatic vagal afferents,[50] have a sensory function is also unknown. With respect to the functional relevance of hepatic metabolic sensors, it is puzzling that a particular change in hepatic vagal afferent activity does not always predict a specific response. Regarding food intake, there are peptides that reduce food intake after peripheral administration, but that markedly increase rather than decrease hepatic vagal afferent nerve activity.[1] Regarding the metabolic responses, alanine and leucine, for instance, both increase hepatic vagal afferent activity, but under the same experimental conditions they have different effects on pancreatic hormone secretion, which appear to be mediated by the hepatic branch of the vagus. These discrepancies raise the critical question of whether sensory afferent vagal fibers in the hepatic branch of the vagus are specific for particular substrates or whether the same fibers respond to different substrates. To answer this question, single unit recordings instead of the usual multiunit recording is required. Single-unit recordings from the hepatic branch of the vagus, in which the origin of the fiber can be visually traced, may also help to verify that the recorded signals are derived from the portal-hepatic area. This is important because about 50% of hepatic afferent fibers do not originate in the liver, but are derived from the duodenum and proximal jejunum.[50]

Another important task is to elucidate how certain substances change afferent nerve activity, and to answer the question of whether or not a sensory function of hepatocytes is required for some of them, which is very likely at least for fatty acid oxidation. This also relates to the question of signal interactions in general, which is largely unresolved. For instance, *c-fos* immunohistochemistry studies revealed that intravenously infused MA and 2DG did not activate exactly the same brain sites, and induction of *fos*-like immunoreactivity in the brain by MA but not 2DG was totally abolished by subdiaphragmatic vagotomy.[60] Moreover, central glucose administration significantly reduced 2DG- but not MA-induced eating,[90] and bilateral destruction of the lateral parabrachial nucleus severely impaired or abolished MA- but not 2DG-induced feeding.[91] All these results indicate that 2DG and MA stimulate eating through at least partially separate central neural pathways and mechanisms.[91] Given that metabolic feedback signals from portal-hepatic sensors are probably involved in the control of gut functions, it is very likely that portal-hepatic and gastrointestinal feedback signals also interact with respect to eating. Such interactions between various metabolic and between metabolic and gastrointestinal controls of eating merit more attention.

All in all, portal-hepatic sensors for metabolites are integrated in feedback control mechanisms of food intake, gut function, and metabolic homeostasis. It is tempting to consider these feedback loops as a functional entity, because they all subserve energy homeostasis. They just influence different steps of food (= energy) processing in the body, such as energy intake and nutrient selection, the temporary storage and progression of the ingested food through the gastrointestinal tract, the absorption of digestive products, and ultimately their metabolic fate. The adequate postabsorptive handling of nutrients by the body appears to require substrate-specific portal-hepatic sensing mechanisms to govern the endocrine and metabolic responses. It is reasonable to assume that the information derived from such substrate-specific sensors somehow feeds back into the mechanisms of nutrient selection. Thus, signals from portal-hepatic sensors for glucose and fructose utilization and fatty acid oxidation apparently influence taste processing and nutrient selection, in addition to any effect such signals probably exert on energy intake. Yet, whether this also applies to signals derived from portal-hepatic sensors for other substrates is presently unknown, and much more needs to be learned about these processes. At present, the available data do not allow for a more comprehensible answer to the question of whether signals derived from portal-hepatic sensors are part of separate control mechanisms for macronutrient selection. To further elucidate the role of portal-hepatic sensors in nutrient selection and to unravel the exact mechanisms involved remain challenging tasks for the future.

REFERENCES

1. Niijima, A., Role played by vagal hepatic afferents form chemical sensors in the hepato-portal region: an electrophysiological study, in *Liver Innervation*, Shimazu, T., Ed., John Libbey, London, 1996, 323
2. Niijima, A., Afferent impulse discharges from glucoreceptors in the liver of the guinea pig, *Ann. N.Y. Acad. Sci.,* 157, 690, 1969.
3. Niijima, A., Glucose-sensitive afferent nerve fibers in the hepatic branch of the vagus nerve in the guinea-pig, *J. Physiol. (London),* 332, 315, 1982.
4. Schmitt, M., Influences of hepatic portal receptors on hypothalamic feeding and satiety centers, *Am. J. Physiol.* 225, 1089, 1973.
5. Russek, M. and Grinstein, S., Coding of metabolic information by hepatic glucoreceptors, in *Neurohumoral Coding of Brain Function*, Myers, R. D. and Drucker-Colin, R. R., Eds., Plenum, New York, 1974, 81.

6. Russek, M., Current status of the hepatostatic theory of food intake control, *Appetite*, 2, 137, 1981.

7. Meyer, A. H. and Scharrer, E., Hyperpolarization of the cell membrane of mouse hepatocytes by metabolizable and nonmetabolizable monosaccharides, *Physiol. Behav.*, 50, 351, 1991.

8. Niijima, A., Glucose-sensitive afferent nerve fibers in the liver and their role in food intake and blood glucose regulation, *J. Auton. Nerv. Syst.*, 9, 207, 1983.

9. Oomura, Y. and Yoshimatsu, H., Neural network of glucose monitoring system, *J. Auton. Nerv. Syst.*, 10, 359, 1984.

10. Russek, M., Demonstration of the influence of an hepatic glucosensitive mechanism on food intake, *Physiol. Behav.*, 5, 1207, 1970.

11. Langhans, W. and Scharrer, E., Metabolic control of eating, *World Rev. Nutr. Diet.*, 70, 1, 1992.

12. Tordoff, M. G. and Friedman, M.I., Hepatic portal glucose infusions decrease food intake and increase food preference, *Am. J. Physiol.*, 251, R192, 1986.

13. Tordoff, M. G. and Friedman, M. I., Hepatic control of feeding: effect of glucose, fructose, and mannitol infusion, *Am. J. Physiol.*, 254, R969, 1988.

14. Tordoff, M. G., Tluczek, J. P., and Friedman, M. I., Effect of hepatic portal glucose concentration on food intake and metabolism, *Am. J. Physiol.*, 257, R1474, 1989.

15. Tordoff, M. G. and Friedman, M.I., Drinking saccharin increases food intake and preference: IV. Cephalic phase and metabolic factors, *Appetite*, 12, 37, 1989.

16. Baird, J. P., Grill, H. J., and Kaplan, J.M., Intake suppression after hepatic portal glucose infusion: all-or-none effect and its temporal threshold, *Am. J. Physiol.*, 272, R1454, 1997.

17. Hariharan, N., Farrelly, D., Hagan, D., Hillyer, D., Arbeeny, C., Sabrah, T., Treloar, A., Brown, K., Kalinowski, S., and Mookhtiar, K., Expression of human hepatic glucokinase in transgenic mice liver results in decreased glucose levels and reduced body weight, *Diabetes*, 46, 11, 1997.

18. Novin, D., VanderWeele, D. A., and Rezek, M., Infusion of 2-deoxy-D-glucose into the hepatic-portal system causes eating: evidence for peripheral glucoreceptors, *Science*, 181, 858, 1973.

19. Delprete, E. and Scharrer, E., Hepatic branch vagotomy attenuates the feeding response to 2-deoxy-D-glucose in rats, *Exp. Physiol.*, 75, 259, 1990.

20. DelPrete, E. and Scharrer, E., Circadian effects of hepatic branch vagotomy on the feeding response to 2-deoxy-D-glucose in rats, *J. Auton. Nerv. Syst.*, 46, 27, 1994.

21. Tordoff, M. G., Novin, D., and Russek, M., Effects of hepatic denervation on the anorexic response to epinephrine, amphetamine, and lithium chloride: a behavioral identification of glucostatic afferents, *J. Comp. Physiol. Psychol.*, 96, 361, 1982.

22. Campfield, L. A. and Smith, F. J., Systemic factors in the control of food intake, in *Neurobiology of Food and Fluid Intake, Handbook of Behavioral Neurobiology*, Vol. 10, Stricker, E. M., Ed., Plenum Press, New York, 1990, 183.

23. Langhans, W. and Scharrer, E., Evidence for a role of the sodium pump of hepatocytes in the control of food intake, *J. Auton. Nerv. Syst.*, 20, 199, 1987.

24. Sakaguchi, T., Control of motor and secretory functions of the stomach by a portal glucose signal, *Neurosci. Biobehav. Rev.*, 19, 469, 1995.

25. Sakaguchi, T., Aono, T., Ohtake, M., and Sandoh, N., Interaction of glucose signals between the nucleus of the vagus nerve and the portal vein area in the regulation of gastric motility in rats, *Brain Res. Bull.*, 33, 469, 1994.

26. Sakaguchi, T. and Ohtake, M., Hepatic glucose signals vagally modulate the cyclicity of gastric motility in rats, *Experientia*, 49, 795, 1993.

27. Niijima, A., Reflex control of the autonomic nervous system activity from the glucose sensors in the liver in normal and midpontine-transected animals, *J. Auton. Nerv. Syst.*, 10, 279, 1984.

28. Adkins-Marshall, B., Pagliassotti, M. J., Asher, J. R., Connolly, C. C., Neal, D. W., Williams, P. E., Myers, S. R., Hendrick, G. K., Adkins, R. B., Jr., and Cherrington, A. D., Role of hepatic nerves in response of liver to intraportal glucose delivery in dogs, *Am. J. Physiol.*, 262, E679, 1992.

29. Moore, M. C. and Cherrington, A.D., Regulation of net hepatic glucose uptake: Interaction of neural and pancreatic mechanisms, *Reprod. Nutr. Dev.*, 36, 399, 1996.

30. Pagliassotti, M. J., Moore, M. C., Neal, D. W., and Cherrington, A. D., Insulin is required for the liver to respond to intraportal glucose delivery in the conscious dog, *Diabetes*, 41, 1247, 1992.

31. Stümpel, F. and Jungermann, K., Sensing by intrahepatic muscarinic nerves of a portal-arterial glucose concentration gradient as a signal for insulin-dependent glucose uptake in the perfused rat liver, *FEBS Lett.*, 406, 119, 1997.

32. Stümpel, F., Kucera, T., Gardemann, A., and Jungermann, K., Acute increase by portal insulin in intestinal glucose absorption via hepatoenteral nerves in the rat, *Gastroenterology*, 110, 1863, 1996.

33. Takahashi, A., Ishimaru, H., Ikarashi, Y., Kishi, E., and Maruyama, Y., Effects of hepatic nerve stimulation on blood glucose and glycogenolysis in rat liver: studies with *in vivo* microdialysis, *J. Auton. Nerv. Syst.*, 61, 181, 1996.

34. Jungermann, K., Püschel, G. P., Gardemann, A., and Stümpel, F., Control of liver metabolism and haemodynamics by hepatic nerves, in *Liver Innervation*, Shimazu, T., Ed., John Libbey, London, 1996, 105.

35. Bernal, R., Hutson, D. G., Dombro, R. S., Livingstone, A., Levi, J. U., and Zeppa, R., A possible hepatic factor in the control of plasma free fatty acid levels, *Metabolism*, 31, 533, 1982.

36. Niijima, A., An electrophysiological study on hepatovisceral reflex: the role played by vagal hepatic afferents from chemosensors in the hepatoportal region, in *Liver and Nervous System*, Häussinger, D. and Jungermann, K., Eds., Kluwer Academic Publishers, Dordrecht, 1998, 159.

37. Russek, M., Hepatic receptors and the neurophysiological mechanisms controlling feeding behavior, *Neurosci. Res.*, 4, 213, 1971.

38. Rezek, M. and Novin, D., Hepatic-portal nutrient infusion: effect on feeding in intact and vagotomized rabbits, *Am. J. Physiol.*, 232, E119, 1977.

39. Niijima, A. and Meguid, M. M., An electrophysiological study on amino acid sensors in the hepato-portal system in the rat, *Obesity Res.*, 3 (Suppl. 5), 741S, 1995.

40. Tanaka, K., Inoue, S., Takamura, Y., Jiang, Z. Y., and Niijima, A., Arginine sensors in the hepato-portal system and their reflex effects on pancreatic efferents in the rat, *Neurosci. Lett.*, 72, 69, 1986.

41. Tanaka, K., Inoue, S., Fujii, T., and Takamura, Y., Enhancement of insulin and glucagon secretion by arginine after hepatic vagotomy, *Neurosci. Lett.*, 72, 74, 1986.

42. Tanaka, K., Inoue, S., Saito, S., Nagase, H., and Takamura, Y., Hepatic vagal amino acid sensors modulate amino acid induced insulin and glucagon secretion in the rat, *J. Auton. Nerv. Syst.*, 42, 225, 1993.

43. Saitou, S., Tanaka, K., Inoue, S., Takamura, Y., and Niijima, A., Glycine sensor in the hepato-portal system and their reflex effects on pancreatic efferents in the rat, *Neurosci. Lett.*, 149, 12, 1993.

44. Orbach, J. and Andrews, W.H., Stimulation of afferent nerve terminals in the perfused rabbit liver by sodium salts of some long-chain fatty acids, *Q. J. Exp. Physiol. Cogn. Med. Sci.*, 58, 267, 1973.

45. Lutz, T. A., Diener, M., and Scharrer, E., Intraportal mercaptoacetate infusion increases afferent activity in the common hepatic vagus branch of the rat, *Am. J. Physiol.*, 273, R442, 1997.

46. Seifter, S. and England, S., Energy metabolism, in *The Liver: Biology and Pathobioogy*, Arias, I. M., Jakoby, W. B., Popper, H., Schachter, D., and Shafritz, D. A., Eds., Raven Press, New York, 1988, 279.

47. Plomp, P. J. A. M., van Roermund, C. W. T., Groen, A. K., Meijer, A. J., and Tager, J. M., Mechanism of the stimulation of respiration by fatty acids in rat liver, *FEBS Lett.*, 193, 243, 1985.

48. Rossi, R., Geronimi, M., Gloor, P., Seebacher, M. C., and Scharrer, E., Hyperpolarization of the cell membrane of mouse hepatocytes by fatty acid oxidation, *Physiol. Behav.*, 57, 509, 1995.

49. Langhans, W., Role of the liver in the metabolic control of eating: what we know — and what we do not know, *Neurosci. Biobehav. Rev.*, 20, 145, 1996.

50. Berthoud, H. R., Kressel, M., and Neuhuber, W. L., An anterograde tracing study of the vagal innervation of rat liver, portal vein and biliary system, *Anat. Embryol. Berl.*, 186, 431, 1992.

51. Scharrer, E., Lutz, T. A., and Rossi, R., Coding of metabolic information by hepatic sensors controlling food intake, in *Liver Innervation*, Shimazu, T., Ed., John Libbey, London, 1996, 381.

52. Horn, C. C. and Friedman, M. I., Methyl palmoxirate increases eating behavior and brain Fos-like immunoreactivity in rats, *Brain Res.*, 781, 8, 1998.

53. Scharrer, E. and Langhans, W., Control of food intake by fatty acid oxidation, *Am. J. Physiol.*, 250, R1003, 1986.

54. Bauché, F., Sabourault, D., Giudicelli, Y., Nordmann, J., and Nordmann, R., 2-Mercaptoacetate administration depresses the beta-oxidation pathway through an inhibition of long-chain acyl-CoA dehydrogenase activity, *Biochem J.*, 196, 803, 1981.

55. Friedman, M. I., Ramirez, I., Bowden, C. R., and Tordoff, M. G., Fuel partitioning and food intake: role for mitochondrial fatty acid transport, *Am. J. Physiol.*, 258, R216, 1990.

56. Langhans, W. and Scharrer, E., Role of fatty acid oxidation in control of meal pattern, *Behav. Neural Biol.*, 47, 7, 1987.

57. Beverly, J. L., Yang, Z. J., and Meguid, M. M., Hepatic vagotomy effects on metabolic challenges during parenteral nutrition in rats, *Am. J. Physiol.*, 266, R646, 1994.

58. Langhans, W. and Scharrer, E., Evidence for a vagally mediated satiety signal derived from hepatic fatty acid oxidation, *J. Auton. Nerv. Syst.*, 18, 13, 1987.

59. Ritter, S. and Taylor, J. S., Capsaicin abolishes lipoprivic but not glucoprivic feeding in rats, *Am. J. Physiol.*, 256, R1232, 1989.

60. Ritter, S. and Taylor, J. S., Vagal sensory neurons are required for lipoprivic but not glucoprivic feeding in rats, *Am. J. Physiol.*, 258, R1395, 1990.

61. Calingasan, N. Y. and Ritter, S., Hypothalamic paraventricular nucleus lesions do not abolish glucoprivic or lipoprivic feeding, *Brain Res.*, 595, 25, 1992.

62. Ritter, S. and Dinh, T. T., 2-mercaptoacetate and 2-deoxy-D-glucose induce fos-like immunoreactivity in rat brain, *Brain Res.*, 641, 111, 1994.

63. Furuse, M., Choi, Y. H., Mabayo, R.T., and Okumura, J., Feeding behavior in rats fed diets containing medium chain triglyceride, *Physiol. Behav.*, 52, 815, 1992.

64. Satabin, P., Auclair, E., Servan, E., Larue-Achagiotis, C., and Guezennec, C. Y., Influence of glucose, medium-chain and long-chain triglyceride gastric loads and forced exercise on food intake and body weight in rats, *Physiol. Behav.*, 50, 147, 1991.

65. Rolls, B. J., Gnizak, N., Summerfelt, A., and Laster, L. J., Food intake in dieters and nondieters after a liquid meal containing medium-chain triglycerides, *Am. J. Clin. Nutr.*, 48, 66, 1988.

66. Stubbs, R. J. and Harbron, C. G., Covert manipulation of the ratio of medium- to long-chain triglycerides in isoenergetically dense diets: Effect on food intake in *ad libitum* feeding men, *Int. J. Obesity*, 20, 435, 1996.

67. Singer, L. K., York, D. A., and Bray, G.A., Macronutrient selection following 2-deoxy-D-glucose and mercaptoacetate administration in rats, *Physiol. Behav.*, 65, 115, 1998

68. Reed, D. R., Tordoff, M. G., and Friedman, M. I., Enhanced acceptance and metabolism of fats by rats fed a high-fat diet, *Am. J. Physiol.*, 261, R1084, 1991.

69. Davidson, T. L., Altizer, A. M., Benoit, S. C., Walls, E. K., and Powley, T. L., Encoding and selective activation of "metabolic memories" in the rat, *Behav. Neurosci.*, 111, 1014, 1997.

70. Scheurink, A. and Ritter, S., Sympathoadrenal responses to glucoprivation and lipoprivation in rats, *Physiol. Behav.*, 53, 995, 1993.

71. vanDijk, G., Scheurink, A., Ritter, S., and Steffens, A., Glucose homeostasis and sympathoadrenal activity in mercaptoacetate-treated rats, *Physiol. Behav.*, 57, 759, 1995.

72. Felig, P., Wahren, J., and Hendler, R., Influence of oral glucose ingestion on splanchnic glucose and gluconeogenic substrate metabolism in man, *Diabetes*, 24, 468, 1975.

73. Langhans, W., Hepatic and intestinal handling of metabolites during feeding in rats, *Physiol. Behav.*, 49, 1203, 1991.

74. Dambach, G. and Friedmann, N., Substrate-induced membrane potential changes in the perfused rat liver, *Biochim. Biophys. Acta*, 367, 366, 1974.

75. Baile, C. A., Zinn, W. M., and Mayer, J., Effects of lactate and other metabolites on food intake of monkeys, *Am. J. Physiol.*, 219, 1606, 1970.

76. Racotta, R. and Russek, M., Food and water intake of rats after intraperitoneal and subcutaneous administration of glucose, glycerol and sodium lactate, *Physiol. Behav.*, 18, 267, 1977.

77. Langhans, W., Damaske, U., and Scharrer, E., Different metabolites might reduce food intake by the mitochondrial generation of reducing equivalents, *Appetite*, 6, 143, 1985.

78. Tordoff, M.G., Ulrich, P.M., and Sandler, F., Flavor preferences and fructose: evidence that the liver detects the unconditioned stimulus for calorie-based learning, *Appetite*, 14, 29, 1990.

79. Langhans, W., Egli, G., and Scharrer, E., Selective hepatic vagotomy eliminates the hypophagic effect of different metabolites, *J. Auton. Nerv. Syst.*, 13, 255, 1985.

80. Nagase, H., Bray, G. A., and York, D. A., Effects of pyruvate and lactate on food intake in rat strains sensitive and resistant to dietary obesity, *Physiol. Behav.*, 59, 555, 1996.

81. Cardin, S., Helie, R., Bergeron, R., Comte-B; van de Werve, G., and Lavoie, J. M., Effect of hepatic portal infusion of pyruvate on pancreatic hormone response during exercise, *Am. J. Physiol.,* 266, R1630, 1994.

82. Lutz, T. A., Niijima, A., and Scharrer, E., Intraportal infusion of 2,5-anhydro-D-mannitol increases afferent activity in the common hepatic vagus branch, *J. Auton. Nerv. Syst.,* 61, 204, 1996.

83. Friedman, M. I., An energy sensor for control of energy intake, *Proc. Nutr. Soc.,* 56, 41, 1997.

84. Scharrer, E., Rossi, R., Sutter, D. A., Seebacher, M. C., Boutellier, S., and Lutz, T. A., Hyperpolarization of hepatocytes by 2,5-AM: Implications for hepatic control of food intake, *Am. J. Physiol.,* 272, R874, 1997.

85. Lutz, T. A., Boutellier, S., and Scharrer, E., Hyperpolarization of the rat hepatocyte membrane by 2,5-anhydro-D-mannitol in vivo, *Life Sci.,* 62, 1427, 1998.

86. Tordoff, M. G., Rawson, N., and Friedman, M. I., 2,5-Anhydro-D-mannitol acts in liver to initiate feeding, *Am. J. Physiol.,* 261, R283, 1991.

87. Ritter, S., Dinh, T. T., and Friedman, M. I., Induction of Fos-like immunoreactivity (Fos-li) and stimulation of feeding by 2,5-anhydro-D-mannitol (2,5-AM) require the vagus nerve, *Brain Res.,* 646, 53, 1994.

88. Kurokawa, M., Sakata, T., Yoshimatsu, H., and Machidori, H., 2,5-Anhydro-D-mannitol — its unique central action on food intake and blood glucose in rats, *Brain Res.,* 566, 270, 1991.

89. Horn, C. C. and Friedman, M. I., Metabolic inhibition increases feeding and brain Fos-like immunoreactivity as a function of diet, *Am. J. Physiol.,* 275, R448, 1998.

90. Singer, L. K. and Ritter, S., Intraventricular glucose blocks feeding induced by 2-deoxy-D-glucose but not mercaptoacetate, *Physiol. Behav.,* 59, 921, 1996.

91. Calingasan, N. Y. and Ritter, S., Lateral parabrachial subnucleus lesions abolish feeding induced by mercaptoacetate but not by 2-deoxy-D-glucose, *Am. J. Physiol.,* 265, R1168, 1993.

22 Glucosensing Neurons in the Central Nervous System

Barry E. Levin, Vanessa H. Routh, and
Ambrose A. Dunn-Meynell

CONTENTS

1 INTRODUCTION

The brain has a vested interest in maintaining its supply of glucose. Along with oxygen, glucose is its primary fuel under physiological conditions.[1] As such, the nervous system has evolved mechanisms for the detection, utilization, and regulation of glucose metabolism in the body. Jean Mayer[2] first proposed the glucostatic hypothesis for the regulation of food intake based on a remarkable and prescient series of predictions about the physiological role of glucose. He reasoned that, since carbohydrate stores are limited, glucose availability would be a likely regulator of short-term energy intake. Mayer and colleagues showed that peripheral glucose utilization was directly tied to ratings of hunger and satiety[3] and predicted, based on Brobeck's lateral hypothalamic (LH) "feeding center" hypothesis,[4] that this area contained the "glucoreceptors" which integrated this response. He further predicted that "the passage of potassium ions into the glucoreceptor cells along with the glucose phosphate represents the point at which effective glucose level is translated into an electric or neural mechanism."[3] Later physiological studies by Oomura et al.[5] and Anand et al.[6] confirmed that the brain did indeed contain neurons which respond to changes in plasma glucose by altering their firing rate. LH neurons decreased and hypothalamic ventromedial nucleus (VMN) neurons increased their firing rates when plasma glucose levels rose. Oomura et al.[7] next showed that direct application of glucose had a similar effect on these neurons. They called neurons which increase their firing rate when glucose increases *glucose responsive* (GR) and those which decrease their firing rate when glucose rises *glucose sensitive* (GS). We now know that GR neurons are regulated by a K^+ channel which is sensitive to glucose availability as Mayer predicted.[3] Here we will also use the term *glucosensing* for any neuron which responds with altered function of any kind to changes in glucose.

Louis-Sylvestre and Le Magnen[8] showed that a 6 to 8% dip in plasma glucose levels preceded the onset of meals in rats by 5 to 6 min. While this observation was in keeping with the glucostatic hypothesis, it is still unclear whether such small changes in plasma glucose levels are either necessary or sufficient to stimulate the onset of feeding or whether glucosensing neurons are even capable of sensing such small changes. This issue aside, it is now well established that some neurons utilize glucose to modulate their rate of cell firing, transmitter or neurohormone release, or receptor binding. This wide range of glucose actions suggests that glucose might play a critical role in the regulation of energy homeostasis by acting directly and indirectly on a number of central systems involved in this process. We propose that the interaction of GS and GR neurons in brain areas such as the LH, VMN, arcuate hypothalamic nucleus (ARC), amygdala, and nucleus tractus solitarius (NTS) provides both a metabolic and behavioral foundation for this regulation of energy homeostasis. While we will emphasize the effect on energy intake, energy expenditure and storage must also be considered since the three are inextricably intertwined. To support this hypothesis, we will ask and attempt to answer three questions: (1) How does the brain sense glucose? (2) Where is glucose sensed? (3) Why does the brain sense glucose?

2   HOW DOES THE BRAIN SENSE GLUCOSE?

Although it is now well established that the brain contains neurons which can sense glucose, it is often not possible to isolate the direct effects of glucose upon these neurons from those resulting from afferent inputs from peripheral glucosensing sites such as hepatic portal glucosensors.[9,10] However, our main focus here will be on the direct effects of glucose on the brain. To act on central neurons, glucose must first cross the blood–brain barrier. It does this by facilitated transport using GLUT1 glucose transporters.[11] Once across the barrier, glucose uptake into most neurons is mediated by the GLUT3 transporter.[12] Depending on the method of measurement, interstitial glucose levels in the brain are estimated at 0.5 to 2.5 mM when plasma levels are in the basal physiological range (4 to 5 mM).[13,14] Interstitial glucose levels follow changes in plasma levels within 10 min under most conditions.[14] Since the K_m for glucose transport of both GLUT1 and GLUT3 are in the micromolar range,[11] both are fully saturated under physiological conditions and are, therefore, not likely to be critical regulators of the rate of glucose entry into the brain or neurons under most physiological conditions. Additionally, astrocytes, whose foot processes form part of the blood–brain barrier,[15] take up glucose and are capable of storing it as an energy buffer in the form of glycogen.[15] They contain GLUT2 transporters which have a K_m in the millimolar range,[11,16] making them potential regulators of the rate of glucose entry into astrocytes. To date, there has been no clear demonstration of GLUT2 in adult neurons although some may contain GLUT4 transporters which are insulin sensitive.[17]

2.1   GLUCOSE-RESPONSIVE NEURONS

Of the two types of glucosensing neurons, the GR neurons are best characterized with regard to potential sensing mechanisms. This is based on their apparent similarity to the pancreatic β-cell which utilizes an ATP-sensitive K$^+$ channel (K_{ATP}) as a GS regulator of insulin release.[18] The K_{ATP} channel consists of a regulatory subunit (the sulfonylurea receptor, SUR) and a pore-forming subunit (K_{IR}6.2).[19] The K_{IR}6.2 conducting pore is a member of the inwardly rectifying K$^+$ channel family.[20,21] It serves as an "energy sensor" to which ATP generated from glucose metabolism is directly bound. This inactivates (closes) the K_{ATP} channel[22] and decreases K$^+$ egress leading to increased intracellular K$^+$ levels, membrane depolarization, entry of Ca^{2+} and insulin release.[19] The SUR is a member of the ATP-binding cassette superfamily.[23] It is essential for the function of the K_{ATP} channel[20] and confers sensitivity to sulfonylureas which are insulin secretogogues by virtue of their binding to the

SUR which inactivates the K_{ATP} channel and depolarizes the cell, releasing insulin.[22] Importantly, a similar K_{ATP} channel is also found in GR neurons where a rise in the intracellular ATP:ADP ratio leads to channel inactivation, membrane depolarization, and increased neuronal activity.[24-27] Neurotransmitters and neuropeptides might potentially influence the activity of the K_{ATP} channel in neurons by activation of intracellular protein kinases to change the phosphorylation state of SUR and/or $K_{IR}6.2$, thus altering K_{ATP} channel activity.[27] In addition, both leptin[28] and insulin[29] activate the K_{ATP} channel causing hyperpolarization of VMN neurons. Thus, the K_{ATP} channel on GR neurons represents a potentially critical point at which metabolic signals from the periphery and endogenous neuromodulators might converge upon neurons involved in the regulation of energy homeostasis.

The pore-forming $K_{IR}6.2$ and both a high- (SUR1) and a low-affinity (SUR2) SUR have been cloned and are present in the brain.[30,31] Since the SUR and $K_{IR}6.2$ form the functional K_{ATP} channel, both binding of radiolabeled sulfonylureas and neurophysiological studies have been used to assess the anatomical distribution and cellular location of the K_{ATP} channel in the brain.[26,30,32,33] In areas such as the VMN or hypothalamic paraventricular nucleus (PVN), GR neurons compose 20 to 40% of the total of sampled neurons.[34,35] These same areas contain a similar proportion of low-affinity sulfonylurea-binding sites (possibly SUR2),[33] which appear to reside predominantly on neuronal cell bodies.[26] The high-affinity binding site (possibly SUR1) appears to be on nerve terminals of GABA[26] and glutamate[36] neurons. At these terminals, both glucose and sulfonylureas can evoke transmitter release independent of neuronal activity.[37] It appears that $K_{IR}6.2$ is the primary form of ATP-sensitive pore-forming unit in the brain. Thus, K_{ATP} channel function is determined by the type and location of the SUR with which it is complexed,[19,32] and it is these SURs which are sensitive to the physiological state of the animal. For example, rats prone to develop diet-induced obesity (obesity-prone) have few if any low-affinity sites in their hypothalamus or amygdala while these sites make up 20 to 45% of total SUR sites in rats resistant to dietary obesity (obesity-resistant) or in fully obese rats.[33] Also, binding to low-affinity sites is highly dependent upon ambient glucose concentrations and this too is dependent upon the weight gain phenotype of the animal.[38,39]

But both SUR binding[26,30,33] and *in situ* hybridization for $K_{IR}6.2$ mRNA[32,40] show a much wider anatomical distribution than expected from physiological studies of GR neurons. This suggests that some neurons contain K_{ATP} channels but do not necessarily sense physiological levels of glucose by altering their firing rate. Since the K_{ATP} channel is responsive to intracellular ATP/ADP levels, anything which severely disrupts ATP formation will activate the channel and hyperpolarize these cells.[25,41] This would serve a neuroprotective role, preventing these neurons from being driven beyond their limits when energy substrate was restricted during anoxia, hypoglycemia or other states associated with massive release of the excitatory amino acid glutamate.[41]

As with the pancreatic β-cell,[42,43] GR neurons may differ from all the others containing the K_{ATP} channel by containing one or more "gatekeepers" for glucose entry and metabolism (see Figure 1). These would govern K_{ATP} channel function by controlling the rate of ATP formation. In the β-cell, glucose entry is regulated by GLUT2[42] and the rate of glucose metabolism by glucokinase.[43] These potentially regulate the rate of ATP formation from glucose by virtue of their K_m for glucose in the plasma glucose (m*M*) range. But it unclear if neurons contain either a transporter or hexokinase with an appropriate K_m for brain glucose levels. While both GLUT2[16,44,45] and glucokinase[44,45] may be present in the brain, it is uncertain whether either is present normally in neurons. Finally, intracellular ATP levels are highly buffered so that a generalized increase in ATP production would be unlikely to affect K_{ATP} channel function. Instead, it has been proposed that glycolytic enzymes (including a high K_m hexokinase like glucokinase) are embedded in the plasma membrane where ATP generated during glycolysis could be directly "tunneled"[46] to the adjacent K_{ATP} channel (see Figure 1). Thus, non-GR neurons containing a K_{ATP} channel would not contain this glucosensing machinery and could not use glucose as a signaling molecule. Instead, they would utilize lactate (provided by astrocytes[15]) or exogenous glucose as substrate for their ongoing metabolic needs. The K_{ATP} channel in these neurons would serve a neuroprotective role when energy substrate became limited.

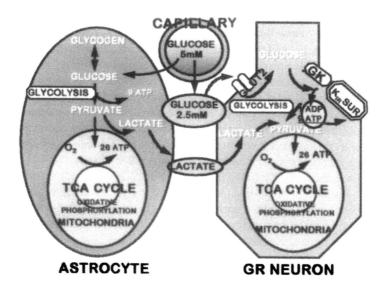

FIGURE 1   Proposed model of the way in which GR neurons use glucose as a signal to control the rate of cell firing. Glucose from the plasma (5 mM) crosses the blood–brain barrier where it is taken up and stored in the form of glycogen in astrocytes. Astrocytes utilize glycogen to produce lactate which is released into the extracellular space and can be taken up and utilized for mitochondrial oxidative phosphorylation in neurons after conversion back to pyruvate. This provides the majority of ATP for neuronal housekeeping functions. Some glucose from the plasma equilibrates with the extracellular fluid at 2 to 2.5 mM where it can be taken up into non-GR neurons via a GLUT3 transporter (not shown). Glycolysis supplements the lactate-derived ATP supplies in these neurons when neuronal activity and energy demands increase. In the GR neuron pictured here, "gatekeepers" in the form of either GLUT2 and/or glucokinase (GK) control the rate of glucose metabolism by virtue of their low mM K_m. In addition, glycolytic enzymes (including GK) are located within the plasmalemma, near the K_{ATP} channel ($K_{IR}6.2$ and SUR1 or 2) where ATP generated by this reaction regulates the activity of the channel. These neurons can use ATP generated by oxidative phosphorylation for ongoing metabolic demands but use glucose specifically as a signaling molecule.

2.2   GLUCOSE-SENSITIVE NEURONS

As with GR neurons, these cells form a subset of resident neurons within a given brain area. Unlike GR neurons, GS neurons increase their firing rates when ambient glucose levels fall.[7,47] An important feature is their location in areas such as the LH[47] and amygdala[48] which are linked to motivational systems involved in food intake.[49,50] Thus, when animals are cued to expect a food reward, GS neurons in these areas "learn" to respond to the conditioned stimulus as well as the actual presentation of the food itself.[49] Since GS neurons directly monitor glucose levels and receive inputs from peripheral glucosensing organs[51] they have the potential to integrate these signals with the affective components of food intake (Figure 2). Relatively little is known about the mechanisms by which glucose regulates the firing rate of GS neurons. While the Na$^+$,K$^+$-ATP pump may be involved,[47] some presumptive GS neurons contain a K$^+$ channel which, unlike the K_{ATP} channel in GR neurons, is inactivated (closed) when glucose levels fall.[52] This would lead to increased firing under conditions of low glucose availability.

2.3   GLUCOSE-MODULATED RECEPTOR BINDING

In addition to regulating neuronal firing and transmitter release, glucose also affects the binding of various ligands to their receptors by an unknown mechanism. For example, binding to α_2-adrenoceptors[53,54] and a set of putative anorectic binding sites in the brain[53] is increased by rising

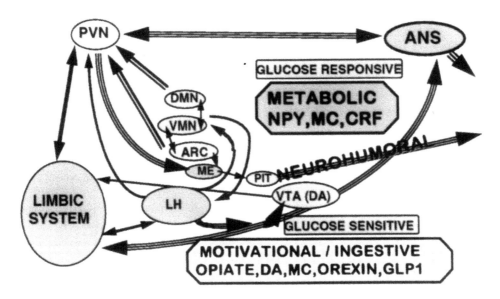

FIGURE 2 Schematic of the interrelationship of some of the forebrain systems involved in the metabolic and motivational aspects of energy homeostasis. This postulates that GR neurons are primarily distributed in metabolic areas and GS neurons function primarily in systems involved in motivation. The hypothalamic DMN, VMN, and ARC nuclei are highly interconnected and receive and integrate multiple metabolic and neural signals from the periphery and other brain areas. ARC and DMN neurons project to the PVN which has major projections to autonomic nervous system (ANS) and neurohumoral (median eminence, ME, and pituitary, PIT) effectors. Neuropeptide Y (NPY), melanocortins (MC: α-melanocyte stimulating hormone and agouti-related peptide), and corticotrophin releasing factor (CRF) are involved in the metabolic processes regulating ingestive behavior, energy expenditure, and storage. Dopamine (DA) neurons in the ventral tegmental area (VTA) project through the LH to areas of the limbic system (medial prefrontal cortex, nucleus accumbens, bed nucleus of the stria terminalis, central and basolateral amygdala) where they interact with opiate, melanocortin, orexin, glucagon-like peptide 1 (GLP1), GABA, and glutamate systems. These pathways receive visceral inputs from the periphery and integrate the affective associations related to ingestion. They are highly integrated with the metabolic systems as well as those involved in motivated behaviors of all sorts.

glucose levels, depending upon the weight gain phenotype of the animal.[54] Additionally, glucose alters binding to opiate receptors,[55] as well as to low-affinity SUR binding sites.[33] Thus, glucose-modulated receptor binding can alter neuronal function in a variety of ways which link it to both the metabolic and motivational components of ingestive behavior.

3   WHERE IS GLUCOSE SENSED?

We have seen how glucose availability can alter cell firing, transmitter release, and receptor binding in those select neurons which contain the appropriate channels, pumps, and receptors to transduce a glucose-related signal. In addition, GR and GS neurons can respond to and integrate a variety of metabolic and affective signals from other brain areas which regulate both the metabolic and motivational aspects of food intake and energy homeostasis. They are also integrated into efferent systems which modulate the acquisition, assimilation, expenditure, and storage of energy (Figure 2). For example, direct infusion of glucose into the carotid artery selectively activates neurons in the PVN, VMN, ARC, and dorsomedial hypothalamic nucleus (DMN). The VMN and DMN are interconnected and the ARC and DMN project to the PVN which projects directly to sympathetic outflow areas in the medulla and spinal cord.[56-58] Carotid glucose infusions both activate a subpopulation of these efferent PVN neurons[57] and stimulate sympathetic activity.[59,60] Again, this activation of neurons and sympathetic outflow is highly dependent upon the weight gain phenotype of the animal. Only obesity-prone rats activate their sympathetic nervous system to such infusions,[60] yet only diet-resistant rats activate neurons in the PVN, VMN, ARC, and DMN to this stimulus (unpublished observation).

GR neurons have been identified electrophysiologically in the VMN,[61,62] PVN,[63] DMN,[64] pituitary,[65] globus pallidus,[64] substantia nigra,[66,67] locus coeruleus,[68] caudal NTS,[69,70] area postrema,[71] and the dorsal horn of the spinal cord.[72] Neurons which express $K_{IR}6.2$ mRNA for the pore-forming unit of the K_{ATP} channel include ARC neuropeptide Y (NPY) and dopamine, substantia nigra dopamine and locus coeruleus norepinephrine, and substantia nigra GABA neurons.[40] In addition, glucose and/or sulfonylureas increase release of GABA in the substantia nigra,[37] growth hormone in anterior pituitary cells,[65] acetylcholine in the striatum,[73-75] and glutamate in motor cortex neuronal synaptosomes.[36]

Induction of intracellular glucopenia increases cFos expression (an index of neuronal activation) in presumptive GS neurons in many of the same areas in which GR neurons have been identified. These include the PVN, LH, DMN, and ARC, plus the central nucleus of the amygdala, lateral parabrachial nucleus, locus coeruleus, dorsal raphe, A1 and C1 catecholamine neurons, and dorsal motor nucleus of the vagus.[76,77] Electrophysiological studies have identified GS neurons in the amygdala,[48] LH,[64,78] globus pallidus,[79] and NTS.[69] Thus, GR and GS neurons reside in a number of brain areas involved in the motivational, neuroendocrine, and metabolic aspects of energy homeostasis. Their integration within these systems gives them the potential of receiving and integrating a multiplicity of metabolic and neural signals relating to energy homeostasis. This leads us to our third question.

4   WHY DOES THE BRAIN SENSE GLUCOSE?

4.1   GLUCOSE AS A MODULATOR OF FEEDING

To date, there is little proof of Mayer's postulate that glucose is a physiological regulator of ingestion,[2] let alone of macronutrient selection. Certainly, the small, transient fall in plasma glucose levels preceding meal initiation suggests a connection.[8] But it is uncertain that such small changes in plasma glucose could possibly be detected by any central or peripheral sensor as a signal for meal initiation. However, marked peripheral and/or central glucoprivation clearly does induce food intake.[80,81] On the other hand, hyperglycemia can produce satiety.[82] Intraportal glucose infusions decrease the future intake of glucose,[83] while intracerebroventricular glucose produces a slight decrease in food intake in lean but not obese Zucker rats.[80] But all of these perturbations of glucose levels are orders of magnitude greater than the 6 to 8% decline in plasma glucose levels which precede meal initiation in rats.[8]

Some insights can be gained by dividing ingestive behavior into motivational and metabolic components (see Figure 2). Since GS neurons respond to affective cues related to ingestion,[84] they may mediate responses to changes in diet palatability and the emotional context of meals as part of the motivational component of ingestion. On the other hand, GR neurons are located largely in brain areas such as the ARC, VMN, DMN, and PVN which are linked to autonomic and neuroendocrine systems which could alter the metabolic drive to eat. To paraphrase the ventromedial hypothalamic "satiety center" and LH "feeding center" concept,[4,85] glucopenia (a metabolic drive) would stimulate food intake by turning off VMN GR neurons, blocking their inhibitory effects on meal initiation. Thus, either inhibition or destruction of VMN GR neurons would dysinhibit the LH feeding center and any of several other feeding-related systems. This would explain the hyperphagia seen after medial hypothalamic lesions.[86] Palatability and the emotional context of meal taking would be encoded by GS neurons in the LH-limbic pathways with links to neuroendocrine systems involved in the metabolic drive to eat. Thus, two parallel but interacting and overlapping glucosensing systems would regulate food intake, one motivational and one metabolic.

It is quite possible that glucose might affect macronutrient selection by acting on one or both of these two systems, but there are no data addressing this issue to date.

4.2 Glucose as a Modulator of Metabolism

Glucose availability can affect the function of systems which control energy homeostasis. But because most of our measures of energy expenditure and/or neuroendocrine function are relatively insensitive, most studies use extremes of glucose levels to induce changes in these systems. Thus, it is unclear whether glucose serves as a physiological regulator or, instead, functions only at these extremes as an emergency system. When glucose levels rise, glucose is stored as glycogen, used for ongoing energy demands, and the excess is expended as heat.[87,88] Thus, hyperglycemia activates the sympathetic nervous system via a centrally-mediated process which does not require insulin.[59,60,89-92] Carotid glucose infusions also increase firing in the sympathetic nerve to interscapular brown adipose tissue,[93] suggesting a role in glucose-induced thermogenesis. Carotid glucose infusions also increase vagal output to the pancreas,[94] and this is mimicked by glucose infusion into the LH but not VMN.[95] Thus, the effects upon energy homeostasis of raising glucose levels varies depending upon route, dose, and site of administration.

On the other hand, a drop in glucose below a critical level needed for cerebral function evokes a centrally mediated counter regulatory response in which the adrenal medulla releases epinephrine and the cortex releases cortisol.[96,97] This response can be initiated by local 2-deoxyglucose infusion into the VMN,[96] while the response to systemic hypoglycemia can be prevented by neurotoxin lesions of the VMN.[98] Since the VMN contains a relatively high proportion of GR and low proportion of GS neurons,[29] this suggests that glucopenia-induced inhibition of VMN GR neurons may underlie the counterregulatory response to glucoprivation.

4.3 Glucose as a Modulator of Body Weight

Numerous studies suggest that glucosensing neurons in the brain play a role in long-term body weight regulation. For example, injections of gold thioglucose, a presumptive toxin for glucosensing neurons, produces obesity.[99,100] Obesity-prone but not obesity-resistant rats activate their sympathetic nervous system following peripheral[89] or central glucose infusions.[101] But only resistant rats activate hypothalamic neurons to central glucose infusions (unpublished observation). In these same areas, obesity-prone rats have reduced low-affinity cell body SUR binding[102] and their VMN neuronal K_{ATP} channel does not function normally.[103] Similarly, genetically obese Zucker rats have abnormal VMN neuronal K_{ATP} channels.[104] Finally, obesity-prone rats have defective glucose-induced upregulation of α_2-adrenoceptor binding in their brains.[54] Thus, aberrant glucosensing appears to be a common feature of genetic and lesion-induced obesity suggesting that this defect plays a role in long-term body weight regulation.

5 GLUCOSENSING NEURONS AS INTEGRATORS OF METABOLIC SIGNALS

We have shown that glucose affects neurons involved in both the metabolic and motivational aspects of energy homeostasis. In fact, glucosensing neurons appear to be integrated into the final common pathways of a host of systems responsive to intrinsic and extrinsic signals relating metabolism and motivational state to neuronal function. Both GR and GS neurons respond to multiple metabolic and neurochemical signals including insulin,[29] leptin,[28] cholecystokinin (CCK),[71] monoamines, oxytocin and vasopressin,[62] GABA,[105] and cytokines.[106] The ARC NPY neurons are a good example of neurons which are linked to the metabolic regulation of energy homeostasis by virtue of their responsiveness to a number of metabolic and neuronal signals.[107-110] They project directly to the PVN which relays their signals to the pituitary and autonomic outflow areas of the brain stem and spinal cord.[56,58] ARC NPY neurons contain the K_{ATP} channel and colocalize agouti-related peptide.[112] They also contain the signal-transducing form of the leptin receptor[111] and may contain insulin receptors since insulin affects the firing of GR neurons.[29] It is likely that these ARC NPY neurons are only one of several neuronal populations which serve as central integrators upon which metabolic and neural signals from the periphery converge to relay critical information to autonomic and neuroendocrine effectors regulating energy homeostasis.

6   CONCLUSION

The brain has evolved multiple mechanisms for sensing, utilizing, and regulating glucose availability. We are at the threshold of our understanding of the underlying mechanisms that allow specialized neurons to perform this task. The brain appears to use glucose as a signal of the energetic status of the body. Glucosensing neurons also receive and integrate other metabolically relevant signals from the periphery as well as inputs from areas involved in the motivational aspects and ingestive behavior. At least at the low and high ends of glucose availability, glucose is involved in the regulation of energy consumption and expenditure. It is still unclear how glucose affects neural function and overall energy homeostasis within the narrower physiological range encountered in the day-to-day regulation of energy balance. Nevertheless, glucosensing neurons are likely to play an important role because they are highly integrated into all of the systems which affect the behavioral and metabolic aspects of energy homeostasis in the body.

REFERENCES

1. Sokoloff, L., Reivich, M., Kennedy, C., DesRosiers, M. H., Patlak, C. S., Pettigrew, O., Sakaruda, O., and Shinohara, M., The [¹⁴C]deoxyglucose method for the measurement of local cerebral glucose utilization: theory, procedure, and normal values in the conscious and anesthetized albino rat, *J. Neurochem.*, 23, 897, 1977.
2. Mayer, J., Glucostatic mechanism of regulation of food intake, *N. Engl. J. Med.*, 249, 13, 1953.
3. Van Itallie, T. B., Beaudoin, R., and Mayer, J., Ateriovenous glucose differences, metabolic hypoglycemia and food intake in man, *J. Clin. Nutr.*, 1, 208, 1953.
4. Brobeck, J. R., Physiology of appetite., in *Overeating, Overweight and Obesity: Proceedings of the Nutrition Symposium*, Boston National Vitamin Foundation, New York, 1953, 36.
5. Oomura, Y., Kimura, K., Ooyama, H., Maeo, T., Iki, M., and Kuniyoshi, N., Reciprocal activities of the ventromedial and lateral hypothalamic area of cats, *Science*, 143, 484, 1964.
6. Anand, B. K., Chhina, G. S., Sharma, K. N., Dua, S., and Singh, B., Activity of single neurons in the hypothalamus feeding centers: effect of glucose, *Am. J. Physiol.*, 207, 1146, 1964.
7. Oomura, Y., Ono, T., Ooyama, H., and Wayner, M.J., Glucose and osmosensitive neurons of the rat hypothalamus, *Nature*, 222, 282, 1969.
8. Louis-Sylvestre, J. and Le Magnen, J., A fall in blood glucose level precedes meal onset in free-feeding rats, *Neurosci. Biobehav. Rev.*, 4(Suppl. 1), 13, 1980.
9. Tordoff, M. G., Tluczek, J. P., and Friedman, M. I., Effect of hepatic portal glucose concentration on food intake and metabolism, *Am. J. Physiol.*, 257, R1474, 1989.
10. Hevener, A. L., Bergman, R. N., and Donovan, C. M., Novel glucosensor for hypoglycemic detection localized to the portal vein, *Diabetes*, 46, 1521, 1997.
11. Vannucci, S. J., Maher, F., and Simpson, I. A., Glucose transporter proteins in brain: delivery of glucose to neurons and glia, *Glia*, 21, 2, 1991.
12. Gerhart, D. Z., Broderius, M. A., Borson, N. D., and Drewes, L. R., Neurons and microvessels express the brain glucose transporter protein GLUT3, *Proc. Natl. Acad. Sci. U.S.A.*, 89, 733, 1992.
13. Fellows, L. K. and Boutelle, M. G., Rapid changes in extracellular glucose levels and blood flow in the striatum of the freely moving rat, *Brain Res.*, 604, 225, 1993.

14. Cummings, S. and Seybold, V., Relationship of alpha-1- and alpha-2-adrenergic-binding sites to regions of the paraventricular nucleus of the hypothalamus containing corticotropin-releasing factor and vasopressin neurons, *Neuroendocrinology*, 47, 523, 1988.

15. Wiesinger, H., Hamprecht, B., and Dringen, R., Metabolic pathways for glucose in astrocytes, *Glia*, 21, 22, 1997.

16. Leloup, C., Arluison, M., Lepetit, N., Cartier, N., Marfaing-Jallat, P., Ferre, P., and Penicaud, L., Glucose transporter 2 (GLUT 2): expression in specific brain nuclei, *Brain Res.*, 638, 221, 1994.

17. Leloup, C., Arluison, M., Kassis, N., Lepetit, N., Cartier, N., Ferre, P., and Penicaud, L., Discrete brain areas express the insulin-responsive glucose transporter GLUT4, *Brain Res. Mol. Brain Res.*, 45, 1996.

18. Gopalakrishnan, M., Janis, R. A., and Triggle, D. J., ATP-sensitive K^+ channels: pharmacologic properties, regulation, and therapeutic potential, *Drug Dev. Res.*, 28, 95, 1993.

19. Trapp, S. and Ashcroft, F. M., A metabolic sensor in action: news from the ATP-sensitive K^+- channel, *News Physiol. Sci.*, 12, 255, 1997.

20. Inagaki, N., Gonoi, T., Clement IV, J. P., Namba, N., Inazawa, J., Gonzalez, G., Aguilar-Bryan, L., Seino, S., and Bryan, J., Reconstitution of IKATP: an inward rectifier subunit plus the sulfonylurea receptor, *Science*, 270, 1166, 1995.

21. Sakura, H., Ammala, C., Smith, P. A., Gribble, P. A., and Ashcroft, F. M., Cloning and functional expression of the cDNA encoding a novel ATP-sensitive potassium channel subunit expressed in pancreatic beta-cells, brain, heart and skeletal muscle, *FEBS Lett.*, 377, 338, 1995.

22. Tucker, S. J., Gribble, F. M., Zhao, C., Trapp, S., and Ashcroft, F. M., Truncation of Kir6.2 produces ATP-sensitive K+ channels in the absence of the sulphonylurea receptor, *Nature*, 387, 179, 1997.

23. Fosset, M., Allard, B., and Lazdunski, M., Coexistence of two classes of glibenclamide-inhibitable ATP-regulated K^+ channels in avian skeletal muscle, *Eur. J. Physiol.*, 431, 117, 1995.

24. Ashford, M. L. J., Boden, P. R., and Treherne, J. M., Tolbutamide excites rat glucoreceptive ventromedial hypothalamic neurones by indirect inhibition of ATP-K+ channels, *Br. J. Pharmacol.*, 101, 531, 1990.

25. Roper, J. and Ashcroft, F. M., Metabolic inhibition and low internal ATP activate K-ATP channels in rat dopaminergic substantia nigra neurones, *Pflugers Arch.*, 430, 44, 1995.

26. Dunn-Meynell, A. A., Routh, V. H., McArdle, J. J., and Levin, B. E., Low affinity sulfonylurea binding sites reside on neuronal cell bodies in the brain, *Brain Res.*, 745, 1, 1997.

27. Routh, V. H., McArdle, J. J., and Levin, B. E., Phosphorylation modulates the activity of the ATP-sensitive K^+ channel in the ventromedial hypothalamic nucleus, Brain Res., 778, 107, 1997.

28. Spanswick, D., Smith, M. A., Groppi, V. E., Logan, S. D., and Ashford, M. L., Leptin inhibits hypothalamic neurons by activation of ATP-sensitive potassium channels, *Nature*, 390, 521, 1997.

29. Routh, V. H., McArdle, J. J., Spanswick, D. C., Levin, B. E., and Ashford, M. L. J., Insulin modulates the activity of glucose responsive neurons in the ventromedial hypothalamic nucleus (VMN), *Abst. Soc. Neurosci.*, 23, 577A, 1997.

30. Mourre, C., Widmann, C., and Lazdunski, M., Sulfonylurea binding sites associated with ATP-regulated K^+ channels in the central nervous system: autoradiographic analysis of their distribution and of their localization in mutant mice cerebellum, *Brain Res.*, 519, 29, 1990.

31. Zini, S., Tremblay, E., Roisin, M.-P., and Ben-Ari, Y., Two binding sites for [^{3}H] glibenclamide in the rat brain, *Brain Res.*, 542, 151, 1991.

32. Karschin, C., Ecke, C., Ashcroft, F. M., and Karschin, A., Overlapping distribution of K_{ATP} channel-forming unit Kir6.2 subunit and the sulfonylurea receptor SUR1 in rodent brain, *FEBS Lett.*, 401, 59, 1997.

33. Levin, B. E. and Dunn-Meynell, A., *In vivo* and *in vitro* regulation of ^{3}H glyburide binding to brain sulfonylurea receptors in obesity-prone and resistant rats by glucose, *Brain Res.*, 776, 146, 1998.

34. Kow, L. M. and Pfaff, D. W., Actions of feeding-relevant agents on hypothalamic glucose-responsive neurons in vitro, *Brain Res. Bull.*, 15, 509, 1985.

35. Minami, T., Oomura, Y., and Sugimori, M., Electrophysiological properties and glucose responsiveness of guinea-pig ventromedial hypothalamic neurones in vitro, *J. Physiol.*, 380, 127, 1986.

36. Lee, K., Dixon, A. K., Rowe, I. C. M., Ashford, M. L. J., and Richardson, P. J., The high-affinity sulfonylurea receptor regulates K_{ATP} channels in nerve terminal of the rat motor cortex, *J. Neurochem.*, 66, 2562, 1996.

37. Amoroso, S., Schmid-Antomarchi, H., Fosset, M., and Lazdunski, M., Glucose, sulfonylureas, and neurotransmitter release: role of ATP-sensitive K⁺ channels, *Science*, 247, 852, 1990.

38. Clement IV, J. P., Kunjilwar, K., Gonzalez, G., Schwanstecher, M., Panten, U., Aguilar-Bryan, L., and Bryan, J., Association and stoichiometry of K(atp) channel subunits, *Neuron*, 18, 827, 1997.

39. Levin, B. E. and Dunn-Meynell, A., Effect of streptozotocin-induced diabetes on rat brain sulfonylurea biding sites, *Brain Res. Bull*, 46, 513, 1998.

40. Dunn-Meynell, A. A., Rawson, N. E., and Levin, B. E., Distribution and phenotype of neurons containing the ATR sensitive K⁺ channel in rat brain, *Brain Res.*, 814, 41, 1998.

41. Jiang, C., Xia, Y., and Haddad, G. G., Role of ATP-sensitive K⁺ channels during anoxia: major differences between rat (newborn and adult) and turtle neurons, *J. Physiol.*, 448, 599, 1992.

42. Schuit, F. C., Is GLUT2 required for glucose sensing? *Diabetologia*, 40,104, 1997.

43. Matchinsky, F. M., A lesson in metabolic regulation inspired by the glucokinase glucose sensor paradigm, *Diabetes*, 45, 223, 1996.

44. Jetton, T. L., Liang, Y., Pettepher, C. C., Zimmerman, E. C., Cox, F. G., Horvath, K., Matschinsky, F. M., and Magnuson, M. A., Analysis of upstream glucokinase promoter activity in transgenic mice and identification of glucokinase in rare neuroendocrine cells in the brain and gut, *J. Biol. Chem.*, 269, 3641, 1994.

45. Navarro, M., de Fonseca, F. R., Alvarez-Buylla, R., Chowen, J. A., Zueco, J. A., Gomez, R., Eng, J., and Blazquez, E., Colocalization of glucagon-like peptide-1 (GLP-1) receptors, glucose transporter GLUT-2, and glucokinase mRNAs in rat hypothalamic cells: evidence for a role of GLP-1 receptor antagonists as an inhibitory signal for food and water intake, *J. Neurochem.*, 67, 1982, 1996.

46. Lynch, R. M., Mejia, R., and Balaban, R. S., Energy transduction at the cell plasmalemma: coupling through membrane associated adenosine triphosphate? *Commun. Mol. Cell. Biophys.*, 5, 151, 1988.

47. Oomura, Y., Ooyama, H., Sugimori, M., Nakamura, T., and Yamada, Y., Glucose inhibition of the glucose-sensitive neurone in the rat lateral hypothalamus, *Nature*, 247, 284, 1974.

48. Nakano, Y., Oomura, Y., Lenard, L., Nishino, H., Aou, S., Yamamoto, T., and Aoyagi, K., Feeding-related activity of glucose- and morphine-sensitive neurons in the monkey amygdala, *Brain Res.*, 399, 167, 1986.

49. Lénárd, L., Oomura, Y., Nakano, Y., Aou, S., and Nishino, H., Influence of acetylcholine on neuronal activity of monkey amygdala during bar press feeding behavior, *Brain Res.*, 500, 359, 1989.

50. Gallagher, M., Graham, P. W., and Holland, P. C., The amygdala central nucleus and appetitive Pavlovian conditioning: lesions impair one class of conditioned behavior, *J. Neurosci.*, 10, 1906, 1990.

51. Adachi, A., Shimizu, N., Oomura, Y., and Kobashi, M., Convergence of heptoportal glucose-sensitive afferent signals to glucose-sensitive units within the nucleus of the solitary tract, *Neurosci. Lett.*, 46, 215, 1984.

52. Ashford, M. L. J., Sturgess, N. J., Trout, N. J., Gardner, N. J., and Hales, C. N., Adenosine-5′-triphosphate-sensitive ion channels in neonatal rat cultured central neurones, *Pflugers Arch.*, 412, 297, 1988.

53. Angel, I. and Taranger, M. A., Coupling between hypothalamic α₂-adrenoceptors and [³H]mazindol binding sites in response to several hyperglycemic stimuli in mice, *Brain Res.*, 490, 367, 1991.

54. Levin, B. E. and Planas, B., Defective glucoregulation of brain α₂-adrenoceptors in obesity-prone rats, *Am. J. Physiol.*, 264, R305, 1993.

55. Brase, D. A., Han, Y.H., and Dewey, W. L., Effects of glucose and diabetes on binding of naloxone and dihydromorphine to opiate receptors in mouse brain, *Diabetes*, 36, 1173, 1987.

56. Bai, F. L., Yamano, M., Shiotani, Y., Emson, P. C., Smith, A. D., Powell, J. F., and Tohyama, M., An arcuato-paraventricular and dorsomedial hypothalamic neuropeptide Y containing system which lacks norepinephrine in the rat, *Brain Res.*, 331, 172, 1985.

57. Dunn-Meynell, A. A., Govek, E., and Levin, B. E., Intracarotid glucose infusions selectively increase Fos-like immunoreactivity in paraventricular, ventromedial and dorsomedial nuclei neurons, *Brain Res.*, 748, 100, 1997.

58. Swanson, L. W. and Kuypers, H. G. J. M., The paraventricular nucleus of the hypothalamus: cytoarchitectonic subdivisions and organization of projections to the pituitary, dorsal vagal complex, and spinal cord as demonstrated by retrograde fluorescence double-labeling methods, *J. Comp. Neurol.*, 194, 555, 1980.

59. Levin, B. E., Glucose increases rat plasma norepinephrine levels by direct action on the brain, *Am. J. Physiol. Regul. Integr. Comp. Physiol.*, 261, R1351, 1991.

60. Levin, B. E., Intracarotid glucose-induced norepinephrine response and the development of diet-induced obesity, *Int. J. Obesity,* 16, 451, 1992.

61. Sellers, A. J., Boden, P. R., and Ashford, M. L. J., Lack of effect of potassium channel openers on ATP-modulated potassium channels recorded from rat ventromedial hypothalamic neurones, *Br. J. Pharmacol.,* 107, 1068, 1992.

62. Kow, L. M. and Pfaff, D. W., Vasopressin excites ventromedial hypothalamic glucose-responsive neurons in vitro, *Physiol. Behav.,* 37, 153, 1986.

63. Kow, L. M. and Pfaff, D. W., Responses of hypothalamic paraventricular neurons *in vitro* to norepi-nephrine and other feeding-relevant agents, *Physiol. Behav.,* 46, 265, 1989.

64. Fukuda, M., Ono, T., Nishino, H., and Sasaki, K., Independent glucose effects on rat hypothalamic neurons: an *in vitro* study, *J. Auton. Nerv. Syst.,* 10, 373, 1984.

65. Bernardi, H., de Weille, J. R., Epelbaum, J., Mourre, C., Amoroso, S., Slama, A., Fosset, M., and Lazdunski, M., ATP-modulated K^+ channels sensitive to antidiabetic sulfonylureas are present in adenohypophysis and are involved in growth hormone release, *Proc. Natl. Acad. Sci. U.S.A.,* 9, 1340, 1993.

66. Hausser, M. A., de Weille, J. R., and Lazdunski, M., Activation by cromakalim of pre- and post-synaptic ATP-sensitive K^+ channels in substantia nigra, *Biochem. Biophys. Res. Commun.,* 174, 909, 1991.

67. Schwanstecher, C. and Panten, U., Tolbutamide- and diazoxide-sensitive K^+ channel in neurons of substantia nigra pars reticulata, *Naunyn-Schmiedebargg Arch. Pharmacol.,* 348, 113, 1993.

68. Illes, P., Sevcik, J., Finta, E. P., Frolich, R., Nieber, K., and Norenberg, W., Modulation of locus coeruleus neurons by extra- and intracellular adenosine 5′-triphosphate, *Brain Res. Bull.,* 35, 513, 1994.

69. Mizuno, Y. and Oomura, Y., Glucose responding neurons in the nucleus tractus solitarius of the rat, *in vitro* study, *Brain Res.,* 307, 109, 1984.

70. Yettefti, K., Orsini, J. C., and Perrin, J., Characteristics of glycemia-sensitive neurons in the nucleus tractus solitarii: possible involvement in nutritional regulation, *Physiol. Behav.,* 61, 93, 1997.

71. Funahashi, M. and Adachi, A., Glucose-responsive neurons exist within the area postrema of the rat: *in vitro* study on the isolated slice preparation, *Br. Res. Bull.,* 32, 531, 1993.

72. Yamashita, S., Park, J. B., Ryu, P. D., Inukai, H., Tanifuji, M., and Murase, K., Possible presence of the ATP-sensitive K^+ channel in isolated spinal dorsal horn neurons of the rat, *Neurosci. Lett.,* 170, 208, 1994.

73. Lee, K., Brownhill, V., and Richardson, P. J., Antidiabetic sulphonylureas stimulate acetylcholine release from striatal cholinergic interneurons through inhibition of K_{ATP} channel activity, *J. Neurochem.,* 69, 1774, 1997.

74. Takahashi, A., Ishimaru, H., Ikarashi, Y., Kishi, E., and Maruyama, Y., Hypothalamic cholinergic activity associated with 2-deoxyglucose-induced hyperglycemia, *Brain Res.,* 734, 116, 1996.

75. Ragozzino, M. E., Unick, K. E., and Gold, P. E., Hippocampal acetylcholine release during memory testing in rats: augmentation by glucose, *Proc. Natl. Acad. Sci. U.S.A.,* 93, 4693, 1996.

76. Ritter, S. and Dinh, T. T., 2-Mercaptoacetate and 2-deoxy-d-glucose induce Fos-like immunoreactivity in rat brain, *Brain Res.,* 641, 111, 1994.

77. Niimi, M., Sato, M., Tamaki, M., Wada, Y., Takahara, J., and Kawanishi, K., Induction of Fos protein in the rat hypothalamus elicited by insulin-induced hypoglycemia, *Neurosci. Res.,* 23, 361, 1995.

78. Orsini, J. C., Armstrong, D. L., and Wayner, M. J., Responses of lateral hypothalamic neurons recorded *in vitro* to moderate changes in glucose concentration, *Brain Res. Bull.,* 29, 503, 1992.

79. Lenard, L., Karadi, Z., Faludi, B., Czurko, A., Niedetzky, C., Vida, I., and Nishino, H., Glucose-sensitive neurons of the globus pallidus: I. Neurochemical characteristics, *Br. Res. Bull.,* 37, 149, 1995.

80. Tsujii, S. and Bray, G. A., Effects of glucose, 2-deoxyglucose, phlorizin, and insulin on food intake of lean and fatty rats, *Am. J. Physiol.,* 258, E476, 1990.

81. Ritter, S., Murnane, S. M., and Ladenheim, E. E., Glucoprivic feeding is impaired by lateral or fourth ventricular alloxan injection, *Am. J. Physiol.,* 243, R312, 1982.

82. Gielkens, H. A. J., Verkijk, M., Lam, W. F., Lamers, C. B. H. W., and Masclee, A. A. M., Effects of hyperglycemia and hyperinsulinemia on satiety in humans, *Metabolism,* 47, 321, 1998.

83. Baird, J. P., Grill, H. J., and Kaplan, J. M., Intake suppression after hepatic portal glucose infusion: all-or-none effect and its temporal threshold, *Am. J. Physiol.,* 272, R1454, 1997.

84. Aou, S., Oomura, Y., Lenard, L., Nishino, H., Inokuchi, A., Minami, T., and Misaki, H.Y., Behavioral significance of monkey hypothalamic glucose-sensitive neurons, *Brain Res.*, 302, 69, 1984.

85. Stellar, E., The physiology of motivation, *Psychol. Rev.*, 5, 5, 1954.

86. Hetherington, A. W. and Ranson, S. W., Hypothalamic lesions and adiposity in the rat, *Anat. Rec.*, 78, 149, 1940.

87. Schutz, Y., Golay, A., Felberg, J. P., and Jequier, E., Decreased glucose-induced thermogenesis after weight loss in obese subjects: a predisposing factor for relapse of obesity? *Am. J. Clin. Nutr.*, 39, 380, 1984.

88. Le Feuvre, R. A., Woods, A. J., Stock, M. J., and Rothwell, N. J., Effects of central injection of glucose on thermogenesis in normal, VMH-lesioned and genetically obese rats, *Brain Res.*, 547, 110, 1991.

89. Levin, B. E. and Sullivan, A. C., Glucose-induced norepinephrine levels and obesity resistance, *Am. J. Physiol.*, 253, R475, 1987.

90. Levin, B. E. and Sullivan, A. C., Glucose-induced sympathetic activation in obesity-prone and resistant rats, *Int. J. Obesity*, 13, 235, 1989.

91. Levin, B. E. and Sullivan, A. C., Glucose, insulin and sympathoadrenal activation, *J. Auton. Nerv. Syst.*, 20, 233, 1987.

92. Tappy, L., Randin, J. P., Felber, J. P., Chiolero, R., Simonsen, D. C., Jequier, E., and DeFronzo, R., Comparison of thermogenic effect of fructose and glucose in normal humans, *Am. J. Physiol.*, 250, E718, 1986.

93. Sakaguchi, T. and Bray, G. A., Ventromedial hypothalamic lesions attenuate responses of sympathetic nerves to carotid arterial infusions of glucose and insulin, *Int. J. Obesity*, 14, 127, 1990.

94. Niijima, A., The effect of glucose on the activity of the adrenal nerve and pancreatic branch of the vagus nerve in the rabbit, *Neurosci. Lett.*, 1, 159, 1975.

95. Niijima, A., Kannan, H., and Yamashita, H., Neural control of blood glucose homeostasis: effect of microinjection of glucose into hypothalamic nuclei on efferent activity of pancreatic branch of vagus nerve in the rat, *Brain Res. Bull.*, 20, 811, 1988.

96. Borg, W. P., Sherwin, R. S., During, M. J., Borg, M. A., and Shulman, G. I., Local ventromedial hypothalamic glucopenia triggers counterregulatory hormone release, *Diabetes*, 44, 180, 1995.

97. Dagogo-Jack, S., Rattarasarn, C., and Cryer, P. E., Reversal of hypoglycemia unawareness, but not defective glucose counterregulation, in IDDM, *Diabetes*, 43, 1426, 1994.

98. Borg, W. P., During, M. J., Sherwin, R. S., Borg, M. A., Brines, M. L., and Shulman, G. I., Ventromedial hypothalamic lesions in rats suppress counterregulatory responses to hypoglycemia, *J. Clin. Invest.*, 93, 1677, 1994.

99. Marks, J. L., Waite, K., Cameron-Smith, D., Blair, S. C., and Cooney, G. J., Effects of gold thioglucose on neuropeptide Y messenger RNA levels in the mouse hypothalamus, *Am. J. Physiol.*, 270, R1208, 1996.

100 Bergen, H. T., Monkman, N., and Mobbs, C. V., Injection with gold thioglucose impairs sensitivity to glucose: evidence that glucose-responsive neurons are important for long-term regulation of body weight, *Brain Res.*, 734, 332, 1996.

101. Smythe, G. A., Bradshaw, J. E., Nicholson, M. V., Grunstein, H. S., and Storlein, L. H., Rapid bidirectional effects of insulin on hypothalamic noradrenergic and serotoninergic neuronal activity in the rat: role in glucose homeostasis, *Endocrinology*, 117, 1590, 1985.

102. Levin, B. E., Brown, K. L., and Dunn-Meynell, A. A., Differential effects of diet and obesity on high and low affinity sulfonylurea binding sites in the rat brain, *Brain Res.*, 739, 293, 1996.

103. Routh, V. H., Levin, B. E., and McArdle, J. J., Defective ATP-sensitive K⁺ (K_{ATP}) channel in ventro-medial hypothalamic nucleus (VMN) of obesity-prone (DIO) rats, *FASEB J.*, 12, A864, 1998.

104. Rowe, I. C., Boden, P. R., and Ashford, M. L., Potassium channel dysfunction in hypothalamic glucose-receptive neurones of obese Zucker rats, *J. Physiol.*, 497, 365, 1996.

105. During, M. J., Leone, P., Davis, K. E., Kerr, D., and Sherwin, R. S., Glucose modulates rat substantia nigra GABA release *in vivo* via ATP-sensitive potassium channels, *J. Clin. Invest.*, 95, 2403, 1995.

106. Kuriyama, K., Hori, T., Mori, T., and Nakashima, T., Actions of interferon and interleukin-1 on the glucose-responsive neurons in the ventromedial hypothalamus, *Br. Res. Bull.*, 24, 803, 1990.

107. Schwartz, M. W., Sipols, A. J., Marks, J. L., Sanacora, G., White, J. D., Scheurink, A., Kahn, S. E., Baskin, D. G., Woods, S. C., Figlewicz, D. P., and Porte, D., Jr., Inhibition of hypothalamic neuropeptide Y gene expression by insulin, *Endocrinology*, 130, 3608, 1992.
108. Giraudo, S. Q., Kotz, C. M., Grace, M. K., Levine, A. S., and Billington, C. J., Rat hypothalamic NPY mRNA and brown fat uncoupling protein mRNA after high-carbohydrate or high-fat diets, *Am. J. Physiol.*, 266, R1578, 1994.
109. Schwartz, M. W., Baskin, D. G., Bukowski, T. R., Kuijper, J. L., Foster, D., Lasser, G., Prunkard, D. E., Porte, D., Jr., Woods, S. C., Seeley, R. J., and Weigle, D. S., Specificity of leptin action on elevated blood glucose levels and hypothalamic neuropeptide Y gene expression in ob/ob mice, *Diabetes*, 45, 531, 1996.
110. Levin, B. E. and Dunn-Meynell, A. A., Dysregulation of arcuate nucleus preproneuropeptide Y mRNA in diet-induced obese rats, *Am. J. Physiol.*, 272, R1365, 1996.
111. Hakansson, M. L., Brown, H., Ghilardi, N., Skoda, R. C., and Meister, B., Leptin receptor immunoreactivity in chemically defined target neurons of the hypothalamus, *J. Neurosci.*, 18, 559, 1998.
112. Hahn, T. M., Breininger, J. F., Baskin, D. G., and Schwartz, M. W., Colocalization of Agouti-related protein and neuropeptide Y in accurate nucleus neurons activated by fasting, *Nature (Neuroscience)*, 1, 271, 1998.

23 Amino Acid Recognition in the Central Nervous System

Dorothy W. Gietzen

CONTENTS

1 INTRODUCTION

1.1 SIGNIFICANCE OF AMINO ACID RECOGNITION

Humans,[1] as well as animals,[2] can detect the presence of protein in their meals. High levels of protein (>40%) can be unpalatable, and may cause metabolic disturbances.[2,3] Very low protein diets are rejected as well; this may be due to the recognition of a decrease in the growth-limiting amino acid, which would limit protein synthesis.[2] At dietary protein levels just below the requirement, but with well-balanced amino acid profiles, animals become hyperphagic, and show increased levels of neuropeptide Y (NPY) in the hypothalamus.[4] The hyperphagia in these animals supports a role for protein in satiety. Indeed, such a role for protein has been widely accepted. The original aminostatic hypothesis arose from observations that humans report decreased hunger that correlates inversely with plasma amino nitrogen levels.[5]

The mechanisms that underlie the recognition of protein and protein quality must act by way of the amino acids resulting from intestinal digestion of protein. Then, after absorption into the plasma and transportation throughout the body in the bloodstream, amino acids can be detected very quickly. Although energy takes priority over amino acid balance in extreme situations,[2] the ability to synthesize protein is essential for survival, and protein synthesis is dependent on a simultaneous supply of all of the 20 precursor amino acids. Because eukaryotes can synthesize some, but not all of these amino acids, the diet must provide the remaining, essential or dietary indispensable amino acids (IAA). Free amino acids are not stored[6] and synthesis of new protein has priority. Thus, in the event of a deficiency of even one of the IAA, the remaining (unusable) amino acids are catabolized and lost, while body proteins are broken down to provide the limiting IAA, resulting in negative nitrogen balance.[7] The literature is replete with behavioral reports of selection for protein and amino acids in the diet in many species. However, no attempt will be made here to provide an in-depth review of that behavioral work.

Not all of the protein sources in the diets of omnivores contain IAA in optimal balance for protein synthesis, but to avoid deficiencies dietary proteins limiting in one or more of the IAA can be combined with other sources having complementary IAA. As an example, meals combining beans (limiting in sulfur-containing amino acids) and rice (limiting in lysine) provide a better balance in the amino acid profile, and thus the IAA precursors for protein synthesis, within those meals. However, because the protein synthetic process requires simultaneous availability of all amino acid precursors, such supplementation has severe time constraints.[8]

While deficiencies of many of the essential nutrients, including IAA,[9,10] are recognized by animals,[9] IAA are second only to sodium in the rapidity with which responses to a deficiency can be seen. For example, studies of IAA choice in the white crowned sparrow have shown that this small, metabolically active bird cannot use two diets containing complementary IAA efficiently if their presentation is separated by as little as 2 h.[11] Therefore, a system that quickly detects IAA levels, particularly deficiencies and their repletion, provides adaptive advantage for the animal.

The vertebrate system for detection of IAA appears to be in the central nervous system. It is the purpose of this chapter to discuss the evidence for specific sites within the central nervous system that (1) recognize deficiencies of IAA, (2) recognize repletion of IAA, and (3) make associations with cues for learned aversions and preferences for dietary sources that induce either deficiency or repletion of IAA.

1.2   Time Course of the Responses to IAA

It is assumed that an organism must detect the level of the limiting IAA before any compensatory responses can take place. The behavioral data are consistent with the following sequence:

1. After eating an IAA deficient or imbalanced diet, first, there is recognition of the metabolic consequence, i.e., an IAA deficiency, within 1 h.[12,13] Evidence of this recognition was seen in rats eating one of two novel diets, either IAA imbalanced or replete. Differences in their rate of eating could be seen during the second half of the first meal (Figure 1).[14] Chicks also can show recognition of an IAA devoid diet in less than 1 h.[15] In the absence of a choice, early recognition of a deficiency can be demonstrated by increased locomotor activity[16] or increased spillage,[17] which can be seen even before the decrease in intake of the diet causing IAA deficiency.[12] Moreover, selection for the better diet can occur in IAA deficiency without food intake depression.[18]

2. In the absence of a choice, rejection of the diet that induces a deficiency is the next step. This is a very robust finding in all omnivores that have been studied.[2] The time to the hypophagic response is inversely correlated with the severity of the deficiency or imbalance.[19]

3. Acquisition of a repleting IAA depends on availability of other food sources in the environment, opportunities for foraging, or the experimental protocol.

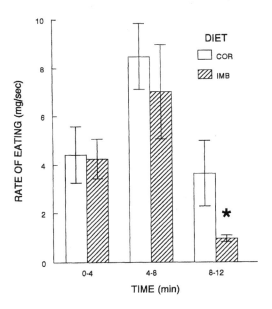

FIGURE 1   Rate of eating during the rats' first meal of either a threonine-imbalanced diet (hatched bars) or a corrected diet (solid bars). The rate of eating did not differ between the diets during the first 8 min. The * over the hatched bar indicates a lower rate of eating in the 8 to 12 min period for the imbalanced diet, $P < 0.05$. (Data are adapted from Gietzen, D. W. and Barrett, J. A. *Appetite*, 29, 396, 1997.)

4. Increased intake of the corrected diet, indicating recognition of repletion, is seen in approximately 30 min. The time course for this step depends on the relative amounts of the limiting AA in the repletion diet.[20]

5. Simson and Booth have shown that repletion must take place within 4 h in order to prevent the development of a conditioned taste aversion to cues associated with histidine deficiency.[13] This appears to be a form of the single-trial long-delay learning described by Garcia et al.[21]

6. Development of a learned preference for cues associated with repletion can take several trials.[22,23] This preference is usually seen only in deficient animals and humans;[1] i.e., it appears to be a need-based preference.[24] However, it is important to note that animals can make appropriate selection in IAA choice trials in the absence of gastrointestinal malaise.[14]

7. Although adaptation does not occur with IAA devoid diets,[2] adaptation to an IAA imbalanced diet does occur over several days, and is based on behavioral changes (such as altered feeding patterns) and increased activity of hepatic IAA catabolic enzymes.[25]

1.3   THE AMINOPRIVIC MODEL

While IAA disproportion can occur in a variety of dietary conditions,[2] a convenient nutritional model is available for studies of rapid deficiency of a single IAA, the amino acid imbalanced diet of A .E. Harper, which has been the subject of several reviews.[2,26-28] In this model, animals are prefed a low-protein (basal) diet that is limiting in an IAA for a few days. They are then given an imbalanced diet, consisting of the basal diet plus an addition of all of the IAA except the one to be limiting, in amounts that individually are nontoxic. For short-term studies, some investigators use a devoid diet, which contains 100% of the requirement for all of the IAA, except the one of which it is devoid. Eating an IAA-imbalanced or -devoid diet causes an influx of the nonlimiting IAA, which stimulates protein synthesis in the liver and other peripheral tissues.[29-31] Increased protein synthesis results in rapid depletion of the limiting IAA in the plasma. The combination of the increased IAA in the plasma and the depletion of the limiting IAA causes increased competition, and a disadvantage for the limiting IAA, at the blood–brain barrier with the result that the level of the limiting IAA is decreased in the brain.[32] This amino acid deficiency in the central nervous system is associated with a rapid recognition and rejection of the imbalanced diet, which is readily reversed by addition of a small amount of the limiting IAA. This has been called "aminoprivic" feeding, because of the similarity to pharmacologically induced glucoprivic and lipoprivic feeding models. The IAA imbalance model has been used widely for tryptophan-depletion studies in humans. Although the prefeeding of low protein is only carried out for 1 to 2 days and the subjects have no prior history of a low-protein or IAA diet, a significant decrease in plasma tryptophan is seen within 5 h.[33]

2   RECOGNITION OF IAA DEFICIENCY

2.1   THE CENTRAL NERVOUS SYSTEM AS A RECOGNITION SITE

Several lines of evidence suggest that IAA are detected in the brain. The first studies implicating the brain were provided when Leung and Rogers[34] infused the limiting IAA into the carotid artery or the jugular vein and found that 4 mg/day of threonine infused into the carotid artery (beginning 4 h before introduction of an imbalanced diet) completely blocked the feeding depression with a threonine-imbalanced diet, while that amount of threonine had no effect on food intake when infused into the jugular vein.[34] These findings were replicated in the bird by Tobin and Boorman.[35] Work in Dr. Harper's laboratory also supports the idea that the brain is involved in the responses to IAA-imbalanced diets. They measured plasma and whole-brain concentrations of the limiting amino acid and found that the limiting IAA decreases more rapidly in brain than in plasma.[32] Thereafter, Tews and colleagues[37,38] concluded, from an elegant series of studies in intact rats and in brain slices, that the decrease of the limiting amino acid in the brain is due to competition at the blood–brain barrier. These studies describe the mechanism for establishing amino acid imbalance in the brain via competition at the transport systems for the various amino acids.[38] As a novel demonstration of the importance of transport systems in the model, Tews et al.[37] also showed that a threonine imbalance can be created by increasing the amounts of dispensable amino acids in the diet, if those amino acids compete with threonine for the transporter that carries threonine across the blood–brain barrier.

Taken together, these studies strongly suggest that there is a role for the brain in recognition of IAA deficiency, but this work does not provide evidence for a particular brain area either functioning as the primary sensor, or involved in learned aspects of the behavioral responses. It should be noted that crucial regional differences may be missed in measurements made in whole brain.

2.1.1   Anatomical Studies

The effects of IAA deficiency are seen in discrete forebrain areas in humans, including the middle frontal gyrus, orbitofrontal cortex, and thalamus, rather than uniformly throughout the brain.[39] Similarly, in their experiments using electrolytic lesions, Leung and Rogers[26-28,40] showed that the rat brain is not uniformly responsive to IAA disproportion. From these lesion studies, it was found that a forebrain area in the rat, the anterior piriform cortex (APC; also called the anterior prepyriform cortex or PPC) may contain the IAA chemosensor because animals with APC lesions do not appear to recognize the imbalanced diet; they continue to eat it at 85 to 100% of their baseline level of food intake,[41] in contrast to control rats that eat an imbalanced diet at only 50% of baseline. This observation has been replicated by Noda and Chikamori[42] in the rat and Firman and Kuenzel[43] in the chicken. Interestingly, this area either overlaps with or is very near the most chemically sensitive area of the brain, the "area tempestas" as determined in studies of seizure threshold with chemical stimulants by Piredda and Gale.[44] Thus, there is a precedent for the APC to respond first, among all the areas in the central nervous system, to a metabolic disruption caused by IAA deficiency.

We have seen that the APC is activated soon after the rat eats an IAA-deficient diet, as indicated by evoked potential recordings in the awake, behaving animal,[45] and in neurochemical measurements (see below). Although we have consistently challenged its role over the past 15 years, we still have not been able to rule out the APC as the site of recognition for IAA.

More recently, the dorsomedial hypothalamus was lesioned electrolytically, resulting in an increase in imbalanced diet intake for the first 3 h of feeding.[46] However, ibotenic acid lesions in the dorsomedial hypothalamus, which destroy glutamate-sensitive cell bodies, have no effect on the initial responses to IAA deficiency. Still, because there are increased FOS-like immunoreactive stained cells in the dorsomedial hypothalamus that are seen only with the imbalanced diet at 2 h,[47] we assume that cell bodies, which do not have glutamate receptors and thus would have survived ibotenic acid lesions, may be involved in the responses to IAA. Alternatively, the dorsomedial hypothalamus may contain fibers of passage that mediate early responses to imbalanced diets. Selective lesions of fiber tracts around the dorsomedial hypothalamus[48] suggest that fibers running anteriorly to the dorsomedial hypothalamus mediate a decrease in imbalanced diet intake that lasts throughout the first experimental day, while tracts going dorsally are important in the earliest period and tracts going laterally and ventrally, although less effective than the anterior fibers, are also involved in longer-term responses. The fibers extending ventrally and laterally can communicate with the lateral hypothalamus and amygdala, which are likely to be associated with the learned responses to imbalanced diets, as described below. The brain sites to which these fibers project will be most interesting in future studies of these interacting brain areas. It is intriguing that the effects of electrolytic lesions of the dorsomedial hypothalamus have their effect so early (0 to 3 h),[46] but that replacement of the limiting IAA into the dorsomedial hypothalamus does not increase intake of the imbalanced diet at that time.[49] In preliminary studies, we have seen a few animals with increased intake of a threonine-imbalanced diet after injections of threonine into the dorsomedial hypothalamus, but not until the 6 to 9 h period (Blevins and Gietzen, unpublished data).

Other potential central nervous system sites for recognition of IAA deficiencies have been suggested, and many sites have been investigated using electrolytic lesions. In 1970, Scharrer et al.[50] lesioned the ventromedial hypothalamus and the lateral hypothalamus, and found that animals with either lesion could still recognize imbalanced diets. Similar results were seen with lesions of the ventromedial hypothalamus by Krauss and Mayer.[51] The ability of olfactory bulbectomized rats to recognize an IAA-imbalanced diet[52] shows that olfaction is not required for IAA recognition. Intravenous (i.v.) administration of an imbalanced mixture, which bypasses the oral taste receptors, causes reduced food intake. Also, behavioral tests indicate that rats do not distinguish an imbalanced diet from the "corrected" control diet by taste.[12] Thus, taste also is unlikely to carry the information of acute IAA deficiency.

Expression of the FOS protein, for which the immediate early gene c-fos codes, can be used to identify brain areas that are activated in various experimental models.[53] Therefore, we evaluated FOS-like immunoreactivity throughout the rat brain at several time points after the initial introduction of either an imbalanced diet or one of two control diets: the basal prefeeding diet or a novel corrected diet.[47,54] These studies showed that the APC was activated by both novel diets at 1 and 1.5 h, but at 2 and 3 h, the APC showed increased FOS-like immunoreactivity selectively after the imbalanced diet only (i.e., not by the corrected diet, which has a similar taste in behavioral tests, containing just enough of the limiting IAA to correct the imbalance). These studies gave the first indication that the dorsomedial hypothalamus might be involved in the responses to IAA, leading to the lesion studies in this area, mentioned above.

The c-fos mapping studies yielded additional evidence that taste does not have a primary role in the recognition of IAA-imbalanced diets. The nucleus of the solitary tract, the first relay in the taste pathway, showed increased FOS-like immunoreactivity for over an hour, but this was similar after introduction of both the imbalanced and the corrected (novel) diets. Thus, the animals' taste pathway was activated by both new diets, but evidence for selective recognition of a difference between the two diets, indicating recognition of IAA in the nucleus of the solitary tract, was not seen.

2.1.2   Neurochemical Systems

To determine neurochemical changes in various brain areas in response to the IAA deficiency induced by ingesting imbalanced diets, the concentrations of amino acids along with monoamine neurotransmitters and their metabolites were measured in 15 brain areas at 2.5 and 3.5 h, and in six brain areas at ½ h intervals after introduction of imbalanced diets. Again, the APC was among the few areas implicated in these studies. Clearly, the limiting IAA is not decreased uniformly throughout the brain after imbalanced diets in rats, or humans, as noted above.[28,55,57] In the rat, the APC consistently shows a decrease in the limiting IAA, but most other brain areas studied do not. Those additional areas showing an occasional decrease in the limiting IAA were the parabrachial nucleus (an important relay in the taste pathway), the anterior cingulate cortex, and the dorsomedial hypothalamus.[58] Still, the one area that consistently shows the most severe reduction of the limiting IAA remains the APC.

What are the next steps in the responses to IAA after recognition of a decreased level of the limiting IAA? One of the first enzymes that the amino acids see in the protein synthetic process are transfer RNA (tRNA) synthetases, which become charged with their cognate amino acids, prior to interacting with the many components in the initiation of protein synthesis. We expected that uncharged tRNA could be the signal carrier in our model, as it is in other amino acid-limited situations, such as mammalian single-cell systems subjected to IAA limitation.[59] Contrary to our expectations, tRNA charging was increased, rather than decreased, in tissue taken 2 h after introduction of the imbalanced diet, both in whole brain and in the APC.[60] Thus, uncharged tRNA does not appear to provide the second step in the recognition process. The increase in charged tRNA suggests that the decrease in the limiting IAA provides some neural signal prior to any shortage of aminoacylated tRNA. Rather, although we do not yet know the intermediate steps, the signal transduction mechanisms activated by a decrease in the limiting IAA appear to influence several neurochemical systems.

Among changes in the neurotransmitters seen in response to a IAA limitation, the catecholamine systems seem to be activated earliest, with the precursor, tyrosine, along with norepinephrine and dopamine, altered within 0.5 h. The neurochemical changes that were measured at 0.5 h after ingestion of a threonine-imbalanced diet, or before food intake behavior was altered on a mild isoleucine-imbalanced diet, are seen in Figure 2A. These results are consistent with activation of the catecholamine systems by 0.5 h, alerting the animal to a potential dietary problem, and are also consistent with the observations of increased spillage and locomotor activity noted above. The changes seen after behavioral rejection of an isoleucine imbalanced diet, or at 1 h after introduction of a threonine-imbalanced diet are seen in Figure 2B. It is tempting to speculate that this early activation of serotonin in the brain is associated with the early decrease in food intake in IAA deficiency.

The role of serotonin in protein selection has been controversial. Fernstrom and Wurtman[61] reported that a pure carbohydrate meal, after an overnight fast, increased tryptophan levels in the rat brain. This led to the proposal of a relationship between serotonin and carbohydrate/protein choices.[62] This proposal was hotly debated for several years; for example, see Booth[63] and Harper and Peters.[64] More recently, serotonin has been shown to have a greater impact on the selection of fat rather than either carbohydrate or protein.[65]

Serotonin appears to act at the type-3 receptor in the APC in aminoprivic feeding. Other systems in the APC that are implicated in the responses to IAA by pharmacologic studies using microinjections of agonists and antagonists into the area include the α-2-noradrenergic receptor, gamma amino butyric acid (GABA), at the $GABA_A$ receptor, the dopamine D2 receptor, NPY, somatostatin, and nitric oxide. The results of these pharmacological studies are summarized in Table 1. *In vitro* studies show that the effect of norepinephrine on the excitatory postsynaptic potentials seen in the APC slice is mediated by the α-2 receptor,[66] supporting the *in vivo* pharmacology studies, at least

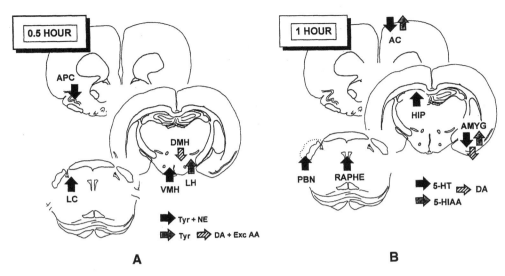

FIGURE 2 (A) Neurochemical changes at 0.5 h after introduction of an imbalanced diet. Abbreviations are Tyr + NE: both tyrosine and norepinephrine were altered in the direction indicated by the arrow, in the specified brain area at this early time; Tyr: tyrosine only was altered; DA + Exc AA: both dopamine and the excitatory amino acids were altered in the DMH: dorsomedial hypothalamus; VMH: ventromedial hypothalamus; LH: lateral hypothalamus; APC: anterior piriform cortex; LC: locus ceruleus, site of norephinephrine cell bodies. (B) Neurochemical changes at 1.0 h after introduction of an imbalanced diet, as in Figure 2A. Arrows indicate changes in 5-HT: serotonin; 5-HIAA: 5-hydroxyindole acetic acid, the major metabolite of serotonin; DA: dopamine. Brain areas are AC: anterior cingulate cortex; PBN: parabrachial nucleus, an important taste relay; RAPHE: raphe nucleus, site of serotonin cell bodies; HIP: hippocampus, a member of the limbic system; AMYG: amygdala, also in the limbic system and important for conditioned taste aversions. (Data are adapted from Gietzen, D.W. et al., *J. Nutr.*, 128, 771, 1998.)

for norepinephrine in that area. Clearly, some signal in response to the decrease in the limiting IAA must have activated the neurons of the APC, resulting in changes in many different neurochemical systems shortly after introduction of an imbalanced or IAA-deficient diet.

3   RECOGNITION OF IAA REPLETION

The paradigm for which the most definitive data are available is one in which the animals are made slightly deficient in an IAA, using the aminoprivic feeding model, and then the limiting IAA is replaced, either systemically by iv injection or feeding a corrected diet, or by microinjection into specific anatomical sites in the brain. Recognition of this repletion is rapid (about 30 min) in animals switched from an imbalanced to a corrected diet[12] and has been measured in a variety of contexts.

3.1   THE APC

Persuasive evidence that the APC recognizes IAA repletion was provided by Beverly and colleagues,[67] who showed that injections of the limiting IAA into the APC (2 nmol of threonine or 1 nmol of isoleucine, depending on the diet) restore feeding of the imbalanced diet to about 80 to 85% of control, baseline feeding levels on the basal diet, whereas animals injected with saline or artificial cerebrospinal fluid (CSF) eat the imbalanced diet at control levels, 50 to 60% of basal baseline intake. These injections are selective for the limiting IAA; i.e., isoleucine has no effect on a threonine-devoid or -imbalanced diet and threonine has no effect when isoleucine is the dietary-limiting IAA. The injections are also anatomically selective, because neither injections 2 mm posterior to the effective site in the APC nor injections into the amygdala have any effect.

TABLE 1
Results of Pharmacological Treatments in the APC on Intake of an IAA Imbalanced Diet

Substance (Receptor)	Effect	Dose	% of BAS	Time of effect (h)	Ref.
Saline or ACSF	–	0.5 µl	50	—	117
Limiting amino acid					
Threonine	↑	2 nmol	75	3–24	117, 118
	↑	2 nmol	75	1.5	49
	–	4 nmol	50	—	67
Isoleucine	↑	1 nmol	75	3–24	117
Nonlimiting amino acid	–	1–4 nmol	50	—	117
Puromycin	–	100 µM	50	—	119
Clonidine (Alpha$_2$ +)	↑	1.5 µg	106	0–3	120
Idazoxan (Alpha$_2$ –)	↓	1.5 µg/0.5 µl	20	0–12	121
Tropisetron (5-HT$_{3/4}$–)	—	0.1–1.0 ng	50	—	122
		0.1–1 nmol	50	—	99
Ondansetron (5-HT$_3$–)	↑	0.5 ng	84	6–24	122
GR113808 (5-HT$_4$–)	↓	1.0 µg	50	0–3	122
Bicuculline (GABA$_A$–)	↑	79 pmol	64	0–6	123
	↑	25 pmol	78	0–3	123
Muscimol (GABA$_A$+)	↓	1–2 nmol	10	0–3	123
(also decreased control diet intake)					
Phaclofen (GABA$_B$–)	–	31 ng/rat	50	—	123
NPY	↓	1–1.5 nmol	30	0–3	124
Somatostatin	↑↓	1 pmol ↑	60	0–3	124
		2 nmol ↓	28	0–3	124
AP5 (NMDA–)	↑	4 nM/side	88	6–9	125
L-NAME (NOS–)	↑	0.5 µg/side	78	0–3	126
Eticlopride (D2–)	↑	0.3 nmol/side	108	0–3	127
Leptin (also decreased control diet intake)	–	0.1–0.25 µg	40	12–24	128

BAS, basal diet baseline, taken as the average of the previous 3 days intake for that time on the basal diet, 50% of BAS is the expected value after control injections, i.e., the null result; ACSF: artificial cerebrospinal fluid, a buffer used for vehicle and control injections; 5-HT: serotonin; GABA: gamma aminobutyric acid; D2: the dopamine-2 receptor; AP5: 2-amino-5-phosphopentanoic acid, antagonist at the glutamate NMDA receptor; NPY: neuropeptide Y; L-NAME: L-N-methylarginine methyl ester, inhibitor of the nitric oxide synthetase (NOS) enzyme, given two times per day for 4 days; +: agonist; –: antagonist; ↑↓: increases or decreases of imbalanced diet intake after injection into the APC.

In an effort to identify cellular mechanisms in the responses to IAA repletion in the APC, we examined changes in intracellular calcium in deficient and replete APC slices. When we added the limiting IAA to deficient (but not to control) APC slices loaded with fura-2, fluorescence ratio measurements indicated an increase in intracellular calcium.[68] This was true with either threonine or lysine as the limiting IAA and was not due to pH changes (Magrum and Gietzen, unpublished data). Thus, an increase in intracellular calcium may be associated with the initial signal leading to alterations in neurotransmitter activity upon repletion of the limiting IAA.

In studies of connections between the APC and other brain areas in the aminoprivic model, Monda et al.,[69] using Beverly's[67] methods, injected threonine into the APC of rats that had been eating a threonine-devoid diet for 5 days. After threonine injections into the APC, intake of a threonine-devoid diet is increased, similar to Beverly's results with an imbalanced diet. Increased neuronal firing in the lateral hypothalamus and decreased neuronal activity in sympathetic fibers to the brown adipose tissue occur approximately 30 min after the injection into the APC. This is

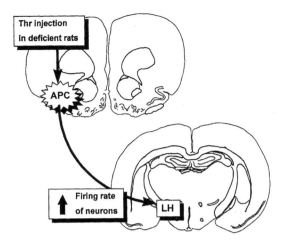

FIGURE 3   Effects of threonine injections in the APC on firing rate of neurons in the lateral hypothalamus, LH. Rats were fed a threonine-devoid diet for 5 days and then given 2 nmol of threonine into the APC. The increased firing rate of the LH neurons was observed 0.5 h after injection. (Data are adapted from Monda, M. et al., *Am J. Physiol.*, 273, R554, 1997.)

the first observation in this model of such a direct connection between the two brain areas (Figure 3) that have been most closely associated with the responses to IAA: the APC, which may house the IAA chemosensor, and the lateral hypothalamic area, which is important in the conditioned responses to repletion of the limiting IAA.[70]

3.2   The Lateral Hypothalamus

Microiontophoretic injections of IAA have been shown to evoke neuronal responses in the lateral hypothalamic area.[71] Also, Panksepp and Booth[72] injected a balanced amino acid solution into an area in the lateral hypothalamus, the dorsal perifornical area, where stimulation induces feeding, and where NPY, a peptide that increases carbohydrate intake, has its predominant effect. Food intake is depressed after injections of amino acid mixtures, but not after saline or glucose into that area.[73] More recently, Ono, Torii, Tabuchi, and colleagues[70,74] used an elegant experimental paradigm in which single neurons are recorded extracellularly in awake, behaving animals. They have clearly demonstrated selective lysine-responsive neurons in the lateral hypothalamus of lysine-deficient animals that had previously learned to recognize auditory cues associated with a lysine solution. The lateral hypothalamus may be involved in recognition of repletion of the limiting IAA in deficient animals, rather than recognizing depletion of the IAA, because the lysine-sensitive neurons only respond to lysine solutions or cues associated with lysine repletion when the animal is in a deficient state. Alternatively, since several days of training are necessary in this model, the role of this area may be associated with the learning phase of the responses to IAA deficiency.

4   LEARNED RESPONSES TO CUES FOR AMINO ACIDS

4.1   Aversions

Meliza et al.[75] found an area in the medial amygdala, which is involved in the conditioned taste aversion that develops after recognition of the acute IAA deficiency. They compared the APC and the medial amygdala using electrolytic lesions, and found that the animals bearing lesions of the amygdala lose their ability to reduce intake of the imbalanced diet, as do APC-lesioned rats, but the amygdala- but not APC-lesioned rats also lose their ability to form a conditioned taste aversion to saccharin after it is paired with an intraperitoneal injection of lithium chloride. The conclusions from these studies were that the APC is involved in the recognition and rejection of an IAA, while the amygdala responds later, being involved in the subsequent conditioned taste aversion. The results of the c-*fos* mapping studies mentioned above[47,54] are consistent with this notion, because the central nucleus of the amygdala showed clear increases in FOS-like immunoreactivity at 3 h, persisting to 6 h, while the APC and dorsomedial hypothalamus were activated earlier, at 2 to 3 h.

Norgren and colleagues[76] obtained results similar to those seen in the amygdala when they lesioned the parabrachial nucleus. The rats with parabrachial lesions eat more of an amino–acid devoid diet, but also lose their ability to show a conditioned aversion. Because the parabrachial nucleus, a secondary relay in the taste pathway, is important in taste aversions, this suggests that the learned aversions to IAA deficiency can be cued to taste.

Some fish have amino acid receptors in their skin and barbels,[77] responding to as little as 10^{-12} M L-amino acids, and they can be trained to behavioral tasks by the introduction of small amounts of amino acids into their water supply. This response is not dependent on olfaction,[78] yet how these selective amino acid receptors relate to central nervous system recognition of IAA in mammals is not known. Nonetheless, there are suggestions that mammals use cranial nerve function in making associations with cues to IAA and proteins in their diets. Torii[79] and colleagues have shown that rats with lesions of the glossopharyngeal nerve do not show a preference for a lysine solution when they are lysine deficient, and somatosensory denervation by partial deafferentation of the trigeminal nerve also renders animals less able to make choices among different protein levels.[80] These denervations likely affect the associations that animals make between the metabolic responses to the diet and taste or oral sensations, and would diminish the cues available for making appropriate choices. An effect of these denervations in the initial detection of IAA deficiency or the level of dietary protein has yet to be shown.

In general, it seems that animals can use a variety of cues to aid their selection of IAA. In addition to taste (see above) and smell,[24] the Tabuchi et al.[70] studies demonstrate that lysine repletion can be cued to an auditory stimulus, and Fromentin et al.[81] have shown learned place preferences associated with IAA deficiency and repletion, as well.

4.2 PERIPHERAL SITES

In studies of peripheral recognition sites, iv administration of a complete IAA mixture leads to full caloric compensation.[82] This level of caloric compensation is not seen with carbohydrate, fat, or mixed infusions,[82] suggesting that gastrointestinal factors are required for compensation with glucose and fatty acids, but not with amino acids. Also, iv administration of imbalanced IAA mixtures leads to the same decrease in food intake[2] as that seen with oral ingestion of imbalanced diets. Therefore, neither the gastrointestinal tract nor taste appear to be necessary for IAA recognition, per se. Yet, several workers have suggested that amino acids are recognized in peripheral sites associated with vagal function, such as the liver,[83] gastrointestinal tract,[84,85] or portal vein.[86] We[87] also have evidence, in the rat, for a role for the vagus in the later (after the first 3 h) feeding responses to IAA-imbalanced diets. Similarly, Torii and colleagues have seen that hepatic branch vagotomy delays the responses in lysine deficiency.[88]

4.2.1 The Vagus Nerve

The C-fiber vagal units in the nodose ganglion that Jeanningros[85] recorded with infusions into the duodenum appear to be selective for various amino acids, but both IAA and dispensable amino acids increase this vagal activity. In addition, many of the amino acid–sensitive hepatic vagal branch units, recorded by Niijima and Meguid,[89] are also responsive to glucose in the portal vein. Ritter's group[84] saw a reduction of sham feeding with phenylalanine infused into the upper small intestine, which is selective for the physiologically relevant L-phenylalanine; the D-form is inactive. The effect of phenylalanine is only partially reduced after vagotomy, whereas the surgery abolishes the responses to maltose and oleate.[84] Also, the responses to maltose and oleate are reduced and the effects of phenylalanine are not altered after capsaicin in the fourth ventricle, near the nucleus of the solitary tract, where vagal afferents terminate.[90] Thus, the vagus clearly responds to amino acids. Yet, like the results of Yox and colleagues,[90] the vagus does not act alone in the aminoprivic model. In our vagotomy studies,[87,91-93] imbalanced diet intake has never been increased to the level achieved with pharmacological antagonists into the APC (see Table 1). Still, vagotomized rats do not reduce their intake of an amino acid-imbalanced diet as drastically as do intact rats, and they do not choose the protein-free diet over the imbalanced diets, as intact rats do. Selective vagal branch ablations[91-93] have shown that the ventral vagal branch, specifically a combination of the hepatic vagal branch and the ventral gastric branch,[91] likely mediate the vagal aspects of the responses to IAA-imbalanced diets. Both afferent and efferent fibers in the vagus are implicated by results after capsaicin is used to abolish most of the vagal afferent fibers,[94] and from pharmacological blockade of vagal efferents.[95,96]

4.2.2 Serotonin

Antagonists of the serotonin₃ receptor also can increase intake of an IAA-imbalanced diet, while having little or no effect on a control, high-carbohydrate diet.[97,98] However, a quaternized form of the serotonin₃ and serotonin₄ receptor antagonist, tropisetron, which does not cross the blood–brain barrier, is as effective as the parent compound.[99] Thus, the effects of serotonin in the aminoprivic model have a peripheral component that involves the serotonin₃ receptor. We have reported that either tropisetron or MDL 72,222, a selective serotonin₃ antagonist, block the conditioned taste aversion to saccharin induced after saccharin is included in the imbalanced diet (Figure 4),[100] suggesting that this peripheral 5-HT₃ system is likely to be associated with the later learned aspects of the responses to IAA, i.e., after 3 h, by which time the recognition phase is past.[101] This is in contrast to the effects of subtypes of the serotonin₁ and serotonin₂ receptor that have effects on carbohydrate[102] or fat[65] intake in the central nervous system.

There also are interactions between the vagus nerve and the serotonin₃/₄ system in the responses to amino acid-imbalanced diets, because a portion of the response to tropisetron is abolished in

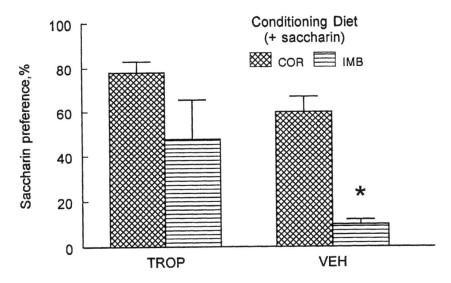

FIGURE 4 Blockade of conditioned taste aversion to saccharin after it had been paired with an isoleucine-imbalanced diet. The conditioning diet was either imbalanced (IMB) or corrected (COR); both conditioning diets contained saccharin. The rats were later tested for saccharin preference by a choice test between the COR diet with or without saccharin. The star * indicates significance at $P < 0.05$. (Data are adapted from Terry-Nathan, V. R. et al., *Am. J. Physiol.*, 268, R1203, 1995.)

vagotomized animals.[87] Meal patterns recorded after either tropisetron or vagotomy[101] indicate that these systems both function after the initial recognition phase of the responses to IAA, i.e., after the first 3 h. These observations have led us to conclude that both the serotoninergic and the vagally mediated aspects of the responses to IAA deficiency are part of the later, learned responses to IAA.

Clear evidence of amygdalar and hypothalamic input to the vagus is abundant.[103] These known anatomical connections allow us to hypothesize that the signals generated in the APC (and likely the dorsomedial hypothalamus) and transmitted to the lateral hypothalamus and amygdala also may be projected to the vagal motor nucleus to stimulate vagal efferents. The vagal stimulus would cause the release of serotonin from enterochromaffin cells in the gastrointestinal tract to act at serotonin$_3$ receptors.[104] Then, along with input from afferent units sensitive to amino acids,[85] a vagal loop could be activated, reinforcing the development of the conditioned aversion.

4.3   PREFERENCES

Preference for a flavor cue associated with protein has been seen after a protein-free meal in humans.[1] Conditioning to either protein or carbohydrate cues is also seen in animals.[105] However, preferences for cues associated with a corrected diet take longer to learn than aversions, in our experience,[22,23] and they were seen only in depleted, not replete, animals (Gietzen, unpublished data), again supporting the idea that these are need-based preferences. Brain sites associated with learned preferences may include those that mediate repletion of an IAA, or those associated with learned aversions, or both, as well as aspects of the reward system.[106]

5   BRAIN AREAS ASSOCIATED WITH ADAPTATION TO AN IMBALANCED DIET

The lesion studies of Leung and Rogers, reviewed in 1987[40] and 1993,[28] showed that several other brain areas are associated with adaptation to the imbalanced diet over several days, because animals with lesions in those areas showed accelerated or diminished adaptation to imbalanced diets. In addition, we lesioned the infralimbic cortex, after noting increased FOS-like immnoreactivity there, both electrolytically and with ibotenic acid.[107,108] No effect was seen during the first day of eating an imbalanced diet after both types of lesions, but the ibotenic acid-lesioned animals (both in APC and infralimbic cortex) showed accelerated adaptation to an IAA-imbalanced diet. Note that there is no adaptation to an IAA-devoid diet, because of damage to vital organs, so these adaptation studies cannot be done with IAA-devoid diets. Nonetheless, after lesions of the area postrema, Leung and Rogers[109] found increased adaptation to a high-protein diet, and no changes in the responses to single IAA deficiencies.[110] No effect on either imbalanced or high-protein intake is seen after lesions of the nucleus of the solitary tract.[28] The area postrema and nucleus of the solitary tract together are important in responses associated with both lipoprivic and glucoprivic feeding,[111] but apparently have no effect on the acute responses to IAA-imbalanced diets in aminoprivic feeding. Also, as noted above, a role for the vagus nerve must be factored into this picture, because lesions of the ventral branches (including at least the hepatic vagal branch and likely the ventral gastric branch) increase adaptation to the imbalanced diet.[87,91-93]

6   SUMMARY OF BRAIN AREAS ASSOCIATED WITH SELECTION OF IAA

We have presented evidence that the APC serves as the IAA chemosensor in the brain. A pair of circuits can be postulated, which could mediate the subsequent steps in the responses to IAA depletion or repletion. In our hypothetical circuit, the APC responds to IAA repletion[67] with a signal recognized in the lateral hypothalamus[69,70] that could project to the nucleus accumbens, a brain area associated with food reward[106] resulting in the development of a preference for the repleting food. There likely are two projections from the APC that would be activated with IAA deficiency in this scheme. First, for the increased locomotor activity[16] and spillage,[17] tract tracing studies show that fibers from the IAA-responsive site in the APC project to the ventral pallidum (Aja, unpublished Ph.D. dissertation). Also, for the hypophagic responses, projections to hypothalamic areas[48] could lead to decreased activity in appetitive circuits, as well as to activation of the vagal loop[94] postulated above. Subsequently, activation of the parabrachial nucleus[76] and amygdala[75] may mediate the aversion to the diet that is associated with the IAA deficiency.

6.1 SEPARATE SYSTEMS FOR IAA/PROTEIN

The APC is a primary olfactory relay, and also is near the gustatory neocortex and the orbitofrontal areas which are activated by IAA deficiency in humans.[39] The APC is sensitive to chemical stimulation and is a short axon away from the taste and other sensory cortices for the somatic and visual aspects of food.[112] This area, which apparently also receives postabsorptive information about the IAA balance of the diet, is in an ideal location to evaluate a new diet and determine its suitability for the long-term maintenance of IAA homeostasis, and thus the precursors for protein synthesis. The dorsomedial hypothalamus is also in a unique situation, in the midst of the hypothalamus, with access to the contents of the CSF via the third ventricle and direct connections with the ventral medial hypothalamus, the paraventricular nucleus and the lateral hypothalamus. The role of the dorsomedial hypothalamus in the determination of body size and composition could be directly related to the ability of the body to maintain the precursors for protein synthesis as well.[113] These two areas, while always recognized as having some role in the control of food intake, have been assigned other primary functions: the APC in various associations with olfactory input and the dorsomedial hypothalamus in regulation of body weight set point. Now, it appears that the APC is the primary chemosensor for IAA; the potential for a similar role for the dorsomedial hypothalamus, or fibers therein, is clear. However, as far as we have been able to determine from our neurochemical and c-*fos* mapping studies, the paraventricular nucleus, which is important in carbohydrate intake[102] may not be involved in the responses to single IAA, although growth hormone-releasing hormone acts there to increase balanced protein intake.[114]

6.2 OVERLAPPING BRAIN SYSTEMS FOR MACRONUTRIENT SELECTION

Parallel systems for the three macronutrients may include the lateral hypothalamus and associated feeding pathways that are activated by repletion of the limiting IAA, stimulating food intake. Our neurochemical measures (see Figure 2A) also support the involvement of the ventromedial hypothalamus in the earliest responses to IAA deficiency. Clearly, the conditioned aversions mediated by the central nucleus of the amygdala, lateral parabrachial nucleus, and vagus nerve have parallels in the aminoprivic, glucoprivic, and lipoprivic models. The central nucleus of the amygdala is essential for lipoprivic feeding, and plays a role in glucoprivic feeding,[115] while fibers in the lateral parabrachial nucleus are important in lipoprivic, but not glucoprivic, feeding.[111] Also, the release of cholecystokinin by protein, but not IAA, in the gastrointestinal tract is also seen with fat.[116]

Clearly, our understanding of the complete neural circuits that recognize IAA and contribute to the selection of protein among the macronutrients is far from complete. Yet, the information reviewed here suggests that the brain areas, which mediate the central nervous system recognition of indispensable amino acids in the diet, comprise very different neural circuits from those for fat and carbohydrate.

ACKNOWLEDGMENTS

Work the author's laboratory presented here was supported by National Institutes of Health Grants: NS33347, DK50347, DK42274, and DK35747 and U.S. Department of Agriculture NRICGP: 97-35200-4477.

REFERENCES

1. Gibson, E. L., Wainwright, C. J., and Booth, D. A., Disguised protein in lunch after low-protein breakfast conditions food-flavor preferences dependent on recent lack of protein intake, *Physiol. Behav.,* 58, 363, 1995.
2. Harper, A. E., Benevenga, N. J., and Wohlhueter, R. M., Effects of ingestion of disproportionate amounts of amino acids, *Physiol. Rev.,* 50, No.3, 428, 1970.
3. Semon, B. A., Leung, P. M. B., Rogers, Q. R., and Gietzen, D. W., Increase in plasma ammonia and amino acids when rats are fed a 44% casein diet, *Physiol. Behav.,* 43, 631, 1988.
4. White, B. D., Dean, R. G., and Martin, R. J., Low protein diets increase neuropeptide Y gene expression in the basomedial hypothalamus of rats, *J. Nutr.,* 124, 1152, 1994.
5. Mellinkoff, S. M., Frankland, M., Boyle, D., and Greipel, M., Relationship between serum amino acid concentration and fluctuations in appetite, *J. Appl. Physiol.,* 8, 535, 1956.
6. Berg, C. P. and Rose, W. C., Tryptophane and growth, *J. Biol. Chem.,* 82, 479, 1929.
7. Munro, H. N., Regulation of body protein metabolism in relation to diet, *Proc. Nutr. Soc.,* 35, 297, 1976.
8. Elman, R., Time factor in retention of nitrogen after intravenous injection of a mixture of amino acids, *Proc. Soc. Exp. Biol. Med.,* 40, 484, 1939.
9. Rozin, P. and Rodgers, W., Novel-diet preferences in vitamin-deficient rats and rats recovered from vitamin deficiency, *J. Comp. Physiol. Psychol.,* 63, 421, 1967.
10. Booth, D. A. and Simson, P. C., Food preferences acquired by association with variations in amino acid nutrition, *Q. J. Exp. Psychol.,* 23, 135, 1971.
11. Murphy, M. E. and Pearcy, S. D., Dietary amino acid complementation as a foraging strategy for wild birds, *Physiol. Behav.,* 53, 689, 1993.
12. Gietzen, D. W., Leung, P. M. B., Castonguay, T. W., Hartman, W. J., and Rogers, Q. R., Time course of food intake and plasma and brain amino acid concentrations in rats fed amino acid-imbalanced or -deficient diets, in *Interaction of the Chemical Senses with Nutrition*, Kare, M. R. and Brand, J. G., Eds., Academic Press, New York, 1986, 415.
13. Simson, P. C. and Booth, D. A., Effect of CS-US interval on the conditioning of odour preferences by amino acid loads, *Physiol. Behav.,* 11, 801, 1973.
14. Gietzen, D. W. and Barrett, J. A., Recognition of amino acid repletion before 45 min in ad lib feeding rats, *Appetite,* 29, 396, 1997 (Abstr.).
15. Picard, M. L., Uzu, G., Dunnington, E. A., and Siegel, P. B., Food intake adjustments of chicks: short term reactions to deficiencies in lysine, methionine and tryptophan, *Br. Poult. Sci.,* 34, 737, 1993.
16. Gietzen, D. W., Le, T., and Carlson, C., A comparison of the behavioral effects of acute tryptophan and threonine deficiencies: differences and similarities, *Eur. Winter Conf. Brain Res.,* 16, P31, 1996 (Abstr.).
17. Rozin, P., Specific aversions as a component of specific hungers, *J. Comp. Physiol. Psychol.,* 64(2), 237, 1967.
18. Hrupka, B. J., Lin, Y. M., Gietzen, D. W., and Rogers, Q. R., Lysine deficiency alters diet selection without depressing food intake in rats, *J. Nutr.,* 129, 424, 1999.
19. Leung, P. M. B., Rogers, Q. R., and Harper, A., Effect of amino acid imbalance on dietary choice in the rat, *J. Nutr.,* 95(3), 483, 1968.
20. Hrupka, B. J., Lin, Y. M., Gietzen, D. W., and Rogers, Q. R., Small changes in essential amino acid concentrations alter diet selection in amino acid-deficient rats, *J. Nutr.,* 127, 777, 1997.
21. Garcia, J., Kimeldorf, D. J., and Koelling, R. A., Conditioned aversion to saccharin resulting from exposure to gamma radiation, *Science,* 122, 157, 1955.
22. Gietzen, D. W., McArthur, L. H., Theisen, J. C., and Rogers, Q. R., Learned preference for the limiting amino acid in rats fed a threonine-deficient diet, *Physiol. Behav.,* 51, 909, 1992.
23. Naito-Hoopes, M., McArthur, L. H., Gietzen, D. W., and Rogers, Q. R., Learned preference and aversion for complete and isoleucine-devoid diets in rats, *Physiol. Behav.,* 53, 485, 1993.
24. Baker, B. J. and Booth, D. A., Genuinely olfactory preferences conditioned by protein repletion, *Appetite,* 13, 223, 1989.
25. Anderson, H. L., Benevenga, N. J., and Harper, A. E., Associations among food and protein intake, serine dehydratase, and plasma amino acids, *Am. J. Physiol.,* 214(5), 1008, 1968.

26. Rogers, Q. R. and Leung, P. M. B., The influence of amino acids on the neuroregulation of food intake, *Federation Proc.*, 32, 1709, 1973.

27. Rogers, Q. R. and Leung, P. M. B., The control of food intake: when and how are amino acids involved? in *The Chemical Senses and Nutrition*, Kare, M. R. and Maller, O., Eds., Academic Press, New York, 1977, 213.

28. Gietzen, D. W., Neural mechanisms in the responses to amino acid deficiency, *J. Nutr.*, 123, 610, 1993.

29. Moja, E. A., Restani, P., Corsini, E., Stacchezzini, M. C., Assereto, R., and Galli, C. L., Cycloheximide blocks the fall of plasma and tissue tryptophan levels after tryptophan-free amino acid mixtures, *Life Sci.*, 49, 1121, 1991.

30. Pronczuk, A. W., Rogers, Q. R., and Munro, H. N., Liver polysome patterns of rats fed amino acid imbalanced diets, *J. Nutr.*, 100, 1249, 1970.

31. Yokogoshi, H., Hayase, K., and Yoshida, A., The quality and quantity of dietary protein affect brain protein synthesis in rats, *J. Nutr.*, 122, 2210, 1992.

32. Peng, Y., Tews, J. K., and Harper, A. E., Amino acid imbalance, protein intake, and changes in rat brain and plasma amino acids, *Am. J. Physiol.*, 222, 314, 1972.

33. Young, S. N., Smith, S. E., Phil, R. O., and Ervin, F. R., Tryptophan depletion causes a rapid lowering of mood in normal males, *Psychopharmacology*, 87, 173, 1985.

34. Leung, P. M.-B. and Rogers, Q. R., Food intake: regulation by plasma amino acid pattern, *Life Sci.*, 8, 1, 1969.

35. Tobin, G. and Boorman, K. N., Carotid or jugular amino acid infusions and food intake in the cockerel, *Br. J. Nutr.*, 41, 157, 1979.

36. Tews, J. K., Good, S. S., and Harper, A. E., Transport of threonine and tryptophan by rat brain slices: relation to other amino acids at concentrations found in plasma, *J. Neurochem.*, 31, 581, 1978.

37. Tews, J. K., Kim, Y.-W. L., and Harper, A. E., Induction of threonine imbalance by dispensable amino acids: relation to competition for amino acid transport into brain, *J. Nutr.*, 109, 304, 1979.

38. Tackman, J. M., Tews, J. K., and Harper, A. E., Dietary disproportions of amino acids in the rat: effects on food intake, plasma and brain amino acids and brain serotonin, *J. Nutr.*, 120, 521, 1990.

39. Bremner, J. D., Innis, R. B., Salomon, R. M., Staib, L. H., Ng, C. K., Miller, H. L., Bronen, R. A., Krystal, J. H., Duncan, J., Rich, D., Price, L. H., Malison, R., Dey, H., Soufer, R., and Charney, D. S., Positron emission tomography measurement of cerebral metabolic correlates of tryptophan depletion-induced depressive relapse, *Arch. Gen. Psychiatr.*, 54, 364, 1997.

40. Leung, P. M. B. and Rogers, Q. R., The effect of amino acids and protein on dietary choice, in *Umami: A Basic Taste*, Kawamura, Y. and Kare, M. R., Eds., Marcel Dekker, New York, 1987, 565.

41. Leung, P. M. B. and Rogers, Q. R., Importance of prepyriform cortex in food-intake response of rats to amino acids, *Am. J. Physiol.*, 221, 929, 1971.

42. Noda, K. and Chikamori, K., Effect of ammonia via prepyriform cortex on regulation of food intake in the rat, *Am. J. Physiol.*, 231, 1263, 1976.

43. Firman, J. D. and Kuenzel, W. J., Neuroanatomical regions of the chick brain involved in monitoring amino acid deficient diets, *Brain Res. Bull.*, 21, 637, 1988.

44. Piredda, S. and Gale, K., A crucial epileptogenic site in the deep prepiriform cortex, *Nature*, 317, 623, 1985.

45. Hasan, Z., Woolley, D. E., and Gietzen, D. W., Responses to indispensable amino acid deficiency and replenishment recorded in the anterior piriform cortex of the behaving rat, *Nutr. Neurosci.*, 1, 373, 1998.

46. Bellinger, L. L., Evans, J. F., and Gietzen, D. W., Dorsomedial hypothalamic lesions alter intake of an imbalanced amino acid diet in rats, *J. Nutr.*, 128, 1213, 1998.

47. Wang, Y., Cummings, S. L., and Gietzen, D. W., Temporal-spatial pattern of c-*Fos* expression in the rat brain in response to indispensable amino acid deficiency. I. The initial recognition phase, *Mol. Brain Res.*, 40, 27, 1996.

48. Evans, J. F., Bellinger, L. L., Tillberg, C. M., and Gietzen, D. W., The effect of dorsomedial hypothalamic (DMN) knife cuts (KC) on intake of an imbalanced amino acid diet (IMB), *FASEB J.*, A854, 1998 (Abstr.).

49. Blevins, J. E., Bellinger, L. L., and Gietzen, D. W., Effects of threonine injection in the dorsomedial hypothalamus (DMH) on intake of an amino acid imbalanced diet in rats, *Int. Cong. Physiol. Food Fluid Intake, Pecs, Hungary, July 5–8*, 52, 1998 (Abstr.).

50. Scharrer, E., Baile, C. A., and Mayer, J., Effect of amino acids and protein on food intake of hyperphagic and recovered aphagic rats, *Am. J. Physiol.*, 218 (2), 400, 1970.

51. Krauss, R. M. and Mayer, J., Influence of protein and amino acids on food intake in the rat, *Am. J. Physiol.*, 209(3), 479, 1968.

52. Leung, P. M. B., Larson, D. M., and Rogers, Q. R., Food intake and preference of olfactory bulbectomized rats fed amino acid imbalanced or deficient diets, *Physiol. Behav.*, 9, 553, 1972.

53. Armstrong, R. C. and Montmini, M. R., Transsynaptic control of gene expression, *Annu. Rev. Neurosci.*, 16, 17, 1993.

54. Wang, Y., Cummings, S. L., and Gietzen, D. W., Temporal-spatial pattern of c-*Fos* expression in the rat brain in response to indispensable amino acid deficiency. II. The learned taste aversion, *Mol. Brain Res.*, 40, 41, 1996.

55. Gietzen, D. W., Leung, P. M. B., and Rogers, Q. R., Norepinephrine and amino acids in prepyriform cortex of rats fed imbalanced amino acid diets, *Physiol. Behav.*, 36, 1071, 1986.

56. Gietzen, D. W., Leung, P. M. B., and Rogers, Q. R., Dietary amino acid imbalance and neurochemical changes in three hypothalamic areas, *Physiol. Behav.*, 46, 503, 1989.

57. Gietzen, D. W., Erecius, L. F., and Rogers, Q. R., Neurochemical changes after imbalanced diets suggest a brain circuit mediating anorectic responses to amino acid deficiency in rats, *J. Nutr.*, 128, 771, 1998.

58. Gietzen, D. W., Erecius, L. F., and Dixon, K. D., Appropriate choice against amino acid deficient diet requires vagal function, *Proc. 7th Multidisciplin. Food Choice Conf.*, 1998 (Abstr.).

59. Rabinovitz, M., The phosphofructokinase-uncharged tRNA interaction in metabolic and cell cycle control: an interpretive review, *Nucleic Acids Symposium Series*, 33, 182, 1995.

60. Hickman, M. A., Kreiter, M. R., Magrum, L. J., and Gietzen, D. W., Increased tRNA charging in response to amino acid imbalanced diets in rats, *Soc. Neurosci. Abstr.*, 24, 193, 1998.

61. Fernstrom, J. D. and Wurtman, R. J., Brain serotonin content: increase following ingestion of carbohydrate diet, *Science*, 174, 1023, 1971.

62. Ashley, D. V. M. and Anderson, G. H., Correlation between the plasma tryptophan to neutral amino acid ratio and protein intake in the self-selecting weanling rat, *J. Nutr.*, 105, 1412, 1975.

63. Booth, D. A., Central dietary "feedback onto nutrient selection": not even a scientific hypothesis, *Appetite*, 8, 195, 1987.

64. Harper, A. E. and Peters, J. C., Protein intake, brain amino acid and serotonin concentrations and protein self-selection, *J. Nutr.*, 119, 677, 1989.

65. Smith, B. K., York, D. A., and Bray, G. A., Chronic d-fenfluramine (d-Fen) administration reduces fat intake and increases carbohydrate intake in rats, *Obesity Res.*, 4, S63, 1996.

66. Rechs, A. J., Horowitz, J. M., and Gietzen, D. W., Alpha-2 adrenoceptors mediate inhibition in layer III of the anterior piriform cortex, submitted.

67. Beverly, J. L., Gietzen, D. W., and Rogers, Q. R., Effect of dietary limiting amino acid in prepyriform cortex on food intake, *Am. J. Physiol.*, 259, R709, 1990.

68. Magrum, L. J., Hickman, M. A., and Gietzen, D. W., Increased intracellular calcium in rat anterior piriform cortex in response to threonine after threonine deprivation, *J. Neurophysiol.*, 81, 1147, 1999.

69. Monda, M., Sullo, A., De Luca, V., Pellicano, M. P., and Viggiano, A., L-Threonine injection into PPC modifies food intake, lateral hypothalamic activity, and sympathetic discharge, *Am. J. Physiol.*, 273, R554, 1997.

70. Tabuchi, E., Ono, T., Nishijo, H., and Torii, K., Amino acid and NaCl appetite, and LHA neuron responses of lysine-deficient rat, *Physiol. Behav.*, 49, 951, 1991.

71. Wayner, M. J., Ono, T., DeYoung, A., and Barone, F. C., Effects of essential amino acids on central neurons, *Pharmacol. Biochem. Behav.*, 3, 85, 1975.

72. Panksepp, J. A. and Booth, D. A., Decreased feeding after injections of amino-acids into the hypothalamus, *Nature (London)*, 233, 341, 1971.

73. Stanley, B. G., Chin, A. S., and Leibowitz, S. F., Feeding and drinking elicited by central injection of neuropeptide Y: evidence for a hypothalamic site(s) of action, *Brain Res. Bull.*, 14, 521, 1985.

74. Torii, K., Yokawa, T., Tabuchi, E., Hawkins, R. L., Mori, M., Kondoh, T., and Ono, T., Recognition of deficient nutrient intake in the brain of rat with L-lysine deficiency monitored by functional magnetic resonance imaging, electrophysiologically and behaviorally, *Amino Acids*, 10, 73, 1996.

75. Meliza, L. L., Leung, P. M. B., and Rogers, Q. R., Effect of anterior prepyriform and medial amygdaloid lesions on acquisition of taste-avoidance and response to dietary amino acid imbalance, *Physiol. Behav.,* 26, 1031, 1981.

76. Norgren, R., Fromentin, G., Feurte, S. and Nicolaidis, S., Parabrachial lesions disrupt responses of rats to amino acid deficient diets, *Proc. Eur. Winter Conf. Brain Res.,* 16, 1996 (Abstr.).

77. Wegert, S. and Caprio, J., Receptor sites for amino acids in the facial taste system of the channel catfish, *J. Comp. Physiol.,* 168, 201, 1991.

78. Valentincic, T. and Caprio, J., Consummatory feeding behavior to amino acids in intact and anosmic channel catfish *Ictalurus punctatus*, *Physiol. Behav.,* 55, 857, 1994.

79. Ninomiya, Y., Kajiura, H., Ishibashi, T., and Imai, Y., Different responsiveness of the chorda tympani and glossopharyngeal nerves to L-lysine in mice, *Chem. Senses,* 19, 617, 1994.

80. Miller, M. G. and Teates, J. F., Acquisition of dietary self-selection in rats with normal and impaired oral sensation, *Physiol. Behav.,* 34, 401, 1985.

81. Fromentin, G., Feurte, S., and Nicolaidis, S., Spatial cues are relevant for learned preferences/aversion shifts due to amino-acid deficiencies, *Appetite,* 30, 223, 1998.

82. Walls, E. K. and Koopmans, H. S., Differential effects of intravenous glucose, amino acids, and lipid on daily food intake in rats, *Am. J. Physiol.,* 262, R225, 1992.

83. Russek, M., Hepatic receptors and the neurophysiological mechanisms controlling feeding behavior, in *Neurosciences Research*, Vol. 4, Ehrenpries, S. and Solnitsky, O. C., Eds., Academic Press, New York, 1971, 213.

84. Yox, D. P., Stokesberry, H., and Ritter, R. C., Vagotomy attenuates suppression of sham feeding induced by intestinal nutrients, *Am. J. Physiol.,* 260, R503, 1991.

85. Jeanningros, R., Vagal unitary responses to intestinal amino acid infusions in the anesthetized cat: a putative signal for protein induced satiety, *Physiol. Behav.,* 28, 9, 1982.

86. Niijima, A. and Meguid, M. M., An electrophysiological study on amino acid sensors in the hepato-portal system in the rat, *Obesity Res.,* 3, 741S, 1995.

87. Washburn, B. S., Jiang, J. C., Cummings, S. L., Dixon, K., and Gietzen, D. W., Anorectic responses to dietary amino acid imbalance: effects of vagotomy and tropisetron, *Am. J. Physiol.,* 266, R1922, 1994.

88. Inoue, M., Funaba, M., Hawkins, R. L., Mori, M., and Torii, K., Effect of continuous infusion of lysine via different routes and hepatic vagotomy on dietary choice in rats, *Physiol. Behav.,* 58, 379, 1995.

89. Niijima, A. and Meguid, M. M., Parenteral nutrients in rat suppresses hepatic vagal afferent signals from portal vein to hypothalamus, *Surgery,* 116, 294, 1994.

90. Yox, D. P., Stokesberry, H., and Ritter, R. C., Fourth ventricular capsaicin attenuates suppression of sham feeding induced by intestinal nutrients, *Am. J. Physiol.,* 260, R681, 1991.

91. Dixon, K. D., Le, T., Bellinger, L. L., and Gietzen, D. W., Selective vagotomy increases dietary intake of an amino acid imbalanced diet (IMB) in rats, *Soc. Neurosci. Abstr.,* 24, 1021, 1998.

92. Vo, B., Pavelka, J., Lucente, J., Williams, F. E., Bellinger, L. L. and Gietzen, D. W., Vagal denervations on intake of an imbalanced amino acid diet (IMB), *FASEB J.,* 10, A224, 1996 (Abstr.).

93. Wiggins, R. L., Williams, F. E., Gietzen, D. W., Norwood, J. K., and Bellinger, L. L., The effect of selective vagotomies (VAGX) on intake of an amino acid imbalanced diet (IAAD), *FASEB J.,* 9, A1003, 1995 (Abstr.).

94. Dixon, K. D., Jones, A. C., and Gietzen, D. W., Capsaicin (CAP) enhanced the effects of tropisetron (Trop) on adaptation to an amino acid imbalanced diet (IMB), *FASEB J.,* 9, A1003, 1995 (Abstr.).

95. Lucente, J., Vo, B., Pavelka, J., Williams, F. E., Bellinger, L. L., and Gietzen, D. W., Effect of blocking sympathetic (SYMP) and vagal (VAG) efferents on intake of an isoleucine imbalanced amino acid diet (IMB), *FASEB J.,* 10, A224, 1996 (Abstr.).

96. Dixon, K. D. and Gietzen, D. W., The effects of vagal efferent blockade and tropisetron (Trop) on amino acid imbalanced diet (IMB) intake, *FASEB J.,* 10, A224, 1996 (Abstr.).

97. Hammer, V. A., Gietzen, D. W., Beverly, J. L., and Rogers, Q. R., Serotonin[3] receptor antagonists block anorectic responses to amino acid imbalance, *Am. J. Physiol.,* 259, R627, 1990.

98. Jiang, J. C. and Gietzen, D. W., Anorectic response to amino acid imbalance: a selective serotonin[3] effect? *Pharmacol. Biochem. Behav.,* 47, 59, 1994.

99. Hrupka, B. J., Gietzen, D. W., and Beverly, J. L., ICS 205-930 and feeding responses to amino acid imbalance: a peripheral effect? *Pharmacol. Biochem. Behav.*, 40, 83, 1991.

100. Terry-Nathan, V. R., Gietzen, D. W., and Rogers, Q. R., Serotonin-3 antagonists block aversion to saccharin in an amino acid imbalanced diet, *Am. J. Physiol.*, 268, R1203, 1995.

101. Erecius, L. F., Dixon, K. D., Jiang, J. C., and Gietzen, D. W., Meal patterns reveal differential effects of vagotomy and tropisetron on responses to indispensable amino acid deficiency in rats, *J. Nutr.*, 126, 1722, 1996.

102. Leibowitz, S. F., Weiss, G. F., Walsh, U. A., and Viswanath, D., Medial hypothalamic serotonin: role in circadian patterns of feeding and macronutrient selection, *Brain Res.*, 503, 132, 1989.

103. Price, J. L. and Amaral, D. G., An autoradiographic study of the projections of the central nucleus of the monkey amygdala, *J. Neurosci.*, 1, 1242, 1981.

104. Costall, B., Kelly, M. E., Naylor, R. J., Onaivi, E. S., and Tyers, M. B., Neuroanatomical sites of action of 5-HT$_3$ receptor agonists and antagonists for alteration of aversive behavior in the mouse, *Br. J. Pharmacol.*, 96, 325, 1989.

105. Perez, C., Ackroff, K., and Sclafani, A., Carbohydrate- and protein-conditioned flavor preferences: effects of nutrient preloads, *Physiol. Behav.*, 59, 467, 1996.

106. Hoebel, B. G., Neuroscience and appetitive behavior research: 25 years, *Appetite*, 29, 119, 1997.

107. Aja, S. M., Gospe, S. M., Jr., and Gietzen, D. W., Ibotenic acid lesions of the anterior piriform cortex accelerate adaptation to amino acid imbalance, *Soc. Neurosci. Abstr.*, 23, 254, 1997.

108. Aja, S. M. and Gietzen, D. W., The infralimbic cortex (ILC) and serotonin-3 receptors (5-HT$_3$-R) in anorectic responses to a threonine imbalanced diet (IMB), *Proc. Soc. Study Ingest. Behav.*, 4, 28, 1998 (Abstr.).

109. Leung, P. M. B., Gietzen, D. W., and Rogers, Q. R., Effect of area postrema lesions on dietary choice of rats fed disproportionate amounts of dietary amino acids, *Soc. Neurosci. Abstr.*, 12, 1556, 1986.

110. Borison, H. L., Borison, R., and McCarthy, L. E., Role of the area postrema in vomiting and related functions, *Fed. Proc.*, 43, 2955, 1984.

111. Calingasan, N. Y. and Ritter, S., Lateral parabrachial subnucleus lesions abolish feeding induced by mercaptoacetate but not by 2-deoxy-D-glucose, *Am. J. Physiol.*, 265, R1168, 1993.

112. Rolls, E. T., The neural control of feeding in primates, in *Neurophysiology of Ingestion*, Booth, D. A., Ed., Pergamon Press, Oxford, 1993, 137.

113. Bernardis, L. L. and Bellinger, L. L., The dorsomedial hypothalamic nucleus revisited: 1986 update, *Brain Res. Rev.*, 12, 321, 1987.

114. Feifel, D. and Vaccarino, F. J., Growth hormone-regulatory peptides (GHRH and somatostatin) and feeding: a model for the integration of central and peripheral function, *Neurosci. Biobehav. Rev.*, 18, 421, 1994.

115. Ritter, S. and Dinh, T. T., 2-Mercaptoacetate and 2-deoxy-D-glucose induce Fos-like immunoreactivity in rat brain, *Brain Res.*, 641, 111, 1994.

116. Burton-Freeman, B., Gietzen, D. W., and Schneeman, B. O., Meal pattern analysis to investigate the satiating potential of fat, carbohydrate, and protein in rats, *Am. J. Physiol.*, 273, R1916, 1997.

117. Beverly, J. L., Gietzen, D. W., and Rogers, Q. R., Threonine concentration in the prepyriform cortex has separate effects on dietary selection and intake of a threonine-imbalanced diet by rats, *J. Nutr.*, 121, 1287, 1991.

118. Beverly, J. L., Hrupka, B. J., Gietzen, D. W., and Rogers, Q. R., Timing and dose of amino acids injected into prepyriform cortex influence food intake, *Physiol. Behav.*, 53, 899, 1993.

119. Beverly, J. L., Gietzen, D. W., and Rogers, Q. R., Protein synthesis in the prepyriform cortex: effects on intake of an amino acid-imbalanced diet by Sprague-Dawley rats, *J. Nutr.*, 121, 754, 1991.

120. Gietzen, D. W. and Beverly, J. L., Clonidine in the prepyriform cortex blocked anorectic response to amino acid imbalance, *Am. J. Physiol.*, 263, R885, 1992.

121. Gietzen, D. W. and Truong, B. G., Noradrenergic involvement in the prepyriform cortex of rats eating amino acid imbalanced diets: further evidence for the alpha-2 receptor, in *Proc. XXXII Cong. Int. Un. Physiol. Sci.*, Glasgow, Aug 1–6, 32, 253, 1993 (Abstr.).

122. Truong, B. G. and Gietzen, D. W., Central 5-HT$_3$ and 5-HT$_4$ receptors influence aminoprivic feeding, Submitted.

123. Truong, B. G. and Gietzen, D. W., γ-Aminobutyric acid (GABA) receptors in anterior piriform cortex (APC) mediate feeding responses in rats, *Soc. Neurosci. Abstr.*, 23, 253, 1997.

124. Cummings, S. L., Truong, B. G., and Gietzen, D. W., Neuropeptide Y and somatostatin in the anterior piriform cortex alter intake of amino acid-deficient diets, *Peptides,* 19, 527, 1998.
125. Gietzen, D. W., Truong, B. G., and Dang, B., Is the NMDA receptor involved in the responses to amino acid imbalanced diets? *Soc. Neurosci. Abstr.,* 20, 1679, 1994.
126. Gietzen, D. W., Diepenbrock, M. T., Truong, B. G., and Nguyen, T. Q.-T., Inhibition of nitric oxide synthase (NOS) alters responses to amino acid deficiency, *FASEB J.,* 7, A646, 1993 (Abstr.).
127. Truong, B. G., Barrett, J. A., and Gietzen, D. W., Dopamine2 (DA2) receptors in the anterior piriform cortex (APC) are involved in mediating the responses to amino acid imbalanced diets (IMB), *Soc. Neurosci. Abstr.,* 24, 2132, 1998.
128. Blevins, J. E., Havel, P. J., Truong, B. G., Aja, S. M., and Gietzen, D. W., Injection of leptin into the anterior piriform cortex (APC) inhibits food intake in rats, *FASEB J.,* 12, A348, 1998 (Abstr.).

Section IV

Neural Integration of Sensory
and Metabolic Information
in the Control of
Macronutrient Selection

24 An Overview of Neural Pathways and Networks Involved in the Control of Food Intake and Selection

Hans-Rudolf Berthoud

CONTENTS

0-8493-2752-0/00/$0.00+$.50
© 2000 by CRC Press LLC

1   INTRODUCTION

Behaviors such as searching for and selection and ingestion of food can ultimately only be expressed through specific actions of the nervous system. Almost a century of intense research has, however, been unable to identify clearly the neural substrate subserving these behavioral functions. Many textbooks still portray a rather simplistic view of hypothalamic feeding and satiety centers. Although there is no doubt that various hypothalamic structures play a key role in feeding behavior and body energy regulation, the hypothalamus is only one piece in the neurological puzzle, and, equally importantly, the hypothalamus is also involved in the control of many other behaviors and homeostatic regulations.[1,2]

Our inability to describe the complete neural circuitry involved in the control of food intake and selection is due in part to confusion about the exact nature of the processes that are controlled and regulated, as well as to a rather casual, not well-defined, use of psychological terms such as appetite, hunger and satiety. For more detailed discussions of terminology, see also Chapter 6 by Booth and Thibault in this volume. As long as operational terms such as hunger, and satiety are used to describe the mere presence or absence of eating, and not the occurrence of specific neurological events, this inability is likely to persist. In the meantime, it would be helpful if we would use operational terms that are better defined than appetite, hunger, and satiety. Table 1 is an attempt to partition the ingestive sequence into a number of such operational units that should make it easier to attach specific neurological events.

It is immediately apparent that we have considerable knowledge of the information that might be used by the nervous system (sensory inputs) and what motor nuclei and muscle groups must execute the behavior (outputs), but we know little about the neurological processing that takes place in between (Figure 1). This chapter is intended to provide some basic information about the neuroanatomical areas and connections that are potentially involved in the overall control of ingestive behavior, energy metabolism, and the regulation of body weight. Such information should facilitate understanding of rationale and interpretation of the various experimental approaches discussed in the chapters of this book. Space limitations do not allow full discussion of the equally important neurochemical dimension of this neural network. The recently published atlas of neuroactive substances and their receptors in the rat is, however, an excellent source of such information.[3] In addition, the neurochemical dimension has also been discussed in a recent neural model of feeding.[4]

2   METHODS USED TO IDENTIFY THE RELEVANT NEURAL CIRCUITS

2.1   IDENTIFICATION OF CONNECTIVITY

Neuronal tracing has become the method of choice to identify neuronal connections between different areas of the nervous system. It has almost completely replaced the use of electrophysiological methods employing stimulating and recording electrodes.

Retrograde tracers such as Fluorogold are taken up by local axonal nerve terminals if injected in their vicinity by either pressure or electrophoresis. The tracers are then transported against the normal direction of propagation of action potentials (retrogradely) to the neuronal cell body. This method works best for longer axons (>500 μm) because the injection artifact does not obscure the specific label. The major pitfall of this method is the possibility that "axons of passage," without terminals in that local injection site, may also take up the tracer, leading to false-positive identification of projection targets.

The use of transneuronal viral tracing (e.g., pseudorabies virus) has been an important addition to the tracing repertoire. These tracers are retrogradely transported to the soma of a neuron where they replicate and then "jump" to synaptic terminals of another neuron.[5,6] They can therefore be used to help define complete multisynaptic pathways.

Anterograde tracing (e.g., with *Phaseolus vulgaris* leucoaglutinine [PHAL], or the carbocyanine dyes DiI and DiA), is used to establish the extent of axonal projections originating from a specific group of neuronal perikarya.

2.2   Functional Mapping

Although demonstration of connectivity does sometimes allow general inferences about the nature of neural signals carried by the traced neurons, it does not link them to specific neuronal activity and physiological functions. In contrast, functional mapping uses physiological or near-physiological stimuli and measures the activity of single neurons or the pattern of neuronal activity throughout the brain. Both invasive and noninvasive methods are in use.

Among the invasive methods, electrophysiological recording from single neurons has the great advantage that more than one type of (physiological) stimulus can be assessed in a given subject and neuron. This method therefore allows us to learn a lot about relatively few neurons, but is not very efficient to assess the "global" response to a given stimulus.

Induction of immediate-early gene (IEG) expression has almost completely replaced the earlier use of 2-deoxy-glucose (2DG) uptake. Fos mapping has a higher anatomical resolution, does not require the use of radioisotopes, and is more economical.[7] However, it cannot totally replace 2DG mapping because this method measures a different parameter of neuronal activation. While 2DG uptake measures local glucose utilization, Fos expression assesses "out-of-the-ordinary" synaptic excitatory activation of individual neurons. The major disadvantage of both methods is that negative results are difficult or impossible to interpret. In the case of 2DG mapping which does not achieve single-neuron resolution, activation of only a few discrete neurons within a larger ensemble does not necessarily increase local glucose uptake. Fos may not be expressed in some neurons, and it is apparently not induced by inhibitory synaptic inputs.[8-10]

As another type of functional mapping, we could consider the use of *in situ* hybridization, immunohistochemistry, and any molecular biochemical identification method on microdissected tissue, in conjunction with functional evidence for a particular transmitter, peptide, or receptor from local injection experiments. A disadvantage that is common to all of these invasive methods (with the exception of single cell recording) is that they do not allow comparison of two different stimuli in the same subject.

Rapid technological progress in the past decade has resulted in powerful new, noninvasive methods (functional magnetic resonance imaging, FMRI, and positron emission tomography, PET) that detect minute changes in local brain metabolism that may be reliable reflections of neuronal activity and can be used in humans.[11,12] A major advantage of these powerful new methods, in addition to their noninvasiveness, is the possibility of performing longitudinal time series and comparing the effects of two or more stimuli in the same subject. Such methods will undoubtedly expand our views on the neural substrate involved in food intake and selection (see below).

3   INPUT (SENSORY) SYSTEMS

Input systems are defined here as systems through which the nervous system receives nutritional/metabolic or other information important in the regulation of energy homeostasis (Figures 1 and 2). There are several classification schemes for such inputs. One distinguishes on the basis of

TABLE 1
List of Some Operational Tasks to Be Carried Out by the Neural Control System

Anticipatory or preparatory phase (to establish readily available "knowledge base" about the internal state of affairs and its relationship to prospectively ingested foods):
- Constantly monitor levels of all important fuels and substrates
- Estimate the potential to interconvert fuels and substrates and to mobilize them from storage through autonomic nervous system and endocrine actions, by monitoring metabolic fluxes and turnover rates
- Keep memorial representations concerning location, abundance, and price of food in general (calories) and specific macro- and micronutrients
- Estimate the impact of prospective ingestion on homeostatic functions and initiate counterregulatory (anticipatory) mechanisms, if necessary
- Use all this stored information to compute value for time horizon and threshold levels for initiation of behavioral action (when food is unexpectedly encountered this information is immediately available for appetitive and consummately phases)

Appetitive phase (to make decision regarding other competing behaviors):
- Make decision to start searching, approaching, and sampling particular food source (if food is not already present)
- Detect sensory qualities of sample
- Compare sensory qualities of sample with expected or remembered values
- Adjust processes under anticipatory phase if necessary

Consummmatory phase (to reach final goal of specific motivated behavior and learn from new foods):
- Make decision about and start ingesting food item A
- Constantly monitor sensory qualities (smell, flavor, taste, texture) of food and amount of "pleasure" derived
- Constantly monitor gastrointestinal and postabsorptive signals and amount of pleasure or aversion derived
- Start storing associations between sensory qualities and postingestive consequences of food A
- Stop ingestion of food item A
- Initiate same cycles for food item B
- Start storing associations between sensory qualities and postingestive consequences of food B
- Stop ingestion and engage in other behavior

Postingestive or sated phase (to acquire associative memorial representations of just-engaged ingestive experience and make transition to other behavior):
- Continue to store associations between postingestive effects and sensory qualities of food A and B
- Use information to recalibrate content of stored information if there was discrepancy between expected and actual value

the anatomical location where the stimulus is transduced into a neural signal. External stimuli include visual, olfactory, and perhaps auditory stimuli. Internal stimuli generated by food after its ingestion can be divided into stimuli generated in each of the segments of the alimentary canal and are sometimes grouped into pregastric (mainly taste), gastric (e.g., distension), and postgastric, or preabsorptive, stimuli. Postabsorptive stimuli include those generated by (1) mucosal nutrient transport mechanisms and the associated release of local hormones (e.g., cholecystokinin, CCK) acting on visceral sensory nerves; (2) nutrients, metabolites, and hormones acting on sensors in the portal hepatic space (e.g., glucose); (3) metabolic processing steps and their potentially associated local messengers and hormones in the liver acting on hepatic sensors (e.g., ATP, glucagon); and (4) metabolites, hormones, and other factors circulating in the blood (or lymph) and activating corresponding sensors directly in the brain (e.g., glucose, amino acids, insulin, and leptin).

It is important to recognize that such a categorization based on the site of neural transduction, while heuristically useful, does not take into consideration that the neural network distal to the transduction site is functioning in essentially the same way. That is, each neuron within the network is more or less sensitive to a variety of chemical signals originating from either other neurons or the bloodstream.

Inputs have also been categorized according to their pro- or antiorexigenic effects. For example, gustatory input from sweet receptors is a positive (reinforcing or feeding-forward) signal that

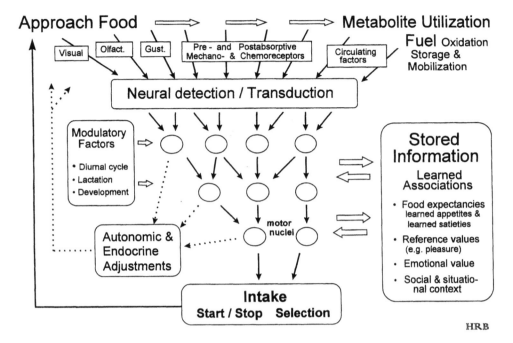

FIGURE 1 Schematic diagram indicating the general flow of information required for neural control of food intake and selection. The essential task for the nervous system is to reduce and integrate the multiple sources of relevant information and instruct the behavioral, autonomic, and endocrine motor systems to initiate appropriate responses. Ultimately, to explain the controls in neurological terms, each processing step as represented by the open circles will have to be identified and characterized.

increases ingestion, while other kinds of stimuli such as CCK and GLP-1 are considered negative feedback or satiety signals that suppress ingestion. However, as with any sensory system, the organism's interpretation of a stimulus depends critically on cognitive processing, so that a previously highly preferred taste can quickly turn into an nonpreferred or aversive one following satiation or food poisoning.

Finally, inputs can be divided into information derived directly from ingested food and its metabolic/hormonal actions or indirectly from prior memorial representations of foods (see Chapter 14 by Davidson et al.). At present, the form and site as well as the recall mechanisms of such food memories are poorly understood.

A change in neural electrical activity is not the only effect produced by sensory input signals. Sensory inputs can also activate second-messenger systems that generate changes in gene expression without immediate changes in the electrical property of the receiving neuron. For example, a particular input or stimulus may increase the expression of a particular receptor in a neuron, and this in turn could make the neuron more sensitive to stimulation by ligands to this receptor. Alternatively, a stimulus might produce morphological changes such as in dendritic geometry and complexity of the recipient neuron. Effects such as these may be particularly common during development or pregnancy, or when major metabolic changes are occurring as during dieting, or during disease or other pathology.

3.1 Information Carried by Visual and Olfactory Cues

In humans and other primates visual cues derived from food are undoubtedly very important for food intake and food choice. However, visual cues cannot directly inform the brain about the nutritional value of a particular food item as, for example, glucoreceptor input does, and can only

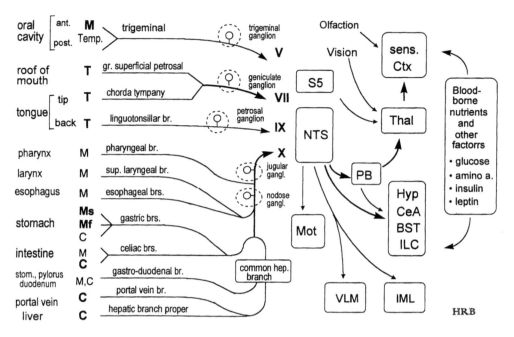

FIGURE 2   Visceral sensory input to the brain. All along the alimentary canal, various mechano- and chemosensors are located that transmit food- and nutrition-related signals via primary visceral afferents in the trigeminal (V), facial (VII), glossopharyngeal (IX), and vagus nerve (X) to the brain stem. The sensory modality transmitted by each nerve branch is indicated by the letters: M = mechanoreceptors (Ms = slowly adapting tension receptor; Mf = fast-adapting touch receptor); T = taste receptors; Temp = temperature sensors; C = chemoreceptors. Blood-borne nutrients and metabolites can also enter the brain directly at various sites. The possibility of dorsal root spinal afferents mediating some nutrient-related signals is not shown. Abbreviations: BST, bed nucleus of stria terminalis; CeA, central nucleus of amygdala; Hyp, hypothalamus; ILC, insular cortex; IML, intermedilateral column of spinal cord (sympathetic preganglionic neurons); Mot, brain stem oral motor nuclei; NTS, nucleus of solitary tract; PB, parabrachial nuclear complex; S5, principal sensory nucleus of trigeminal nerve; sens. Ctx., sensory cortex; Thal., thalamus; VLM, ventrolateral medulla.

gain nutritional salience by associative processes through learning (see also Chapter 6 by Booth and Thibault, this volume). It is not surprising then, that visual cues are first extensively processed by cortical structures before entering feeding-relevant circuits. It has been clearly demonstrated that highly processed visual information available in the cortical visual area IV is a rich source of input to other specialized cortical and cortical-like areas such as the orbitofrontal cortex and amygdala of the macaque monkey, where it is integrated with information from the olfactory and gustatory systems (see Chapter 17 by Rolls).

Olfactory input is equally important in primates and humans, and probably more so in rodents. Because olfactory receptors are chemoreceptors, they have the potential to code nutritional value directly or innately, just like the smell of certain pheromones can produce innate, unlearned responses. This might suggest that the smell of perhaps amino acids or proteins in certain food items may innately label an item as edible.[13] There is, however, not much experimental evidence that the olfactory system codes directly for different macro- or micronutrients or for energy density. Similar to visual cues, olfactory cues probably achieve nutritional salience mainly through associative learning. In contrast to the gustatory system (see below), the olfactory system seems to have a greater ratio of specialist vs. generalist receptor neurons. In other words, there are more narrowly tuned receptor neurons in the olfactory system. The significance of this difference for the control of ingestive behavior is not clear. The pathways and relays for olfactory perception are organized differently than for other sensory systems. From the olfactory bulb, information directly reaches various cortical areas including the olfactory tubercle, piriform, dorsal peduncular and entorhinal cortex. Information from the major primary olfactory cortex, the piriform cortex, is then relayed to polymodal association areas such as the orbitofrontal and insular cortex as well as the amygdala, where it is integrated with gustatory, visceral, and somatosensory input, and from the entorhinal cortex to the hippocampal formation.

3.2   GUSTATORY AND TRIGEMINAL INPUT

Gustatory input via taste receptor cells on the tongue and palate is considered most important for guiding food intake and selection. In addition, textural characteristics and the temperature of foods are detected by trigeminal mechanosensors and thermal sensors on the tongue and throughout the entire oral cavity (see Figure 2). Taste receptor physiology is in a fluid state, with recently recognized specific tastes for fatty acids, umami, complex carbohydrates, and water added to the classical four taste qualities — sweet, salty, sour, and bitter. Those four classical taste modalities have been interpreted as mainly representing innate detectors for acceptable food (sweet), for dangerous or toxic foods (bitter and sour), and for special needs (salt, water), providing little other information about the macro- or micronutrient composition or energy density (see also discussion in Chapter 6 by Booth and Thibault). If this view were completely accepted, gustatory input, like visual and olfactory inputs, could only serve as cues for such macronutrient information by learned associations. Given the newer findings of fat, umami, and complex carbohydrate (polycose) taste, however, it is entirely possible that gustatory input has some capacity to code for macronutrient content directly in an unlearned fashion.

A more detailed description of central taste pathways in rats and primates is provided in Chapter 17 by Rolls. Briefly, from the second-order taste neurons in the rostral part of the solitary nucleus, taste information is relayed in a largely ipsilateral fashion to the parabrachial nuclei in the pons (third-order taste neurons). From the parabrachial nucleus, taste information takes two routes, one quite discrete via the thalamic taste area (ventral posteromedial nucleus, VPM), to the gustatory cortex (agranular or dysgranular insular cortex), the other more diffuse, to the amygdala, hypothalamus, and other ventral forebrain areas.[14] Input from trigeminal mechanosensors and temperature sensors is first relayed through the trigeminal sensory nucleus (S5) and the solitary nucleus, and then closely parallels the flow of gustatory information to the pons and forebrain.[15]

3.3   GASTROINTESTINAL AND PORTAL-HEPATIC SENSORY INFORMATION

The various sensory signals generated by food passing through the gastrointestinal tract, and linked to metabolic processing in the liver are well covered by other chapters in this book (Chapter 18 by Covasa and Ritter; Chapter 19 by Tso and Fukagawa; Chapter 21 by Langhans), and will be discussed only briefly here. Figure 2 depicts the important role played by vagal primary afferent neurons in the transmission of many of these signals to the brain. The gastric distension signal is thought to play a major role in the satiation process, but in itself cannot provide information about the chemical composition of the ingesta. However, if combined with signals from chemosensors, it has the potential to transmit quantitatively how much of a given macronutrient was ingested.

There is evidence for chemosensors in the small intestine, as well as in the portal vein and liver. Sensitivity to representatives of all three classes of macronutrients has been reported, in particular to glucose and other hexoses, various amino acids, and fatty acids. Together, these sensors have the potential to measure directly the qualitative and quantitative macronutrient composition of ingested food, but this has not yet been experimentally demonstrated.

It has become clear that gastrointestinal and hepatic nutrient detection is typically not a one step transduction mechanism. In the case of intestinal fat detection, there seems to be a cascade of events involving CCK and its receptors, at least two CCK-releasing factors, glucagon-like peptide (GLP-1) and its receptors, chylomicron formation, apolipoprotein (ApoAIV), and procolipase/enterostatin. If these messengers have differential sensitivities to different forms of fat such as chain length and degree of saturation, then the ensemble could be very efficient in telling the brain what kind of fat and how much was ingested. Together with similar signaling cascades for protein and carbohydrates, the gastrointestinal tract and liver could be very efficient in the nutritional encoding of sensory neural signals (see Chapter 21 by Langhans).

As depicted in Figure 2, many of these signals are carried to the brain by vagal primary afferent neurons terminating in the solitary nucleus of the medulla. However, there is also some evidence for transmission of such signals by dorsal root spinal afferents. Besides mechanical and chemical signals directly from the gastrointestinal tract, it is very likely that mechanoreceptors and pain receptors in the peritoneal wall and skin, as well as in visceral omenta contribute to the general perception of fullness after a large meal. Furthermore, metabolites and hormonal signals generated by interaction of nutrients in the gastrointestinal tract, pancreas, and liver can potentially directly affect the brain if appropriate receptors and transport mechanisms in the blood–brain barrier are present (see Chapter 22 by Levin et al., Chapter 23 by Gietzen, and Chapter 30 by Van Dijk et al.).

3.4 SIGNALS FROM STORED NUTRIENTS

With the discovery of leptin, the existence and importance of direct signals from the major site of stored energy to the brain has become clear. Leptin receptors (long and short form) are found in the hypothalamus (see Section 4.3) and other forebrain areas such as thalamus, cerebellum, and substantia nigra,[16] as well as in several peripheral locations in viscera such as the liver, pancreas, adrenal gland, and gastrointestinal tract. It has been proposed that leptin modulates CCK-sensitive vagal afferent fibers innervating the gastrointestinal tract.[17] Adipose tissue may release additional signaling factors besides leptin,[18] and it may send sensory information to the brain via dorsal root spinal afferent neurons.[19] An extensive body of work has also identified the pancreatic hormone insulin, whose average plasma levels closely track the level of adiposity, as a possible adiposity signal used by the brain (see Chapter 30 by Van Dijk et al.). However, its signaling characteristics are clearly different from that of leptin, because it is primarily released by ingested carbohydrates, and thus has a very distinctive short-term plasma concentration profile. Also, insulin is not able to prevent the development of hyperphagia and obesity in the absence of leptin or its receptor, as in the ob/ob and db/db mouse or fa/fa Zucker rat.

The other functional energy store is hepatic glycogen. Although it has long been hypothesized that hepatic glycogen generates a signal to the brain by affecting hepatocyte membrane potential and vagal sensory neurons,[20] this has not yet received general acceptance. In the light of the leptin discovery it is interesting to speculate that an analogous hormonal signal is generated by hepatic glycogen or some concomitant of it (see Chapter 21 by Langhans).

3.5 INFORMATION DERIVED THROUGH RECALL FROM STORED MEMORIES

Information about familiar foods can also be stored as memorial representations in the brain (see Chapter 6 by Booth and Thibault; Chapter 7 by Sclafani; and Chapter 14 by Davidson et al.). The issue of memory and its neuroanatomical substrate is discussed later in this chapter. Input from memory is defined as information retrieved from memorial representations in the brain. There is evidence that a particular internal state selectively facilitates recall of certain memorial representations. It has been reported that rats, for example, attend more to a stimulus that has previously been paired with oily food when fatty acid oxidation is inhibited by the administration of β-mercaptoacetate, and attend more to a different stimulus paired with glucose when glucose utilization is inhibited by central administration of 2DG (see Chapter 14 by Davidson et al.).

4 INTEGRATIVE SYSTEMS

To divide the overall neural substrate involved in food intake and selection into sensory, motor, and integrative systems is obviously arbitrary and is done more for heuristic reasons and out of ignorance of the ultimate functional neuroanatomical description. Whenever we do not fully under-stand a neurological system we use terms like integrator or network.

If integration is defined as combining two or more independent (sensory) signals by a given formula or transfer function into a single new signal, then it already takes place within both the input (sensory) and output (motor) systems. For example, most vagal primary afferent neurons in the rat are polymodal, integrating mechanical and certain chemical (CCK, leptin) signals.[17,21]

4.1 WHAT INFORMATION NEEDS TO BE INTEGRATED?

For the organism to make a (conscious or unconscious) decision about what food to eat and how much to ingest in a given meal, there are many types of information it should consider.

1. Information about the availability of nutrients and energy from various *endogenous* sources:
   a. Immediately available (e.g., blood glucose, hepatic ATP, signaling via intestinal hor-mones);
   b. Available with delay (e.g., gastric fill, adipose tissue mass/leptin, insulin).
2. Information about the availability and characteristics of *exogenous* (to be ingested) nutrients:
   - *Unlearned* (food is present)
     a. Direct smell, taste, and texture and its hedonic value;
     b. Novelty (neophobia);
   - *Learned* (memory retrieval triggered by external or internal signals, or thought)
     c. Amount of pleasure to be derived;
     d. Potential harmfulness or toxicity (e.g., conditioned aversion);
     e. Location, abundance and cost of obtaining;
     f. Energy density and nutrient composition;
     g. Social (and scientific) acceptability of the food;
   Difference between remembered and actual characteristics.
3. Information about the level of energy and specific nutrient demand to be anticipated in the postingestive period (e.g., time of day, physical work, lactation, cold environment):
   a. Total energy demand;
   b. Specific nutrients/metabolites in specific tissues.
4. Information about the urgency of other competing behaviors (e.g., sleep, sexual behavior, and defense, including defensive anorexic effect of gastrointestinal or systemic infection, or seasonal activities such as migration and hibernation).

4.2 GENERAL CONCEPTS OF INTEGRATIVE NEURAL NETWORKS

4.2.1. Parallel vs. Linear Processing

It is increasingly recognized that the brain takes advantage of parallel processing to deal more rapidly and efficiently with the great number of requisite computational tasks.[22-24] Although there is not yet experimental evidence for this type of processing within the neural network involved in the control of food intake and selection, it would be an extremely powerful way to integrate rapidly the vastly different types of information into behavioral action. I therefore make the speculative proposition to view the "feeding network" as consisting of a central processor unit with sensory inputs and motor outputs, and several parallel loops as depicted in Figure 3. Each of the parallel loops has the capacity to modify the way the central processor unit handles the sensory input and transfers it to a meaningful output. In analogy to computer technology it could be said that activation of each loop changes the software that runs the central processor, and thus the sensory-motor transfer function.

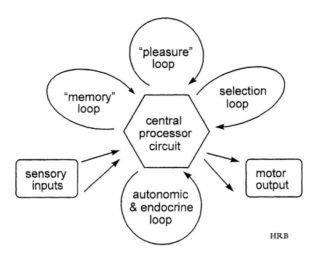

FIGURE 3 Parallel processing model for neural control of food intake and selection. Information within the central processor circuit has essentially simultaneous access to the various "loops." The "memory loop" allows sensory information associated with an actual event to (1) be compared with memories from earlier experience and (2) be used to form new memorial representations. The "pleasure loop" estimates the potential amount of pleasure to be derived from engaging in the actual behavior. The "selection loop" determines a hierarchy among all the competing behaviors and selects the most beneficial behavior. Finally, the "autonomic and endocrine loop" evaluates the possibility to use and, if necessary, engages the autonomic and endocrine systems in order to solve the homeostatic imbalance.

In a "memory loop," relevant stored information is retrieved if any of the sensory cues matches a property of any stored association. The "pleasure loop" assesses the amount of pleasure or displeasure derived from the sensory input, and the "selection or competition loop" evaluates the appropriateness of engaging in this or other competing behaviors. Emotional context and conditioning such as in conditioned taste aversion could be seen as an additional loop. Alternatively, it could be embedded in either the memory loop, the pleasure loop, or both.

4.2.2.   Phylogenetically Older vs. Newer Layers of Processing

Anatomical identification of the neural network subserving food intake control may be complicated by the fact that evolutionary pressure has emphasized different control functions and systems in a way that is neither linear nor modular and makes it, therefore, difficult to analyze experimentally. For example, at the anatomical level, some phylogenetically older systems may still have the connections with the rest of the system but not be functionally important. At the neurochemical level, phylogenetically old signaling molecules may be used for completely new functions.

It has been proposed that in lower vertebrates the major function of a primitive vagal sensory-paraventricular system was to detect and expel any invader (toxin, antigen) from the alimentary canal by activating a somatic motor reflex.[25] This "ancient" function still exists in modern vertebrates in the form of the vomiting reflex, the innate acceptance/rejection reflexes to the basic taste stimuli (see Section 3.2), and the recently discovered vagal involvement in immune defense. According to the hypothesis, additional components have then been added to this primitive pathway during evolution. One of these add-ons may be the more modern taste system that makes use of combinatorial processing, connects with forebrain reward systems, and is in a position to discriminate at least some of the nutritional components of a food. Another add-on may be the vagal efferent system, which allows not only somatic but also visceral-secretory reflexes to be generated. The important point is that this system appears homogeneous at first inspection, but may be composed of functionally quite different parts.

4.2.3   Forebrain vs. Brain Stem Controversy: Indirect or Top-Down Control

The general organization of the nervous system can be viewed as essentially hierarchical, with certain lower or vegetative functions and areas being controlled or modulated by progressively higher levels of integration. In the field of food intake control, the controversy regarding the respective roles of the brain stem or hindbrain vs. the diencephalon or forebrain has been particularly stimulating, peaking in the use of midbrain decerebrate rats as an experimental model. Because such rats do not lose the ability to exhibit the typical acceptance and rejection reflexes to sweet and bitter tastants, respectively, and if satiated by gastric feeding let nutritive solutions dribble out of their mouth if infused through an intraoral fistula, it has been suggested that the caudal brain stem contains the basic machinery for the display of ingestive responses including satiation.[26] Furthermore, these basic brain stem controls have been termed direct controls in distinction to indirect controls effected by inputs from descending forebrain connections.[27] In contrast to this "bottom-up" interpretation, it could be argued that the primary event in any display of ingestive behavior (start, stop, maintain, and suppress) is a conscious or unconscious decision made somewhere in the forebrain, which then leads to "loading" the appropriate software that dictates the rules of ingestive behavior, including brain stem functions.[4] If viewed in the light of this "top-down" organization, the designation "indirect controls" for the forebrain functions seems not quite appropriate. However, the two brain regions do not have to be played against each other, but rather the high degree of interconnectedness and functional interplay should be emphasized. The strong reciprocal anatomical connections between the relevant forebrain and brain stem areas have been amply documented, and the functional importance of such connections is starting to be fully appreciated by recent demonstrations that feeding induced by hypothalamic infusion of neuropeptide Y (NPY) is blocked by opioid receptor antagonism at the level of the nucleus of solitary tract (NTS),[28] and that MC-4 receptor activation-induced suppression of food intake can be obtained not only in the hypothalamus, but also in the NTS[29] and parabrachial nucleus (personal communication by F. Koegler).

I suggest that certain areas of the brain stem together with other major areas of the forebrain should be considered as a central processing unit, within which key information is liberally circulating and available for parallel processing in any of the constituent and their associated specialized areas (Figure 4). In the brain stem these core areas may include the nucleus of the solitary tract, the parabrachial complex, and parts of the reticular formation directly related to these two nuclei. In the diencephalon most parts of the hypothalamus, and in the telencephalic areas of

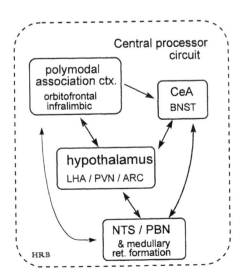

FIGURE 4   Suggested central processor circuit for the control of food intake. The four general areas are the most heavily interconnected areas with clearly demonstrated relevance for ingestive control.

the polymodal and supramodal association cortices as well as the extended amygdala, should be included. It has to be stressed that this concept is at best speculative, and that there is room to add or subtract some areas from the list.

4.3   HYPOTHALAMUS

4.3.1. Overview of Hypothalamic Organization

More than 40 histologically distinct nuclei and areas form the mammalian hypothalamus, and many of these nuclei have been further subdivided into subnuclei.[30] A general, schematic anatomical and functional map of the entire hypothalamus is depicted in Figure 5. One of the broad generalizations that could be made concerns the medial-to-lateral zonation. The most medial or periventricular zone is mainly concerned with the detection of blood-borne (and possibly CSF-borne) signals and the organization and control of endocrine responses. The medial zone is primarily composed of large nuclei such as the dorsomedial and ventromedial nuclei, the anterior hypothalamic area, and the medial preoptic nucleus, all of which receive various sensory inputs, interconnect heavily with the rest of the hypothalamus, and are involved in the organization of adaptive behaviors. The lateral zone has long been characterized by a lack of clearly definable cell groups and it is traversed by the medial forebrain bundle and the fornix. More recently, however, some very distinct and neurochemically defined cell groups with relevance for food intake have been described. The lateral zone has an extensive intra- and exrahypothalamic communication system and could be viewed as the interface between more medial hypothalamic areas with the somatic and autonomic motor systems that seems to be involved in mediating general arousal, sensory sensitization, and sensorimotor coordination associated with the expression of motivated behaviors such as feeding.[30,31]

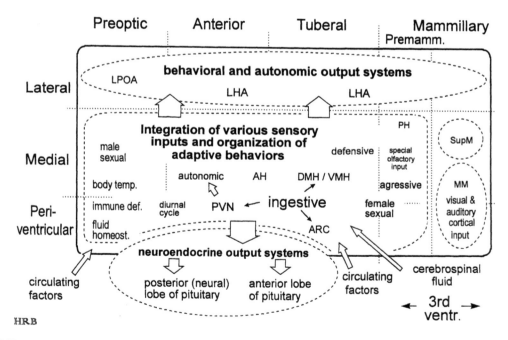

FIGURE 5   Gross "geographic" and functional map of the rat hypothalamus. The map is based on R. B. Simerly, "Anatomical Substrates of Hypothalamic Integration,"[30] and other anatomical descriptions of hypothalamic organization. The separation into behavioral and functional domains is not as specific as indicated, with considerable overlap. For abbreviations see legend to Figure 6.

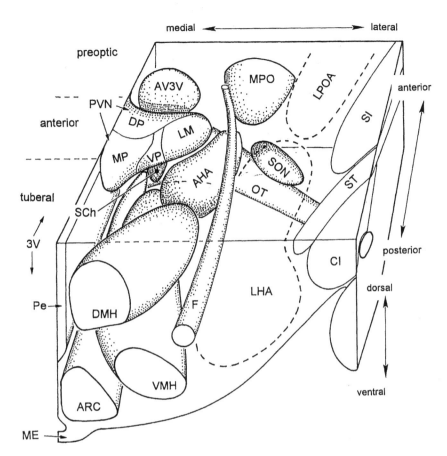

FIGURE 6   Three-dimensional view from dorsal and caudal of major hypothalamic nuclei in the right hemisphere of rat hypothalamus. Abbreviations: AHA, anterior hypothalamic area; ARC, arcuate nucleus; AV3V, anterioventral area of 3rd ventricle; CI, capsula interna; DP, dorsal parvocellular subnucleus of paraventricular nucleus; DMN, dosomedial nucleus; F, fornix; LHA, lateral hypothalamic area; LM, lateral magnocellular subnucleus of paraventricular nucleus; LPOA, lateral preoptic area; MP, medial parvocellular PVN; MPO, medial preoptic area; OT, optic tract; SCh, suprachiasmatic nucleus; SON, supraoptic nucleus; SI, substantia inomminata; ST, subthalamic nucleus; VMN, ventromedial nucleus; VP, ventral parvocellular subnucleus of paraventricular nucleus.

The paraventricular nucleus of the hypothalamus represents a microcosm within the hypothalamus, in that various subnuclei interface with all three effector systems, the endocrine (magnocellular groups), autonomic and behavioral (parvocellular groups). A three-dimensional view of a restricted area and a limited number of nuclei of the rat hypothalamus is depicted in Figure 6.

4.3.2   Connections of Some Important Hypothalamic Nuclei

Rapid progress has been made over the last few years in identifying hypothalamic neuron populations that contain specific neurotransmitters, receptors, and other factors of crucial importance in feeding behavior and the development of obesity (Figure 7). Some of these neurochemically identified connections are considered in the following discussion for each of the key nuclei. This area of investigation is very fluid, and some of the neurochemical and anatomical details of this hypothalamic network will likely change.

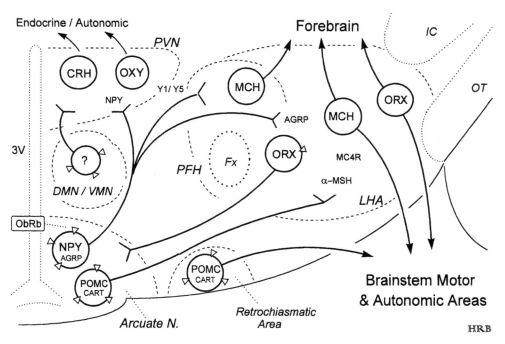

FIGURE 7   Some neurochemically identified connections and projections in the rat hypothalamus that have significant effects on ingestive behavior and body adiposity. A medial (arcuate nucleus, retrochiasmatic area, dorsomedial, DMN, and ventromedial nucleus, VMN, not shown) group of neurons contains leptin receptors (ObRb or long form, triangles). Neurons in the ARC containing mRNA for NPY and AGRP project to the PVN and perifornical/lateral hypothalamus (PFH). A different population of neurons in the ARC and retrochiasmatic area contain mRNA for POMC and CART project to the lateral hypothalamus as well as to autonomic and motor areas of brain stem and spinal cord. In the LHA, one population of neurons contains MCH and another population contains orexin-A, with both ascending (e.g., cortex, amygdala, accumbens) and descending autonomic and motor projections to the brain stem and spinal cord. The lines do not follow actual pathways and bifurcations do not necessarily imply collateral pathways. IC, internal capsule; OT, optic tract.

4.3.2.1   Lateral hypothalamic area (LHA) (including perifornical area)

As mentioned above, the LHA is the most extensively interconnected area of the hypothalamus, allowing it to modulate many different functions, spanning the entire spectrum from cognitive to autonomic. The LHA or zone is a large and heterogeneous area with several distinct nuclear groups, and each of these subnuclei or areas will, in the future, undoubtedly have to be investigated separately (see Reference 30 for a more detailed discussion).

As shown in Figure 8, the LHA has vast efferent projections to the entire cortical mantle including the hippocampal formation, extended amygdala, basal ganglia and thalamus, the midbrain and pons, the brain stem and spinal cord, as well as most other nuclei of the hypothalamus. These projections have mainly been established using retrograde tracer injections into the various projection targets resulting in labeled perikarya in the LHA, and erroneous colabeling of fibers of passage is not a problem. More recently, many of these projections have been confirmed on the basis of immunohistochemical studies using antibodies to peptide neurotransmitters which are exclusively produced in LHA neurons, such as melanin-concentrating hormone (MCH) and orexin.[57] One population of LHA neurons contains mRNA and stains immunohistochemically for MCH[32], a peptide that elicits food intake when injected intracerebroventricularly,[33] (Figure 7). A separate population contains the peptide orexin-A, another peptide that elicits food intake if injected locally.[34]

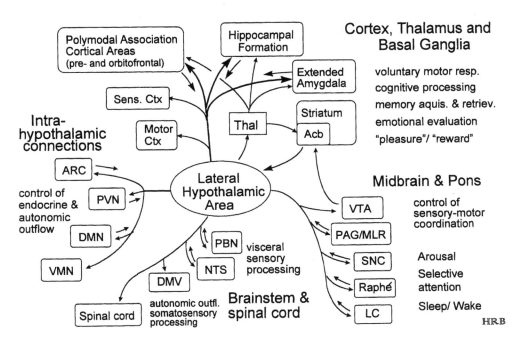

FIGURE 8   Projections and afferent inputs of LHA. Efferent projections are mainly based on retrograde tracing from respective targets as described by Simerly[30] and others. Afferents are based on anterograde and retrograde tracing, and are thus less reliable, because of the problem with axons of passage (medial forebrain bundle) in the lateral hypothalamus. Bifurcations do not necessarily imply axon collaterals. For abbreviations see legends to Figures 2, 6, and 10.

Together with the dorsomedial and mammillary nuclei, the lateral hypothalamic area provides the largest nonthalamic input to the cerebral cortex,[35] supporting the concept that some cortical areas should be included in the core processing circuit for food intake control (see Figure 4). Many of these projections arise from MCH-containing neurons.

Afferents to the lateral hypothalamic area are more difficult to assess, because locally injected retrograde tracers are easily taken up not only by axon terminals but also by fibers of passage running in the extensive medial forebrain bundle. By using anterograde tracing, afferents have been demonstrated to originate from parts of the amygdala, hippocampus, and the ventral striatum/nucleus accumbens area. Also, most of the connections to brain stem and midbrain areas are reciprocal. Within the hypothalamus the lateral zone has reciprocal connections to the arcuate (some containing orexin) and paraventricular nuclei, and efferent projections to the dorsomedial, ventromedial, and anterior hypothalamic nuclei but avoids most of the periventricular nuclei except the paraventricular nucleus. The perifornical area of the LHA receives substantial NPYergic input from the arcuate nucleus (see Figure 7). The strongest feeding response to NPY can be elicited by local injection into the perifornical area.[36]

4.3.2.2   Paraventricular nucleus of the hypothalamus (PVN)

The major afferents to the PVN are provided by the ventrolateral medulla (A1), nucleus solitarius (A2), and locus coeruleus (A6) noradrenergic cell groups, brain stem adrenalin-containing cell groups (C1–C3), by the median preoptic nucleus, and subfornical organ, by other hypothalamic nuclei such as the anteroventral periventricular (atrial natriuretic peptide), arcuate (NPY), dosomedial (galanin), and lateral hypothalamus, by the bed nucleus of the stria terminalis and the central and medial amygdaloid nuclei.[37,38] Many of the NPY-containing projections from the arcuate nucleus also contain agouti-related protein (AGRP) which acts as a competitive antagonist at the MC-3 and MC-4 melanocortin receptor subtypes, shown to potently modulate food intake.[39]

The major nonendocrine efferent projections of the parvocellular paraventricular nucleus are directed to autonomic preganglionic and related nuclei in the brain stem and spinal cord, including the dorsal vagal motor nucleus, ventrolateral medulla (A1), solitary nucleus, area postrema, periaqueductal gray, dorsal raphé, locus coeruleus, parabrachial and pedunculopontine nuclei.[40-42] In addition, the paraventricular nucleus of the hypothalamus also projects to di- and telencephalic regions such as the thalamic paraventricular nucleus and to the medial and central amygdaloid nuclei,[43] allowing it to modulate sensitivity to visceral sensory inputs and their emotional responses. Finally, the paraventricular nucleus provides inputs to most other hypothalamic areas, including dorsomedial, ventromedial, and arcuate hypothalamic nuclei, the anterior hypothalamus, and the perifornical area in the lateral hypothalamus.[41,43,44]

4.3.2.3 Dorsomedial hypothalamic nucleus (DMN)

The DMN of the hypothalamus, like the lateral hypothalamic area and paraventricular nucleus, also receives afferent input from the brain stem, but considerably less and mainly from the periaqueductual gray, parabrachial nucleus, and the ventrolateral rather than the dorsal medulla.[45] The major inputs to this nucleus are provided by most other hypothalamic nuclei, including the ventromedial, dorsomedial, and paraventricular nuclei. There are few inputs from other parts of the forebrain, including the ventral subiculum, prefrontal cortex, and extended amygdala (BST).[45] The efferent projections of the dorsomedial hypothalamus are mostly intrahypothalamic, directed toward other members of the medial zone including the ventromedial nucleus and anterior hypothalamus, the periventricular zone with notably the paraventricular nucleus.[46,47]

In addition to neural inputs, many DMN neurons contain the long-form leptin receptor (ObRb) and some of them have been demonstrated to project to the PVN.[48]

4.3.2.4 Ventromedial hypothalamic nucleus (VMN)

The VMN has been further divided into four distinct subnuclei, with each differing considerably in its cytoarchitecture, neurochemistry, and connectivity.[30] In general, the VMN receives substantial inputs from various parts of the amygdala, the lateral hypothalamus, and from most nuclei of the medial hypothalamus. There is also input from the parabrachial and solitary nuclei in the brain stem.[49] Many neurons, particularly in the dorsal portion of the nucleus, contain leptin receptors and some of them project to an area just ventral to the PVN.[48]

The pattern of efferent projections of the VMN is different from the other hypothalamic areas discussed above, in that it has few direct descending projections to autonomic preganglionic nuclei. Instead, it projects heavily to the periaqueductal gray and other brain stem reticular nuclei. It has massive projections to other hypothalamic medial zone nuclei, avoiding periventricular and lateral zones.[40,50] In the forebrain, the zona incerta, the midline thalamic nuclei, the extended amygdala, the nucleus accumbens, and various areas of the prefrontal and polymodal association cortex receive some input from the ventromedial nucleus.[50]

4.3.2.5 Arcuate nucleus (ARC)

The connections of the ARC are mostly intrahypothalamic. The strongest inputs are from other periventricular areas including the paraventricular nucleus, the medial zone nuclei such as the median preoptic nucleus,[51,52] and the lateral zone, in particular an orexin-containing projection from the LHA.[53,54] Among the sparse extrahypothalamic inputs are those from the extended amygdala and several brain stem sites.[55] However, the most important inputs to this nucleus are hormonal, with both leptin and insulin acting via specific receptors found on many neurons.[16,56] One such population of leptin receptor-bearing neurons contains mRNA for NPY, often coexpresses AGRP, and projects heavily to most periventricular zone nuclei, including the PVN, as well as the perifornical area of the lateral zone. Another distinct population of leptin receptor-bearing ARC neurons coexpresses proopiomelanocortin (POMC) and cocaine-amphetamine-related transcript (CART), and projects to MCH, orexin, and other LHA neurons.[57] There is also a small but distinct population of leptin receptor-bearing POMC/CART neurons located at the base of the medial hypothalamus, the retrochiasmatic area, some of which project very prominently to autonomic and motor areas in the spinal cord and brain stem[57] (see Figure 7).

4.4   CORTEX

The largest differences in both architecture and nomenclature among species exist for the cortical areas, and it is not possible here to provide detailed information. The cortical areas which appear to be of most interest to the control of food intake and selection are those that contain highly processed polymodal or supramodal sensory information, and present an "integrated view" of the outside and inside world of the organism. Obviously, for such information to get to these "higher" cortical areas, the individual sensory (visual, gustatory, and tactile) signals have to be channeled through the respective relay stations in the thalamus and sensory cortex, and the olfactory signals through the olfactory bulb and piriform cortex. Also, the primary motor cortex is obviously involved in any voluntary motor act necessary for procurement and ingestion of food (see below). From the discussion above, it is also clear that the greater lateral hypothalamic zone provides the largest nonthalamic input to the entire cortical mantle, suggesting that organizing any motivated behavior involves hypothalamic modulation at any step of cortical processing. Nevertheless, the key areas seem to be the polymodal association areas of the prefrontal and limbic cortex. In the rat the prefrontal cortex is the rostral part of the transition zone between iso- and allocortex, including the anterior cingulate, agranular insular, and orbitofrontal areas. The infralimic cortex is sometimes also included. The most direct evidence for a critical involvement of this area in food intake and selection comes from studies in macaque monkeys (see Chapter 17 by Rolls, this volume), and from PET studies in humans, showing that largest areas of neural activation differentially affected by hunger, satiation, and taste stimuli occurred in the cortex, including the prefrontal and orbito-frontal areas.[58,59] In addition, studies using amino acid–imbalanced diets in rats have shown that the anterior piriform cortex located in immediate proximity to the agranular insular and orbitofrontal cortex contains sensors for essential amino acids.

4.5   AMYGDALA

The amygdalar complex consists of over 20 distinct nuclear groups, and the interested reader should consult the more extensive descriptions of its anatomy and connections.[60] The concept of the extended amygdala considers the central and medial parts of the amygdala proper, together with the bed nucleus of the stria terminalis and two bands of interconnecting cell groups, as well as the caudalmost portions of the nucleus accumbens shell as a functional unit, and distinguishes it from the more cortical-like lateral and basal amygdaloid nuclei.[60] Clearly, the extended amygdala including the central and medial nuclei has received the most attention in the field of food intake control, both because of direct effects on ingestive behavior following stimulation or ablation of this system[61] and because of its connectivity with other relevant brain structures. There are four major categories of afferents to the extended amygdala. First, it is a major termination site for descending projections from the agranular insular cortex and from other cortical areas via relays in the lateral and basal amygdaloid nuclei, and receives thus some of the same highly processed sensory information as do polymodal cortical association areas such as the orbitofrontal cortex (see Chapter 17 by Rolls, this volume). Second, it receives substantial visceral sensory information from the solitary and parabrachial nuclei in the brain stem. Third, it receives input from the paraventricular nucleus and lateral hypothalamus. Finally, there is input from the shell of the nucleus accumbens, which possibly links the hedonic or rewarding to the emotional aspects of a particular food stimulus.

Projections of the central and medial amygdaloid nuclei and their related extended amygdala limbs include the paraventricular, dorsomedial, ventromedial, arcuate, preoptic, and premammillary nuclei, and the lateral hypothalamus; several parts of the olfactory sensory pathways, the hippoc-ampus, nucleus accumbens, and medial thalamus, the ventral tegmental area and substantia nigra pars compacta, the periaqueductal gray and dorsal raphé nucleus in the midbrain; as well as the parabrachial and solitary nuclei, the vagal motor nucleus, the ventrolateral medulla; and spinal cord.[62] Although most of these latter, descending projections are direct, some use the lateral, dorsomedial, and ventromedial hypothalamus as a relay. The basomedial amygdaloid nuclei, in addition, project heavily to cortical areas including the insular, piriform, and infralimbic prefrontal cortex.[63] Because of these extensive connections, the extended amygdala is considered here as another integral part of the central processing unit (see Figure 4).

4.6   "MEMORY LOOP"

The human brain is capable of storing detailed representations of the entire food space, just as it stores representations of other external and internal daily activities and events. Although the rat neocortex is considerably smaller, basic cortical and subcortical anatomy and connections relevant to memory functions are very similar to those in humans, and behavior is guided by the same behavioral principles other than explicit verbal cognitive functions.

The organism presumably uses all of its senses to generate a large number of associations. Given the many sensory input channels described above, with their qualitative, quantitative, and temporal differences, even food items that are very similar produce a specific signature of sensory consequences in the broadest sense. Taste is just one, and may not be the most important, of the distinguishing features between two foods (see discussion in Chapter 6 by Booth and Thibault). Theoretically, any two or more salient sensory features of a particular food item (or pure macronutrient) for which the organism has sensors can be acquired in an associative memorial representation. Such associative learning enables the organism to use past experience for the management of daily tasks such as feeding. By linking internal signals of metabolic repletion or depletion with external cues, it provides a mechanism to recognize the nutritive value of a food item before it is ingested.

The contribution of associative learning and classical conditioning in the control of food selection is beginning to be more widely appreciated, although it has been propagated by many investigators earlier in this century. However, behavioral analyses (see Chapter 6 by Booth and Thibault; Chapter 14 by Davidson et al.; Chapter 7 by Sclafani) are clearly ahead of anatomical and neurochemical investigations. Except for a few very recent reports,[64,65] we simply do not know in any specific fashion how nutritional-ingestive associations are formed or acquired, where memorial representations are stored, and how they are retrieved. We can only resort to the field of cognitive neurosciences to obtain a very general idea about possible pathways and structures involved.[66-69]

4.6.1   Declarative or Explicit Memory Involving
the Hippocampal Formation

According to Cohen and Eichenbaum,[70] the key features of declarative memory are that it is the outcome of processing operations, it is "promiscuously accessible to various processing systems," and "it can be expressed flexibly, and is capable of being used in even completely novel contexts." Furthermore, the medial temporal lobe of the primate diencephalon, including entorhinal, perirhinal, and parahippocampal cortices and the hippocampal formation, together with various cortical association areas, are considered the key structures involved in declarative memory. A schematic diagram of these areas together with the major connections to other areas relevant for food intake control are depicted in Figure 9.

4.6.2   Nondeclarative or Implicit Memory: Classical Conditioning,
Procedural Learning, and Nonassociative Learning

In distinction to declarative memory, (1) procedural memory involves the tuning and modifying of the particular processors engaged during training, (2) its representations are dedicated to the modified processors, unavailable to other processors, (3) its representations are fundamentally individual,

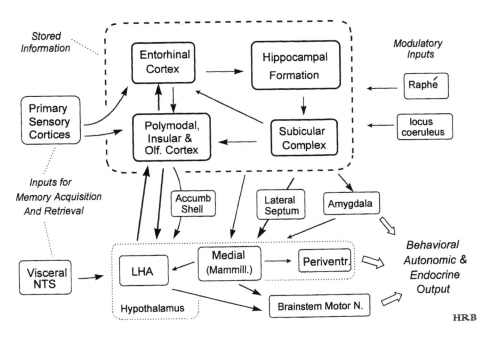

FIGURE 9 Simplified schematic diagram of potential involvement of hippocampus and associated structures in memory functions related to the control of food intake and selection. General inflow of sensory information is on the left, outflow through the septum, hypothalamus, and amygdala on the lower right. Arrows in and out of the stippled boxes represent information flow to/from more than one of the structures in the box; for example, projections from the hippocampus via the lateral septum to the hypothalamus are topographically organized, with specific areas in the hippocampal/subicular complex projecting via specific relays in the lateral septum to specific targets in the lateral, medial, and periventricular hypothalamic zones.

and (4) memory can be expressed only inflexibly, only in a repetition of the original processing situation.[70] This type of learning and memory may or may not involve the hippocampal formation. There is considerable experimental evidence to implicate the amygdala in classical fear conditioning and conditioned taste aversion, and the cerebellum in classical conditioning of specific motor responses such as in eye blink conditioning. Other processors that are likely involved in learning of motor skills and habits include the striatum, thalamus, and cerebellum. It is interesting in this context that recent studies monitoring changes in regional cerebral blood flow as a measurement of neural activity found significant changes in all of these structures when "hunger" was induced by 36 h of food deprivation, or when a mixed gustatory stimulus was presented in hungry subjects,[58,59] and that the long form of the leptin receptor (ObRb) is found not only in the hypothalamus, but also the thalamus and cerebellum.[16] These findings are consistent with the view that internal signals reflecting relative depletion/repletion may be used to retrieve or activate specific mechanisms in these sensory-motor processing and programming units (and/or retrieve specific memories from hippocampal circuits). Finally, nonassociative learning could take place at the basic reflex level in the medulla and spinal cord, where evidence for cellular adaptive responses such as long-term potentiation has recently been found.[71]

4.7 "PLEASURE LOOP"

Eating is typically linked with pleasure. In fact, emphasis on pleasure in eating, as manifest in the more and more sophisticated and ethnically diverse restaurants and the ever-present television food advertisements, make it suspect as one of the driving forces and causes of overeating and obesity in Western society. Behavioral studies have come to the conclusion that humans and animals develop behavioral strategies to maximize the amount of pleasure to be derived.[72,73] The question is what are the neural substrates of pleasure or food reward. Berridge[74] has argued that reward contains distinguishable psychological and functional components, with different underlying neural substrates. His term "liking" is the pleasure derived from palatable food, and is associated with opioid and GABA/benzodazepine systems in the brain stem gustatory relay nuclei and the ventral pallidum, while "wanting" represents the appetitive/incentive motivation component, associated with the mesotelencephalic dopamine system, including nucleus accumbens and prefrontal cortex. However, there seems to be complete disagreement on whether pleasure/liking and reward/wanting are conscious functions. While for Cabanac[73] pleasure or joy is a sign of a useful conscious event, Berridge thinks that both components can exist without subjective (conscious) awareness. Mogenson et al.[75] were the first to investigate the role of the ventral striatum/nucleus accumbens and its dopamine modulation in the control of ingestive behaviors. The anatomical[76,77] as well as neurochemical and functional behavior of this system in relation to ingestive behavior have since been well-examined aspects[78-80] (see also Chapter 26 by Glass et al., this volume). The most important afferent inputs and efferent projections of the nucleus accumbens are depicted in Figure 10.

5   OUTPUT SYSTEMS

The output system for food intake control and selection, in the narrower sense, only includes the skeletal motor system,[81] including final common pathways for locomotion and oropharyngeal motor control.[82] In a broader sense, however, the autonomic and enteric nervous systems and the pituitary-endocrine axis should also be considered, because they significantly modulate food handling as

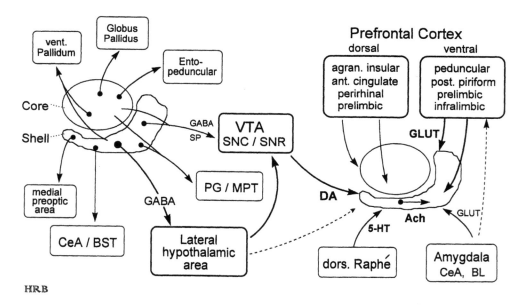

FIGURE 10   Major afferents (right) and efferents (left) of the nucleus accumbens, suspected to be involved in reward and pleasure functions. The shell area is different from the rest of the ventral striatum in that it has prominent projections to the lateral hypothalamus, extended amygdala (CeA and BST), and the medial preoptic area. The shell receives massive glutamatergic input from mainly ventral prefrontal association cortex areas, and dopaminergic input from the ventral tegmental/substantia nigra (pars compacta) area. Output related to food intake and selection is thought to be organized via GABA-ergic neurons projecting to the lateral hypothalamic area, and via the central nucleus of the amygdala. Abbreviations: Ach, cholinergic interneurons; BL, basolateral amygdala; BST, bed nucleus of the stria terminalis; CeA, central nucleus of amygdala; DA, dopaminergic neurons; Glut, glutamatergic neurons; MPT, mesopontine tegmental area; PG, periaqueductal gray; SNC, substantia nigra pars compacta; SNR, substantia nigra pars reticularis; SP, substance P; VTA, ventral tegmental area (A10); 5-HT, serotonin-containing neurons.

well as metabolic processing, oxidation, and storage, and thus codetermine the level of potential signals on the sensory side of the regulatory loop (see Figure 1).

As for the sensory systems, we know the most distal elements of these output or motor systems quite well, but have a limited view of the neural organization that enables access to them or, in other words, translates cognitive, emotional, and motivational processes into behavioral action. Increased use of transsynaptic viral neuronal tracing, which delineates not only motor neuron pools innervating particular muscles and glands, but also premotor and higher-order motor neurons, will undoubtedly help further define various output systems.

5.1   LOCOMOTOR AND OROMOTOR CONTROL: THE TRANSLATION OF MOTIVATION INTO ACTION

The basic organization of the motor system can be best appreciated if it is divided into four components, the final motoneuron pool, the premotor interneurons forming the basic motor system, the somatic motor system, and the emotional motor system.[81] Although both the somatic and emotional motor systems can directly access motoneurons, many of their effects are of modulatory character via the basic premotor interneurons.

Analysis of the behavioral motor output system is complicated by the fact that integrative and motor steps cannot be clearly separated. This is because behavior is typically displayed as a temporal sequence of a number of steps or elements, and each step has to be cleared independently through the parallel processing loops of memory, pleasure, and emotion, as well as neuroendocine and autonomic adjustments. To achieve this, motor programming is intimately linked to and originates within these integrative processing areas such as the prefrontal association cortex, lateral hypothalamus, hippocampus, nucleus accumbens, septum, and central amygdala (Figure 11). More-detailed

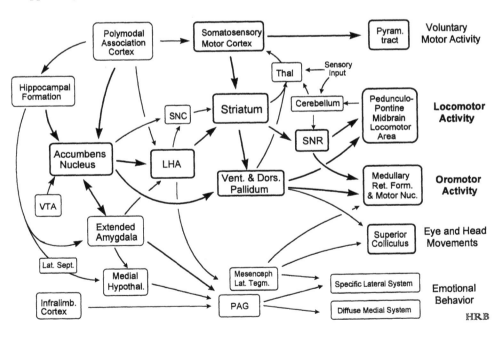

FIGURE 11   Schematic diagram showing major pathways involved in the organization of behavioral output during the various phases and modes of feeding behavior. General flow of information is from left to right. Although the emotional motor system (bottom) is not involved in normal feeding behavior, it is also shown because of its central origin in the hypothalamus and limbic system, related to ingestive behavior. Abbreviations: Medullary RF, medullary reticular formation; PAG, peiaqueductal gray; Lat Sept., lateral septum; Lat. Tegm., lateral midbrain tegmentum; for other abbreviations see legend to Figure 9.

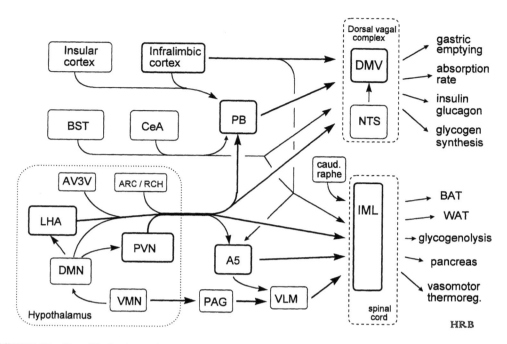

FIGURE 12 Simplified schematic diagram of major structures and pathways involved in the organization of autonomic outflow related to feeding behavior and nutritional metabolic regulation. Connections to preganglionic neurons of lumbo-sacral outflow is not indicated. Bifurcations do not necessarily imply collateral pathways. Abbreviations: A5, noradrenergic cell group in ventrolateral pons; DMV, dorsal motor nucleus of vagus; IML, intermediolateral column of spinal cord; RCH, retrochiasmatic area of ventral hypothalamus; for other abbreviations see legend to Figure 7.

"software programs" of motor patterns are then accessed in the striatum. Finally, specific pools of premotor and motorneurons are then activated, under constant sensory-motor feedback control at appropriate levels.

5.2   CONTROL OF AUTONOMIC AND ENDOCRINE OUTFLOW

Behavior is not the only option to restore homeostasis. Temporary deficits in immediately available fuel, for example, can be rapidly restored by increased absorption from the gastrointestinal tract, or mobilization from stores, or decreased utilization. These functions can be controlled by autonomic and endocrine outflow. The neuroendocrine system has been briefly introduced in the sections above and in-depth reviews can be found elsewhere.[83]

The central neural control of autonomic outflow has been discussed in numerous excellent reviews[84] and a highly schematic diagram is shown in Figure 12. A restricted number of brain areas have direct, monosynaptic connections to autonomic preganglionic neurons in the medulla and spinal cord. A more inclusive group of cortical and diencephalic/hypothalamic structures can also talk to these preganglionic motor neurons via one or more relay stations such as the parabrachial nuclear complex, the A5 region in the pons, and the ventrolateral medullary (C1) area and caudal raphe nuclei in the caudal brain stem.

6   CONCLUSIONS

1. The neural network responsible for the control of energy intake and body weight is far from being comprehensively identified. Progress has been recently made on a few key

areas such as the hypothalamus and brain stem by identifying critical pathways and their neurochemical codes. However, considering the multiple operational tasks and neurological processing steps necessary for the complete expression of ingestive behavior, the neural network is likely to be a distributed system, involving most, if not all, areas of the nervous system.

2. If the neural system controlling energy intake in general is ill-defined, we know even less about possible separate systems controlling ingestion of protein, carbohydrate, and fat. Clearly, the visual, olfactory, gustatory, and visceral sensory systems together with blood-borne internal signals are in a position to detect the sensory characteristics as well as the energetic and metabolic consequences of different foods, including their different macro- and micronutrient composition. The question is whether and how the brain uses this sensory information to control or regulate macronutrient selection.

3. Evidence for separate neural control systems is almost exclusively based on findings of altered macronutrient selection following local brain injection of peptides or neurotransmitter analogues. This simple and pragmatic approach may well lead to the development of pharmacologic therapeutic strategies to change macronutrient selection in humans. However, it is not suited to demonstrate the existence and functional characteristics of real physiological control systems. For example, injection of a given peptide into a rat's brain may reduce ingestion of a dry, powdery casein diet because it interferes with saliva secretion. Although this effect may be therapeutically useful, no one would conclude that the peptide is part of a protein regulatory system. Therefore, the demonstration that anatomical, pharmacological, and genetic manipulations can selectively alter intake of specific macronutrients in itself is not necessarily evidence for, and does not delineate, a control system.

4. In order to identify the anatomical pathways and neurochemical characteristics of systems specifically involved in the control of particular macro- or micronutrients, behavioral analysis has to evolve beyond simply measuring the amount of a particular food ingested in a given time. The ingestive behavior sequence including its preparatory and postingestive phases, as well as its learning aspects, has to be broken down into a series of operational steps, tasks, or units that can be brought under experimental control, and that allow neuroanatomical and neurochemical modeling as well as measurement and/or intervention.

5. If separate systems exist for each macronutrient, they are likely to differ in relatively few areas with larger areas of overlap. It is also possible that the same basic neural substrate is used in a time-sharing or multiplexed fashion, or with colocalized transmitters/modulators encoding for the different macronutrients.

ACKNOWLEDGMENTS

I would like to thank Drs. Steve Woods, Larry Swanson, David York, and Joel Elmquist for critically reading and providing constructive suggestions on earlier versions of this manuscript, and Patricia James for secretarial help. Research by the author is supported by the National Institute of Diabetes and Digestive and Kidney Diseases, Grant No. 47348; and the Pennington Biomedical Research Foundation.

REFERENCES

1. Swanson, L. W. and Mogenson, G. J., Neural mechanisms for the functional coupling of autonomic, endocrine and somatomotor responses in adaptive behavior, *Brain Res.,* 228, 1, 1981.
2. Risold, P. Y., Thompson, R. H., and Swanson, L. W., The structural organization of connections between hypothalamus and cerebral cortex, *Brain Res. Brain Res. Rev.,* 24, 197, 1997.

3. Tohyama, M. and Takatsuji, K., *Atlas of Neuroactive Substances and Their Receptors in the Rat*, Oxford University Press, Oxford, U.K., 1998.

4. Somerville, E. M. and Clifton, P. G., Neurochemical interactions in the control of ingestive behavior, in *Drug Receptor Subtypes and Ingestive Behavior*, Cooper, S. J. and Clifton, P. G., Eds., Academic Press, London, 1995, chap. 17, 369.

5. Card, J. P., Rinaman, L., Schwaber, J. S., Miselis, R. R., Whealy, M. E., Robbins, A. K., and Enquist, L. W., Neurotropic properties of pseudorabies virus: uptake and transneuronal passage in the rat central nervous system, *J. Neurosci.*, 10, 1974, 1990.

6. Sams, J. M., Jansen, A. S., Mettenleiter, T. C., and Loewy, A. D., Pseudorabies virus mutants as transneuronal markers, *Brain Res.*, 687, 182, 1995.

7. Morgan, J. I. and Curran, T., Stimulus-transcription coupling in the nervous system: involvement of the inducible proto-oncogenes fos and jun, *Annu. Rev. Neurosci.*, 14, 421, 1991.

8. Chaudhuri, A., Neural activity mapping with inducible transcription factors, *Neuroreport*, 8, 5, 1997.

9. Herrera, D. G. and Robertson, H. A., Activation of c-*fos* in the brain, *Prog. Neurobiol.*, 50, 83, 1996.

10. Dragunow, M., A role for immediate-early transcription factors in learning and memory, *Behav. Genet.*, 26, 293, 1996.

11. Levin, J. M., Ross, M. H., and Renshaw, P. F., Clinical applications of functional MRI in neuropsychiatry, *J. Neuropsych. Clin. Neurosci.*, 7, 511, 1995.

12. Aine, C. J., A conceptual overview and critique of functional neuroimaging techniques in humans: I. MRI/FMRI and PET, *Crit. Rev. Neurobiol.*, 9, 229, 1995.

13. Friedrich, R. W. and Korsching, S. I., Chemotopic, combinatorial, and noncombinatorial odorant representations in the olfactory bulb revealed using a voltage-sensitive axon tracer, *J. Neurosci.*, 18, 9977, 1998.

14. Norgren, R., Gustatory system, in *The Rat Nervous System*, Paxinos, G., Ed., Academic Press, San Diego, 1995, chap. 29, 751.

15. Waite, M. E. and Tracey, D.J., Trigeminal sensory system, in *The Rat Nervous System*, Paxinos, G., Ed., Academic Press, San Diego, 1995, chap. 27, 705.

16. Elmquist, J. K., Bjorback, C., Ahima, R. S., Flier, J. S., and Saper, C. B., Distributions of leptin receptor mRNA isoforms in the rat brain, *J. Comp. Neurol.*, 395, 535, 1998.

17. Wang, Y. H., Tache, Y., Scheibel, A. B., Go, V. L. W., and Wei, J. Y., Two types of leptin-responsive gastric vagal afferent terminals: an *in vitro* single-unit study in rats, *Am. J. Physiol.*, 273, R833, 1997.

18. Weigle, D. S., Hutson, A. M., Kramer, J. M., Fallon, M. G., Lehner, J. M., Lok, S., and Kuijper, J. L., Leptin does not fully account for the satiety activity of adipose tissue-conditioned medium, *Am. J. Physiol.*, 275, R976, 1998.

19. Bartness, T. J. and Bamshad, M., innervation of mammalian white adipose tissue: implications for the regulation of total body fat, *Am J. Physiol.*, 275, R1399, 1998.

20. Russek, M., Current status of the hepatostatic theory of food intake control, *Appetite*, 2, 137, 1981.

21. Schwartz, G. J. and Moran, T. H., Sub-diaphragmatic vagal afferent integration of meal-related gastrointestinal signals, *Neurosci., Biobehav. Rev.*, 20, 47, 1996.

22. Sabbatini, R. M., Using neural networks for professing biologic signals, *MD Comput.*, 13, 165, 1996.

23. Juergens, E. and Eckhorn, R., Parallel processing by a homogeneous group of coupled model neurons can enhance, reduce and generate signal correlations, *Biol. Cybern.*, 76, 217, 1997.

24. Kauer, J. S., Contributions of topography and parallel processing to odor coding in the vertebrate olfactory pathway, *Trends Neurosci.*, 14, 79, 1991.

25. Andrews, P. L. R. and Lawes I. N. C., A protective role for vagal afferents: an hypothesis, in *Neuroanatomy and Physiology of Abdominal Vagal Afferents*, Ritter, S., Ritter, R. C., and Barnes, C. D., Eds., CRC Press, Boca Raton, FL, 1992, chap. 12, 279.

26. Grill, H. J. and Kaplan, J. M., Caudal brain stem participates in the distributed neural control of feeding, in *Handbook of Behavioral Neurobiology*, Vol. 10, *Neurobiology of Food and Fluid Intake*, Stricker, E. M., Ed., Plenum Press, New York, 1990, chap. 6, 125.

27. Smith, G. P., The direct and indirect controls of meal size, Neurosci. Biobehav. Rev., 20, 41, 1996.

28. Kotz, C. M., Briggs, J. E., Grace, M. K., Levine, A. S., and Billington, C. J., Diverence of the feeding and thermogenic pathways influenced by NPY in the hypothalamic PVN of the rat, *Am. J. Physiol.*, 275, 271, 1998.

29. Grill, H. J., Ginsberg, A., Seeley, R. J., and Kaplan, J., Brainstem application of melanocortin receptor ligands produces long-lasting effects on feeding and body weight, *J. Neurosci.*, 18, 10128, 1998.

30. Simerly, R. B., Anatomical substrates of hypothalamic integration, in *The Rat Nervous System*, Paxinos, G., Ed., Academic Press, San Diego, 1995, chap. 17, 353.

31. Swanson, L. W., The neural basis of motivated behavior, *Acta Morphol. Neerl. Scand.*, 26, 165, 1988.

32. Skofitsch, G., Jacobowitz, D. M., and Zamir, N., Immunohistochemical localization of a melanin concentrating hormone-like peptide in the rat brain, *Brain Res. Bull.*, 15, 635, 1985.

33. Tritos, N. A., Vicent, D., Gillette, J., Ludwig, D. S., Flier, J. S., and Maratos-Flier, E., Functional interactions between melanin-concentrating hormone, neuropeptide Y, and anorectic neuropeptides in the rat hypothalamus, *Diabetes*, 47, 1687, 1998.

34. Sakurai, T., Amemiya, A., and Yanagisawa, M., Orexins and orexin receptors; a family of hypothalamic neuropeptides and G protein-coupled receptors that regulate feeding behavior, *Cell*, 20, 573, 1998.

35. Saper, C. B., Organization of cerebral cortical afferent systems in the rat. II. Hypothalamocortical projections, *J. Comp. Neurol.*, 237, 21, 1985.

36. Stanley, B. G., Magdalin, W., Seirafi, A, Nguyen, M. M., and Leibowitz, S. F., The perifornical area: the major focus of (a) patchily distributed hypothalamic neuropeptide y-sensitive feeding system(s), *Brain Res.*, 604, 304, 1993.

37. Sawchenko, P. E. and Swanson, L. W., The organization of noradrenergic pathways from the brain stem to the paraventricular and supraoptic nuclei in the rat, *Brain Res. Rev.*, 4, 275, 1982.

38. Sawchenko, P. E. and Swanson, L. W., The organization of forebrain afferents to the paraventricular and supraoptic nuclei of the rat, *J. Comp. Neurol.*, 218, 121, 1983.

39. Fan, W., Boston, B. A., Kesterton, R. A., Hruby, V. J., and Cone, R. A., Role of melanocortinergic neurons in feeding and the agouti obesity syndrome, *Nature*, 385, 165, 1997.

40. Saper, C. B., Loewy, A. D., Swanson, L. W., and Cowan, C. M., Direct hypothalamo-autonomic connections, *Brain Res.*, 117, 305, 1976.

41. Luiten, P. G., ter Horst, G. J., and Steffens, A. B., The hypothalamus, intrinsic connections and outflow pathways to the endocrine system in relation to the control of feeding and metabolism, *Prog. Neurobiol.*, 28, 1, 1987.

42. Hosoya, Y., Siguiura, Y., Okado, N., Loewy, A. D., and Kohno, K., Descending input from the hypothalamic paraventricular nucleus to sympathetic preganglionic neurons in the rat, *Exp. Brain Res.* 85, 10, 1991.

43. Larsen, P. J., Moller, M., and Mikkelsen, J. D., Efferent projections from the periventricular and medial parvicellular subnuclei of the hypothalamic paraventricular nucleus to circumventricular organs of the rat: a Phaseolus vulgaris-leucoagglutinin (PHA-L) tracing study, *J. Comp. Neurol.*, 306, 462, 1991.

44. ter Horst, G. J. and Luiten, P. G., Phaseolus vulgaris leuco-agglutinin tracing of intrahypothalamic connections of the lateral, ventromedial, dorsomedial and paraventricular hypothalamic nuclei in the rat, *Brain Res. Bull.*, 18, 191, 1987.

45. Thompson, R. H. and Swanson, L. W., Organization of inputs to the dorsomedial nucleus of the hypothalamus: a reexamination with Fluorogold and PHAL in the rat, *Brain Res. Brain Res. Rev.*, 27, 89, 1998.

46. ter Horst, G. J. and Luiten, P.G., The projections of the dorsomedial hypothalamic nucleus in the rat, *Brain Res. Bull.*, 16, 231, 1986.

47. Thompson, R. H., Canteras, N. S., and Swanson, L. W., Organization of projections from the dorsomedial nucleus of the hypothalamus: a PHA-L study in the rat, *J. Comp. Neurol.*, 376, 143, 1996.

48. Elmquist, J. K., Ahima, R. S., Elias, C. F., Flier, J. S., and Saper, C. B., Leptin activates distinct projections from the dorsomedial and ventromedial hypothalamic nuclei, *Proc. Natl. Acad. Sci. U.S.A.*, 95, 741, 1998.

49. Fulwiler, C. E. and Saper, C. B., Cholecystokinin-immunoreactive innervation of the ventromedial hypothalamus in the rat: possible substrate for autonomic regulation of feeding, *Neurosci. Lett.*, 53, 289, 1985.

50. Canteras, N. S., Simerly, R. B., and Swanson, L. W., Organization of projections from the ventromedial nucleus of the hypothalamus: a *Phaseolus vulgaris*-leucoagglutinin study in the rat, *J. Comp. Neurol.*, 348, 41, 1994.

51. Sawchenko, P. E., Swanson, L.W., Steinbusch, H. W., and Verhofstad, A. A., The distribution and cells of origin of serotonergic inputs to the paraventricular and supraaoptic nuclei of the rat, *Brain Res.*, 277, 355, 1983.

52. Sawchenko, P. E. and Swanson, L. W., The organization of forebrain afferents to the paraventricular and supraoptic nuclei of the rat, *J. Comp. Neurol.*, 218, 121, 1983.

53. Horvath, T. L., Diano, S., and van den Pol, A. N., Hypocretin (Orexin)-containig neurons make synaptic contact with arcuate nucleus NPY- and POMC-producing cells that express leptin receptors in rodent and primate; novel hypothalamic circuit involved in energy homeostasis, *Soc. Neurosci. Abstr.*, 24, 12, 1998.

54. Peyron, C., Tighe, D. K., van den Pol, A. N., de Lecea, L., Heller, H. C., Sutcliffe, J. G., and Kidluff, T. S., Neurons containing hypocretin (orexin) project to multiple neuronal systems, *J. Neurosci.*, 18, 9996, 1998.

55. Ricardo, J. A. and Koh, E. T., Anatomical evidence of direct projections from the nucleus of the solitary tract to the hypothalamus, amygdala, and other forebrain structures in the rat, *Brain Res.*, 153, 1, 1978.

56. Håkansson, M.-L., Brown, H., Ghilardi, N., Skoda, R. C., and Meister, B., Leptin receptor immunoreactivity in chemically defined target neurons of the hypothalamus, *J. Neurosci.*, 18, 559, 1998.

57. Elias, C. F., Saper, C. B., Maratos-Flier, E., Tritos, N. A., Lee, C., Kelly, J., Tatro, J. B., Hoffman, G. E., Ollmann, M. M., Barsh, G. S., Sakurai, T., Yanagishawa, M., and Elmquist, J. K., Chemically defined projections linking the mediobasal hypothalamus and the lateral hypothalamic area, *J. Comp. Neurol.*, 402, 442, 1988.

58. Tataranni, P. A., Chen, K., Uecker, A., and Ravussin, E., Identification of regions of the human brain involved in hunger and satiation using positron emission tomography, *Int. J. Obesity*, 22, S34, 1998.

59. Gauthier, J.-F., Tataranni, P. A., and Reiman, E. M., Neuroanatomical correlates of taste in fasted humans using positron emission tomography, *Int. J. Obesity*, 24, S34, 1998.

60. Alheid, G. F., de Olmos, J. S., and Beltramino, C. A., Amygdala and extended amygdala, in *The Rat Nervous System*, Paxinos, G., Ed., Academic Press, San Diego, 1995, chap. 22, 495.

61. King, B. M., Kass, J. M., Cadieux, N. L., Sam, H., Neville, K. L., and Arceneaux, E. R., Hyperphagia and obesity in female rats with temporal lobe lesions, *Physiol. Behav.*, 54, 759, 1993.

62. Canteras, N. S., Simerly, R. B., and Swanson, L.W., Organization of projection from the medial nucleus of the amygdala: a PHAL study in the rat, *J. Comp. Neurol.*, 360, 213, 1995.

63. Petrovich, G. D., Risold, P. Y., and Swanson, L. W., Organization of projections from the basomedial nucleus of the amygdala: a PHAL study in the rat, *J. Comp. Neurol.*, 374, 387, 1996.

64. Berman, D. E., Hasvi, S., Rosenblum, K., Seger, R., and Dudai, Y., Specific and differential activation of mitogen-activated protein kinase cascades by unfamiliar taste in the insular cortex of the behaving rat, *J. Neurosci.*, 18, 10037, 1998.

65. Alvarez, P. and Eichenbaum, H., Representations of odors and their reward values in orbitofrontal cortex of the rat change over time and are independent of the hippocampus, *Soc. Neurosci. Abstr.*, 24, 1424, 1998.

66. Schmajuk, N. A., *Animal Learning and Cognition (Problems in the Behavioral Sciences*, Vol. 16), Cambridge University Press, Cambridge, 1997, chap. 12, 241.

67. Schimamura, A. P., Forms of memory: issues and directions, in *Brain Organization and Memory, Cells Systems and Circuits*, McGaugh, J. L., Weinberger, M. L., and Lynch, G., Eds., Oxford University Press, Oxford, 1990, chap. 8, 159.

68. Kohonen, T., Notes on neural computing and associative memory, in *Brain Organization and Memory, Cells Systems and Circuits*, McGaugh, J. L., Weinberger, M. L., and Lynch, G., Eds., Oxford University Press, Oxford, 1990, chap. 16, 323.

69. Freeman, W. J. and Skarda, C. A., Representations: who needs them? in *Brain Organization and Memory, Cells Systems and Circuits*, McGaugh, J. L., Weinberger, M. L., and Lynch, G., Eds., Oxford University Press, Oxford, 1990, chap. 19, 375.

70. Cohen, N. J. and Eichenbaum H., *Memory, Amnesia, and the Hippocampal System*, MIT Press, Cambridge, MA, 1994, chap. 3, 55.

71. Alford, S., Zompa, I., and Dubuc, R., Long-term potentiation of glutamatergic pathways in the lamprey brain stem, *J. Neurosci.*, 15, 7528, 1995.

72. Balasko, M. and Cabanac, M., Motivational conflict among water need, palatability, and cold discomfort in rats, *Physiol. Behav.*, 65, 35, 1998.

73. Cabanac, M., On the origin of consciousness, a postulate and its corollary, *Neurosci. Biobehav. Rev.*, 20, 33, 1996.

74. Berridge, K. C., Food reward: brain substrates of wanting and liking, *Neurosci. Biobehav. Rev.*, 20, 1, 1996.

75. Mogenson, G. J., Jones, D. L., and Yim, C. Y., From motivation to action: functional interface between the limbic system and the motor system, *Prog. Neurobiol.*, 14, 69, 1980.

76. Groenewegen, H. J., Berendse, H. W., Meredith, G. E., Haber, S. N., Voorn, P., Wolters, J. G., and Lohman, A. H., Functional anatomy of the ventral, limbic system-innervated striatum, in *The Mesolimbic Dopamine System: From Motivation to Action*, Wilner, P. and Scheel-Krueger, J., Eds., John Wiley & Sons, Chichester, 1991, chap. 2, 19.

77. Scheel-Krueger, J. and Wilner, P., The mesolimbic system: principles of operation, in *The Mesolimbic Dopamine System: From Motivation to Action*, Wilner, P. and Scheel-Krueger, J., Eds., John Wiley & Sons, Chichester, 1991, chap. 22, 559.

78. Hoebel, B. G., Rada, P., Mark, G. P., and Hernandez, L., The power of integrative peptides to reinforce behavior by releasing dopamine, *Ann. N.Y. Acad. Sci.*, 739, 36, 1994.

79. Maldonado-Irizarry, C. S., Swanson, C. J., and Kelley, A. E., Glutamate receptors in the nucleus accumbens shell control feeding behavior via the lateral hypothalamus, *J. Neurosci.*, 15, 6779, 1995.

80. Stratford, T. R. and Kelley, A. E., GABA in the nucleus accumbens shell participates in the central regulation of feeding behavior, *J. Neurosci.*, 17, 4434, 1997.

81. Holstege, G., The basic, somatic, and emotional components of the motor system in mammals, in *The Rat Nervous System*, Paxinos, G., Ed., Academic Press, San Diego, 1995, chap. 8, 137.

82. Travers, J. B., Oromotor nuclei, in *The Rat Nervous System*, Paxinos, G., Ed., Academic Press, San Diego, 1995, chap. 130, 239.

83. Armstrong, W. E., Hypothalamic supraoptic and paaventricular nuclei, in *The Rat Nervous System*, Paxinos, G., Ed., Academic Press, San Diego, 1995, chap. 18, 377.

84. Saper, C. B., Central autonomic system, in *The Rat Nervous System*, Paxinos, G., Ed., Academic Press, San Diego, 1995, chap. 7, 107.

25 Macronutrients and Brain Peptides: What They Do and How They Respond

Sarah F. Leibowitz

CONTENTS

ABSTRACT

This chapter reviews evidence suggesting that different neurochemicals in the brain are involved in controlling the ingestion and metabolism of specific macronutrients. With a focus on the two hypothalamic peptide systems, neuropeptide Y (NPY) and galanin (GAL), evidence is described which demonstrates how these peptides have different behavioral and physiological effects. They are also differentially responsive to feedback signals from circulating steroids, nutrient metabolism, and diet consumption. They can be further distinguished by their relation to natural biological rhythms and developmental patterns. The neuroanatomical substrates involved in these actions of NPY and GAL are also distinct. The neurocircuit mediating the NPY actions originates in the arcuate nucleus (ARC) and terminates in the medial portion of the paraventricular nucleus (mPVN). The GAL-containing neurons, in contrast, are concentrated in the antero-parvocellular PVN (aPVN), in addition to the medial preoptic area, which contribute to local GAL innervation as well as projections to the median eminence (ME). With recent evidence indicating that consumption of a high-carbohydrate diet stimulates NPY gene expression in the ARC, it is proposed that the NPY ARC-mPVN system is more closely related to patterns of carbohydrate ingestion and carbohydrate utilization, channeling nutrients toward the synthesis of fat. It is activated in close association with the adrenal steroid, corticosterone. The GAL aPVN-ME system, in contrast, is more closely associated with patterns of fat consumption and signals related to fat oxidation. The expression and production of this peptide is stimulated by the consumption of a high-fat diet. Moreover, the GAL system is most active at times, during the middle of the feeding cycle, the proestrous period or at puberty in females, when fat consumption and body fat, along with the gonadal steroids, are naturally rising. It is through these mechanisms of carbohydrate and fat balance that these two peptides are believed to contribute, differentially, to the overall process of body weight regulation that involves both the synthesis and deposition of body fat.

KEY WORDS: *neuropeptide Y, galanin, hypothalamus, feeding behavior, carbohydrate, fat, body weight*

1   INTRODUCTION

Over the past decade, it has become evident that there exist mechanisms that govern the partitioning of specific macronutrients in the body. These mechanisms work through metabolic responses, which channel food into either protein, lipid, or carbohydrate stores. They may also involve the behavioral process of food ingestion, affecting the selection of or preference for specific macronutrients. The brain very likely plays an important role in these functions, balancing intake and metabolism of the macronutrients, with the ultimate goal of maintaining or achieving energy balance. In this process, it appears to recruit the help of circulating hormones, in part, by controlling their release into the blood. These endocrine substances, long known to have a myriad of metabolic actions in the periphery, are additionally found to have potent feedback effects on the brain.

To perform these functions, the brain requires the participation of multiple neurochemicals and neurocircuits. It has become clear that these neurochemicals can have differential effects, on behavioral as well as metabolic and endocrine processes, that are geared toward the partitioning of particular macronutrients. They may also be differentially responsive to the feedback effects of various hormones. Most intriguing is the recent evidence that neurochemicals in the brain can also respond differently, and rapidly, to the ingestion of specific macronutrients.

This evidence underscores the importance of feedback loops between the brain and body, as well as between the brain and the external environment. That is, in addition to the distinct effects of the neurochemicals on the intake and metabolism of specific macronutrients, there are the circulating hormones, metabolic fuels, and neural signals affected by these processes that return to the brain to control the activity of its neurochemicals. Thus, the peptide-synthesizing neurons that control the release of particular hormones or the intake of particular nutrients may, themselves, be responsive to the specific feedback signals produced by these hormones or dietary nutrients. A thorough investigation of these regulatory feedback loops, which may involve changes in peptide gene expression, peptide synthesis or release, and peptide binding to its receptor site, is essential for a full understanding of the physiological role of these peptides in the control of nutrient intake and metabolism.

Analyses of these feedback effects on the neurochemical systems of the brain have provided the clearest evidence for functional specificity of the different neurochemicals in controlling nutrient balance. This concept, however, receives further support from evidence showing natural fluctuations, in nutrient intake, metabolism, and hormone secretion, that occur across biological rhythms as well as developmental stages, in conjunction with changes in brain neurochemical activity. This temporal association of important traits very likely reflects the differential contribution of neurochemical systems to different states of nutrient balance. It is these biological rhythms, together with environmental influences, that contribute to the life-long process of body weight regulation.

The evidence supporting these ideas is only just emerging in the literature, still tentative but strongly suggestive of brain mechanisms controlling nutrient balance, both intake and metabolism. It is clear that multiple neurochemicals in the brain, both the classical neurotransmitters as well as peptide neuromodulators, are functional in different aspects of this process. These substances act through both inhibitory and stimulatory mechanisms. Examples of the inhibitory neurotransmitters which reduce food intake and body weight are the monoamines, serotonin and dopamine, and the peptide, enterostatin, which differentially affect the consumption of fat and carbohydrate.[1,2]

In contrast, there are two additional peptide systems, neuropeptide Y (NPY) and galanin (GAL), which are similarly geared toward enhancing food ingestion and weight gain but have markedly different effects in relation to carbohydrate and fat diets. These neuropeptide systems, the focus of this chapter, illustrate most clearly how the brain operates through diverse mechanisms that monitor, modulate, and coordinate complex physiological processes to maintain nutrient balance. These neuropeptide systems are anatomically distinct and can be distinguished by differences in their behavioral and physiological actions. They also differ in their responsiveness to the feedback actions of circulating hormones, metabolic signals, and ingested nutrients. Further, they show different patterns of endogenous activity across natural physiological states or stages associated with shifts in nutrient ingestion or metabolism.

These peptides, together with their receptor sites, exist in high concentrations in the hypothalamus.[3-6] However, within this structure, they perform their functions through different nuclei, cell groups, and neural projections.[2,7,8] For NPY, the neurocircuit mediating its actions originates in the arcuate nucleus (ARC), which has a dense population of NPY-synthesizing cells, and it terminates in the medial portion of the paraventricular nucleus (mPVN), which is densely innervated by NPY projections from the ARC.[9,10] The GAL-containing neurons involved in nutrient balance, in contrast, are concentrated in the anterior parvocellular region of the PVN (aPVN), in addition to the medial preoptic area (MPOA) rostral to the PVN which exhibits functional relationships in females.[11,12] Through interconnecting circuits and local innervation, these cell groups contribute to the dense GAL projection that courses to the PVN as well as to the median eminence (ME) in the most basal region of the hypothalamus just above the pituitary.[13-15]

Through these different anatomical substrates, the two peptides are believed to have distinct functions in controlling the body's carbohydrate and fat stores.[16] Under different conditions and in different physiological states, the NPY system is found to be more closely linked to processes of carbohydrate ingestion and carbohydrate utilization. The GAL system, in contrast, is strongly associated with patterns of fat consumption and fat oxidation. The variety of evidence supporting these proposals are summarized in detail below. It is through these mechanisms of carbohydrate and fat balance that these two peptides are believed to contribute to the overall process of body weight regulation that involves both the synthesis and accumulation of body fat.

2   PEPTIDE INJECTIONS AND NUTRIENT INGESTION

When administered into the medial hypothalamus, NPY and GAL both stimulate food consumption in satiated rats, with NPY producing a somewhat stronger response than GAL.[17-22] To understand this feeding response further, investigations have been conducted to determine whether these two peptides have differential effects on the ingestion of the macronutrients, carbohydrate, fat and protein. Tests that allow rats to select freely their macronutrient diets provide some evidence for differences between NPY and GAL. These peptides, however, are not selective in their effects on nutrient choice, and the differential patterns they do produce can be significantly altered by the rat's own natural preferences for the macronutrients.

The initial report on NPY and macronutrient choice[23] demonstrated that NPY preferentially stimulates carbohydrate ingestion. This enhancement of carbohydrate intake has similarly been described in reports from other laboratories, in a two-diet as well as three-diet choice paradigm.[24-26] It is further supported by the finding that the magnitude of the NPY feeding response is markedly attenuated when the carbohydrate diet is removed and only fat and protein remain.[23] Moreover, the rat's responsiveness to NPY is strongest at the onset of the natural feeding cycle, when carbohydrate is the naturally preferred macronutrient.[21,27,28]

Whereas these findings support a possible link between NPY and the intake of carbohydrate more than fat or protein, there is further evidence that baseline nutrient preference is an important factor in this peptide effect.[26] That is, when the food intake scores of each rat are adjusted for their individual preferences, the NPY-induced feeding response involves an increase in both carbohydrate and fat intake, although the carbohydrate response still remains significantly larger. In rats that naturally prefer fat,[29] NPY potentiates the intake of this macronutrient, although to a lesser extent than carbohydrate. It can also produce a significant feeding response in rats on a high-fat diet.[20] Thus, this evidence with different feeding paradigms generally indicates that NPY can stimulate a relative preference for carbohydrate; however, it also enhances fat intake, particularly under conditions when this nutrient is naturally preferred.

While these issues of selectivity and baseline preferences similarly apply to studies with GAL, the evidence obtained to date suggests that this peptide and NPY differ in their overall impact on nutrient consumption. The two original reports on GAL, in rats given a choice of macronutrient diets, demonstrated that this peptide potentiates the intake of both fat and carbohydrate,[21,30] in contrast to NPY which, under the same test conditions, stimulates predominantly carbohydrate intake.[21,23] Thus, in a diet-choice condition, GAL fails to exhibit selectivity in its effect on these two macronutrients, although the consumption of protein is clearly not affected by this peptide. Similar results have been obtained in more recent publications, showing that GAL potentiates the intake of either fat or carbohydrate in rats with different nutrient preferences,[31,32] and that GAL has no impact on the rats' choice between a fat-rich and laboratory chow diet.[33]

Although GAL lacks nutrient specificity in a choice paradigm, studies in a single-diet condition reveal further differences between this peptide and NPY and suggest a closer relation of GAL to dietary fat. This relationship was originally suggested by the finding that the magnitude of the GAL feeding response is markedly reduced when fat is removed from the diet and only carbohydrate and protein remain.[21] In more recent investigations, injection of GAL, in contrast to NPY, has been shown to produce a stronger, more prolonged feeding response in subgroups[34] or strains[20] of rats that naturally prefer fat. Moreover, GAL unlike NPY is found to have a greater effect in rats tested on a high-fat compared to a low-fat diet.[35]

Also suggestive of a relationship between GAL and fat intake are results obtained with the intestinal peptide, enterostatin, and also the opioid peptides. Enterostatin is synthesized in proportion to the amount of fat ingested, and it selectively reduces the consumption of this macronutrient.[36,37] Injection of this pentapeptide strongly attenuates the GAL-stimulated feeding response on a high-fat diet, while producing little change in the feeding elicited by NPY.[38] This effect, with GAL and enterostatin interacting on a high-fat diet, is not detected in rats examined with GAL and enterostatin on a high-carbohydrate diet. Further, in studies of the opioids, a close relationship between these peptides and GAL has been postulated. Both mu and kappa receptor stimulants have a potent and selective stimulatory effect on the consumption of fat.[26,39,40] It has been demonstrated[41] that GAL-stimulated feeding of a high-fat diet is completely antagonized by a mu opioid receptor blocker. This has led to the proposal that GAL functions through a pathway that involves mu opioid receptors, which may be involved in the control of fat intake.

Thus, taken together, the evidence supports a possible relationship between the peptides and the enhancement of nutrient preferences, although the effects of NPY and GAL are not selective and are compromised by natural baseline patterns. In a recent study,[42] a different approach was used to seek further information on these issues of peptide–nutrient relationships. In this report, tests with peptide injections were performed in animals that show little preference for the different macronutrients and naturally choose a balanced diet. The subjects in this experiment ($n = 18$) all consumed a daily diet consisting of approximately 45% carbohydrate (ranging from 35 to 55%), 30% fat (20 to 35%), and 25% protein (18 to 32%), and they were examined within a relatively short (3-week) period during which their baseline preferences remained relatively stable. In these rats with chronic cannulae aimed at the PVN, four peptides were tested, given in counterbalanced order with saline vehicle. Their impact on nutrient consumption, with three pure macronutrient diets available, was examined at the start of the feeding cycle. These peptides included GAL (300 pmol) and NPY (100 pmol) and, for comparison, the opioid agonists, DAMGO (3 nmol) and dynorphin A (DYNA) (3 nmol), and food intake measurements were taken at 90 min for NPY, GAL, and DAMGO and at 3 h for dynorphin A.

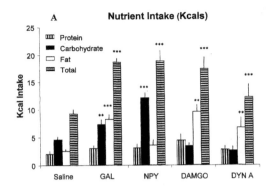
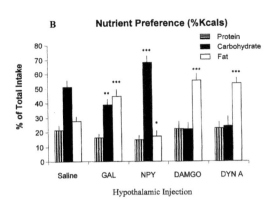

FIGURE 1   Nutrient intake (A) and nutrient preference (B) following PVN peptide injections (0.3 µl). Saline, GAL (300 pmol), NPY (100 pmol), DAMGO (3 nmol), or DYN A (3 nmol) were tested, on different days, in the same set of rats ($N = 18$). Macronutrient intake was measured over a 90-min period for GAL, NPY, and DAMGO and over a 180-min period for DYN A. $*p < 0.05$, $**p < 0.01$; $***p < 0.001$ compared to saline scores.

In these rats with no strong dietary preferences, clear differences between the peptides are observed (Figure 1A). When injected with NPY, the rats exhibit an increase in the consumption of carbohydrate, having little impact on fat intake. This contrasts with GAL, which increases both fat and carbohydrate intake. These patterns, consistent with evidence described above, are seen whether the scores are calculated in kilocalories (Figure 1A) or in grams. The strongest differences between the peptides, however, are evident when the data are presented in terms of percent of total diet or relative preference (Figure 1B). After GAL injection, the rats' preference for fat is significantly increased from 28 to 45% ($p < 0.05$), whereas it is actually reduced to 19% ($p < 0.05$) by injection of NPY. Thus, when tested in the same set of rats, these two peptides produce very different response patterns, with a preferential effect on fat and carbohydrate intake, respectively. Whereas the results are clear, this study will need to be repeated in other laboratories that have previously obtained divergent results from this laboratory, possibly due, in part, to their use of different types of diets, test times and procedures, and rat strains.[32,35]

From these findings, however, the conclusion is clear, that hypothalamic peptides have distinct and differential effects on nutrient consumption. Whereas GAL is least specific in a choice paradigm, the contrasting effects of NPY and the opioids with the three nutrients available are obvious. In these same rats within the same test week, a carbohydrate feeding response is evident with NPY in stark contrast to the selective effect of the opioids on fat consumption (see Figure 1A). This opioid effect, previously described with systemic injections,[26,35,43] is confirmed here with PVN injections. The specificity of the actions of these peptides is additionally supported by the failure of each peptide to stimulate the consumption of protein (Figure 1). This is in distinct contrast to another hypothalamic peptide, growth hormone-releasing factor, which preferentially enhances the ingestion of this macronutrient.[44] Thus, the concept of peptideinduced changes in nutrient preferences is strongly supported by these results.

3   CIRCULATING HORMONES, HYPOTHALAMIC PEPTIDES, AND NUTRIENT INTAKE

In understanding the differential roles of the peptides in maintaining nutrient balance, it is also important to assess their impact on circulating hormones, in addition to nutrient ingestion. These hormones are known to have effects on the metabolism as well as consumption of the macronutrients,[45-48] and their levels in the blood are differentially affected by the ingestion of these macronutrients.[49-52] Moreover, their administration has potent effects on the synthesis and action of endogenous peptides in the brain that affect nutrient consumption.[13,16,53]

Thus, it is significant that the peptides NPY and GAL, in addition to their distinct behavioral profiles, are found to have different effects on hormone release. This is most clearly seen in the case of corticosterone (CORT), insulin, and vasopressin. Whereas blood levels of these hormones are enhanced by hypothalamic injection of NPY,[54,55] they are reduced by injection of GAL.[56-59] In the reproductive system, these hypothalamic peptides exert opposite effects on mating behavior,[60-63] as well as on the release of luteinizing hormone (LH), specifically in animals with low endogenous steroid levels.[13,53] Divergent effects of the peptides can also be seen with measurements of growth hormone and prolactin release.[62,64-66]

In addition to these patterns of release, there is evidence that the peptide systems differ in the nature of the feedback signals that modulate them and also in the anatomical sites that are responsive to hormone action. This is most evident with CORT and the gonadal steroid, estrogen (E_2), which differentially affect gene expression and peptide production in the NPY and GAL systems. In the case of CORT, all aspects of the ARC-PVN NPY projection system, at the level of the cell body, terminal and receptor site, are strongly positively responsive to this steroid in the blood.[46,67-71] The NPY system has a dense concentration of glucocorticoid (type II) receptors that mediate this feedback regulation by CORT, in addition to its permissive role in the stimulatory effect of NYP on carbohydrate intake as well as body weight.[46,72-75] This is in contrast to the GAL neurons, receptors, and feeding projection in the PVN, which can function at low levels of CORT or even independently of this adrenal steroid and that are unresponsive to or negatively affected by CORT.[76,77] The GAL-containing neurons in the PVN exhibit little glucocorticoid-receptor immunoreactivity,[75] and their expression of the GAL gene or production of the peptide, in addition to the stimulatory effect of GAL on fat ingestion, is unaffected or reduced by CORT replacement in adrenalectomized rats.[30,72,76] This relative independence of the PVN GAL system in relation to circulating CORT and the strong dependence of NPY on this steroid may be reflective of the physiological actions of these peptides which as described above include the inhibition and stimulation, respectively, of CORT release.

Another steroid, E_2, also has differential effects on the GAL and NPY systems. There is considerable evidence revealing a very potent effect of this steroid on GAL gene expression and production in the hypothalamus, specifically the MPOA, aPVN, and ME, as well as in the anterior pituitary.[13,78-81] This is in contrast to the inhibitory effect of E_2 on NPY gene expression in the ARC, as well as on NPY release in the PVN.[82-84] This differential responsiveness of GAL and NPY neurons to E_2 may reflect their different effects on LH release.[13,53] It may also be explained by the differential concentration of the glucocorticoid and estrogen receptors, in the area of the MPOA and aPVN GAL neurons compared with the NPY neurons in the ARC.[85,86]

The significance of these differential patterns can be more fully appreciated in relation to the effects of nutrient consumption on circulating hormone levels, in addition to the hypothalamic peptides. Whereas this relationship will be discussed in greater detail below, it is noteworthy that dietary carbohydrate and fat have differential effects on circulating levels of these steroids.[49-52] Specifically, levels of CORT are found to be highest in adult rats maintained on a high-carbohydrate diet (65% carbohydrate), compared with those on a moderate- (45%) or low- (15%) carbohydrate diet.[49] In contrast, additional findings indicate that E_2 levels are actually increased by the consumption of a high-fat diet.[51,80]

Thus, in addition to underscoring the differences between the two peptides in their relation to the steroids, these results suggest that the hormones may play an intermediary role in the relationship of the hypothalamic peptides to dietary nutrients. Whereas distinct hormone–peptide relationships exist for the NPY and GAL systems, it may be noted that the production of these peptides in the hypothalamus, while differentially affected by the steroids, are similarly controlled by two other hormones, namely, insulin and the adipose tissue hormone, leptin. These hormones, with known inhibitory effects on nutrient intake and body weight,[47,48,87] reduce NPY expression and levels in the ARC and PVN, as well as GAL mRNA and peptide immunoreactivity in the PVN, ARC, and ME.[88-91]

4   NUTRIENT METABOLISM AND HYPOTHALAMIC PEPTIDES

Further differences between GAL and NPY can be found in studies of their relationship to the rats' nutritional state. In response to food restriction or deprivation, rats exhibit a marked increase in NPY content, NPY-positive immunoreactivity, peptide release, or mRNA levels.[92-94] This response is anatomically localized, seen most consistently and dramatically in the NPY projection from the ARC to PVN. While there are fewer studies on GAL, the available evidence indicates that GAL-synthesizing neurons are considerably less responsive than NPY neurons to food deprivation.[95,96] Whereas a small increase in GAL mRNA and peptide immunoreactivity is seen in the PVN after food deprivation,[97] chronic food restriction causes a decline in GAL mRNA in the ARC.[93] A further difference between these peptides is seen in investigations of dietary zinc deficiency, accompanied by reduced food intake and body weight, which produces a decrease in hypothalamic GAL mRNA levels in association with a rise in NPY gene expression.[98]

At this point, one can only speculate as to the endocrine or metabolic signal(s) to which these hypothalamic peptide systems are responding in food-deprived animals. It is likely that both the hormones and metabolic fuels play a role in these relationships. The importance of the hormones is supported by evidence showing that the deprivation-induced rise in NPY gene expression and peptide levels, normally associated with an increase in CORT[46] and decline in insulin,[47] is attenuated or reversed by adrenalectomy or insulin administration.[99,100] This deprivation effect is restored by high doses of CORT replacement,[99] although it appears independent of physiological CORT levels.[101] Whereas a decline in insulin and leptin in food-deprived rats may contribute to the deprivation-induced rise in NPY mRNA in the ARC as well as GAL in the PVN, this leaves unexplained why deprivation can enhance NPY production in diabetic[102] or leptin-deficient[103] animals.

Whereas there is relatively little information on the metabolic effects of NPY and GAL, the available evidence demonstrates that hypothalamic NPY and GAL injection can both reduce energy expenditure[104,105] and sympathetic nervous system activity.[106,107] However, NPY but not GAL has been shown to alter substrate utilization, stimulating the metabolism of carbohydrate.[104,105] The possibility that this effect of NPY may be due to an increase in locomotor activity,[108] however, needs further investigation.

More information on a potential relationship between the peptides and nutrient metabolism is provided by investigations showing that local changes in nutrients or their metabolism may have potent and direct effects on peptide gene expression.[97] The evidence suggests that NPY neurons, specifically in the ARC, may be more responsive to alterations in carbohydrate metabolism, whereas GAL is affected by signals of fat oxidation.[97,109,110] Specifically, in the ARC, NPY levels are increased by administration of 2-deoxy-D-glucose,[109,110] which blocks glucose utilization.[111] This compound stimulates a feeding response, which is antagonized by injection of an NPY antibody and is associated with a preferential increase in carbohydrate intake.[112,113] This agent, in contrast, has no impact on hypothalamic GAL.[97]

Conversely, whereas NPY is unaffected by another metabolic inhibitor, mercaptoacetate (MA), which blocks fatty acid oxidation,[114] the production of GAL is reduced by this compound specifically in the PVN.[97] This suppressive effect on GAL is accompanied by an MA-induced reduction in fat consumption, which may be a consequence of the animals' inability to metabolize lipids.[97] A compensatory feeding response after MA injection, which involves a stronger increase in protein consumption relative to carbohydrate intake,[97,115] is found to be suppressed by a GAL receptor antagonist.[116]

The significance of these findings with metabolic inhibitors can be more fully appreciated in relation to studies of the effects of diet consumption on nutrient metabolism and, in turn, on peptide production in the hypothalamus. The evidence is clear in showing the differential effects of fat-rich and carbohydrate-rich diets on nutrient metabolism. In particular, the consumption of fat enhances the oxidation of fatty acids, whereas carbohydrate ingestion potentiates the utilization of carbohydrates.[117-119] Given that alterations in nutrient metabolism differentially affect NPY and

GAL, the possibility exists that nutrient intake itself also affects brain peptide activity. This will be discussed in the next section.

5   DIETARY NUTRIENTS AND ENDOGENOUS HYPOTHALAMIC PEPTIDES

The evidence described above demonstrates that NPY and GAL injection have potent as well as differential effects on the behavioral processes of carbohydrate or fat ingestion and the physiological and endocrine processes involved in nutrient metabolism. There are additional results showing that the hormones and metabolic signals, in turn, have impact on the activity of the endogenous peptide systems that further differentiates their potential functions. The question to be addressed here is whether nutrient ingestion itself can alter or be linked to the endogenous peptides. Specifically, are endogenous NPY and GAL related to an animal's natural appetite for carbohydrate or fat? Can experimentally imposed dietary manipulations differentially affect peptide gene expression in the hypothalamus? The evidence suggests that they can and, once again, shows that the peptide systems can be readily differentiated, both anatomically and physiologically, in their relation to the macronutrients.

The first studies to demonstrate this relationship took advantage of the finding that Sprague-Dawley rats exhibit considerable individual differences in their appetite for the macronutrients, carbohydrate and fat. Approximately 50% of the population shows a strong preference for carbohydrate (>45% of total diet), while an additional 35% consume relatively large amounts of fat (>30%) and, consequently, exhibit greater fat deposition and body weight gain.[27,28,49,120,121] Analyses of NPY and GAL in these rats with distinct preferences for the two nutrients indicate that they can be related to these natural preferences, in a highly brain site-specific and nutrient-specific manner.[12,122-124]

With NPY, a strong positive correlation exists between the rats' daily intake of carbohydrate and their peptide levels specifically in the ARC and mPVN.[123] This contrasts with GAL, which in the PVN is positively related to spontaneous fat ingestion but inversely related to carbohydrate intake.[122] These differential patterns have been replicated and expanded in recent studies[12,124] which have more clearly defined the GAL projection related to dietary fat. This projection is found to originate in the aPVN, where GAL-synthesizing neurons are dense and relatively small in size, and it projects to the ME, its external zone, where GAL terminals are highly concentrated. This aPVN-ME GAL system contrasts with the magnocellular GAL-containing neurons, concentrated in the central portion of the PVN <0.5 mm caudal to the aPVN cell group, which project to the internal zone of the ME on their course to the posterior pituitary[13,14] and appear unrelated to dietary fat.[12] Whereas GAL in neurons of the ARC also show little change in relation to fat intake,[12] one additional site, the MPOA, is found to exhibit greater peptide production in relation to higher fat intake, although only in female subjects.[11,79,80]

These relationships between the peptides and either carbohydrate or fat consumption, in subjects expressing their natural selection patterns, are also detected in rats with a single diet available, a paradigm that provides no opportunity for a spontaneous choice of macronutrients.[12,124] This indicates that the nutrient–peptide relationships reflect, in part, the impact of the diet itself on the peptide projections (Figure 2). Specifically, in rats maintained on a high-carbohydrate diet, NPY gene expression in the ARC and peptide immunoreactivity in the ARC and PVN are greatly enhanced[124] (Figure 2A). This effect is strongest when dietary carbohydrate increases from 45 to 65% of the total diet. In contrast, GAL mRNA in the aPVN and peptide levels in the aPVN and ME are increased by the consumption of a high-fat diet[12] (Figure 2B). The precise amount of fat is also important in this response. Whereas a 30% fat diet compared to a 10% fat diet stimulates GAL gene expression but not peptide production, a further rise in dietary fat above 30% increases the synthesis of GAL, as well as its release.[12] The anatomical specificity of this effect is underscored, once again, by the finding that GAL neurons in the ARC and MPOA of male rats are unaffected

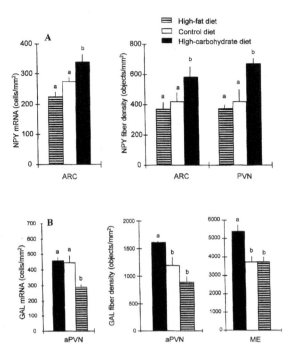

FIGURE 2 Neuropeptide Y (A), and GAL (B) mRNA and peptide-ir after consumption of a high-carbohydrate, control, or high-fat diet for a 4-week period. Shared alphabet letters refer to groups that are not significantly different, at $p < 0.05$. Abbreviations: ARC, arcuate nucleus; PVN, paraventricular nucleus; aPVN, anterior portion of the PVN; ME, median eminence. ([A] Modified from Wang, J., Akabayashi, A., Dourmashkin, J., Yu, H., Alexander, J. T., Chae, H. J., and Leibowitz, S. F., Neuropeptide Y in relation to carbohydrate intake, corticosterone and dietary obesity, *Brain Res.*, 802, 75, 1998. With permission from Elsevier Science; [B] modified from Leibowitz, S. F., Akabayashi, A., and Wang, J., Obesity on a high-fat diet: role of hypothalamic galanin in neurons of the anterior paraventricular nucleus projecting to the median eminence, *J. Neurosci.*, 18, 2709, 1998. With permission from Elsevier Science.)

by the consumption of fat.[12] Only in the female is MPOA GAL stimulated in rats on a high-fat diet.[11,79,80]

6   POSITIVE NUTRIENT–PEPTIDE–NUTRIENT FEEDBACK LOOP

The above investigations provide initial support for the concept of a positive feedback loop between the hypothalamic peptides and dietary nutrients. The strongest evidence comes from the diet studies with measurements of endogenous peptides, which clearly relate fat intake to the production of GAL and carbohydrate ingestion to the production of NPY. The other side of the loop, suggested by studies with injected peptides affecting nutrient ingestion, is less definitive but at least consistent with results summarized above. Thus, in these positive feedback loops, GAL and NPY are both proposed to stimulate the ingestion of a macronutrient which further enhances the endogenous activity of the peptides.

Further studies are clearly needed to understand the physiological functions of such positive interactions and also to elaborate on the inhibitory signals, such as the monoamines,[2] that are needed to control them. However, it is well known that the introduction of fat-rich or sweet, carbohydrate-rich diets induces hyperphagia and increases meal size,[28,49,125] an effect attributed, in part, to the reduced satiating capacity of fat[28,125] as well as to the enhancement in diet palatability associated with sucrose as well as fat.[126,127]

In light of the evidence reviewed above, the possibility exists that these behavioral effects of the imbalanced diets are mediated, in part, by the positive feedback loops that may exist between the dietary macronutrients and the hypothalamic peptides, specifically GAL neurons in the PVN or NPY neurons in the ARC. This idea is supported by the finding that the range of fat critical to the hyperphagia, above 30%, is the same as that needed to potentiate GAL peptide synthesis and release in the PVN, along with GAL peptide immunoreactivity in the ME.[12] Also, the greatest increase in NPY production in the ARC is seen when dietary carbohydrate rises from 40 to 65% of total diet,[124] and in response to the consumption of glucose as well.[128] In recent experiments, fat-rich or carbohydrate-rich meals, ingested over a 2-h interval, are found to affect these endogenous peptides, in precisely the same manner as that described above under chronic diet conditions.[128,129] Thus, the specific relationships between the peptides and dietary nutrients are, once again, confirmed and are demonstrated to occur rapidly, possibly reflecting a role of the peptides in controlling the size of nutrient-specific meals.

7   HYPOTHALAMIC PEPTIDES AND NUTRIENT INGESTION ACROSS PHYSIOLOGICAL STATES AND STAGES

It is now clear that the peptide systems in the hypothalamus can produce specific behavioral and physiological responses that, through feedback interactions, help to balance the ingestion and metabolism of macronutrients. Through these feedback loops, the peptides function in a distinct manner, in relation to different macronutrients and in an anatomically specific manner. This evidence provides essential information for research efforts designed to determine whether these neurochemical systems in the brain are, in fact, functionally active under natural conditions. Studies of biological rhythms and different developmental stages reveal clear, temporal associations between natural shifts in the brain peptides, circulating hormones, and the behavioral process of nutrient ingestion. They can be detected in relation to the 24-h diurnal cycle, the 4-day female cycle, as well as different stages of development. Once again, endogenous NPY and GAL are found to be differentially related to natural shifts in carbohydrate and fat intake, respectively.

Briefly, across the light/dark cycle, there exists a clear diurnal rhythm of nutrient intake.[27,28,130] Subsequent to a 12-h period of little eating resulting in low carbohydrate stores, the early hours of the feeding cycle are characterized by a strong preference for carbohydrate and, then, followed by a rise in fat consumption 3 to 4 h later. In association with these behavioral patterns are diurnal rhythms in the hypothalamic peptides that are distinct and anatomically localized. For NPY, a rise in mRNA is evident in the ARC a few hours before feeding onset, followed by increased peptide synthesis and peak peptide levels in the mPVN, at the time when carbohydrate ingestion is strongest.[27,68,131] A very different pattern is seen for endogenous GAL, which exhibits a distinctive pattern only in the PVN. This is characterized by a single peptide peak toward the middle of the nocturnal feeding cycle, between the 3rd and 6th hours,[132] at the same time when there is a rise in spontaneous fat-rich meals.[27]

There is further evidence for a shift in eating behavior and nutrient preference across the female cycle.[11,45,133] In studies of rats on the macronutrient diets,[11] a significant rise in preference for fat is evident during the proestrous period, followed by a marked decline in fat as well as caloric intake during the estrous stage. This pattern may be attributed, in part, to a reliable increase in GAL during proestrous, followed by a decline during estrous, specifically in the MPOA, aPVN, and ME.[11,13,134] These are the same sites as those found to be responsive to dietary fat in a single-diet feeding paradigm.[12,79] In fat-preferring compared to carbohydrate-preferring female rats, GAL levels are significantly higher in these three areas.[11] This is in contrast to NPY levels in the ARC and PVN, which are reliably lower in high-fat eaters, similar to results obtained in male rats.[123,124]

Significant shifts in GAL can also be detected across developmental stages, in association with changes in endocrine, behavioral, and physiological responses. Most notable is the natural burst in fat intake as well as body fat that is evident around puberty.[135,136] In addition to GAL mRNA in gonadotropin-releasing hormone neurons,[137] levels of this peptide, in the whole hypothalamus and particularly in the PVN, MPOA, and ME, show a sharp rise around puberty.[138,139] This increase in GAL production in these areas contrasts with the pattern observed with NPY, which shows little change in the ARC and PVN at this time and rises only in the MPOA.[139,140] Moreover, levels of GAL in the PVN, MPOA, and ME are greater in rats that prefer fat, and their rise in the PVN and ME occurs earlier in female rats compared with males, presumably reflecting the excitatory effects of E_2 as well as progesterone on GAL gene expression.[11,13,79,80,139]

8 FUNCTIONAL SIGNIFICANCE OF RELATIONSHIP BETWEEN HYPOTHALAMIC PEPTIDES AND NUTRIENT BALANCE UNDER PHYSIOLOGICAL CONDITIONS

The evidence described above demonstrates strong effects of hypothalamic peptide injections on behavioral, physiological, and endocrine systems involved in nutrient balance. They reveal clear feedback effects of circulating hormones, macronutrient diets, and metabolic signals on peptide gene expression and production in hypothalamic neurons. In addition to these causal relationships, there is evidence for temporal associations between the brain peptides and various physiological and behavioral traits. In each case, endogenous activity of the brain peptides can be linked to an animal's natural appetite for carbohydrate or fat as it shifts naturally across biological rhythms or during development.

These studies of temporal associations and causal relationships yield a concordance of evidence linking the hypothalamic peptide systems to specific physiological and behavioral traits. They argue for differential activity of the two peptide systems across biological rhythms that can be related to specific dietary nutrients and circulating hormones. Together, this evidence provides the basis for formulating specific working hypotheses concerning the potential physiological functions and mechanisms of action of the peptides.

From the evidence summarized above, it is suggested that the NPY ARC-PVN system has a prominent role in maintaining carbohydrate balance, through processes of ingestion and metabolism. Hormones that are released by NPY have prominent effects on carbohydrate metabolism and, in turn, have potent regulatory effects on NPY gene expression. Through its preferential effect on the ingestion of carbohydrate, NPY can provide the necessary substrates for restoring nutrient balance. In this process, NPY works in close association with the adrenal steroid, CORT, in an effort to maintain circulating glucose levels and, when there is sufficient substrate availability, enhance the synthesis of body fat.

It is further proposed that GAL neurons in the PVN that project to the ME have a particular function in controlling fat ingestion. They may, in turn, be responsive to shifts in fat oxidation as well as fat consumption. The studies of natural rhythms and feeding patterns demonstrate a close relation between GAL production in the PVN and natural patterns of fat intake. The specific states when the GAL PVN-ME system may be active include the middle period of the natural feeding cycle, the proestrous period, and the time of puberty, when natural appetite for fat and endogenous GAL production are markedly increased. These functions of GAL may be performed with the help of the gonadal steroids, E_2 and PROG, which potently enhance GAL production. In each of these states, this relationship between GAL and fat intake may contribute to longterm patterns of body fat accumulation.

ACKNOWLEDGMENTS

Some of the research described in this chapter has been supported by U.S. Public Health Service Grant MH 43422. I am indebted to Ms. Jesline Alexander for her help in the preparation of this manuscript.

REFERENCES

1. Lin, L. and York, D. A., Enterostatin actions in the amygdala and PVN to suppress feeding in the rat, *Peptides*, 18, 1341, 1997.
2. Leibowitz, S. F. and Hoebel, B. G., Behavioral Neuroscience and Obesity, in *The Handbook of Obesity*, G. Bray, C. Bouchard, and P. T. James, Eds., Marcel Dekker, New York, 1997, 313.
3. Melander, T., Hokfelt, T., and Rokaeus, A., Distribution of galanin-like immunoreactivity in the rat central nervous system, *J. Comp. Neurol.*, 248, 475, 1986.
4. Bonnefond, C., Palacios, J. M., Probst, A., and Mengod, G., Distribution of galanin mRNA containing cells and galanin receptor binding sites in human and rat hypothalamus, *Eur. J. Neurosci.*, 2, 629, 1997.
5. Dumont, Y., Martel, J. C., Fournier, A., St-Pierre, S., and Quirion, R., Neuropeptide Y and neuropeptide Y receptor subtypes in brain and peripheral tissues, *Prog. Neurobiol.*, 38, 125, 1992.
6. Wahlestedt, C., Ekman, R., and Widerlov, E., Neuropeptide Y (NPY) and the central nervous system: distribution effects and possible relationship to neurological and psychiatric disorders, *Prog. Neuro-Psychopharmacol. Biol. Psych.*, 13, 31, 1989.
7. Ojeda, S. R., Andrews, W. W., Advis, J. P., and White, S. S., Recent advances in the endocrinology of puberty, *Endocr. Rev.*, 1, 228, 1980.
8. Wilding, J. P., Metabolic actions of neuropeptide Y and their relevance to obesity, *Biochem. Soc. Trans.*, 24, 576, 1996.
9. Bai, F. L., Yamano, M., Shiotani, Y., Emson, P. C., Smith, A. D., Powell, J. F., and Tohyama, M., An arcuato-paraventricular and -dorsomedial hypothalamic neuropeptide Y-containing system which lacks noradrenaline in the rat, *Brain Res.*, 331, 172, 1985.
10. Sawchenko, P. E., Swanson, L. W., Grzanna, R., Howe, P. R., Bloom, S. R., and Polak, J. M., Colocalization of neuropeptide Y immunoreactivity in brain stem catecholaminergic neurons that project to the paraventricular nucleus of the hypothalamus, *J Comp. Neurol.*, 241, 138, 1985.
11. Leibowitz, S. F., Akabayashi, A., Alexander, J. T., and Wang, J., Gonadal steroids and hypothalamic galanin and neuropeptide Y: role in eating behavior and body weight control in female rats, *Endocrinology*, 139, 1771, 1998.
12. Leibowitz, S. F., Akabayashi, A., and Wang, J., Obesity on a high-fat diet: role of hypothalamic galanin in neurons of the anterior paraventricular nucleus projecting to the median eminence. *J. Neurosci.*, 18, 2709, 1998.
13. Merchenthaler, I., Lopez, F. J., and Negro-Vilar, A., Anatomy and physiology of central galanin-containing pathways, *Prog. Neurobiol.*, 40, 711, 1993.
14. Palkovits, M., Rokaeus, A., Antoni, F. A., and Kiss, A., Galanin in the hypothalamo-hypophyseal system, *Neuroendocrinology*, 46, 417, 1987.
15. Levin, M. C., Sawchenko, P. E., Howe, P. R., Bloom, S. R., and Polak, J. M., Organization of galanin-immunoreactive inputs to the paraventricular nucleus with special reference to their relationship to catecholaminergic afferents, *J. Comp. Neurol.*, 261, 562, 1987.
16. Leibowitz, S. F., Brain peptides and obesity: pharmacologic treatment, *Obesity Res.*, 3 (Suppl. 4), 573S, 1995.
17. Kyrkouli, S. E., Stanley, B. G., and Leibowitz, S. F., Galanin: stimulation of feeding induced by medial hypothalamic injection of this novel peptide, *Eur. J. Pharmacol.*, 122, 159, 1986.
18. Crawley, J. N., Austin, M. C., Fiske, S. M., Martin, B., Consolo, S., Berthold, M., Langel, U., Fisone, G., and Bartfai, T., Activity of centrally administered galanin fragments on stimulation of feeding behavior and on galanin receptor binding in the rat hypothalamus, *J. Neurosci.*, 10, 3695, 1990.
19. Stanley, B. G. and Leibowitz, S. F., Neuropeptide Y: stimulation of feeding and drinking by injection into the paraventricular nucleus, *Life Sci.*, 35, 2635, 1984.
20. Lin, L., York, D. A., and Bray, G. A., Comparison of Osborne-Mendel and S5B/PL strains of rat: central effects of galanin, NPY, beta-casomorphin and CRH on intake of high-fat and low-fat diets, *Obesity Res.*, 4, 117, 1996.
21. Tempel, D. L. and Leibowitz, S. F., Diurnal variations in the feeding responses to norepinephrine, neuropeptide Y and galanin in the PVN, *Brain Res. Bull.*, 25, 821, 1990.
22. Kalra, S. P., Dube, M. G., Sahu, A., Phelps, C. P., and Kalra, P. S., Neuropeptide Y secretion increases in the paraventricular nucleus in association with increased appetite for food, *Proc. Natl. Acad. Sci. U.S.A.*, 88, 10931, 1991.

23. Stanley, B. G., Daniel, D. R., Chin, A. S., and Leibowitz, S. F., Paraventricular nucleus injections of peptide YY and neuropeptide Y preferentially enhance carbohydrate ingestion, *Peptides,* 6, 1205, 1985.

24. Glass, M. J., Cleary, J. P., Billington, C. J., and Levine, A. S., Role of carbohydrate type on diet selection in neuropeptide Y-stimulated rats, *Am. J Physiol.,* 273, R2040, 1997.

25. Morley, J. E., Levine, A. S., Gosnell, B. A., Kneip, J., and Grace, M., Effect of neuropeptide Y on ingestive behaviors in the rat, *Am. J. Physiol.,* 252, R599, 1987.

26. Welch, C. C., Grace, M. K., Billington, C. J., and Levine, A. S., Preference and diet type affect macronutrient selection after morphine, NPY, norepinephrine, and deprivation, *Am. J. Physiol.,* 266, R426, 1994.

27. Shor-Posner, G., Ian, C., Brennan, G., Cohn, T., Moy, H., Ning, A., and Leibowitz, S. F., Self-selecting albino rats exhibit differential preferences for pure macronutrient diets: characterization of three subpopulations, *Physiol. Behav.,* 50, 1187, 1991.

28. Shor-Posner, G., Brennan, G., Ian, C., Jasaitis, R., Madhu, K., and Leibowitz, S. F., Meal patterns of macronutrient intake in rats with particular dietary preferences, *Am. J. Physiol.,* 266, R1395, 1994.

29. Stanley, B. G., Anderson, K. C., Grayson, M. H., and Leibowitz, S. F., Repeated hypothalamic stimulation with neuropeptide Y increases daily carbohydrate and fat intake and body weight gain in female rats, *Physiol. Behav.,* 46, 173, 1989.

30. Tempel, D. L., Leibowitz, K. J., and Leibowitz, S. F., Effects of PVN galanin on macronutrient selection, *Peptides,* 9, 309, 1988.

31. Smith, B. K., York, D. A., and Bray, G. A., Effects of dietary preference and galanin administration in the paraventricular or amygdaloid nucleus on diet self-selection, *Brain Res. Bull.,* 39, 149, 1996.

32. Smith, B. K., Berthoud, H. R., York, D. A., and Bray, G. A., Differential effects of baseline macronutrient preferences on macronutrient selection after galanin, NPY, and an overnight fast, *Peptides,* 18, 207, 1997.

33. Corwin, R. L., Rowe, P. M., and Crawley, J. N., Galanin and the galanin antagonist M40 do not change fat intake in a fat-chow choice paradigm in rats, *Am. J. Physiol.,* 269, R511, 1995.

34. Leibowitz, S. F. and Kim, T., Impact of a galanin antagonist on exogenous galanin and natural patterns of fat ingestion, *Brain Res.,* 599, 148, 1992.

35. Barton, C., Lin, L., York, D. A., and Bray, G. A., Differential effects of enterostatin, galanin and opioids on high-fat diet consumption, *Brain Res.,* 702, 55, 1995.

36. Okada, S., York, D. A., Bray, G. A., and Erlanson-Albertsson, C., Enterostatin (Val-Pro-Asp-Pro-Arg), the activation peptide of procolipase, selectively reduces fat intake, *Physiol. Behav.,* 49, 1185, 1991.

37. Erlanson-Albertsson, C. and York, D., Enterostatin — a peptide regulating fat intake, *Obesity Res.,* 5, 360, 1997.

38. Lin, L., Gehlert, D. R., York, D. A., and Bray, G. A., Effect of enterostatin on the feeding response to galanin and NPY, *Obesity Res.,* 1 (3), 1993.

39. Shor-Posner, G., Azar, A. P., Filart, R., Tempel, D., and Leibowitz, S. F., Morphine-stimulated feeding: analysis of macronutrient selection and paraventricular nucleus lesions, *Pharmacol. Biochem. Behav.,* 24, 931, 1986.

40. Ookuma, K., Barton, C., York, D. A., and Bray, G. A., Effect of enterostatin and kappa-opioids on macronutrient selection and consumption, *Peptides,* 18, 785, 1997.

41. Barton, C., York, D. A., and Bray, G. A., Opioid receptor subtype control of galanin-induced feeding, *Peptides,* 17, 237, 1996.

42. Chae, H. J., Hoebel, B. G., Tempel, D. L., Paredes, M., and Leibowitz, S. F., Neuropeptide-Y, galanin and opiate agonists have differential effects on nutrient ingestion, *Soc. Neurosci. Abstr.,* 21, 696, 1995.

43. Shor-Posner, G., Azar, A. P., Filart, R., Tempel, D., and Leibowitz, S. F., Morphine-stimulated feeding: analysis of macronutrient selection and paraventricular nucleus lesions, *Pharmacol. Biochem. Behav.,* 24, 931, 1986.

44. Dickson, P. R. and Vaccarino, F. J., GRF-induced feeding: evidence for protein selectivity and opiate involvement, *Peptides,* 15, 1343, 1994.

45. Wade, G. N., Schneider, J. E., and Li, H., Control of fertility by metabolic cues, *Am. J. Physiol.,* 270, 1, 1996.

46. Tempel, D. L. and Leibowitz, S. F., Adrenal steroid receptors: interactions with brain neuropeptide systems in relation to nutrient intake and metabolism, *J. Neuroendocrinol.,* 6, 479, 1994.

47. Woods, S. C., Chavez, M., Park, C. R., Riedy, C., Kaiyala, K., Richardson, R. D., Figlewicz, D. P., Schwartz, M. W., Porte, D., Jr., and Seeley, R. J., The evaluation of insulin as a metabolic signal influencing behavior via the brain, *Neurosci. Biobehav. Rev.,* 20, 139, 1996.

48. Bray, G. A. and York, D. A., Clinical review 90: leptin and clinical medicine: a new piece in the puzzle of obesity [published erratum appears in *J. Clin. Endocrinol. Metab.*, 82(11), 3878, 1997], *J. Clin. Endocrinol. Metab.*, 82, 2771, 1997.

49. Wang, J., Alexander, J. T., Zheng, P., Yu, H., Dourmashkin, J., and Leibowitz, S. F., Behavioral and endocrine traits of obesity-prone and obesity-resistant rats on macronutrient diets, *Am. J. Physiol.*, 274, E1057, 1998.

50. Boivin, A. and Deshaies, Y., Dietary rat models in which the development of hypertriglyceridemia and that of insulin resistance are dissociated, *Metabolism*, 44, 1540, 1995.

51. Hilakivi-Clarke, L., Cho, E., and Onojafe, I., High-fat diet induces aggressive behavior in male mice and rats, *Life Sci.*, 58, 1653, 1996.

52. Hilakivi-Clarke, L., Onojafe, I., Raygada, M., Cho, E., Clarke, R., and Lippman, M. E., Breast cancer risk in rats fed a diet high in n-6 polyunsaturated fatty acids during pregnancy, *J. Natl. Cancer Inst.*, 88, 1821, 1996.

53. Kalra, S. P. and Kalra, P. S., Nutritional infertility: the role of the interconnected hypothalamic neuropeptide Y-galanin-opioid network, *Front. Neuroendocrinol.*, 17, 371, 1996.

54. Sainsbury, A., Rohner-Jeanrenaud, F., Grouzmann, E., and Jeanrenaud, B., Acute intracerebroventricular administration of neuropeptide Y stimulates corticosterone output and feeding but not insulin output in normal rats, *Neuroendocrinology*, 63, 318, 1996.

55. Leibowitz, S. F., Sladek, C., Spencer, L., and Tempel, D., Neuropeptide Y, epinephrine and norepinephrine in the paraventricular nucleus: stimulation of feeding and the release of corticosterone, vasopressin and glucose, *Brain Res. Bull.*, 21, 905, 1988.

56. Pierroz, D. D., Catzeflis, C., Aebi, A. C., Rivier, J. E., and Aubert, M. L., Chronic administration of neuropeptide Y into the lateral ventricle inhibits both the pituitary-testicular axis and growth hormone and insulin-like growth factor I secretion in intact adult male rats, *Endocrinology*, 137, 3, 1996.

57. Tempel, D. L. and Leibowitz, S. F., Galanin inhibits insulin and corticosterone release after injection into the PVN, *Brain Res.*, 536, 353, 1990.

58. Hooi, S. C., Maiter, D. M., Martin, J. B., and Koenig, J. I., Galaninergic mechanisms are involved in the regulation of corticotropin and thyrotropin secretion in the rat, *Endocrinology*, 127, 2281, 1990.

59. Kondo, K., Murase, T., Otake, K., Ito, M., and Oiso, Y., Centrally administered galanin inhibits osmotically stimulated arginine vasopressin release in conscious rats, *Neurosci. Lett.*, 128, 245, 1991.

60. Bloch, G. J., Butler, P. C., Kohlert, J. G., and Bloch, D. A., Microinjection of galanin into the medial preoptic nucleus facilitates copulatory behavior in the male rat, *Physiol. Behav.*, 54, 615, 1993.

61. Bloch, G. J., Butler, P. C., and Kohlert, J. G., Galanin microinjected into the medial preoptic nucleus facilitates female- and male-typical sexual behaviors in the female rat, *Physiol. Behav.*, 59, 1147, 1996.

62. Catzeflis, C., Pierroz, D. D., Rohner-Jeanrenaud, F., Rivier, J. E., Sizonenko, P. C., and Aubert, M. L., Neuropeptide Y administered chronically into the lateral ventricle profoundly inhibits both the gonadotropic and the somatotropic axis in intact adult female rats, *Endocrinology*, 132, 224, 1993.

63. Clark, J. T., Kalra, P. S., and Kalra, S. P., Neuropeptide Y stimulates feeding but inhibits sexual behavior in rats, *Endocrinology*, 117, 2435, 1985.

64. Murakami, Y., Kato, Y., Koshiyama, H., Inoue, T., Yanaihara, N., and Imura, H., Galanin stimulates growth hormone (GH) secretion via GH-releasing factor (GRF) in conscious rats, *Eur. J. Pharmacol.*, 136, 415, 1987.

65. Koshiyama, H., Kato, Y., Inoue, T., Murakami, Y., Ishikawa, Y., Yanaihara, N., and Imura, H., Central galanin stimulates pituitary prolactin secretion in rats: possible involvement of hypothalamic vasoactive intestinal polypeptide, *Neurosci. Lett.*, 75, 49, 1987.

66. Harfstrand, A., Eneroth, P., Agnati, L., and Fuxe, K., Further studies on the effects of central administration of neuropeptide Y on neuroendocrine function in the male rat: relationship to hypothalamic catecholamines [published erratum appears in *Regul. Pept.* 17(5), 300, 1987], *Regul. Pept.*, 17, 167, 1987.

67. Akabayashi, A., Watanabe, Y., Wahlestedt, C., McEwen, B. S., Paez, X., and Leibowitz, S. F., Hypothalamic neuropeptide Y, its gene expression and receptor activity: relation to circulating corticosterone in adrenalectomized rats, *Brain Res.*, 665, 201, 1994.

68. Akabayashi, A., Levin, N., Paez, X., Alexander, J. T., and Leibowitz, S. F., Hypothalamic neuropeptide Y and its gene expression: relation to light/dark cycle and circulating corticosterone, *Mol. Cell. Neurosci.*, 5, 210, 1994.

69. Dallman, M. F., Akana, S. F., Strack, A. M., Hanson, E. S., and Sebastian, R. J., The neural network that regulates energy balance is responsive to glucocorticoids and insulin and also regulates HPA axis responsivity at a site proximal to CRF neurons, *Ann. N.Y. Acad. Sci.*, 771, 730, 1995.

70. White, B. D., Dean, R. G., Edwards, G. L., and Martin, R. J., Type II corticosteroid receptor stimulation increases NPY gene expression in basomedial hypothalamus of rats, *Am. J. Physiol.*, 266, R1523, 1994.

71. Larsen, P. J., Jessop, D. S., Chowdrey, H. S., Lightman, S. L., and Mikkelsen, J. D., Chronic administration of glucocorticoids directly upregulates prepro-neuropeptide Y and Y1-receptor mRNA levels in the arcuate nucleus of the rat, *J. Neuroendocrinol.*, 6, 153, 1994.

72. Tempel, D. L. and Leibowitz, S. F., Glucocorticoid receptors in PVN: interactions with NE, NPY, and Gal in relation to feeding, *Am. J. Physiol.*, 265, E794, 1993.

73. Sainsbury, A., Cusin, I., Rohner-Jeanrenaud, F., and Jeanrenaud, B., Adrenalectomy prevents the obesity syndrome produced by chronic central neuropeptide Y infusion in normal rats, *Diabetes*, 46, 209, 1997.

74. Harfstrand, A., Cintra, A., Fuxe, K., Aronsson, M., Wikstrom, A. C., Okret, S., Gustafsson, J. A., and Agnati, L. F., Regional differences in glucocorticoid receptor immunoreactivity among neuropeptide Y immunoreactive neurons of the rat brain, *Acta Physiol. Scand.*, 135, 3, 1989.

75. Cintra, A., Fuxe, K., Solfrini, V., Agnati, L. F., Tinner, B., Wikstrom, A. C., Staines, W., Okret, S., and Gustafsson, J. A., Central peptidergic neurons as targets for glucocorticoid action. Evidence for the presence of glucocorticoid receptor immunoreactivity in various types of classes of peptidergic neurons, *J. Steroid Biochem. Mol. Biol.*, 40, 93, 1991.

76. Akabayashi, A., Watanabe, Y., Gabriel, S. M., Chae, H. J., and Leibowitz, S. F., Hypothalamic galanin-like immunoreactivity and its gene expression in relation to circulating corticosterone, *Mol. Brain Res.*, 25, 305, 1994.

77. Hedlund, P. B., Koenig, J. I., and Fuxe, K., Adrenalectomy alters discrete galanin mRNA levels in the hypothalamus and mesencephalon of the rat, *Neurosci. Lett.*, 170, 77, 1994.

78. Kaplan, L. M., Gabriel, S. M., Koenig, J. I., Sunday, M. E., Spindel, E. R., Martin, J. B., and Chin, W. W., Galanin is an estrogen-inducible, secretory product of the rat anterior pituitary, *Proc. Natl. Acad. Sci. U.S.A.*, 85, 7408, 1988.

79. Leibowitz, S. F., Wang, J., Dourmashkin, J. T., and Yu, H., Galanin gene espression in medial preoptic nucleus of female rats: relation to high-fat diet and gonadal steroids, *Soc. Neurosci. Abstr.* 23, 1075, 1997.

80. Leibowitz, S. F., Xuereb, M., Alexander, J. T., and Wang, J., Body weight regulation in the female rat: role of steroid-responsive and dietary-fat responsive galanin neurons in the medial preoptic area, *Soc. Neurosci. Abstr.* 24, 12, 1998.

81. Levin, M. C. and Sawchenko, P. E., Neuropeptide co-expression in the magnocellular neurosecretory system of the female rat: evidence for differential modulation by estrogen, *Neuroscience*, 54, 1001, 1993.

82. Bonavera, J. J., Dube, M. G., Kalra, P. S., and Kalra, S. P., Anorectic effects of estrogen may be mediated by decreased neuropeptide-Y release in the hypothalamic paraventricular nucleus, *Endocrinology*, 134, 2367, 1994.

83. Shimizu, H., Ohtani, K., Kato, Y., Tanaka, Y., and Mori, M., Estrogen increases hypothalamic neuropeptide Y (NPY) mRNA expression in ovariectomized obese rat, *Neurosci. Lett.*, 204, 81, 1996.

84. Crowley, W. R., Tessel, R. E., O'Donohue, T. L., Adler, B. A., and Kalra, S. P., Effects of ovarian hormones on the concentrations of immunoreactive neuropeptide Y in discrete brain regions of the female rat: correlation with serum luteinizing hormone (LH) and median eminence LH-releasing hormone, *Endocrinology*, 117, 1151, 1985.

85. Fuxe, K., Cintra, A., Agnati, L. F., Harfstrand, A., Wikstrom, A. C., Okret, S., Zoli, M., Miller, L. S., Greene, J. L., and Gustafsson, J. A., Studies on the cellular localization and distribution of glucocorticoid receptor and estrogen receptor immunoreactivity in the central nervous system of the rat and their relationship to the monoaminergic and peptidergic neurons of the brain, *J. Steroid Biochem.*, 27, 159, 1987.

86. Simerly, R. B., Chang, C., Muramatsu, M., and Swanson, L. W., Distribution of androgen and estrogen receptor mRNA-containing cells in the rat brain: an in situ hybridization study, *J. Comp. Neurol.*, 294, 76, 1990.

87. Chavez, M., Riedy, C. A., van Dijk, G., and Woods, S. C., Central insulin and macronutrient intake in the rat, *Am. J. Physiol.*, 271, R727, 1996.

88. Sahu, A., Evidence suggesting that galanin (GAL), melanin-concentrating hormone (MCH), neurotensin (NT), proopiomelanocortin (POMC) and neuropeptide Y (NPY) are targets of leptin signaling in the hypothalamus, *Endocrinology,* 139, 795, 1998.

89. Leibowitz, S. F. and Wang, J., Circulating leptin: specific effects on brain peptides involved in eating and body weight regulation, *Obesity Res.,* 4, 1S, 1996.

90. Tang, C., Akabayashi, A., Manitiu, A., and Leibowitz, S. F., Hypothalamic galanin gene expression and peptide levels in relation to circulating insulin: possible role in energy balance, *Neuroendocrinology,* 65, 265, 1997.

91. Wang, J. and Leibowitz, K. L., Central insulin inhibits hypothalamic galanin and neuropeptide Y gene expression and peptide release in intact rats, *Brain Res.,* 777, 231, 1997.

92. Dube, M. G., Sahu, A., Kalra, P. S., and Kalra, S. P., Neuropeptide Y release is elevated from the microdissected paraventricular nucleus of food-deprived rats: an *in vitro* study, *Endocrinology,* 131, 684, 1992.

93. Brady, L. S., Smith, M. A., Gold, P. W., and Herkenham, M., Altered expression of hypothalamic neuropeptide mRNAs in food-restricted and food-deprived rats, *Neuroendocrinology,* 52, 441, 1990.

94. Schwartz, M. W., Sipols, A. J., Grubin, C. E., and Baskin, D. G., Differential effect of fasting on hypothalamic expression of genes encoding neuropeptide Y, galanin, and glutamic acid decarboxylase, *Brain Res. Bull.,* 31, 361, 1993.

95. Barker-Gibb, M. L. and Clarke, I. J., Increased galanin and neuropeptide-Y immunoreactivity within the hypothalamus of ovariectomised ewes following a prolonged period of reduced body weight is associated with changes in plasma growth hormone but not gonadotropin levels, *Neuroendocrinology,* 64, 194, 1996.

96. Brogan, R. S., Fife, S. K., Conley, L. K., Giustina, A., and Wehrenberg, W. B., Effects of food deprivation on the GH axis: immunocytochemical and molecular analysis, *Neuroendocrinology,* 65, 129, 1997.

97. Wang, J., Akabayashi, A., Yu, H., Dourmashkin, J., Silva, I., Lighter, J., and Leibowitz, S. F., Hypothalamic galanin: control by signals of fat metabolism, *Brain Res.,* 804, 7, 1998.

98. Selvais, P. L., Labuche, C., Nguyen, X. N., Ketelslegers, J. M., Denef, J. F., and Maiter, D. M., Cyclic feeding behaviour and changes in hypothalamic galanin and neuropeptide Y gene expression induced by zinc deficiency in the rat, *J. Neuroendocrinol.,* 9, 55, 1997.

99. Ponsalle, P., Srivastava, L., Unt, R., and White, J. D., Glucocorticoids are required for food-deprivation induced increases in hypothalamic neuropeptide-Y expression, *J. Neuroendocrinol.,* 4, 585, 1992.

100. Malabu, U. H., McCarthy, H. D., McKibbin, P. E., and Williams, G., Peripheral insulin administration attenuates the increase in neuropeptide Y concentrations in the hypothalamic arcuate nucleus of fasted rats, *Peptides,* 13, 1097, 1992.

101. Hanson, E. S., Levin, N., and Dallman, M. F., Elevated corticosterone is not required for the rapid induction of neuropeptide Y gene expression by an overnight fast, *Endocrinology,* 138, 1041, 1997.

102. McKibbin, P. E., McCarthy, H. D., Shaw, P., and Williams, G., Insulin deficiency is a specific stimulus to hypothalamic neuropeptide Y: a comparison of the effects of insulin replacement and food restriction in streptozocin-diabetic rats, *Peptides,* 13, 721, 1992.

103. Schwartz, M. W., Marks, J. L., Sipols, A. J., Baskin, D. G., Woods, S. C., Kahn, S. E., and Porte, D., Jr., Central insulin administration reduces neuropeptide Y mRNA expression in the arcuate nucleus of food-deprived lean (Fa/Fa) but not obese (fa/fa) Zucker rats, *Endocrinology,* 128, 2645, 1991.

104. Menendez, J. A., Atrens, D. M., and Leibowitz, S. F., Metabolic effects of galanin injections into the paraventricular nucleus of the hypothalamus, *Peptides,* 13, 323, 1992.

105. Menendez, J. A., McGregor, I. S., Healey, P. A., Atrens, D. M., and Leibowitz, S. F., Metabolic effects of neuropeptide Y injections into the paraventricular nucleus of the hypothalamus, *Brain Res.,* 516, 8, 1990.

106. Nagase, H., Bray, G. A., and York, D. A., Effect of galanin and enterostatin on sympathetic nerve activity to interscapular brown adipose tissue, *Brain Res.,* 709, 44, 1996.

107. Billington, C. J., Briggs, J. E., Harker, S., Grace, M., and Levine, A. S., Neuropeptide Y in hypothalamic paraventricular nucleus: a center coordinating energy metabolism, *Am. J. Physiol.,* 266, R1765, 1994.

108. Ruffin, M. P., Even, P. C., El-Ghissassi, M., and Nicolaidis, S., Metabolic action of neuropeptide Y in relation to its effect on feeding, *Physiol. Behav.,* 62, 1259, 1997.

109. Akabayashi, A., Zaia, C. T., Silva, I., Chae, H. J., and Leibowitz, S. F., Neuropeptide Y in the arcuate nucleus is modulated by alterations in glucose utilization, *Brain Res.,* 621, 343, 1993.

110. Minami, S., Kamegai, J., Sugihara, H., Suzuki, N., Higuchi, H., and Wakabayashi, I., Central glucoprivation evoked by administration of 2-deoxy-D-glucose induces expression of the c-*fos* gene in a subpopulation of neuropeptide Y neurons in the rat hypothalamus, *Mol. Brain Res.,* 33, 305, 1995.

111. Brown, J., Effects of 2-deoxy-D-glucose on carbohydrate metabolism: review of the literature and studies in the rat, *Metabolism,* 11, 1098, 1962.

112. Kanarek, R. B., Marks-Kaufman, R., Ruthazer, R., and Gualtieri, L., Increased carbohydrate consumption by rats as a function of 2-deoxy-D-glucose administration, *Pharmacol. Biochem. Behav.,* 18, 47, 1983.

113. He, B., White, B. D., Edwards, G. L., and Martin, R. J., Neuropeptide Y antibody attenuates 2-deoxy-D-glucose induced feeding in rats, *Brain Res.,* 781, 348, 1998.

114. Bauche, F., Sabourault, D., Giudicelli, Y., Nordmann, J., and Nordmann, R., Inhibition *in vitro* of acyl-CoA dehydrogenases by 2-mercaptoacetate in rat liver mitochondria, *Biochem. J.,* 215, 457, 1983.

115. Singer, L. K., York, D. A., and Bray, G. A., Macronutrient selection following 2-deoxy-D-glucose and mercaptoacetate administration in rats, *Physiol. Behav.,* 65, 115, 1998.

116. Koegler, F. H. and Ritter, S., Feeding induced by pharmacological blockade of fatty acid metabolism is selectively attenuated by hindbrain injections of the galanin receptor antagonist, M40, *Obesity Res.,* 4, 329, 1996.

117. Flatt, J. P., The difference in the storage capacities for carbohydrate and for fat, and its implications in the regulation of body weight, *Ann. N.Y. Acad. Sci.,* 499, 104, 1987.

118. Thomas, C. D., Peters, J. C., Reed, G. W., Abumrad, N. N., Sun, M., and Hill, J. O., Nutrient balance and energy expenditure during ad libitum feeding of high-fat and high-carbohydrate diets in humans, *Am. J. Clin. Nutr.,* 55, 934, 1992.

119. Wang, S. W., Wang, M., Grossman, B. M., and Martin, R. J., Effects of dietary fat on food intake and brain uptake and oxidation of fatty acids, *Physiol. Behav.,* 56, 517, 1994.

120. Cook, C. B., Shawar, L., Thompson, H., and Prasad, C., Caloric intake and weight gain of rats depends on endogenous fat preference, *Physiol. Behav.,* 61, 743, 1997.

121. Larue-Achagiotis, C., Martin, C., Verger, P., and Louis-Sylvestre, J., Dietary self-selection vs. complete diet: body weight gain and meal pattern in rats, *Physiol. Behav.,* 51, 995, 1992.

122. Akabayashi, A., Koenig, J. I., Watanabe, Y., Alexander, J. T., and Leibowitz, S. F., Galanin-containing neurons in the paraventricular nucleus: a neurochemical marker for fat ingestion and body weight gain, *Proc. Natl. Acad. Sci. U.S.A.,* 91, 10375, 1994.

123. Jhanwar-Uniyal, M., Beck, B., Jhanwar, Y. S., Burlet, C., and Leibowitz, S. F., Neuropeptide Y projection from arcuate nucleus to parvocellular division of paraventricular nucleus: specific relation to the ingestion of carbohydrate, *Brain Res.,* 631, 97, 1993.

124. Wang, J., Akabayashi, A., Dourmashkin, J., Yu, H., Alexander, J. T., Chae, H. J., and Leibowitz, S. F, Neuropeptide Y in relation to carbohydrate intake, corticosterone and dietary obesity, *Brain Res.,* 802, 75, 1998.

125. Warwick, Z. S. and Schiffman, S. S., Role of dietary fat in calorie intake and weight gain, *Neurosci. Biobehav. Rev.,* 16, 585, 1992.

126. Rolls, B. J. and Shide, D. J., The influence of dietary fat on food intake and body weight [published erratum appears in *Nutr. Rev.* 51(1), 31, 1993], *Nutr. Rev.,* 50, 283, 1992.

127. Sclafani, A., Dietary obesity, in *Obesity: Theory and Therapy,* Stunkard, A. J. and Wadden, T. A., Eds., Raven Press, New York, 1993, 125.

128. Dourmashkin, J., Yun, R., Wang, J., Bedrin, Z. U., Zheng, P., and Leibowitz, S. F., Hypothalamic neuropeptide Y (NPY) projection from the arcuate to paraventricular nucleus (ARC-PVN) is activated by the consumption of meals rich in carbohydrate and is associated with changes in circulating glucose, *Soc. Neurosci. Abstr.,* 24, 12, 1998.

129. Castellanos, L., Dourmashkin, J., Yun, R., Kendzior, J., Chang, G. Q., and Leibowitz, S. F., Hypothalamic neurons expressing galanin are stimulated by meals rich in fat and by third ventricular injections of fatty acids, *Soc. Neurosci. Abstr.* 24, 447, 1998.

130. Armstrong, S., A chronometric approach to the study of feeding behavior, *Neurosci. Biobehav. Rev.,* 4, 27, 1980.

131. Jhanwar-Uniyal, M., Beck, B., Burlet, C., and Leibowitz, S. F., Diurnal rhythm of neuropeptide Y-like immunoreactivity in the suprachiasmatic, arcuate and paraventricular nuclei and other hypothalamic sites, *Brain Res.*, 536, 331, 1990.

132. Akabayashi, A., Zaia, C. T., Koenig, J. I., Gabriel, S. M., Silva, I., and Leibowitz, S. F., Diurnal rhythm of galanin-like immunoreactivity in the paraventricular and suprachiasmatic nuclei and other hypothalamic areas, *Peptides,* 15, 1437, 1994.

133. Geiselman, P. J., Martin, J. R., Vanderweele, D. A., and Novin, D., Dietary self-selection in cycling and neonatally ovariectomized rats, *Appetite,* 2, 87, 1981.

134. Marks, D. L., Smith, M. S., Vrontakis, M., Clifton, D. K., and Steiner, R. A., Regulation of galanin gene expression in gonadotropin-releasing hormone neurons during the estrous cycle of the rat, *Endocrinology,* 132, 1836, 1993.

135. Leibowitz, S. F., Lucas, D. J., Leibowitz, K. L., and Jhanwar, Y. S., Developmental patterns of macronutrient intake in female and male rats from weaning to maturity, *Physiol. Behav.,* 50, 1167, 1991.

136. Blum, W. F., Englaro, P., Hanitsch, S., Juul, A., Hertel, N. T., Muller, J., Skakkebaek, N. E., Heiman, M. L., Birkett, M., Attanasio, A. M., Kiess, W., and Rascher, W., Plasma leptin levels in healthy children and adolescents: dependence on body mass index, body fat mass, gender, pubertal stage, and testosterone, *J. Clin. Endocrinol. Metab.,* 82, 2904, 1997.

137. Rossmanith, W. G., Marks, D. L., Clifton, D. K., and Steiner, R. A., Induction of galanin gene expression in gonadotropin-releasing hormone neurons with puberty in the rat, *Endocrinology,* 135, 1401, 1994.

138. Gabriel, S. M., Kaplan, L. M., Martin, J. B., and Koenig, J. I., Tissue-specific sex differences in galanin-like immunoreactivity and galanin mRNA during development in the rat, *Peptides,* 10, 369, 1989.

139. Alexander, J. T., Akabayashi, A., Gabriel, S. M., Thomas, B. E., and Leibowitz, S. F. Galanin and neuropeptide Y immunoreactivity in brain nuclei of female and male rats in relation to puberty, *Soc. Neurosci. Abstr.,* 20, 99, 1994.

140. Sutton, S. W., Mitsugi, N., Plotsky, P. M., and Sarkar, D. K., Neuropeptide Y (NPY): a possible role in the initiation of puberty, *Endocrinology,* 123, 2152, 1988.

26 Opioids, Food Reward, and Macronutrient Selection

Michael J. Glass, Charles J. Billington, and Allen S. Levine

CONTENTS

1 INTRODUCTION

Many of the most burdensome health problems plaguing the industrialized world are linked to increased weight gain, likely related to increased consumption of high-fat energy-dense foods. Endogenous opioids in the brain have been implicated in this phenomenon, particularly as they may be associated with the experience of pleasure. Many have plausibly hypothesized that much overeating is in part due to the rewards of eating.

The endogenous opioid system is composed of a family of three peptides; endorphins — enkephalins, and dynorphins — which interact with three receptor types — mu, delta, and kappa.[1] More recent data add a novel ligand and receptor referred to as orphanin FQ and orphan (ORL1), respectively.[2] A converging body of behavioral, pharmacological, and anatomical data supports the hypothesis that the endogenous opioids regulate feeding behavior.[3-6] For example, there is a close correspondence between brain sites of opioid peptide and receptor synthesis[7,8] and areas that have been implicated in feeding. Support for a structure–function relationship between the brain opioid system and food intake is demonstrated by evidence showing that consumption is altered by injection of opioid agonists[5] and antagonists[6] into areas of the brain thought to mediate feeding behavior. These include the paraventricular nucleus (PVN), nucleus accumbens, nucleus of the solitary tract (NTS), parabrachial nucleus, ventral tegmental area, and amygdala.[5,6] In addition, food intake[9-11] and food restriction or deprivation[12,13] are each associated with changes in opioid peptide levels, and mRNA expression and opioid receptor binding in several of these brain regions implicated in food intake.[14,15] This body of data indicates that manipulations affecting brain opioid transmission alter food consumption, and that food consumption alters levels of brain opioids.

2   THE OPIOID SYSTEM AND FAT INGESTION

One of the earliest hypotheses regarding the role of opioids on food intake suggested that this system selectively modulated consumption of fat. Several decades after Curt Richter' s systematic studies on nutrient self-selection there was a resurgence of studies using his methodology to evaluate the effects of biogenic amines and peptides on macronutrient or diet selection patterns.[16-19] Marks-Kaufman and Kanarek[16-18] reported that acute or chronic morphine stimulated fat intake, while naloxone[19] or naltrexone[20] reduced fat consumption in animals provided with separate sources of fat, carbohydrate, or protein. In addition to these studies, Romsos and co-workers[21] reported that the preferential kappa opioid receptor agonists ketocyclazocine or butorphanol tartrate selectively increased intake of a high-fat diet when offered along with a high-carbohydrate diet.[21]

However, the interpretation that the opioid system is involved in fat consumption is suspect for several reasons. Morphine was typically administered under conditions of 6-h schedule feeding, a procedure which itself may impact nutrient selection since both food deprivation and restriction have been reputed to stimulate intake of fat.[22,23] Fat intake is elevated only at later time points after morphine administration (Figure 1), or in response to a secondary injection regimen that permits acclimation to presumed sedative effects. Further, there is a marked reduction in carbohydrate

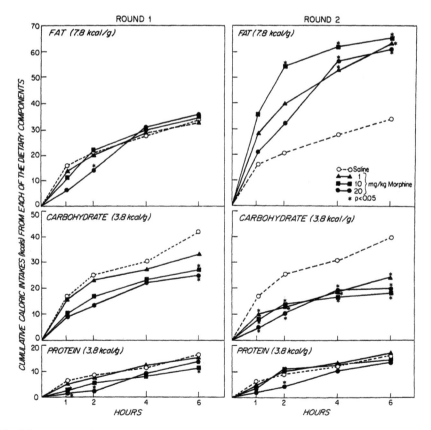

FIGURE 1   Mean cumulative caloric intakes across a 6-h feeding period of fat, carbohydrate, and protein, following (1) saline and the first round of 1.0, 10.0, and 20.0 mg/kg morphine sulfate injections and (2) saline and the second round of 1.0, 10.0, and 20.0 mg/kg morphine sulfate injections. Significantly different from saline injections: *$p < 0.05$. (From Marks-Kaufman, R. and Kanarek, R. B., *Pharmacol. Biochem. Behav.*, 12, 427, 1980. With permission.)

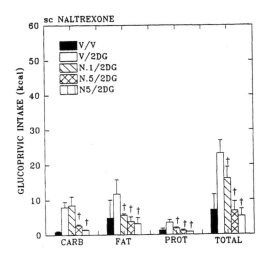

FIGURE 2   Alterations in macronutrient intake (2 h, kcal, ±S.E.M.) following subcutaneous administration of naltrexone in rats pretreated with 2DG (200 mg/kg, i.p.). Significant differences were observed for carbohydrate intake (0.5 h: $F = 5.90$, $p < 0.0005$; 2 h: $F = 15.57$, $p < 0.0001$; 4 h: $F = 14.96$, $p < 0.0001$), fat intake (0.5 h: $F = 4.14$, $p < 0.006$; 2 h: $F = 5.02$, $p < 0.002$; 4 h: $F = 3.85$, $p < 0.008$), protein intake (0.5 h: $F = 5.87$, $p < 0.0006$; 2 h: $F = 6.31$, $p < 0.0003$; 4 h: $F = 6.77$, $p < 0.0002$), and total intake (0.5 h: $F = 10.05$, $p < 0.0001$; 2 h: $F = 18.05$, $p < 0.0001$; 4 h: $F = 16.25$, $p < 0.0001$). 2DG significantly increased all forms of macronutrient intake as well as total intake over the entire time course in this and subsequent groups. In this and all subsequent glucoprivation figures, significant differences (Turkey comparisons, $p < 0.05$) following antagonist treatment relative to vehicle (Veh)-2DG treatment are denoted by crosses. (From Koch, J. E. and Bodnar, R. J., Selective alterations in macronutrient intake of food-deprived or glucoprivic rats by centrally-administered opioid receptor subtype antagonists in rats, *Brain Res.*, 657, 191, 1994. With permission from Elsevier Science.)

consumption (Figure 1). In addition to these issues, empirical reports from Leibowitz's laboratory suggested that morphine increased protein intake.[24,25]

Opioid antagonists have also been shown to suppress macronutrient intake nonselectively. Koch and Bodnar[26] reported that peripherally administered naltrexone decreased intake of fat, carbohydrate and protein in rats stimulated to eat by 2-deoxy-D-glucose (2DG), and decreased consumption of both carbohydrate and fat in food-deprived animals (Figure 2).[26]

Studies in which morphine was given repeatedly indicate that experience with this opiate modulates its effects on nutrient selection.[17] When non-food-restricted rats were given a single daily injection (40 mg/kg) of morphine for 22 days, fat intake was elevated relative to controls only on the last 5 days, while carbohydrate intake was suppressed throughout most of the treatment, and protein intake was not affected.[27] Similarly, in food-restricted rats given a single daily morphine injection (10 mg/kg), fat intake was elevated only on the last 5 days, while carbohydrate intake was suppressed throughout the treatment.[18]

3   THE ROLE OF BASELINE NUTRIENT PREFERENCE

Evans and Vaccarino[28] deprived rats of either a carbohydrate or a protein diet, and then provided a choice between the two. Morphine was associated with increased consumption of the deprived nutrient. Gosnell and co-workers[29] attempted to determine the relationship between baseline diet preferences and morphine-stimulated nutrient intake. Rats were provided with both high-carbohydrate and high-fat diets, and were classified as carbohydrate, fat, or intermediate preferrers based on daily intake patterns. The effect of morphine upon nutrient selection was then evaluated in all groups. At 4 h postinjection, morphine (10 mg/kg) stimulated intake of the preferred nutrient in fat or carbohydrate preferring groups, while nonsignificantly increasing fat and carbohydrate in the intermediate group (Figure 3). These data demonstrate the potential importance of baseline preferences in morphine-stimulated nutrient intake.

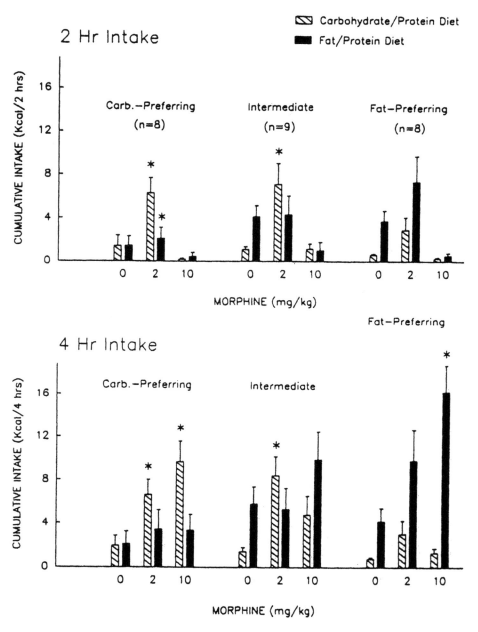

FIGURE 3 Cumulative intakes of carbohydrate/protein and fat/protein diets after injections of morphine 0, 2, or 10 mg/kg. Rats were divided into carbohydrate-preferring, intermediate, and fat-preferring groups on the basis of average daily intakes of the two diets; all rats were tested with the 0, 2, and 10 mg/kg doses. Asterisks indicate significant differences from the corresponding control condition (0 mg/kg) (Dunnett's procedure, one-tailed, $p < 0.05$). (From Gosnell, B. A., Krahn, D. D., and Majchrzak, M. J., The effects of morphine on diet selection are dependent upon baseline diet preferences, *Pharmacol. Biochem. Behav.*, 37, 210, 1990. With permission from Elsevier Science.)

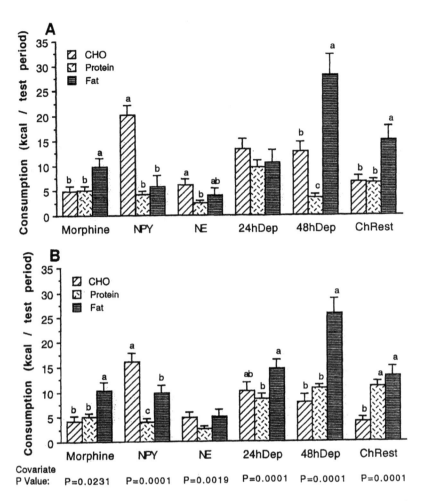

FIGURE 4 Effect of morphine, NPY, NE, 24hDep, 48hDep, and ChRest on macronutrient selection (regimen I). Intake of carbohydrate (CHO), protein, and fat diets is depicted as observed values (A) and as kilocalories adjusted for covariate, baseline preference (B). Note that test period varied (morphine 6 h, NPY 4 h, NE 2 h, 24hDep 4 h, 48hDep 4 h, and ChRest 4 h). Means within an experimental procedure with no common superscript differ significantly ($p < 0.05$). (From Welch, C. C., Grace, M. K., Billington, C. J., and Levine, A. S., *Am. J. Physiol.*, 266, R429, 1994. With permission.)

In a further demonstration of the impact of baseline preferences, Welch et al.[30] investigated the effect of food deprivation and various orexigenic agents on nutrient selection in animals receiving individual macronutrients or complete diets high in fat or carbohydrate. Morphine, 48-h food deprivation, and chronic food restriction (80% of *ad libitum* intake) all increased fat selection relative to carbohydrate and protein, while neuropeptide Y (NPY) and norepinephrine (NE) increased carbohydrate selection relative to fat and protein (Figure 4). Analysis of covariance, using baseline preference (calculated as the mean of 3 days of consistent intake) as the covariate, showed baseline selection patterns were significantly altered for certain treatments. For example, fat intake was significantly greater than carbohydrate and protein in chronically food-restricted rats; however, the difference between fat and protein disappears when accounting for preference (Figure 4). In all cases baseline preference (the covariate) was significant, indicating that macronutrient preference is an important contributor to drug-stimulated diet selection (Figure 4).

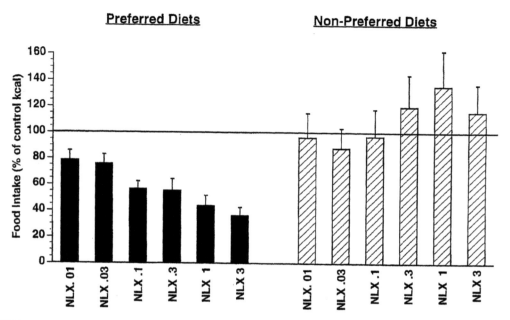

FIGURE 5   Effect of naloxone (NLX) on 24-h deprivation-induced intake of preferred and nonpreferred diets. Data are expressed as %kcal ingested by the control (0 mg/kg naloxone) group. Filled bars indicate significant differences ($p < 0.05$) compared with control (100%) values. (From Glass, M. J., Grace, M., Cleary, J. P., Billington, C. J., and Levine, A. S., *Am. J. Physiol.*, 271, R219, 1996. With permission.)

Differences in diet preferences could potentially explain anomalies in the literature. For example, in the Welch study the results without covariate analysis indicate no specific effect upon nutrient selection in 24-h food-deprived animals, which is contrary to some reports indicating that 24-h food deprivation increases fat intake.[31] However, when baseline preference is taken into account, fat intake was significantly elevated relative to carbohydrate and protein (Figure 4). It is possible that discrepancies in particular treatment effects may result from confounding nutrient preferences between animals.

Baseline preferences have been shown to affect naloxone suppression of diet intake in 24-h food-deprived rats.[32] Rats were provided with a choice between a high-fat and high-carbohydrate diet, and baseline preferences were determined as the mean of 3 days of consistent intake. The effects of acute administration of naloxone were then determined. Intake from preferred and nonpreferred diets were pooled into separate categories and analyzed. At doses as low as 0.01 mg/kg, naloxone significantly suppressed intake of the preferred diet relative to the nonpreferred diet (Figure 5), indicating the potent effect of this agent on consumption of preferred foods.

4   OPIOIDS AND FOOD REWARD

Evidence derived from nutrient or diet selection methods suggests that opioid agonists or antagonists affect consumption of the preferred diet or nutrient, rather than selectively influencing fat intake. This interpretation of nutrient/diet selection studies seems consistent with certain data derived from examination of the relationship between the opioid system and food reward. This evidence suggests that opioids play a special role in feeding behavior by influencing processes involved in the integration of internal and external food-related stimuli. Rewarding processes have been thought to play a role in sustaining the activity of behavioral systems which maintain feeding behavior once it has begun and may be associated with the pleasure derived from eating, an area in which opioids might have some influence. Long associated with the pleasant qualities of drugs of abuse, the opioids in some brain sites may mediate pleasure derived from food, and maintain feeding once initiated. Evidence supporting this idea includes the ability of opioid antagonists such as naloxone or naltrexone to inhibit maintenance rather than initiation of feeding, enhanced opioid antagonist potency in reducing intake of palatable foods, and heightened effects in situations which combine metabolic and hedonic events.

In an observational analysis of meal patterns in rats, Kirkham and Blundell[33] showed that naloxone failed to reduce initiation of food intake, while hastening the development of satiety. Similarly, naltrexone had little effect on latency or goal box running speed in animals running a straight-alley maze for reinforcement, while decreasing the amount of food eaten in the goal box after some contact with food.[34] In an operant version of the previous study, rats worked for food pellets by pressing a lever 80 times (fixed ratio 80 [FR80], initiation phase) to obtain the first pellet, while each subsequent pellet was earned by 10 lever presses (FR10, maintenance phase). Naloxone had no effect on the effort required to earn the first pellet, but inhibited pressing after some food had been earned. Beczkowska et al.[35,36] conducted a series of studies employing an analysis of fluid consumption over 5 min intervals for a 1-h period. Both general and selective opioid antagonists were administered, and the overall pattern was consistent with the results found in the above operant studies; that is, only the maintenance phase of fluid consumption was affected.

In a related vein, naloxone is less effective in reducing responding on the progressive ratio schedule (PR), a test purportedly measuring the motivational value of food and other reinforcers.[37] Subjects on the PR must emit an ever-increasing number of responses to earn successive food pellets. The final ratio completed is termed the break point, which indicates the motivational strength of the reinforcer. The effect of naloxone on break point for various sucrose concentrations declined as the sucrose concentration increased. With 10% sucrose, naloxone failed to reduce the break point even at a dose of 10 mg/kg (Figure 6). On the other hand, naloxone robustly reduced consumption of 10% sucrose when given free access to it from a sipper tube (Figure 7). A similar pattern was observed in genetically obese Zucker rats, where naloxone was more effective in reducing free consumption of food pellets compared with responding on a PR-3 schedule of reinforcement.[66] These results indicate that the efficacy of naloxone in reducing motivated behavior can be overcome by potent reinforcers, while it strongly affects free consumption of foods.

One potential explanation for the seeming disparity of naloxone's effects on free feeding compared with working for foods is that only small amounts of food are ingested during operant tests. There may be a relationship between naloxone effects and postingestive processes leading to early satiety. This hypothesis would have to contend with evidence showing that opioid antagonists reduce sham feeding, which minimizes postabsorptive events. Opioid antagonists reduce sucrose sham feeding in rats,[38] and attenuation of sucrose sham feeding in response to naloxone,[38] beta-funaltrexamine (FNA) (mu-receptor antagonist), or nor-binaltorphimine (BNI) (kappa-receptor antagonist)[39] resembles dilution of sucrose concentration. Opioid antagonists also decrease intake of noncaloric solutions such as saccharin[40] and sodium chloride.[41] However, as oral factors have been implicated in regulation of intake of familiar foods in real and sham-fed animals, the evidence may point to opioid involvement in processes whereby oral signals regulate food intake. Overall, the results of tests of break points and food intake represent an important dissociation of the effects of naloxone on feeding behavior, although the reasons for this are unclear.

Reduced efficacy of opioid antagonist administration in rats performing for small amounts of sucrose or sweet pellets presents an interesting contrast to studies showing that naloxone has a more robust effect on free consumption of sweet foods. Opioid antagonists inhibit intake of high fat, sucrose, and cafeteria diets, and robustly reduce sweet ingestates compared with nonsweet foods.[42,43] Levine et al.[44] tested rats under a series of energy conditions including acute 24- and 48-h food deprivation, chronic food restriction, and a restricted access feeding schedule. Rats were provided either plain or sweet chow and food intake measured in response to naloxone treatment. During acute deprivation, both regular and sweet chow were significantly suppressed, but reductions in sweet chow were more robust (Figure 8). However, during chronic restriction only sweet chow

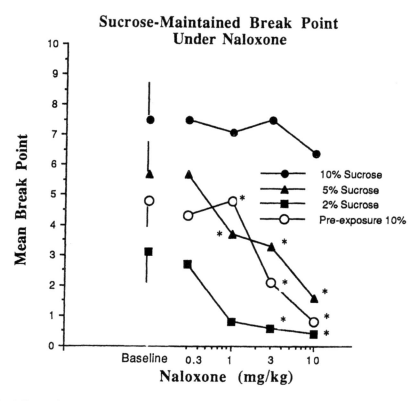

FIGURE 6 Effects of naloxone dose on break points under different sucrose concentrations and following 5 h of free access in the home cage. Baseline values are derived from the data under saline injections. Lines associated with baseline points are equal to 1 S.E.M. Asterisks indicate significant differences from break points under saline control injections. (From Cleary, J. P., Weldon, D. T., O'Hare, E., Billington, C. J., and Levine, A. S., *Psychopharmacology*, 126, 110, 1996. With permission.)

ingestion was significantly suppressed (Figure 8). These results indicate that the effects of naloxone are most pronounced under conditions that combine chronic metabolic challenge with sweet foods.

Changes in the brain opioid system occur in response to consumption of sweet and high-fat foods. Rats fed a fat-sucrose diet display increased hypothalamic levels of proDynorphin mRNA and peptide in the arcuate nucleus, and paraventricular nucleus respectively.[11] Further, dynorphin binds preferentially to the kappa receptor, and intrahypothalamic injections of the selective kappa receptor antagonist nor-BNI inhibits sucrose drinking.[45] This set of findings suggests that a functional relationship exists between hypothalamic opioids and intake of palatable food intake.

In contrast to palatable diet studies, energy restriction decreases opioid mRNA and peptide levels[12,13] in the arcuate and paraventricular nuclei, and alters opioid receptor binding.[14,15] Such changes could reflect arcuate opioid involvement in energy regulation or may be related to an integration of both energy and reward. In humans food pleasantness ratings are elevated in food-deprived people.[46] Food-restricted rats consume more of a sucrose diet relative to a cornstarch diet, consistent with an interaction between energy deprivation and sweet food consumption.[47] Changes in rat brain opioid receptor binding occur in areas which have been implicated in both regulatory feeding and taste.[14,15] Such changes may play a role in hedonic reactions to foods responding to metabolic challenge,[48,49] and increased sweet food consumption relative to nonsweet foods by food-restricted rats.

Thus far, the data reviewed in this section indicate that naloxone may reduce palatable food consumption by affecting taste discrimination. To test this hypothesis rats were trained to discriminate 10% sucrose from water in a two-lever operant chamber.[50] Sucrose was chosen because of

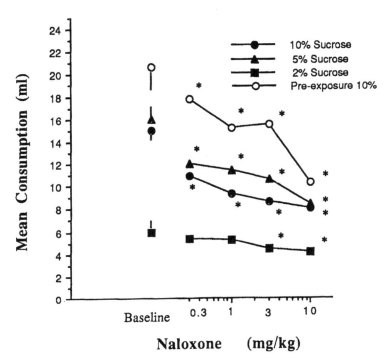

FIGURE 7   Effects of naloxone on sucrose consumption, during the first 0.5 h of unrestricted access, under 2, 5, and 10% sucrose concentrations. Baseline values are derived from the data under saline injections. Lines associated with baseline points are equal to 1 S.E.M. Asterisks indicate significant differences from consumption under saline control injections. (From Cleary, J. P., Weldon, D. T., O'Hare, E., Billington, C. J., and Levine, A. S., *Psychopharmacology*, 126, 110, 1996. With permission.)

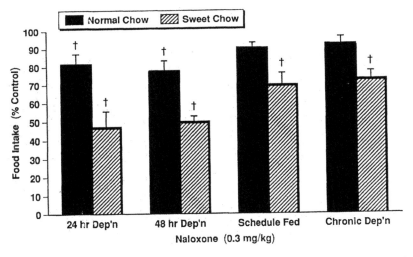

FIGURE 8   Effect of naloxone (0.3 mg/kg) on normal and sweet chow intake (%control) in food-deprived and schedule-fed rats. $†P < 0.05$ compared with vehicle-injected group (0 mg/kg). (From Levine, A. S., Weldon, D. T., Grace, M., Cleary, J. P., and Billington, C. J., *Am. J. Physiol.*, 268, R250, 1995. With permission.)

TABLE 1

Effect of Naloxone on Sucrose Concentration Discrimination at 10% Sucrose Training Sucrose Concentration (n = 5) and 5% Sucrose Training Concentration (n = 6)

Naloxone Dose (mg/kg)	Sucrose Concentration (%)						
	10	1.0	0.5	0.1	0.05	0.01	0.005
	10% sucrose training concentration						
0	100 ± 0	80 ± 13	80 ± 13	60 ± 16	50 ± 16	30 ± 13	20 ± 13
0.1	100 ± 0	100 ± 0	80 ± 13	50 ± 16	30 ± 15	40 ± 16	30 ± 15
0.3	100 ± 0	90 ± 10	80 ± 13	50 ± 16	50 ± 16	30 ± 15	30 ± 15
1.0	100 ± 0	100 ± 0	70 ± 15	70 ± 15	40 ± 16	20 ± 13	20 ± 13
3.0	100 ± 0	90 ± 10	60 ± 16	60 ± 16	30 ± 15	30 ± 15	40 ± 16
	5% Sucrose Training Concentration						
0	100 ± 0	100 ± 0	83 ± 11	77 ± 13	58 ± 14	24 ± 13	33 ± 14
0.1	100 ± 0	100 ± 0	77 ± 13	66 ± 14	49 ± 15	24 ± 13	24 ± 13
0.3	100 ± 0	100 ± 0	77 ± 13	83 ± 11	66 ± 14	41 ± 14	24 ± 13
1.0	100 ± 0	100 ± 0	74 ± 13	74 ± 13	49 ± 15	24 ± 13	24 ± 13
3.0	100 ± 0	100 ± 0	83 ± 11	74 ± 13	49 ± 15	33 ± 14	8 ± 13

From O'Hare, E., Cleary, J. P., Billington, C. J., and Levine, A. S., *Psychopharmacology*, 129, 289, 1997. With permission.

the potent ability of naloxone to reduce intake of this nutrient. The rats learned to press one lever upon receipt of 10% sucrose and the other lever when presented water. After rats displayed consistent discrimination between 10% sucrose and water, they were then tested under a series of sucrose concentrations (0.005 to 1.0%) to generate a generalization curve. As expected, the ability to discriminate sucrose was a function of sucrose concentration (Table 1). Naloxone failed to affect sucrose discrimination (Table 1). These results were not a function of the training concentration of sucrose, because in another group trained under 5% sucrose, naloxone again failed to alter sucrose discrimination (Table 1). These data suggest that taste perception per se is not affected by opiates, at least in the case of sucrose.

If opioids do not affect sweet taste perception, then postsensory processing may be modulated. Opioid antagonists have been shown to play a role in the hedonic response to solutions, as measured by the taste reactivity test. This evidence has been derived by examing the patterns of stereotypic orofacial responses emitted by rats in reaction to passive oral infusions of sucrose, quinine, or sucrose–quinine solutions.[49,51] Rats display what have been termed ingestive reactions (i.e., lateral tongue protrusion, paw licks) in response to passive oral infusions of solutions that are consumed readily by rats and which taste sweet to humans, such as sucrose. Rats also display what are termed aversive reactions (i.e., headshakes, forelimb flails) in response to oral infusions of solutions that are avoided by rats and which taste bitter to humans, such as quinine.[49,51] A hedonic response can be thought of as representing a hypothetical neural state, which reflects the integration of interoceptive signals related to metabolic state, exeroceptive signals from the environment, and experience.[49,51] Integration of these factors under particular conditions determines the hedonic reaction to stimuli.[49,51] Naltrexone reduces only ingestive reactions in response to an infusion of sucrose,[52] but does not modify ingestive or aversive reactions resulting from infusion of a quinine solution.[52] Such data appear to be consistent with reports that naltrexone reduces food pleasantness ratings in humans.[53]

One of the most notable of the anorectic effects of naloxone is its ability to preferentially reduce consumption of "sweet" carbohydrates in chronically restricted rats. For example, food intake was reduced in both cornstarch- and sucrose-fed rats fed *ad libitum* (Figure 9), while sucrose but not

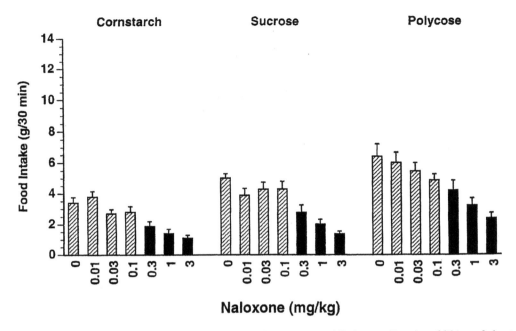

FIGURE 9 Effect of naloxone on intake of cornstarch, sucrose, and Polycose diets in *ad libitum*-fed rats. Filled bars indicate significant ($p < 0.05$) differences compared with the control group (0 mg/kg naloxone). (From Weldon, D. T., O'Hare, E., Cleary, J., Billington, C. J., and Levine, A. S., *Am. J. Physiol.*, 270, R1184, 1996. With permission.)

cornstarch intake was reduced in chronically restricted rats (85% free-fed body weight) (Figure 10). The interaction between carbohydrate type and food restriction was further investigated using an automated meal pattern analysis. A minimum amount of food ingestion or time appeared necessary before naloxone suppressed food intake independent of carbohydrate type (unpublished observations). Perhaps opioids released during an encounter with palatable food in a "hungry" rat results in meal maintenance. Opioid receptor blockade with naloxone would then interrupt meal maintenance.

5   WHAT DOES IT ALL MEAN?

The anorectic profile of the effects of opioid antagonists on feeding includes decreased feeding duration and reduction of hedonic reactions by rats, without influencing sweet taste perception. Naloxone or naltrexone effects appear similar to the psychological phenomenon of alliesthesia, a term used to denote changes in hedonic ratings by humans in response to metabolic and consummatory manipulations.[46] Caloric loads at the start of testing decrease pleasantness ratings for foods in humans,[46] and decrease hedonic reactions[48] and meal size[54] in rats. Thus, opioid antagonists seem to affect feeding similar to administered nutrients. It is unclear whether each respective treatment acts by common or dissimilar mechanisms. Attenuation of hedonic responses and early food intake termination by naloxone or naltrexone may reflect modulation of systems relating food-associated (both taste and postingestional) and interoceptive (metabolic) stimuli.

One problem raised by the pattern of effects induced by opioid antagonists is that hedonic reactions are affected rather early in testing (as early as 1 min, although stronger effects occur at around 10 min[52]), yet these agents do not alter food intake initiation. If palatability is a crucial feeding determinant, as has been traditionally hypothesized, then evidence reviewed here suggests that early palatability reduction is not sufficient to manifest in reduced food consumption. However, this interpretation must be tempered by differences in testing conditions between these studies (i.e., deprivation state, time of testing, strain and age of animals, etc.). A more important issue concerns

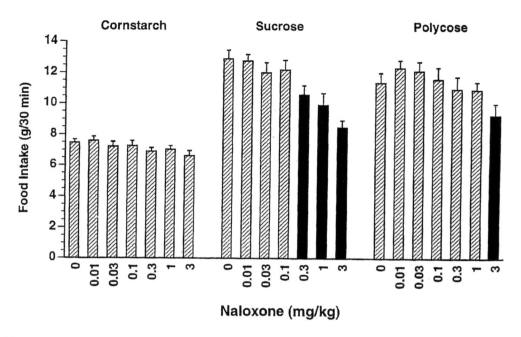

FIGURE 10   Effect of naloxone on intake of cornstarch, sucrose, and Polycose diets in food-restricted rats. Filled bars indicate significant ($p < 0.05$) differences compared with the control group (0 mg/kg naloxone). (From Weldon, D. T., O'Hare, E., Cleary, J., Billington, C. J., and Levine, A. S., *Am. J. Physiol.*, 270, R1186, 1996. With permission.)

experimental paradigms employed, which encompass a wide range of procedures, each of which may reflect distinct underlying processes. For example, food consumption tests measure a conflation of behaviors including approach, manipulation, and mastication, as well as conditional and unconditional responses to the environment and to food-related stimuli. The taste reactivity test measures classes of orofacial patterns in response to tastants within the context of greater experimental control. The variable measured by each test may reflect unrelated processes or differential sensitivity to drug effects.

It may be plausible to relate the findings from nutrient selection studies to the ideas developed here. Evidence indicates that opioids affect consumption of preferred nutrients. Food preference may reflect differences in hedonic responses to each. Opioids may modulate hedonic reactions to relevant stimuli, such as nutrients and diets. Diminished reward could account for decreases in preferred food consumption.

6   OPIOID INTERACTIONS WITH BRAIN APPETITE CIRCUITRY

Opioids in key brain sites may play a specific role in energy balance control circuitry by modulating state and sensory processes associated with affect and feeding maintenance. It may be useful to integrate these actions with other features of the distributed neural mechanisms that regulate food intake and energy balance. One working model that has received empirical support hypothesizes that a critical locus of energy balance coordination lies within an ARC–PVN (arcuate nucleus–paraventricular nucleus of the hypothalamus) circuit.[55] Within this circuitry NPY may be a crucial neuroregulatory signal that initiates a cascade of events which controls energy regulation, including lowered brown fat metabolism, increased white fat storage, as well as increased energy input through food intake.[56] Part of this coordinative action may engage other circuitry including PVN-sympathetic (SNS) pathways related to alterations in peripheral metabolism, the nucleus of the solitary tract (NTS), and the amygdala. The NTS and amygdala also receive sensory information from the taste system and gastrointestinal (GI) tract.[57] Whether the NTS and amygdala are directly activated by NPY action in the PVN enacted by sensory stimulation, energy changes, or a combination of both is unclear. Opioids appear to participate in food and energy control circuitry linking the hypothalamic PVN, the NTS, and amygdala. Small doses of naltrexone in the NTS block changes in food intake, but not energy metabolism elicited by NPY given into the PVN.[58] Larger doses of naltrexone in NTS blocks both food intake and changes in peripheral energy metabolism. Naltrexone administered into the amygdala also attenuates food consumption stimulated by NPY given into the PVN[59] and patterns of c-FOS in the amygdala induced by PVN NPY are affected by peripheral naloxone treatment.[60] The effect of opioid antagonists administered into these regions on nutrient selection and food reward needs to be evaluated. Whether or not this hypothetical neural system accurately represents feeding circuitry will only be revealed by future investigation.

7 MACRONUTRIENT CLASSIFICATION: A FURTHER COMPLICATION

Within this chapter, and in the literature generally, nutrients have been discussed with little concern for their differences. While the designations fat and carbohydrate are useful taxonomic devices, there is much variation in the properties of nutrients within a class. For example, carbohydrates differ in terms of their component saccharides (i.e., glucose, fructose), length of saccharide chain (mono-, di-, poly-), and in the case of starches the degree of branching of glucose polymers (amylose, amylopectin). Carbohydrates differ in various metabolic parameters including cephalic and postcephalic insulin release, rate of absorption, uptake and utilization by tissue, and lipogenic capacity.[61-64] Further, carbohydrates differ in their sensory characteristics. Mono- and disaccharides taste sweet to humans, while starches do not, and this distinction may also be made by other species such as the rat.[64,65] These animals display different preferences for sugar or starches in solution[64] or when presented in diet form,[47] and a conditioned taste aversion to starch does not generalize to simple sugars.[64] These results suggest that there may be different receptor systems for simple and complex carbohydrates. Similar differences in lipids and proteins exist. For example, lipids can be liquid or solid at room temperature due to the degree of saturation. Highly unsaturated lipids can become rancid resulting in off-flavors. Proteins can be complete or incomplete; that is, they may have all essential amino acids or be limiting in one or more essential amino acids. To further complicate matters, macronutrient texture and taste can be altered by cooking and processing.

Despite these differences rarely have subtypes of macronutrients been considered as important variables in their own right within nutrient selection studies. This fact raises questions concerning the degree of generalizability of nutrient regulation studies and raises the possibility that the effects of a treatment may be restricted to the particular class of carbohydrate, protein, or fat. Perhaps sweetness or crispness are more important than the macronutrient type. For example, we noted that ventricular NPY administration resulted in preference for carbohydrate over fat when the carbohydrate source was sucrose and the fat source was vegetable shortening. However, when offered a choice between cornstarch and vegetable shortening NPY-injected rats ate more of the vegetable shortening.[67]

8 CONCLUSION

Nutrient selection patterns represent a complex behavioral process which is likely the outcome of diverse neural circuitry related to metabolic, endocrine, and behavioral systems, each of which can be influenced by environmental factors. While the opioid system has been implicated in nutrient intake, this evidence has been largely based on a limited set of experimental situations which include the implementation of classic self-selection methodology, particular nutrient regimens, and peripheral administration of agents. The latter experimental constraint is crucial as the opioid system is widely distributed within the brain, and plays a neuromodulatory role within diverse circuits related to a variety of functions. Given this behavioral, neurochemical, and anatomical intricacy, it is certainly questionable whether opioids would be expected to affect consumption of fat across different neural sites, peptide and receptor types, and different testing conditions. Indeed, the somewhat confusing results in this literature may very well be the result of the complex nature of the system and the blunt tools that have, heretofore, been used to measure its workings. In order to address these issues expansion of methodology is needed, which includes administration of drugs into different brain regions, manipulation of specific receptor subtypes, variation of nutrient source as an important independent variable, and use of behavioral tests which tap into important learning and affective processes related to nutrient intake such as nutrient conditioning and taste reactivity.

While the evidence seems consistent with a role for opioids in food reward, a more accurate description of its role in feeding behavior, macronutrient selection, and feeding pathways awaits inquiry. It does, however, seem unlikely that opioids result in the selection of one macronutrient independent of taste and texture. Sucrose and cornstarch may be related in the test tube, but not in the palate.

REFERENCES

1. Akil, H., Watson, S. J., Young, E., Lewis, M. E., Khachaturian, H., and Walker, J. M., Endogenous opioids: biology and function, *Annu. Rev. Neurosci.*, 7, 223, 1984.
2. Rainer, K., Reinscheid, R. K., Nothacker, H. P., Bourson, A., Ardati, A., Henningsen, R. A., Bunzow, J. R., Grandy, D. K., Langen, H., Monsma, F. J., and Civelli, O., Orphanin FQ: a neuropeptide that activates an opioidlike G protein-coupled receptor, *Science*, 270, 792, 1995.
3. Morley, J. E., Levine, A. S., Yim, G. K., and Lowy, M. T., Opioid modulation of appetite, *Neurosci. Biobehav. Rev.*, 7, 281, 1983.
4. Levine, A. S. and Billington, C. J., Opioids. Are they regulators of feeding? *Ann. N.Y. Acad. Sci.*, 575, 209, 1989.
5. Gosnell, B. A. and Levine, A. S., Stimulation of ingestive behavior by preferential and selective opioid agonists, in *Drug Receptor Subtypes and Ingestive Behavior*, Cooper, S. J. and Clifton, P. G., Eds., Academic Press, London, 1996, 147.
6. Bodnar, R. J. Opioid receptor subtype antagonists and ingestion, in *Drug Receptor Subtypes and Ingestive Behavior*, Cooper, S. J. and Clifton, P. G., Eds., Academic Press, London, 1996, 127.
7. Mansour, A., Khachaturian, H., Lewis, M. E., Akil, H., and Watson, S. J., Anatomy of CNS opioid receptors, *Trends Neurosci.*, 11, 308, 1988.
8. Mansour, A., Fox, C. A., Akil, H., and Watson, S. J., Opioid-receptor mRNA expression in the rat CNS: anatomical and functional implications, *Trends Neurosci.*, 18, 22, 1995.
9. Dum, J., Gramsch, C., and Herz, A., Activation of hypothalamic beta-endorphin pools by reward induced by highly palatable food, *Pharmacol. Biochem. Behav.*, 18, 443, 1983.
10. Brennan, G., Bachus, S. E., and Jhanwar-Uniyal, M., Involvement of gene expression for hypothalamic dynorphin in dietary preference, *J. Neurosci.*, 20, 818, 1994.
11. Welch, C. C., Kim, E., Grace, M. K., Billington, C. J., and Levine, A. S., Palatability-induced hyperphagia increases hypothalamic dynorphin peptide and mRNA levels, *Brain Res.*, 721, 126, 1996.
12. Brady, L. S., Smith, M. A., Gold, P. W., and Herkenham, M., Altered expression of hypothalamic neuropeptide mRNA's in food-restricted and food-deprived rats, *Neuroendocrinology*, 52, 441, 1990.
13. Kim, E., Welch, C. C., Grace, M. K., Billington, C. J., and Levine, A. S., Chronic food restriction and acute food deprivation decrease mRNA levels of opioid peptides in arcuate nucleus, *Am. J. Physiol.*, 270, R1019, 1996.
14. Wolinsky, T. D., Carr, K. D., Hiller, J. M., and Simon, E. J., Effects of chronic food restriction on mu and kappa opioid binding in rat forebrain: a quantitative autoradiographic study, *Brain Res.*, 656, 274, 1994.

15. Wolinsky, T. D., Carr, K.D., Hiller, J. M., and Simon, E. J., Chronic food restriction alters mu and kappa opioid binding in the parabrachial nucleus of the rat: a quantitative autoradiographic study, *Brain Res.*, 706, 333, 1996.

16. Marks-Kaufman, R. and Kanarek, R. B., Morphine selectively influences macronutrient intake in the rat, *Pharmacol. Biochem. Behav.*, 12, 427, 1980.

17. Marks-Kaufman, R., Increased fat consumption induced by morphine administration in rats, *Pharmacol. Biochem. Behav.*, 16, 949, 1982.

18. Marks-Kaufman, R. and Kanarek, R. B., Diet selection following a chronic morphine and naloxone regimine. *Pharmacol. Biochem. Behav.*, 35, 665, 1990.

19. Marks-Kaufman, R. and Kanarek, R. B., Modifications in nutrient selection induced by naloxone in the rat, *Psychopharmacology*, 74, 321, 1981.

20. Marks-Kaufman, R., Plager, A., and Kanarek, R. B., Central and peripheral contributions of endogenous opioid systems to nutrient selection in rats, *Psychopharmacology*, 85, 414, 1985.

21. Romsos, D. R., Gosnell, B. A., Morley, J. E., and Levine, A. S., Effects of kappa opiate agonists, cholecystokinin, bombesin on intake of diets varying in carbohydrate-to-fat ratio in rats, *J. Nutr.*, 117, 976, 1987.

22. Piquard, F., Schaefer, A., and Haberey, P., Influence of fasting and protein deprivation on food self-selection in the rat, *Physiol. Behav.*, 20, 771, 1978.

23. Bligh, M. E., DeStefano, M. B., Kramlik, S. K., Douglass, L. W., Dubuc, P., and Castonguay, T. W., Adrenal modulation of the enhanced fat intake subsequent to fasting, *Physiol. Behav.*, 48, 373, 1990.

24. Bhakthavatsalam, P. and Leibowitz, S. F., Morphine-elicited feeding: diurnal rhythm, circulating corticosterone and macronutrient selection, *Pharmacol. Biochem. Behav.*, 24, 911, 1986.

25. Shor-Posner, G., Azar, A. P., Filart, R., Tempel, D., and Leibowitz, S. F., Morphine-stimulated feeding: analysis of macronutrient selection and paraventricular nucleus lesions, *Pharmacol. Biochem. Behav.*, 24, 931, 1986.

26. Koch, J. E. and Bodnar, R. J., Selective alterations in macronutrient intake of food-deprived or glucoprivic rats by centrally-administered opioid receptor subtype antagonists in rats, *Brain Res.*, 657, 191, 1994.

27. Ottaviani, R. and Riley, A. L., Effect of chronic morphine administration on the self-selection of macronutrients in the rat, *Nutr. Behav.*, 2, 27, 1984.

28. Evans, K. R. and Vaccarino, F. J., Amphetamine-and morphine-induced feeding: evidence for involvement of reward mechanisms, *Neurosci. Biobehav. Rev.*, 14, 2, 1990.

29. Gosnell, B. A., Krahn, D. D., and Majchrzak, M. J., The effects of morphine on diet selection are dependent upon baseline diet preferences, *Pharmacol. Biochem. Behav.*, 37, 207, 1990.

30. Welch, C. C., Grace, M. K., Billington, C. J., and Levine, A. S., Preference and diet type affect macronutrient selection after morphine, NPY, norepinephrine, and deprivation, *Am. J. Physiol.*, 266, R426, 1994.

31. Schutz, H. G. and Pilgrim, F. J., Changes in the self-selection pattern for purified dietary components by rats after starvation, *J. Comp. Physiol. Psychol.*, 47, 444, 1954.

32. Glass, M. J., Grace, M., Cleary, J. P., Billington, C. J., and Levine, A. S., Potency of naloxone's anorectic effect in rats is dependent on diet preference, *Am. J. Physiol.*, 271, R217, 1996.

33. Kirkham, T. C. and Blundell, J. E., Dual action of naloxone on feeding revealed by behavioral analysis: separate effects on initiation and termination of eating, *Appetite*, 5, 45, 1984.

34. Kirkham, T. C. and Blundell, J. E., Effects of naloxone and naltrexone on the development of satiation measured in the runway: comparisons with d-amphetamine and d-fenfluramine, *Pharmacol. Biochem. Behav.*, 25, 123, 1986.

35. Beczkowska, I. W., Bowen, W. D., and Bodnar, R. J., Central opioid receptor subtype antagonists differentially alter sucrose and deprivation-induced water intake, *Brain Res.*, 589, 291, 1992.

36. Beczkowska, I. W., Koch, J. E., Bostock, M. E., Leibowitz, S. F., and Bodnar, R. J., Central opioid receptor subtype antagonists differentially reduce intake of saccharin and maltose dextrin solutions in rats, *Brain Res.*, 618, 261, 1993.

37. Cleary, J. P., Weldon, D. T., O'Hare, E., Billington, C. J., and Levine, A. S., Naloxone effects on sucrose-motivated behavior, *Psychopharmacology*, 126, 110, 1996.

38. Kirkham, T. C. and Cooper, S. J., Naloxone attenuation of sham feeding is modified by manipulation of sucrose concentration, *Physiol. Behav.*, 44, 491, 1988.

39. Leventhal, L., Kirkham, T. C., Cole, J. L., and Bodnar, R. J., Selective actions of central mu and kappa opioid antagonists upon sucrose intake in sham-feeding rats, *Brain Res.*, 685, 205, 1995.
40. Cooper, S. J., Effects of opiate agonists and antagonists on fluid intake and saccharin choice in the rat, *Neuropharmacology*, 22, 323, 1983.
41. Bodnar, R. J., Glass, M. J., and Koch, J. E., Analysis of central opioid receptor subtype antagonism of hypotonic and hypertonic saline intake in water-deprived rats, *Brain Res. Bull.*, 36, 293, 1995.
42. Levine, A. S., Murray, S. S., Kneip, J., Grace, M., and Morley, J. E., Flavor enhances the antidipsogenic effect of naloxone, *Physiol. Behav.*, 28, 23, 1982.
43. Giraudo, S. Q., Grace, M. K., Welch, C. C., Billington, C. J., and Levine, A. S., Naloxone's anorectic effect is dependent upon the relative palatability of food, *Pharmacol. Biochem. Behav.*, 46, 917, 1993.
44. Levine, A. S., Weldon, D. T., Grace, M., Cleary, J. P., and Billington, C. J., Naloxone blocks that portion of feeding driven by sweet taste in food-restricted rats, *Am. J. Physiol.*, 268, R248, 1995.
45. Koch, J. E., Glass, M. J., Cooper, M. L., and Bodnar, R. J., Alterations in deprivation, glucoprivic and sucrose intake following general, mu and kappa opioid antagonists in the hypothalamic paraventricular nucleus of rats, *Neuroscience*, 66, 951, 1995.
46. Cabanac, M., Physiological role of pleasure, *Science*, 173, 1103, 1971.
47. Weldon, D. T., O'Hare, E., Cleary, J., Billington, C. J., and Levine, A. S., Effect of naloxone on intake of cornstarch, sucrose, and polycose diets in restricted and non-restricted rats, *Am. J. Physiol.*, p. R1183, 1996.
48. Berridge, K. C., Modulation of taste by affect by hunger, caloric satiety, and sensory-specific satiety in the rat, *Appetite*, 16, 103, 1991.
49. Berridge, K. C., Food reward: brain substrates of wanting and liking, *Neurosci. Biobehav. Rev.*, 20, 1, 1996.
50. O'Hare, E., Cleary, J. P., Billington, C. J., and Levine, A. S., Naloxone administration following operant training of sucrose/water discrimination in the rat, *Psychopharmacology*, 129, 289, 1997.
51. Grill, H. J. and Berridge, K.C., Taste reactivity as a measure of the neural control of palatability, *Prog. Psychobiol. Physiol. Psychol.*, 11, 1, 1985.
52. Parker, L. A., Maier, S., Rennie, M., and Crebolder, J., Morphine- and naltrexone-induced modification of palatability: analysis by the taste reactivity test, *Behav. Neurosci.*, 106, 999, 1992.
53. Fantino, M., Hosotte, J., and Apfelbaum, M., An opioid antagonist naltrexone reduces the preference for sucrose in humans, *Am. J. Physiol.*, 251, R91, 1986.
54. Hodos, W., Progressive ratio as a measure of reward strength, *Science*, 134, 943, 1961.
55. Billington, C. J. and Levine, A. S., Hypothalamic neuropeptide Y regulation of feeding and energy metabolism, *Curr. Opin. Neurobiol.*, 2, 847, 1992.
56. Levine, A. S. and Billington, C. J., Why do we eat? A neural systems approach [Review, 139 refs], *Annu. Rev. Nutr.*, 17, 597, 1997.
57. Kotz, C. M., Grace, M. K., Briggs, J., Levine, A. S., and Billington, C. J., Effects of opioid antagonists naloxone and naltrexone on neuropeptide Y-induced feeding and brown fat thermogenesis in the rat: neural site of action, *J. Clin. Invest.*, 96, 163, 1995.
58. Kotz, C. M., Briggs, J., Grace, M., Levine, A. S., and Billington, C. J., Divergence of the feeding and thermogenic pathways influenced by neuropeptide Y in the hypothalamic paraventricular nucleus of the rat, *Am. J. Physiol.*, 275, R471, 1998.
59. Giraudo, S. Q., Billington, C. J., and Levine, A. S., Effects of the opioid antagonist naltrexone on feeding induced by DAMGO in the central nucleus of the amygdala and in the paraventricular nucleus in the rat, *Brain Res.*, 782, 18, 1998.
60. Pomonis, J. D., Levine, A. S., and Billington, C. J., Interaction of the hypothalamic paraventricular nucleus and central nucleus of the amygdala in naloxone blockade of neuropeptide Y-induced feeding revealed by c-fos expression, *J. Neurosci.*, 17, 5175, 1997.
61. Reiser, S. and Hallfrisch, J., Insulin sensitivity and adipose tissue weight of rats fed starch or sucrose diets ad libitum or in meals, *J. Nutr.*, 107, 147, 1977.
62. Reiser, S. and Lewis, C. G., Effect of the type of dietary carbohydrate on small intestine function, *Proc. Biochem. Pharmacol.*, 21, 135, 1986.
63. Jenkins, D. J. A., Jenkins, A. L., Wolver, T. M. S., Thompson, L. H., and Rao, A. V., Simple and complex carbohydrates, *Nutr. Rev.*, 44, 44, 1986.

64. Sclafani, A., Carbohydrate taste, appetite, and obesity: an overview, *Neurosci. Biobehav. Rev.*, 11, 131, 1987.

65. Sclafani, A., The hedonics of sugar and starch, in *The Hedonics of Taste,* Bolles, R. C., Ed., Erlbaum Associates, Mahway, NJ, 1991, 59.

66. Glass, M. J., O'Hare, E., Cleary, J. P., Billington, C. J., and Levine, A. S., The effect of naloxone on food-motivated behavior in the obese Zucker rat, *Psychopharmacology,* 141, 378, 1999.

67. Glass, M. J., Cleary, J. P., Billington, C. J., and Levine, A. S., The role of carbohydrate type upon diet selection in neuropeptide Y stimulated rats, *Am. J. Physiol.,* 273, R2040, 1997.

27 Serotonin (5-HT) and Serotoninergic Receptors in the Regulation of Macronutrient Intake

*Jason C.G. Halford, Brenda K. Smith,
and John E. Blundell*

CONTENTS

ABSTRACT

There exists a close link between nutritional intake and serotonin 5-HT (5-hydroxytryptamine) activity. Manipulations of 5-HT functioning influence eating behaviour, appetite, and the response to various nutritional challenges. Conversely, nutritional manipulations or restrictions also appear to alter the sensitivity of this 5-HT system. Of the 14 5-HT receptor subtypes currently identified, 5-HT_{1B} and 5-HT_{2C} receptors are believed to mediate the 5-HT-induced hypophagia associated with the process of satiety. Much interest has been generated by the possible effects of 5-HT drugs on food choice and macronutrient selection. The response of the 5-HT system to carbohydrate intake and its possible role in controlling carbohydrate intake have been much investigated. Despite a large number of publications, the results of both animal and human studies into 5-HT-induced carbohydrate intake suppression are inconclusive, possibly because of the variety of differing methodologies employed. The data from early dietary choice paradigms are particularly contradictory. Long-term dietary-induced obesity studies in animals using a variety of procedures suggest 5-HT manipulation does effectively reduce the intake of dietary fat. d-Fenfluramine specifically can inhibit consumption of a variety of weight-inducing high-fat diets in rodents. Moreover, in a series of more recent dietary choice studies 5-HT and 5-HT drugs administered in the hypothalamic area have been shown to suppress fat intake selectively. As in animal research, much human research has concentrated on the role of 5-HT in controlling carbohydrate consumption. However, in humans 5-HT drugs such as d-fenfluramine have been shown to suppress the consumption of palatable high-fat foods such as snacks and in some cases can selectively reduce fat consumption. With a new generation of drugs which selectively activate the 5-HT_{1B} and 5-HT_{2C} satiety receptors, any nutrient-selective effects of the operation of the 5-HT system may become more apparent. Drugs that curtail the overconsumption of high-fat energy-dense foods would be particularly useful in treating obesity and in reducing the risks of dietary fat-related diseases.

1   INTRODUCTION

The interlinked phenomenon of the increasing consumption of high-fat diets and the continuing rise in the prevalence of obesity has become a key health concern as we start the new millennium. The occurrence of obesity appears to be rising in both developed and developing countries. This increase in population body weight has been linked to dramatic changes in our food source, primarily the availability of high-fat energy-dense food stuffs (as well as the reduction in our level of daily activity which compounds the effect). A number of large-scale surveys in various countries have demonstrated a positive relationship between the proportion of the diet containing fat and degree of body weight gain.[1,2] Moreover, reduction in obesity appears to be associated with decrease in the amount of fat consumed.

Until its recent withdrawal, the serotoninergic drug d-fenfluramine (and its less pharmacologically specific predecessor fenfluramine) was one of the most effective treatments for obesity available.[3] Evidence from a large number of differing countries over a 15-year period supports the assertion that d-fenfluramine (and fenfluramine) treatment significantly reduces body weight, and sustains this lower body weight whilst treatment continues. The primary action of d-fenfluramine is to reduce the total daily calorie consumption specifically by both reducing meal size and eliminating snacking behaviour (see later). As d-fenfluramine is still considered an effective anti-obesity drug, and dietary fat intake is linked to obesity, does the effect of this serotoninergic drugs on energy consumption include an effect on fat intake? Drugs that specifically reduced fat intake would be key in treating obesity.

Serotoninergic drugs such as d-fenfluramine, which stimulate the release of neuronal serotonin, can be utilised experimentally to assess if enhanced 5-HT function selectively reduces the intake of specific macronutrients. The hypophagic effects of serotoninergic drugs which increase synaptic 5-HT levels such as d-fenfluramine (5-HT releaser and reuptake inhibitor), or fluoxetine and sertraline (selective serotonin reuptake inhibitors, SSRIs) have been well documented in both humans and animals.[3] Animal data and some preliminary human data on the effect of selective 5-HT receptor agonists is also now available. This chapter restricts itself in the main to studies in which at least some measurement of the macronutrient composition of the food intake is detailed. More general reviews of 5-HT and food intake are available elsewhere.[3]

2   5-HT AND FOOD INTAKE

The monoamine neurotransmitter, serotonin (5-HT, 5-hydroxytryptamine) was first specifically linked to the control of food intake, and of feeding behaviour, more than 20 years ago.[4] 5-HT is a key central nervous system (CNS) transmitter and the product of a metabolic process of only two steps. Neuronal 5-HT is synthesised from the essential amino acid tryptophan. Ingested tryptophan enters the CNS from the blood plasma across the blood–brain barrier via a process of active transport. In the neuronal cytoplasm tryptophan is hydroxylased to produce 5-hydroxytryptophan (5-HTP). 5-HTP is itself converted to 5-HT and stored in dense vesicles on the presynaptic membrane ready for release. The rate of 5-HT synthesis is key to one of the theories of how 5-HT modulates macronutrient selection. A review of early pharmacological interventions into this process of 5-HT synthesis and release indicated a role for serotonin in the control of food intake. Analysis of feeding behaviour during these 5-HT interventions further indicated that the functioning of the 5-HT system may be critical in the body's natural control of food intake.[4]

Behavioural analysis of 5-HT-induced reductions of food intake suggested that 5-HT acted to enhance the process of satiation and to promote the state of satiety (to naturally bring food intake to an end). Thus, it was proposed that serotoninergic functioning may be an integral part of the negative feedback aspect of the body's appetite regulation system, terminating the consumption of food.[4] This raised two questions: How did food consumed, in terms of its energy content and its differing macronutrient content affect CNS 5-HT? And, in turn, how did 5-HT modulate subsequent food intake, in terms of total energy consumed and macronutrient composition of food subsequently consumed? Much of the resultant research concentrated on the ability of macronutrients to alter the availability of the 5-HT precursor tryptophan differentially in the brain and how this affected subsequent food intake.

3   5-HT RECEPTORS

Neuronal 5-HT released into the synaptic cleft stimulates 5-HT postsynaptic receptors. Early pharmacological interventions which used 5-HT-active compounds such as fenfluramine to increase indirectly the stimulation of postsynaptic 5-HT receptors resulted in decreased food intake. Since then, large advances in the discovery and identification of novel 5-HT receptors have taken place. Cloning and radioligand techniques have allowed the subdivision of 5-HT receptors into 14 distinct subtypes.[5] These subtypes are 5-HT_{1A}, $5\text{-HT}_{1B(r\,and\,h)}$,* 5-HT_{1D}, 5-HT_{1E},** 5-HT_{1F}, 5-HT_{2A}, 5-HT_{2B}, 5-HT_{2C},*** 5-HT_3, 5-HT_4, $5\text{-HT}_{5\alpha}$, $5\text{-HT}_{5\beta}$, 5-HT_6, and 5-HT_7. The ongoing identification of new 5-HT receptor subtypes has in turn led to research to determine which of these 5-HT receptors are involved in the reduction of food intake via the processes of satiety. The functional significance of many of these new 5-HT receptors still remains to be determined.

Early pharmacological studies used drugs such as fenfluramine (and later d-fenfluramine) and fluoxetine which both increased synaptic levels of 5-HT by either increasing its release from and/or blocking its reuptake back into the presynaptic membrane, thus increasing 5-HT function. Research has progressed with the availability of more-selective 5-HT agonists and antagonists. Briefly, summarising a large body of subsequent pharmacological research, the 5-HT receptor subtypes most directly implicated in feeding control are 5-HT_{1A}, 5-HT_{1B}, and 5-HT_{2C}.[3] It is the postsynaptic 5-HT_{1B} and the 5-HT_{2C} receptors that are generally believed to be involved in the 5-HT satiety system. 5-HT_{1A} receptors are believed to act as autoreceptors to limit 5-HT functioning. In certain conditions activation of 5-HT_{1A} receptors results in an increase in food intake. Any distinct 5-HT receptor subtype involved in a macronutrient-specific satiety effect remains to be determined. However, there is no current evidence that any specific 5-HT receptor subtype mediates the intake of a specific macronutrient.

4   5-HT, DIET COMPOSITION, AND FOOD CHOICE

Studies of hypothalamic 5-HT suggest it is indeed sensitive to distinct macronutrients, and may be integral in the systems regulating macronutrient intake. It has been proposed that neuronal 5-HT functions as a sensor of the plasma amino acid ratio.[6,7] This proposal was based on the observation of a relationship between the T/LNAA (tryptophan/large neutral amino acids) plasma ratio and the relative proportions of dietary carbohydrate and protein consumed. Increased carbohydrate intake raises the T/LNAA ratio.[8] The resulting increase in available tryptophan increases 5-HT synthesis and release. Therefore, CNS 5-HT function is directly linked to the nutritional status of the body, and can consequently alter subsequent feeding behaviour to maintain energy balance.

---

* r and h = rodent and human.
** ht = functional significance remains to be determined.
*** Previously termed 5-HT_{1C}.

The key to this system is the ability of ingested food to alter tryptophan levels in the brain which in turn has a potent effect on neuronal 5-HT synthesis.[9] To summarise, the 5-HT synthesis rate is sensitive to even modest changes in brain tryptophan levels. The availability of tryptophan to the brain is itself determined by the ability of differing macronutrients circling in the bloodstream to modify tryptophan transport across the blood–brain barrier. Tryptophan has to compete with other LNAAs for transport into the brain. The LNAAs, like tryptophan, come from postabsorptive dietary protein in blood plasma. As transport across the blood–brain barrier is limited, proportional increases in the levels of LNAAs available for transport results in a direct reduction in the uptake of tryptophan. However, postabsorptive carbohydrate induces insulin secretion, which has the effect of decreasing the level of available LNAAs (by stimulating demand for LNAAs in the periphery). This increases the proportion of tryptophan which can cross the blood–brain barrier. Tryptophan levels in the brain then increase resulting in an increase in the rate of neuronal 5-HT synthesis. Protein-rich meals containing both tryptophan and other LNAA do not alter the net T/LNAA ratio and cause no change to 5-HT synthesis rate. Carbohydrate-rich meals increase the net T/LNAA ratio (via insulin action) and cause an increase in 5-HT synthesis.[9]

Given that it is well established that neuronal 5-HT levels directly influence subsequent food intake, the operation of a system which affects the synthesis of 5-HT should have direct implications on consequent total food intake. First, with regard to protein intake, consumption of a protein source containing a balanced range of amino acids (equal tryptophan and LNAAs) should have no net effect on the T/LNAA ratio and therefore not affect the 5-HT synthesis rate. The only exceptions to this would be either in the case of diets artificially high or low in tryptophan content which would drastically alter the net T/LNAA ratio. Second, with regard to carbohydrate intake, diets high in carbohydrate increase the net T/LNAA ratio. This should increase 5-HT functioning and so reduce subsequent total food intake. This has provided experimenters with a theoretical basis to test the link between 5-HT, dietary carbohydrate, and total subsequent food intake. Some researchers have also proposed a key role for the T/LNAA 5-HT mechanism in explaining how the macronutrient composition of previous food intake could in turn determine subsequent macronutrient selection (see later).

The relative strength or contribution of this system has been questioned.[10,11] The observed degree of protein intake regulation could not be achieved by the action of the T/LNAA alone. However, along with a 5-HT–vagal link with the peripheral satiety factor cholecystokinin (CCK), and with other directly acting factors such as enterostatin (sensitive to fat intake), the T/LNAA ratio is one theoretical way in which macronutrient consumption could modulate the direction of feeding behaviour. The T/LNAA, along with other inputs could affect both 5-HT synthesis and/or its release. Thus, the 5-HT system may be open to differing degrees of modulation by all the macronutrients.

Indeed, evidence indicates that carbohydrate and protein action on the T/LNAA ratio is altered in patients with anorexia nervosa[12] and obesity.[13] Glucose preloading does not correct this abnormality. In normal subjects this carbohydrate manipulation would alter the T/LNAA ratio. This demonstrates that in two groups of people who display differing and abnormal eating behaviour, the functioning of the T/LNAA 5-HT food intake control system appears to be significantly altered. Brewerton[14] reviewed evidence of general 5-HT dysregulation in subjects with eating and weight disorders (specifically those suffering from bulimia nervosa). Brewerton noted some clinical features of eating disorders, such as feeding disturbance, depression, and impassivity all indicated abnormal or reduced 5-HT function. In these people increases in available tryptophan in the brain may be somehow unable to reverse this decrease in 5-HT functioning.

Moreover, psychobiological stressors such as binge eating (see later) and/or dieting 'may perturb and interact with a vulnerable 5-HT system.' Jimerson et al.[15] suggested individual differences in 5-HT functioning may be a dispositional risk factor for abnormal eating behaviour. Thus, aberrations of normal nutritional status could have a potent effect on 5-HT functioning, or vice versa. Indeed, 5-HT levels, especially in women,[16,17] do appear to be sensitive to dieting. The prolactin response to tryptophan is elevated in women following 3 weeks adherence to a low-calorie diet (LCD) (1000 kcal/day).[18] The LCD also leads to a fall in plasma tryptophan, decreasing the T/LNAA ratio.[19] Prolactin response to d-fenfluramine or mCPP (m-chlorophenylpiperazine) (5-HT_{2C} agonist) in healthy female subjects showed that moderate dieting also causes 5-HT_{2C} receptor supersensitivity.[17,20] This evidence collectively suggests a sensitive connection between nutritional status and 5-HT metabolism, and a strong relationship between abnormal eating behaviour and 5-HT dysregulation.

5   DIET SELECTION IN ANIMAL STUDIES

The relationship between 5-HT and nutrient intake has been conceptualised as a system in which the postabsorptive action of carbohydrate acts on CNS 5-HT synthesis via tryptophan uptake into the brain. This system has provided a basis for numerous studies on the action of acute and chronic doses of 5-HT drugs on subsequent macronutrient selection in animals. In these studies animals are presented with a choice of differing diets varying in macronutrient composition (diet selection paradigms). Wurtman and Wurtman[21,22] performed the first early studies to determine if fenfluramine and other serotoninergic drugs available at that time selectively reduced the intake of specific macronutrients (protein and carbohydrate) as well as total food intake.

In one of their first diet selection paradigm studies Wurtman and Wurtman[21] discovered that fenfluramine reduced total food intake in rats but did not reduce the total intake of protein (in the form of casein) once the protein consumed from both tested diets was calculated (see Table 1). Even when animals' total food intake was reduced, the animals appeared to be preserving ('or sparing') their protein intake by reducing (or 'selectively suppressing') their intake of other macro-nutrients (specifically carbohydrate). In a second study they confirmed that fenfluramine reduced total caloric intake and the intake of carbohydrate, but again did not affect protein intake.[22] Consequently, it seemed that fenfluramine, which increased 5-HT release, reduced the total food intake of these animals by specifically reducing carbohydrate intake, whilst protein intake was defended or 'spared.'

The Wurtmans combined their pharmacological data with the existing data on the effect diet had on various biochemical measures of the T/LNAA ratio and 5-HT synthesis and proposed a 5-HT carbohydrate intake control loop. Basically as before, the consumption of carbohydrate increases the synthesis of 5-HT by increasing tryptophan availability in the CNS. Increase in neuronal 5-HT in turn specifically inhibits the consumption of carbohydrates. Conversely, reduction in carbohydrate intake results in less brain tryptophan and reduced 5-HT functioning. This in turn leads to carbohydrate consumption. The exact mechanisms underlying this section of the feedback loop were less well defined. However, this early research, whether it ultimately proved correct, provided the first testable theory of how 5-HT functioning could alter subsequent food choice. It predicted how 5-HT could specifically alter macronutrient intake, rather than just how previous consumed macronutrients could alter 5-HT functioning. However, did these early studies actually prove that the 5-HT drugs employed (mainly fenfluramine) were selectively suppressing carbohydrate?

Unfortunately, diet selection paradigm studies are fraught with methodological problems which in turn result in a lack of definitive results. In the previous two-diet-choice paradigm, the proportion of protein and carbohydrate had been varied, but fat levels were held constant, precluding the study of 5-HT effects on fat intake. Similar studies have continued to demonstrate 5-HT-induced selective reduction in carbohydrate intake[23-25] using various 5-HT-acting agents. Kim and Wurtman[23] demonstrated that both fenfluramine and fluoxetine specifically suppressed the intake of a diet which was very low in protein (5%) and high in carbohydrate (40%) compared with a diet high in both macronutrients (45% protein, 40% carbohydrate). Thus, it seems the protein content of a test diet would have to be low before the selective, 5-HT-induced suppression of carbohydrate intake could be observed. Similarly, Luo and Li[24] noted that acute doses of d-fenfluramine, fluoxetine, and the selective $5\text{-HT}_{1A/1B}$ receptor agonist RU-24969 specifically reduced the intake of high-carbohydrate,

TABLE 1
Effects of 5-HT Manipulation on Macronutrient Selection in Animals

Study	Drug(s)	Diet(s)	Results
Wurtman and Wurtman (21)	Fenfluramine (2.5 mg/kg i.p.) Fluoxetine (5–10 mg/kg i.p.) Sprague-Dawley rats	Choice between two diets: High CHO/low protein and low CHO/high protein; fat held constant	Fenfluramine reduced total caloric intake but did not reduce protein intake; protein intake was defended by a selective reduction in CHO consumption; fluoxetine produced similar but weaker effects
Wurtman and Wurtman (22)	MK-212 (1.5, 3.0 mg/kg i.p.) Fenfluramine (2.5 mg/kg i.p.) Sprague-Dawley rats	Eight diets given in pairs; varying in protein and CHO content	In drug conditions animals consumed less calories; the intake of high-carbohydrate diets was reduced the most
Moses and Wurtman (90)	MK-212 (1.5 mg/kg i.p.) d-Fenfluramine (2.0 mg/kg i.p.) Male Sprague-Dawley rats	Diets high or low in CHO (75% vs. 25% CHO); diets differed in fat content: 5 or 27% by weight	Both drugs reduced only the intake of the high CHO diets
Kim and Wurtman (23)	dl-Fenfluramine (3 and 6 mg/kg i.p.) Fluoxetine (7.5 and 15 mg/kg i.p.) CGS 10686 B (2.5–5 mg/kg i.p.) Male Sprague-Dawley rats	Diets high or low in protein (5 or 45%) while CHO held constant (40%); fat was varied (2.7% in high protein and 23.9% in low protein)	CGS 10686 B induced a marked decrease in the consumption of the low protein (high fat) diet; fenfluramine and fluoxetine produced similar effects; the effects of fluoxetine were less marked
Luo and Li (24)	d-Fenfluramine (0.5, 1.0, 1.5 mg/kg i.p) Fluoxetine (2, 3, 4 mg/kg i.p.) RU-24969 (1, 2, 3 mg/kg i.p.) Male Wistar rats	A choice of high CHO (78.5%)/low protein (5%) or low CHO (38.5%)/high protein (45%) diets	At midrange doses all drugs selectively suppressed the intake of the high-carbohydrate diet
Luo and Li (25)	d-Fenfluramine (1.5 mg/kg i.p.) Fluoxetine (3 mg/kg i.p.) RU-24969 (1,5 mg/kg i.p.) (Chronic — 6 days) Male Wistar rats	High CHO/low protein (78.5% CHO/5% protein) or (73.5% CHO/10% protein) Low CHO/high protein (38.5% CHO/45% protein) or (23.5% CHO/60% protein)	All drugs selectively suppressed the intake of the high-carbohydrate diet on the first and on all subsequent days days of the study.
Li and Luo (91)	Buspirone (0.6 mg/kg s.c.) Male Wistar Rats	Same as Luo and Li (25)	Buspirone selectively increased the consumption of CHO
Lawton and Blundell (26)	d-Fenfluramine (1, 2, and 3 mg/kg i.p.) Male Lister Hooded rats	Pelleted or hydrated standard diet offered with either sucrose or polycose as a pure powder or dissolved in water	
Lawton and Blundell (27)	d-Fenfluramine (1, 2 and 3 mg/kg i.p.) Fluoxetine (5, 10 and 15 mg/kg i.p.) mCPP (1, 2, and 4 mg/kg i.p.) RU-24969 (0.3, 1, and 2 mg/kg i.p.) MK-212 (0.3, 1.5, and 3 mg/kg i.p.) DOI (0.72, 1.43, and 2.86 mg/kg i.p.) Male Lister Hooded rats	Pelleted or hydrated standard diet offered with either sucrose or polycose as a pure powder or dissolved in water	All drugs reduced the intake of both diet and CHO supplement; however, fluoxetine and DOI exerted stronger effects on polycose intake

TABLE 1 (continued)
Effects of 5-HT Manipulation on Macronutrient Selection in Animals

Study	Drug(s)	Diet(s)	Results
Orthen-Gambill and Kanarek (29)	Fenfluramine (1.5, 3.0, 6.0 mg/kg i.p.) Female Sprague-Dawley rats	Offered three macronutrients (CHO, protein, and fat) or a single, standard diet	Fenfluramine reduced total caloric intake and under certain conditions selectively reduced both fat and protein intake
Kanarek and Duskbin (30)	5-HT (2-6 mg/kg i.p.) Male Sprague-Dawley rats	Offered three macronutrients (CHO, protein, and fat) or a single, standard diet	5-HT reduced total caloric intake and selectively suppressed fat intake in the selection paradigm
Kanarek et al. (92)	dl-Fenfluramine (1.5, 3, and 6 mg/kg) Dosed for 6 days Male Sprague-Dawley rats	Offered a single high CHO (65% CHO by kcal) or a single high fat (65% by kcal) diet	Dose-related reduction in food intake in animals offered either diet; a greater reduction was observed in rats fed high-fat diet
White et al. (93)	Diet supplemented with tryptophan Male Sprague-Dawley rats	Offered three macronutrients (CHO, protein, and fat)	Tryptophan loading of diet resulted in a selective decrease in CHO intake, but over time maintained total caloric intake by increasing fat intake
Heisler et al. (32)	Fluoxetine (5 or 10 mg/kg) Acute and chronic (28 days) Male Long Evans rats	Offered three macronutrients (CHO, protein, and fat)	In acute phase, fluoxetine decreased fat and protein intake in a dose-dependent manner with a weaker effect on CHO consumption; chronic fluoxetine decreased weight gain over the 28-day period probably due to its continuing suppression of fat and protein intake
Leibowitz et al. (36)	5-HT into PVN region and other hypothalamic sites Male Sprague-Dawley rats	Offered three macronutrients (CHO, protein, and fat)	5-HT infusion into PVN and VMH led to a selective decrease in CHO consumption
Leibowitz et al. (37)	5-HT into PVN region d-Norfenfluramine into PVN Fluoxetine into PVN Male Sprague-Dawley rats	Offered three macronutrients (CHO, protein, and fat)	5-HT infusion into PVN selectively decreased CHO intake at start of dark cycle; similar changes in macronutrient selection were produced by both d-norfenfluramine and fluoxetine
Weiss et al. (38)	d-Norfenfluramine into PVN d-Fenfluramine (0.06, .013, 0.25, 0.5, and 1 mg/kg i.p.) Male Sprague-Dawley rats	Offered three macronutrients (CHO, protein, and fat)	d-Norfenfluramine infusion into PVN selectively decreased CHO intake at start of dark cycle; similar changes were produced by the lowest doses of d-fenfluramine; at high doses, the intake of all macronutrients was suppressed

TABLE 1 (continued)
Effects of 5-HT Manipulation on Macronutrient Selection in Animals

Study	Drug(s)	Diet(s)	Results
Weiss et al. (39)	Fluoxetine into PVN region or peripherally (0.6–10 mg/kg i.p.) Male Sprague-Dawley rats	Offered three macronutrients (CHO, protein, and fat)	Fluoxetine selectively decreased CHO intake in the first hour of the dark cycle only
Smith et al. (33)	d-Fenfluramine (1.5 mg/kg i.p.) over 12 days Male Sprague-Dawley rats	Offered three macronutrients (CHO, protein, and fat); experiment repeated in animals classified as either CHO or fat preferrers	d-Fenfluramine suppressed both absolute (kcal) and relative (% of total energy) fat intake; this effect was also observed independent of baseline macronutrient preference; if anything, CHO intake was enhanced with drug treatment
Smith et al. (34)	d-Norfenfluramine (3V) d-Norfenfluramine (PVN region) 5-HT (PVN region) Male Sprague-Dawley rats	Offered three macronutrients (CHO, protein, and fat); in final study, responses of fat-preferring and CHO-preferring rats were compared	d-Norfenfluramine into the 3V selectively suppressed fat intake; d-norfenfluramine in the PVN selectively suppressed fat and protein intake; 5-HT in the PVN selectively decreased fat and protein intake; the effect of 5-HT was more robust in fat-preferring rats
Blundell and Hill (44)	d-Fenfluramine (approx. 2–4 mg/kg in drinking water) over 90 days Female Lister Hooded rats	Offered standard diet and a palatable supplement of either pure fat or pure CHO	d-Fenfluramine suppressed food intake and body weight gain in both diet supplement options

Note: CHO: carbohydrate; PVN: paraventricular nucleus.

low-protein test meals. When these drugs were given chronically, the combination of 5-HT drug and high-carbohydrate, low-protein diet had the most potent effect on food intake and body weight.

However, other two-choice studies have shown that 5-HT drugs can reduce the intake of all diet options offered except those high in carbohydrate.[26-28] In a series of studies Lawton and Blundell[26-28] demonstrated that carbohydrate suppression only occurred under certain experimental conditions. Under other conditions selective protein supression and carbohydrate sparing could be reliably demonstrate. In replication studies, one of the most robust selective carbohydrate-sparing effects could be demonstrated if the carbohydrate supplement was offered in a liquid rather than a powder form (Halford and Blundell, unpublished data). Therefore, contextual variables, such as palatability, quality of the test diets, and form/type of macronutrient chosen, amongst other things, could have a marked effect on any observed results.[26-28] In summary, two-diet-choice selective paradigms demonstrate that 5-HT suppression of food intake is not dependent on that food containing large amounts of carbohydrate. Additionally, low-carbohydrate diets do not prevent 5-HT-induced hypophagia. This would not be the case if food choice were solely (or predominately) determined by the operation of a postulated 5-HT carbohydrate intake control loop.

Three-choice-diet selection paradigms consist of diets containing largely one macronutrient, or alternatively pure macronutrients. This fuller choice paradigm should allow a clearer identification of 5-HT macronutrient-specific effects. However, these designs are not entirely free of the effects of contexual variables either. Othen-Gambill and Kanarek[29] either gave rats a standard balanced diet or provided them with three separate macronutrient diets. In these studies, after the differing caloric densities of the diets had been controlled for, it was found that fenfluramine did produce some reduction in carbohydrate intake but also a sizeable reduction in fat intake. Similar results were observed after 5-HT administration.[30] The observed lack of drug suppression of protein intake (protein-sparing effect) seen in the two-diet paradigm appeared less robust in the three-diet version. In some experimental situations carbohydrate intake was even selectively spared at the expense of other macronutrients. Thus, the differing paradigm methodology produced strikingly different results. The results of these studies do not appear to support the dominant operation of a specific 5-HT carbohydrate feedback loop either. Moreover, in certain situations 5-HT-releasing drugs selectively reduced the intake of fat, the macronutrient excluded from the two-diet paradigm.

Other evidence also supports a selective action of *systemically* administered 5-HT or 5-HT receptor agonists to suppress fat intake in studies designed to allow a concurrent evaluation of the consumption of individual fat, carbohydrate, and protein diets.[30-34] It is well known that treatment with fluoxetine, a serotonin reuptake inhibitor, results in decreased food intake and weight loss. When the effect of chronic, daily fluoxetine injections on percent of calories from macronutrients was examined in rats across a 28-day study, proportional fat intake was decreased during the drug phase and proportional carbohydrate consumption was increased whilst there was no change in the percent of protein.[32] Similarly, in a study of the effect of chronic d-fenfluramine treatment on macronutrient selection, d-fenfluramine reduced proportional fat intake from 61 to 47% of total energy compared with saline-treated controls[33] (Figure 1). Notably, carbohydrate intake was increased from 28 to 41% although this difference did not reach statistical significance, and protein intake was not altered. The results of these two studies suggest that serotonin agonists can induce a pharmacological modification of appetite in which fat calories are exchanged, at least in part, for carbohydrate calories. Furthermore, it appears that this particular serotoninergic effect (enhancement of carbohydrate intake) may only be expressed in experimental designs employing chronic treatment regimens. For example, the acute effect of a single dose of fluoxetine[32] or fenfluramine[31] was to reduce the consumption of all three macronutrients, although the suppression of carbohydrate and protein intakes was less robust. Certainly, chronic administration of serotoninergic agents has clinical applications for obesity. Moreover, these chronic effects on voluntary diet composition may reveal key information regarding the role of biological feedback in the serotonergic control of feeding.[32]

One explanation offered for the novel findings of serotonergic effects on fat consumption was the possible influence of baseline dietary intakes, or that serotonin decreased intake of the most preferred diet. In fact, baseline fat intakes were noted to be high in some studies.[29,30,32] However, this hypothesis was tested in an experiment that characterised animals by initial fat or carbohydrate preference. It was found that daily injections of low-dose dexfenfluramine to both fat- and carbohydrate-preferring rats suppressed the intake of fat and not carbohydrate[33] (Figure 2). Compared with the pretreatment period, fat-preferring rats reduced their daily fat intake from 62 to 53% of

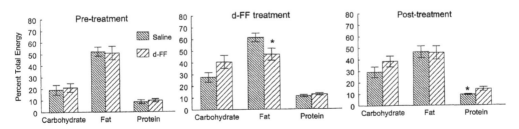

FIGURE 1 Effect of 12 days of d-fenfluramine (d-FF, $n = 7$) 1.5 mg/kg per day or saline ($n = 7$) on cumulative macronutrient intakes expressed as percent of total energy and averaged across each time period: pre-treatment, d-FF treatment, and post-treatment. *$P < 0.05$ compared to saline. (From Smith, B. K., York, D. A., and Bray, G. A., Chronic *d*-fenfluramine treatment reduces fat intake independent of macronutrient preference, *Pharmacol. Biochem. Behav.,* 60, 105, 1998. With permission from Elsevier Science.)

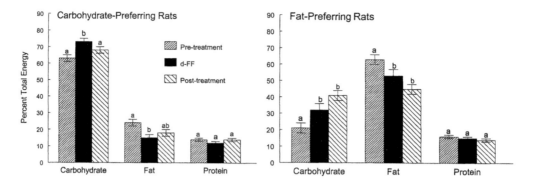

FIGURE 2 Effect of 12 days of d-fenfluramine (d-FF) 1.5 mg/kg per day on cumulative macronutrient intakes (expressed as percent of total intake) of carbohydrate ($n = 12$) and fat preferring ($n = 12$) rats. Within each macronutrient diet, means without common superscripts differ significantly between time periods, $P < 0.05$. (From Smith, B. K., York, D. A., and Bray, G. A., Chronic *d*-fenfluramine treatment reduces fat intake independent of macronutrient preference, *Pharmacol. Biochem. Behav.*, 60, 105, 1998. With permission from Elsevier Science.)

total energy and the low baseline fat intake of carbohydrate-preferring rats was further reduced by d-fenfluramine treatment (from 24 to 15% of total energy). Remarkably, proportional carbohydrate intake was increased with d-fenfluramine administration in both preference groups (Figure 2). These changes in percent fat and carbohydrate consumption during d-fenfluramine treatment returned to pretreatment levels in carbohydrate-preferring rats, but not in fat-preferring rats. The reduction in fat intake observed in both preference groups treated with d-fenfluramine indicates that the anorexic effect of the drug is not simply to suppress intake of the preferred macronutrient diet. Thus, it appears that the serotonin system exerts its effects on macronutrient selection in a manner independent of the mechanism(s) responsible for dietary preferences.

Few laboratories have investigated the effects of *centrally* administered serotoninergic drugs on macronutrient self-selection in animals. However, previous studies employing the three-choice diet selection paradigm have shown uniformly that stimulation of serotonin activity, through microinjection of 5-HT or 5-HT agonists into the paraventricular nucleus of the hypothalamus, leads to the selective suppression of carbohydrate consumption with no change in the consumption of protein or fat.[31,35-39] One exception was the report by Max et al.[40] showing that the combined injection of fenfluramine and norfenfluramine into the PVN decreased both fat and carbohydrate intake. In these studies, the 5-HT drug was administered to nondeprived rats at the end of the light period and the modifications in food intake and macronutrient selection were measured during the first 1 to 2 h after injection, which coincides with the early dark period of spontaneous feeding.

Despite similar experimental conditions and diet selection paradigms, the results from a recent series of experiments[34] stand in contrast to earlier work and to the once prevailing concept that drugs that increase serotonergic activity selectively reduce carbohydrate intake. Rather, a clear reduction in fat intake accompanied in some cases by an inhibition of protein intake has been demonstrated in response to intrahypothalamic administration of d-norfenfluramine or 5-HT when given in doses that decreased total caloric intake by 40 to 50% at 2 h after injection (Figure 3). In another study, the dose effects of 5-HT on macronutrient intake were examined in rats that were grouped by baseline fat or carbohydrate preference and infused into the PVN with 5-HT (0.3 to 300 nmol) 30 min before onset of the dark period.[34] 5-HT administration led to a dose-related reduction in fat intake at 60 min in fat-preferring rats (Figure 4). In contrast, all doses tested were generally ineffective in suppressing carbohydrate or protein intake except that the highest dose of 5-HT inhibited both carbohydrate and protein intake in fat-preferring rats. These data show that increased serotonergic activity in the PVN leads to a reduction in fat intake independent of

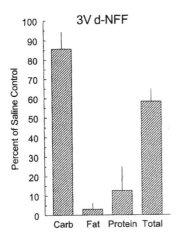
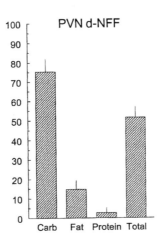

FIGURE 3   Effects of d-norfenfluramine (d-NFF) or saline administered into the third cereboventricle (3V: 416 nmol; $n = 6$) or into the paraventricular nucleus (PVN; 208 nmol; $n = 11$) of the hypothalamus on 2 h food intake of nondeprived rats injected immediately before lights out. Data are expressed as percent of saline control and presented as mean ±S.E.M. For two animals that consumed more than 100% carbohydrate after PVN injection of d-NFF, values were capped at 100%.

macronutrient preference. The suppressed intake of all three macronutrient diets observed in fat-preferring rats suggests an interaction of serotonin receptors in PVN with baseline fat consumption.

Possible reasons for the disparities in results across studies include differences in diet composition. One apparent difference between the most recent studies and earlier serotonin literature is the fat source: plant vs. animal, respectively. It is possible that the amount of saturated fat in the diet may differentially affect feeding responses to exogenously administered serotonin or its receptor agonists. As an example, fenfluramine was shown to have a greater anorectic effect in rats fed diets containing beef tallow compared to those fed corn oil,[41] indicating a possible interaction of dietary fat source and serotonin on feeding behaviour.

Drug dose also may contribute to contrasting study results. For example, the doses of d-norfenfluramine tested in previous studies were low (3 to 50 nmol) and only the highest dose led to a decrease in total caloric intake[38] compared with that used by Smith et al.[33] (208 nmol). Regarding the effect of PVN 5-HT infusion, the range of doses previously reported to inhibit carbohydrate intake (2.5 to 20 nmol)[35-37] has been replicated. Thus, a range of 5-HT doses (0.3 to 300 nmol) infused into the PVN led to a dose-related decrease in fat intake regardless of the animals' baseline macronutrient intakes (Figure 4). In contrast, there were no dose effects of 5-HT on carbohydrate or protein intake in either preference group. However, in fat-preferring rats, the highest dose of 5-HT reduced intake of all three macronutrient diets. These results combined demonstrate a selective effect of exogenous serotonergic drugs in the hypothalamus to reduce fat rather than carbohydrate intake and suggest that higher baseline fat intake enhances responsivity to serotonergic drugs.

Therefore, in contrast to previous research, more recent studies employing three-choice diet selection paradigms indicate that increasing serotonergic neurotransmission by microinjection of 5-HT or norfenfluramine directly into the PVN results primarily in the reduction of fat and protein intake.[34] In particular, the data also demonstrate that only in fat-preferring rats does hypothalamic 5-HT injection lead to a suppression of carbohydrate intake. That the highest dose of 5-HT suppressed intake of all three macronutrient diets in fat-preferring rats suggests an interaction of the PVN 5-HT feeding system with baseline dietary fat consumption. Thus, the later results from studies of centrally administered serotonin agonists are consistent with those obtained using peripheral administration.

Some chronic 5-HT studies use the phenomenon of dietary-induced obesity[42] based on procedures such as cafeteria feeding. In the cafeteria-diet paradigm animals are offered a choice between a standard lab chow control or the laboratory chow control or the laboratory chow together with one or two highly palatable composite foods.[43] This palatable diet alternative often takes the form of human snack foods and so is consequently very high in fat. Long-term exposure to the cafeteria choice leads to dietary-induced obesity. An alternative version of this chronic diet supplement paradigm involves offering laboratory chow supplemented with another highly preferred diet predominately consisting of fat or sugar. The animals are offered purer versions of the macronutrients along with their normal diet. Blundell and Hill[10] noted that animals additionally offered these purer supplement diets, like the cafeteria diets, also demonstrated an increased daily intake, meal size, and induced obesity. In this supplement paradigm, rats chronically treated with d-fenfluramine and offered a fat-supplemented diet do not display the dietary-induced obesity seen in the nondrug control animals. d-Fenfluramine reduced rodent body weight gain over time, as well as the consumption of both fat supplement and standard diets.[44,45]

The results from these long-term choice studies are supported by data from long-term one-diet (or no-choice) studies. In these studies separate groups of animals are maintained on diets differing in macronutrient composition. The high-fat diets (50 to 60% fat) employed in such studies, like the supplement or cafeteria diets, readily induce obesity in rats (compared with a 4 to 5% low-fat control). In Osborne-Mendel rats, which normally feed readily on high-fat diets and gain weight, chronic treatment with d-fenfluramine reduced food intake and prevented weight gain with no sign of developing tolerance to the drug effects in either measure.[46] Therefore, exposure to high-fat diets (containing low carbohydrate) does not weaken the hypophagic effects of d-fenfluramine.[44]

The results of these chronic studies (cafeteria, supplement, or high-fat maintenance diets) appear to be consistent with some of the data from the two-diet choice, and much of the data from the three-diet choice selection paradigms. This is true for studies in which drugs were given either acutely or chronically, and either administered in the periphery or directly into the CNS. 5-HT manipulation does not necessarily result in selective reduction of carbohydrate intake over other macronutrients. Moreover, in a variety of experimental situations 5-HT manipulation does appear to have distinct effects specifically on fat consumption. The most recent data included in this chapter confirm this.[33,34] The 5-HT-induced reduction in the consumption of high-fat diets, demonstrated predominately by the use of the drugs fenfuramine and d-fenfluramine, could be of therapeutic value in obesity treatment if demonstrated in humans.[47]

6   FOOD CHOICE IN HUMANS

As with animal studies, human research is affected by similar range of analogous methodological issues.[11] The proposed relationship between 5-HT synthesis and protein–carbohydrate intake has also influenced greatly the design of human food-choice experiments. Human studies also initially tested protein and carbohydrate differences, in which fat intake (which was high in both options) was again held constant. In these studies subjects were offered food either high in carbohydrate or in protein. As with animal studies, the fat level was generally held constant. Notably, although fat was held constant in these studies fat formed a significant, if not the major macronutrient component of the test foods offered. For example, some snacks offered in these studies contained 51% fat. Some of the test meals employed contained an average of 55% fat. The net effect of 5-HT manipulation in these studies was a greater reduction in carbohydrate than in protein intake. However, if 5-HT manipulation reduced carbohydrate intake to a greater degree than protein intake, the greatest reduction in intake in either choice involved an obligatory reduction in fat.[48,49] Therefore, 5-HT drugs could be considered to reduce meals high in carbohydrate *and* fat selectively. In summary, it is possible the intake of either high-carbohydrate or high-protein items may be preferentially suppressed when they are combined with high proportions of fat in the form of a snack. Any inconsistent effects of d-fenfluramine on protein and carbohydrate intake in humans could instead be interpreted as a consistent 5-HT-induced suppression of fat intake.

Additionally, it should be noted that in these studies[48,49] and in others (cited in Table 2), the most dramatic effect of d-fenfluramine on human feeding behaviour is a reduction in snacking, a significant risk factor for weight gain. In fact in many of these studies the subjects were selected were normal or overweight self-defined carbohydrate cravers with a distinct snacking habit. In the normal diet (rather than in laboratory test meals) snack food usually contains much larger amounts of fat than normal meals. Drewnowski[50] notes that the fat content of snacks can range from 40 to 60% and some snack items, which are often employed in laboratory-based studies, can contain as much as 81% fat. Any effect of d-fenfluramine on snacking will obviously involve an obligatory reduction in fat intake. However, do 5-HT drugs directly and selectively reduce fat intake?

Free-selection designs incorporating foods differing in all macronutrients have been employed to investigate the direct effects of 5-HT manipulation on fat intake. Hill and Blundell[51] using a variety of food items found that d-fenfluramine-induced hypophagic in humans was not macronutrient specific. Foltin et al.[52] examined the effects of fenfluramine on either low- or high-carbohydrate lunches, or a free-selection buffet lunch. The total food intake after drug administration was analysed in each of the three lunch conditions. Fenfluramine appeared to be most effective in reducing the intake of food given in the diet condition manipulated to be the lowest in carbohydrate. Thus, drug action was greatest on the diet containing foods with proportionally higher levels of fat and protein. However, in the free selection condition which allow subjects to express their appetite naturally, no drug-induced carbohydrate-selective effects were observed.

In similar short-term studies the SSRI fluoxetine also did not give rise to any selective macronutrient effect.[53] Fluoxetine (like fenfluramine) has also been shown to lack specific effects on the intake of diets consisting of food items selected by the experimenter to produce differing levels of carbohydrate intake.[54] These results are consistent with those of a previous study which demonstrated that fluoxetine had no differential effects on macronutrient intake in a test meal.[55] Collectively, these studies were unable to show evidence of selective effect on the intake of certain macronutrients by drugs which increased synaptic 5-HT. There was no evidence of a selective suppression of carbohydrate intake or protein sparing. Additionally, Walsh et al.[56] showed that the 5-HT$_{2C}$ receptor agonist mCPP also failed to produce a differential effect on macronutrient intake during a test meal.

However, in the Pilj et al.[55] study, fluoxetine reduced snacking significantly which resulted in a net decrease in carbohydrate consumption. This could be interpreted as evidence supporting a selective 5-HT effect on carbohydrate intake. However, in a normal diet snack foods contain the highest levels of dietary fat and so constitute the greatest risk factor for weight gain and eventual obesity. Again, the 5-HT drug effect on snacking behaviour could lead to an obligatory reduction in fat intake in a real-life situation.

Other studies have found that 5-HT drugs do selectively reduce the consumption of a specific macronutrient. The 5-HT active drugs d-fenfluramine, and a novel 5-HT$_{1B/1D}$ agonist sumatriptan have in some studies produced a selective reduction in fat intake.[57,58] Goodall et al.[57] found that d-fenfluramine selectively reduced fat intake when subjects were offered a variety of food options from a food dispenser similar to a vending machine. This d-fenfluramine-induced selective suppression of fat intake was reversed by the administration of the selective 5-HT$_2$ antagonist ritanserin. The effect of d-fenfluramine on fat intake appeared to be mediated by postsynaptic 5-HT$_2$ receptors, probably the 5-HT$_{2C}$ feeding receptor identified by pharmacological studies.

Boeles et al.[58] found that the 5-HT$_{1B/1D}$ receptor agonist sumatriptan decreased food intake in normal-weight women offered a buffet-style lunch. The most potent effect of sumatriptan was on fat intake. Carbohydrate intake was not significantly reduced by this 5-HT$_{1B/1D}$ antagonist. Collectively, these three studies would appear to suggest that in humans 5-HT manipulation by either increasing synaptic 5-HT or by directly activating 5-HT receptors results in a selective suppression

TABLE 2
Effects of 5-HT Drugs on Macronutrient Selection in Humans

Study	Drug and Subjects	Diet	Results
Wurtman et al. (94)	dl-Fenfluramine (60 mg and 45 mg) or l-tryptophan (2.4 g) in capsules 24 obese subjects (21 female) Classified as CHO cravers	Subjecs given fixed meals over 3 weeks and offered a range of snacks rich in protein or CHO	Fenfluramine reduced CHO snacking in some subjects; effects of tryptophan less pronounced
Wurtman et al. (48)	d-Fenfluramine 30 mg (two 15-mg tablets) 20 obese (16 women) Classified as CHO cravers	Subjects offered high CHO and high protein foods; fat held constant at 55.5% in meals, 51.7% in snacks	d-Fenfluramine reduced snack and meal intake, particularly CHO snacks
Wurtman et al. (49)	d-Fenfluramine (as above) 51 obese (41 females) CHO cravers and noncravers	As above	d-Fenfluramine reduced snacking and mealtime CHO intake in the cravers; drug effects on snacking in the noncravers were weaker; the drug did not reduce mealtime CHO or protein intake in noncravers
Wurtman et al. (95)	d-Fenfluramine (15 mg twice daily) Fluoxetine (20 mg three times daily) 12-week period with 4-day test phases 64 obese women Classified as CHO cravers	Diets high or low in protein (5 or 45%) while CHO held constant (40%); fat was varied (2.7% in high protein and 23.9% in low protein)	Both drugs reduced subject weight over 12 weeks and reduced CHO consumption from snacks
Blundell and Hill (10)	d-Fenfluramine 30 mg (two 15-mg tablets) 8 obese women	High CHO (63% by kcal) or high protein (54% by kcal) lunch	Lower levels of food intake were observed in the high protein, d-fenfluramine condition
Lawton et al. (53)	Fluoxetine 60 mg orally 13 obese women	High CHO (52% by kcal) or high fat (55% by kcal) lunch	Fluoxetine equally suppressed high CHO and high fat lunch intakes
Foltin et al. (52)	Fenfluramine 20 or 40 mg (orally) 5 males, 4 females of normal weight	High CHO or low CHO lunch, or self-selection lunch	Fenfluramine was most effective in reducing food intake in the low CHO lunch condition
Foltin et al. (54)	Fluoxetine 40 mg (orally) 10 males, 1 female of normal weight	Diets manipulated to be high in CHO, high in fat, or regular	Fluoxetine reduced food intake; no selective macronutrient effects were observed
Pilj et al. (55)	Fluoxetine 60 mg (orally) 23 obese women; 11 received drug, 12 received placebo	No diet prescribed; food diaries kept on 4 test days during 14-day period	Fluoxetine reduced total food intake and induced weight loss; fluoxetine did not affect macronutrient intakes of meals, although consumption of CHO snacks was reduced
Walsh et al. (56)	mCPP 0.4 mg/kg orally 12 normal-weight women	Test meal buffet lunch consisting of foods differing in macronutrient content	mCPP reduced total food intake during the test meal; the intake of all three macronutrients was reduced equally
Goodall et al. (57)	d-Fenfluramine 30 mg orally in two 15-mg tablets 12 normal-weight men	Subjects selected food from an automated dispensing machine	d-Fenfluramine reduced total energy intake and selectively suppressed fat intake

TABLE 2 (continued)
Effects of 5-HT Drugs on Macronutrient Selection in Humans

Study	Drug and Subjects	Diet	Results
Boeles et al. (58)	Sumatriptan 6 mg s.c. 15 normal-weight women	Test meal buffet lunch similar to Walsh et al. (56)	Sumatriptan reduced total food intake of the test meal; the drug selectively suppressed the intake of fat and protein
Poppitt et al. (61)	d-Fenfluramine 15 mg twice daily 6 moderately obese women	Provided with either high or low levels of dietary fat	d-Fenfluramine had the most potent effect on reducing caloric intake in the high-fat diet condition
Green et al. (62)	d-Fenfluramine 15 mg twice daily 15 obese women	On test days, subjects offered high fat or high CHO meal and snack items	d-Fenfluramine reduced meal and snack intake by 14.5 and 17.5% respectively; d-fenfluramine induced the largest reduction in total caloric intake in subjects offered high fat/sweet foods
Cangiano et al. (60)	5-HTP 750 mg per day for 2 weeks 25 overweight diabetics	Food intake and macronutrient selection assessed using food diaries	5-HTP decreased energy intake by reducing the intake of fat and CHO

Note: CHO: carbohydrate

of fat intake. This effect does not appear to be dependent on a single 5-HT receptor subtype. Rather, a number of 5-HT_1 and 5-HT_2 receptor subtypes may be involved. Whether it is the 5-HT_{1B} or 5-HT_{2C} receptors alone that modulate 5-HT-induced reduction in fat intake remains to be determined.

In these acute laboratory studies, when subjects were allowed to choose from a variety of food items, and not given just fixed manipulative test meals, macronutrient selective effects were observed. This also allowed experimenters to determine the difference between 'real' and obligatory changes in macronutrient consumption. As with the animal studies, a selective reduction in fat intake can only be demonstrated when the subjects are given the opportunity within the experimental paradigm to express it fully. In human studies this means providing a large number of normal/familiar (and palatable) foods which differ both in macronutrient content (by all three macronutrients) and their snack/meal context such as in a buffet-type test meal or food dispenser/vending machine studies.

As with the animal studies, support in the human literature for 5-HT-induced suppression of fat intake has been provided by long-term clinical studies of 5-HT antiobesity compounds. Lafreniere et al.[59] found that chronic administration of d-fenfluramine to obese subjects selectively reduced their fat intake over the entire length of the study. At the end of 3 months the energy intake of the d-fenfluramine group was 16% lower than the placebo control, a 13% reduction in energy from meal intake, and a 23% reduction in energy from snacks. This energy reduction was characterised by a decrease in the percent of energy as fat from 34 to 30%, a reduction of about 25% in total fat consumption. These obese individuals, given d-fenfluramine chronically, displayed a selective and robust avoidance of high-fat food. So 5-HT manipulation did produce a selective reduction in fat intake. Such effects are produced by 5-HT precursors as well as d-fenfluramine. Cangiano et al.[60] gave overweight people with diabetes the serotonin precursor 5-HTP over a period of 2 weeks. Differences in food intake were assessed using self-report food diaries. 5-HTP reduced both carbohydrate and fat intake, as well as body weight (although underreporting of food intake, particularly of snacks, is a problem with such studies).

Laboratory-based studies measuring energy expenditure as well as energy intake in the obese[61] have supported the Lafreniere et al. data[59] data.[61] Subjects in a high-control laboratory setting were provided with either low-fat (high-carbohydrate) or high-fat diets. d-Fenfluramine produced the greatest effect on energy balance (energy intake minus energy expenditure) in the high-fat diet condition. In addition, Green et al.[62] noted that in obese females d-fenfluramine induced the largest reduction in daily energy intake when they were offered high-fat sweet snacks, rather than sweet carbohydrate-equivalent snacks offered. As obese women show a preference for sweet-tasting high-fat foods[63,64] the selective effect of d-fenfluramine on sweet fat snacks may have important implications for treating obesity.

7   5-HT: INTERACTION WITH OTHER MACRONUTRIENT-SPECIFIC SATIETY SIGNALS

The results of several studies point to a role of 5-HT in mediating the satiety effect of peripherally administered CCK. CCK is secreted by the intestinal mucosa in response to the presence of food and, in particular, to ingested fat. CCK appears to be involved in the physiological adaptation to dietary fat consumption because plasma CCK levels in response to an intraduodenal fat infusion are higher in rats adapted to a high-fat diet than the levels observed in rats fed a low-fat diet.[65] CCK receptors on vagal afferent fibers (CCK_A) are thought to connect via the nucleus tractus solilarius (NTS) to a 5-HT satiety mechanism in the region of the paraventricular nucleus[34,66,67] which is innervated by serotonergic neurons originating in the raphe nuclei. Anorexia induced by CCK-8 is reversed by systemic treatment with a number of 5-HT antagonists.[68,69] These effects could be additive because metergoline administration alone increases food intake. However, a dose of metergoline that did not increase food intake above baseline was shown to attenuate satiety induced by CCK-8.[70] Earlier studies documented the release of medial and lateral hypothalamic 5-HT in response to feeding.[71] More recent neurochemical evidence obtained through microdialysis has now confirmed that extracellular 5-HT in the lateral hypothalamus is increased after peripheral injection of anorectic doses of CCK-8.[72]

5-HT may also mediate the macronutrient-specific effects of CCK on satiety. By using a dietary self-selection paradigm containing isocaloric protein, carbohydrate, or fat diets, intraperitoneal injection of CCK-8 decreased the consumption of fat and protein intake,[73] an observation that is consistent with the effects of 5-HT and 5-HT agonists when administered centrally or peripherally.[30,32-34] Thus, it may be proposed that a hypothalamic serotonergic system inducing satiety that is specific for fat, and in some cases protein, is linked to a pathway stimulated by peripheral CCK induction.

5-HT receptor agonists stimulate the hypothalamic pituitary adrenal (HPA) axis and affect energy balance. The possibility that 5-HT and corticotropin-releasing hormone (CRH) pathways interact in the regulation of feeding and energy expenditure is implicated by both the anorectic effects of CRH and evidence for synaptic connections of serotonergic fibers from the raphe nucleus with CRH-containing neurons in the paraventricular nucleus.[74] Acute administration of the 5-HT-releasing drug dexfenfluramine increases the concentration of CRH in the hypothalamus.[75] Thus, the anorexic effect of dexfenfluramine may involve CRH, a peptide that was shown to inhibit specifically fat, not carbohydrate, intake when given by the intracerebroventricular route in rats adapted to the three-choice macronutrient diet protocol.[76] However, central administration of the α-helical CRH antagonist failed to prevent the hypophagic effects of the 5-HT agonists RU-24969 (1A/1B) or DOI (2A/2C),[77] an observation that appears to argue against a role for CRH in the central anorectic effect of 5-HT. Nonetheless, both RU-24969 and DOI disrupt the behavioral satiety sequence (BSS) by inducing hyperactivity,[78] suggesting that these drugs produce hypophagia by a different pharmacological mechanism than that of other drugs that increase synaptic 5-HT activity, i.e., through the activation of 5-HT_{2A} receptors.[3] For example, the $5\text{-HT}_{1B/2C}$ agonists mCPP and

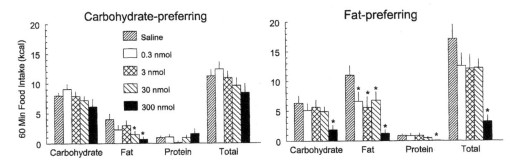

FIGURE 4   Effect of 5-HT infused into the paraventricular nucleus (PVN) of the hypothalamus of carbohydrate ($n = 11$) and fat-preferring rats on mean ($\pm$S.E.M.) kcal macronutrient intake at 60 min after dark onset. *$P < 0.05$. Bonferroni tests, compared to saline control (0 nmol).

d-fenfluramine, and the 5-HT$_{1B}$ agonist CP-94,253 all preserve the BSS.[78] Whether α-helical CRH will block the hypophagia or the fat-specific satiety induced by these additional serotonergic agonists remains to be tested.

Another hypothalamic peptide system that appears to interact with serotonin is neuropeptide Y (NPY). Several studies have shown that serotonergic drugs block the potent feeding behaviour induced by NPY in the PVN.[79-80] Antagonising serotonin receptors stimulates feeding and results in increased NPY mRNA and NPY secretion in the hypothalamus.[81] The feasibility of reciprocal effects of NPY on 5-HT is less clear, e.g., during feeding induced by NPY perfusion of the hypothalamus, 5-HT release and turnover may be decreased[82] or unchanged.[83]

It is clear that serotonergic drugs can decrease the consumption and selection of high-fat diets, as well as reduce daily fat intake (see Figures 1 through 4). Recent developments allow for the possibility that activation of specific 5-HT receptor subtypes may differentially shift the consumption/selection of macronutrient diets through an interaction with the NPY system. 5-HT$_{1B}$ agonists infused into the PVN result in hypophagia.[67,84] 5-HT$_{1B}$ receptors in the hypothalamus are located presynaptically[85] where they have been shown to inhibit neurotransmitter release.[86] As previously suggested,[87] some of the 5-HT$_{1B}$ receptors found in the PVN may be located on presynaptic NPY axon terminals originating from the arcuate nucleus, which contains a high density of 5-HT$_{1B}$ receptor mRNA.[88] Thus, it may be hypothesised that the stimulation of 5-HT$_{1B}$ receptors located on these terminals inhibits NPY release.[89] Given the evidence that hypothalamic NPY preferentially stimulates carbohydrate consumption,[89] the activation of 5-HT$_{1B}$ receptors in the PVN might result in a satiating effect on carbohydrate feeding. Relevant to this proposition, microinfusion of the selective 5-HT$_{1B}$ agonist CP-93,129 into the PVN decreased 60 min consumption of carbohydrate by 77%, with no effects on fat or protein intake (B. Smith, personal communication). In contrast, local application of d-norfenfluramine, the active metabolite of d-fenfluramine thought to activate postsynaptic 5-HT$_{2C}$ receptors in the PVN region, specifically decreased fat and protein intake but did not decrease carbohydrate feeding[34] (see Figure 3). The different effects on macronutrient selection produced by these 5-HT agonists suggest a possible role of receptor subtypes in macronutrient-specific satiety.

8   5-HT AND THE RISK FACTORS FOR OVEREATING

Many humans, possibly a majority, demonstrate a vulnerability to gain weight. One reason for this is the apparent ease with which certain people develop a positive energy balance. The risk factors that provoke the occurrence of a positive energy balance reside within the biological system, in the environment (food supply itself), and at the interface between food and the biological system. Some of the most potent risk factors that generate unwanted overconsumption (energy intake rising above energy expenditure) have been described elsewhere.[1,2] Foods that are energy dense and high in fat content provide a substantial risk for the overconsumption of energy and eventually obesity. Such risky foods include what are often termed convenience or snack food items. Drugs that could specifically target the intake of these foods would be useful in both treating and possibly preventing the development of obesity in vulnerable populations.

In conclusion, the effect of 5-HT manipulation on total food intake and satiety has been well documented. This has generated much interest in the possible effects of 5-HT drugs on food choice and macronutrient selection. A role for 5-HT in controlling carbohydrate intake was initially postulated, but more recent evidence indicates that 5-HT drugs could be used to control fat intake. In animals, drugs such as d-fenfluramine can inhibit the consumption of weight-inducing high-fat diets. Work by Smith et al.[33,34] demonstrates that 5-HT and 5-HT drugs administered in the hypo-thalamic area can suppress fat intake selectively. Recent human studies have also shown that 5-HT drugs suppress the consumption of palatable high-fat foods such as snacks and, in some cases, can selectively reduce fat consumption. The precise role of $5-HT_{1B}$ and $5-HT_{2C}$ receptors in these macronutrient-selective, 5-HT satiety effects are not yet known. If direct activation of these receptors does curtail the overconsumption of high-fat, energy-dense foods these drugs may be useful in treating obesity and in reducing the disease risks of dietary fat-related diseases.

REFERENCES

1. Blundell, J. E. and MacDiarmid, J. I., Passive overconsumption: fat intake and short-term energy balance, *Ann. N.Y. Acad. Sci.,* 827, 392, 1997.
2. Golay, A. and Bobbioni, E., The role of dietary fat in obesity, *Int. J. Obesity,* 21, s2 1997.
3. Blundell, J. E. and Halford, J. C. G., Serotonin and appetite regulation: implications for the pharma-cological treatment of obesity, *CNS Drugs,* 9, 473, 1998.
4. Blundell, J. E., Is there a role for serotonin (5-hydroxytryptamine) in feeding? *Int. J. Obesity,* 1, 15, 1977.
5. Hoyer, D. and Martin, G., 5-HT receptor classification and nomenclature: towards a harmonisation with the human genome, *Neuropharmacology,* 36, 419, 1997.
6. Fernstrom, J. D. and Wurtman, R. J., Control of brain 5-HT content by dietary carbohydrates, in *Serotonin and Behaviour,* Barchas, J. and Usdin, E., Eds., Academic Press, New York, 1973, 121.
7. Wurtman, R. J. and Fernstrom, J. D., Effects of diet on brain neurotransmitters, *Nutr. Rev.,* 32, 193, 1974.
8. Teff, K. L., Young, S. N., and Blundell, J. E., The effect of protein or carbohydrate breakfasts on subsequent plasma amino acid levels, satiety and nutrient selection in normal males, *Pharmacol. Biochem. Behav.,* 34, 410, 1989.
9. Fernstrom, J. D., Food-induced changes in brain serotonin synthesis: is there a relationship to appetite for specific macronutrients? *Appetite,* 8, 163, 1987.
10. Blundell, J. E. and Hill, A. J., On the mechanism of action of dexfenfluramine: effect on alliesthesia and appetite motivation in lean and obese subjects, *Cln. Neuropharmacol.,* 11(s), 121, 1988.
11. Blundell, J. E. and Lawton, C. L., Serotonin and dietary fat intake: effects of dexfenfluramine, *Met. Clin. Exp.,* 44, 33, 1995.
12. Pirke, K. M., Schweiger, U., and Laessle, R. G., Effects on diet composition on affective state in anorexia nervosa and bulimia, *Clin. Neuropharmacol.,* 9, 561, 1986.
13. Ashley, D. V. M., Fleury, M. O., Golay, A., Maeder, E., and Leathwood, P. D., Evidence for diminished brain 5-HT biosynthesis in obese diabetic humans, *Am. J. Clin. Nutr.,* 42, 1240, 1985.
14. Brewerton, T. D., Toward a unified theory of serotonin dysregualtion in eating related disorders, *Psychoneuroendocrinology,* 20, 561, 1995.
15. Jimerson, D. C., Wolfe, B. E., Metzger, E. D., Finkelstein, D. M., Cooper, T. B., and Levine, J. M., Decreased serotonin function in bulimia nervosa, *Arch. Gen. Psychiatr.,* 54, 529, 1997.
16. Anderson, I. M., Parry-Billins, M., Newsholme, E. A., Fairburn, C. G., and Cowen, P. J., Dieting reduced plasma tryptophan and alters brain 5-HT functioning in women, *Psychol. Med.,* 20, 785, 1990.

17. Walsh, A. E., Oldman, A. D., Franklin, M., Fairburn, C. G., and Cowen, P. J., Dieting decreases plasma tryptophan and increases the prolactin response to d-fenfluramine in women but not men, *J. Affective Disorders,* 33, 89, 1995.

18. Goodwin, G. M., Fairburn, C. G., and Cowen, P. J., Dieting changes 5-HT function in women but not in men, implications for the aetiology of anorexia nervosa? *Psychol. Med.,* 17, 839, 1987.

19. Goodwin, G. M., Cowen, P. J., Fairburn, C. J., Parry-Billings, M., Calder, P. C., and Newsholme, E. A., Plasma concentrations of tryptophan and dieting, *Br. Med. J.,* 17, 839, 1990.

20. Cowen, P. J., Clifford, E. M., Walsh A. E., Williams, C., and Fairburn, C. G., Moderate dieting causes 5-HT$_{2C}$ supersensitization, *Psychol. Med.,* 26, 1156, 1996.

21. Wurtman, J. J. and Wurtman, R. J., Fenfluramine and fluoxetine spare protein consumption while suppressing caloric intake by rats, *Science,* 198, 1178, 1977.

22. Wurtman, J. J. and Wurtman, R. J., Drugs that enhance central serotoninergic transmission diminish elective carbohydrate consumption by rats, *Life Sci.,* 24, 895, 1979.

23. Kim, S. and Wurtman, R. J., Selective effects of CGS 10686B, dl-fenfluramine or fluoxetine on nutrient selection, *Physiol. Behav.,* 42, 319, 1988.

24. Luo, S. and Li, E. T. S., Food intake and selection pattern of rats treated with dexfenfluramine, fluoxetine and RU-24969, *Brain Res. Bull.,* 24, 729, 1990.

25. Luo, S. and Li, E. T. S., Effects of repeated administration of serotonergic agonists on diet selection and body weight in rats, *Pharmacol. Biochem. Behav.,* 38, 495, 1991.

26. Lawton, C. L. and Blundell, J. E., The effects of d-fenfluramine on intake of carbohydrate supplements is influenced by the hydration of test diets, *Physiol. Behav.,* 53, 517, 1992.

27. Lawton, C. L. and Blundell, J. E., 5-HT and carbohydrate suppression — effects of 5-HT anatagonist on the action of d-fenfluramine DOI, *Pharmacol. Biochem. Behav.,* 46, 349, 1993.

28. Lawton, C. L. and Blundell, J. E., 5-HT manipulation and dietary choice: variable carbohydrate (polycose) suppression demonstrated only under specific experimental conditions, *Psychopharmacology,* 112, 375, 1993.

29. Orthen-Gambill, N. and Kanarek, R. B., Differential effects of amphetamine and fenfluramine on dietary self-selection in rats, *Pharmacol. Biochem. Behav.,* 16, 303, 1982.

30. Kanarek, R. B. and Dushkin, H., Peripheral serotonin administration selectively reduces fat intake in rats, *Pharmacol. Biochem. Behav.,* 31, 113, 1988.

31. Shor-Posner, G., Grinker, J. A., Marinescu, C., Brown, O., and Leibowitz, S. F., Hypothalamic serotonin in the control of meal patterns and macronutrient selection, *Brain Res. Bull.,* 17, 663, 1986.

32. Heisler, L. K., Kananek, R. B., and Gerstien, A., Fluoxetine decreases fat and protein intake but not carbohydrate intake in male rats, *Pharmacol. Biochem. Behav.,* 58, 767, 1997.

33. Smith, B. K., York, D. A., and Bray, G. A., Chronic *d*-fenfluramine treatment reduces fat intake independent of macronutrient preference, *Pharmacol. Biochem. Behav.,* 60, 105, 1998.

34. Smith, B. K., York, D. A., and Bray, G. A., Activation of hypothalamic serotonin receptors reduced intake of dietary fat and protein but not carbohydrate, *Am. J. Physiol.,* 277, 1999.

35. Leibowitz, S. F., Weiss, G. F., Walsh, U. A., and Viswanath, D., Medial hypothalamic serotonin: role in circadian patterns of feeding and macronutrient selection, *Brain Res.,* 503, 132, 1989.

36. Leibowitz, S. F., Weiss, G. F., and Suh, J. S., Medical hypothalamic nuclie mediate serotonins inhibitory effect of feeding behaviour, *Pharmacol. Biochem. Behav.,* 37, 735, 1990.

37. Leibowitz, S. F., Alexander, J. T., Cheung, W. K., and Weiss, G. F., Effects of serotonin and serotonin blocker metergoline on meal patterns macronutrient selection, *Pharmacol. Biochem. Behav.,* 45, 185, 1993.

38. Weiss, G. F., Rogacki, N., Fueg, A., Buchen, D., and Leibowitz, S. F., Impact of hypothalamic d-norfenfluramine and peripheral d-fenfluramine on macronutrient intake in the rat, *Brain Res. Bull.,* 25, 849, 1990.

39. Weiss, G. F., Rogacki, N., Fueg, A., Buchen, D., Suh, J. S., Wong, D. T., and Leibowitz, S. F., Effect of hypothalamic and peripheral fluoxetine injection on natural patterns of macronutrient intake in the rat, *Psychopharmacology,* 105, 467, 1991.

40. Max, J. P., Thystere, P., Chapleur-Chateau, M., Burlet, A., Nicolas, J. P., and Burlet, C., Hypothalamic neuropeptides could mediate the anorectic effects of fenfluramine, *NeuroReport,* 5, 1925, 1994.

41. Mullen, B. J. and Martin, R. J., The effect of dietary fat on diet selection may involve central serotonin, *Am. J. Physiol.,* 263, R559, 1992.

42. Sclafani, A., Animal models of obesity: classification and characterisation, *Int. J. Obesity,* 8, 491, 1984.
43. Rogers, P. J. and Blundell, J. E., Meal patterns and food selection during the development of obesity in rats fed a cafeteria diet, *Neurosci. Biobehav. Rev.,* 8, 441, 1984.
44. Blundell, J. E. and Hill, A. J., Do serotoninergic drugs decrease energy intake by reducing fat or carbohydrate intake? Effects of d-fenfluramine with supplemented weight increase diets, *Pharmacol. Biochem. Behav.,* 31, 773, 1989.
45. Prats, E., Monfar, M., Castell, J., Iglesias, R., and Allemany, M., Energy intake of rats fed a cafeteria diet, *Physiol. Behav.,* 45, 2263, 1989.
46. Fisler, J. S., Underberger, S. J., York, D. A., and Bray, G. A., d-Fenfluramine in a rat model of dietary fat-induced obesity, *Pharmacol. Biochem. Behav.,* 45, 487, 1993.
47. Blundell, J. E., Lawton, C. L., and Halford, J. C. G., Serotonin, eating behaviour and fat intake, *Obesity Res.,* 3(4s), 471, 1995.
48. Wurtman, J. J., Wurtman, R. J., and Marks, S., d-Fenfluramine selectively suppresses carbohydrate snacking by obese subjects, *Int. J. Eating Dis.,* 4, 89, 1985.
49. Wurtman, J. J., Wurtman, R. J., Reynolds, S., Fenfluramine suppresses snack intake among carbohydrate cravers but not among noncarbohydrate cravers, *Int. J. Eating Dis.,* 6, 687, 1987.
50. Drewnowski, A. Changes in mood after carbohydrate consumption, *Am. J. Clin. Nutr.,* 46, 703, 1987.
51. Hill, A. J. and Blundell, J. E., Model system for investigating the actions of anorectic drugs: effects of d-fenfluramine on food intake, nutrient selection, food preference, meal patterns, hunger and satiety in human subjects, in *Advances in the Biosciences,* Pergamon Press, Oxford, 1986, 377.
52. Foltin, R. W., Haney, M., Comer, S., and Fischman, M. W., Effect of fenfluramine on food intake, mood, and performance of humans living in a residential laboratory, *Physiol. Behav.,* 59, 295 1996.
53. Lawton, C. L., Wales, J. K., Hill, A. J., and Blundell, J. E., Serotoninergic manipulation, meal-induced satiety and eating patterns, *Obesity Res.,* 3, 345, 1995.
54. Foltin, R. W., Haney, M., Comer, S., and Fischman, M. W., Effects of fluoxetine on food intake of humans living in a residential laboratory, *Appetite,* 27, 165, 1996.
55. Pilj, H., Koppeschaar, H. P. F., Willekens, F. L. A., de Kamp, I. O., Veldhuis, H. D., and Meinder, A. E., Effect of serotonin re-uptake inhibition by fluoxetine on body weight and spontaneous food choice in obesity, *Int. J. Obesity,* 15, 237, 1991.
56. Walsh, A. E., Smith, K. A., Oldman, A. D., Williams, C., Goodall, E. M., and Cowen, P. J., m-Chlorophenylpiperazine decrease food intake in a test meal, *Psychopharmacology,* 116, 120, 1994.
57. Goodall, E. M., Cowen, P. J., Franklin, M., and Silverstone, T., Ritanserin attenuates anorectic endocrine and thermic responses to d-fenfluramine in human volunteers, *Psychopharmacology,* 112, 461, 1993.
58. Boeles, S., Williams, C., Campling, G. M., Goodall, E. M., and Cowen, P. J., Sumatriptan decreases food intake and increases plasma growth hormone in healthy women, *Psychopharmacology,* 129, 179, 1997.
59. Lafreniere, F., Lambert, J., Rasio, E., and Serri, O., Effects of dexfenfluramine treatment on body weight and postprandial thermogenise in obese subjects. A double blind placebo-controlled study, *Int. J. Obesity,* 17, 25, 1993.
60. Cangiano, C., Laviano, A., Del Ben, M., Preziosa, I., Angelico, F., Cascino, A., and Rossi Fanelli, F. Effects of oral 5-hydroxy-tryptophan on energy intake and macronutrient selection in non-insulin dependent diabetic patients, *Int. J. Obesity,* 22, 648, 1998.
61. Poppitt, S. D., Murgatroyed, P. R., Tainsh, K. R., and Prentice, A. M., The effect of dexfenfluramine on energy and macronutrient balance of obese women on high fat and low-fat diets, *Int. J. Obesity,* 21(s), 197, 1997.
62. Green, S., Lawton, C. L., Wales, J. K., and Blundell, J. E., Risk factors for overeating: dex-fenfluramine suppresses the intake of sweet high fat or carbohydrate food in obese women, *Int. J. Obesity,* 21(s), s64, 1997.
63. Drewnowski, A., Kurth, C., Holden-Wiltse, J., and Saari, J., Food preference in human obesity: carbohydrates verses fat, *Appetite,* 18, 207, 1992.
64. Drewnowski, A., Why do we like fat? *J. Am. Diet. Assoc.,* 97, s58, 1997.
65. Spannagel, A. W., Nakano, I., Tawil, T., Chey, W. Y., Liddle, R. A., and Green, G. M., Adaptation to fat markedly increases pancreatic secretory response to intraduodenal fat in rats, *Am. J. Physiol.,* 270, G128, 1996.

66. Leibowitz, S. F., Weiss, G. F., and Shor-Posner, G., Hypothalamic serotonin: pharmacological, biochemical, and behavioral analyses of its feeding-suppressive action, *Clin. Neuropharmacol.*, 11(S1), S51, 1988.
67. Hutson, P. H., Donohoe, T. P., and Curzon, G., Infusion of the 5-hydroxytryptamine agonists RU24969 and TFMPP into the paraventricular nucleus of the hypothalamus causes hypophagia, *Psychopharmacology*, 95, 550, 1988.
68. Grignaschi, G., Mantelli, B., Fracasso, C., Anelli, M., Caccia, S., and Samanin, R., Reciprocal interaction of 5-hydroxytryptamine and cholecystokinin in the control of feeding patterns in rats, *Br. J. Pharmacol.*, 109, 491, 1993.
69. Poeschla, B., Gibbs, J., Simansky, K. J., Greenberg, D., and Smith, G. P., Cholecystokinin-induced satiety depends on activation of 5-HT1C receptors, *Am. J. Physiol.*, 264, R62–R64, 1993.
70. Stallone, D., Nicolaidis, S., and Gibbs, J., Cholecystokinin-induced anorexia depends on serotoninergic function, *Am. J. Physiol.*, 256, R1138, 1989.
71. Schwartz, D. H., Hernandez, L., and Hoebel, B. G., Serotonin release in lateral and medial hypothalamus during feeding and its anticipation, *Brain Res. Bull.*, 25, 797, 1990.
72. Voigt, J.-P., Sohr, R., and Fink, H., CCK-8S facilitates 5-HT release in the rat hypothalamus, *Pharmacol. Biochem. Behav.*, 59, 179, 1998.
73. McCoy, J. G., Stump, B., and Avery, D. D., Intake of individual macronutrients following IP injections of BBS and CCK in rats, *Peptides*, 11, 221, 1990.
74. Liposits, Z., Phelix, C., and Paull, W. K., Synaptic interaction of serotonergic axons and corticotropin releasing factor (CRF) synthesizing neurons in the hypothalamic paraventricular nucleus of the rat, *Histochemistry*, 86, 541, 1987.
75. Holmes, M. C., Di Renzo, G., Beckford, U., Gillham, B., and Jones, M. T., Role of serotonin in the control of secretion of corticotrophin releasing factor, *J. Endocrinol.*, 93, 151, 1982.
76. Lin, L., York, D., and Bray, G., Acute effects of intracerebroventricular corticotropin releasing hormone (CRH) on macronutrient selection, *Int. J. Obesity*, 16 (Suppl. 1), 52, 1992.
77. Bovetto, S., Rouillard, C., and Richard, D., Role of CRH in the effects of 5-HT-receptor agonists on food intake and metabolic rate, *Am. J. Physiol.*, 271, R1231, 1996.
78. Halford, J. C., Wanninayake, S. C., and Blundell, J. E., Behavioral satiety sequence (BSS) for the diagnosis of drug action on food intake, *Pharmacol. Biochem. Behav.*, 61, 159, 1998.
79. Dryden, S., Frankish, H. M., Wang, Q., Pickavance, L., and Williams, G., The serotonergic agent fluoxetine reduces neuropeptide Y levels and neuropeptide Y secretion in the hypothalamus of lean and obese rats, *Neuroscience*, 72, 557, 1996.
80. Dryden, S., Wang, Q., Frankish, H. M., and Williams, G., Differential effects of the 5-HT1B/2C receptor agonist mCPP and the 5-HT1A agonist flesinoxan on hypothalamic neuropeptide Y in the rat, *Peptides*, 6, 943, 1996.
81. Dryden, S., Wang, Q., Frankish, H. M., Pickavance, L., and Williams, G., The serotonin (5-HT) antagonist methysergide increases neuropeptide Y (NPY) synthesis and secretion in the hypothalamus of the rat, *Brain Res.*, 13, 12, 1995.
82. Shimizu, H. and Bray, G.A., Effects of neuropeptide Y on norepinephrine and serotonin metabolism in rat hypothalamus in vivo, *Brain Res. Bull.*, 22, 945, 1989.
83. Myers, R. D., Lankford, M. F., and Paez, X., Norepinephrine, dopamine, and 5-HT release from perfused hypothalamus of the rat during feeding induced by neuropeptide Y, *Neurochem. Res.*, 17, 1123, 1992.
84. Macor, J. E., Burkhart, C. A., Heym, J. M., Ives, J. L., Lebel, L. A., et al., 3-(1,2,5,6-Tetrahydropyrid-4-yl)pyrrolo[3,2-*b*]pyrid-5-one: a potent and selective serotonin (5-HT$_{1B}$) agonist and rotationally restricted phenolic analogue of 5-methoxy-3-(1,2,5,6-tetrahydropyrid-4-yl)indole, *J. Med. Chem.*, 33, 2087, 1990.
85. Frankfurt, M., Mendelson, S. D., McKittrick, C. R., and McEwen, B. S., Alterations of serotonin receptor binding in the hypothalamus following acute denervation, *Brain Res.*, 601, 349, 1993.
86. Hoyer, D. and Middlemiss, D. N., Species differences in the pharmacology of terminal 5-HT autoreceptors in mammalian brain, *Trends Pharmacol. Sci.*, 10, 130, 1989.
87. Lucas, J. J., Yamamoto, A., Scearce-Levie, K., Saudou, F., and Hen, R., Absence of fenfluramine-induced anorexia and reduced c-*fos* induction in the hypothalamus and central amygdaloid complex of serotonin 1B receptor knock-out mice, *J. Neurosci.*, 18, 5537, 1998.

88. Bruinvels, A. T., Landwehrmeyer, B., Gustafson, E. L., Durkin, M. M., Mengod, G., Branchek, D., Hoyer, T. A., and Palacios, J. M., Localization of 5-HT$_{1B}$, 5-HT$_{1Da}$, 5-HT$_{1E}$ and 5-HT$_{1F}$ receptor messenger RNA in rodent and primate brain, *Neuropharmacology*, 33, 367, 1994.
89. Stanley, B. G., Daniel, D. R., Chin, A. S., and Leibowitz, S. F., Paraventricular nucleus injections of peptide YY and neuropeptide Y preferentially enhance carbohydrate ingestion, *Peptides*, 6, 1205, 1985.
90. Moses, P. L. and Wurtman, R. J., The ability of certain anorexic drugs to suppress food consumption depends on the nutrient compostion of the test diet, *Life Sci.*, 35, 1297, 1984.
91. Li, E. T. S. and Luo, S. Q., Buspirone-induced carbohydrate feeding in not influenced by the route of administration and nutritional status, *Brain Res. Bull.*, 30, 547, 1993.
92. Kanarek, R. B., Glick, A. L., and Marks-Kaufman, R., Dietary influences on the acute effects of anorectic drugs, *Physiol. Behav.*, 49, 149, 1991.
93. White, P. J., Cybulski, K. A., Primus, R., Johnson, D. F., Collier, G. H., and Wagner G. C., Changes in macronutrient selection as a function of dietary tryptophan, *Physiol. Behav.*, 43, 73, 1988.
94. Wurtman, J. J., Wurtman, R. J., Growdon, J. H., Henry, P., Lipscomb, A., and Zeisel, S. H., Carbohydrate craving in obese people: suppression by treatments affecting serotoninergic transmission, *Int. J. Eating Dis.*, 1, 2, 1981.
95. Wurtman, J. J., Wurtman, R. J., Berry, E., Gleason, R., Goldberg, H., McDermott, J., Kahne, M., and Tsay, R., Dexfenfluramine, fluoxetine, and weight less among female carbohydrate cravers, *Neuropsychopharmacology*, 9, 201, 1993.
96. Hill, A. J. and Blundell, J. E., Sensitivity of the appetite control systm in obese subjects to nutritional and serotoninergic challenges, *Int. J. Obesity*, 14, 219, 1990.

28 Effects of Pure Macronutrient Diets on 5-HT Release in the Rat Hypothalamus: Relationship to Insulin Secretion and Possible Mechanism for Feedback Control of Fat and Carbohydrate Ingestion

Martine Orosco, Kyriaki Gerozissis, and Stylianos Nicolaïdis

CONTENTS

1 5-HT AND MACRONUTRIENT SELECTION

The involvement of brain 5-HT in the control of food intake is well established and has been extensively reviewed[1,2] (see also Chapter 27 by Halford et al. in this volume). Numerous pharmacological manipulations using precursors, agonists, or transmission enhancers have contributed to evidence consistent with a role for serotonin in promoting satiety. In search of the site of action, studies have mainly concentrated on the hypothalamus, especially the paraventricular (PVN) and the ventromedial (VMN) nuclei,[3] although 5-HT transmission in the brain stem and in the gastrointestinal tract is also involved in the control of food intake.[4,5] The introduction of *in vivo* techniques brought about important improvements in the knowledge of the physiological role of 5-HT in food intake regulation. Using microdialysis in the PVN and VMH in freely moving rats and simultaneous recording of spontaneous feeding behavior, we measured the levels of serotonin before, during, and after a spontaneously occurring meal. We observed increases in 5-HT and in its metabolite, 5-hydroxyindolacetic acid (5-HIAA) as soon as the beginning of the meal.[6] We ascribed these changes to the release of 5-HT as a signal of meal termination in line with the role of the amine in satiety and particularly in the satiation process.

For many years, specific intakes with regard to carbohydrates and proteins were especially investigated because they both affect 5-HT synthesis in a way that was proposed to underlie specific appetites and specific satieties.[7-9] Until recently, the studies on the effects of serotonergic compounds focused exclusively on carbohydrate and protein ingestion and described a selective reduction in carbohydrate intake.[7,10,11] The relation between serotonin and dietary fat was less investigated, although indirectly suggested.[12,13] However, recent evidence, using dexfenfluramine, indicates that enhancement of serotonin activity may also, and even more selectively, reduce fat consumption.[14,15] Such a reduction of fat intake by serotonin had been observed previously. However, a peripheral mechanism was involved.[16]

2   CHANGES OF HYPOTHALAMIC 5-HT RELEASE IN RESPONSE TO THE INGESTION OF SPECIFIC NUTRIENTS

The method of injection of neuroactive substances directly into the brain has a nonphysiological component, which can make functional interpretation difficult. Such injections do not exactly mimic the spatial distribution, temporal profile, and concentration of naturally released transmitters. The technique of microdialysis is a useful tool because it allows one to follow changes of local transmitter release from a circumscribed brain area throughout the display of specific behaviors, including ingestive behavior. We used the microdialysis technique to measure 5-HT release into the interstitial space in response to the ingestion of pure macronutrients.[16a] We placed the tip of the probe into an area approximately in the middle between the paraventricular and ventromedial nuclei. Because of the geometry of the probe, interstitial fluid from any of the neighboring areas in the medial hypothalamus, including the paraventricular, ventromedial, dorsomedial, anterior, and periventricular nuclei, could potentially enter the probe. Animals that had previously been familiarized with the pure macronutrient diets were food deprived overnight. On test days, the probe was inserted, and after the collection of four baseline samples (30 µl, 15 min, 2 µl/min), access to the food was allowed.

Ingestion of a carbohydrate meal (85% starch, 15% sucrose, without vitamins and minerals) resulted in an immediate increase in hypothalamic 5-HT release, that reached significance in the third sample (30 to 45 min, +157%), peaked between 45 and 75 min (+200%), and returned to baseline within about 2 h after the start of ingestion (Figure 1).

In stark contrast, ingestion of pure fat (lard) was accompanied by a decrease of hypothalamic 5-HT release which reached significance in the 15 to 30 min sample (–42%), remained for about 90 min and then returned to control levels. Finally, ingestion of pure protein (casein) also resulted in a significant decrease of 5-HT of a similar magnitude and time course as with fat ingestion (Figure 1).

We then calculated correlation coefficients between the amount of a specific macronutrient ingested as well as the total time spent eating and the responses of 5-HT. The responses were estimated both from their peak value and from the mean of peri- and postprandial changes. The only significant correlation was found between the mean decrease in 5-HT and the ingested amount of lard ($r^2 = 0.864$, $p < 0.01$). No significant correlations were found with the carbohydrate or protein meals. Also, no significant correlations were found between the changes in 5-HT and the total time spent eating.

Therefore, ingestion of pure carbohydrate, fat, or protein produced vastly different and opposite changes in extracellular hypothalamic 5-HT concentration, most likely the result of changes of 5-HT release from serotonergic nerve terminals. Does this support a role for serotonin in the control of specific macronutrient intake? As reviewed above and discussed in greater detail in Chapter 27 by Halford et al., experiments using local injection of various serotonergic agonists and antagonists suggest that in the hypothalamus, 5-HT acts to suppress fat intake selectively. Suppression of 5-HT release by eating fat, as found in our study, would thus tend to stimulate, rather than suppress,

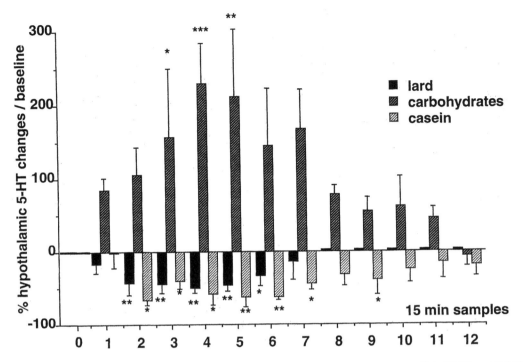

FIGURE 1   Changes in 5-HT levels in 15 min microdialysate samples from rostromedial (VMH and PVN) hypothalamus during and after a meal of lard, carbohydrates, or casein. Results are expressed as mean % change from basal levels preceding the meal ± SEM. *$p < 0.05$, **$p < 0.01$, ***$p < 0.001$, compared to basal levels.

further fat intake. Therefore, the results of our study are not consistent with the concept of a simple negative feedback control of fat intake via hypothalamic 5-HT release. The results rather suggest a positive feedback relationship with fat ingestion causing decreased hypothalamic 5-HT release and a possible disinhibition of fat appetite. Such a positive feedback or feed-forward relationship could be responsible for maintaining fat intake within a meal. Smith et al.[15] has observed that the dose–response curve for intracerebral injection of 5-HT to suppress fat intake is shifted to the left in fat-preferring as compared with carbohydrate-preferring rats. Such an increased sensitivity to exogenous 5-HT could be due to chronically lowered hypothalamic 5-HT release in fat-preferring rats, as predicted from our microdialysis results.

With respect to carbohydrate, the massive and rapid increase in 5-HT release during and following ingestion of a pure carbohydrate meal may be taken as evidence for hypothalamic 5-HT playing a role as negative feedback signal to bring the meal to an end (satiation), and the long duration of its increase to maintain satiety during the ensuing intermeal interval. Although this explanation is consistent with Fernstrom's view, it is not consistent with experiments demonstrating selective effects of hypothalamic 5-HT transmission on fat but not carbohydrate intake.[14-16] In view of these latter reports, increases of hypothalamic 5-HT-induced by carbohydrate could be interpreted as a carbohydrate specific feed-forward mechanism with concomitant suppression of fat appetite.

3   EFFECTS OF INSULIN ON FOOD SELECTION

Among the peptide hormones, insulin is also well recognized to play a role in the central control of food intake and body weight, and its possible effects on macronutrient selection are discussed in detail in Chapter 30 by van Dijk et al. The first investigation to show a direct effect of insulin on central structures used slow and prolonged bilateral insulin infusion into the medial hypothalamus of the rat.[17] These infusions produced hypophagia and a permanent reduction in body weight. Similar effects of intracerebroventricular and hypothalamic infusions have been reported in both the rat and the baboon.[18,19] The transport of insulin into the brain, and the presence of insulin receptors, especially in the hypothalamus, is well established.[20-23] In microdialysis investigations, we assessed immunoreactive insulin (IRI) using a sensitive radioimmunoassay (RIA) in the dialysates from the VMN and PVN, while freely behaving rats were ingesting a meal of laboratory chow.[24,24a] The same rats were equipped with a jugular catheter for remote blood sampling. In these dynamic and parallel measurements of both hypothalamic and circulating insulin, the chow meal induced not only the expected plasma insulin increase, but also a clear-cut increase in extracellular hypothalamic insulin.[25,25a]

Brain insulin has not been much studied with regard to macronutrient ingestion so far, while at the peripheral level, insulin secretion is known to be nutrient dependent.[26] Data on the effects of centrally administered insulin are still confusing. On the one hand, insulin loses its capacity in reducing feeding when the proportion of fat in the diet increases, suggesting that central insulin is ineffective in reducing fat intake.[27] On the other hand, when a choice of the three macronutrients is offered, insulin administered by i.c.v. route or in the arcuate nucleus selectively decreases fat and augments carbohydrate consumption.[28]

4   EFFECTS OF SPECIFIC NUTRIENTS ON PERIPHERAL AND HYPOTHALAMIC INSULIN

The carbohydrate meal induced the expected large increases in both plasma and hypothalamic IRI levels (Figures 2 and 3). Extracellular hypothalamic IRI concentrations were increased in the first 30-min sample, and the increase reached statistical significance in the second 30-min sample of the meal and lasted beyond the duration of the meal. In contrast, the casein meal, while producing a delayed but significant increase in plasma IRI levels, did not induce any significant change in

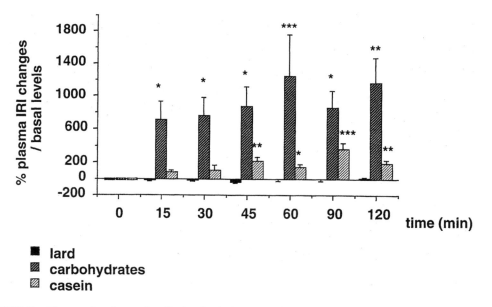

FIGURE 2 Changes in plasma insulin levels during and after a meal of lard, carbohydrates, or casein. Results are expressed as mean % change from basal levels preceding the meal ± SEM. *$p < 0.05$, **$p < 0.01$, ***$p < 0.001$, compared to basal levels.

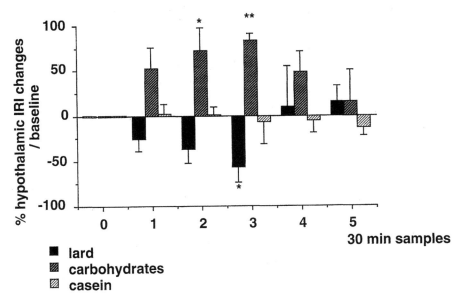

FIGURE 3 Changes in immunoreactive insulin levels in 30 min microdialysate samples from rostromedial (VMH and PVN) hypothalamus during and after a meal of lard, carbohydrates, or casein. Results are expressed as mean % change from basal levels preceding the meal ± SEM. $*p < 0.05$, $**p < 0.01$, compared to basal levels.

hypothalamic IRI levels at any time, and the lard meal, while not changing plasma levels, significantly lowered hypothalamic IRI levels (Figure 3). Therefore, with both the protein and fat meals, brain insulin did not closely track peripheral insulin levels, raising the possibility that temporary changes in insulin transporter activity or insulin degradation, or both, may be differentially affected by these diets.

Finally, because insulin action on the hypothalamus has been portrayed as an important modulator of food intake, let us consider its potential to control macronutrient selection selectively. As with 5-HT, ingestion of pure macronutrient meals produced distinctly different effects on hypothalamic insulin. As expected, a pure carbohydrate meal produced a large increase of hypothalamic IRI. Under the assumption that insulin plays a role as a negative-feedback signal for carbohydrate intake, it could be expected that local injection of insulin specifically suppresses carbohydrate intake. This is, however, not what has been reported. Insulin injected directly into the arcuate nucleus of the hypothalamus, but not the PVN or third ventricle, selectively suppressed fat consumption without an effect on carbohydrate consumption in a choice situation.[28] Therefore, at first glance, our microdialysis results measuring hypothalamic IRI are not consistent with a negative-feedback model (but see discussion in Chapter 30 by van Djik et al.).

In contrast to carbohydrate ingestion, hypothalamic IRI concentration, just like 5-HT release, was significantly decreased following ingestion of a pure fat meal (Figure 3). Therefore, similar theoretical arguments as developed above for 5-HT can be made with respect to hypothalamic IRI playing a role as a negative- or positive-feedback signal for fat intake.

5   RELATIONSHIP BETWEEN CHANGES IN HYPOTHALAMIC 5-HT, INSULIN, AND THE SATIATING EFFECT OF SINGLE-MACRONUTRIENT DIETS

Since the central action of both serotonin and insulin as inhibitors of feeding is well recognized and recently, their involvement in specific inhibition of fat intake was suggested, and because both serotonin and insulin levels in the medial hypothalamus increase in response to a meal, a potential interaction between the indoleamine and the peptide in the control of food intake and selection was indicated. We may ask the question whether 5-HT release is mediated by changes in hypothalamic insulin. The completely parallel but reciprocal changes induced by carbohydrate and fat may suggest some interdependency of the two signals. However, because the pure-protein meal significantly decreased 5-HT release in the face of modestly changed peripheral and unchanged hypothalamic IRI levels, these two anorectic signals seem to operate independently. It is difficult to obtain clarification of this issue from the literature. Extensive investigations on the possible interactions between 5-HT and insulin have led to contradictory results, because of the multiplicity of the routes of administration and of the parameters measured. Peripheral administration of insulin by itself was reported to enhance 5-HT synthesis,[29-31] but also to reduce its release.[32] Central administration of insulin also resulted in conflicting data on 5-HT changes.[33-35] On the other hand, 5-HT agonists or antagonists acted in various ways on pancreatic insulin secretion, either enhancing[36] or inhibiting it.[37] A more recent work instead reports an activation of insulin secretion in response to 5-HT.[38]

As we have seen above, some of the changes in hypothalamic 5-HT and insulin may reflect peripheral modifications. This is the case for the changes in 5-HT levels following the carbohydrate and protein meals, in agreement with the classical mechanism proposed by Fernstrom and Wurtman[9,39] and also for the changes in hypothalamic insulin after carbohydrate ingestion in agreement with a transport of peripheral insulin into the brain.[40,41] In that case, the smaller and shorter 5-HT increase observed during a chow meal[6] may be due to the fact that the carbohydrate effect is blunted by the influence of the other nutrients. However, this theory cannot be applied to hypothalamic insulin, the increase of which is still higher in response to chow than to carbohydrates. Thus, the mechanisms cannot be the same in all cases and cannot be only of metabolic origin. Furthermore, if the mechanisms of central 5-HT and IRI changes were all peripheral in origin, the effects of both chow and carbohydrate meals should be postabsorptive and thus should appear later after the beginning of ingestion but not as soon as the first 15 min of the meal. Even the decreases observed during the protein meals (5-HT) and during the lard meals (5-HT and insulin) occur too early to follow only a primary peripheral mechanism.

Thus, other mechanisms should exist, probably directly at the central level, at least during the first minutes of the meals. Although the nature of these mechanisms is not known, they may underlie the release of 5-HT and increased transport of insulin into the hypothalamus caused directly by the ingestion of a meal to induce short-term satiation, as we proposed in our previous work.[6,25] Then, subsequent mechanisms involving a peripheral primary site would be at the origin of longer-term satiety. For these reasons, some consideration should be given to the satiating capacity of our pure macronutrient test meals. Because both 5-HT and IRI are considered as anorexigenic agents, it is tempting to correlate the level of their release with the corresponding level of satiety. To this end, chow was presented after the rats had ingested each macronutrient test meal to satiation. The delay of initiation of the chow meal as well as its volume were considered: the more satiated the animal, the longer the delay for the initiation of the subsequent chow meal and the less its volume. It happened that following ingestion of the three macronutrients, all the rats immediately consumed a meal of chow without any latency. Consequently, the satiating potency of the nutrients could not be classified on this basis. In terms of the amount of chow consumed by the three groups, rats that ate fat to satiation consumed the largest amount of chow, and rats that became satiated on the pure carbohydrate meal consumed the smallest amount of chow, with the protein group being intermediate. In other words, eating a large meal of pure fat but not pure carbohydrate leaves rats still hungry for their chow maintenance diet. Because chow is mostly carbohydrate, one explanation of this finding is the concept of sensory-specific satiety.

6   CONCLUSIONS

There are at least two neurochemical systems in the hypothalamus, insulin and 5-HT, that are similarly affected by the intake of two different macronutrients. Carbohydrate ingestion increases while fat ingestion decreases hypothalamic concentration of both substances. When injected directly into the brain the two substances also seem to act similarly on macronutrient selection, by selectively reducing fat intake. We have, therefore, asked the question whether hypothalamic 5-HT and insulin are part of a unique cascade of signals, but additional experiments are necessary to answer the question. Furthermore, the facts that (1) fat ingestion decreases and carbohydrate ingestion increases hypothalamic 5-HT and IRI (present results) and (2) both exogenous 5-HT and insulin tend to decrease fat consumption selectively are not consistent with a simple negative-feedback model for macronutrient self-selection via these substances. Future experiments using the microdialysis technique will have to inspect different anatomical locations, taking into account the differential distribution and involvement of 5-HT receptor subtypes as well as other neurochemical systems such as NPY and proopio-melanocoutin (POMC)/α-melanocyte-stimulating hormone (α-MSH). It may also be of interest to take into consideration the individual macronutrient preference, and compare hypothalamic transmitter dynamics in lean, obese, and obesity-prone rats.

REFERENCES

1. Blundell, J. E. and Halford, J. C. G., Serotonin and appetite regulation. Implications for the pharmacological treatment of obesity, *CNS Drugs*, 9, 473, 1998.
2. Leibowitz, S. F. and Shor-Posner, G., Brain serotonin and eating behavior, *Appetite*, 7, 1, 1986.
3. Leibowitz, S. F., Neurochemical systems of the hypothalamus, in control of feeding and drinking behavior and water-electrolyte excretion, in *Handbook of the Hypothalamus,* Morgane, P. J. and Panksepp, J., Eds., Marcel Dekker, New York, 1980, 299.
4. Simansky, K. J., Serotonergic control of the organization of feeding and satiety, *Behav. Brain Res.,* 73, 37, 1996.
5. Kaplan, J .M., Song, S., and Grill, H. J., Serotonin receptors in the caudal brain stem are necessary and sufficient for the anorectic effect of peripherally administered mCPP, *Psychopharmacology* (Berlin), 137, 43, 1998.
6. Orosco, M. and Nicolaïdis, S., Spontaneous feeding-related monoaminergic changes in the rostromedial hypothalamus revealed by microdialysis, *Physiol. Behav.,* 52, 1015, 1992.
7. Fernstrom, J. D., Dietary effects on brain serotonin synthesis: relationship to appetite regulation, *Am. J. Clin. Nutr.*, 42, 1072, 1985.
8. Fernstrom, J. D. and Wurtman, R. J., Brain serotonin content: increase following ingestion of carbohydrate diet, *Science,* 174, 197, 1971.
9. Fernstrom, J. D. and Wurtman, R. J., Brain serotonin content: physiological regulation by plasma neutral amino acids, *Science,* 178, 414, 1972.
10. Wurtman, J. J. and Wurtman, R. J., Fenfluramine and fluoxetine spare protein consumption while suppressing caloric intake by rats, *Science,* 198, 1178, 1977.
11. Wurtman, J. J. and Wurtman, R. J., Drugs that enhance serotoninergic transmission diminish elective carbohydrate consumption by rats, *Life Sci.*, 24, 895, 1979.
12. Mullen, B. J. and Martin, R. J., The effect of dietary fat on diet selection may involve central serotonin, *Am. J. Physiol.*, 32, R559, 1992.
13. Uemura, K. and Young, J. B., Effects of fat feeding on epinephrine secretion in the rat, *Am. J. Physiol.,* 36, R1329, 1994.
14. Smith, B. K., York, D. A., and Bray, G. A., Chronic d-fenfluramine treatment reduces fat intake and increases carbohydrate intake in rats, *Pharmacol. Biochem. Behav.,* 18, 207, 1997.
15. Smith, B. K., York, D. A., and Bray, G. A., Effects of intrahypothalamic serotonin or serotonin receptor agonists on macronutrient selection, *Am. J. Physiol.*, in press.
16. Kanarek, R. B. and Dushkin, H., Peripheral serotonin administration selectively reduces fat intake in rats, *Pharmacol. Biochem. Behav.*, 31, 113, 1988.
16a. Rouch, C., Nicolaïdis, S., and Orosco, M., Determination, using microdialysis, of hypothalamic serotonin variations in response to different macronutrients, *Physiol. Behav.,* 65, 653, 1999.
17. Nicolaïdis, S., Mécanisme nerveux de l'équilibre énergétique, *Journ. Annu. Diabetol. Hotel* (Paris), 1, 152, 1978.
18. Mc Gowan, M. K., Andrews, K. M., Kelly, J., and Grossman, S. P., Effects of chronic intrahypothalamic infusion of insulin on food intake and diurnal meal patterning in the rat, *Behav. Neurosci.*, 104, 371, 1990.

19. Woods, S. C., Lotter, E. C., McKay, D., and Porte, D., Jr., Chronic intracerebroventricular infusion of insulin reduces food intake and body weight in baboons, *Nature*, 282, 503, 1979.

20. Havrankova, J. M., Schmechel, D., Roth, J., and Brownstein, M., Identification of insulin in rat brain, *Proc. Natl. Acad. Sci. U.S.A.*, 75, 5737, 1978.

21. Baskin, D. G., Porte, D. J., Guest, K., and Dorsa, D. M., Regional concentrations of insulin in the rat brain, *Endocrinology*, 112, 898, 1983.

22. Le Roith, D., Rojeski, M., and Roth, J., Insulin receptors in brain and other tissues: similarities and differences, *Neurochem. Int.*, 12, 419, 1988.

23. Unger, J. W., Livingston, J. N., and Moss, A. M., Insulin receptors in the central nervous system: localization, signaling mechanisms and functional aspects, *Prog. Neurobiol.*, 36, 343, 1991.

24. Gerozissis, K., Orosco, M., Rouch, C., and Nicolaïdis, S., Basal and hyperinsulinemia-induced immunoreactive insulin changes in lean and genetically obese Zucker rats revealed by microdialysis, *Brain Res.*, 611, 258, 1993.

24a. Gerozissis, K., Orosco, M., Rouch, C., and Nicolaïdis, S., Insulin responses to a fat meal in hypothalamic microdialysates and plasma, *Physiol. Behav.*, 62, 767, 1997.

25. Orosco, M., Gerozissis, K., Rouch, C., and Nicolaïdis, S., Feeding-related immunoreactive insulin changes in the PVN-VMH revealed by microdialysis, *Brain Res.*, 671, 149, 1995.

25a. Gerozissis, K., Rouch, C., Nicolaïdis, S., and Orosco, M., Brain insulin response to feeding in the rat is both macronutrient and area specific, *Physiol. Behav.*, 65, 271, 1998.

26. Nuttal, F. Q. and Gannon, M. C., Plasma glucose and insulin response to macronutrients in nondiabetic and NIDDM subjects, *Diabetes Care*, 14, 824, 1991.

27. Chavez, M., Riedy, C. A., Van Dijk, G., and Woods, S. C., Central insulin and macronutrient intake in the rat, *Am. J. Physiol.*, 271, R72, 1996.

28. Van Dijk, G., de Groote, C., Chavez, M., Van der Werf, Y., Steffens, A. B., and Strubbe, J. H., Insulin in the arcuate nucleus of the hypothalamus reduces fat consumption in rats, *Brain Res.*, 777,147, 1997.

29. Mackenzie, R. G. and Trulson, M. E., Effects of insulin and steptozotocin-induced diabetes on brain tryptophan and serotonin metabolism in rats, *J. Neurochem.*, 30, 205, 1978.

30. Shimizu, H. and Bray, G. A., Effects of insulin on hypothalamic monoamine metabolism, *Brain Res.*, 510, 251, 1990.

31. Orosco, M., Rouch, C., Gripois, D., Blouquit, M. F., Roffi, J., Jacquot, C., and Cohen, Y., Effects of insulin on brain monoamine metabolism in the Zucker rat: influence of genotype and age, *Psychoneuroendocrinology*, 16, 537, 1991.

32. Orosco, M. and Nicolaïdis, S., Insulin and glucose-induced changes in feeding and medial hypothalamic monoamines revealed by microdialysis, *Brain Res. Bull.*, 33, 289, 1994.

33. Mackenzie, R. G. and Trulson, M. E., Does insulin act directly on the brain to increase tryptophan levels, *J. Neurochem.*, 30, 1205, 1978.

34. Minano, F. J., Peinado, J. M., and Myers, R. D., Profile of NE, DA and 5-HT activity shifts in medial hypothalamus perfused by 2DG and insulin in the sated or fasted rat, *Brain Res. Bull.*, 22, 695, 1989.

35. Myers, R. D., Peinado, J. M., and Minano, F. J., Monoamine transmitter activity in lateral hypothalamus during its perfusion with insulin and 2DG in sated and fasted rats, *Physiol. Behav.*, 44, 633, 1988.

36. Telib, M., Raptis, S., Schröder, K. E., and Pfeiffer, E. F., Serotonin and insulin release in vitro, *Diabetologia*, 4, 253, 1968.

37. Feldman, J. M. and Lebovitz, H. E., Mechanism of epinephrine and serotonin inhibition of insulin release in the golden hamster in vitro, *Diabetes*, 19, 480, 1970.

38. Peschke, E., Peschke, D., Hammer, T., and Csernus, V., Influence of melatonin and serotonin on glucose-stimulated insulin release from perifused rat pancreatic islets in vitro, *J. Pineal Res.*, 23, 156, 1997.

39. Fernstrom, J. D. and Wurtman, R. J., Elevation of plasma tryptophan by insulin in the rat, *Metabolism*, 21, 337, 1972.

40. Schwartz, M. W., Bergman, R. N. Kahn, S. E., Taborsky, G. J., Fisher, L. D., Sipols, A. J., Woods, S. C., Steil, G. M., and Porte, D. J., Evidence for entry of plasma insulin into cerebrospinal fluid through an intermediate compartment in dogs. Quantitative aspects and implications for transport, *J. Clin. Invest.*, 88, 1272, 1991.

41. Schwartz, M. W., Figlewicz, D. P., Baskin, D. G ., Woods, S. C., and Porte, D., Jr., Insulin in the brain: a hormonal regulator of energy balance, *Endocr. Rev.*, 13, 387, 1992.

29 Quantitative and Macronutrient-Related Regulation of Hypothalamic Neuropeptide Y, Galanin, and Neurotensin

Bernard Beck

CONTENTS

1   INTRODUCTION

For more than 20 years now, neuropeptides present in the brain have appeared to constitute a special class of neuromodulators of feeding behavior in addition to the classical neurotransmitters such as catecholamines or serotonin.[1] These neuropeptides are predominantly found in the hypothalamus. They act through complex networks involving several nuclei or areas. They are present in the arcuate (ARC), ventromedian (VMN), dorsomedian (DMN), paraventricular (PVN), and suprachiasmatic (SCN) nuclei as well as in the lateral hypothalamus, the main areas participating in the food intake regulation. They are also sensible to information transiting through the limbic system and the brain stem which are relays for peripheral and cognitive events.

The orexigenic peptides are less numerous than the inhibitory peptides. The most important are neuropeptide Y (NPY) and galanin[1] as well as the new class formed by the orexins.[2] NPY is one of the most abundant peptides found in the brain.[3] In the hypothalamus, this 36-AA peptide is synthetized in neurons of the ARC which project to the PVN to form a pure peptidergic pathway.[4] In the PVN, it is also present in fibers which originate in the brain stem where it is colocalized with catecholamines.[5,6] It strongly stimulates food intake even in satiated rats (reviewed in

Reference 7). It preferentially stimulates carbohydrate intake[7,8] (see also Chapter 25 by Leibowitz in this book) and to a lesser extent fat intake. Its elevated concentrations in the Zucker rat contribute to the development of overeating and obesity observed in this strain of rat.[9,10] Galanin, a 29-AA peptide, is synthetized in neurons of the PVN as well as in the ARC and other brain areas.[11] It stimulates food intake less strongly than NPY.[12] Several studies have shown that it may be linked to fat ingestion.[13-15] Other experiments do not confirm this unique effect on fat intake[16,17] and/or suggested that individual food preferences might play a role in the feeding response to galanin injection.[18] Galanin is also augmented in the PVN of the obese Zucker rat,[19] but contrary to NPY[8,20] it does not induce sustained hyperphagia and obesity when it is chronically injected in brain ventricles of normal rats.[24]

The main inhibitory neuropeptides are cholecystokinin, corticotropin-releasing hormone, bombesin, and neurotensin.[1,22] Neurotensin is a smaller peptide (13 AA) which inhibits food intake when it is injected in brain ventricles or in different hypothalamic or extrahypothalamic nuclei (reviewed in Reference 22). It has no specific effect on the intake of the different macronutrients even if fats are a strong stimulus for its peripheral release.[23] It is also involved in the peptidergic dysregulation observed in the hyperphagic Zucker rat.[24,25]

All of these peptides interact one with the others to trigger or to block feeding behavior. In a counterregulatory mechanism, ingested food can influence the central levels of these peptides to achieve a fine regulation of the energy balance. The aim of this chapter is to focus on the role of food on hypothalamic NPY, galanin, and neurotensin on both a qualitative and quantitative point of view.

2 QUANTITATIVE ASPECTS

The quantitative aspects were studied with different experimental protocols. Food was either totally withdrawn for variable periods of time or given in a limited quantity for several days. In some cases, food restriction induced larger effects on neuropeptides than absolute fasting.[26]

2.1 NEUROPEPTIDE Y

2.1.1 NPY and Fasting

Several experiments have shown that fasting could affect NPY at each stage of its metabolism: mRNA expression, peptide content, and release. After 48 h of food deprivation with water available, NPY concentration is significantly increased in the ARC of Sprague-Dawley and Long-Evans rats.[27,28] The increase is still larger after 72 or 96 h of food withdrawal. This increased peptide content is paralleled by a significant augmentation of NPY mRNA expression in the ARC.[26,29-32] Similar results are observed in mice and hamsters.[33,34] These increases in NPY synthesis in the ARC are associated with important augmentation of NPY content and innervation in the PVN.[27,28,35] No changes are observed either in other hypothalamic nuclei, such as the DMN and VMN[27,28] or in the brain stem,[30,31,33] and, as NPY injection in the PVN is very potent for stimulating intake,[7] these experiments suggested that the arcuate to paraventricular (AP) axis could be the main pathway for the regulation of food intake by NPY. This is confirmed by the measurement of the NPY release in the PVN by the push–pull perfusion technique in different conditions. First, in fasted animals, this release is markedly greater than in satiated rats.[36-38] It returns to normal levels after refeeding. Second, other experiments have shown that increased release induced by hyperosmotic stimulation is associated with feeding events.[39] Moreover, increased release and content are measured at the dark/light transition when the first meals are ingested by the rats.[37,40]

The duration of the fast or refeeding is a very important factor for the NPY variation. Increased NPY mRNA expression is obvious after 24 h of deprivation but a trend (+40%) is already observed after an overnight fast.[29] A 6-h refeeding period is sufficient to observe a return to normal levels in the ARC[28] but not in the PVN where 24 h is necessary.[27] This pattern suggests the existence of an active process for stimulating food intake until the total restoration of energy stores is achieved. In obese Zucker rats, the very high levels of NPY in the ARC and PVN[9] corresponding to increased synthesis[41] and release[37,42] were not further enhanced by fasting.[43] They already correspond to a fasting situation in the lean rat and could be at the origin of their sustained food intake whatever the period of the light/dark cycle.

2.1.2   NPY and Food Restriction

The effects of food restriction on the NPY system were generally measured after longer periods of time than in fasting experiments (2 to 6 weeks). When the level of restriction is about 50% of the daily intake of control rats (10 to 15 g/day), the variations of NPY mRNA expression in the ARC are similar to that induced by fast and even larger.[26] Peptide content increases similarly to levels corresponding to those induced by intense exercise.[44] At lower levels of restriction (20% during 2 weeks), NPY mRNA expression is still increased but the peptide content is not altered.[45] In rats with a daily feeding restriction (2 h/day), NPY release is augmented before a meal and decreased after meal supply.[38] In Zucker rats, food restriction is also very potent on the NPY system. It induces a large increase in NPY mRNA expression in lean as well as in obese rats.[46] This increased expression associated with stable content suggests that the axonal transport of NPY from ARC to PVN is increased. The association with the increased release could be at the origin of the overeating observed when food is again available.

2.2   GALANIN AND NEUROTENSIN

The studies on galanin and neurotensin are less numerous than for NPY. Neither 24 h or 48 h fasting has a significant effect on preprogalanin mRNA expression measured in the ARC and DMN. This is observed either in the Wistar rat[32] or in the Sprague-Dawley rat.[30] In the lean as well as in obese Zucker rat, fasting for 48 h has also no significant effect.[19] A longer food deprivation (4 days) induces a slight decrease in galanin mRNA, whereas food restriction has no effect.[26]

For neurotensin, fasting (2 or 4 days) has no effect in normal lean rats either on peptide expression or content.[24,47] However, in obese rats, a 45% decrease in neurotensin levels is observed in the VMN.[46] As suggested above, this decrease might contribute with other peptides to the greater food intake observed in these rats when food is again available.

3   QUALITATIVE ASPECTS

Carbohydrates and fats are the two macronutrients that have the most impact on hypothalamic neuropeptides. They have short-term and long-term effects on body weight regulation. Each one can induce obesity with or without hyperphagia. These effects depend on the nature and form of each macronutrient as well as of age and duration of ingestion (reviewed in References 48 and 49).

Three factors besides the nature of each macronutrient and the nutritional status need to be taken into account when examining and comparing the data relative to the effects of macronutrients on hypothalamic neuropeptides: the time of brain sampling during the light/dark cycle, the age of the animals ingesting the different diets, and the duration of the experiments with these unbalanced diets. Time of brain sampling is important because each peptide varies according to a specific circadian cycle for each area.[14,40] It can also correspond to specific periods for feeding, such as, for rats, relative feeding inactivity at the beginning of the light phase or very active feeding activity at the beginning of the dark cycle. The age of the animals could also influence the results as all peptidergic systems do not reach their maturity at the same moment of life. The NPY system is quite totally operational very early (at least at weaning), whereas galanin and neurotensin systems continue to evolve until adulthood and might remain more receptive to diet changes.[10,50] Finally,

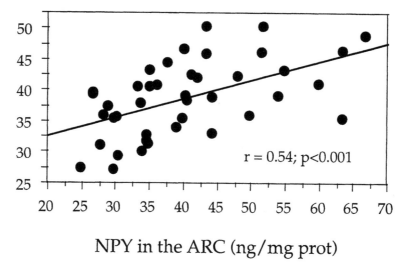

NPY in the PVN (ng/mg prot)

NPY in the ARC (ng/mg prot)

FIGURE 1 Functionality of the AP axis for the regulation of food intake by NPY, as demonstrated by the good correlation existing between NPY concentration (in ng/mg protein) in each area in adult Long-Evans rats ingesting different diets (control, HF, HC) for 14 weeks.

the duration of the experiments could be a major factor for the more or less good adaptation of the different systems to dietary changes.

3.1   NEUROPEPTIDE Y

As previously described, NPY preferentially stimulates carbohydrate intake.[7,8] Carbohydrates in turn can regulate the orexigenic effects of NPY. Intravenous injection of glucose in rats can diminish the intake induced by intracerebroventricular injection of NPY.[51] This is not observed with fructose suggesting that glucose utilization by the brain is an important factor for local hypothalamic regulation.[52] As for fasting, the AP axis appears to be the main pathway involved in this regulation by macronutrients. This is clearly shown by the parallel evolution of NPY levels in the ARC and PVN. There was indeed a significant correlation ($r = 0.54$, $p < 0.0002$) between NPY in these two areas in rats which are given diets differing in carbohydrate and fat content over a 14-week period (Figure 1). This correlation is also verified in rats with clear dietary preferences either for fats or for carbohydrates ($r = 0.45$, $p < 0.01$; Figure 2).

As previously underlined, time of brain sampling is important for NPY. For example, when compared with fat-preferring rats, carbohydrate-preferring rats are characterized by higher NPY levels in the PVN at the dark onset[53] and lower NPY levels in the same area during the light phase.[54] This suggests that not only absolute levels but also dynamic evolution of the peptide during the nycthemere could be a factor for modifying the food choice of the animals.

The effects of the different macronutrients are not exactly the same when they are ingested on short or long term. At short term (2 weeks), when increasing the carbohydrate-to-fat ratio of the diet given to the rats, there is an inverse relationship with the NPY concentrations of the parvocellular part of the PVN ($r = -0.44$; $p < 0.005$). The NPY concentrations decrease when the carbohydrate content of the diet augments. No significant variations of the peptide content are found in the ARC,[55,56] whereas mRNA expression is decreased by a high-fat (HF) diet.[57] At long term, when the rats are fed on either HF diet or high-carbohydrate (HC) diets from weaning until the age of 4 months, the decrease of NPY in the PVN with the HC diet is confirmed in the parvocellular part

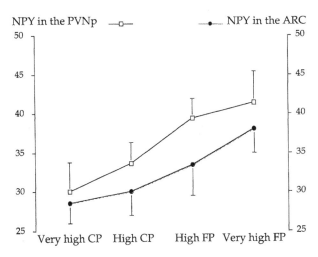

FIGURE 2 Parallel variations of NPY concentrations (in ng/mg protein) in the ARC and paraventricular nuclei in adult Long-Evans rats with marked dietary preferences PVN from very HC preference (CP) to very HF preference (FP).

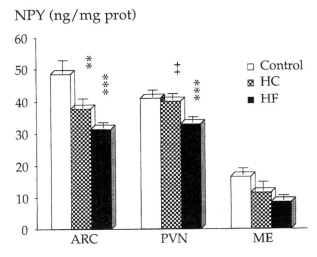

FIGURE 3 NPY levels (in ng/mg protein) measured during the first third of the light period in the ARC and PVN nuclei and in the median eminence (ME) of Long-Evans rats fed for 14 weeks on either a control well-balanced diet, an HC, or an HF diet. The HC diet consisted in a high starch diet plus a 25% sucrose solution to drink. $**p < 0.02$ vs. control; $***p < 0.005$ vs. control; $++p < 0.02$ vs. HF.

of the nucleus and also visible in the ARC.[58] Moreover, there is a further control by the HF diet, and NPY concentration in the HF diet-fed rats is smaller than in rats ingesting the well-balanced diet and even smaller than in rats with the HC diet in both ARC and PVN (Figure 3). This diminution of NPY appears necessary to decrease the consumption of the HF diet and therefore to limit the energy intake brought by this diet of high caloric density.

Other factors such as diet palatability might also play a role in NPY regulation. Increased NPY concentrations in the ARC and stability of NPY expression in the whole hypothalamus were indeed measured in rats fed on highly palatable sweetened diet which did not differ in carbohydrate content from rat chow.[59] The factors responsible for such changes are not known.

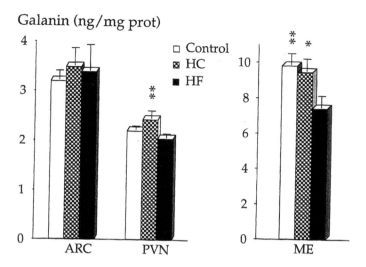

FIGURE 4 Galanin concentrations (in ng/mg protein) measured during the first third of the light period in the ARC and PVN nuclei and in the median eminence (ME) of Long-Evans rats fed on either a control well-balanced diet, an HC diet or an HF diet for 14 weeks. The HC diet consisted of a high-starch diet plus a 25% sucrose solution to drink. $*p < 0.05$ vs. HF diet; $**p < 0.01$ vs. HF diet.

3.2   GALANIN AND NEUROTENSIN

Both peptides are more or less related to fat intake.[13-18,23] As for NPY, time of brain sampling is a critical factor for the effect of an HF diet on galanin concentration in the PVN and median eminence. After 4 weeks of ingestion of this diet, galanin mRNA and peptide are higher than after the ingestion of an HC diet[15] at dark onset.

On the other hand, at the beginning of the light phase, after a similar treatment length, galanin gene expression in the PVN of obese rats fed on HC diet were higher than those of obese rats on the HF diet.[60] We confirmed this results after feeding the rats on HF diet for 14 weeks (Figure 4). In all three studies, there was no variation at all in the ARC but once again there were opposite variations of galanin concentrations in the median eminence depending on the time of the light/dark cycle (Figure 4).[15]

The Brattleboro rat has a natural preference for fat-rich diet and is characterized by higher galanin mRNA expression in the PVN. This is associated with lower peptide content in the same area suggesting the existence of an increased turnover of the peptide.[61]

For neurotensin, ingestion of an HF diet for a short or long period induces a decrease in peptide levels in the PVN[62,63] when compared with the ingestion of an HC diet. Diminutions are also observed in other hypothalamic areas such as the lateral hypothalamus at short term or the DMN at long term. However, no significant variation can be observed in extrahypothalamic areas such as the ventral tegmental area where neurotensin can also inhibit food intake.[22] This indicates a relative specificity of the hypothalamic areas for the relationship between neurotensin and macronutrients.

4   CONCLUSION

It results from the current literature that there is a complex regulatory loop between hypothalamic neuropeptides and ingested food. The stimulatory peptides induce feeding behavior when receiving a signal of energy depletion and their action is then controlled by the type and quantity of food ingested. This action is performed either directly by inhibiting the orexigenic peptides or by stimulating the inhibitory peptides. This regulation is active on numerous neuropeptides and pathways.

The balance existing between these peptides could drive the qualitative aspects of feeding behavior and could lead to specific macronutrient choice. It is not fixed in time and can vary during the nycthemere to adapt the energy intake to energy needs. The environmental factors were not taken into account in this chapter. Some data concerning the variations of corticotropin-releasing hormone, the stress-related hormone, can be found in some references of the cited literature. These factors also play a very important role in addition to the primary action of food components.

ACKNOWLEDGMENTS

The author thanks his colleagues and more particularly Dr. A. Burlet and Dr. A. Stricker-Krongrad for their active collaboration in the cited experiments. He also thanks F. Bergerot, B. Fernette, and F. Giannangeli for their excellent technical assistance and C. Habert for preparing this manuscript. Several experiments were supported by the Institut Benjamin Delessert (Paris). This constant help is gratefully acknowledged.

REFERENCES

1. Morley, J. E., Neuropeptide regulation of appetite and weight., *Endocr. Rev.,* 8, 256, 1987.
2. Sakurai, T., Amemiya, A., Ishii, M., Matsuzaki, I., Chemelli, R. M., Tanaka, H., Williams, S. C., Richardson, J. A., Kozlowski, G. P., Wilson, S., Arch, J. R. S., Buckingham, R. E., Haynes, A. C., Carr, S. A., Annan, R. S., McNulty, D. E., Liu, W. S., Terrett, J. A., Elshourbagy, N. A., Bergsma, D. J., and Yanagisawa, M., Orexins and orexin receptors: a family of hypothalamic neuropeptides and G protein-coupled receptors that regulate feeding behavior, *Cell,* 92, 573, 1998.
3. Chronwall, B. M., Di Maggio, D. A., Massari, V. J., Pickel, V. M., Ruggiero, D. A., and O'Donohue, T. L., The anatomy of neuropeptide Y-containing neurons in rat brain, *Neuroscience,* 15, 1159, 1985.
4. Bai, F. L., Yamano, M., Shiotani, Y., Emson, P. C., Smith, A. D., Powell, J. F., and Toyama, M., An arcuato-paraventricular and -dorsomedial hypothalamic neuropeptide Y-containing system which lacks noradrenaline in the rat, *Brain Res.,* 331, 172, 1985.
5. Sawchenko, P. E., Swanson, L. W., Grzanna, R., Howe, P. R. C., Bloom, S. R., and Polak, J. M., Colocalization of neuropeptide Y immunoreactivity in brain stem catecholaminergic neurons that project to the paraventricular nucleus of the hypothalamus, *J. Comp. Neurol.,* 241, 138, 1985.
6. Sahu, A., Kalra, S. P., Crowley, W. R., and Kalra, P. S., Evidence that NPY-containing neurons in the brain stem project into selected hypothalamic nuclei: implication in feeding behavior, *Brain. Res.,* 457, 376, 1988.
7. Stanley, B. G., Neuropeptide Y in multiple hypothalamic sites controls eating behavior, endocrine, and autonomic systems for body energy balance, in *Biology of Neuropeptide Y and Related Peptides,* Colmers, W. F. and Wahlestedt, C., Eds., Humana Press, Totowa, NJ, 1993, 457.
8. Beck, B., Stricker-Krongrad, A., Nicolas, J. P., and Burlet, C., Chronic and continuous intracerebroventricular infusion of neuropeptide Y in Long-Evans rats mimics the feeding behaviour of obese Zucker rats, *Int. J. Obesity,* 16, 295, 1992.
9. Beck, B., Burlet, A., Nicolas, J. P., and Burlet, C., Hypothalamic neuropeptide Y (NPY) in obese Zucker rats: implications in feeding and sexual behaviors, *Physiol. Behav.,* 47, 449, 1990.
10. Beck, B., Burlet, A., Bazin, R., Nicolas, J. P., and Burlet, C., Elevated neuropeptide-Y in the arcuate nucleus of young obese Zucker rats may contribute to the development of their overeating, *J. Nutr.,* 123, 1168, 1993.
11. Skofitsch, G. and Jacobowitz, D. M., Galanin in the central nervous system: a review, in *Current Aspects of the Neurosciences,* Vol. 1, Osborne, N. N., Eds., Macmillan, New York, 1990, 1.
12. Kyrkouli, S. E., Stanley, B. G., Seirafi, R. D., and Leibowitz, S. F., Stimulation of feeding by galanin — anatomical localization and behavioral specificity of this peptides effects in the brain, *Peptides,* 11, 995, 1990.
13. Tempel, D. L., Leibowitz, K. J., and Leibowitz, S. F., Effects of PVN galanin on macronutrient selection, *Peptides,* 9, 309, 1988.

14. Akabayashi, A., Koenig, J. I., Watanabe, Y., Alexander, J. T., and Leibowitz, S. F., Galanin-containing neurons in the paraventricular nucleus: a neurochemical marker for fat ingestion and body weight gain, *Proc. Natl. Acad. Sci. U.S.A.,* 91, 10375, 1994.

15. Leibowitz, S. F., Akabayashi, A., and Wang, J. A., Obesity on a high-fat diet: role of hypothalamic galanin in neurons of the anterior paraventricular nucleus projecting to the median eminence, *J. Neurosci.,* 18, 2709, 1998.

16. Corwin, R. L., Rowe, P. M., and Crawley, J. N., Galanin and the galanin antagonist M40 do not change fat intake in a fat-chow choice paradigm in rats, *Am. J. Physiol.,* 38, R511, 1995.

17. Lin, L., York, D. A., and Bray, G. A., Comparison of Osborne-Mendel and S5B/PL strains of rat: central effects of galanin, NPY, beta-casomorphin and CRH on intake of high-fat and low-fat diets, *Obesity Res.,* 4, 117, 1996.

18. Smith, B. K., York, D. A., and Bray, G. A., Effects of dietary preference and galanin administration in the paraventricular or amygdaloid nucleus on diet self-selection, *Brain Res. Bull.,* 39, 149, 1996.

19. Beck, B., Burlet, A., Nicolas, J. P., and Burlet, C., Galanin in the hypothalamus of fed and fasted lean and obese Zucker rats, *Brain Res.,* 623, 124, 1993.

20. Stanley, B. G., Anderson, K. C., Grayson, M. H., and Leibowitz, S. F., Repeated hypothalamic stimulation with neuropeptide Y increases daily carbohydrate and fat intake and body weight gain in female rats, *Physiol. Behav.,* 46, 173, 1989.

21. Smith, B. K., York, D. A., and Bray, G. A., Chronic cerebroventricular galanin does not induce sustained hyperphagia or obesity, *Peptides,* 15, 1267, 1994.

22. Beck, B., Cholecystokinin, neurotensin and corticotropin-releasing factor — 3 important anorexic peptides, *Ann. Endocrinol.,* 53, 44, 1992.

23. Rosell, S. and Rokaeus, A., The effect of ingestion of amino-acids, glucose and fat on circulating neurotensin-like immunoreactivity (NTLI) in man, *Acta Physiol. Scand.,* 107, 263, 1979.

24. Beck, B., Burlet, A., Nicolas, J. P., and Burlet, C., Neurotensin in microdissected brain nuclei and in the pituitary of the lean and obese Zucker rats, *Neuropeptides,* 13, 1, 1989.

25. Beck, B., Burlet, A., Nicolas, J. P., and Burlet, C., Hyperphagia in obesity is associated with a central peptidergic dysregulation in rats, *J. Nutr.,* 120, 806, 1990.

26. Brady, L. S., Smith, M. A., Gold, P. W., and Herkenham, M., Altered expression of hypothalamic neuropeptide messenger RNAs in food-restricted and food-deprived Rats, *Neuroendocrinology,* 52, 441, 1990.

27. Sahu, A., Kalra, P. S., and Kalra, S. P., Food deprivation and ingestion induce reciprocal changes in neuropeptide Y concentrations in the paraventricular nucleus, *Peptides,* 9, 83, 1988.

28. Beck, B., Jhanwar-Uniyal, M., Burlet, A., Chapleur-Chateau, M., Leibowitz, S. F., and Burlet, C., Rapid and localized alterations of neuropeptide Y in discrete hypothalamic nuclei with feeding status, *Brain Res.,* 528, 245, 1990.

29. White, J. D. and Kershaw, M., Increased hypothalamic neuropeptide Y expression following food deprivation, *Mol. Cell. Neurosci.,* 1, 41, 1990.

30. O'Shea, R. D. and Gundlach, A. L., Preproneuropeptide-Y messenger ribonucleic acid in the hypothalamic arcuate nucleus of the rat is increased by food deprivation or dehydration, *J. Neuroendocrinol.,* 3, 11, 1991.

31. Marks, J. L., Li, M., Schwartz, M., Porte, D., and Baskin, D. G., Effect of fasting on regional levels of neuropeptide-Y messenger RNA and insulin receptors in the rat hypothalamus — an autoradiographic study, *Mol. Cell. Neurosci.,* 3, 199, 1992.

32. Schwartz, M. W., Sipols, A. J., Grubin, C. E., and Baskin, D. G., Differential effect of fasting on hypothalamic expression of genes encoding neuropeptide-Y, galanin, and glutamic acid decarboxylase, *Brain Res. Bull.,* 31, 361, 1993.

33. Chua, S. C., Leibel, R. L., and Hirsch, J., Food Deprivation and age modulate neuropeptide gene expression in the murine hypothalamus and adrenal gland, *Mol. Brain Res.,* 9, 95, 1991.

34. Mercer, J. G., Lawrence, C. B., Beck, B., Burlet, A., Atkinson, T., and Barrett, P., Hypothalamic NPY and prepro-NPY mRNA in Djungarian hamsters: effects of food deprivation and photoperiod, *Am. J. Physiol.,* 38, R1099, 1995.

35. Calza, L., Giardino, L., Battistini, N., Zanni, M., Galetti, S., Protopapa, F., and Velardo, A., Increase of neuropeptide Y-like immunoreactivity in the paraventricular nucleus of fasting rats, *Neurosci. Lett.,* 104, 99, 1989.

36. Kalra, S. P., Dube, M. G., Sahu, A., Phelps, C. P., and Kalra, P. S., Neuropeptide-Y secretion increases in the paraventricular nucleus in association with increased appetite for food, *Proc. Natl. Acad. Sci. U.S.A.*, 88, 10931, 1991.
37. Stricker-Krongrad, A., Kozak, R., Burlet, C., Nicolas, J. P., and Beck, B., Physiological regulation of hypothalamic neuropeptide Y release in lean and obese rats, *Am. J. Physiol.*, 42, R2112, 1997.
38. Yoshihara, T., Honma, S., and Honma, K., Effects of restricted daily feeding on neuropeptide Y release in the rat paraventricular nucleus, *Am. J. Physiol.*, 33, E589, 1996.
39. Stricker-Krongrad, A., Barbanel, G., Beck, B., Burlet, A., Nicolas, J. P., and Burlet, C., K⁺-stimulated neuropeptide-Y release into the paraventricular nucleus and relation to feeding behavior in free-moving rats, *Neuropeptides*, 24, 307, 1993.
40. Jhanwar-Uniyal, M., Beck, B., Burlet, C., and Leibowitz, S. F., Diurnal rhythm of neuropeptide Y-like immunoreactivity in the suprachiasmatic, arcuate and paraventricular nuclei and other hypothalamic sites, *Brain Res.*, 536, 331, 1990.
41. Sanacora, G., Kershaw, M., Finkelstein, J. A., and White, J. D., Increased hypothalamic content of preproneuropeptide Y messenger ribonucleic acid in genetically obese Zucker rats and its regulation by food deprivation, *Endocrinology*, 127, 730, 1990.
42. Dryden, S., Pickavance, L., Frankish, H. M., and Williams, G., Increased neuropeptide Y secretion in the hypothalamic paraventricular nucleus of obese (fa/fa) Zucker rats, *Brain Res.*, 690, 185, 1995.
43. Beck, B., Burlet, A., Nicolas, J. P., and Burlet, C., Unexpected regulation of hypothalamic neuropeptide-Y by food deprivation and refeeding in the Zucker rat, *Life Sci.*, 50, 923, 1992.
44. Lewis, D. E., Shellard, L., Koeslag, D. G., Boer, D. E., McCarthy, H. D., McKibbin, P. E., Russell, J. C., and Williams, G., Intense exercise and food restriction cause similar hypothalamic neuropeptide-Y increases in rats, *Am. J. Physiol.*, 264, E279, 1993.
45. Wilding, J. P. H., Ajala, M. O., Lambert, P. D., and Bloom, S. R., Additive effects of lactation and food restriction to increase hypothalamic neuropeptide Y mRNA in rats, *J. Endocrinol.*, 152, 365, 1997.
46. Pesonen, U., Huupponen, R., Rouru, J., and Koulu, M., Hypothalamic neuropeptide expression after food restriction in Zucker rats — evidence of persistent neuropeptide-Y gene activation, *Mol. Brain Res.*, 16, 255, 1992.
47. Beck, B., Nicolas, J. P., and Burlet, C., Neurotensin decreases with fasting in the ventromedian nucleus of obese Zucker rats, *Metabolism*, 44, 972, 1995.
48. Sclafani, A., Carbohydrate-induced hyperphagia and obesity in the rat: effects of saccharide type, form, and taste, *Neurosci. Biobehav. Rev.*, 11, 155, 1987.
49. Warwick, Z. S. and Schiffman, S. S., Role of dietary fat in calorie intake and weight gain, *Neurosci. Biobehav. Rev.*, 16, 585, 1992.
50. Beck, B., Burlet, A., Bazin, R., Nicolas, J. P., and Burlet, C., Early modification of neuropeptide-Y but not of neurotensin in the suprachiasmatic nucleus of the obese Zucker rat, *Neurosci. Lett.*, 136, 185, 1992.
51. Rowland, N. E., Peripheral and central satiety factors in neuropeptide Y-induced feeding in rats, *Peptides*, 9, 989, 1988.
52. Akabayashi, A., Zaia, C. T. B. V., Silva, I., Chae, H. J., and Leibowitz, S. F., Neuropeptide-Y in the arcuate nucleus is modulated by alterations in glucose utilization, *Brain Res.*, 621, 343, 1993.
53. Jhanwar-Uniyal, M., Beck, B., Jhanwar, Y. S., Burlet, C., and Leibowitz, S. F., Neuropeptide Y projection from arcuate nucleus to parvocellular division of paraventricular nucleus — specific relation to the ingestion of carbohydrate, *Brain Res.*, 631, 97, 1993.
54. Stricker-Krongrad, A., Beck, B., Burlet, A., Nicolas, J. P., and Burlet, C., Dietary preference for carbohydrate or fat is related to neuropeptide Y variations in a specific hypothalamic network, *Int. J. Obesity*, 18 (Suppl. 2), 102, 1994.
55. Beck, B., Stricker-Krongrad, A., Burlet, A., Nicolas, J. P., and Burlet, C., Influence of diet composition on food intake and hypothalamic neuropeptide Y (NPY) in the rat, *Neuropeptides*, 17, 197, 1990.
56. Beck, B., Stricker-Krongrad, A., Burlet, A., Nicolas, J. P., and Burlet, C., Specific hypothalamic neuropeptide-Y variation with diet parameters in rats with food choice, *Neuroreport*, 3, 571, 1992.
57. Giraudo, S. Q., Kotz, C. M., Grace, M. K., Levine, A. S., and Billington, C. J., Rat hypothalamic NPY mRNA and brown fat uncoupling protein mRNA after high-carbohydrate or high-fat diets, *Am. J. Physiol.*, 266, R1578, 1994.

58. Beck, B., Stricker-Krongrad, A., Burlet, A., Max, J. P., Musse, N., Nicolas, J. P., and Burlet, C., Macronutrient type independently of energy intake modulates hypothalamic neuropeptide Y in Long-Evans rats, *Brain Res. Bull.*, 34, 85, 1994.

59. Wilding, J. P. H., Gilbey, S. G., Mannan, M., Aslam, N., Ghatei, M. A., and Bloom, S. R., Increased neuropeptide-Y content in individual hypothalamic nuclei, but not neuropeptide-Y messenger RNA, in diet-induced obesity in rats, *J. Endocrinol.*, 132, 299, 1992.

60. Mercer, J. G., Lawrence, C. B., and Atkinson, T., Regulation of galanin gene expression in the hypothalamic paraventricular nucleus of the obese Zucker rat by manipulation of dietary macronutrients, *Mol. Brain Res.*, 43, 202, 1996.

61. Burlet, A., Odorisis, M., Beck, B., Max, J. P., Fernette, B., Angel, E., Nicolas, J. P., and Burlet, C., Hypothalamic expression of galanin varies with the preferential consumption of fat, *Soc. Neurosci. Abstr.*, 22, 1685, 1996.

62. Beck, B., Stricker-Krongrad, A., Burlet, A., Nicolas, J. P., and Burlet, C., Changes in hypothalamic neurotensin concentrations and food intake in rats fed a high-fat diet, *Int. J. Obesity*, 16, 361, 1992.

63. Beck, B., Burlet, A., Nicolas, J. P., and Burlet, C., Opposite influence of carbohydrates and fat on hypothalamic neurotensin in Long-Evans rats, *Life Sci.*, 59, 349, 1996.

30 Adiposity Signals and Macronutrient Selection

Gertjan van Dijk, Mark Chavez, Christine A. Riedy, and Stephen C. Woods

CONTENTS

1 INTRODUCTION

The brain is a major controller of the amount of fat stored in the body. Through its control of energy intake (food intake) and energy expenditure (exercise and metabolic rate), it is able to balance the energy equation and thereby maintain stable fat stores over prolonged intervals. In order to accomplish this, the brain requires continuous information about how well it is doing; i.e., the brain must have a means of knowing how much fat is actually stored at any point of time. Further, for regulation by the brain to be successful, this knowledge, once acquired, must be wired neuronally to systems that control energy intake and expenditure. The molecules that convey information regarding the amount of fat stored in the body at any point of time are called adiposity signals.

2 ADIPOSITY SIGNALS

Over the past few years tremendous new knowledge has been gained in our understanding of the means by which the size of the fat stores is signaled to the brain.[1] At least two hormones appear to be key. Leptin is a recently described peptide hormone secreted from adipocytes.[2] The amount of leptin secreted throughout the day is positively proportional to the amount of fat stored in the body.[3-6] Insulin is a peptide hormone secreted from the pancreatic B cells in response to glucose and other stimuli, and it has a major role in controlling glucose and lipid utilization and storage.[7] Importantly, the amount of insulin secreted is also positively proportional to the size of the body's fat stores. More obese animals and humans secrete proportionately more insulin than lean individuals during basal conditions as well as in response to identical meals or glucose challenges.[8] Hence, the amount of insulin secreted during meals and throughout the day is proportional to body fat such that plasma insulin is a reliable indicator of adiposity.[9] Therefore, both plasma leptin levels and plasma insulin levels are excellent indicators of the amount of stored fat.

However, there are fundamental differences between insulin and leptin with regard to their potential role as signals to the brain indicating the amount of fat stored in adipose tissue. For one, insulin secretion is adjusted in response to every acute change of metabolism.[7] It increases during meals or when glucose is elevated for some other reason, and it decreases during stress and exercise.[7] Leptin secretion does not track these acute changes of metabolism, but its overall level each day is nonetheless directly proportional to the size of the fat stores.[5] Hence, insulin levels reflect the interaction of ongoing metabolic processes and body adiposity, whereas leptin levels reflect adiposity more directly. Second, insulin secretion is more closely related to the amount of visceral white adipose tissue,[10,11] whereas leptin secretion is more closely related to total fat mass and especially to subcutaneous fat.[12,13] This is a potentially very important difference with regard to the message each hormone conveys to the brain since visceral fat is a much greater risk factor for the metabolic complications associated with obesity than is subcutaneous fat. In particular, elevated visceral fat is associated with an increased incidence of insulin resistance, type 2 diabetes mellitus, hypertension, cardiovascular disease, and certain cancers. Hence, while the circulating level of leptin and of insulin each conveys specific information about the distribution of fat, the combination of the two additionally conveys information about the total fat mass of the body.

2.1   ADIPOSITY SIGNALS AND THE BRAIN

Because both leptin and insulin levels are so well correlated with adiposity, in order to account for the ability of the brain to maintain a relatively constant size of the adipose mass, one needs only hypothesize that the brain is able to monitor plasma insulin and/or leptin. Although conceptually simple, there are problems with such a hypothesis. Both insulin and leptin are large peptides that cannot easily penetrate the tight junctions that comprise the blood–brain barrier. This potential problem was eliminated when it was discovered, initially for insulin[14-16] and more recently for leptin,[17] that hormone-specific transport mechanisms exist that pass these compounds from the plasma into the brain interstitial fluid. Receptors for both insulin and leptin exist on brain capillary endothelial cells, and the transcytotic transport of both peptides from the plasma into the brain is saturable and receptor mediated.[18-20] The bottom line is that when insulin or leptin becomes elevated in the plasma, more is transported into the brain and a greater "adiposity" signal is therefore present to influence the regulation of energy homeostasis.

Receptors for insulin and for leptin exist on neurons in discrete areas of the brain.[1] In support of a major role of these two hormones in energy homeostasis, neurons in hypothalamic areas important in the regulation of energy homeostasis have been found to express specific receptors for them. Consistent with this, when the levels of either peptide are elevated locally in the third cerebral ventricle or in the nearby ventral hypothalamus, animals eat less food, increase their energy expenditure, and reduce their body weight.[21-25] They therefore behave as if they are "overweight," and the appropriate response is to stimulate catabolic pathways while simultaneously eating less food. Conversely, when antibodies to insulin are administered into the hypothalamus, rats eat more food and gain weight.[23,26] Although antibodies to leptin have not been administered into the brain, there are animals that do not synthesize endogenous leptin (ob/ob mice) and which are hyperphagic and obese.[27] Consistent with this, animals that are genetically leptin resistant (db/db mice and fa/fa rats) are also hyperphagic and obese.[27,28] Analogously, animals which do not synthesize insulin are also hyperphagic. They do not become obese because insulin is a necessary factor for adipocytes to store fat. When leptin is administered into the brain of hyperphagic leptin-deficient animals,[29] or when insulin is administered into the brain of hyperphagic insulin-deficient animals,[30] the hyperphagia is eliminated. In addition to effects on food intake, leptin and insulin also affect metabolism. When leptin is administered into the brains of experimental animals, there is a selective reduction of respiratory quotient (RQ) and body fat, with lean body mass being spared.[31,32] Likewise, when insulin is administered into the brain, there is also a reduction of the RQ indicating that the body is oxidizing relatively more fat.[33] An obvious paradox based on all of these observations is that obese individuals (including humans) have elevated insulin and leptin throughout the body, including within the brain. If the same levels of insulin and leptin were present in the brains of lean animals, they would reduce their food intake and lose weight. Hence, obesity is associated with an apparent central as well as systemic insulin and leptin "resistance." Nonetheless, obese individuals regulate their (elevated) level of adiposity with the same rigor as do lean individuals. This suggests that perhaps achieving very high concentrations of insulin and/or leptin in the brain might be an effective weight-loss strategy, in much the same way that patients with insulin-resistant type 2 diabetes can be treated with systemic insulin.

2.2   ADIPOSITY SIGNALS AND MACRONUTRIENT INTAKE

Taken together, considerable evidence exists supporting the hypothesis that insulin and leptin provide the brain with key information regarding the size of the body's fat stores. When the levels of either hormone are manipulated locally within the brain, there are predictable changes of the amount of food consumed as well as of energy expenditure.[1] While these effects have been studied extensively and have been consistently found by many groups, very few experiments have considered the interaction between dietary macronutrients and the efficacy of leptin and insulin to regulate food intake and body weight. With respect to this interaction, two issues can be raised, and these comprise the main subject of this chapter. The first refers to the vast majority of experiments assessing leptin and insulin effects on food intake in animals maintained on standard laboratory chow. Laboratory chow tends to be a high-carbohydrate/low-fat source of calories and therefore may provide a biased assessment of the actions of insulin or leptin on food intake and body weight. Only a handful of experiments have used alternative diets in these experiments. The second issue concerns whether leptin or insulin, acting in the central nervous system (CNS), regulates food intake by emphasizing one or another macronutrient. From a teleological point of view, it might be reasoned that if leptin and insulin are signals that bias the brain to reduce food intake and body weight, their administration into the CNS would lead to a selective reduction of fat consumption. It is reasonable that reduced dietary fat intake would be particularly effective for reducing total body fat. This is because, of all the macronutrients, dietary fat most efficiently contributes to fat deposition.

2.3   EFFICACY OF ADIPOSITY SIGNALS IN RATS MAINTAINED ON DIETS
OF DIFFERENT FAT CONTENT

In a pioneering experiment, Arase et al.[34] maintained rats on diets containing either low (10%) or high (62%) fat content, and then administered insulin (10 mU/day) or its vehicle into the third cerebral ventricle. They found that rats on the low-fat (and therefore high-carbohydrate) diet reduced their food intake and body weight similarly to what is generally observed when rats are maintained on laboratory chow. However, the rats on the high-fat (and therefore low-carbohydrate) diet had no change of food intake or body weight when given the same amount of insulin. Arase et al.[34] interpreted their findings to indicate that the elevated fatty acid oxidation of the rats adapted to the high-fat diet rendered their brains relatively resistant to the anorexic action of insulin. In a replication and extension of the Arase et al. work, we maintained different groups of rats on one of four diets containing 7, 22, 39, or 54% calories as fat and then infused them intraventricularly with insulin (10 mU/day) or its vehicle.[35] As seen in Figure 1, the ability of third-ventricular insulin to reduce food intake decreased as the percent of fat in the diet increased. Body weight changes paralleled changes of food intake. In fact, only the rats maintained on the two diets with the lowest fat content reduced their food intake and body weight.

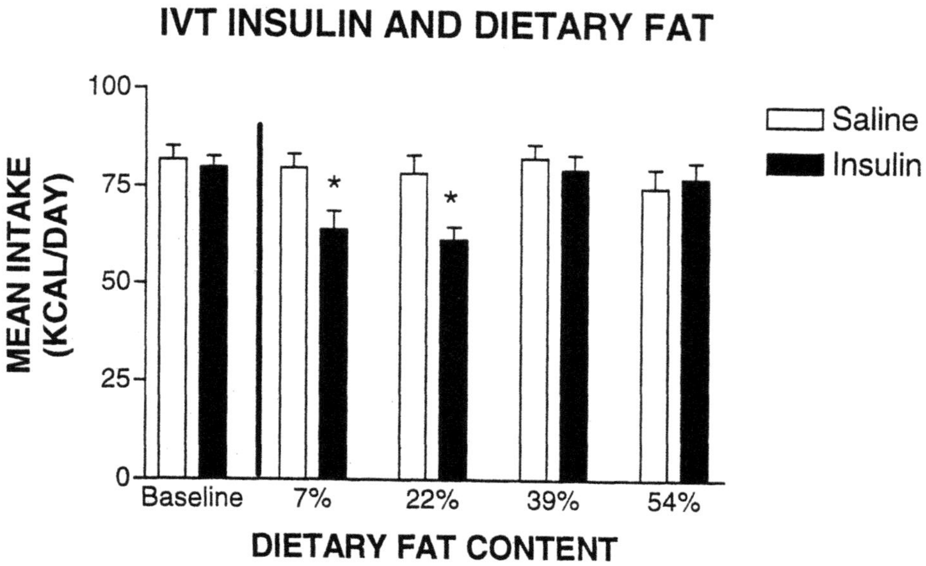

IVT INSULIN AND DIETARY FAT

FIGURE 1 Mean food intake (+ SEM) of rats maintained on 7, 22, 39, or 54% of dietary calories as fat. The baseline values are the mean of the final 2 days without an infusion. The other values are the mean of the final 4 days of a 6-day period during which rats received third cerebroventricular infusions (IVT) of either insulin (10 mU/day) or saline. $*p < 0.05$. (From Chavez, M. et al., *Am. J. Physiol.*, 271, R727, 1996.)

In an analogous experiment (C. A. Riedy, G. van Dijk, and S. C. Woods, in preparation), we investigated the ability of central leptin to reduce food intake in rats that were maintained on a low-fat/high-carbohydrate (AIN-76, with 52% cornstarch by weight) or high-fat/low-carbohydrate diet (AIN-76, with 10.4% cornstarch). Animals that had adapted to these diets for 2 weeks received bolus infusions of leptin (1 µg/day) into the third cerebral ventricle on three consecutive days, 1 h prior to the onset of the dark cycle. Relative to vehicle treatment, leptin reduced both food intake and body weight. However, its effect was equivalent for the two dietary groups over the 3-day period (see Figure 2), suggesting that dietary fat or carbohydrate content may not be an important variable for the catabolic activity of leptin in the brain.

Thus, there appears to be a fundamental difference between the anorexic effects of insulin and leptin and their interaction with dietary fat. Rats consuming 39% or more dietary fat appear "blind" to central insulin, whereas (at least preliminarily) central leptin reduces food intake and body weight irrespective of the fat content of the diet. An explanation that is consistent with both experiments is based on the finding that, as discussed above, whereas both insulin and leptin are secreted in direct proportion to adiposity, insulin secretion increases during individual meals while leptin secretion does not. Insulin entry into the brain is consequently also increased during meals.[36-38] The prandial and immediately postprandial intervals are periods of increased glucose oxidation and reduced hepatic glucose production. These two phenomena, combined, could potentially result in hypoglycemia if the supply of readily available glucose from the gut were to be attenuated. Hypoglycemia, in turn, would put the brain at risk since, unlike the rest of the body, the brain is largely dependent on glucose as a source of energy. Hence, an increase in the level or activity of insulin in the brain of a magnitude sufficient to reduce food intake could potentially reduce available glucose as well. The point is that the action of insulin to reduce food intake might be manifest only so long as a critical level of carbohydrate availability and oxidation can be maintained. If carbohydrate availability has the possibility of decreasing to too low a level, as might occur on a high-fat (thus low-carbohydrate) diet, the neuronal circuitry would not respond to elevated insulin. This blunting or inhibition would not occur with increases in brain leptin, since elevated levels of leptin do not have a large impact on blood glucose levels.[39,40] When the amount of carbohydrate

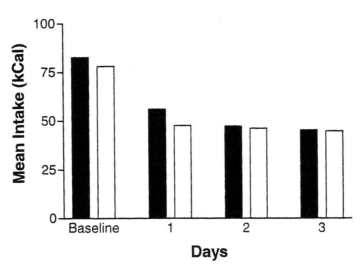

FIGURE 2   Mean food intake of rats maintained on a high-carbohydrate/low-fat diet (solid bars) or a high-fat/low-carbohydrate diet (open bars). The baseline values are the mean of the 5 vehicle-administration days which immediately preceded days on which leptin was administered into the third cerebral ventricle (1 µg/rat/day). Leptin caused a significant decrease of intake in both groups relative to the baseline. Animals administered vehicle on the same days did not change their intakes reliably from that on the baseline days (data not shown). (From C. A. Riedy, G. van Dijk, and S. C. Woods, in preparation.)

in the diet is high, a reduction of overall caloric intake would not necessarily jeopardize the stream of newly absorbed glucose into the blood. Hence, the anorexic effect of insulin, but not of leptin, would be anticipated to be sensitive to the carbohydrate content of the diet, and that is exactly what has been observed.

2.4   ADIPOSITY SIGNALS AND MACRONUTRIENT SELECTION

There are several implications of the data and interpretation of experiments in which the relative proportions of dietary fat and carbohydrate were manipulated. One important implication is that brain insulin should not reduce food intake, when doing so would be at the expense of a reduction of available carbohydrates. In a test of this, we assessed the effects of CNS insulin administration on macronutrient selection.[35] Rats had a choice of pure macronutrients that they could select from three separate jars in their home cages. After their baseline intakes were stable (around 10 days), they were chronically infused with insulin (6 mU/day) into the third cerebral ventricle for 7 days. This dose of insulin reliably reduced caloric intake and body weight (by 6% over the 7 days) relative to vehicle infusion. Further analysis revealed that there was a selective reduction (a decrease of 41%) of dietary fat intake with no discernible effect on carbohydrate or protein intake (Figure 3). The percentage of calories consumed as fat by these animals during the control period before the start of insulin infusion was around 43%. Based on the results of rats maintained on fixed diets with differential proportions of dietary fat as discussed above (see Figure 1), the average rat was therefore consuming a relatively high-fat diet and might have been predicted to be insensitive to the anorexic effect of insulin. However, since the rats could choose their own macronutrient mix, when given insulin into the brain they were able to reduce their total caloric consumption while maintaining a stable carbohydrate balance in the latter experiment. Comparable experiments have not been reported for leptin infusions.

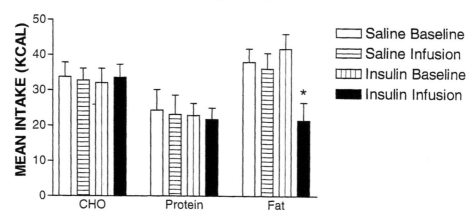

FIGURE 3 Mean (+ SEM) food intake of rats with a choice of three macronutrients in their home cages. Baseline values are the average of the final 2 days of the preinfusion period. Infusion values are the average of the final 4 days of a 6-day period during which saline or insulin (6 mU/day) was infused into the third cerebral ventricle. $*p < 0.05$. (From Chavez, M. et al., *Am. J. Physiol.*, 271, R727, 1996.)

3   BRAIN LOCI RESPONSIVE TO ADIPOSITY SIGNALS

In an attempt to localize the site(s) where insulin affects macronutrient selection, rats received bilateral cannulas aimed at either the arcuate (ARC) or the paraventricular (PVH) nuclei of the hypothalamus.[41] ARC neurons contain insulin-binding sites,[42,43] and many of them which are important in the control of food intake and energy homeostasis project directly to the PVH.[44] The rats were then given a choice of three diets consisting of powdered chow enriched by 25 to 30% with protein, fat, or carbohydrate. Local infusion of a small dose of insulin (2 µU) into the ARC at the beginning of the dark cycle caused a 44% reduction in the consumption of fat relative to the effect of the vehicle. This very low dose of insulin was not sufficient to reduce total food intake, perhaps because there were nonsignificant increases in carbohydrate and protein consumption.

In contrast to what occurred in the ARC, the same dose of insulin infused into the PVH (or a fourfold higher dose infused into the third cerebral ventricle) had no effect on diet selection. Thus, consistent with other data suggesting that the ARC is a major site of the action of insulin to reduce food intake,[1,44] the ARC appears to be an important site for determining the effect of insulin on macronutrient intake as well. Further, the data from the infusions into the ARC are consistent with the hypothesis that insulin reduces caloric intake only when glucose homeostasis is ensured. Results reported by Oomura[45] are consistent with an interaction of central insulin interacting with central glucose availability. He demonstrated that neurons in the ventromedial hypothalamus (VMH, another site in close proximity to the third cerebral ventricle) increase their firing rate to iontophoretically applied insulin. Lowering the background level of glucose, and thus perhaps reducing glucose oxidation locally, caused a pronounced reduction of firing rate of these neurons to applied insulin. Thus, the data of Oomura are consistent with the hypothesis that insulin, presumably at the level of the VMH, causes reduced food intake provided a sufficient level of carbohydrate oxidation is maintained.

4   SUMMARY

The infusion of insulin into either the third cerebral ventricle (or the ARC nuclei of the hypothalamus) reduces caloric intake so long as sufficient carbohydrate intake is maintained. This can be accomplished by having the maintenance diet contain a high proportion of carbohydrate (as occurs with most commercial laboratory chows for rats) or else by letting the animals choose their proportions of macronutrients. In the latter instance, the central infusion of insulin causes a selective reduction of fat intake (sparing both carbohydrate and protein). While still preliminary, the data with leptin indicate that there may be no comparable interaction with dietary macronutrient content.

ACKNOWLEDGMENTS

Research described in this chapter was supported in part by National Institutes of Health Grant DK-17844 and the Dutch Diabetes Foundation.

REFERENCES

1. S. C. Woods, R. J. Seeley, D. J. Porte, and M. W. Schwartz, *Science*, 280, 1378, 1998.
2. Y. Zhang, R. Proenca, M. Maffie, M. Barone, L. Leopold, and J. M. Friedman, *Nature*, 372, 425, 1994.
3. R. C. Frederich, A. Hamann, S. Anderson, B. Lollmann, B. B. Lowell, and J. S. Flier, *Nat. Med.*, 1, 1311, 1995.
4. M. Maffei, J. Halaas, E. Rayussin, R. E. Pratley, G. M. Lee, Y. Zhang, H. Fei, S. Kim, R. Lallone, S. Ranganathan, P. A. Kern, and J. M. Friedman, *Nat. Med.*, 1, 1155, 1995.
5. R. V. Considine, M. K. Sinha, M. L. Heiman, A. Kriaucinas, T. W. Stephens, M. R. Nyce, J. P. Ohannesian, C. C. Marco, L. J. McKee, T. L. Bauer, and J. F. Caro, *N. Engl. J. Med.*, 334, 292, 1996.
6. G. P. McGregor, J. F. Desaga, K. Ehlenz, A. Fischer, F. Heese, A. Hegele, C. Lammer, C. Peiser, and R. E. Lang, *Endocrinology*, 137, 1501, 1996.
7. S. C. Woods, G. J. Taborsky, Jr, and D. Porte, Jr., in *CNS Control of Nutrient Homeostasis,* American Physiological Society, Bethesda, 1986.
8. J. D. Bagdade, E. L. Bierman, and D. Porte, Jr., *J. Clin. Invest.*, 46, 1549, 1967.
9. B. D. Polonsky, E. Given, and V. Carter, *J. Clin. Invest.*, 81, 442, 1988.
10. M. Cigolini, J. C. Seidell, G. Targher, J. P. Deslypere, B. M. Ellsinger, J. Charzewska, A. Cruz, and P. Bjorntorp, *Metabolism*, 44, 35, 1995.
11. D. J. Galanis, S. T. McGarvey, J. Sobal, L. Bausserman, and P. D. Levinson, *Int. J. Obesity Relat. Metab. Disorders*, 19, 731, 1995.
12. A. Dua, M. I. Hennes, R. G. Hoffman, D. L. Maas, G. R. Krakower, G. E. Sonnenberg, and A. H. Kissebah, *Diabetes*, 45, 1635, 1996.
13. H. Masuzaki, Y. Ogawa, N. Isse, N. Satoh, T. Okazaki, M. Shigemoto, K. Mori, N. Tamura, K. Hosoda, Y. Yoshimasa, H. Jingami, T. Kawada, and K. Nakao, *Diabetes*, 44, 855, 1995.
14. G. Baura, D. Foster, D. Porte, Jr., S. E. Kahn, R. N. Bergman, C. Cobelli, and M. W. Schwartz, *J. Clin. Invest.*, 92, 1824, 1993.
15. M. W. Schwartz, A. J. Sipols, S. E. Kahn, D. P. Lattemann, G. J. Taborsky, Jr., R. N. Bergman, S. C. Woods, and D. Porte, Jr., *Am. J. Physiol.*, 259, E378, 1990.
16. M. W. Schwartz, R. N. Bergman, S. E. Kahn, G. J. Taborsky, Jr., L. D. Fisher, A. J. Sipols, S. C. Woods, G. M. Steil, and D. Porte, Jr., *J. Clin. Invest.*, 88, 1272, 1991.
17. W. A. Banks, A. J. Kastin, and W. E. A. Huang, *Peptides*, 17, 305, 1996.
18. P. L. Golden, T. J. Maccagnan, and W. M. Pardridge, *J. Clin. Invest.*, 99, 14, 1997.
19. H. J. Frank and W. M. Pardridge, *Diabetes*, 30, 757, 1981.
20. H. J. Frank, T. Jankovic-Vokes, W. M. Pardridge, and W. L. Morris, *Diabetes*, 34, 728, 1985.
21. D. J. Brief and J. D. Davis, *Brain Res. Bull.*, 12, 571, 1984.
22. M. K. McGowan, K. M. Andrews, J. Kelly, and S. P. Grossman, *Behav. Neurosci.*, 104, 373, 1990.
23. M. K. McGowan, K. M. Andrews, and S. P. Grossman, *Physiol. Behav.*, 51(4), 753, 1992.
24. C. R. Plata-Salaman, Y. Oomura, and N. Shimizu, *Physiol. Behav.*, 37, 717, 1986.
25. S. C. Woods, L. J. Stein, L. D. McKay, and D. Porte, Jr., *Nature*, 282, 503, 1979.
26. J. H. Strubbe and C. G. Mein, *Physiol. Behav.*, 19, 309, 1977.
27. D. L. Coleman, *Diabetologia*, 14, 141, 1978.
28. G. A. Bray and D. A. York, *Physiol. Rev.*, 51, 598, 1971.

29. L. A. Campfield, F. J. Smith, Y. Gulsez, R. Devos, and P. Burn, *Science*, 269, 546, 1995.

30. A. J. Sipols, D. G. Baskin, and M. W. Schwartz, *Diabetes*, 44, 147, 1995.

31. G. Chen, K. Koyama, X. Yuan, Y. Lee, Y. T. Zhou, R. O'Doherty, C. B. Newgard, and R. H. Unger, *Proc. Natl. Acad. Sci. U.S.A.*, 93, 14795, 1996.

32. N. Levin, C. Nelson, A. Gurney, R. Vandlen, and F. DeSauvage, *Proc. Natl. Acad. Sci. U.S.A.*, 93, 1726, 1996.

33. C. Park, M. Chavez, and S. C. Woods, *Soc. Neurosci.*, Abstr., 939, 1992.

34. K. Arase, J. S. Fisler, N. S. Shargill, D. A. York, and G. A. Bray, *Am. J. Physiol.*, 255, R974, 1988.

35. M. Chavez, C. A. Riedy, G. van Dijk, and S. C. Woods, *Am. J. Physiol.*, 271, R727, 1996.

36. M. Orosco, K. Gerozissis, C. Rouch, and S. Nicolaidis, *Brain Res.*, 671, 149, 1995.

37. K. Gerozissis, M. Orosco, C. Rouch, and S. Nicolaidis, *Physiol. Behav.*, 62, 767, 1997.

38. A. B. Steffens, A. J. Scheurink, D. Porte, Jr., and S. C. Woods, *Am. J. Physiol.*, 255, R200, 1988.

39. S. Dagogo-Jack, C. Fanelli, D. Paramore, and M. Landt, *Diabetes*, 45, 695, 1996.

40. J. W. Kolaczynski, R. V. Considine, J. Ohannesian, C. Marco, I. Opentanova, M. R. Nyce, and J. F. Caro, *Diabetes*, 45, 1511, 1996.

41. G. van Dijk, C. de Groote, M. Chavez, Y. van der Werf, A. B. Steffens, and J. H. Strubbe, *Brain Res.*, 777, 147, 1997.

42. D. G. Baskin, J. L. Marks, M. W. Schwartz, D. P. Figewicz, S. C. Woods, and D. Porte, Jr., in *Insulin and Insulin Receptors in the Brain in Relation to Food Intake and Body Weight*, H. Lehnert, R. Murison, H. Weiner, D. Hellhammer, and J. Beyer, Eds., Hogrefe & Huber Publ., Stuttgart, 1993.

43. E. S. Corp, S. C. Woods, D. Porte, Jr., D. M. Dorsa, D. P. Figlewicz, and D. G. Baskin, *Neurosci. Lett.*, 70, 17, 1986.

44. M. W. Schwartz, D. P. Figlewicz, D. G. Baskin, S. C. Woods, and D. Porte, Jr., *Endocr. Rev.*, 13, 387, 1992.

45. Y. Oomura, *Adv. Metab. Disorders*, 10, 31, 1983.

31 Stress and Macronutrient Selection

Ruth B. S. Harris, Leigh Anne Howell,
Tiffany D. Mitchell, Bradley D. Youngblood,
David A. York, and Donna H. Ryan

CONTENTS

1   INTRODUCTION

Stress has been described as an event that leads to overactivation of the body's normal activational systems.[1] It is generally accepted that the corticotrophin-releasing factor (CRF) system, which includes the neuropeptides CRF and urocortin (UCN), at least four subtypes of CRF receptors, and CRF binding protein,[2] is responsible for initiating and coordinating neurological, behavioral, endocrine, and immunological responses to stress.[3,4] CRF is a 41-amino acid peptide present in many areas of the brain and in peripheral tissues. The highest concentrations are found in the paraventricular nucleus of the hypothalamus (PVN), where it regulates release of proopiomelanocortin (POMC)-derived proteins from the pituitary, including adrenocorticotropin hormone (ACTH) and β-endorphin.[5,6] UCN is a 40-amino acid peptide that has 45% homology to CRF.[7] UCN mRNA has been identified in a number of brain areas, including the midbrain, hypothalamus, and pituitary,[8] and the protein is widely distributed in both brain and peripheral tissues with high concentrations present in the pituitary.[9] Centrally infused UCN has similar bioactivities to CRF, inhibiting food intake,[10] increasing ACTH release,[7] and inducing anxiety-type behaviors.[11] UCN binds to CRF binding protein[7] and both CRF_1 and CRF_2 receptors, having a 40-fold higher affinity than CRF for CRF_2 receptors.[7,12] UCN immunoreactivity is localized in brain areas that contain high levels of CRF_2 receptors, including the ventromedial and supraoptic nuclei of the hypothalamus.[7] These observations lead to the conclusion that, although CRF and UCN are capable of inducing responses typical of stress, they may have distinct functions based on protein and receptor distribution.[7]

In conditions of stress, stimulation of the CRF system results in activation of the sympathetic nervous system, and the hypothalamic–pituitary–adrenal (HPA) axis, release of inflammatory cytokines, inhibition of release of growth hormone and leutinizing hormone, hypotension, abnormal gastrointestinal function, and disrupted feeding behavior.[3,13] Changes in central catecholaminergic, serotonergic, opioid, and cholinergic activity provide the basis for behavioral responses that include anxiety, depression, deficits in motor performance, impaired acquisition and memory, and decreased socialization.[2,3]

2   STRESS AND FOOD INTAKE

Central administration of CRF inhibits food intake of rats[14]; however, the effect of stress on food intake is dependent upon the duration of the stressor and whether the stress is physical or psychological. Very mild stressors, such as pinch,[15] a brief period of restraint or brief handling of an animal,[16] will stimulate feeding and drinking during the 30 min following stress. The response is reinforcing, as the amount of food consumed increases with repeated exposure to mild stress[16,17] and can be blocked by opioid antagonists,[18] dopamine antagonists,[19] or CRF. The response is exaggerated by low doses of CRF receptor antagonist.[17] It has been proposed that very mild stress promotes food consumption due to an increased responsivity to environmental stimuli[20] but that CRF inhibits evoked feeding when excessive food intake has the potential to compromise other behaviors.[17]

More extreme stressors inhibit food intake. An hour of immobilization or restraint[21] inhibits intake during the hour following stress, a response that is prevented by antagonism of serotonin receptors in the PVN[22] or is partially reversed by infusions of a CRF receptor antagonist (αhCRF$_{9\text{-}41}$) into the lateral cerebral ventricle.[21,23] We have shown that a single 3-h restraint stress, with rats confined in Plexiglas tubes, inhibits food intake during the subsequent 24 h and causes weight loss that remains uncorrected for at least 10 days.[24] The effect is exaggerated by applying the stress early in the day, compared with late in the light phase, possibly because of an association with circadian release of corticosterone. The sustained reduction in body weight of stressed rats may, at least partially, be attributed to the absence of any rebound hyperphagia to compensate for hypophagia immediately following stress. Therefore, the mechanisms that normally stimulate feeding during recovery from physical stressors, such as food restriction or cold exposure,[25,26] are not activated by negative energy balance and weight loss associated with exposure to a psychological stressor.

The effect of repeated exposure to stress on food intake has produced discrepant results. Repeated immobilization of rats for 2.5 h/day for 7 days inhibited daily food intake by 25 to 30% on all days of immobilization.[27] In an experiment designed to determine the effect of type and duration of stress on food intake, Marti et al.[28] exposed rats to handling, 1 h of restraint, or 1 h of immobilization each day for 27 days. After 7 and 27 days food intake and body weight were significantly reduced in restrained and immobilized rats, compared with undisturbed controls. Other time points were not reported. Repeated immobilization for 2 h a day for 14 days caused a significant inhibition of food intake, reported for days 1, 10, and 14.[28] The degree of inhibition was greater on day 1 than either of the later time points and it is possible that there was some recovery of intake, especially considering the reduced body size of the stressed rats. Similarly, Krahn et al.[29] showed that 2 h of restraint on each of 5 days caused significant reductions in food intake, measured during the 2 h after the end of stress, on all 5 days but that the response was greater on the first than the last day of stress. Intake measured for 12 or 24 h after restraint was not significantly different from that of controls even after the first stressor. In contrast, Haleem and Parveen[30] reported that, although a single exposure to 1 h of restraint inhibited food intake and growth of rats, this response was no longer apparent after five daily 2-h periods of restraint, suggesting that the animals had adapted to the stress. In a model of repeated restraint used in our laboratory, adult rats are exposed to 3 h of restraint on 3 consecutive days. This protocol causes a significant inhibition of 24-h food intake, that is corrected within a few days, but a sustained reduction in body weight. The rats do not demonstrate any poststress hyperphagia and, although they gain weight after the end of stress, they do not reach the weight of nonrestrained rats even 40 days after the end of the repeated restraint.[31]

Although the majority of studies examining the effects of stress on food intake of rats have been performed with either restraint or immobilization stress, inhibition of intake is a generalized response to psychological stressors. Meerlo et al.[32] exposed Roman high- and low-avoidance rats, two lines with different abilities to cope with stress, to a single social defeat and followed food intake and body weight of the rats for 20 days after the stress. Social defeat is an extreme psychological stress in which an experimental animal is placed in the home cage of an aggressive male rat. This stress inhibited food intake of both the high- and low-avoidance rats, but the response was greater in the more stress-responsive, high-avoidance rats. Food intake returned to control levels within approximately 10 days, but neither line of animals demonstrated a compensatory hyperphagia, resulting in a sustained reduction in body weight, compared with control animals. These results suggest that exposure to psychological stress causes a resetting of some aspect of homeostasis that lowers body weight. Measurements of body composition indicate that all of the weight loss in male rats during stress is lean tissue, but there are compensatory adjustments in fat and protein deposition such that, by 5 days poststress, the difference in weight between stressed and control rats consists of both lean and fat tissue. These observations suggest that stress causes animals to maintain a smaller body size, rather than a different body composition, than control animals.[33]

A majority of studies examining the effect of stress on food intake have used rats maintaining a normal body weight. However, there are some data to indicate that the degree of hypophagia induced by stress is dependent upon the energy balance status of the animal at the time it is exposed to stress. Rats that have been food restricted and trained to obtain food in a fixed ratio operant paradigm will reduce appetitive behavior in a fixed ratio paradigm (FR5) immediately following a short exposure to restraint combined with partial water immersion, but do not reduce intake of freely available food unless the stress is increased to induce a severe disruption of motor function.[34] More convincing evidence has been provided by Lennie et al.[35] Rats were overfed to gain weight or were food restricted to lose weight prior to turpentine injection. Following injection, overweight rats showed a protracted hypophagia and greater weight loss than normal-weight rats, whereas food-restricted rats increased food intake and gained weight following the stress. Similarly, we have observed greater effects of repeated restraint on body weight of rats fed a high-fat diet than those fed a low-fat diet.[33] These observations demonstrate that the effect of stress on food intake is dependent upon the size of body energy stores of the animals at the time of stress, and imply that the response is regulated to prevent excessive tissue catabolism.[35]

All of these studies, determining the poststress food intake of rats, have been conducted with mixed diets, excluding the possibility of determining whether stress inhibits appetite for all nutrients, or whether the decline in food intake results from an avoidance of a specific macronutrient present in a mixed diet.

3   GLUCOCORTICOIDS AND MACRONUTRIENT INTAKE

As described above, the behavioral and physiological responses to stress are orchestrated by activation of the CRF system. Several studies have examined the effect of either CRF infusion or modulation of peripheral corticosterone on macronutrient selection in rats. Lin et al.[36] adapted rats to macronutrient selection diets prior to placement of cannulas in the lateral ventricle. All of the rats highly preferred the fat diet over protein or carbohydrate. After 10 days of recovery from surgery, macronutrient selection was measured for 1 h following an overnight fast. Infusion of 1 μg of CRF into the lateral ventricle inhibited fat intake, whereas 5 μg inhibited intake of both fat and protein but had no effect on carbohydrate, compared with intake of rats infused with saline (see Figure 1A). Infusion of 1 μg CRF into the third ventricle significantly inhibited fat and protein intake but had no effect on carbohydrate intake, which was very low (<0.5 g/h) even in control rats (see Figure 1B). These results confirm that hypothalamic CRF inhibits food intake and demonstrate

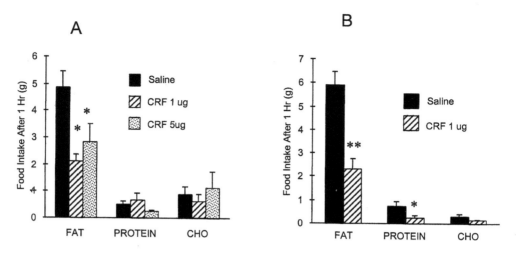

FIGURE 1 The effects of CRF on the selection of fat protein and carbohydrate (CHO). Rats were adapted to a three-choice diet prior to testing. Values represent means + SEM for 1 h of testing measured after an overnight fast. (A) CRF (1 or 5 μg) was infused into the lateral ventricle. Two-way analysis of variance showed a significant effect of CRF on fat intake (*$p < 0.05$). (B) CRF (1 μg) was infused into the third ventricle. Two-way analysis of variance showed a significant effect of CRF on fat (**$p < 0.01$) and protein (*$p < 0.05$) intake.

a macronutrient-specific response. As the greatest effect of CRF was on fat intake, the most preferred macronutrient of the rats, it would be of interest to repeat the study with rats preferring carbohydrate, separating the effects of CRF on intake of a specific energy source compared with intake of a preferred food.

Adrenal glucocorticoids, released by stress-associated activation of the HPA axis, also influence macronutrient selection. Corticosterone is essential for maintenance of normal food intake and body weight[37] but can have anabolic or catabolic actions, depending upon circulating concentrations. Low, or basal, concentrations of corticosterone bind to high-affinity Type I receptors with an anabolic outcome, whereas higher concentrations activate both Type I and Type II receptors leading to a state of catabolism.[37] Adrenalectomy of rats allowed to select from three macronutrient diets almost totally abolishes fat intake, reduces carbohydrate intake, and increases protein intake.[38] Infusion of a Type I receptor agonist returns macronutrient preference to normal.[38] Similar responses were demonstrated by Castonguay[39] who showed a dose-dependent restoration of a normal fat intake in adrenalectomized lean and obese Zucker rats infused with corticosterone. In adrenalectomized Sprague-Dawley rats, which had a reduced energy intake and consumed a reduced proportion of calories as fat and protein compared with sham-operated controls, corticosterone replacement restored fat but not protein intake. Neither adrenalectomy nor corticosterone replacement influenced the proportion of calories consumed as carbohydrate.[40] The results from these studies led to the conclusion that corticosterone was responsible for driving fat preference.[39]

Prasad et al.[41] also examined the effect of corticosterone on fat preference, supplementing corticosterone levels in intact rats rather than replacing corticosterone in adrenalectomized animals. In nonmanipulated rats there was a positive correlation between basal urinary corticosterone excretion and fat preference, an inverse relationship with protein preference and no significant relationship between corticosterone and carbohydrate intake, in agreement with the data from Castonguay.[39] An acute injection of corticosterone increased intake of all macronutrients, but when rats with a low preference for fat were treated chronically with corticosterone, presented in drinking water for 3 weeks, there was a substantial increase in the proportion of calories consumed as fat and a decrease in carbohydrate intake. This suggests that chronic high levels of corticosterone promote fat consumption, the opposite to that expected if the response is mediated by high-affinity Type I receptors.[37]

Promotion of fat intake by corticosterone initially appears contrary to inhibition of fat intake by central infusion of CRF,[36] which would stimulate glucocorticoid release. However, as corticosterone downregulates CRF receptor mRNA expression[42,43] and activation of the HPA axis,[44] it is possible that acute CRF infusions have the opposite effect to chronic corticosterone infusion. The contrasting results may also be explained by experimental design as as the effects of corticosterone on macronutrient intake were measured over 24 h,[40] whereas the effect of CRF was measured during the hour following an acute injection.

In contrast to these studies, work from the Leibowitz[45] laboratory suggests that corticosterone is responsible for ensuring intake of all three macronutrient and for promoting carbohydrate intake during the start of the dark cycle, a time when endogenous corticosterone concentrations are at their diurnal peak. Adrenalectomy decreased total energy intake by inhibiting consumption of fat, protein, and carbohydrate during the dark cycle. A single dose of corticosterone restored fat and protein intake to normal levels but caused a significant increase in the proportion of energy consumed as carbohydrate during the early hours of the dark period[45] a time when normal rats consume a relatively large proportion of their daily carbohydrate intake.[46] This response was not observed when corticosterone was replaced by chronic infusion.[46] The difference in response to acute treatment with corticosterone, compared with chronic infusion, suggests that acute and chronic activation of the HPA axis have differential effects on macronutrient selection, thus one would also anticipate different effects of acute and chronic stress on macronutrient preference.

4   STRESS AND MACRONUTRIENT SELECTION

Very few studies have examined the effect of stress on macronutrient selection. Based on the differential effects of mild, severe, acute, repeated, or chronic stress on food intake, one would expect macronutrient preference also to be determined by the type and intensity of the stressor. There is a substantial literature on stress inhibiting consumption of sucrose or saccharin solution[47] which is used as a behavioral measure of stress-induced anhedonia and depression.[48,49] However, there has been recent controversy over whether the change in consumption of these preferred solutions is secondary to the reduced body weight[50,51] or the hydration state[52,53] of stressed animals, rather than a specific behavioral response to stress.

One study investigated the effects of mild tail-pinch stress on energy and macronutrient intake of rats but failed to find an effect on any aspect of intake.[54] Young rats were subjected to 24 days of tail pinch which did not change energy intake or macronutrient selection but inhibited longitudinal growth, suggesting that the intensity of the stressor had exceeded that which promotes feeding. In an experiment with older animals, rats were fed one of four isocaloric experimental diets.[55] The control diet contained 21% protein, 64% carbohydrate (CHO), and 5% fat. The three other diets were high carbohydrate (19% protein, 0.5% fat, 74% CHO), high protein (31% protein, 5% fat, 53% CHO), or high fat (19% protein, 23% fat, 22% CHO). Baseline intakes of the diets were determined during a 3-h period following a 12-h fast. The rats were then exposed either to food deprivation for an increasing length of time (24 or 48 h) or 12 h food deprivation followed by a 10-min swim in cold water. Food deprivation increased the 3-h intake of all diets with the smallest response in rats on control diet and the greatest response in those fed high-fat diet. The 10-min cold swim also increased food intake over baseline with the smallest response in rats fed control diet (154%) or high-carbohydrate diet (174%), and a much greater response in those fed high-protein (310%) or high-fat diet (423%). Pretreatment of the rats with naloxone (1 mg/kg) prior to the cold swim reduced food intake to approximately half of baseline levels for all dietary treatments. The results from this study confirm that stress-induced eating is mediated by the opioid system, but it is not clear whether the exaggerated hyperphagia in rats fed high-protein or high-fat diets was due to the palatability or the composition of the diets. However, evidence that morphine promotes fat intake in rats given macronutrient selection diets[56] suggests that mild stress should lead to a preferential consumption of fat if stress-induced eating is mediated by the opioid system.

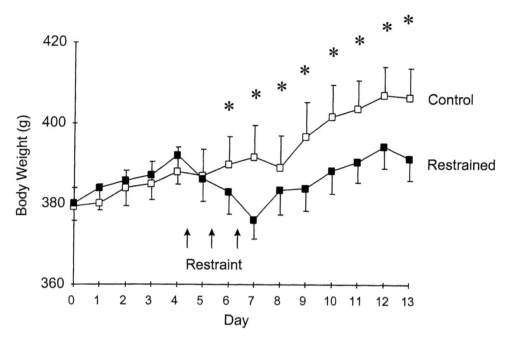

FIGURE 2 Daily body weights of repeatedly restrained or control male Sprague-Dawley rats offered free access to three liquid diets each containing one macronutrient. Data are means ± SEM for 10 rats. Stressed rats were restrained for 3 h/day on each of the days indicated by an arrow. There were significant differences in body weights of the two groups of rats from the second day of stress (day 6 of the experiment) until the end of the experiment, determined by repeated-measures analysis of variance and indicated by asterisks (Stress NS, Day $p < 0.0001$, Interaction $p < 0.004$) and *post hoc* calculation of least-significant difference ($p < 0.05$).

We have conducted two studies to determine the effect of more severe stressors on macronutrient selection. In the first study macronutrient intake of rats exposed to repeated restraint stress was examined. Male Sprague-Dawley rats were adapted to the simultaneous presentation of three isocaloric liquid diets each deriving 100% of energy from one macronutrient. All of the diets contained adequate levels of micronutrients and were flavored with vanillin (Dyets, Inc., Bethlehem, PA). Once a stable 24-h baseline period had been established, half of the rats were exposed to 3 h of restraint on 3 days. As can be seen in Figure 2, rats exposed to repeated restraint lost weight, compared with controls, and did not return to control weight within 4 days of the end of stress. In this study there were no significant differences in energy intakes of the two groups of rats during the three experimental periods (prestress, stress, and poststress), possibly because they slected a low-fat diet (<10% kcal).[33] The energy intakes of both restrained and control rats decreased during the period of restraint, as shown in Table 1, but the percent energy obtained from each macronutrient was different between the groups, as shown in Figure 3. Restrained rats consumed less carbohydrate and more protein during the stress period than during the pre- or the poststress period. There was no effect on macronutrient selection of controls. Throughout the experiment, all of the rats consumed the largest proportion of energy from carbohydrate (>70% kcal) and the least amount from fat (<10%) and it is possible that different responses would have been found in rats that consumed a greater proportion of their energy as fat.

The increased consumption of protein by restrained rats is particularly interesting as we have found that all of the weight loss experienced during the 3 days of stress is lean body mass, with no change in body fat content.[31,33] An increase in protein intake has also been reported for rats recovering from the physical stress of a 13-day fast, a period long enough to exhaust fat stores and initiate protein catabolism.[57] Following a protein-sparing 6-day fast, Wistar rats, a fat-preferring

TABLE 1
Energy Intake of Control and Restrained Rats before, during, and after Repeated Restraint

	Prestress	Stress	Poststress
Controls	113 ± 5^A	96 ± 3^{BC}	105 ± 2^{AC}
Restrained	121 ± 4^A	91 ± 5^B	104 ± 2^C

Data are means $\pm$ SEM for 11 rats. Values that do not share a common superscript are significantly different, determined by repeated measures analysis of variance and *post hoc* determination of least-significant difference at $p < 0.05$.

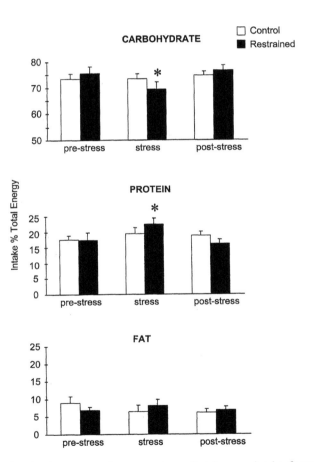

FIGURE 3 Macronutrient intake, expressed as percentage of total energy intake, for rats exposed to repeated restraint. Intakes are the average for 4 baseline days (prestress), 3 days of restraint (stress), or 4 days immediately following the end of stress (poststress). Data are means + SEM for groups of 10 rats. An asterisk indicates a significant difference between the stress period and the pre- and poststress periods for the restrained animals, determined by repeated-measures analysis of variance and *post hoc* calculation of least-significant difference ($p < 0.05$). Restraint caused a small increase in the proportion of energy consumed as protein and a small decrease in the proportion consumed as carbohydrate.

strain, increased energy intake by exclusively increasing fat consumption. Those that were recovering from the extended, 13-day fast initially increased only fat intake but, after 5 days, increased the proportion of calories consumed as protein, presumably as a means of restoring lean tissue mass. These two studies suggest that rats will increase protein intake if body energy stores are replete but lean mass has been lost. In the restrained rats larger changes in preference may have been observed if intake had been monitored at intervals during the day, as we have previously found that the period of greatest difference in intake of restrained and control rats is at the start of the dark period, independent of the time of day that stress is applied,[24] and the data from Kumar and Leibowitz[45] indicate that this is also the time at which acute administration of corticosterone influences macronutrient selection.

In a second study with Wistar rats exposed to the chronic stress of rapid eye movement (REM) sleep deprivation, we found different changes in macronutrient selection from those demonstrated by the restrained rats. Rats were deprived of REM sleep by housing them on a small platform over water. There were two control groups: cage controls, housed in normal cages, and tank controls, housed on larger platforms over water. The rats were offered the same three liquid diets as described for the restraint study and a stable 4-day baseline period was followed by 96 h of sleep deprivation, with intake and body weight recorded daily. Animals were also tested for spatial memory in a Morris water maze; therefore, all rats were subjected to varying degrees of stress. Controls were made to swim in the behavioral test, tank controls were housed in a stressful environment and made to swim, and sleep-deprived rats were housed in a stressful environment, were made to swim, and were unable to sleep for 96 h. Table 2 shows that cage control rats gained weight during the study, whereas both tank controls and sleep-deprived rats lost weight. Despite this weight loss there was no effect of treatment on energy intake for any group. Macronutrient preference of the rats is shown in Figure 4, comparing percent calories consumed as each macronutrient during baseline conditions or on each day of sleep deprivation. Although repeated measures analysis of variance indicated a significant effect of treatment on percent energy consumed as carbohydrate ($p < 0.02$)

TABLE 2
Body Weights and Energy Intakes of Wistar Rats Subjected to 96 h of Sleep Deprivation

	Baseline	Day 1	Day 2	Day 3	Day 4
Body Weights					
Cage controls	328 ± 11^A	331 ± 11^A	335 ± 10^{AB}	336 ± 11^{AB}	341 ± 11^A
Tank controls	330 ± 11^A	313 ± 12^B	311 ± 11^{BC}	303 ± 11^C	296 ± 10^D
Sleep deprived	328 ± 9^A	314 ± 12^B	313 ± 10^B	302 ± 9^C	293 ± 8^D
Energy Intakes					
Cage controls	113 ± 6	125 ± 8	120 ± 8	117 ± 10	120 ± 9
Tank controls	116 ± 4	109 ± 5	110 ± 6	112 ± 6	124 ± 5
Sleep deprived	112 ± 5	131 ± 8	119 ± 6	119 ± 7	129 ± 5

Data are means ± SEM for groups of 8 rats. Baseline is the average for the 2 days prior to the onset of stress. Days 1 through 4 indicate days of sleep deprivation. Repeated-measures analysis of variance indicated significant effects of time and a significant interruption between treatment and time ($p < 0.0001$) on body weight. Superscripts for body weights indicate significant differences ($p < 0.05$) within a treatment group. The cage control rats weighed significantly ($p < 0.05$) more than the tank control or sleep-deprived rats on all 4 days of sleep deprivation. There were no significant effects of time or treatment on energy intakes of the rats.

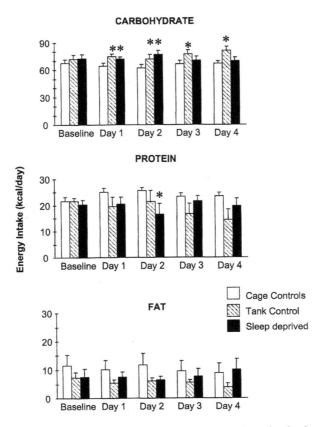

FIGURE 4   Macronutrient intake of cage control, tank control, and sleep-deprived rats during baseline and sleep deprivation. Data are the mean + SEM for groups of 8 rats during baseline period (average of 2 days immediately before the start of sleep deprivation) and each of 4 days of sleep deprivation. An asterisk indicates a significant difference between an experimental group and control animals on that particular day, determined by repeated-measures analysis of variance and *post hoc* calculation of least-significant difference ($p < 0.05$).

and protein ($p < 0.03$), there were few differences between treatment groups determined by *post hoc* calculation of least-significant difference ($p < 0.05$). Carbohydrate preference of tank controls was higher than that of cage controls on all days of stress. Sleep-deprived rats had a higher carbohydrate preference than cage controls only on the first 2 days of stress. This diet was highly preferred by all the animals, representing approximately 70% of total energy intake throughout the study. There was no significant effect of time on macronutrient selection within any of the three groups.

The results from this study are different from those of the previous experiment in that carbohydrate intake decreased following 3 h of restraint, whereas it increased in animals during chronic stress. In either situation the changes in preference were small in relation to the metabolic response to stress, especially in sleep-deprived rats that lost 10% body weight during the 4-day experimental period. Both the tank controls and sleep-deprived rats were chronically stressed, as indicated by serum corticosterone concentrations and thymus weight (see Table 3), but neither groups increased food or calorie intake and there was no evidence for development of a preference for fat, as would be expected based on studies with corticosterone replacement or supplementation in rats.[40,41] The minimal changes in macronutrient selection of stressed rats emphasizes that stress-induced end point behaviors result from activation of multiple systems, rather than the response to activation of a single pathway, such as the HPA axis.

TABLE 3
End-Point Measures in Rats Exposed to 4 Days
of REM Sleep Deprivation

	Corticosterone (ng/ml)	Adrenal Wt. (mg)	Thymus Wt. (mg)
Cage controls	33 ± 4[A]	52 ± 4	487 ± 31[A]
Tank controls	295 ± 125[B]	61 ± 10	366 ± 37[B]
Sleep deprived	341 ± 73[B]	60 ± 4	329 ± 29[B]

Data are means ± SEM for groups of 8 rats. Tissues were collected and serum corticosterone was measured (RIA: ICN Biomedicals, CA) on trunk blood at the end of 96 h of sleep deprivation. Superscripts indicate significant differences between treatment groups, determined by one-way analysis of variance and *post hoc* calculation of least-significant difference ($p < 0.05$).

5 STRESS AND ALCOHOL INTAKE

There is a well-established relationship between stress and alcohol consumption.[58] Alcohol intake increases following exposure to stress and alcohol decreases the response to stress. Although it may not be appropriate to consider alcohol as a macronutrient comparable to fat, protein, or carbohydrate,[59] it can represent a significant caloric intake in animals given access to an ethanol solution.[60]

Animals with high basal levels of corticosterone show a greater preference for 6% v/v ethanol solution over water than do rats with low basal levels of corticosterone,[61] and adrenalectomized rats show no preference for ethanol but preference is restored by corticosterone, but not aldosterone, replacement.[62,63] However, corticosterone alone is not enough to induce a preference for alcohol as Fhalke et al.[64] demonstrated that inhibition of corticosterone production abolished alcohol preference in alcohol-preferring rats but had no effect in nonpreferring rats. Alcohol consumption changes central concentrations of a number of neurotransmitters including dopamine, norepinephrine, opioids and γ-aminobutyric acid, all of which are associated with reward and positive reinforcement.[65] Consistent with this, strains of mice that have a high preference for alcohol have low endogenous levels of dopamine in the brain and preference for alcohol can be inhibited by increasing brain dopamine concentrations.[66] Alcohol-preferring rats have low baseline levels of β-endorphin and show large increases in response to alcohol consumption. Alcohol acutely increases opioid receptor binding, but extended exposure to alcohol reduces receptor activity. The role of opioid receptors in the short-term positive-reinforcing effect of alcohol consumption has been demonstrated by studies in which alcohol consumption is inhibited by morphine administration or is increased during morphine withdrawal.[67]

Acute administration of ethanol increases circulating catecholamine concentrations[68] and activates the HPA axis by stimulating CRF release[69]; however, it also reduces the catecholamine response to acute restraint stress.[68] Prenatal exposure to alcohol has a prolonged effect in rats, blunting the HPA response to cytokines until puberty, after which the animals become hyperresponsive.[70] Alcohol consumption during exposure to a stressor is often decreased. Rats offered 0.2% saccharin and 10% ethanol/0.2% saccharin in a two-bottle preference test while they were immobilized or subjected to isolation stress decreased their intake of alcohol but not saccharin solution during the period of stress.[71] Alcohol consumption is also reported to have been decreased during overcrowding.[72] Rats offered 0.1% saccharin and 10% ethanol/0.1% saccharin during 2 weeks of exposure to unpredictable periods of immobilization or isolation had a higher preference for alcohol than controls, as the two groups consumed similar amounts of alcohol but stressed rats had lower saccharin and total fluid intakes. This pattern of response was maintained during the 3 weeks following the end of stress.[73] The difference in response between studies may be explained by whether or not the stress is uncontrollable[74] and by the baseline alcohol preference of different animals, as cold exposure combined with immobilization stress inhibited ethanol intake of ethanol-preferring rats but increased consumption in rats with a low preference for alcohol.[75] Increased alcohol consumption during a poststress period is a more consistent finding in that rats increase alcohol intake following uncontrollable but not controllable stress.[76-79] This response appears to be dependent upon ACTH, as it is abolished by hypophysectomy, but not by adrenalectomy, and can be induced by repeated administration of ACTH.[80]

6   CONCLUSIONS

There is little information available concerning the effects of stress on macronutrient selection, although there is a large literature on stress and alcohol consumption. Although alcohol may be considered a nutrient based on its caloric content, the interaction between stress and alcohol intake does not represent the relationship between stress and other preferred macronutrients. Severe stressors stimulate alcohol consumption, in part due to changes in opioid receptor affinity, whereas only very mild stress will increase food intake in association with increased release of endogenous opioids. Severe stressors suppress food intake in response to the activation of multiple pathways. A relationship between aspects of the CRF system and macronutrient selection has been established, but this relationship may not dominate in conditions of stress. Further studies are needed to clarify the effect of stress on macronutrient selection; however, the type and duration of stress and baseline macronutrient preference are factors that need to be considered when interpreting the results.

REFERENCES

1. Hennessy, J. and Levine, S., Stress, arousal, and the pituitary-adrenal system: a psychoendocrine hypothesis, in *Progress in Psychobiology and Psychobiological Psychology*, 8th ed., Sprague, J. M. and Epstein, A.N., Eds., Academic Press, New York, 1979, 133.
2. Turnbull, A. V. and Rivier, C., Corticotropin-releasing factor (CRF) and endocrine responses to stress: CRF receptors, binding protein, and related peptides, *Proc. Soc. Exp. Biol. Med.*, 215, 1, 1997.
3. Koob, G. F., Heinrichs, S. C., Pich, E. M., Menzaghi, F., Baldwin, H., Miczek, K., and Britton, K. T., The role of corticotropin-releasing factor in behavioural responses to stress, *Ciba Found. Symp.*, 172, 277, 1993.
4. Dunn, A. J. and Berridge, C. W., Physiological and behavioral responses to corticotropin-releasing factor administration: is CRF a mediator of anxiety or stress responses? *Brain Res. Rev.*, 15, 71, 1990.
5. Vale, W., Rivier, C., Brown, M. R., Spiess, J., Koob, G., Swanson, L., Bilezikjian, L., Bloom, F., and Rivier, J., Chemical and biological characterization of corticotropin releasing factor, *Recent Prog. Horm. Res.*, 39, 245, 1983.
6. Rivier, C., Brownstein, M., Spiess, J., Rivier, J., and Vale, W., In vivo corticotropin-releasing factor-induced secretion of adrenocorticotropin, beta-endorphin, and corticosterone, *Endocrinology*, 110, 272, 1982.
7. Vaughan, J., Donaldson, C., Bittencourt, J., Perrin, M. H., Lewis, K., Sutton, S., Chan, R., Turnbull, A. V., Lovejoy, D., and Rivier, C., Urocortin, a mammalian neuropeptide related to fish urotensin I and to corticotropin-releasing factor, *Nature*, 378, 287, 1995.
8. Wong, M. L., al-Shekhlee, A., Bongiorno, P. B., Esposito, A., Khatri, P., Sternberg, E. M., Gold, P. W., and Licinio, J., Localization of urocortin messenger RNA in rat brain and pituitary, *Mol. Psychiatry*, 1, 307, 1996.
9. Oki, Y., Iwabuchi, M., Masuzawa, M., Watanabe, F., Ozawa, M., Iino, K., Tominaga, T., and Yoshimi, T., Distribution and concentration of urocortin, and effect of adrenalectomy on its content in rat hypothalamus, *Life Sci.*, 62, 807, 1998.

10. Spina, M., Merlo-Pich, E., Chan, R. K., Basso, A. M., Rivier, J., Vale, W., and Koob, G. F., Appetite-suppressing effects of urocortin, a CRF-related neuropeptide, *Science,* 273, 1561, 1996.

11. Moreau, J. L., Kilpatrick, G., and Jenck, F., Urocortin, a novel neuropeptide with anxiogenic-like properties, *NeuroReport,* 8, 1697, 1997.

12. Donaldson, C. J., Sutton, S. W., Perrin, M. H., Corrigan, A. Z., Lewis, K. A., Rivier, J. E., Vaughan, J. M., and Vale, W. W., Cloning and characterization of human urocortin, *Endocrinology,* 137, 2167, 1996.

13. Owens, M. J. and Nemeroff, C. B., Physiology and pharmacology of corticotropin-releasing factor, *Pharmacol. Rev.,* 43, 425, 1991.

14. Hotta, M., Shibasaki, T., Yamauchi, N., Ohno, H., Benoit, R., Ling, N., and Demura, H., The effects of chronic central administration of corticotropin-releasing factor on food intake, body weight, and hypothalamic-pituitary-adrenocortical hormones, *Life Sci.,* 48, 1483, 1991.

15. Levine, A. S. and Morley, J. E., Tail pinch-induced eating: is it the tail or the pinch? *Physiol. Behav.,* 28, 565, 1982.

16. Badiani, A., Jakob, A., Rodaros, D., and Stewart, J., Sensitization of stress-induced feeding in rats repeatedly exposed to brief restraint: the role of corticosterone, *Brain Res.,* 710, 35, 1996.

17. Heinrichs, S. C., Cole, B. J., Pich, E. M., Menzaghi, F., Koob, G. F., and Hauger, R. L., Endogenous corticotropin-releasing factor modulates feeding induced by neuropeptide Y or a tail-pinch stressor, *Peptides,* 13, 879, 1992.

18. Morley, J. E. and Levine, A. S., Stress-induced eating is mediated through endogenous opiates, *Science,* 209, 1259, 1980.

19. Larson, A. A. and Kondzielski, M. H., Serotonin-induced gnawing in rats: comparison with tail pinch-induced gnawing, *Pharmacol. Biochem. Behav.,* 16, 407, 1982.

20. Morley, J. E., Levine, A. S., and Rowland, N. E., Minireview. Stress induced eating, *Life Sci.,* 32, 2169, 1983.

21. Krahn, D. D., Gosnell, B.A., Grace, M., and Levine, A. S., CRF antagonist partially reverses CRF- and stress-induced effects on feeding, *Brain Res. Bull.,* 17, 285, 1986.

22. Grignaschi, G., Mantelli, B., and Samanin, R., The hypophagic effect of restraint stress in rats can be mediated by 5-HT2 receptors in the paraventricular nucleus of the hypothalamus, *Neurosci. Lett.,* 152, 103, 1993.

23. Shibasaki, T., Yamauchi, N., Kato, Y., Masuda, A., Imaki, T., Hotta, M., Demura, H., Oono, H., Ling, N., and Shizume, K., Involvement of corticotropin-releasing factor in restraint stress-induced anorexia and reversion of the anorexia by somatostatin in the rat, *Life Sci.,* 43, 1103, 1988.

24. Rybkin, I. I., Zhou, Y., Volaufova, J., Smagin, G. N., Ryan, D.H., and Harris, R. B., Effect of restraint stress on food intake and body weight is determined by time of day, *Am. J. Physiol.,* 273, R1612, 1997.

25. Harris, R. B., Kasser, T. R., and Martin, R. J., Dynamics of recovery of body composition after overfeeding, food restriction or starvation of mature female rats, *J. Nutr.,* 116, 2536, 1986.

26. Bing, C., Frankish, H. M., Pickavance, L., Wang, Q., Hopkins, D. F., Stock, M. J., and Williams, G., Hyperphagia in cold-exposed rats is accompanied by decreased plasma leptin but unchanged hypothalamic NPY, *Am. J. Physiol.,* 274, R62, 1998.

27. Michajlovskij N., Lichardus B., Kvetnansky R., and Ponec J., Effect of acute and repeated immobilization stress on food and water intake, urine output and vasopressin changes in rats, *Endocr. Exp.,* 22, 143, 1988.

28. Marti, O., Marti, J., and Armario, A., Effects of chronic stress on food intake in rats: influence of stressor intensity and duration of daily exposure, *Physiol. Behav.* 55, 747, 1994.

29. Krahn, D. D., Gosnell, B. A., and Majchrzak, M. J., The anorectic effects of CRH and restraint stress decrease with repeated exposures, *Biol. Psychiatry,* 27, 1094, 1990.

30. Haleem, D. J. and Parveen T., Brain regional serotonin synthesis following adaptation to repeated restraint, *NeuroReport,* 5, 1785, 1994.

31. Harris, R. B. S, Zhou, J., Youngblood, B., Rybkin, I., Smagin, G., and Ryan, D., The effect of repeated restraint stress on body weight and body composition of rats fed low and high-fat diets, *Am. J. Physiol.,* 275, R1928, 1998.

32. Meerlo, P., Overkamp, G. J., and Koolhaas, J. M., Behavioural and physiological consequences of a single social defeat in Roman high- and low-avoidance rats, *Psychoneuroendocrinology,* 22, 155, 1997.

33. Zhou, J., Yan, X., Ryan, D., and Harris, R., Sustained effects of repeated restraint stress on muscle and adipocyte metabolism in high-fat fed rats, *Am. J. Physiol.*, in press.

34. Youngblood, B., Ryan, D., and Harris, R., Appetitive operant behavior and free-feeding in rats exposed to acute stress, *Physiol. Behav.*, 62, 827, 1997.

35. Lennie, T. A., McCarthy, D. O., and Keesey, R. E., Body energy status and the metabolic response to acute inflammation, *Am. J. Physiol.*, 269, R1024, 1995.

36. Lin, L., York, D., and Bray, G., Acute effects of intracerebroventricular corticotropin releasing hormone (CRH) on macronutrient selection, *Int. J. Obesity*, 16, (Suppl. 1), 52, Abstr. P207, 1992.

37. Devenport, L., Knehans, A., Sundstrom, A., and Thomas, T., Corticosterone's dual metabolic actions, *Life Sci.*, 45, 1389, 1989.

38. Devenport, L., Knehans, A., Thomas, T., and Sundstrom, A., Macronutrient intake and utilization by rats: interactions with type I adrenocorticoid receptor stimulation, *Am. J. Physiol.*, 260, R73, 1990.

39. Castonguay, T. W., Glucocorticoids as modulators in the control of feeding. *Brain Res. Bull.*, 27, 423, 1991.

40. Bligh, M. E., Douglass, L. W., and Castonguay, T. W., Corticosterone modulation of dietary selection patterns, *Physiol. Behav.*, 53, 975 1993.

41. Prasad, C., delaHoussaye, A. J., Prasad, A., and Mizuma, H., Augmentation of dietary fat preference by chronic, but not acute, hypercorticosteronemia, *Life Sci.*, 56, 1361, 1995.

42. Pozzoli, G., Bilezikjian, L. M., Perrin, M. H., Blount, A. L., and Vale, W. W., Corticotropin-releasing factor (CRF) and glucocorticoids modulate the expression of type 1 CRF receptor messenger ribonucleic acid in rat anterior pituitary cell cultures, *Endocrinology*, 137, 65, 1996.

43. Iredale, P. A. and Duman, R. S., Glucocorticoid regulation of corticotropin-releasing factor1 receptor expression in pituitary-derived AtT-20 cells, *Mol. Pharmacol.*, 51, 794, 1997.

44. Bradbury, M. J., Akana, S. F., Cascio, C. S., Levin, N., Jacobson, L., and Dallman, M. F., Regulation of basal ACTH secretion by corticosterone is mediated by both type I (MR) and type II (GR) receptors in rat brain, *J. Steroid Biochem. Mol. Biol.*, 40, 133, 1991.

45. Kumar, B. A. and Leibowitz, S. F., Impact of acute corticosterone administration on feeding and macronutrient self-selection patterns, *Am. J. Physiol.*, 254, R222, 1988.

46. Kumar, B. A., Papamichael, M., and Leibowitz, S. F., Feeding and macronutrient selection patterns in rats: adrenalectomy and chronic corticosterone replacement, *Physiol. Behav.*, 42, 581, 1988.

47. Willner, P., Towell, A., Sampson, D., Sophokleous, S., and Muscat, R., Reduction of sucrose preference by chronic unpredictable mild stress, and its restoration by a tricyclic antidepressant, *Psychopharmacology*, 93, 358, 1987.

48. Plaznik, A., Stefanski, R., and Kostowski, W., Restraint stress-induced changes in saccharin preference: the effect of antidepressive treatment and diazepam, *Pharmacol. Biochem. Behav.*, 33, 755, 1989.

49. Papp, M., Willner, P., and Muscat, R., An animal model of anhedonia: attenuation of sucrose consumption and place preference conditioning by chronic unpredictable mild stress, *Psychopharmacology*, 104, 255, 1991.

50. Matthews, K., Forbes, N., and Reid, I. C., Sucrose consumption as an hedonic measure following chronic unpredictable mild stress, *Physiol. Behav.*, 57, 241, 1995.

51. Forbes, N. F., Stewart, C. A., Matthews, K., and Reid, I. C., Chronic mild stress and sucrose consumption: validity as a model of depression, *Physiol. Behav.*, 60, 1481, 1996.

52. Hatcher, J. P., Bell, D. J., Reed, T. J., and Hagan, J. J., Chronic mild stress-induced reductions in saccharin intake depend upon feeding status, *J. Psychopharmacol.*, 11, 331, 1997.

53. Harris, R. B. S., Zhou, J., Youngblood, B. D., Smagin, G. N., and Ryan, D. H., Failure to change exploration or saccharin preference in rats exposed to chronic mild stress, *Physiol. Behav.*, 63, 91, 1998.

54. Bernardis, L. L. and Bellinger, L. L., Dorsomedial hypothalamic hypophagia: self-selection of diets and macronutrients, efficiency of food utilization, "stress eating," response to high-protein diet and circulating substrate concentrations, *Appetite*, 2, 103, 1981.

55. Vaswani, K., Tejwani, G. A., and Mousa, S., Stress induced differential intake of various diets and water by rat: the role of the opiate system, *Life Sci.*, 32, 1983, 1983.

56. Welch, C. C., Grace, M. K., Billington, C. J., and Levine, A. S., Preference and diet type affect macronutrient selection after morphine, NPY, norepinephrine, and deprivation, *Am. J. Physiol.*, 266, R426, 1994.

57. Thouzeau, C., Le Maho, Y., and Larue-Achagiotis, C., Refeeding in fasted rats: dietary self-selection according to metabolic status, *Physiol. Behav.,* 58,1051, 1995.

58. Powers, R. J. and Kutash, I. L., Stress and alcohol, *Int. J. Addictions,* 20, 461, 1985.

59. Gill, K., Amit, Z., and Smith, B. R., Alcohol as a food: a commentary on Richter, *Physiol. Behav.,* 60, 1485, 1996.

60. Weisinger, R. S., Denton, D. A., and Osborne, P. G., Voluntary ethanol intake of individually or pair-housed rats: effect of ACTH or dexamethasone treatment, *Pharmacol. Biochem. Behav.,* 33, 335, 1989.

61. Prasad, C. and Prasad, A., A relationship between increased voluntary alcohol preference and basal hypercorticosteronemia associated with an attenuated rise in corticosterone output during stress, *Alcohol,* 12, 59, 1995.

62. Fahlke, C., Engel, J.A., Eriksson, C.J., Hard, E., and Soderpalm, B., Involvement of corticosterone in the modulation of ethanol consumption in the rat, *Alcohol,* 11, 195, 1994.

63. Lamblin, F. and De Witte, P., Adrenalectomy prevents the development of alcohol preference in male rats, *Alcohol,* 13, 233, 1996.

64. Fahlke, C., Hard, E., Thomasson, R., Engel, J. A., and Hansen, S., Metyrapone-induced suppression of corticosterone synthesis reduces ethanol consumption in high-preferring rats, *Pharmacol. Biochem. Behav.,* 48, 977, 1994.

65. De Witte, P., The role of neurotransmitters in alcohol dependence: animal research, *Alcohol Alcoholism,* 31, 13, 1996.

66. George, S. R., Fan, T., Ng, G. Y., Jung, S. Y., and Naranjo, C. A., Low endogenous dopamine function in brain predisposes to high alcohol preference and consumption: reversal by increasing synaptic dopamine, *J. Pharmacol. Exp. Ther.,* 273, 373, 1995.

67. Ulm, R. R., Volpicelli, J. R., and Volpicelli, L. A., Opiates and alcohol self-administration in animals, *J. Clin. Psychol.,* 56, 5, 1995.

68. Livezey, G. T., Balabkins, N., and Vogel, W. H., The effect of ethanol (alcohol) and stress on plasma catecholamine levels in individual female and male rats, *Neuropsychologia,* 17, 193, 1987.

69. Rivier, C., Bruhn, T., and Vale, W., Effect of ethanol on the hypothalamic-pituitary-adrenal axis in the rat: role of corticotropin-releasing factor (CRF), *J. Pharmacol. Exp. Ther.,* 229, 127, 1984.

70. Lee, S. and Rivier, C., Gender differences in the effect of prenatal alcohol exposure on the hypothalamic-pituitary-adrenal axis response to immune signals, *Psychoneuroendocrinology,* 21, 145, 1996.

71. Sprague, J. E. and Maickel, R. P., Effects of stress and ebiratide (Hoe-427) on free-choice ethanol consumption: comparison of Lewis and Sprague-Dawley rats, *Life Sci.,* 55, 873, 1994.

72. Hannon, R. and Donlon-Bantz, K., Effect of housing density on alcohol consumption by rats, *J. Stud. Alcohol,* 37, 1556, 1976.

73. Nash, J. F., Jr. and Maickel, R. P., Stress-induced consumption of ethanol by rats, *Life Sci.,* 37, 757, 1985.

74. Volpicelli, J. R., Uncontrollable events and alcohol drinking, *Br. J. Addiction,* 82, 381, 1987.

75. Rockman, G. E., Hall, A., Hong, J., and Glavin, G. B., Unpredictable cold-immobilization stress effects on voluntary ethanol consumption in rats, *Life Sci.,* 40, 1245, 1987.

76. Volpicelli, J. R. and Ulm, R. R., The influence of control over appetitive and aversive events on alcohol preference in rats, *Alcohol,* 7, 133, 1990.

77. Boyd, T. L., Callen, E. J., and House, W. J., The effects of post-stress exposure to alcohol upon the development of alcohol consumption in rats, *Behav. Res. Ther.,* 27, 35, 1989.

78. Caplan, M. A. and Puglisi K., Stress and conflict conditions leading to and maintaining voluntary alcohol consumption in rats, *Pharmacol. Biochem. Behav.,* 24, 271, 1986

79. Mills, K. C., Bean, J. W., and Hutcheson, J. S., Shock induced ethanol consumption in rats, *Pharmacol. Biochem. Behav.,* 6, 107, 1997.

80. Nash, J. F., Jr. and Maickel, R. P., The role of the hypothalamic-pituitary-adrenocortical axis in post-stress induced ethanol consumption by rats, *Prog. Neuro-Psychopharmacol. Biol. Psychiatry,* 12, 653, 1988.

Section V

Conclusions

32 Neural and Metabolic Control of Macronutrient Selection: Consensus and Controversy

Randy J. Seeley and Hans-Rudolf Berthoud

CONTENTS

1 INTRODUCTION: THE PROBLEM

The ingestion of substances from the environment, for all practical purposes, represents the sole method by which animals obtain the necessary energy, nutrients, and vitamins to sustain life. As such, ingestive behavior has evolved in an extremely complex web of competing needs for the maintenance of numerous homeostatic systems in the body. One way to conceptualize the substances that the animal consumes is by their macronutrient content, i.e., the amount of protein, fat, and carbohydrate a particular food contains. The central question for this book is how appropriate is that conceptualization and how much do animals use the macronutrient content of foods to guide their food selection. The importance of the answer to this question is difficult to underestimate. Given the clear association between dietary fat and numerous serious diseases such as obesity, diabetes, heart disease, and some cancers, choosing the appropriate intervention is critical to promoting healthier diets. For example, if dietary fat content is tightly regulated, dietary interventions to reduce fat content will ultimately be unsuccessful but physiological interventions might be extremely effective. However, if fat intake is not regulated, dietary interventions would be the therapeutic strategy of choice.

Stricker, in the first chapter of this book, sets a very high standard for what a tightly regulated homeostatic system would look like. First is the argument that only variables critical for organismal function (life), such as body fluid osmolality (water intake) or sodium concentration (salt appetite), need to be under tight homeostatic control. Second, deficits in these variables generate discrete signals that are translated by the brain and other tissues into physiological responses that restore the deviation. In the case of behavioral reactions such as ingestion, the brain is an obligatory site for processing and it has to generate a specific appetite for the depleted variable (nutrient).

2   THE EXTREME POSITIONS

Along the spectrum of possibilities about the regulation of macronutrient selection there are two extreme positions one could take. First, it is possible that each macronutrient has a completely separate and dedicated control system that tightly regulates intake to match ongoing utilization and/or oxidation of that macronutrient. The other extreme position is that the macronutrient content of the diet is simply of no relevance whatsoever to the animal's food selection. We shall first try to stake out these two extreme positions and the evidence that one might garner to support each of those extreme positions.

2.1   THREE SEPARATE AND DEDICATED CONTROL SYSTEMS FOR EACH MACRONUTRIENT

The first position would be that the selection of different macronutrients in the diet is determined separately by the storage and utilization of each macronutrient. Such a system would keep each macronutrient in "balance" so that intake exactly matched loss of each macronutrient to utilization (either for energy or for building tissue in the case of proteins). In essence, this position holds that each macronutrient has its own separate control system (Figures 1 and 2) that would include mechanisms for detection of relative macronutrient depletion/repletion, the ability to generate specific error signals related to a specifc macronutrient, and the ability to identify and ingest separate macronutrients from the environment to correct these error signals and achieve long-term balance.

Of course, not even the strongest advocate for separate systems would expect that these systems operate by only using negative feedback signals during individual meals or even over the course of a single day. Just as it is well recognized that total energy intake can fluctuate widely in the short term, we would expect that separate control systems for each macronutrient would also be subject to short-term variability. However, this extreme view would suggest that in the longer term, intake of each macronutrient would be kept in a relatively narrow range, and would be defended against shortage or surplus.

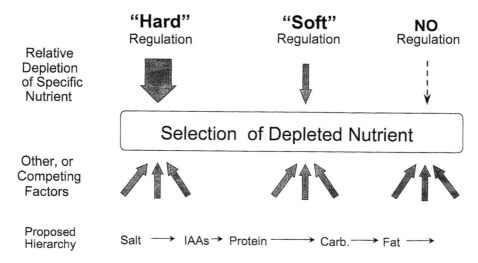

FIGURE 1   Proposed hierarchy for regulatory weight of specific nutrient depletion vs. other competing factors in selection process.

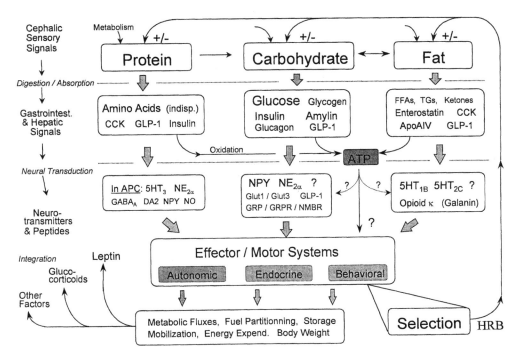

FIGURE 2 Schematic diagram showing general flow of information and chemical character of messengers potentially involved in macronutrient-specific regulatory system of food intake. Signals carrying macronutrient-specific information are generated by the cephalic and gastrointestinal/hepatic phases during ingestion, and by metabolic processes. These signals are transduced and integrated into neuronal activity, which determines current and future food selection as well as autonomic and endocrine activity. The possibility that non-(macronutrient) specific information from hepatic energy production is involved in the control of food intake is also indicated. Abbreviations for messengers and their receptors: GRP, gastrin releasing peptide (bombesin); NMBR, neuromedin-B receptor; CCK, cholecystokinin; GLP-1, glucagon-like peptide-1; 5HT, serotonin; NE, norepinephrine; GABA, gamma-aminobutyric acid; DA, dopamine; NPY, neuropeptide Y; NO, nitric oxide; Glut 1/Glut 3, glucose transporters 1 and 3. Other abbreviations: APC, anterior piriform cortex.

Several lines of evidence can be used to support such a position. First, it is clear that there are optimal levels of macronutrients that produce optimal functioning for the body and hence there could have been evolutionary pressure in favor of omnivore organisms capable of regulating macronutrients separately. In particular, there should have been strong pressure for omnivores to ensure that adequate levels of protein are consumed. Without adequate access to proteins, animals cannot grow or rebuild tissue after injury. Hence, it appears that protein intake can be separately regulated. Paul Rozin[1] showed that when given separate access to a protein source, rats would accurately compensate for dilution of that protein source by increased intake such that protein intake was conserved. Simpson and Raubenheimer (Chapter 4) also provide compelling data that when rats are given choices between carbohydrate and protein that their behavior carefully follows the prediction for regulating carbohydrate and protein. Additionally, Gietzen (Chapter 23) and her colleagues have elegantly described a neural system that can locally detect amino acid imbalances and produce potent changes in ingestive behavior.

For fat and carbohydrate balance, arguments can also be made for their separate regulation. First, it is clear that carbohydrates and fats are processed by very different biochemical pathways and are specifically used as preferred or obligatory fuels by different organs. Such distinct pathways at least provide an opportunity for the utilization of each macronutrient to be coded separately and for that information to be utilized separately by the central nervous system.

Second, the body has many opportunities to detect carbohydrates and fats during their journey through the alimentary canal and beyond. Besides direct sensors for glucose and fatty acids at multiple locations along this journey, the extended gastrointestinal tract has additional coding mechanisms, in that the two macronutrients differentially release signaling peptides such as CCK, GLP-1, enterostatin, insulin, and amylin, which, in turn, have direct effects on the brain. In work going back to that of Oomura and carried forward by Levin and colleagues (Chapter 3), specific types of neurons in the CNS alter their firing rate very sensitively to ambient levels of glucose and ongoing glucose utilization within those neurons. It has also been convincingly shown that specific metabolic blockade of either glucose or fatty acid utilization activates separate neural pathways and differentially changes subsequent macronutrient choice (Chapter 13 by Ritter et al.).

Third, rats can associate the different sensory qualities and/or metabolic consquences of carbohydrates and fats with other cues in the environment through processes of learning, and they apparently use such metabolic memories to guide future food choice. Davidson and colleagues (Chapter 14) have shown that animals can respond to compounds which selectively alter carbohydrate or fat metabolism by increasing responding to the carbohydrate- or fat-associated stimuli in their environment.

Finally, in at least some paradigms exogenous administration of peptides such as NPY, opioids, and enterostatin, as well as neurotransmitters such as 5-HT and NE can produce relatively selective changes in either carbohydrate or fat ingestion (Chapters 20 and 25).

Thus, it is quite clear that the brain does have access to sensory information specific to each macronutrient, although this fact in itself does not prove that the brain uses the information to control intake.

2.2   NO SEPARATE REGULATION OF MACRONUTRIENT INTAKE

The opposite extreme position would be that the macronutrient content of the diet is not of any relevance in determining the animal's ongoing choice among the foods available in its environment. Such a position is bolstered by teleological arguments. First, mammals have a liver that allows them a tremendous amount of flexibility in being able to convert available substrates into the substrates needed by the body under a variety of conditions. Consequently, the need to select the appropriate macronutrients from the environment is not absolute (Chapter 2). Stricker (Chapter 1) even suggests that there is no need for the generation of a hunger signal for energy in general, let alone for macronutrients such as carbohydrate and fat, because bodily cells usually are provided with nourishment adequate for metabolism, and food deprivation does not cause a real shortage of the circulating energy supply. According to his view there are no excitatory signals for hunger (appetite) other than those generated by the palatability of the food, and selection would therefore be solely determined by the sensory qualities of different foods (macronutrients). In addition, Friedman and colleagues (Chapter 2) have provided a number of studies that point to common end points of substrate utilization (such as ATP levels in the liver) rather than the separate pathways associated with each macronutrient as being critical to determining food intake (Figure 1). As reviewed by Ackroff (Chapter 16), a variety of fat dilution experiments fail to provide much evidence of compensatory increases in fat intake.

Next is the powerful data described by Galef (Chapter 3). It is clear that under some situations, rodents with adequate access to protein will fail to select a diet with sufficient protein to maintain proper growth or, even worse, to maintain life. The fact that animals will actually die of protein malnourishment even while having ample access to protein in their environment is a salient failure of the animal to select the appropriate macronutrients from its environment. Hence, some argument can be made that animals neither need nor use feedback from the body to guide selection of macronutrients in their diet.

3   "SOFT" REGULATION OF MACRONUTRIENT INTAKE

The real question for this book is how to come to terms with what seem disparate sets of data pointing at very different conclusions about whether or not animals use the macronutrient content of food to guide their food selection. On its face, it would appear that both extreme positions about the nature of this regulatory system would be untenable. Consequently, we would like to propose a hypothesis that we hope could explain a larger proportion of the available data.

Homeostatic regulatory systems lie on a continuum based on two factors: first, how tightly regulated a system must be in order to maintain adequate internal conditions for sustaining life and, second, the flexibility of the body to adopt alternative strategies in order to maintain those internal conditions. For example, the need to maintain adequate oxygen supply for cells to metabolize fuels is very high on this continuum. If oxygen is not available from the environment, within minutes the levels of oxygen in the body fall outside that compatible with life. The body does have numerous defense responses that extend the time that the body can sustain itself, but none of those is adequate if the supply of oxygen into the body is not quickly restored. Such a system where the cost of not maintaining appropriate levels of a parameter results in death and the body has only a limited number of responses to ameliorate the effects of the deficit we term a "hard" regulatory system. A system where the cost of not defending a parameter is not as high (either death results only after much longer time frames or not at all) and where the body has more available strategies to defend perturbations of the regulated parameter is a "soft" regulatory system. So homeostatic systems can be considered for their place on this continuum from hard to soft (Figure 1).

The second concept we would like to introduce is that homeostatic regulatory systems are constantly in competition and that the animal must continuously make compromises in regulatory systems. For example, under normal conditions, blood glucose levels are defended very accurately and the body has numerous systems that contribute to this accurate regulation. However, maintaining adequate levels of blood glucose over the long run requires that animals consume foods from the environment. The consumption of those foods in and of itself represents a challenge to this system since, typically, glucose levels rise considerably after a meal. Hence, to keep glucose levels from falling in the long run, animals must compromise and accept higher-than-ideal glucose levels in the short run.[2]

Not only do animals have to compromise between short- and long-term regulation within a single homeostatic system, but it is clear that animals must make compromises between the competing needs of different homeostatic systems. An obvious example is that many air-breathing species feed on sources of energy that are found beneath the surface of the water. To maintain adequate energy substrates, these animals are forced to risk adequate oxygen supplies by going below the surface of the water. The animal has clearly compromised. Note, however, that there are serious limits to the compromise. Maintenance of adequate oxygen supplies demands that at some point the animal break off the search for food and return to the surface to replenish levels of oxygen. Barring unforeseen circumstances, it seems unlikely that animals would willingly drown while chasing food. So when push comes to shove, softer homeostatic regulatory systems must give way to harder systems. Hence, where a system falls on this continuum should dictate much about how the animal chooses to compromise the competing needs and influences of different homeostatic systems.

The concept that we would like to forward here is that the separate regulation of macronutrient intake does exist but that it is quite soft on the continuum of other regulatory systems. That is to say, that for omnivores under most circumstances the need to select particular macronutrients is small. First, picking the appropriate mix of macronutrients from the environment can clearly increase the efficiency of the body but rarely does it represent a life-threatening decision (with the possible exception of essential amino and fatty acids). The primary reason for this is the alternative strategies the body has to deal with a mismatch between intake of a particular macronutrient and utilization of that macronutrient. For example, most tissues (with the brain and liver being notable exceptions) can utilize either fats or carbohydrates for fuel. Additionally, the liver is capable of substantial conversions of the available fuels into the most utilizable fuels.

It is important to note that while these strategies typically prevent the choice of macronutrients from being critical to the maintenance of life, the ongoing need by specific tissues for specific sources of fuel do put some pressure on the organism to select appropriate macronutrient mixes from the environment. For example, during weight maintenance, an argument can be made that the amount of protein, carbohydrate and fat oxidized should be equal to the amounts of the three macronutrients ingested (see Chapter 11 by Flatt). If the respiratory quotient is not equal to the food quotient, or in other words, if fat oxidation is not commensurate with fat intake, the adipose depot will change and body weight cannot remain stable. Therefore, it could be argued that systems which tend to keep body weight stable exert some pressure on macronutrient choice. Furthermore, in particular, many of the conversions by the liver are energetically expensive. Hence, there is a clear advantage of selecting the right mix of macronutrients from the environment such that these wasteful conversions can be avoided. We propose to classify this system as a relatively soft regulatory system.

The consequence of being a soft regulatory system is that when it comes in conflict with the needs of a harder regulatory system, the animal's choice should be to sacrifice the needs of the soft regulatory system to serve the more-pressing needs of the harder regulatory system. This brings us to the clear conflict in much of the literature concerning whether animals use information about macronutrients to select diets in their environment. To find evidence of such a soft regulatory system, experiments must be done carefully so as not to put the animal's choices in conflict with the needs of other homeostatic systems. As a consequence, methodological details of various experiments are likely to have large and substantive influences on whether individual experiments do or do not support the concept of macronutrient selection.

4   METHODOLOGICAL ISSUES AND EVIDENCE FOR SOFT REGULATION

4.1   ANIMAL DATA

Considerable debate has been generated by the use of a three-choice test where animals have access to a pure source of each of the three macronutrients. While this method has generated significant data in support of the concept of macronutrient selection as well as the neural underpinnings of such regulation, it represents an artificial situation not encountered in animals' natural habitats. It is also clear that the basal diet that animals are maintained upon and their initial preferences can dramatically alter the results of these experiments. Such outcomes are consistent with a soft regulatory system. For example, enterostatin only decreases fat intake when on a high-fat maintenance diet and when the overall energy balance of the animal has not been compromised.[7] Hence only when macronutrient selection is not in contest with caloric regulation can clear evidence of its existence be seen.

This brings us to another interpretive issue associated with the three-choice method. When rats are given insulin into the third ventricle, it produces a very selective decrease in fat intake. However, as van Dijk and colleagues (Chapter 30) point out, their belief is that what animals really are doing is protecting carbohydrate and protein intake while decreasing overall caloric intake. The only choice the animal can then make to decrease caloric intake is to decrease fat intake. Rather than insulin acting in the brain to control fat intake, it may simply provide a signal about caloric intake. This is consistent with the notion introduced by Friedman that these changes in intake in a three-choice test can be thought of as either macronutrient selection or "macronutrient rejection" (Chapter 2).

In this volume, two approaches are introduced that are complimentary to the standard three-choice test. The first comes from Simpson and Raubenheimer (Chapter 4) in which very specific predictions can be made about how animals will choose food sources if they are regulating various contents of the food. This is a powerful conceptual model that can be applied to understanding the physiological control systems of macronutrient intake. The second is the classical and instrumental learning paradigms described by Davidson and colleagues (Chapter 14). The advantage (and dis-advantage) of these paradigms is that animals produce responding in the absence of any macronu-trient access. Hence, it provides some insight into the signals to which the animal is paying attention without the animal being able to use direct feedback from the food consumed to guide its behavior.

This brings us to another important methodological issue: learning and plasticity. Davidson's data imply that animals are processing information related to macronutrient content of foods and using that information to guide behavior in the future. Certainly a soft regulatory system is expected to be importantly shaped by an animal's experience. Hence, it is important to understand an animal's previous experience before putting it into the standard three-choice test since its history can clearly impact on its choice. While Galef (Chapter 3) points out that rats can die of protein malnourishment even in the face of adequate protein sources, he also points out that this is a very rare occurrence if the animal has had previous experience with the protein source or has had access to a demonstrator rat that was already consuming adequate amounts of the protein source. Therefore, given the right opportunity, rats clearly learn to regulate their protein intake.

4.2   HUMAN DATA

No less difficult than the animal data is the data on humans. Moran and Rolls (Chapter 8) provide data that indicate that over the course of a single meal, there is little compensation for altered macronutrient content of a preload. Such data argue against any short-term compensatory mecha-nism which guides food choice to maintain a constant mixture of macronutrients. However, there is at least some data for a subtle long-term component of macronutrient selection in human beings.

de Castro (Chapter 5) and colleagues have shown that there is a significant heritable factor to the overall macronutrient intake pattern in humans as reported by an intake diary. Additionally, there is a subtle but reliable effect for intake of a particular macronutrient on a given day to influence intake of the macronutrient 2 days hence. Consistent with this, Hill and colleagues[3] have shown that when fat was covertly substituted by the nonabsorbable fat-substitute olestra in the diet of humans for 14 days, there was a small but significant increase in spontaneous fat intake. Again there would appear to be two messages from such data. First, methodological differences result in very different answers to the basic questions associated with macronutrient regulation. Second, the evidence would point to neither extreme position about macronutrient selection, but rather point toward a subtle but demonstrable regulation of macronutrient content of the food. Hence, such data would appear to be consistent with the hypothesis that macronutrients are regulated but only in a "soft" manner. Finally, with humans it would also appear that learning plays an important role in macronutrient selection as would be implied by a soft regulatory system. Booth and Thiebault (Chapter 6) describe compelling evidence that carbohydrate regulation in humans is importantly dependent on learning.

5   POTENTIAL FOR THERAPEUTIC INTERVENTION

The hypothesis that macronutrient content of the diet is regulated but in a soft manner has interesting implications for the potential for therapeutic interventions to alter the macronutrient mix in the diet. First, it would imply that either dietary or physiological interventions could be mildly suc-cessful. In the case of dietary interventions, the systems that regulate macronutrient intake are likely to compensate and provide some pressure to ameliorate the dietary change. However, the ability of the system to compensate for such changes is limited and likely to be incomplete. In the case of physiological interventions that alter the physiological systems responsible for this regulation, they can be effective but only to the point at which these systems begin to interfere with other regulatory systems. Consequently, the amount of change that can be expected from such a manipulation is relatively small. Second, the "softness" of this system does have an advantage. For either dietary or physiological intervention, it will be difficult to push this system around so much as to create deleterious side effects. Hence, if all we would like to do is push the diet from 35% calories from fat to 30%, it may be possible.

6   CONCLUSIONS

Both of our opinions have evolved considerably as we have put together this book. We hope that this volume will provide a significant resource for people interested in the issue of how macronutrients affect food intake. It is clear, however, that much work still needs to be done. Among the areas for which we appear to have the least knowledge is how the taste system might code different macronutrients. The taste system provides major chemosensory input that importantly guides other aspects of ingestive behavior, but its role in guiding macronutrient decisions remains unclear.

Second, a major methodological conclusion is that simply providing an animal or person with a choice of two or three macronutrients is not an effective strategy for coming to conclusions on fat, carbohydrate, and protein. This is an important point made by both Booth and Thiebault as well as Friedman in their chapters. To show that it is the macronutrient that is being regulated, one must vary the source or form of that macronutrient and show that similar results are obtained. A major emphasis of work in this field should address these important methodological issues such that confidence in their generality can be increased. While such parametric analyses are tedious, they are necessary. Both Simpson and Davidson provide some excellent examples of novel methodologies that can powerfully dissect issues associated with macronutrient regulation that are not easily obtained from simpler two- or three choice paradigms.

Finally, we feel continued emphasis on what animals and people learn about the macronutrient composition of food is required. Another aspect of soft regulatory systems is that they are less likely to be hard-wired by evolution and more likely a product of learning over the lifetime of the animal. Consequently, understanding how animals process information about the macronutrient content consumed in the past and how that information is used to influence future food selection choices is a necessary component of understanding the regulation of macronutrients.

REFERENCES

1. Rozin, P., *J. Comp. Physiol. Psychol.,* 65, 23, 1968.
2. Woods, S. C., *Psychol. Rev.,* 98, 488, 1991.
3. Hill, J. O., Seagle, H. M., Johnson, S. L., Smith, S., Reed, G. W., Tran, Z. V., Cooper, D., Stone, M., and Peters, J. C., *Am. J. Clin. Nutr.,* 67, 1178, 1998.

Index

A

AA, *see* Amino acid

Absorption, inhibitors of, *see* Fat substitutes, effects of inhibitors of absorption and

Acetoacetate, 139, 141

Adenosine monophosphate (AMP), 137, 300

Adenosine triphosphate (ATP), 131, 191, 311, 364
    availability of for cellular metabolism, 177
    synthesis, 14, 315

Adipose tissue, 157, 167

Adiposity signals, macronutrient selection and, 465
    adiposity signals and brain, 466
    adiposity signals and macronutrient intake, 467
    adiposity signals and macronutrient selection, 469
    brain loci responsive to adiposity signals, 470
    efficacy of adiposity signals in rats maintained on diets of different fat content, 467

Adrenalectomized (ADX) rats, 219, 228

Adrenalectomy, 313

ADX rats, *see* Adrenalectomized rats

Afferent nerve activity, 319

Agouti-related protein (AGRP), 375

AGRP, *see* Agouti-related protein

Alanine, 133, 140, 313, 314

Albumin, 144

Alcohol, 168, 482

Aldosterone, 6

Alliesthesia, 257

2,5-AM, *see* 2,5-Anhydro-D-mannitol

Aminergic transmitters, 295

Amino acid (AA), 37, 145
    branched-chain, 132
    catabolism, 134

    growth-limiting, 339
    imbalances, 491
    indispensable, *see* Indispensable amino acids
    infusion, 99
    large neutral, 427
    learned responses to cues for, 347
    metabolism, tissue-specific, 138
    mixtures, 96
    oxidation, 143, 150
    sensors, 314

Amino acid recognition, in central nervous system, 339
    aminoprivic model, 341
    brain areas associated with adaptation to imbalanced diet, 350
    brain areas associated with selection of IAA, 350
        overlapping brain systems for macronutrient selection, 351
        separate systems for IAA/protein, 351

learned responses to cues for amino acids, 347
    aversions, 347
    peripheral sites, 348
    preferences, 350
    recognition of IAA deficiency, 342
    recognition of IAA repletion, 345
        APC, 345
        lateral hypothalamus, 347
    significance of amino acid recognition, 339
    time course of responses to IAA, 340

Aminoprivic feeding, 342

Aminoprivic model, 341, 346

AMP, *see* Adenosine monophosphate

Amphetamine, 76, 78

Amylopectin, 419

Amylose, 419

Angiotensin, 5

2,5-Anhydro-D-mannitol (2,5-AM), 189, 318

Animal models, *see* Nutrient preloads, effects of on subsequent macronutrient selection in animal models

Anterior piriform cortex (APC), 342

Antiobesity compounds, 439

APC, *see* Anterior piriform cortex

Apo A-IV, *see* Apolipoprotein A-IV

Apolipoprotein A-IV (apo A-IV), 280, *see also* Fat absorption, role of lymphatic apolipoprotein A-IV and, in regulation of food intake
    action, site of, 282
    food intake-inhibitory effects of, 286
    synthesis, regulation of, 287

Appetite, 19, 63, 67, 176
    behavior, 211, 213
    carbohydrate-specific, 74
    effect of macronutrient balance/oxidation on, 181
    for fatty substances, 234
    learned, 38, 39
    memory-specific control of, 259
    sodium, 20

Appetites, homeostatic systems and specific, 3
    analogy to hunger, 6
    analogy to NaCl appetite, 6
    thirst, 3
        angiotensin, 5
        osmoregulation, 4
        volume regulation, 4

Appetite, for protein, 19
    appetites for other nutrients, 21
    behavioral homeostasis, 19
    cafeteria-feeding experiments and protein intake, 22
    caveat, 25
    common sense and evolutionary theory, 21
    dietary self-selection, 21

Milton Keynes UK
Ingram Content Group UK Ltd.
UKHW052025071024
449327UK00027B/2429

9 780367 399351